Student's Solutions Manual
Part II

Thomas/Finney

Calculus and Analytic Geometry

9TH Edition

Maurice D. Weir
Naval Postgraduate School

Addison-Wesley Publishing Company

Reading, Massachusetts • Menlo Park, California • New York
Don Mills, Ontario • Harlow United Kingdom • Amsterdam • Bonn
Sydney • Singapore • Tokyo • Madrid • San Juan • Milan • Paris

Reproduced by Addison-Wesley from camera-ready copy supplied by the author.

ISBN 0-201-53180-1
8 9 10-CRS-99

PREFACE TO THE STUDENT

This Student's Solutions Manual contains the solutions to all of the odd-numbered exercises in the 9th Edition of CALCULUS AND ANALYTIC GEOMETRY by Ross L. Finney and George B. Thomas, Jr., excluding the Computer Algebra System (CAS) exercises. We have worked each solution to ensure that it

- conforms exactly to the methods, procedures and steps presented in the text
- is mathematically correct
- includes all of the steps necessary so you can follow the logical argument and algebra
- includes a graph or figure whenever called for by the exercise
- is formatted in an appropriate style to aid you in its understanding

How to use a solution's manual

- solve the assigned problem yourself
- if you get stuck along the way, refer to the solution in the manual as an aid but continue to solve the problem on your own
- if you cannot continue, reread the textbook section, or work through that section in the Student Study Guide, or consult your instructor
- after solving the problem (if odd-numbered), carefully compare your solution procedure to the one in the manual
- if your answer is correct but your solution procedure seems to differ from the one in the manual, and you are unsure your method is correct, consult your instructor
- if your answer is incorrect and you cannot find your error, consult your instructor

TABLE OF CONTENTS

CHAPTER 8 INFINITE SERIES

8.1 LIMITS OF SEQUENCES OF NUMBERS

1. $a_1 = \frac{1-1}{1^2} = 0$, $a_2 = \frac{1-2}{2^2} = -\frac{1}{4}$, $a_3 = \frac{1-3}{3^2} = -\frac{2}{9}$, $a_4 = \frac{1-4}{4^2} = -\frac{3}{16}$

3. $a_1 = \frac{(-1)^2}{2-1} = 1$, $a_2 = \frac{(-1)^3}{4-1} = -\frac{1}{3}$, $a_3 = \frac{(-1)^4}{6-1} = \frac{1}{5}$, $a_4 = \frac{(-1)^5}{8-1} = -\frac{1}{7}$

5. $a_1 = \frac{2}{2^2} = \frac{1}{2}$, $a_2 = \frac{2^2}{2^3} = \frac{1}{2}$, $a_3 = \frac{2^3}{2^4} = \frac{1}{2}$, $a_4 = \frac{2^4}{2^5} = \frac{1}{2}$

7. $a_1 = 1$, $a_2 = 1 + \frac{1}{2} = \frac{3}{2}$, $a_3 = \frac{3}{2} + \frac{1}{2^2} = \frac{7}{4}$, $a_4 = \frac{7}{4} + \frac{1}{2^3} = \frac{15}{8}$, $a_5 = \frac{15}{8} + \frac{1}{2^4} = \frac{31}{16}$, $a_6 = \frac{63}{32}$,

$a_7 = \frac{127}{64}$, $a_8 = \frac{255}{128}$, $a_9 = \frac{511}{256}$, $a_{10} = \frac{1023}{512}$

9. $a_1 = 2$, $a_2 = \frac{(-1)^2(2)}{2} = 1$, $a_3 = \frac{(-1)^3(1)}{2} = -\frac{1}{2}$, $a_4 = \frac{(-1)^4\left(-\frac{1}{2}\right)}{2} = -\frac{1}{4}$, $a_5 = \frac{(-1)^5\left(-\frac{1}{4}\right)}{2} = \frac{1}{8}$,

$a_6 = \frac{1}{16}$, $a_7 = -\frac{1}{32}$, $a_8 = -\frac{1}{64}$, $a_9 = \frac{1}{128}$, $a_{10} = \frac{1}{256}$

11. $a_1 = 1$, $a_2 = 1$, $a_3 = 1 + 1 = 2$, $a_4 = 2 + 1 = 3$, $a_5 = 3 + 2 = 5$, $a_6 = 8$, $a_7 = 13$, $a_8 = 21$, $a_9 = 34$, $a_{10} = 55$

13. $a_n = (-1)^{n+1}$, $n = 1, 2, \ldots$

15. $a_n = (-1)^{n+1}n^2$, $n = 1, 2, \ldots$

17. $a_n = n^2 - 1$, $n = 1, 2, \ldots$

19. $a_n = 4n - 3$, $n = 1, 2, \ldots$

21. $a_n = \frac{1 + (-1)^{n+1}}{2}$, $n = 1, 2, \ldots$

23. $\left|\sqrt[n]{0.5} - 1\right| < 10^{-3} \Rightarrow -\frac{1}{1000} < \left(\frac{1}{2}\right)^{1/n} - 1 < \frac{1}{1000} \Rightarrow \left(\frac{999}{1000}\right)^n < \frac{1}{2} < \left(\frac{1001}{1000}\right)^n \Rightarrow n > \frac{\ln\left(\frac{1}{2}\right)}{\ln\left(\frac{999}{1000}\right)} \Rightarrow n > 692.8$

$\Rightarrow N = 692$; $a_n = \left(\frac{1}{2}\right)^{1/n}$ and $\lim\limits_{n\to\infty} a_n = 1$

25. $(0.9)^n < 10^{-3} \Rightarrow n \ln(0.9) < -3 \ln 10 \Rightarrow n > \frac{-3 \ln 10}{\ln(0.9)} \approx 65.54 \Rightarrow N = 65$; $a_n = \left(\frac{9}{10}\right)^n$ and $\lim\limits_{n\to\infty} a_n = 0$

27. (a) $f(x) = x^2 - a \Rightarrow f'(x) = 2x \Rightarrow x_{n+1} = x_n - \frac{x_n^2 - a}{2x_n} \Rightarrow x_{n+1} = \frac{2x_n^2 - \left(x_n^2 - a\right)}{2x_n} = \frac{x_n^2 + a}{2x_n} = \frac{\left(x_n + \frac{a}{x_n}\right)}{2}$

(b) $x_1 = 2$, $x_2 = 1.75$, $x_3 = 1.732142857$, $x_4 = 1.73205081$, $x_5 = 1.732050808$; we are finding the positive number where $x^2 - 3 = 0$; that is, where $x^2 = 3$, $x > 0$, or where $x = \sqrt{3}$.

29. $x_1 = 1$, $x_2 = 1 + \cos(1) = 1.540302306$, $x_3 = 1.540302306 + \cos(1 + \cos(1)) = 1.570791601$, $x_4 = 1.570791601 + \cos(1.570791601) = 1.570796327 = \frac{\pi}{2}$ to 9 decimal places. After a few steps, the $\text{arc}(x_{n-1})$ and line segment $\cos(x_{n-1})$ are nearly the same as the quarter circle.

31. $a_{n+1} \ge a_n \Rightarrow \frac{3(n+1)+1}{(n+1)+1} > \frac{3n+1}{n+1} \Rightarrow \frac{3n+4}{n+2} > \frac{3n+1}{n+1} \Rightarrow 3n^2 + 3n + 4n + 4 > 3n^2 + 6n + n + 2$ $\Rightarrow 4 > 2$; the steps are reversible so the sequence is nondecreasing; $\frac{3n+1}{n+1} < 3 \Rightarrow 3n + 1 < 3n + 3$ $\Rightarrow 1 < 3$; the steps are reversible so the sequence is bounded above by 3

33. $a_{n+1} \le a_n \Rightarrow \frac{2^{n+1}3^{n+1}}{(n+1)!} \le \frac{2^n 3^n}{n!} \Rightarrow \frac{2^{n+1}3^{n+1}}{2^n 3^n} \le \frac{(n+1)!}{n!} \Rightarrow 2 \cdot 3 \le n + 1$ which is true for $n \ge 5$; the steps are reversible so the sequence is decreasing after a_5, but it is not nondecreasing for all its terms; $a_1 = 6$, $a_2 = 18$, $a_3 = 36$, $a_4 = 54$, $a_5 = \frac{324}{5} = 64.8 \Rightarrow$ the sequence is bounded from above by 64.8

35. $a_n = 1 - \frac{1}{n}$ converges because $\frac{1}{n} \to 0$ by Example 2; also it is a nondecreasing sequence bounded above by 1

37. $a_n = \frac{2^n - 1}{2^n} = 1 - \frac{1}{2^n}$ and $0 < \frac{1}{2^n} < \frac{1}{n}$; since $\frac{1}{n} \to 0$ (by Example 2) $\Rightarrow \frac{1}{2^n} \to 0$, the sequence converges; also it is a nondecreasing sequence bounded above by 1

39. $a_n = ((-1)^n + 1)\left(\frac{n+1}{n}\right)$ diverges because $a_n = 0$ for n odd, while for n even $a_n = 2\left(1 + \frac{1}{n}\right)$ converges to 2; it diverges by definition of divergence

41. If $\{a_n\}$ is nonincreasing with lower bound M, then $\{-a_n\}$ is a nondecreasing sequence with upper bound $-M$. By Theorem 1, $\{-a_n\}$ converges and hence $\{a_n\}$ converges. If $\{a_n\}$ has no lower bound, then $\{-a_n\}$ has no upper bound and therefore diverges. Hence, $\{a_n\}$ also diverges.

43. $a_n \ge a_{n+1} \Leftrightarrow \frac{1+\sqrt{2n}}{\sqrt{n}} \ge \frac{1+\sqrt{2(n+1)}}{\sqrt{n+1}} \Leftrightarrow \sqrt{n+1} + \sqrt{2n^2+2n} \ge \sqrt{n} + \sqrt{2n^2+2n} \Leftrightarrow \sqrt{n+1} \ge \sqrt{n}$

and $\frac{1+\sqrt{2n}}{\sqrt{n}} \ge \sqrt{2}$; thus the sequence is nonincreasing and bounded below by $\sqrt{2} \Rightarrow$ it converges

45. $\frac{4^{n+1} + 3^n}{4^n} = 4 + \left(\frac{3}{4}\right)^n$ so $a_n \ge a_{n+1} \Leftrightarrow 4 + \left(\frac{3}{4}\right)^n \ge 4 + \left(\frac{3}{4}\right)^{n+1} \Leftrightarrow \left(\frac{3}{4}\right)^n \ge \left(\frac{3}{4}\right)^{n+1} \Leftrightarrow 1 \ge \frac{3}{4}$ and $4 + \left(\frac{3}{4}\right)^n \ge 4$; thus the sequence is nonincreasing and bounded below by $4 \Rightarrow$ it converges

47. Let $0 < M < 1$ and let N be an integer greater than $\frac{M}{1-M}$. Then $n > N \Rightarrow n > \frac{M}{1-M} \Rightarrow n - nM > M$

$\Rightarrow n > M + nM \Rightarrow n > M(n+1) \Rightarrow \frac{n}{n+1} > M$.

49. The sequence $a_n = 1 + \frac{(-1)^n}{2}$ is the sequence $\frac{1}{2}, \frac{3}{2}, \frac{1}{2}, \frac{3}{2}, \ldots$. This sequence is bounded above by $\frac{3}{2}$, but it clearly does not converge, by definition of convergence.

51. Given an $\epsilon > 0$, by definition of convergence there corresponds an N such that for all $n > N$, $|L_1 - a_n| < \epsilon$ and $|L_2 - a_n| < \epsilon$. Now $|L_2 - L_1| = |L_2 - a_n + a_n - L_1| \le |L_2 - a_n| + |a_n - L_1| < \epsilon + \epsilon = 2\epsilon$. $|L_2 - L_1| < 2\epsilon$ says that the difference between two fixed values is smaller than any positive number 2ϵ. The only nonnegative number smaller than every positive number is 0, so $|L_1 - L_2| = 0$ or $L_1 = L_2$.

53. $a_{2k} \to L \Leftrightarrow$ given an $\epsilon > 0$ there corresponds an N_1 such that $\left[2k > N_1 \Rightarrow |a_{2k} - L| < \epsilon\right]$. Similarly, $a_{2k+1} \to L \Leftrightarrow \left[2k+1 > N_2 \Rightarrow |a_{2k+1} - L| < \epsilon\right]$. Let $N = \max\{N_1, N_2\}$. Then $n > N \Rightarrow |a_n - L| < \epsilon$ whether n is even or odd, and hence $a_n \to L$.

8.2 THEOREMS FOR CALCULATING LIMITS OF SEQUENCES

1. $\lim\limits_{n\to\infty} 2 + (0.1)^n = 2 \Rightarrow$ converges (Table 8.1, #4)

3. $\lim\limits_{n\to\infty} \frac{1-2n}{1+2n} = \lim\limits_{n\to\infty} \frac{\left(\frac{1}{n}\right) - 2}{\left(\frac{1}{n}\right) + 2} = \lim\limits_{n\to\infty} \frac{-2}{2} = -1 \Rightarrow$ converges

5. $\lim\limits_{n\to\infty} \frac{1-5n^4}{n^4+8n^3} = \lim\limits_{n\to\infty} \frac{\left(\frac{1}{n^4}\right) - 5}{1 + \left(\frac{8}{n}\right)} = -5 \Rightarrow$ converges

7. $\lim\limits_{n\to\infty} \frac{n^2-2n+1}{n-1} = \lim\limits_{n\to\infty} \frac{(n-1)(n-1)}{n-1} = \lim\limits_{n\to\infty} (n-1) = \infty \Rightarrow$ diverges

9. $\lim\limits_{n\to\infty} (1 + (-1)^n)$ does not exist $\Rightarrow$ diverges

11. $\lim\limits_{n\to\infty} \left(\frac{n+1}{2n}\right)\left(1 - \frac{1}{n}\right) = \lim\limits_{n\to\infty} \left(\frac{1}{2} + \frac{1}{2n}\right)\left(1 - \frac{1}{n}\right) = \frac{1}{2} \Rightarrow$ converges

13. $\lim\limits_{n\to\infty} \frac{(-1)^{n+1}}{2n-1} = 0 \Rightarrow$ converges

15. $\lim\limits_{n\to\infty} \sqrt{\frac{2n}{n+1}} = \sqrt{\lim\limits_{n\to\infty} \frac{2n}{n+1}} = \sqrt{\lim\limits_{n\to\infty} \left(\frac{2}{1+\frac{1}{n}}\right)} = \sqrt{2} \Rightarrow$ converges

17. $\lim\limits_{n\to\infty} \sin\left(\frac{\pi}{2} + \frac{1}{n}\right) = \sin\left(\lim\limits_{n\to\infty} \left(\frac{\pi}{2} + \frac{1}{n}\right)\right) = \sin \frac{\pi}{2} = 1 \Rightarrow$ converges

19. $\lim\limits_{n\to\infty} \frac{\sin n}{n} = 0$ because $-\frac{1}{n} \le \frac{\sin n}{n} \le \frac{1}{n} \Rightarrow$ converges by the Sandwich Theorem for sequences

21. $\lim\limits_{n\to\infty} \frac{n}{2^n} = \lim\limits_{n\to\infty} \frac{1}{2^n \ln 2} = 0 \Rightarrow$ converges (using l'Hôpital's rule)

23. $\lim\limits_{n\to\infty} \frac{\ln(n+1)}{\sqrt{n}} = \lim\limits_{n\to\infty} \frac{\left(\frac{1}{n+1}\right)}{\left(\frac{1}{2\sqrt{n}}\right)} = \lim\limits_{n\to\infty} \frac{2\sqrt{n}}{n+1} = \lim\limits_{n\to\infty} \frac{\left(\frac{2}{\sqrt{n}}\right)}{1+\left(\frac{1}{n}\right)} = 0 \Rightarrow$ converges

25. $\lim\limits_{n\to\infty} 8^{1/n} = 1 \Rightarrow$ converges (Table 8.1, #3)

27. $\lim\limits_{n\to\infty} \left(1+\frac{7}{n}\right)^n = e^7 \Rightarrow$ converges (Table 8.1, #5)

29. $\lim\limits_{n\to\infty} \sqrt[n]{10n} = \lim\limits_{n\to\infty} 10^{1/n} \cdot n^{1/n} = 1 \cdot 1 = 1 \Rightarrow$ converges (Table 8.1, #3 and #2)

31. $\lim\limits_{n\to\infty} \left(\frac{3}{n}\right)^{1/n} = \frac{\lim\limits_{n\to\infty} 3^{1/n}}{\lim\limits_{n\to\infty} n^{1/n}} = \frac{1}{1} = 1 \Rightarrow$ converges (Table 8.1, #3 and #2)

33. $\lim\limits_{n\to\infty} \frac{\ln n}{n^{1/n}} = \frac{\lim\limits_{n\to\infty} \ln n}{\lim\limits_{n\to\infty} n^{1/n}} = \frac{\infty}{1} = \infty \Rightarrow$ diverges (Table 8.1, #2)

35. $\lim\limits_{n\to\infty} \sqrt[n]{4^n n} = \lim\limits_{n\to\infty} 4\sqrt[n]{n} = 4 \cdot 1 = 4 \Rightarrow$ converges (Table 8.1, #2)

37. $\lim\limits_{n\to\infty} \frac{n!}{n^n} = \lim\limits_{n\to\infty} \frac{1 \cdot 2 \cdot 3 \cdots (n-1)(n)}{n \cdot n \cdot n \cdots n \cdot n} \le \lim\limits_{n\to\infty} \left(\frac{1}{n}\right) = 0$ and $\frac{n!}{n^n} \ge 0 \Rightarrow \lim\limits_{n\to\infty} \frac{n!}{n^n} = 0 \Rightarrow$ converges

39. $\lim\limits_{n\to\infty} \frac{n!}{10^{6n}} = \lim\limits_{n\to\infty} \frac{1}{\left(\frac{(10^6)^n}{n!}\right)} = \infty \Rightarrow$ diverges (Table 8.1, #6)

41. $\lim\limits_{n\to\infty} \left(\frac{1}{n}\right)^{1/(\ln n)} = \lim\limits_{n\to\infty} \exp\left(\frac{1}{\ln n} \ln\left(\frac{1}{n}\right)\right) = \lim\limits_{n\to\infty} \exp\left(\frac{\ln 1 - \ln n}{\ln n}\right) = e^{-1} \Rightarrow$ converges

43. $\lim\limits_{n\to\infty} \left(\frac{3n+1}{3n-1}\right)^n = \lim\limits_{n\to\infty} \exp\left(n \ln\left(\frac{3n+1}{3n-1}\right)\right) = \lim\limits_{n\to\infty} \exp\left(\frac{\ln(3n+1) - \ln(3n-1)}{\frac{1}{n}}\right)$

$= \lim\limits_{n\to\infty} \exp\left(\frac{\frac{3}{3n+1} - \frac{3}{3n-1}}{\left(-\frac{1}{n^2}\right)}\right) = \lim\limits_{n\to\infty} \exp\left(\frac{6n^2}{(3n+1)(3n-1)}\right) = \exp\left(\frac{6}{9}\right) = e^{2/3} \Rightarrow$ converges

45. $\lim\limits_{n\to\infty} \left(\frac{x^n}{2n+1}\right)^{1/n} = \lim\limits_{n\to\infty} x\left(\frac{1}{2n+1}\right)^{1/n} = x \lim\limits_{n\to\infty} \exp\left(\frac{1}{n} \ln\left(\frac{1}{2n+1}\right)\right) = x \lim\limits_{n\to\infty} \exp\left(\frac{-\ln(2n+1)}{n}\right)$

$= x \lim\limits_{n\to\infty} \exp\left(\frac{-2}{2n+1}\right) = xe^0 = x,\ x > 0 \Rightarrow$ converges

47. $\lim\limits_{n\to\infty} \frac{3^n \cdot 6^n}{2^{-n} \cdot n!} = \lim\limits_{n\to\infty} \frac{36^n}{n!} = 0 \Rightarrow$ converges (Table 8.1, #6)

49. $\lim\limits_{n\to\infty} \tanh n = \lim\limits_{n\to\infty} \frac{e^n - e^{-n}}{e^n + e^{-n}} = \lim\limits_{n\to\infty} \frac{e^{2n} - 1}{e^{2n} + 1} = \lim\limits_{n\to\infty} \frac{2e^{2n}}{2e^{2n}} = \lim\limits_{n\to\infty} 1 = 1 \Rightarrow$ converges

51. $\lim\limits_{n\to\infty} \frac{n^2 \sin\left(\frac{1}{n}\right)}{2n-1} = \lim\limits_{n\to\infty} \frac{\sin\left(\frac{1}{n}\right)}{\left(\frac{2}{n} - \frac{1}{n^2}\right)} = \lim\limits_{n\to\infty} \frac{-\left(\cos\left(\frac{1}{n}\right)\right)\left(\frac{1}{n^2}\right)}{\left(-\frac{2}{n^2} + \frac{2}{n^3}\right)} = \lim\limits_{n\to\infty} \frac{-\cos\left(\frac{1}{n}\right)}{-2 + \left(\frac{2}{n}\right)} = \frac{1}{2} \Rightarrow$ converges

53. $\lim\limits_{n\to\infty} \tan^{-1} n = \frac{\pi}{2} \Rightarrow$ converges

55. $\lim\limits_{n\to\infty} \left(\frac{1}{3}\right)^n + \frac{1}{\sqrt{2^n}} = \lim\limits_{n\to\infty} \left(\left(\frac{1}{3}\right)^n + \left(\frac{1}{\sqrt{2}}\right)^n\right) = 0 \Rightarrow$ converges (Table 8.1, #4)

57. $\lim\limits_{n\to\infty} \frac{(\ln n)^{200}}{n} = \lim\limits_{n\to\infty} \frac{200\,(\ln n)^{199}}{n} = \lim\limits_{n\to\infty} \frac{200 \cdot 199\,(\ln n)^{198}}{n} = \ldots = \lim\limits_{n\to\infty} \frac{200!}{n} = 0 \Rightarrow$ converges

59. $\lim\limits_{n\to\infty} \left(n - \sqrt{n^2 - n}\right) = \lim\limits_{n\to\infty} \left(n - \sqrt{n^2 - n}\right)\left(\frac{n + \sqrt{n^2 - n}}{n + \sqrt{n^2 - n}}\right) = \lim\limits_{n\to\infty} \frac{n}{n + \sqrt{n^2 - n}} = \lim\limits_{n\to\infty} \frac{1}{1 + \sqrt{1 - \frac{1}{n}}}$

$= \frac{1}{2} \Rightarrow$ converges

61. $\lim\limits_{n\to\infty} \frac{1}{n} \int_1^n \frac{1}{x}\,dx = \lim\limits_{n\to\infty} \frac{\ln n}{n} = \lim\limits_{n\to\infty} \frac{1}{n} = 0 \Rightarrow$ converges (Table 8.1, #1)

63. $1, 1, 2, 4, 8, 16, 32, \ldots = 1, 2^0, 2^1, 2^2, 2^3, 2^4, 2^5, \ldots \Rightarrow x_1 = 1$ and $x_n = 2^{n-2}$ for $n \geq 2$

65. (a) $f(x) = x^2 - 2$; the sequence converges to $1.414213562 \approx \sqrt{2}$

(b) $f(x) = \tan(x) - 1$; the sequence converges to $0.7853981635 \approx \frac{\pi}{4}$

(c) $f(x) = e^x$; the sequence $1, 0, -1, -2, -3, -4, -5, \ldots$ diverges

67. (a) If $a = 2n + 1$, then $b = \left\lfloor \frac{a^2}{2} \right\rfloor = \left\lfloor \frac{4n^2 + 4n + 1}{2} \right\rfloor = \left\lfloor 2n^2 + 2n + \frac{1}{2} \right\rfloor = 2n^2 + 2n$, $c = \left\lceil \frac{a^2}{2} \right\rceil = \left\lceil 2n^2 + 2n + \frac{1}{2} \right\rceil$

$= 2n^2 + 2n + 1$ and $a^2 + b^2 = (2n+1)^2 + \left(2n^2 + 2n\right)^2 = 4n^2 + 4n + 1 + 4n^4 + 8n^3 + 4n^2$

$= 4n^4 + 8n^3 + 8n^2 + 4n + 1 = \left(2n^2 + 2n + 1\right)^2 = c^2.$

(b) $\lim\limits_{a\to\infty} \frac{\left\lfloor \frac{a^2}{2} \right\rfloor}{\left\lceil \frac{a^2}{2} \right\rceil} = \lim\limits_{a\to\infty} \frac{2n^2 + 2n}{2n^2 + 2n + 1} = 1$ or $\lim\limits_{a\to\infty} \frac{\left\lfloor \frac{a^2}{2} \right\rfloor}{\left\lceil \frac{a^2}{2} \right\rceil} = \lim\limits_{a\to\infty} \sin\theta = \lim\limits_{\theta\to\pi/2} \sin\theta = 1$

69. (a) $\lim\limits_{n\to\infty} \frac{\ln n}{n^c} = \lim\limits_{n\to\infty} \frac{\left(\frac{1}{n}\right)}{cn^{c-1}} = \lim\limits_{n\to\infty} \frac{1}{cn^c} = 0$

(b) For all $\epsilon > 0$, there exists an $N = e^{-(\ln \epsilon)/c}$ such that $n > e^{-(\ln \epsilon)/c} \Rightarrow \ln n > -\frac{\ln \epsilon}{c} \Rightarrow \ln n^c > \ln\left(\frac{1}{\epsilon}\right)$

$\Rightarrow n^c > \frac{1}{\epsilon} \Rightarrow \frac{1}{n^c} < \epsilon \Rightarrow \left|\frac{1}{n^c} - 0\right| < \epsilon \Rightarrow \lim\limits_{n\to\infty} \frac{1}{n^c} = 0$

71. $\lim_{n\to\infty} n^{1/n} = \lim_{n\to\infty} \exp\left(\frac{1}{n}\ln n\right) = \lim_{n\to\infty} \exp\left(\frac{1}{n}\right) = e^0 = 1$

73. Assume the hypotheses of the theorem and let ϵ be a positive number. For all ϵ there exists a N_1 such that when $n > N_1$ then $|a_n - L| < \epsilon \Rightarrow -\epsilon < a_n - L < \epsilon \Rightarrow L - \epsilon < a_n$, and there exists a N_2 such that when $n > N_2$ then $|c_n - L| < \epsilon \Rightarrow -\epsilon < c_n - L < \epsilon \Rightarrow c_n < L + \epsilon$. If $n > \max\{N_1, N_2\}$, then $L - \epsilon < a_n \le b_n \le c_n < L + \epsilon \Rightarrow |b_n - L| < \epsilon \Rightarrow \lim_{n\to\infty} b_n = L$.

75. $g(x) = \sqrt{x}$; $2 \to 1.00000132$ in 20 iterations; $.1 \to 0.9999956$ in 20 iterations; a root is 1

77. $g(x) = -\cos x$; $x_0 = .1 \to 0.73908456$ in 35 iterations

79. $g(x) = 0.1 + \sin x$; $x_0 = -2 \to 0.853748068$ in 43 iterations

81. $x_0 = \text{initial guess} > 0 \Rightarrow x_1 = \sqrt{x_0} = (x_0)^{1/2} \Rightarrow x_2 = \sqrt{x_0^{\,1/2}} = x_0^{\,1/4}, \ldots \Rightarrow x_n = x_0^{\,1/(2n)} \Rightarrow x_n \to 1$ as $n \to \infty$

83. $g(x) = 2x + 3 \Rightarrow g^{-1}(x) = \frac{x-3}{2}$ and when the iterative method is applied to $g^{-1}(x)$ we have $x_0 = 2$ $\to -2.99999881$ in 23 iterations $\Rightarrow -3$ is the fixed point

8.3 INFINITE SERIES

1. $s_n = \frac{a(1-r^n)}{(1-r)} = \frac{2\left(1-\left(\frac{1}{3}\right)^n\right)}{1-\left(\frac{1}{3}\right)} \Rightarrow \lim_{n\to\infty} s_n = \frac{2}{1-\left(\frac{1}{3}\right)} = 3$

3. $s_n = \frac{a(1-r^n)}{(1-r)} = \frac{1-\left(-\frac{1}{2}\right)^n}{1-\left(-\frac{1}{2}\right)} \Rightarrow \lim_{n\to\infty} s_n = \frac{1}{\left(\frac{3}{2}\right)} = \frac{2}{3}$

5. $\frac{1}{(n+1)(n+2)} = \frac{1}{n+1} - \frac{1}{n+2} \Rightarrow s_n = \left(\frac{1}{2} - \frac{1}{3}\right) + \left(\frac{1}{3} - \frac{1}{4}\right) + \ldots + \left(\frac{1}{n+1} - \frac{1}{n+2}\right) = \frac{1}{2} - \frac{1}{n+2} \Rightarrow \lim_{n\to\infty} s_n = \frac{1}{2}$

7. $1 - \frac{1}{4} + \frac{1}{16} - \frac{1}{64} + \ldots$, the sum of this geometric series is $\frac{1}{1-\left(-\frac{1}{4}\right)} = \frac{1}{1+\left(\frac{1}{4}\right)} = \frac{4}{5}$

9. $\frac{7}{4} + \frac{7}{16} + \frac{7}{64} + \ldots$, the sum of this geometric series is $\frac{\left(\frac{7}{4}\right)}{1-\left(\frac{1}{4}\right)} = \frac{7}{3}$

11. $(5+1) + \left(\frac{5}{2} + \frac{1}{3}\right) + \left(\frac{5}{4} + \frac{1}{9}\right) + \left(\frac{5}{8} + \frac{1}{27}\right) + \ldots$, is the sum of two geometric series; the sum is $\frac{5}{1-\left(\frac{1}{2}\right)} + \frac{1}{1-\left(\frac{1}{3}\right)} = 10 + \frac{3}{2} = \frac{23}{2}$

13. $(1+1)+\left(\frac{1}{2}-\frac{1}{5}\right)+\left(\frac{1}{4}+\frac{1}{25}\right)+\left(\frac{1}{8}-\frac{1}{125}\right)+\ldots$, is the sum of two geometric series; the sum is

$$\frac{1}{1-\left(\frac{1}{2}\right)}+\frac{1}{1+\left(\frac{1}{5}\right)}=2+\frac{5}{6}=\frac{17}{6}$$

15. $\frac{4}{(4n-3)(4n+1)}=\frac{1}{4n-3}-\frac{1}{4n+1} \Rightarrow s_n=\left(1-\frac{1}{5}\right)+\left(\frac{1}{5}-\frac{1}{9}\right)+\left(\frac{1}{9}-\frac{1}{13}\right)+\ldots+\left(\frac{1}{4n-7}-\frac{1}{4n-3}\right)$

$$+\left(\frac{1}{4n-3}-\frac{1}{4n+1}\right)=1-\frac{1}{4n+1} \Rightarrow \lim_{n\to\infty} s_n=\lim_{n\to\infty}\left(1-\frac{1}{4n+1}\right)=1$$

17. $\frac{40n}{(2n-1)^2(2n+1)^2}=\frac{A}{(2n-1)}+\frac{B}{(2n-1)^2}+\frac{C}{(2n+1)}+\frac{D}{(2n+1)^2}$

$$=\frac{A(2n-1)(2n+1)^2+B(2n+1)^2+C(2n+1)(2n-1)^2+D(2n-1)^2}{(2n-1)^2(2n+1)^2}$$

$$\Rightarrow A(2n-1)(2n+1)^2+B(2n+1)^2+C(2n+1)(2n-1)^2+D(2n-1)^2=40n$$

$$\Rightarrow A\left(8n^3+4n^2-2n-1\right)+B\left(4n^2+4n+1\right)+C\left(8n^3-4n^2-2n+1\right)=D\left(4n^2-4n+1\right)=40n$$

$$\Rightarrow (8A+8C)n^3+(4A+4B-4C+4D)n^2+(-2A+4B-2C-4D)n+(-A+B+C+D)=40n$$

$$\Rightarrow \begin{cases} 8A+8C=0 \\ 4A+4B-4C+4D=0 \\ -2A+4B-2C-4D=40 \\ -A+B+C+D=0 \end{cases} \Rightarrow \begin{cases} 8A+8C=0 \\ A+B-C+D=0 \\ -A+2B-C-2D=20 \\ -A+B+C+D=0 \end{cases} \Rightarrow \begin{cases} B+D=0 \\ 2B-2D=20 \end{cases} \Rightarrow 4B=20 \Rightarrow B=5 \text{ and}$$

$$D=-5 \Rightarrow \begin{cases} A+C=0 \\ -A+5+C-5=0 \end{cases} \Rightarrow C=0 \text{ and } A=0. \text{ Hence, } \sum_{n=1}^{k}\left[\frac{40n}{(2n-1)^2(2n+1)^2}\right]$$

$$=5\sum_{n=1}^{k}\left[\frac{1}{(2n-1)^2}-\frac{1}{(2n+1)^2}\right]=5\left(\frac{1}{1}-\frac{1}{9}+\frac{1}{9}-\frac{1}{25}+\frac{1}{25}-\ldots-\frac{1}{(2(k-1)+1)^2}+\frac{1}{(2k-1)^2}-\frac{1}{(2k+1)^2}\right)$$

$$=5\left(1-\frac{1}{(2k+1)^2}\right) \Rightarrow \text{the sum is } \lim_{n\to\infty} 5\left(1-\frac{1}{(2k+1)^2}\right)=5$$

19. $s_n=\left(1-\frac{1}{\sqrt{2}}\right)+\left(\frac{1}{\sqrt{2}}-\frac{1}{\sqrt{3}}\right)+\left(\frac{1}{\sqrt{3}}-\frac{1}{\sqrt{4}}\right)+\ldots+\left(\frac{1}{\sqrt{n-1}}+\frac{1}{\sqrt{n}}\right)+\left(\frac{1}{\sqrt{n}}-\frac{1}{\sqrt{n+1}}\right)=1-\frac{1}{\sqrt{n+1}}$

$$\Rightarrow \lim_{n\to\infty} s_n=\lim_{n\to\infty}\left(1-\frac{1}{\sqrt{n+1}}\right)=1$$

21. $s_n=\left(\frac{1}{\ln 3}-\frac{1}{\ln 2}\right)+\left(\frac{1}{\ln 4}-\frac{1}{\ln 3}\right)+\left(\frac{1}{\ln 5}-\frac{1}{\ln 4}\right)+\ldots+\left(\frac{1}{\ln(n+1)}-\frac{1}{\ln n}\right)+\left(\frac{1}{\ln(n+2)}-\frac{1}{\ln(n+1)}\right)$

$$=-\frac{1}{\ln 2}+\frac{1}{\ln(n+2)} \Rightarrow \lim_{n\to\infty} s_n=-\frac{1}{\ln 2}$$

23. convergent geometric series with sum $\frac{1}{1-\left(\frac{1}{\sqrt{2}}\right)}=\frac{\sqrt{2}}{\sqrt{2}-1}=2+\sqrt{2}$

25. convergent geometric series with sum $\frac{\left(\frac{3}{2}\right)}{1-\left(-\frac{1}{2}\right)} = 1$

27. $\lim_{n\to\infty} \cos(n\pi) = \lim_{n\to\infty} (-1)^n \neq 0 \Rightarrow$ diverges

29. convergent geometric series with sum $\frac{1}{1-\left(\frac{1}{e^2}\right)} = \frac{e^2}{e^2-1}$

31. convergent geometric series with sum $\frac{2}{1-\left(\frac{1}{10}\right)} - 2 = \frac{20}{9} - \frac{18}{9} = \frac{2}{9}$

33. difference of two geometric series with sum $\frac{1}{1-\left(\frac{2}{3}\right)} - \frac{1}{1-\left(\frac{1}{3}\right)} = 3 - \frac{3}{2} = \frac{3}{2}$

35. $\lim_{n\to\infty} \frac{n!}{1000^n} = \infty \neq 0 \Rightarrow$ diverges

37. $\sum_{n=1}^{\infty} \ln\left(\frac{n}{n+1}\right) = \sum_{n=1}^{\infty} \left[\ln(n) - \ln(n+1)\right] \Rightarrow s_n = \left[\ln(1) - \ln(2)\right] + \left[\ln(2) - \ln(3)\right] + \left[\ln(3) - \ln(4)\right] + \ldots$
$+ \left[\ln(n-1) - \ln(n)\right] + \left[\ln(n) - \ln(n+1)\right] = \ln(1) - \ln(n+1) = -\ln(n+1) \Rightarrow \lim_{n\to\infty} s_n = -\infty, \Rightarrow$ diverges

39. convergent geometric series with sum $\frac{1}{1-\left(\frac{e}{\pi}\right)} = \frac{\pi}{\pi - e}$

41. $\sum_{n=0}^{\infty} (-1)^n x^n = \sum_{n=0}^{\infty} (-x)^n$; $a = 1$, $r = -x$; converges to $\frac{1}{1-(-x)} = \frac{1}{1+x}$ for $|x| < 1$

43. $a = 3$, $r = \frac{x-1}{2}$; converges to $\frac{3}{1-\left(\frac{x-1}{2}\right)} = \frac{6}{3-x}$ for $-1 < \frac{x-1}{2} < 1$ or $-1 < x < 3$

45. $a = 1$, $r = 2x$; converges to $\frac{1}{1-2x}$ for $|2x| < 1$ or $|x| < \frac{1}{2}$

47. $a = 1$, $r = -(x+1)^n$; converges to $\frac{1}{1+(x+1)} = \frac{1}{2+x}$ for $|x+1| < 1$ or $-2 < x < 0$

49. $a = 1$, $r = \sin x$; converges to $\frac{1}{1-\sin x}$ for $x \neq (2k+1)\frac{\pi}{2}$, k an integer

51. $0.\overline{23} = \sum_{n=0}^{\infty} \frac{23}{100}\left(\frac{1}{10^2}\right)^n = \frac{\left(\frac{23}{100}\right)}{1-\left(\frac{1}{100}\right)} = \frac{23}{99}$

53. $0.\overline{7} = \sum_{n=0}^{\infty} \frac{7}{10}\left(\frac{1}{10}\right)^n = \frac{\left(\frac{7}{10}\right)}{1-\left(\frac{1}{10}\right)} = \frac{7}{9}$

55. $0.0\overline{6} = \sum_{n=0}^{\infty} \left(\frac{1}{10}\right)\left(\frac{6}{10}\right)\left(\frac{1}{10}\right)^n = \frac{\left(\frac{6}{100}\right)}{1-\left(\frac{1}{10}\right)} = \frac{6}{90} = \frac{1}{15}$

57. $1.24\overline{123} = \frac{124}{100} + \sum_{n=0}^{\infty} \frac{123}{10^5}\left(\frac{1}{10^3}\right)^n = \frac{124}{100} + \frac{\left(\frac{123}{10^5}\right)}{1-\left(\frac{1}{10^3}\right)} = \frac{124}{100} + \frac{123}{10^5 - 10^2} = \frac{124}{100} + \frac{123}{99{,}900} = \frac{123{,}753}{99{,}900} = \frac{41{,}251}{33{,}300}$

59. (a) $\sum_{n=-2}^{\infty} \frac{1}{(n+4)(n+5)}$ (b) $\sum_{n=0}^{\infty} \frac{1}{(n+2)(n+3)}$ (c) $\sum_{n=5}^{\infty} \frac{1}{(n-3)(n-2)}$

61. (a) one example is $\frac{1}{2} + \frac{1}{4} + \frac{1}{8} + \frac{1}{16} + \ldots = \frac{\left(\frac{1}{2}\right)}{1-\left(\frac{1}{2}\right)} = 1$

(b) one example is $-\frac{3}{2} - \frac{3}{4} - \frac{3}{8} - \frac{3}{16} - \ldots = \frac{\left(-\frac{3}{2}\right)}{1-\left(\frac{1}{2}\right)} = -3$

(c) one example is $1 - \frac{1}{2} - \frac{1}{4} - \frac{1}{8} - \frac{1}{16} - \ldots$; the series $\frac{k}{2} + \frac{k}{4} + \frac{k}{8} + \ldots = \frac{\left(\frac{k}{2}\right)}{1-\left(\frac{1}{2}\right)} = k$ where k is any positive or negative number.

63. Let $a_n = b_n = \left(\frac{1}{2}\right)^n$. Then $\sum_{n=1}^{\infty} a_n = \sum_{n=1}^{\infty} b_n = \sum_{n=1}^{\infty} \left(\frac{1}{2}\right)^n = 1$, while $\sum_{n=1}^{\infty} \left(\frac{a_n}{b_n}\right) = \sum_{n=1}^{\infty} (1)$ diverges.

65. Let $a_n = \left(\frac{1}{4}\right)^n$ and $b_n = \left(\frac{1}{2}\right)^n$. Then $A = \sum_{n=1}^{\infty} a_n = \frac{1}{3}$, $B = \sum_{n=1}^{\infty} b_n = 1$ and $\sum_{n=1}^{\infty} \left(\frac{a_n}{b_n}\right) = \sum_{n=1}^{\infty} \left(\frac{1}{2}\right)^n = 1 \neq \frac{A}{B}$.

67. Since the sum of a finite number of terms is finite, adding or subtracting a finite number of terms from a series that diverges does not change the divergence of the series.

69. (a) $\frac{2}{1-r} = 5 \Rightarrow \frac{2}{5} = 1 - r \Rightarrow r = \frac{3}{5}$; $2 + 2\left(\frac{3}{5}\right) + 2\left(\frac{3}{5}\right)^2 + \ldots$

(b) $\frac{\left(\frac{13}{2}\right)}{1-r} = 5 \Rightarrow \frac{13}{10} = 1 - r \Rightarrow r = -\frac{3}{10}$; $\frac{13}{2} - \frac{13}{2}\left(\frac{3}{10}\right) + \frac{13}{2}\left(\frac{3}{10}\right)^2 - \frac{13}{2}\left(\frac{3}{10}\right)^3 + \ldots$

71. $s_n = 1 + 2r + r^2 + 2r^3 + r^4 + 2r^5 + \ldots + r^{2n} + 2r^{2n+1}$, $n = 0, 1, \ldots$

$\Rightarrow s_n = \left(1 + r^2 + r^4 + \ldots + r^{2n}\right) + \left(2r + 2r^3 + 2r^5 + \ldots + 2r^{2n+1}\right) \Rightarrow \lim_{n\to\infty} s_n = \frac{1}{1-r^2} + \frac{2r}{1-r^2}$

$= \frac{1+2r}{1-r^2}$, if $|r^2| < 1$ or $|r| < 1$

73. distance $= 4 + 2\left[(4)\left(\frac{3}{4}\right) + (4)\left(\frac{3}{4}\right)^2 + \ldots\right] = 4 + 2\left(\frac{3}{1-\left(\frac{3}{4}\right)}\right) = 28$ m

75. area $= 2^2 + \left(\sqrt{2}\right)^2 + (1)^2 + \left(\frac{1}{\sqrt{2}}\right)^2 + \ldots = 4 + 2 + 1 + \frac{1}{2} + \ldots = \frac{4}{1-\frac{1}{2}} = 8$ m^2

77. (a) $L_1 = 3,\ L_2 = 3\left(\frac{4}{3}\right),\ L_3 = 3\left(\frac{4}{3}\right)^2, \ldots,\ L_n = 3\left(\frac{4}{3}\right)^{n-1} \Rightarrow \lim\limits_{n\to\infty} L_n = \lim\limits_{n\to\infty} 3\left(\frac{4}{3}\right)^{n-1} = \infty$

(b) $A_1 = \frac{1}{2}(1)\left(\frac{\sqrt{3}}{2}\right) = \frac{\sqrt{3}}{4},\ A_2 = A_1 + 3\left(\frac{1}{2}\right)\left(\frac{1}{3}\right)\left(\frac{\sqrt{3}}{6}\right) = \frac{\sqrt{3}}{4} + \frac{\sqrt{3}}{12},\ A_3 = A_2 + 12\left(\frac{1}{2}\right)\left(\frac{1}{9}\right)\left(\frac{\sqrt{3}}{18}\right)$

$= \frac{\sqrt{3}}{4} + \frac{\sqrt{3}}{12} + \frac{\sqrt{3}}{27},\ A_4 = A_3 + 48\left(\frac{1}{2}\right)\left(\frac{1}{27}\right)\left(\frac{\sqrt{3}}{54}\right), \ldots,\ A_n = \frac{\sqrt{3}}{4} + \frac{27\sqrt{3}}{64}\left(\frac{4}{9}\right)^2 + \frac{27\sqrt{3}}{64}\left(\frac{4}{9}\right)^3 + \ldots$

$$= \frac{\sqrt{3}}{4} + \sum_{n=2}^{\infty} \frac{27\sqrt{3}}{64}\left(\frac{4}{9}\right)^n = \frac{\sqrt{3}}{4} + \frac{\left(\frac{27\sqrt{3}}{64}\right)\left(\frac{4}{9}\right)^2}{1-\left(\frac{4}{9}\right)} = \frac{\sqrt{3}}{4} + \frac{\left(\frac{27\sqrt{3}}{64}\right)\left(\frac{16}{9}\right)}{9-4} = \frac{\sqrt{3}}{4} + \frac{3\sqrt{3}}{4\cdot 5} = \frac{5\sqrt{3}+3\sqrt{3}}{20} = \frac{2\sqrt{3}}{5}$$

8.4 THE INTEGRAL TEST FOR SERIES OF NONNEGATIVE TERMS

1. converges; a geometric series with $r = \frac{1}{10} < 1$

3. diverges; by the nth-Term Test for Divergence, $\lim\limits_{n\to\infty} \frac{n}{n+1} = 1 \neq 0$

5. diverges; $\sum\limits_{n=1}^{\infty} \frac{3}{\sqrt{n}} = 3 \sum\limits_{n=1}^{\infty} \frac{1}{\sqrt{n}}$, which is a divergent p-series

7. converges; a geometric series with $r = \frac{1}{8} < 1$

9. diverges by the Integral Test: $\int_2^n \frac{\ln x}{x}\,dx = \frac{1}{2}\left(\ln^2 n - \ln 2\right) \Rightarrow \int_2^\infty \frac{\ln x}{x}\,dx \to \infty$

11. converges; a geometric series with $r = \frac{2}{3} < 1$

13. diverges; $\sum\limits_{n=0}^{\infty} \frac{-2}{n+1} = -2 \sum\limits_{n=0}^{\infty} \frac{1}{n+1}$, which diverges by the Integral Test

15. diverges; $\lim\limits_{n\to\infty} a_n = \lim\limits_{n\to\infty} \frac{2^n}{n+1} = \lim\limits_{n\to\infty} \frac{2^n \ln 2}{1} = \infty \neq 0$

17. diverges; $\lim\limits_{n\to\infty} \frac{\sqrt{n}}{\ln n} = \lim\limits_{n\to\infty} \frac{\left(\frac{1}{2\sqrt{n}}\right)}{\left(\frac{1}{n}\right)} = \lim\limits_{n\to\infty} \frac{\sqrt{n}}{2} = \infty \neq 0$

19. diverges; a geometric series with $r = \frac{1}{\ln 2} \approx 1.44 > 1$

21. converges by the Integral Test: $\int_3^\infty \frac{\left(\frac{1}{x}\right)}{(\ln x)\sqrt{(\ln x)^2 - 1}}\,dx;\ \begin{bmatrix} u = \ln x \\ du = \frac{1}{x}\,dx \end{bmatrix} \to \int_{\ln 3}^{\infty} \frac{1}{u\sqrt{u^2-1}}\,du$

$= \lim\limits_{b\to\infty} \left[\sec^{-1} |u|\right]_{\ln 3}^{b} = \lim\limits_{b\to\infty} \left[\sec^{-1} b - \sec^{-1}(\ln 3)\right] = \lim\limits_{b\to\infty} \left[\cos^{-1}\left(\frac{1}{b}\right) - \sec^{-1}(\ln 3)\right]$

$= \cos^{-1}(0) - \sec^{-1}(\ln 3) = \frac{\pi}{2} - \sec^{-1}(\ln 3) \approx 1.1439$

23. diverges by the nth-Term Test for divergence; $\lim_{n\to\infty} n \sin\left(\frac{1}{n}\right) = \lim_{n\to\infty} \frac{\sin\left(\frac{1}{n}\right)}{\left(\frac{1}{n}\right)} = \lim_{x\to 0} \frac{\sin x}{x} = 1 \neq 0$

25. converges by the Integral Test: $\int_1^\infty \frac{e^x}{1+e^{2x}}\,dx; \begin{bmatrix} u = e^x \\ du = e^x\,dx \end{bmatrix} \to \int_e^\infty \frac{1}{1+u^2}\,du = \lim_{n\to\infty} \left[\tan^{-1} u\right]_e^b$

$= \lim_{b\to\infty} \left(\tan^{-1} b - \tan^{-1} e\right) = \frac{\pi}{2} - \tan^{-1} e \approx 0.35$

27. converges by the Integral Test: $\int_1^\infty \frac{8\tan^{-1} x}{1+x^2}\,dx; \begin{bmatrix} u = \tan^{-1} x \\ du = \frac{dx}{1+x^2} \end{bmatrix} \to \int_{\pi/4}^{\pi/2} 8u\,du = \left[4u^2\right]_{\pi/4}^{\pi/2} = 4\left(\frac{\pi^2}{4} - \frac{\pi^2}{16}\right) = \frac{3\pi^2}{4}$

29. converges by the Integral Test: $\int_1^\infty \operatorname{sech} x\,dx = 2 \lim_{b\to\infty} \int_1^b \frac{e^x}{1+(e^x)^2}\,dx = 2 \lim_{b\to\infty} \left[\tan^{-1} e^x\right]_1^b$

$= 2 \lim_{b\to\infty} \left(\tan^{-1} e^b - \tan^{-1} e\right) = \pi - 2\tan^{-1} e$

31. $\int_1^\infty \left(\frac{a}{x+2} - \frac{1}{x+4}\right) dx = \lim_{b\to\infty} \left[a \ln|x+2| - \ln|x+4|\right]_1^b = \lim_{b\to\infty} \ln \frac{(b+2)^a}{b+4} - \ln\left(\frac{3^a}{5}\right);$

$\lim_{b\to\infty} \frac{(b+2)^a}{b+4} = a \lim_{b\to\infty} (b+2)^{a-1} = \begin{cases} \infty, & a > 1 \\ 1, & a = 1 \end{cases}$ $\Rightarrow$ the series converges to $\ln\left(\frac{5}{3}\right)$ if $a = 1$ and diverges to ∞ if $a > 1$. If $a < 1$, the terms of the series eventually become negative and the Integral Test does not apply. From that point on, however, the series behaves like a negative multiple of the harmonic series, and so it diverges.

33. (a)

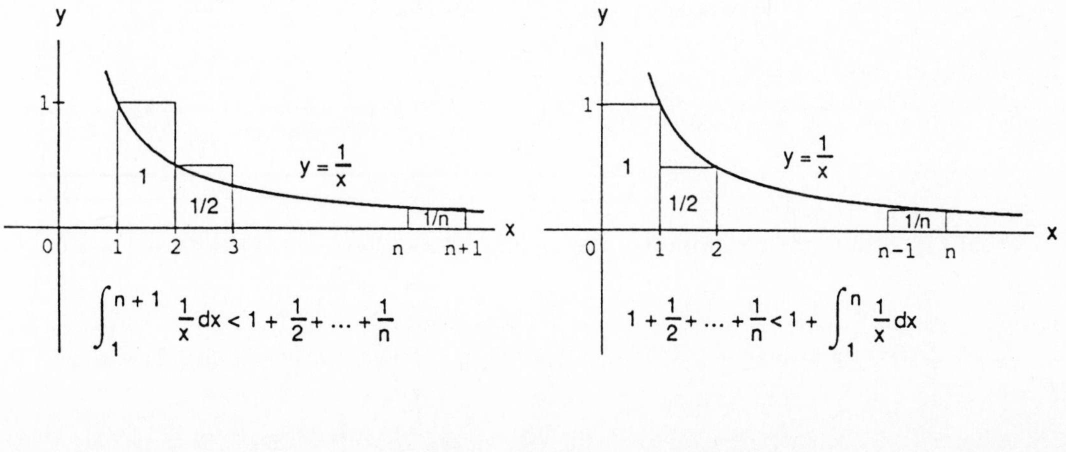

(b) There are $(13)(365)(24)(60)(60)(10^9)$ seconds in 13 billion years; by part (a) $s_n \le 1 + \ln n$ where $n = (13)(365)(24)(60)(60)(10^9) \Rightarrow s_n \le 1 + \ln\left((13)(365)(24)(60)(60)(10^9)\right)$
$= 1 + \ln(13) + \ln(365) + \ln(24) + 2\ln(60) + 9\ln(10) \approx 41.55$

35. Yes. If $\sum_{n=1}^{\infty} a_n$ is a divergent series of positive numbers, then $\left(\frac{1}{2}\right)\sum_{n=1}^{\infty} a_n = \sum_{n=1}^{\infty}\left(\frac{a_n}{2}\right)$ also diverges and $\frac{a_n}{2} < a_n$. There is no "smallest" divergent series of positive numbers: for any divergent series $\sum_{n=1}^{\infty} a_n$ of positive numbers $\sum_{n=1}^{\infty}\left(\frac{a_n}{2}\right)$ has smaller terms and still diverges.

37. Let $A_n = \sum_{k=1}^{n} a_k$ and $B_n = \sum_{k=1}^{n} 2^k a_{(2^k)}$, where $\{a_k\}$ is a nonincreasing sequence of positive terms converging to 0. Note that $\{A_n\}$ and $\{B_n\}$ are nondecreasing sequences of positive terms. Now,

$$B_n = 2a_2 + 4a_4 + 8a_8 + \ldots + 2^n a_{(2^n)} = 2a_2 + (2a_4 + 2a_4) + (2a_8 + 2a_8 + 2a_8 + 2a_8) + \ldots$$
$$+ \underbrace{\left(2a_{(2^n)} + 2a_{(2^n)} + \ldots + 2a_{(2^n)}\right)}_{2^{n-1}\text{ terms}} \le 2a_1 + 2a_2 + (2a_3 + 2a_4) + (2a_5 + 2a_6 + 2a_7 + 2a_8) + \ldots$$
$$+ \left(2a_{(2^{n-1})} + 2a_{(2^{n-1}+1)} + \ldots + 2a_{(2^n)}\right) = 2A_{(2^n)} \le 2\sum_{k=1}^{\infty} a_k.$$

Therefore if $\sum a_k$ converges, then $\{B_n\}$ is bounded above $\Rightarrow \sum 2^k a_{(2^k)}$ converges. Conversely,

$$A_n = a_1 + (a_2 + a_3) + (a_4 + a_5 + a_6 + a_7) + \ldots + a_n < a_1 + 2a_2 + 4a_4 + \ldots + 2^n a_{(2^n)} = a_1 + B_n < a_1 + \sum_{k=1}^{\infty} 2^k a_{(2^k)}.$$

Therefore, if $\sum_{k=1}^{\infty} 2^k a_{(2^k)}$ converges, then $\{A_n\}$ is bounded above and hence converges.

39. (a) $\int_2^{\infty} \frac{dx}{x(\ln x)^p};\ \left[\begin{array}{l} u = \ln x \\ du = \frac{dx}{x} \end{array}\right] \to \int_{\ln 2}^{\infty} u^{-p}\,du = \lim_{b\to\infty}\left[\frac{u^{-p+1}}{-p+1}\right]_{\ln 2}^{b} = \lim_{b\to\infty}\left(\frac{1}{1-p}\right)\left[b^{-p+1} - (\ln 2)^{-p+1}\right]$

$= \begin{cases} \frac{1}{p-1}(\ln 2)^{-p+1}, & p > 1 \\ \infty, & p < 1 \end{cases} \Rightarrow$ the improper integral converges if $p > 1$ and diverges if $p < 1$. For $p = 1$: $\int_2^{\infty} \frac{dx}{x\ln x} = \lim_{b\to\infty}\left[\ln(\ln x)\right]_2^b = \lim_{b\to\infty}\left[\ln(\ln b) - \ln(\ln 2)\right] = \infty$, so the improper integral diverges if $p = 1$.

(b) Since the series and the integral converge or diverge together, $\sum_{n=2}^{\infty} \frac{1}{n(\ln n)^p}$ converges if and only if $p > 1$.

41. (a) From Fig. 8.13 in the text with $f(x) = \frac{1}{x}$ and $a_k = \frac{1}{k}$, we have $\int_1^{n+1} \frac{1}{x}\,dx \le 1 + \frac{1}{2} + \frac{1}{3} + \ldots + \frac{1}{n}$

$\le 1 + \int_1^n f(x)\,dx \Rightarrow \ln(n+1) \le 1 + \frac{1}{2} + \frac{1}{3} + \ldots + \frac{1}{n} \le 1 + \ln n \Rightarrow 0 \le \ln(n+1) - \ln n$

$\le \left(1+\frac{1}{2}+\frac{1}{3}+\ldots+\frac{1}{n}\right)-\ln n \le 1$. Therefore the sequence $\left\{\left(1+\frac{1}{2}+\frac{1}{3}+\ldots+\frac{1}{n}\right)-\ln n\right\}$ is bounded above by 1 and below by 0.

(b) From the graph in Fig. 8.13(a) with $f(x)=\frac{1}{x}$, $\frac{1}{n+1} < \int_n^{n+1} \frac{1}{x}\,dx = \ln(n+1)-\ln n$

$\Rightarrow 0 > \frac{1}{n+1} - [\ln(n+1)-\ln n] = \left(1+\frac{1}{2}+\frac{1}{3}+\ldots+\frac{1}{n+1}-\ln(n+1)\right)-\left(1+\frac{1}{2}+\frac{1}{3}+\ldots+\frac{1}{n}-\ln n\right)$.

If we define $a_n = 1+\frac{1}{2}=\frac{1}{3}+\frac{1}{n}-\ln n$, then $0 > a_{n+1}-a_n \Rightarrow a_{n+1} < a_n \Rightarrow \{a_n\}$ is a decreasing sequence of nonnegative terms.

8.5 COMPARISON TESTS FOR SERIES OF NONNEGATIVE TERMS

1. diverges by the Limit Comparison Test (part 1) when compared with $\sum_{n=1}^{\infty} \frac{1}{\sqrt{n}}$, a divergent p-series:

$$\lim_{n\to\infty} \frac{\left(\frac{1}{2\sqrt{n}+\sqrt[3]{n}}\right)}{\left(\frac{1}{\sqrt{n}}\right)} = \lim_{n\to\infty} \frac{\sqrt{n}}{2\sqrt{n}+\sqrt[3]{n}} = \lim_{n\to\infty}\left(\frac{1}{2+n^{-1/6}}\right) = \frac{1}{2}$$

3. converges by the Direct Comparison Test; $\frac{\sin^2 n}{2^n} \le \frac{1}{2^n}$, which is the nth term of a convergent geometric series

5. diverges since $\lim_{n\to\infty} \frac{2n}{3n-1} = \frac{2}{3} \ne 0$

7. converges by the Direct Comparison Test; $\left(\frac{n}{3n+1}\right)^n < \left(\frac{n}{3n}\right)^n < \left(\frac{1}{3}\right)^n$, the nth term of a convergent geometric series

9. diverges by the Direct Comparison Test; $n > \ln n \Rightarrow \ln n > \ln\ln n \Rightarrow \frac{1}{\ln n} < \frac{1}{\ln(\ln n)}$ and the series $\sum_{n=3}^{\infty} \frac{1}{n}$ diverges

11. converges by the Limit Comparison Test (part 2) when compared with $\sum_{n=1}^{\infty} \frac{1}{n^2}$, a convergent p-series:

$$\lim_{n\to\infty} \frac{\left[\frac{(\ln n)^2}{n^3}\right]}{\left(\frac{1}{n^2}\right)} = \lim_{n\to\infty} \frac{(\ln n)^2}{n} = \lim_{n\to\infty} \frac{2(\ln n)\left(\frac{1}{n}\right)}{1} = 2\lim_{n\to\infty} \frac{\ln n}{n} = 0 \qquad \text{(Table 8.1)}$$

13. diverges by the Limit Comparison Test (part 3) with $\frac{1}{n}$, the nth term of the divergent harmonic series:

$$\lim_{n\to\infty} \frac{\left[\frac{1}{\sqrt{n}\ln n}\right]}{\left(\frac{1}{n}\right)} = \lim_{n\to\infty} \frac{\sqrt{n}}{\ln n} = \lim_{n\to\infty} \frac{\left(\frac{1}{2\sqrt{n}}\right)}{\left(\frac{1}{n}\right)} = \lim_{n\to\infty} \frac{\sqrt{n}}{2} = \infty$$

15. diverges by the Limit Comparison Test (part 3) with $\frac{1}{n}$, the nth term of the divergent harmonic series:

$$\lim_{n\to\infty} \frac{\left(\frac{1}{1+\ln n}\right)}{\left(\frac{1}{n}\right)} = \lim_{n\to\infty} \frac{n}{1+\ln n} = \lim_{n\to\infty} \frac{1}{\left(\frac{1}{n}\right)} = \lim_{n\to\infty} n = \infty$$

17. diverges by the Integral Test: $\int_2^\infty \frac{\ln(x+1)}{x+1}\,dx = \int_{\ln 3}^\infty u\,du = \lim_{b\to\infty}\left[\frac{1}{2}u^2\right]_{\ln 3}^b = \lim_{b\to\infty} \frac{1}{2}\left(b^2 - \ln^2 3\right) = \infty$

19. converges by the Direct Comparison Test with $\frac{1}{n^{3/2}}$, the nth term of a convergent p-series: $n^2 - 1 > n$ for $n \ge 2 \Rightarrow n^2\left(n^2 - 1\right) > n^3 \Rightarrow n\sqrt{n^2-1} > n^{3/2} \Rightarrow \frac{1}{n^{3/2}} > \frac{1}{n\sqrt{n^2-1}}$

21. converges because $\sum_{n=1}^{\infty} \frac{1-n}{n2^n} = \sum_{n=1}^{\infty} \frac{1}{n2^n} + \sum_{n=1}^{\infty} \frac{-1}{2^n}$ which is the sum of two convergent series: $\sum_{n=1}^{\infty} \frac{1}{n2^n}$ converges by the Direct Comparison Test since $\frac{1}{n2^n} < \frac{1}{2^n}$, and $\sum_{n=1}^{\infty} \frac{-1}{2^n}$ is a convergent geometric series

23. converges by the Direct Comparison Test: $\frac{1}{3^{n-1}+1} < \frac{1}{3^{n-1}}$, which is the nth term of a convergent geometric series

25. diverges by the Limit Comparison Test (part 1) with $\frac{1}{n}$, the nth term of the divergent harmonic series:

$$\lim_{n\to\infty} \frac{\left(\sin\frac{1}{n}\right)}{\left(\frac{1}{n}\right)} = \lim_{x\to 0} \frac{\sin x}{x} = 1$$

27. converges by the Limit Comparison Test (part 1) with $\frac{1}{n^2}$, the nth term of a convergent p-series:

$$\lim_{n\to\infty} \frac{\left(\frac{10n+1}{n(n+1)(n+2)}\right)}{\left(\frac{1}{n^2}\right)} = \lim_{n\to\infty} \frac{10n^2+n}{n^2+3n+2} = \lim_{n\to\infty} \frac{20n+1}{2n+3} = \lim_{n\to\infty} \frac{20}{2} = 10$$

29. converges by the Direct Comparison Test: $\frac{\tan^{-1} n}{n^{1.1}} < \frac{\frac{\pi}{2}}{n^{1.1}}$ and $\sum_{n=1}^{\infty} \frac{\frac{\pi}{2}}{n^{1.1}} = \frac{\pi}{2}\sum_{n=1}^{\infty} \frac{1}{n^{1.1}}$ is the product of a convergent p-series and a nonzero constant

31. converges by the Limit Comparison Test (part 1) with $\frac{1}{n^2}$: $\lim_{n\to\infty} \frac{\left(\frac{\coth n}{n^2}\right)}{\left(\frac{1}{n^2}\right)} = \lim_{n\to\infty} \coth n = \lim_{n\to\infty} \frac{e^n + e^{-n}}{e^n - e^{-n}}$
$= \lim_{n\to\infty} \frac{1+e^{-2n}}{1-e^{-2n}} = 1$

33. diverges: $f(x) = \sqrt[x]{x} \Rightarrow f'(x) > 0$ when $1 < x < e$ and $f'(x) < 0$ when $x > e \Rightarrow \sqrt[e]{e} > \sqrt[n]{n}$ for all $n \geq 3$; also $3^e > e \Rightarrow 3 > \sqrt[e]{e}$. Consequently, $3n > n\sqrt[n]{n} \Rightarrow \frac{1}{3n} < \frac{1}{n\sqrt[n]{n}} \Rightarrow \sum_{n=1}^{\infty} \frac{1}{n\sqrt[n]{n}}$ diverges by the Direct Comparison Test

35. $\frac{1}{1+2+3+\ldots+n} = \frac{1}{\left(\frac{n(n+1)}{2}\right)} = \frac{2}{n(n+1)} \leq \frac{1}{n^2} \Rightarrow$ the series converges by the Direct Comparison Test

37. (a) If $\lim_{n\to\infty} \frac{a_n}{b_n} = 0$, then there exists an integer N such that for all $n > N$, $\left|\frac{a_n}{b_n} - 0\right| < 1 \Rightarrow -1 < \frac{a_n}{b_n} < 1$ $\Rightarrow a_n < b_n$. Thus, if $\sum b_n$ converges, then $\sum a_n$ converges by the Direct Comparison Test.

(b) If $\lim_{n\to\infty} \frac{a_n}{b_n} = \infty$, then there exists an integer N such that for all $n > N$, $\frac{a_n}{b_n} > 1 \Rightarrow a_n > b_n$. Thus, if $\sum b_n$ diverges, then $\sum a_n$ diverges by the Direct Comparison Test.

39. $\lim_{n\to\infty} \frac{a_n}{b_n} = \infty \Rightarrow$ there exists an integer N such that for all $n > N$, $\frac{a_n}{b_n} > 1 \Rightarrow a_n > b_n$. If $\sum a_n$ converges, then $\sum b_n$ converges by the Direct Comparison Test

8.6 THE RATIO AND ROOT TESTS FOR SERIES WITH NONNEGATIVE TERMS

1. converges by the Ratio Test: $\lim_{n\to\infty} \frac{a_{n+1}}{a_n} = \lim_{n\to\infty} \frac{\left[\frac{(n+1)^{\sqrt{2}}}{2^{n+1}}\right]}{\left[\frac{n^{\sqrt{2}}}{2^n}\right]} = \lim_{n\to\infty} \frac{(n+1)^{\sqrt{2}}}{2^{n+1}} \cdot \frac{2^n}{n^{\sqrt{2}}}$ $= \lim_{n\to\infty} \left(1+\frac{1}{n}\right)^{\sqrt{2}} \left(\frac{1}{2}\right) = \frac{1}{2} < 1$

3. diverges by the Ratio Test: $\lim_{n\to\infty} \frac{a_{n+1}}{a_n} = \lim_{n\to\infty} \frac{\left(\frac{(n+1)!}{e^{n+1}}\right)}{\left(\frac{n!}{e^n}\right)} = \lim_{n\to\infty} \frac{(n+1)!}{e^{n+1}} \cdot \frac{e^n}{n!} = \lim_{n\to\infty} \frac{n+1}{e} = \infty$

5. converges by the Ratio Test: $\lim_{n\to\infty} \frac{a_{n+1}}{a_n} = \lim_{n\to\infty} \frac{\left(\frac{(n+1)^{10}}{10^{n+1}}\right)}{\left(\frac{n^{10}}{10^n}\right)} = \lim_{n\to\infty} \frac{(n+1)^{10}}{10^{n+1}} \cdot \frac{10^n}{n^{10}} = \lim_{n\to\infty} \left(1+\frac{1}{n}\right)^{10} \left(\frac{1}{10}\right)$ $= \frac{1}{10} < 1$

7. converges by the Direct Comparison Test: $\frac{2+(-1)^n}{(1.25)^n} = \left(\frac{4}{5}\right)^n [2+(-1)^n] \leq \left(\frac{4}{5}\right)^n (3)$

9. diverges; $\lim_{n\to\infty} a_n = \lim_{n\to\infty} \left(1-\frac{3}{n}\right)^n = \lim_{n\to\infty} \left(1+\frac{-3}{n}\right)^n = e^{-3} \approx 0.05 \neq 0$

11. converges by the Direct Comparison Test: $\frac{\ln n}{n^3} < \frac{n}{n^3} = \frac{1}{n^2}$ for $n \geq 2$

13. diverges by the Direct Comparison Test: $\frac{1}{n} - \frac{1}{n^2} = \frac{n-1}{n^2} > \frac{1}{2}\left(\frac{1}{n}\right)$ for $n > 2$

15. diverges by the Direct Comparison Test: $\frac{\ln n}{n} > \frac{1}{n}$ for $n \geq 3$

17. converges by the Ratio Test: $\lim_{n\to\infty} \frac{a_{n+1}}{a_n} = \lim_{n\to\infty} \frac{(n+2)(n+3)}{(n+1)!} \cdot \frac{n!}{(n+1)(n+2)} = 0 < 1$

19. converges by the Ratio Test: $\lim_{n\to\infty} \frac{a_{n+1}}{a_n} = \lim_{n\to\infty} \frac{(n+4)!}{3!\,(n+1)!\,3^{n+1}} \cdot \frac{3!\,n!\,3^n}{(n+3)!} = \lim_{n\to\infty} \frac{n+4}{3(n+1)} = \frac{1}{3} < 1$

21. converges by the Ratio Test: $\lim_{n\to\infty} \frac{a_{n+1}}{a_n} = \lim_{n\to\infty} \frac{(n+1)!}{(2n+3)!} \cdot \frac{(2n+1)!}{n!} = \lim_{n\to\infty} \frac{n+1}{(2n+3)(2n+2)} = 0 < 1$

23. converges by the Root Test: $\lim_{n\to\infty} \sqrt[n]{a_n} = \lim_{n\to\infty} \sqrt[n]{\frac{n}{(\ln n)^n}} = \lim_{n\to\infty} \frac{\sqrt[n]{n}}{\ln n} = \lim_{n\to\infty} \frac{1}{\ln n} = 0 < 1$

25. converges by the Direct Comparison Test: $\frac{n!\,\ln n}{n(n+2)!} = \frac{\ln n}{n(n+1)(n+2)} < \frac{n}{n(n+1)(n+2)} = \frac{1}{(n+1)(n+2)} < \frac{1}{n^2}$

which is the nth-term of a convergent p-series

27. converges by the Ratio Test: $\lim_{n\to\infty} \frac{a_{n+1}}{a_n} = \lim_{n\to\infty} \frac{\left(\frac{1+\sin n}{n}\right)a_n}{a_n} = 0 < 1$

29. diverges by the Ratio Test: $\lim_{n\to\infty} \frac{a_{n+1}}{a_n} = \lim_{n\to\infty} \frac{\left(\frac{3n-1}{2n+1}\right)a_n}{a_n} = \lim_{n\to\infty} \frac{3n-1}{2n+1} = \frac{3}{2} > 1$

31. converges by the Ratio Test: $\lim_{n\to\infty} \frac{a_{n+1}}{a_n} = \lim_{n\to\infty} \frac{\left(\frac{2}{n}\right)a_n}{a_n} = \lim_{n\to\infty} \frac{2}{n} = 0 < 1$

33. converges by the Ratio Test: $\lim_{n\to\infty} \frac{a_{n+1}}{a_n} = \lim_{n\to\infty} \frac{\left(\frac{1+\ln n}{n}\right)a_n}{a_n} = \lim_{n\to\infty} \frac{1+\ln n}{n} = \lim_{n\to\infty} \frac{1}{n} = 0 < 1$

35. diverges by the nth-Term Test: $a_1 = \frac{1}{3}$, $a_2 = \sqrt[2]{\frac{1}{3}}$, $a_3 = \sqrt[3]{\sqrt[2]{\frac{1}{3}}} = \sqrt[6]{\frac{1}{3}}$, $a_4 = \sqrt[4]{\sqrt[3]{\sqrt[2]{\frac{1}{3}}}} = \sqrt[4!]{\frac{1}{3}}, \ldots,$

$a_n = \sqrt[n!]{\frac{1}{3}} \Rightarrow \lim_{n\to\infty} a_n = 1$ because $\left\{\sqrt[n!]{\frac{1}{3}}\right\}$ is a subsequence of $\left\{\sqrt[n]{\frac{1}{3}}\right\}$ whose limit is 1 by Table 8.1

37. converges by the Ratio Test: $\lim\limits_{n\to\infty} \frac{a_{n+1}}{a_n} = \lim\limits_{n\to\infty} \frac{2^{n+1}(n+1)!(n+1)!}{(2n+2)!} \cdot \frac{(2n)!}{2^n n!\, n!} = \lim\limits_{n\to\infty} \frac{2(n+1)(n+1)}{(2n+2)(2n+1)}$

$= \lim\limits_{n\to\infty} \frac{n+1}{2n+1} = \frac{1}{2} < 1$

39. diverges by the Root Test: $\lim\limits_{n\to\infty} \sqrt[n]{a_n} = \lim\limits_{n\to\infty} \sqrt[n]{\frac{(n!)^n}{(n^n)^2}} = \lim\limits_{n\to\infty} \frac{n!}{n^2} = \infty > 1$

41. converges by the Root Test: $\lim\limits_{n\to\infty} \sqrt[n]{a_n} = \lim\limits_{n\to\infty} \sqrt[n]{\frac{n^n}{2^{n^2}}} = \lim\limits_{n\to\infty} \frac{n}{2^n} = \lim\limits_{n\to\infty} \frac{1}{2^n \ln 2} = 0 < 1$

43. converges by the Ratio Test: $\lim\limits_{n\to\infty} \frac{a_{n+1}}{a_n} = \lim\limits_{n\to\infty} \frac{1\cdot 3\cdot \cdots \cdot(2n-1)(2n+1)}{4^{n+1}2^{n+1}(n+1)!} \cdot \frac{4^n\, 2^n\, n!}{1\cdot 3\cdot \cdots \cdot (2n-1)}$

$= \lim\limits_{n\to\infty} \frac{2n+1}{(4\cdot 2)(n+1)} = \frac{1}{4} < 1$

45. Ratio: $\lim\limits_{n\to\infty} \frac{a_{n+1}}{a_n} = \lim\limits_{n\to\infty} \frac{1}{(n+1)^p} \cdot \frac{n^p}{1} = \lim\limits_{n\to\infty} \left(\frac{n}{n+1}\right)^p = 1^p = 1 \Rightarrow$ no conclusion

Root: $\lim\limits_{n\to\infty} \sqrt[n]{a_n} = \lim\limits_{n\to\infty} \sqrt[n]{\frac{1}{n^p}} = \lim\limits_{n\to\infty} \frac{1}{\left(\sqrt[n]{n}\right)^p} = \frac{1}{(1)^p} = 1 \Rightarrow$ no conclusion

47. $a_n \le \frac{n}{2^n}$ for every n and the series $\sum\limits_{n=1}^{\infty} \frac{n}{2^n}$ converges by the Ratio Test since $\lim\limits_{n\to\infty} \frac{(n+1)}{2^{n+1}} \cdot \frac{2^n}{n} = \frac{1}{2} < 1$

$\Rightarrow \sum\limits_{n=1}^{\infty} a_n$ converges by the Direct Comparison Test

8.7 ALTERNATING SERIES, ABSOLUTE AND CONDITIONAL CONVERGENCE

1. converges absolutely $\Rightarrow$ converges by the Absolute Convergence Test since $\sum\limits_{n=1}^{\infty} |a_n| = \sum\limits_{n=1}^{\infty} \frac{1}{n^2}$ which is a convergent p-series

3. diverges by the nth-Term Test since for $n > 10 \Rightarrow \frac{n}{10} > 1 \Rightarrow \lim\limits_{n\to\infty} \left(\frac{n}{10}\right)^n \ne 0 \Rightarrow \sum\limits_{n=1}^{\infty} (-1)^{n+1}\left(\frac{n}{10}\right)^n$ diverges

5. converges by the Alternating Series Test because $f(x) = \ln x$ is an increasing function of $x \Rightarrow \frac{1}{\ln x}$ is decreasing

$\Rightarrow u_n \ge u_{n+1}$ for $n \ge 1$; also $u_n \ge 0$ for $n \ge 1$ and $\lim\limits_{n\to\infty} \frac{1}{\ln n} = 0$

7. diverges by the nth-Term Test since $\lim\limits_{n\to\infty} \frac{\ln n}{\ln n^2} = \lim\limits_{n\to\infty} \frac{\ln n}{2 \ln n} = \lim\limits_{n\to\infty} \frac{1}{2} = \frac{1}{2} \ne 0$

9. converges by the Alternating Series Test since $f(x) = \frac{\sqrt{x}+1}{x+1} \Rightarrow f'(x) = \frac{1 - x - 2\sqrt{x}}{2\sqrt{x}(x+1)^2} < 0 \Rightarrow f(x)$ is decreasing

$\Rightarrow u_n \ge u_{n+1}$; also $u_n \ge 0$ for $n \ge 1$ and $\lim\limits_{n\to\infty} u_n = \lim\limits_{n\to\infty} \frac{\sqrt{n}+1}{n+1} = 0$

11. converges absolutely since $\sum_{n=1}^{\infty} |a_n| = \sum_{n=1}^{\infty} \left(\frac{1}{10}\right)^n$ a convergent geometric series

13. converges conditionally since $\frac{1}{\sqrt{n}} > \frac{1}{\sqrt{n+1}} > 0$ and $\lim_{n\to\infty} \frac{1}{\sqrt{n}} = 0 \Rightarrow$ convergence; but $\sum_{n=1}^{\infty} |a_n| = \sum_{n=1}^{\infty} \frac{1}{n^{1/2}}$ is a divergent p-series

15. converges absolutely since $\sum_{n=1}^{\infty} |a_n| = \sum_{n=1}^{\infty} \frac{n}{n^3+1}$ and $\frac{n}{n^3+1} < \frac{1}{n^2}$ which is the nth-term of a converging p-series

17. converges conditionally since $\frac{1}{n+3} > \frac{1}{(n+1)+3} > 0$ and $\lim_{n\to\infty} \frac{1}{n+3} = 0 \Rightarrow$ convergence; but $\sum_{n=1}^{\infty} |a_n|$ $= \sum_{n=1}^{\infty} \frac{1}{n+3}$ diverges because $\frac{1}{n+3} \geq \frac{1}{4n}$ and $\sum_{n=1}^{\infty} \frac{1}{n}$ is a divergent series

19. diverges by the nth-Term Test since $\lim_{n\to\infty} \frac{3+n}{5+n} = 1 \neq 0$

21. converges conditionally since $f(x) = \frac{1}{x^2} + \frac{1}{x} \Rightarrow f'(x) = -\left(\frac{2}{x^3} + \frac{1}{x^2}\right) < 0 \Rightarrow f(x)$ is decreasing and hence $u_n > u_{n+1} > 0$ for $n \geq 1$ and $\lim_{n\to\infty} \left(\frac{1}{n^2} + \frac{1}{n}\right) = 0 \Rightarrow$ convergence; but $\sum_{n=1}^{\infty} |a_n| = \sum_{n=1}^{\infty} \frac{1+n}{n^2}$ $= \sum_{n=1}^{\infty} \frac{1}{n^2} + \sum_{n=1}^{\infty} \frac{1}{n}$ is the sum of a convergent and divergent series, and hence diverges

23. converges absolutely by the Ratio Test: $\lim_{n\to\infty} \left(\frac{u_{n+1}}{u_n}\right) = \lim_{n\to\infty} \left[\frac{(n+1)^2\left(\frac{2}{3}\right)^{n+1}}{n^2\left(\frac{2}{3}\right)^n}\right] = \frac{2}{3} < 1$

25. converges absolutely by the Integral Test since $\int_1^{\infty} (\tan^{-1} x)\left(\frac{1}{1+x^2}\right) dx = \lim_{b\to\infty} \left[\frac{(\tan^{-1} x)^2}{2}\right]_1^b$ $= \lim_{b\to\infty} \left[(\tan^{-1} b)^2 - (\tan^{-1} 1)^2\right] = \frac{1}{2}\left[\left(\frac{\pi}{2}\right)^2 - \left(\frac{\pi}{4}\right)^2\right] = \frac{3\pi^2}{32}$

27. diverges by the nth-Term Test since $\lim_{n\to\infty} \frac{n}{n+1} = 1 \neq 0$

29. converges absolutely by the Ratio Test: $\lim_{n\to\infty} \left(\frac{u_{n+1}}{u_n}\right) = \lim_{n\to\infty} \frac{(100)^{n+1}}{(n+1)!} \cdot \frac{n!}{(100)^n} = \lim_{n\to\infty} \frac{100}{n+1} = 0 < 1$

31. converges absolutely by the Direct Comparison Test since $\sum_{n=1}^{\infty} |a_n| = \sum_{n=1}^{\infty} \frac{1}{n^2+2n+1}$ and $\frac{1}{n^2+2n+1} < \frac{1}{n^2}$ which is the nth-term of a convergent p-series

33. converges absolutely since $\sum_{n=1}^{\infty} |a_n| = \sum_{n=1}^{\infty} \left|\frac{(-1)^n}{n\sqrt{n}}\right| = \sum_{n=1}^{\infty} \frac{1}{n^{3/2}}$ is a convergent p-series

35. converges absolutely by the Root Test: $\lim_{n\to\infty} \sqrt[n]{|a_n|} = \lim_{n\to\infty} \left(\frac{(n+1)^n}{(2n)^n}\right)^{1/n} = \lim_{n\to\infty} \frac{n+1}{2n} = \frac{1}{2}$

37. diverges by the nth-Term Test since $\lim_{n\to\infty} |a_n| = \lim_{n\to\infty} \frac{(2n)!}{2^n n!\, n} = \lim_{n\to\infty} \frac{(n+1)(n+2)\cdots(2n)}{2^n n}$

$= \lim_{n\to\infty} \frac{(n+1)(n+2)\cdots(n+(n-1))}{2^{n-1}} > \lim_{n\to\infty} \left(\frac{n+1}{2}\right)^{n-1} = \infty \neq 0$

39. converges conditionally since $\frac{\sqrt{n+1}-\sqrt{n}}{1} \cdot \frac{\sqrt{n+1}+\sqrt{n}}{\sqrt{n+1}+\sqrt{n}} = \frac{1}{\sqrt{n+1}+\sqrt{n}}$ and $\left\{\frac{1}{\sqrt{n+1}+\sqrt{n}}\right\}$ is a decreasing sequence of positive terms which converges to $0 \Rightarrow \sum_{n=1}^{\infty} \frac{(-1)^n}{\sqrt{n+1}+\sqrt{n}}$ converges; but $n > \frac{1}{3} \Rightarrow 3n > 1$ $\Rightarrow 4n > n+1 \Rightarrow 2\sqrt{n} > \sqrt{n+1} \Rightarrow 3\sqrt{n} > \sqrt{n+1}+\sqrt{n} \Rightarrow \frac{1}{3\sqrt{n}} < \frac{1}{\sqrt{n+1}+\sqrt{n}} \Rightarrow \sum_{n=1}^{\infty} \frac{1}{\sqrt{n+1}+\sqrt{n}}$ diverges by the Direct Comparison Test

41. diverges by the nth-Term Test since $\lim_{n\to\infty} \left(\sqrt{n+\sqrt{n}}-\sqrt{n}\right) = \lim_{n\to\infty} \left[\left(\sqrt{n+\sqrt{n}}-\sqrt{n}\right)\left(\frac{\sqrt{n+\sqrt{n}}+\sqrt{n}}{\sqrt{n+\sqrt{n}}+\sqrt{n}}\right)\right]$

$= \lim_{n\to\infty} \frac{\sqrt{n}}{\sqrt{n+\sqrt{n}}+\sqrt{n}} = \lim_{n\to\infty} \frac{1}{\sqrt{1+\frac{1}{\sqrt{n}}}+1} = \frac{1}{2} \neq 0$

43. converges absolutely by the Direct Comparison Test since $\operatorname{sech}(n) = \frac{2}{e^n + e^{-n}} = \frac{2e^n}{e^{2n}+1} < \frac{2e^n}{e^{2n}} = \frac{2}{e^n}$ which is the nth term of a convergent geometric series

45. $|\text{error}| < \left|(-1)^6\left(\frac{1}{5}\right)\right| = 0.2$

47. $|\text{error}| < \left|(-1)^6 \frac{(0.01)^5}{5}\right| = 2 \times 10^{-11}$

49. $\frac{1}{(2n)!} < \frac{5}{10^6} \Rightarrow (2n)! > \frac{10^6}{5} = 200{,}000 \Rightarrow n \geq 5 \Rightarrow 1 - \frac{1}{2!} + \frac{1}{4!} - \frac{1}{6!} + \frac{1}{8!} \approx 0.54030$

51. (a) $a_n \geq a_{n+1}$ fails since $\frac{1}{3} < \frac{1}{2}$

(b) Since $\sum_{n=1}^{\infty} |a_n| = \sum_{n=1}^{\infty} \left[\left(\frac{1}{3}\right)^n + \left(\frac{1}{2}\right)^n\right] = \sum_{n=1}^{\infty} \left(\frac{1}{3}\right)^n + \sum_{n=1}^{\infty} \left(\frac{1}{2}\right)^n$ is the sum of two absolutely convergent series, we can rearrange the terms of the original series to find its sum:

$\left(\frac{1}{3}+\frac{1}{9}+\frac{1}{27}+\ldots\right) - \left(\frac{1}{2}+\frac{1}{4}+\frac{1}{8}+\ldots\right) = \frac{\left(\frac{1}{3}\right)}{1-\left(\frac{1}{3}\right)} - \frac{\left(\frac{1}{2}\right)}{1-\left(\frac{1}{2}\right)} = \frac{1}{2} - 1 = -\frac{1}{2}$

53. The unused terms are $\sum_{j=n+1}^{\infty} (-1)^{j+1}a_j = (-1)^{n+1}(a_{n+1} - a_{n+2}) + (-1)^{n+3}(a_{n+3} - a_{n+4}) + \ldots$
$= (-1)^{n+1}[(a_{n+1} - a_{n+2}) + (a_{n+3} - a_{n+4}) + \ldots]$. Each grouped term is positive, so the remainder has the same sign as $(-1)^{n+1}$, which is the sign of the first unused term.

55. Using the Direct Comparison Test, since $|a_n| \geq a_n$ and $\sum_{n=1}^{\infty} a_n$ diverges we must have that $\sum_{n=1}^{\infty} |a_n|$ diverges.

57. (a) $\sum_{n=1}^{\infty} |a_n + b_n|$ converges by the Direct Comparison Test since $|a_n + b_n| \leq |a_n| + |b_n|$ and hence $\sum_{n=1}^{\infty} (a_n + b_n)$ converges absolutely

(b) $\sum_{n=1}^{\infty} |b_n|$ converges $\Rightarrow \sum_{n=1}^{\infty} -b_n$ converges absolutely; since $\sum_{n=1}^{\infty} a_n$ converges absolutely and $\sum_{n=1}^{\infty} -b_n$ converges absolutely, we have $\sum_{n=1}^{\infty} [a_n + (-b_n)] = \sum_{n=1}^{\infty} (a_n - b_n)$ converges absolutely by part (a)

(c) $\sum_{n=1}^{\infty} |a_n|$ converges $\Rightarrow |k| \sum_{n=1}^{\infty} |a_n| = \sum_{n=1}^{\infty} |ka_n|$ converges $\Rightarrow \sum_{n=1}^{\infty} ka_n$ converges absolutely

59. $s_1 = -\frac{1}{2}$, $s_2 = -\frac{1}{2} + 1 = \frac{1}{2}$,

$s_3 = -\frac{1}{2} + 1 - \frac{1}{4} - \frac{1}{6} - \frac{1}{8} - \frac{1}{10} - \frac{1}{12} - \frac{1}{14} - \frac{1}{16} - \frac{1}{18} - \frac{1}{20} - \frac{1}{22} \approx -0.5099$,

$s_4 = s_3 + \frac{1}{3} \approx -0.1766$,

$s_5 = s_4 - \frac{1}{24} - \frac{1}{26} - \frac{1}{28} - \frac{1}{30} - \frac{1}{32} - \frac{1}{34} - \frac{1}{36} - \frac{1}{38} - \frac{1}{40} - \frac{1}{42} - \frac{1}{44} \approx -0.512$,

$s_6 = s_5 + \frac{1}{5} \approx -0.312$,

$s_7 = s_6 - \frac{1}{46} - \frac{1}{48} - \frac{1}{50} - \frac{1}{52} - \frac{1}{54} - \frac{1}{56} - \frac{1}{58} - \frac{1}{60} - \frac{1}{62} - \frac{1}{64} - \frac{1}{66} \approx -0.51106$

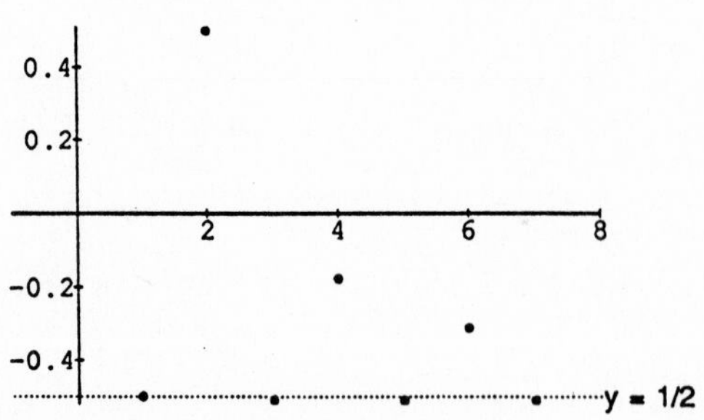

61. (a) If $\sum_{n=1}^{\infty} |a_n|$ converges, then $\sum_{n=1}^{\infty} a_n$ converges and $\frac{1}{2}\sum_{n=1}^{\infty} a_n + \frac{1}{2}\sum_{n=1}^{\infty} |a_n| = \sum_{n=1}^{\infty} \frac{a_n + |a_n|}{2}$ converges where $b_n = \frac{a_n + |a_n|}{2} = \begin{cases} a_n, & \text{if } a_n \geq 0 \\ 0, & \text{if } a_n < 0 \end{cases}$.

(b) If $\sum_{n=1}^{\infty} |a_n|$ converges, then $\sum_{n=1}^{\infty} a_n$ converges and $\frac{1}{2}\sum_{n=1}^{\infty} a_n - \frac{1}{2}\sum_{n=1}^{\infty} |a_n| = \sum_{n=1}^{\infty} \frac{a_n - |a_n|}{2}$ converges where $c_n = \frac{a_n - |a_n|}{2} = \begin{cases} 0, & \text{if } a_n \geq 0 \\ a_n, & \text{if } a_n < 0 \end{cases}$.

63. Here is an example figure when $N = 5$. Notice that $u_3 > u_2 > u_1$ and $u_3 > u_5 > u_4$, but $u_n \geq u_{n+1}$ for $n \geq 5$.

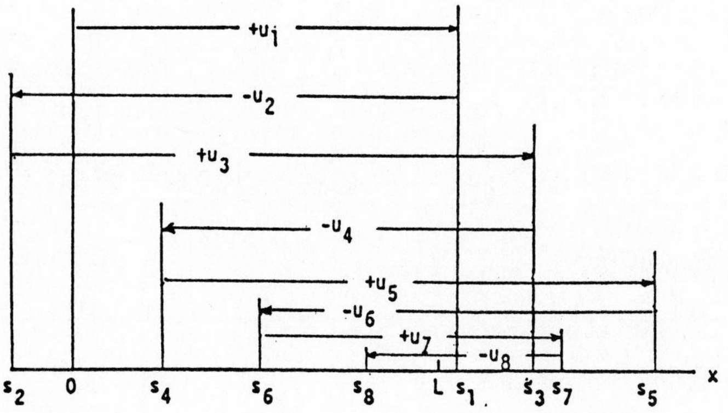

8.8 POWER SERIES

1. $\lim_{n\to\infty} \left|\frac{u_{n+1}}{u_n}\right| < 1 \Rightarrow \lim_{n\to\infty} \left|\frac{x^{n+1}}{x^n}\right| < 1 \Rightarrow |x| < 1 \Rightarrow -1 < x < 1$; when $x = -1$ we have $\sum_{n=1}^{\infty} (-1)^n$, a divergent series; when $x = 1$ we have $\sum_{n=1}^{\infty} 1$, a divergent series

   (a) the radius is 1; the interval of convergence is $-1 < x < 1$

   (b) the interval of absolute convergence is $-1 < x < 1$

   (c) there are no values for which the series converges conditionally

3. $\lim_{n\to\infty} \left|\frac{u_{n+1}}{u_n}\right| < 1 \Rightarrow \lim_{n\to\infty} \left|\frac{(4x+1)^{n+1}}{(4x+1)^n}\right| < 1 \Rightarrow |4x+1| < 1 \Rightarrow -1 < 4x+1 < 1 \Rightarrow -\frac{1}{2} < x < 0$; when $x = -\frac{1}{2}$ we have $\sum_{n=1}^{\infty} (-1)^n(-1)^n = \sum_{n=1}^{\infty} (-1)^{2n} = \sum_{n=1}^{\infty} 1^n$, a divergent series; when $x = 0$ we have $\sum_{n=1}^{\infty} (-1)^n(1)^n = \sum_{n=1}^{\infty} (-1)^n$, a divergent series

   (a) the radius is $\frac{1}{4}$; the interval of convergence is $-\frac{1}{2} < x < 0$

   (b) the interval of absolute convergence is $-\frac{1}{2} < x < 0$

   (c) there are no values for which the series converges conditionally

5. $\lim_{n\to\infty} \left|\frac{u_{n+1}}{u_n}\right| < 1 \Rightarrow \lim_{n\to\infty} \left|\frac{(x-2)^{n+1}}{10^{n+1}} \cdot \frac{10^n}{(x-2)^n}\right| < 1 \Rightarrow \frac{|x-2|}{10} < 1 \Rightarrow |x-2| < 10 \Rightarrow -10 < x-2 < 10$

$\Rightarrow -8 < x < 12$; when $x = -8$ we have $\sum_{n=1}^{\infty} (-1)^n$, a divergent series; when $x = 12$ we have $\sum_{n=1}^{\infty} 1$, a divergent series

(a) the radius is 10; the interval of convergence is $-8 < x < 12$

(b) the interval of absolute convergence is $-8 < x < 12$

(c) there are no values for which the series converges conditionally

7. $\lim_{n\to\infty} \left|\frac{u_{n+1}}{u_n}\right| < 1 \Rightarrow \lim_{n\to\infty} \left|\frac{(n+1)x^{n+1}}{(n+3)} \cdot \frac{(n+2)}{nx^n}\right| < 1 \Rightarrow |x| \lim_{n\to\infty} \frac{(n+1)(n+2)}{(n+3)(n)} < 1 \Rightarrow |x| < 1$

$\Rightarrow -1 < x < 1$; when $x = -1$ we have $\sum_{n=1}^{\infty} (-1)^n \frac{n}{n+2}$, a divergent series by the nth-term Test; when $x = 1$ we have $\sum_{n=1}^{\infty} \frac{n}{n+2}$, a divergent series

(a) the radius is 1; the interval of convergence is $-1 < x < 1$

(b) the interval of absolute convergence is $-1 < x < 1$

(c) there are no values for which the series converges conditionally

9. $\lim_{n\to\infty} \left|\frac{u_{n+1}}{u_n}\right| < 1 \Rightarrow \lim_{n\to\infty} \left|\frac{x^{n+1}}{(n+1)\sqrt{n+1}\,3^{n+1}} \cdot \frac{n\sqrt{n}\,3^n}{x^n}\right| < 1 \Rightarrow \frac{|x|}{3}\left(\lim_{n\to\infty} \frac{n}{n+1}\right)\left(\sqrt{\lim_{n\to\infty} \frac{n}{n+1}}\right) < 1$

$\Rightarrow \frac{|x|}{3}(1)(1) < 1 \Rightarrow |x| < 3 \Rightarrow -3 < x < 3$; when $x = -3$ we have $\sum_{n=1}^{\infty} \frac{(-1)^n}{n^{3/2}}$, an absolutely convergent series; when $x = 3$ we have $\sum_{n=1}^{\infty} \frac{1}{n^{3/2}}$, a convergent p-series

(a) the radius is 3; the interval of convergence is $-3 \le x \le 3$

(b) the interval of absolute convergence is $-3 \le x \le 3$

(c) there are no values for which the series converges conditionally

11. $\lim_{n\to\infty} \left|\frac{u_{n+1}}{u_n}\right| < 1 \Rightarrow \lim_{n\to\infty} \left|\frac{x^{n+1}}{(n+1)!} \cdot \frac{n!}{x^n}\right| < 1 \Rightarrow |x| \lim_{n\to\infty} \left(\frac{1}{n+1}\right) < 1$ for all x

(a) the radius is ∞; the series converges for all x

(b) the series converges absolutely for all x

(c) there are no values for which the series converges conditionally

13. $\lim_{n\to\infty} \left|\frac{u_{n+1}}{u_n}\right| < 1 \Rightarrow \lim_{n\to\infty} \left|\frac{x^{2n+3}}{(n+1)!} \cdot \frac{n!}{x^{2n+1}}\right| < 1 \Rightarrow x^2 \lim_{n\to\infty} \left(\frac{1}{n+1}\right) < 1$ for all x

(a) the radius is ∞; the series converges for all x

(b) the series converges absolutely for all x

(c) there are no values for which the series converges conditionally

15. $\lim_{n\to\infty} \left|\frac{u_{n+1}}{u_n}\right| < 1 \Rightarrow \lim_{n\to\infty} \left|\frac{x^{n+1}}{\sqrt{(n+1)^2+3}} \cdot \frac{\sqrt{n^2+3}}{x^n}\right| < 1 \Rightarrow |x| \sqrt{\lim_{n\to\infty} \frac{n^2+3}{n^2+2n+4}} < 1 \Rightarrow |x| < 1$

$\Rightarrow -1 < x < 1$; when $x = -1$ we have $\sum_{n=1}^{\infty} \frac{(-1)^n}{\sqrt{n^2+3}}$, a conditionally convergent series; when $x = 1$ we have $\sum_{n=1}^{\infty} \frac{1}{n^2+3}$, a divergent series

(a) the radius is 1; the interval of convergence is $-1 \le x < 1$

(b) the interval of absolute convergence is $-1 < x < 1$

(c) the series converges conditionally at $x = -1$

17. $\lim_{n\to\infty} \left|\frac{u_{n+1}}{u_n}\right| < 1 \Rightarrow \lim_{n\to\infty} \left|\frac{(n+1)(x+3)^{n+1}}{5^{n+1}} \cdot \frac{5^n}{n(x+3)^n}\right| < 1 \Rightarrow \frac{|x+3|}{5} \lim_{n\to\infty} \left(\frac{n+1}{n}\right) < 1 \Rightarrow \frac{|x+3|}{5} < 1$

$\Rightarrow |x+3| < 5 \Rightarrow -5 < x+3 < 5 \Rightarrow -8 < x < 2$; when $x = -8$ we have $\sum_{n=1}^{\infty} \frac{n(-5)^n}{5^n} = \sum_{n=1}^{\infty} (-1)^n n$, a divergent series; when $x = 2$ we have $\sum_{n=1}^{\infty} \frac{n5^n}{5^n} = \sum_{n=1}^{\infty} n$, a divergent series

(a) the radius is 5; the interval of convergence is $-8 < x < 2$

(b) the interval of absolute convergence is $-8 < x < 2$

(c) there are no values for which the series converges conditionally

19. $\lim_{n\to\infty} \left|\frac{u_{n+1}}{u_n}\right| < 1 \Rightarrow \lim_{n\to\infty} \left|\frac{\sqrt{n+1}\,x^{n+1}}{3^{n+1}} \cdot \frac{3^n}{\sqrt{n}\,x^n}\right| < 1 \Rightarrow \frac{|x|}{3} \sqrt{\lim_{n\to\infty} \left(\frac{n+1}{n}\right)} < 1 \Rightarrow \frac{|x|}{3} < 1 \Rightarrow |x| < 3$

$\Rightarrow -3 < x < 3$; when $x = -3$ we have $\sum_{n=1}^{\infty} (-1)^n \sqrt{n}$, a divergent series; when $x = 3$ we have $\sum_{n=1}^{\infty} \sqrt{n}$, a divergent series

(a) the radius is 3; the interval of convergence is $-3 < x < 3$

(b) the interval of absolute convergence is $-3 < x < 3$

(c) there are no values for which the series converges conditionally

21. $\lim_{n\to\infty} \left|\frac{u_{n+1}}{u_n}\right| < 1 \Rightarrow \lim_{n\to\infty} \left|\frac{\left(1+\frac{1}{n+1}\right)^{n+1} x^{n+1}}{\left(1+\frac{1}{n}\right)^n x^n}\right| < 1 \Rightarrow |x| \left(\frac{\lim_{t\to\infty} \left(1+\frac{1}{t}\right)^t}{\lim_{n\to\infty} \left(1+\frac{1}{n}\right)^n}\right) < 1 \Rightarrow |x|\left(\frac{e}{e}\right) < 1 \Rightarrow |x| < 1$

$\Rightarrow -1 < x < 1$; when $x = -1$ we have $\sum_{n=1}^{\infty} (-1)^n \left(1+\frac{1}{n}\right)^n$, a divergent series by the nth-Term Test since $\lim_{n\to\infty} \left(1+\frac{1}{n}\right)^n = e \ne 0$; when $x = 1$ we have $\sum_{n=1}^{\infty} \left(1+\frac{1}{n}\right)^n$, a divergent series

(a) the radius is 1; the interval of convergence is $-1 < x < 1$

(b) the interval of absolute convergence is $-1 < x < 1$

(c) there are no values for which the series converges conditionally

23. $\lim_{n\to\infty} \left|\frac{u_{n+1}}{u_n}\right| < 1 \Rightarrow \lim_{n\to\infty} \left|\frac{(n+1)^{n+1}x^{n+1}}{n^n x^n}\right| < 1 \Rightarrow |x| \left(\lim_{n\to\infty} \left(1+\frac{1}{n}\right)^n\right)\left(\lim_{n\to\infty} (n+1)\right) < 1$

$\Rightarrow e|x| \lim_{n\to\infty} (n+1) < 1 \Rightarrow$ only $x = 0$ satisfies this inequality

(a) the radius is 0; the series converges only for $x = 0$

(b) the series converges absolutely only for $x = 0$

(c) there are no values for which the series converges conditionally

25. $\lim_{n\to\infty} \left|\frac{u_{n+1}}{u_n}\right| < 1 \Rightarrow \lim_{n\to\infty} \left|\frac{(x+2)^{n+1}}{(n+1)\,2^{n+1}} \cdot \frac{n2^n}{(x+2)^n}\right| < 1 \Rightarrow \frac{|x+2|}{2} \lim_{n\to\infty} \left(\frac{n}{n+1}\right) < 1 \Rightarrow \frac{|x+2|}{2} < 1 \Rightarrow |x+2| < 2$

$\Rightarrow -2 < x+2 < 2 \Rightarrow -4 < x < 0$; when $x = -4$ we have $\sum_{n=1}^{\infty} \frac{-1}{n}$, a divergent series; when $x = 0$ we have $\sum_{n=1}^{\infty} \frac{(-1)^{n+1}}{n}$, the alternating harmonic series which converges conditionally

(a) the radius is 2; the interval of convergence is $-4 < x \le 0$

(b) the interval of absolute convergence is $-4 < x < 0$

(c) the series converges conditionally at $x = 0$

27. $\lim_{n\to\infty} \left|\frac{u_{n+1}}{u_n}\right| < 1 \Rightarrow \lim_{n\to\infty} \left|\frac{x^{n+1}}{(n+1)(\ln(n+1))^2} \cdot \frac{n(\ln n)^2}{x^n}\right| < 1 \Rightarrow |x| \left(\lim_{n\to\infty} \frac{n}{n+1}\right)\left(\lim_{n\to\infty} \frac{\ln n}{\ln(n+1)}\right)^2 < 1$

$\Rightarrow |x|(1)\left(\lim_{n\to\infty} \frac{\left(\frac{1}{n}\right)}{\left(\frac{1}{n+1}\right)}\right)^2 < 1 \Rightarrow |x| \left(\lim_{n\to\infty} \frac{n+1}{n}\right)^2 < 1 \Rightarrow |x| < 1 \Rightarrow -1 < x < 1$; when $x = -1$ we have $\sum_{n=1}^{\infty} \frac{(-1)^n}{n(\ln n)^2}$ which converges absolutely; when $x = 1$ we have $\sum_{n=1}^{\infty} \frac{1}{n(\ln n)^2}$ which converges

(a) the radius is 1; the interval of convergence is $-1 \le 1 \le 1$

(b) the interval of absolute convergence is $-1 \le x \le 1$

(c) there are no values for which the series converges conditionally

29. $\lim_{n\to\infty} \left|\frac{u_{n+1}}{u_n}\right| < 1 \Rightarrow \lim_{n\to\infty} \left|\frac{(4x-5)^{2n+3}}{(n+1)^{3/2}} \cdot \frac{n^{3/2}}{(4x-5)^{2n+1}}\right| < 1 \Rightarrow (4x-5)^2 \left(\lim_{n\to\infty} \frac{n}{n+1}\right)^{3/2} < 1 \Rightarrow (4x-5)^2 < 1$

$\Rightarrow |4x-5| < 1 \Rightarrow -1 < 4x-5 < 1 \Rightarrow 1 < x < \frac{3}{2}$; when $x = 1$ we have $\sum_{n=1}^{\infty} \frac{(-1)^{2n+1}}{n^{3/2}} = \sum_{n=1}^{\infty} \frac{-1}{n^{3/2}}$ which is absolutely convergent; when $x = \frac{3}{2}$ we have $\sum_{n=1}^{\infty} \frac{(1)^{2n+1}}{n^{3/2}}$, a convergent p-series

(a) the radius is $\frac{1}{4}$; the interval of convergence is $1 \le x \le \frac{3}{2}$

(b) the interval of absolute convergence is $1 \le x \le \frac{3}{2}$

(c) there are no values for which the series converges conditionally

31. $\lim_{n\to\infty}\left|\frac{u_{n+1}}{u_n}\right|<1 \Rightarrow \lim_{n\to\infty}\left|\frac{(x+\pi)^{n+1}}{\sqrt{n+1}}\cdot\frac{\sqrt{n}}{(x+\pi)^n}\right|<1 \Rightarrow |x+\pi|\lim_{n\to\infty}\left|\sqrt{\frac{n}{n+1}}\right|<1$

$\Rightarrow |x+\pi|\sqrt{\lim_{n\to\infty}\left(\frac{n}{n+1}\right)}<1 \Rightarrow |x+\pi|<1 \Rightarrow -1<x+\pi<1 \Rightarrow -1-\pi<x<1-\pi;$

when $x=-1-\pi$ we have $\sum_{n=1}^{\infty}\frac{(-1)^n}{\sqrt{n}}=\sum_{n=1}^{\infty}\frac{(-1)^n}{n^{1/2}}$, a conditionally convergent series; when $x=1-\pi$ we have $\sum_{n=1}^{\infty}\frac{1^n}{\sqrt{n}}=\sum_{n=1}^{\infty}\frac{1}{n^{1/2}}$, a divergent p-series

(a) the radius is 1; the interval of convergence is $(-1-\pi)\le x<(1-\pi)$

(b) the interval of absolute convergence is $-1-\pi<x<1-\pi$

(c) the series converges conditionally at $x=-1-\pi$

33. $\lim_{n\to\infty}\left|\frac{u_{n+1}}{u_n}\right|<1 \Rightarrow \lim_{n\to\infty}\left|\frac{(x-1)^{2n+2}}{4^{n+1}}\cdot\frac{4^n}{(x-1)^{2n}}\right|<1 \Rightarrow \frac{(x-1)^2}{4}\lim_{n\to\infty}|1|<1 \Rightarrow (x-1)^2<4 \Rightarrow |x-1|<2$

$\Rightarrow -2<x-1<2 \Rightarrow -1<x<3$; at $x=-1$ we have $\sum_{n=0}^{\infty}\frac{(-2)^{2n}}{4^n}=\sum_{n=0}^{\infty}\frac{4^n}{4^n}=\sum_{n=0}^{\infty}1$, which diverges; at $x=3$ we have $\sum_{n=0}^{\infty}\frac{2^{2n}}{4^n}=\sum_{n=0}^{\infty}\frac{4^n}{4^n}=\sum_{n=0}^{\infty}1$, a divergent series; the interval of convergence is $-1<x<3$; the series $\sum_{n=0}^{\infty}\frac{(x-1)^{2n}}{4^n}=\sum_{n=0}^{\infty}\left(\left(\frac{x-1}{2}\right)^2\right)^n$ is a convergent geometric series when $-1<x<3$ and the sum is

$$\frac{1}{1-\left(\frac{x-1}{2}\right)^2}=\frac{1}{\left[\frac{4-(x-1)^2}{4}\right]}=\frac{4}{4-x^2+2x-1}=\frac{4}{3+2x-x^2}$$

35. $\lim_{n\to\infty}\left|\frac{u_{n+1}}{u_n}\right|<1 \Rightarrow \lim_{n\to\infty}\left|\frac{\left(\sqrt{x}-2\right)^{n+1}}{2^{n+1}}\cdot\frac{2^n}{\left(\sqrt{x}-2\right)^n}\right|<1 \Rightarrow \left|\sqrt{x}-2\right|<2 \Rightarrow -2<\sqrt{x}-2<2 \Rightarrow 0<\sqrt{x}<4$

$\Rightarrow 0<x<16$; when $x=0$ we have $\sum_{n=0}^{\infty}(-1)^n$, a divergent series; when $x=16$ we have $\sum_{n=0}^{\infty}(1)^n$, a divergent series; the interval of convergence is $0<x<16$; the series $\sum_{n=0}^{\infty}\left(\frac{\sqrt{x}-2}{2}\right)^n$ is a convergent geometric series when $0<x<16$ and its sum is $\frac{1}{1-\left(\frac{\sqrt{x}-2}{2}\right)}=\frac{1}{\left(\frac{2-\sqrt{x}+2}{2}\right)}=\frac{2}{4-\sqrt{x}}$

37. $\lim_{n\to\infty}\left|\frac{u_{n+1}}{u_n}\right|<1 \Rightarrow \lim_{n\to\infty}\left|\left(\frac{x^2+1}{3}\right)^{n+1}\cdot\left(\frac{3}{x^2+1}\right)^n\right|<1 \Rightarrow \frac{(x^2+1)}{3}\lim_{n\to\infty}|1|<1 \Rightarrow \frac{x^2+1}{3}<1 \Rightarrow x^2<2$

$\Rightarrow |x|<\sqrt{2} \Rightarrow -\sqrt{2}<x<\sqrt{2}$; at $x=\pm\sqrt{2}$ we have $\sum_{n=0}^{\infty}(1)^n$ which diverges; the interval of convergence is $-\sqrt{2}<x<\sqrt{2}$; the series $\sum_{n=0}^{\infty}\left(\frac{x^2+1}{3}\right)^n$ is a convergent geometric series when $-\sqrt{2}<x<\sqrt{2}$ and its sum is

$$\frac{1}{1-\left(\frac{x^2+1}{3}\right)}=\frac{1}{\left(\frac{3-x^2-1}{3}\right)}=\frac{3}{2-x^2}$$

39. $\lim_{n\to\infty} \left|\frac{(x-3)^{n+1}}{2^{n+1}} \cdot \frac{2^n}{(x-3)^n}\right| < 1 \Rightarrow |x-3| < 2 \Rightarrow 1 < x < 5$; when $x = 1$ we have $\sum_{n=1}^{\infty} (1)^n$ which diverges; when $x = 5$ we have $\sum_{n=1}^{\infty} (-1)^n$ which also diverges; the interval of convergence is $1 < x < 5$; the sum of this convergent geometric series is $\frac{1}{1+\left(\frac{x-3}{2}\right)} = \frac{2}{x-1}$. If $f(x) = 1 - \frac{1}{2}(x-3) + \frac{1}{4}(x-3)^2 + \ldots + \left(-\frac{1}{2}\right)^n (x-3)^n + \ldots$ $= \frac{2}{x-1}$ then $f'(x) = -\frac{1}{2} + \frac{1}{2}(x-3) + \ldots + \left(-\frac{1}{2}\right)^n n(x-3)^{n-1} + \ldots$ is convergent when $1 < x < 5$, and diverges when $x = 1$ or 5. The sum for $f'(x)$ is $\frac{-2}{(x-1)^2}$, the derivative of $\frac{2}{x-1}$.

41. (a) Differentiate the series for sin x to get $\cos x = 1 - \frac{3x^2}{3!} + \frac{5x^4}{5!} - \frac{7x^6}{7!} + \frac{9x^8}{9!} - \frac{11x^{10}}{11!} + \ldots$ $= 1 - \frac{x^2}{2!} + \frac{x^4}{4!} - \frac{x^6}{6!} + \frac{x^8}{8!} - \frac{x^{10}}{10!} + \ldots$. The series converges for all values of x since

$$\lim_{n\to\infty} \left|\frac{x^{n+1}}{(n+1)!} \cdot \frac{n!}{x^n}\right| = |x| \lim_{n\to\infty} \left(\frac{1}{n+1}\right) = 0 < 1 \text{ for all } x$$

(b) $\sin 2x = 2x - \frac{2^3x^3}{3!} + \frac{2^5x^5}{5!} - \frac{2^7x^7}{7!} + \frac{2^9x^9}{9!} - \frac{2^{11}x^{11}}{11!} + \ldots = 2x - \frac{8x^3}{3!} + \frac{32x^5}{5!} - \frac{128x^7}{7!} + \frac{512x^9}{9!} - \frac{2048x^{11}}{11!} + \ldots$

(c) $2 \sin x \cos x = 2\Big[(0\cdot 1) + (0\cdot 0 + 1\cdot 1)x + \left(0\cdot\frac{-1}{2} + 1\cdot 0 + 0\cdot 1\right)x^2 + \left(0\cdot 0 - 1\cdot\frac{1}{2} + 0\cdot 0 - 1\cdot\frac{1}{3!}\right)x^3$

$+\left(0\cdot\frac{1}{4!} + 1\cdot 0 - 0\cdot\frac{1}{2} - 0\cdot\frac{1}{3!} + 0\cdot 1\right)x^4 + \left(0\cdot 0 + 1\cdot\frac{1}{4!} + 0\cdot 0 + \frac{1}{2}\cdot\frac{1}{3!} + 0\cdot 0 + 1\cdot\frac{1}{5!}\right)x^5$

$+\left(0\cdot\frac{1}{6!} + 1\cdot 0 + 0\cdot\frac{1}{4!} + 0\cdot\frac{1}{3!} + 0\cdot\frac{1}{2} + 0\cdot\frac{1}{5!} + 0\cdot 1\right)x^6 + \ldots\Big] = 2\left[x - \frac{4x^3}{3!} + \frac{16x^5}{5!} - \ldots\right]$

$= 2x - \frac{2^3x^3}{3!} + \frac{2^5x^5}{5!} - \frac{2^7x^7}{7!} + \frac{2^9x^9}{9!} - \frac{2^{11}x^{11}}{11!} + \ldots$

43. (a) $\ln|\sec x| + C = \int \tan x\, dx = \int \left(x + \frac{x^3}{3} + \frac{2x^5}{15} + \frac{17x^7}{315} + \frac{62x^9}{2835} + \ldots\right) dx$

$= \frac{x^2}{2} + \frac{x^4}{12} + \frac{x^6}{45} + \frac{17x^8}{2520} + \frac{31x^{10}}{14{,}175} + \ldots + C$; $x = 0 \Rightarrow C = 0 \Rightarrow \ln|\sec x| = \frac{x^2}{2} + \frac{x^4}{12} + \frac{x^6}{45} + \frac{17x^8}{2520} + \frac{31x^{10}}{14{,}175} + \ldots$, converges when $-\frac{\pi}{2} < x < \frac{\pi}{2}$

(b) $\sec^2 x = \frac{d(\tan x)}{dx} = \frac{d}{dx}\left(x + \frac{x^3}{3} + \frac{2x^5}{15} + \frac{17x^7}{315} + \frac{62x^9}{2835} + \ldots\right) = 1 + x^2 + \frac{2x^4}{3} + \frac{17x^6}{45} + \frac{62x^8}{315} + \ldots$, converges when $-\frac{\pi}{2} < x < \frac{\pi}{2}$

(c) $\sec^2 x = (\sec x)(\sec x) = \left(1 + \frac{x^2}{2} + \frac{5x^4}{24} + \frac{61x^6}{720} + \ldots\right)\left(1 + \frac{x^2}{2} + \frac{5x^4}{24} + \frac{61x^6}{720} + \ldots\right)$

$= 1 + \left(\frac{1}{2} + \frac{1}{2}\right)x^2 + \left(\frac{5}{24} + \frac{1}{4} + \frac{5}{24}\right)x^4 + \left(\frac{61}{720} + \frac{5}{48} + \frac{5}{48} + \frac{61}{720}\right)x^6 + \ldots$

$= 1 + x^2 + \frac{2x^4}{3} + \frac{17x^6}{45} + \frac{62x^8}{315} + \ldots,\ -\frac{\pi}{2} < x < \frac{\pi}{2}$

45. (a) If $f(x) = \sum_{n=0}^{\infty} a_n x^n$, then $f^{(k)}(x) = \sum_{n=k}^{\infty} n(n-1)(n-2)\cdots(n-(k-1))\, a_n x^{n-k}$ and $f^{(k)}(0) = k!a_k$

$\Rightarrow a_k = \frac{f^{(k)}(0)}{k!}$; likewise if $f(x) = \sum_{n=0}^{\infty} b_n x^n$, then $b_k = \frac{f^{(k)}(0)}{k!} \Rightarrow a_k = b_k$ for every nonnegative integer k

(b) If $f(x) = \sum_{n=0}^{\infty} a_n x^n = 0$ for all x, then $f^{(k)}(x) = 0$ for all x $\Rightarrow$ from part (a) that $a_k = 0$ for every nonnegative integer k

47. The series $\sum_{n=1}^{\infty} \frac{x^n}{n}$ converges conditionally at the left-hand endpoint of its interval of convergence $[-1, 1]$; the series $\sum_{n=1}^{\infty} \frac{x^n}{(n^2)}$ converges absolutely at the left-hand endpoint of its interval of convergence $[-1, 1]$

8.9 TAYLOR AND MACLAURIN SERIES

1. $f(x) = \ln x$, $f'(x) = \frac{1}{x}$, $f''(x) = -\frac{1}{x^2}$, $f'''(x) = \frac{2}{x^3}$; $f(1) = \ln 1 = 0$, $f'(1) = 1$, $f''(1) = -1$, $f'''(1) = 2 \Rightarrow P_0(x) = 0$, $P_1(x) = (x-1)$, $P_2(x) = (x-1) - \frac{1}{2}(x-1)^2$, $P_3(x) = (x-1) - \frac{1}{2}(x-1)^2 + \frac{1}{3}(x-1)^3$

3. $f(x) = \frac{1}{x} = x^{-1}$, $f'(x) = -x^{-2}$, $f''(x) = 2x^{-3}$, $f'''(x) = -6x^{-4}$; $f(2) = \frac{1}{2}$, $f'(2) = -\frac{1}{4}$, $f''(2) = \frac{1}{4}$, $f'''(x) = -\frac{3}{8}$ $\Rightarrow P_0(x) = \frac{1}{2}$, $P_1(x) = \frac{1}{2} - \frac{1}{4}(x-2)$, $P_2(x) = \frac{1}{2} - \frac{1}{4}(x-2) + \frac{1}{8}(x-2)^2$, $P_3(x) = \frac{1}{2} - \frac{1}{4}(x-2) + \frac{1}{8}(x-2)^2 - \frac{1}{16}(x-2)^3$

5. $f(x) = \sin x$, $f'(x) = \cos x$, $f''(x) = -\sin x$, $f'''(x) = -\cos x$; $f\left(\frac{\pi}{4}\right) = \sin \frac{\pi}{4} = \frac{\sqrt{2}}{2}$, $f'\left(\frac{\pi}{4}\right) = \cos \frac{\pi}{4} = \frac{\sqrt{2}}{2}$, $f''\left(\frac{\pi}{4}\right) = -\sin \frac{\pi}{4} = -\frac{\sqrt{2}}{2}$, $f'''\left(\frac{\pi}{4}\right) = -\cos \frac{\pi}{4} = -\frac{\sqrt{2}}{2} \Rightarrow P_0 = \frac{\sqrt{2}}{2}$, $P_1(x) = \frac{\sqrt{2}}{2} + \frac{\sqrt{2}}{2}\left(x - \frac{\pi}{4}\right)$, $P_2(x) = \frac{\sqrt{2}}{2} + \frac{\sqrt{2}}{2}\left(x - \frac{\pi}{4}\right) - \frac{\sqrt{2}}{4}\left(x - \frac{\pi}{4}\right)^2$, $P_3(x) = \frac{\sqrt{2}}{2} + \frac{\sqrt{2}}{2}\left(x - \frac{\pi}{4}\right) - \frac{\sqrt{2}}{4}\left(x - \frac{\pi}{4}\right)^2 - \frac{\sqrt{2}}{12}\left(x - \frac{\pi}{4}\right)^3$

7. $f(x) = \sqrt{x} = x^{1/2}$, $f'(x) = \left(\frac{1}{2}\right)x^{-1/2}$, $f''(x) = \left(-\frac{1}{4}\right)x^{-3/2}$, $f'''(x) = \left(\frac{3}{8}\right)x^{-5/2}$; $f(4) = \sqrt{4} = 2$, $f'(4) = \left(\frac{1}{2}\right)4^{-1/2} = \frac{1}{4}$, $f''(4) = \left(-\frac{1}{4}\right)4^{-3/2} = -\frac{1}{32}$, $f'''(4) = \left(\frac{3}{8}\right)4^{-5/2} = \frac{3}{256} \Rightarrow P_0(x) = 2$, $P_1(x) = 2 + \frac{1}{4}(x-4)$, $P_2(x) = 2 + \frac{1}{4}(x-4) - \frac{1}{64}(x-4)^2$, $P_3(x) = 2 + \frac{1}{4}(x-4) - \frac{1}{64}(x-4)^2 + \frac{1}{512}(x-4)^3$

9. $e^x = \sum_{n=0}^{\infty} \frac{x^n}{n!} \Rightarrow e^{-x} = \sum_{n=0}^{\infty} \frac{(-x)^n}{n!} = 1 - x + \frac{x^2}{2!} - \frac{x^3}{3!} + \frac{x^4}{4!} - \ldots$

11. $f(x) = (1+x)^{-1} \Rightarrow f'(x) = -(1+x)^{-2}$, $f''(x) = 2(1+x)^{-3}$, $f'''(x) = -3!(1+x)^{-4} \Rightarrow \ldots f^{(k)}(x)$
$= (-1)^k k!(1+x)^{-k-1}$; $f(0) = 1$, $f'(0) = -1$, $f''(0) = 2$, $f'''(0) = -3!, \ldots, f^{(k)}(0) = (-1)^k k!$
$\Rightarrow \frac{1}{1+x} = 1 - x + x^2 - x^3 + \ldots = \sum_{n=0}^{\infty} (-x)^n = \sum_{n=0}^{\infty} (-1)^n x^n$

13. $\sin x = \sum_{n=0}^{\infty} \frac{(-1)^n x^{2n+1}}{(2n+1)!} \Rightarrow \sin 3x = \sum_{n=0}^{\infty} \frac{(-1)^n (3x)^{2n+1}}{(2n+1)!} = \sum_{n=0}^{\infty} \frac{(-1)^n 3^{2n+1} x^{2n+1}}{(2n+1)!} = 3x - \frac{3^3 x^3}{3!} + \frac{3^5 x^5}{5!} - \ldots$

15. $7 \cos(-x) = 7 \cos x = 7 \sum_{n=0}^{\infty} \frac{(-1)^n x^{2n}}{(2n)!} = 7 - \frac{7x^2}{2!} + \frac{7x^4}{4!} - \frac{7x^6}{6!} + \ldots$, since the cosine is an even function

17. $\cosh x = \frac{e^x + e^{-x}}{2} = \frac{1}{2}\left[\left(1 + x^2 + \frac{x^2}{2!} + \frac{x^3}{3!} + \frac{x^4}{4!} + \ldots\right) + \left(1 - x + \frac{x^2}{2!} - \frac{x^3}{3!} + \frac{x^4}{4!} - \ldots\right)\right] = 1 + \frac{x^2}{2!} + \frac{x^4}{4!} + \frac{x^6}{6!} + \ldots$
$= \sum_{n=0}^{\infty} \frac{x^{2n}}{(2n)!}$

19. $f(x) = x^4 - 2x^3 - 5x + 4 \Rightarrow f'(x) = 4x^3 - 6x^2 - 5$, $f''(x) = 12x^2 - 12x$, $f'''(x) = 24x - 12$, $f^{(4)}(x) = 24$
$\Rightarrow f^{(n)}(x) = 0$ if $n \ge 5$; $f(0) = 4$, $f'(0) = -5$, $f''(0) = 0$, $f'''(0) = -12$, $f^{(4)}(0) = 24$, $f^{(n)}(0) = 0$ if $n \ge 5$
$\Rightarrow x^4 - 2x^3 - 5x + 4 = 4 - 5x - \frac{12}{3!}x^3 + \frac{24}{4!}x^4 = x^4 - 2x^3 - 5x + 4$ itself

21. $f(x) = x^3 - 2x + 4 \Rightarrow f'(x) = 3x^2 - 2$, $f''(x) = 6x$, $f'''(x) = 6 \Rightarrow f^{(n)}(x) = 0$ if $n \ge 4$; $f(2) = 8$, $f'(2) = 10$,
$f''(2) = 12$, $f'''(2) = 6$, $f^{(n)}(2) = 0$ if $n \ge 4 \Rightarrow x^3 - 2x + 4 = 8 + 10(x-2) + \frac{12}{2!}(x-2)^2 + \frac{6}{3!}(x-2)^3$
$= 8 + 10(x-2) + 6(x-2)^2 + (x-2)^3$

23. $f(x) = x^4 + x^2 + 1 \Rightarrow f'(x) = 4x^3 + 2x$, $f''(x) = 12x^2 + 2$, $f'''(x) = 24x$, $f^{(4)}(x) = 24$, $f^{(n)}(x) = 0$ if $n \ge 5$;
$f(-2) = 21$, $f'(-2) = -36$, $f''(-2) = 50$, $f'''(-2) = -48$, $f^{(4)}(-2) = 24$, $f^{(n)}(-2) = 0$ if $n \ge 5 \Rightarrow x^4 + x^2 + 1$
$= 21 - 36(x+2) + \frac{50}{2!}(x+2)^2 - \frac{48}{3!}(x+2)^3 + \frac{24}{4!}(x+2)^4 = 21 - 36(x+2) + 25(x+2)^2 - 8(x+2)^3 + (x+2)^4$

25. $f(x) = x^{-2} \Rightarrow f'(x) = -2x^{-3}$, $f''(x) = 3!\,x^{-4}$, $f'''(x) = -4!\,x^{-5} \Rightarrow f^{(n)}(x) = (-1)^n(n+1)!\,x^{-n-2}$;
$f(1) = 1$, $f'(1) = -2$, $f''(1) = 3!$, $f'''(1) = -4!$, $f^{(n)}(1) = (-1)^n(n+1)! \Rightarrow \frac{1}{x^2}$
$= 1 - 2(x-1) + 3(x-1)^2 - 4(x-1)^3 + \ldots = \sum_{n=0}^{\infty} (-1)^n(n+1)(x-1)^n$

27. $f(x) = e^x \Rightarrow f'(x) = e^x$, $f''(x) = e^x \Rightarrow f^{(n)}(x) = e^x$; $f(2) = e^2$, $f'(2) = e^2, \ldots f^{(n)}(2) = e^2$
$\Rightarrow e^x = e^2 + e^2(x-2) + \frac{e^2}{2}(x-2)^2 + \frac{e^3}{3!}(x-2)^3 + \ldots = \sum_{n=0}^{\infty} \frac{e^2}{n!}(x-2)^n$

29. If $e^x = \sum_{n=0}^{\infty} \frac{f^{(n)}(a)}{n!}(x-a)^n$ and $f(x) = e^x$, we have $f^{(n)}(a) = e^a$ f or all $n = 0, 1, 2, 3, \ldots$
$\Rightarrow e^x = e^a\left[\frac{(x-a)^0}{0!} + \frac{(x-a)^1}{1!} + \frac{(x-a)^2}{2!} + \ldots\right] = e^a\left[1 + (x-a) + \frac{(x-a)^2}{2!} + \ldots\right]$ at $x = a$

31. $f(x) = f(a) + f'(a)(x-a) + \frac{f''(a)}{2}(x-a)^2 + \frac{f'''(a)}{3!}(x-a)^3 + \ldots \Rightarrow f'(x)$

$= f'(a) + f''(a)(x-a) + \frac{f'''(a)}{3!}3(x-a)^2 + \ldots \Rightarrow f''(x) = f''(a) + f'''(a)(x-a) + \frac{f^{(4)}(a)}{4!}4 \cdot 3(x-a)^2 + \ldots$

$\Rightarrow f^{(n)}(x) = f^{(n)}(a) + f^{(n+1)}(a)(x-a) + \frac{f^{(n+2)}(a)}{2}(x-a)^2 + \ldots$

$\Rightarrow f(a) = f(a) + 0,\ f'(a) = f'(a) + 0,\ \ldots,\ f^{(n)}(a) = f^{(n)}(a) + 0$

33. $f(x) = \ln(\cos x) \Rightarrow f'(x) = -\tan x$ and $f''(x) = -\sec^2 x$; $f(0) = 0$, $f'(0) = 0$, $f''(0) = -1$

$\Rightarrow L(x) = 0$ and $Q(x) = -\frac{x^2}{2}$

35. $f(x) = \left(1-x^2\right)^{-1/2} \Rightarrow f'(x) = x\left(1-x^2\right)^{-3/2}$ and $f''(x) = \left(1-x^2\right)^{-3/2} + 3x^2\left(1-x^2\right)^{-5/2}$; $f(0) = 1$, $f'(0) = 0$, $f''(0) = 1 \Rightarrow L(x) = 1$ and $Q(x) = 1 = \frac{x^2}{2}$

37. $f(x) = \sin x \Rightarrow f'(x) = \cos x$ and $f''(x) = -\sin x$; $f(0) = 0$, $f'(0) = 1$, $f''(0) = 0 \Rightarrow L(x) = x$ and $Q(x) = x$

8.10 CONVERGENCE OF TAYLOR SERIES; ERROR ESTIMATES

1. $e^x = 1 + x + \frac{x^2}{2!} + \ldots = \sum_{n=0}^{\infty} \frac{x^n}{n!} \Rightarrow e^{-5x} = 1 + (-5x) + \frac{(-5x)^2}{2!} + \ldots = 1 - 5x + \frac{5^2x^2}{2!} - \frac{5^3x^3}{3!} + \ldots = \sum_{n=0}^{\infty} \frac{(-1)^n 5^n x^n}{n!}$

3. $\sin x = x - \frac{x^3}{3!} + \frac{x^5}{5!} - \ldots = \sum_{n=0}^{\infty} \frac{(-1)^n x^{2n+1}}{(2n+1)!} \Rightarrow 5\sin(-x) = 5\left[(-x) - \frac{(-x)^3}{3!} + \frac{(-x)^5}{5!} - \ldots\right]$

$= \sum_{n=0}^{\infty} \frac{5(-1)^{n+1} x^{2n+1}}{(2n+1)!}$

5. $\cos x = \sum_{n=0}^{\infty} \frac{(-1)^n x^{2n}}{(2n)!} \Rightarrow \cos\sqrt{x} = \sum_{n=0}^{\infty} \frac{(-1)^n \left(x^{1/2}\right)^{2n}}{(2n)!} = \sum_{n=0}^{\infty} \frac{(-1)^n x^n}{(2n)!} = 1 - \frac{x}{2!} + \frac{x^2}{4!} - \frac{x^3}{6!} + \ldots$

7. $e^x = \sum_{n=0}^{\infty} \frac{x^n}{n!} \Rightarrow xe^x = x\left(\sum_{n=0}^{\infty} \frac{x^n}{n!}\right) = \sum_{n=0}^{\infty} \frac{x^{n+1}}{n!} = x + x^2 + \frac{x^3}{2!} + \frac{x^4}{3!} + \frac{x^5}{4!} + \ldots$

9. $\cos x = \sum_{n=0}^{\infty} \frac{(-1)^n x^{2n}}{(2n)!} \Rightarrow \frac{x^2}{2} - 1 + \cos x = \frac{x^2}{2} - 1 + \sum_{n=0}^{\infty} \frac{(-1)^n x^{2n}}{(2n)!} = \frac{x^2}{2} - 1 + 1 - \frac{x^2}{2} + \frac{x^4}{4!} - \frac{x^6}{6!} + \frac{x^8}{8!} - \frac{x^{10}}{10!} + \ldots$

$= \frac{x^4}{4!} - \frac{x^6}{6!} + \frac{x^8}{8!} - \frac{x^{10}}{10!} + \ldots = \sum_{n=2}^{\infty} \frac{(-1)^n x^{2n}}{(2n)!}$

11. $\cos x = \sum_{n=0}^{\infty} \frac{(-1)^n x^{2n}}{(2n)!} \Rightarrow x\cos \pi x = x\sum_{n=0}^{\infty} \frac{(-1)^n (\pi x)^{2n}}{(2n)!} = \sum_{n=0}^{\infty} \frac{(-1)^n \pi^{2n} x^{2n+1}}{(2n)!} = x - \frac{\pi^2 x^3}{2!} + \frac{\pi^4 x^5}{4!} - \frac{\pi^6 x^7}{6!} + \ldots$

13. $\cos^2 x = \frac{1}{2} + \frac{\cos 2x}{2} = \frac{1}{2} + \frac{1}{2}\sum_{n=0}^{\infty} \frac{(-1)^n (2x)^{2n}}{(2n)!} = \frac{1}{2} + \frac{1}{2}\left[1 - \frac{(2x)^2}{2!} + \frac{(2x)^4}{4!} - \frac{(2x)^6}{6!} + \frac{(2x)^8}{8!} - \ldots\right]$

$= 1 - \frac{(2x)^2}{2 \cdot 2!} + \frac{(2x)^4}{2 \cdot 4!} - \frac{(2x)^6}{2 \cdot 6!} + \frac{(2x)^8}{2 \cdot 8!} - \ldots = 1 + \sum_{n=1}^{\infty} \frac{(-1)^n (2x)^{2n}}{2 \cdot (2n)!}$

15. $\frac{x^2}{1-2x} = x^2\left(\frac{1}{1-2x}\right) = x^2 \sum_{n=0}^{\infty} (2x)^n = \sum_{n=0}^{\infty} 2^n x^{n+2} = x^2 + 2x^3 + 2^2x^4 + 2^3x^5 + \ldots$

17. $\frac{1}{1-x} = \sum_{n=0}^{\infty} x^n = 1 + x + x^2 + x^3 + \ldots \Rightarrow \frac{d}{dx}\left(\frac{1}{1-x}\right) = \frac{1}{(1-x)^2} = 1 + 2x + 3x^2 + \ldots = \sum_{n=1}^{\infty} nx^{n-1}$

$= \sum_{n=0}^{\infty} (n+1)x^n$

19. By the Alternating Series Estimation Theorem, the error is less than $\frac{|x|^5}{5!} \Rightarrow |x|^5 < (5!)(5 \times 10^{-4})$

$\Rightarrow |x|^5 < 600 \times 10^{-4} \Rightarrow |x| < \sqrt[5]{6 \times 10^{-2}} \approx 0.56968$

21. If $\sin x = x$ and $|x| < 10^{-3}$, then the $|\text{error}| = |R_2(x)| = \left|\frac{-\cos c}{3!}x^3\right| < \frac{(10^{-3})^3}{3!} \approx 1.67 \times 10^{-10}$, where c is between 0 and x. The Alternating Series Estimation Theorem says $R_2(x)$ has the same sign as $-\frac{x^3}{3!}$. Moreover, $x < \sin x \Rightarrow 0 < \sin x - x = R_2(x) \Rightarrow x < 0 \Rightarrow -10^{-3} < x < 0$.

23. $|R_2(x)| = \left|\frac{e^c x^3}{3!}\right| < \frac{3^{(0.1)}(0.1)^3}{3!} < 1.87 \times 10^{-5}$, where c is between 0 and x

25. $|R_4(x)| < \left|\frac{\cosh c}{5!}x^5\right| = \left|\frac{e^c + e^{-c}}{2}\frac{x^5}{5!}\right| < \frac{1.65 + \frac{1}{1.65}}{2} \cdot \frac{(0.5)^5}{5!} = (1.3)\frac{(0.5)^5}{5!} \approx 0.000293653$

27. $|R_1| = \left|\frac{1}{(1+c)^2}\frac{x^2}{2!}\right| < \frac{x^2}{2} = \left|\frac{x}{2}\right||x| < .01\,|x| = (1\%)\,|x| \Rightarrow \left|\frac{x}{2}\right| < .01 \Rightarrow 0 < |x| < .02$

29. (a) $\sin x = x - \frac{x^3}{3!} + \frac{x^5}{5!} - \frac{x^7}{7!} + \ldots \Rightarrow \frac{\sin x}{x} = 1 - \frac{x^2}{3!} + \frac{x^4}{5!} - \frac{x^6}{7!} + \ldots$, $s_1 = 1$ and $s_2 = 1 - \frac{x^2}{6}$; if L is the sum of the series representing $\frac{\sin x}{x}$, then by the Alternating Series Estimation Theorem, $L - s_1 = \frac{\sin x}{x} - 1 < 0$ and $L - s_2 = \frac{\sin x}{x} - \left(1 - \frac{x^2}{6}\right) > 0$. Therefore $1 - \frac{x^2}{6} < \frac{\sin x}{x} < 1$

(b) The graph of $y = \frac{\sin x}{x}$, $x \neq 0$, is bounded below by the graph of $y = 1 - \frac{x^2}{6}$ and above by the graph of $y = 1$ as derived in part (a).

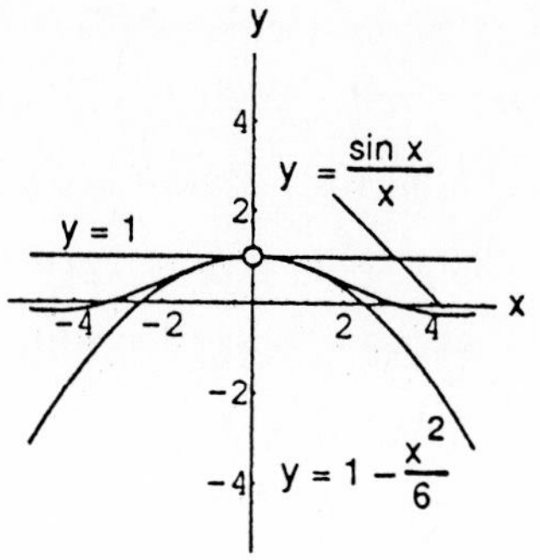

31. $\sin x$ when $x = 0.1$; the sum is $\sin(0.1) \approx 0.099833416$

33. $\tan^{-1} x$ when $x = \frac{\pi}{3}$; the sum is $\tan^{-1}\left(\frac{\pi}{3}\right) = \sqrt{3} \approx 0.808448$

35. $e^x \sin x = 0 + x + x^2 + x^3\left(-\frac{1}{3!} + \frac{1}{2!}\right) + x^4\left(-\frac{1}{3!} + \frac{1}{3!}\right) + x^5\left(\frac{1}{5!} - \frac{1}{2!}\frac{1}{3!} + \frac{1}{4!}\right) + x^6\left(\frac{1}{5!} - \frac{1}{3!}\frac{1}{3!} + \frac{1}{5!}\right) + \ldots$
$= x + x^2 + \frac{1}{3}x^3 = \frac{1}{30}x^5 - \frac{1}{90}x^6 + \ldots$

37. $\sin^2 x = \left(\frac{1 - \cos 2x}{2}\right) = \frac{1}{2} - \frac{1}{2}\cos 2x = \frac{1}{2} - \frac{1}{2}\left(1 - \frac{(2x)^2}{2!} + \frac{(2x)^4}{4!} - \frac{(2x)^6}{6!} + \ldots\right) = \frac{2x^2}{2!} - \frac{2^3x^4}{4!} + \frac{2^5x^6}{6!} - \ldots$
$\Rightarrow \frac{d}{dx}(\sin^2 x) = \frac{d}{dx}\left(\frac{2x^2}{2!} - \frac{2^3x^4}{4!} + \frac{2^5x^6}{6!} - \ldots\right) = 2x - \frac{(2x)^3}{3!} + \frac{(2x)^5}{5!} - \frac{(2x)^7}{7!} + \ldots \Rightarrow 2 \sin x \cos x$
$= 2x - \frac{(2x)^3}{3!} + \frac{(2x)^5}{5!} - \frac{(2x)^7}{7!} + \ldots = \sin 2x$, which checks

39. A special case of Taylor's Formula is $f(x) = f(a) + f'(c)(x - a)$. Let $x = b$ and this becomes $f(b) - f(a) = f'(c)(b - a)$, the Mean Value Theorem

41. (a) $f'' \leq 0$, $f'(a) = 0$ and $x = a$ interior to the interval $I \Rightarrow f(x) - f(a) = \frac{f''(c_2)}{2}(x - a)^2 \leq 0$ throughout I
$\Rightarrow f(x) \leq f(a)$ throughout I $\Rightarrow$ f has a local maximum at $x = a$

(b) similar reasoning gives $f(x) - f(a) = \frac{f''(c_2)}{2}(x - a)^2 \geq 0$ throughout I $\Rightarrow f(x) \geq f(a)$ throughout I $\Rightarrow$ f has a local minimum at $x = a$

43. (a) $f(x) = (1 + x)^k \Rightarrow f'(x) = k(1 + x)^{k-1} \Rightarrow f''(x) = k(k - 1)(1 + x)^{k-2}$; $f(0) = 1$, $f'(0) = k$, and $f''(0) = k(k - 1)$
$\Rightarrow Q(x) = 1 + kx + \frac{k(k - 1)}{2}x^2$

(b) $|R_2(x)| = \left|\frac{3 \cdot 2 \cdot 1}{3!}x^3\right| < \frac{1}{100} \Rightarrow |x^3| < \frac{1}{100} \Rightarrow 0 < x < \frac{1}{100^{1/3}}$ or $0 < x < .21544$

45. If $f(x) = \sum_{n=0}^{\infty} a_n x^n$, then $f^{(k)}(x) = \sum_{n=k}^{\infty} n(n - 1)(n - 2)\cdots(n - k + 1)a_n x^{n-k}$ and $f^{(k)}(0) = k!\, a_k$
$\Rightarrow a_k = \frac{f^{(k)}(0)}{k!}$ for k a nonnegative integer. Therefore, the coefficients of f(x) are identical with the corresponding coefficients in the Maclaurin series of f(x) and the statement follows.

47. (a) Suppose f(x) is a continuous periodic function with period p. Let x_0 be an arbitrary real number. Then f assumes a minimum m_1 and a maximum m_2 in the interval $[x_0, x_0 + p]$; i.e., $m_1 \le f(x) \le m_2$ for all x in $[x_0, x_0 + p]$. Since f is periodic it has exactly the same values on all other intervals $[x_0 + p, x_0 + 2p]$, $[x_0 + 2p, x_0 + 3p]$, ..., and $[x_0 - p, x_0]$, $[x_0 - 2p, x_0 - p]$, ..., and so forth. That is, for all real numbers $-\infty < x < \infty$ we have $m_1 \le f(x) \le m_2$. Now choose $M = \max\{|m_1|, |m_2|\}$. Then $-M \le -|m_1| \le m_1 \le f(x) \le m_2 \le |m_2| \le M \Rightarrow |f(x)| \le M$ for all x.

(b) The dominate term in the nth order Taylor polynomial generated by cos x about x = a is $\frac{\sin(a)}{n!}(x-a)^n$ or $\frac{\cos(a)}{n!}(x-a)^n$. In both cases, as |x| increases the absolute value of these dominate terms tends to ∞, causing the graph of $P_n(x)$ to move away from cos x.

49. (a) $e^{-i\pi} = \cos(-\pi) + i\sin(-\pi) = -1 + i(0) = -1$

(b) $e^{i\pi/4} = \cos\left(\frac{\pi}{4}\right) + i\sin\left(\frac{\pi}{4}\right) = \frac{1}{\sqrt{2}} + \frac{i}{\sqrt{2}} = \left(\frac{1}{\sqrt{2}}\right)(1+i)$

(c) $e^{-i\pi/2} = \cos\left(-\frac{\pi}{2}\right) + i\sin\left(-\frac{\pi}{2}\right) = 0 + i(-1) = -i$

51. $e^x = 1 + x + \frac{x^2}{2!} + \frac{x^3}{3!} + \frac{x^4}{4!} + \ldots \Rightarrow e^{i\theta} = 1 + i\theta + \frac{(i\theta)^2}{2!} + \frac{(i\theta)^3}{3!} + \frac{(i\theta)^4}{4!} + \ldots$ and

$$e^{-i\theta} = 1 - i\theta + \frac{(-i\theta)^2}{2!} + \frac{(-i\theta)^3}{3!} + \frac{(-i\theta)^4}{4!} + \ldots = 1 - i\theta + \frac{(i\theta)^2}{2!} - \frac{(i\theta)^3}{3!} + \frac{(i\theta)^4}{4!} - \ldots$$

$$\Rightarrow \frac{e^{i\theta} + e^{-i\theta}}{2} = \frac{\left(1 + i\theta + \frac{(i\theta)^2}{2!} + \frac{(i\theta)^3}{3!} + \frac{(i\theta)^4}{4!} + \ldots\right) + \left(1 - i\theta + \frac{(i\theta)^2}{2!} - \frac{(i\theta)^3}{3!} + \frac{(i\theta)^4}{4!} - \ldots\right)}{2}$$

$$= 1 - \frac{\theta^2}{2!} + \frac{\theta^4}{4!} - \frac{\theta^6}{6!} + \ldots = \cos\theta;$$

$$\frac{e^{i\theta} - e^{-i\theta}}{2} = \frac{\left(1 + i\theta + \frac{(i\theta)^2}{2!} + \frac{(i\theta)^3}{3!} + \frac{(i\theta)^4}{4!} + \ldots\right) - \left(1 - i\theta + \frac{(i\theta)^2}{2!} - \frac{(i\theta)^3}{3!} + \frac{(i\theta)^4}{4!} - \ldots\right)}{2}$$

$$= \theta - \frac{\theta^3}{3!} + \frac{\theta^5}{5!} - \frac{\theta^7}{7!} + \ldots = \sin\theta$$

53. $e^x \sin x = \left(1 + x + \frac{x^2}{2!} + \frac{x^3}{3!} + \frac{x^4}{4!} + \ldots\right)\left(x - \frac{x^3}{3!} + \frac{x^5}{5!} - \frac{x^7}{7!} + \ldots\right)$

$= (1)x + (1)x^2 + \left(-\frac{1}{6} + \frac{1}{2}\right)x^3 + \left(-\frac{1}{6} + \frac{1}{6}\right)x^4 + \left(\frac{1}{120} - \frac{1}{12} + \frac{1}{24}\right)x^5 + \ldots = x + x^2 + \frac{1}{3}x^3 - \frac{1}{30}x^5 + \ldots;$

$e^x \cdot e^{ix} = e^{(1+i)x} = e^x(\cos x + i\sin x) = e^x\cos x + i(e^x\sin x) \Rightarrow e^x\sin x$ is the series of the imaginary part of $e^{(1+i)x}$ which we calculate next; $e^{(1+i)x} = \sum_{n=0}^{\infty} \frac{(x+ix)^n}{n!} = 1 + (x+ix) + \frac{(x+ix)^2}{2!} + \frac{(x+ix)^3}{3!} + \frac{(x+ix)^4}{4!} + \ldots$

$= 1 + x + ix + \frac{1}{2!}(2ix^2) + \frac{1}{3!}(2ix^3 - 2x^3) + \frac{1}{4!}(-4x^4) + \frac{1}{5!}(-4x^5 - 4ix^5) + \frac{1}{6!}(-8ix^6) + \ldots \Rightarrow$ the imaginary part of $e^{(1+i)x}$ is $x + \frac{2}{2!}x^2 + \frac{2}{3!}x^3 - \frac{4}{5!}x^5 - \frac{8}{6!}x^6 + \ldots = x + x^2 + \frac{1}{3}x^3 - \frac{1}{30}x^5 - \frac{1}{90}x^6 + \ldots$ in agreement with our product calculation

55. (a) $e^{i\theta_1}e^{i\theta_2} = (\cos\theta_1 + i\sin\theta_1)(\cos\theta_2 + i\sin\theta_2) = (\cos\theta_1\cos\theta_2 - \sin\theta_1\sin\theta_2) + i(\sin\theta_1\cos\theta_2 + \sin\theta_2\cos\theta_1)$

$= \cos(\theta_1+\theta_2) + i\sin(\theta_1+\theta_2) = e^{i(\theta_1+\theta_2)}$

(b) $e^{-i\theta} = \cos(-\theta) + i\sin(-\theta) = \cos\theta - i\sin\theta = (\cos\theta - i\sin\theta)\left(\frac{\cos\theta + i\sin\theta}{\cos\theta + i\sin\theta}\right) = \frac{1}{\cos\theta + i\sin\theta} = \frac{1}{e^{i\theta}}$

8.11 APPLICATIONS OF POWER SERIES

1. $(1+x)^{1/2} = 1 + \frac{1}{2}x + \frac{\left(\frac{1}{2}\right)\left(-\frac{1}{2}\right)x^2}{2!} + \frac{\left(\frac{1}{2}\right)\left(-\frac{1}{2}\right)\left(-\frac{3}{2}\right)x^3}{3!} + \ldots = 1 + \frac{1}{2}x - \frac{1}{8}x^2 + \frac{1}{16}x^3 - \ldots$

3. $(1-x)^{-1/2} = 1 - \frac{1}{2}(-x) + \frac{\left(-\frac{1}{2}\right)\left(-\frac{3}{2}\right)(-x)^2}{2!} + \frac{\left(-\frac{1}{2}\right)\left(-\frac{3}{2}\right)\left(-\frac{5}{2}\right)(-x)^3}{3!} + \ldots = 1 + \frac{1}{2}x + \frac{3}{8}x^2 + \frac{5}{16}x^3 + \ldots$

5. $\left(1+\frac{x}{2}\right)^{-2} = 1 - 2\left(\frac{x}{2}\right) + \frac{(-2)(-3)\left(\frac{x}{2}\right)^2}{2!} + \frac{(-2)(-3)(-4)\left(\frac{x}{2}\right)^3}{3!} + \ldots = 1 - x + \frac{3}{4}x^2 - \frac{1}{2}x^3$

7. $\left(1+x^3\right)^{-1/2} = 1 - \frac{1}{2}x^3 + \frac{\left(-\frac{1}{2}\right)\left(-\frac{3}{2}\right)\left(x^3\right)^2}{2!} + \frac{\left(-\frac{1}{2}\right)\left(-\frac{3}{2}\right)\left(-\frac{5}{2}\right)\left(x^3\right)^3}{3!} + \ldots = 1 - \frac{1}{2}x^3 + \frac{3}{8}x^6 - \frac{5}{16}x^9 + \ldots$

9. $\left(1+\frac{1}{x}\right)^{1/2} = 1 + \frac{1}{2}\left(\frac{1}{x}\right) + \frac{\left(\frac{1}{2}\right)\left(-\frac{1}{2}\right)\left(\frac{1}{x}\right)^2}{2!} + \frac{\left(\frac{1}{2}\right)\left(-\frac{1}{2}\right)\left(-\frac{3}{2}\right)\left(\frac{1}{x}\right)^3}{3!} + \ldots = 1 + \frac{1}{2x} - \frac{1}{8x^2} + \frac{1}{16x^3}$

11. $(1+x)^4 = 1 + 4x + \frac{(4)(3)x^2}{2!} + \frac{(4)(3)(2)x^3}{3!} + \frac{(4)(3)(2)x^4}{4!} = 1 + 4x + 6x^2 + 4x^3 + x^4$

13. $(1-2x)^3 = 1 + 3(-2x) + \frac{(3)(2)(-2x)^2}{2!} + \frac{(3)(2)(1)(-2x)^3}{3!} = 1 - 6x + 12x^2 - 8x^3$

15. Assume the solution has the form $y = a_0 + a_1x + a_2x^2 + \ldots + a_{n-1}x^{n-1} + a_nx^n + \ldots$

$\Rightarrow \frac{dy}{dx} = a_1 + 2a_2x + \ldots + na_nx^{n-1} + \ldots$

$\Rightarrow \frac{dy}{dx} + y = (a_1 + a_0) + (2a_2 + a_1)x + (3a_3 + a_2)x^2 + \ldots + (na_n + a_{n-1})x^{n-1} + \ldots = 0$

$\Rightarrow a_1 + a_0 = 0,\ 2a_2 + a_1 = 0,\ 3a_3 + a_2 = 0$ and in general $na_n + a_{n-1} = 0$. Since $y = 1$ when $x = 0$ we have $a_0 = 1$. Therefore $a_1 = -1$, $a_2 = \frac{-a_1}{2\cdot 1} = \frac{1}{2}$, $a_3 = \frac{-a_2}{3} = -\frac{1}{3\cdot 2}$, …, $a_n = \frac{-a_{n-1}}{n} = \frac{(-1)^n}{n!}$

$\Rightarrow y = 1 - x + \frac{1}{2}x^2 - \frac{1}{3!}x^3 + \ldots + \frac{(-1)^n}{n!}x^n + \ldots = \sum_{n=0}^{\infty} \frac{(-1)^n x^n}{n!} = e^{-x}$

17. Assume the solution has the form $y = a_0 + a_1x + a_2x^2 + \ldots + a_{n-1}x^{n-1} + a_nx^n + \ldots$

$\Rightarrow \frac{dy}{dx} = a_1 + 2a_2x + \ldots + na_nx^{n-1} + \ldots$

$\Rightarrow \frac{dy}{dx} - y = (a_1 - a_0) + (2a_2 - a_1)x + (3a_3 - a_2)x^2 + \ldots + (na_n - a_{n-1})x^{n-1} + \ldots = 1$

$\Rightarrow a_1 - a_0 = 1$, $2a_2 - a_1 = 0$, $3a_3 - a_2 = 0$ and in general $na_n - a_{n-1} = 0$. Since $y = 0$ when $x = 0$ we have $a_0 = 0$. Therefore $a_1 = 1$, $a_2 = \frac{a_1}{2} = \frac{1}{2}$, $a_3 = \frac{a_2}{3} = \frac{1}{3 \cdot 2}$, $a_4 = \frac{a_3}{4} = \frac{1}{4 \cdot 3 \cdot 2}$, $\ldots$, $a_n = \frac{a_{n-1}}{n} = \frac{1}{n!}$

$\Rightarrow y = 0 + 1x + \frac{1}{2}x^2 + \frac{1}{3 \cdot 2}x^3 + \frac{1}{4 \cdot 3 \cdot 2}x^4 + \ldots + \frac{1}{n!}x^n + \ldots$

$= \left(1 + 1x + \frac{1}{2}x^2 + \frac{1}{3 \cdot 2}x^3 + \frac{1}{4 \cdot 3 \cdot 2}x^4 + \ldots + \frac{1}{n!}x^n + \ldots\right) - 1 = \sum_{n=0}^{\infty} \frac{x^n}{n!} - 1 = e^x - 1$

19. Assume the solution has the form $y = a_0 + a_1x + a_2x^2 + \ldots + a_{n-1}x^{n-1} + a_nx^n + \ldots$

$\Rightarrow \frac{dy}{dx} = a_1 + 2a_2x + \ldots + na_nx^{n-1} + \ldots$

$\Rightarrow \frac{dy}{dx} - y = (a_1 - a_0) + (2a_2 - a_1)x + (3a_3 - a_2)x^2 + \ldots + (na_n - a_{n-1})x^{n-1} + \ldots = x$

$\Rightarrow a_1 - a_0 = 0$, $2a_2 - a_1 = 1$, $3a_3 - a_2 = 0$ and in general $na_n - a_{n-1} = 0$. Since $y = 0$ when $x = 0$ we have $a_0 = 0$. Therefore $a_1 = 0$, $a_2 = \frac{1 + a_1}{2} = \frac{1}{2}$, $a_3 = \frac{a_2}{3} = \frac{1}{3 \cdot 2}$, $a_4 = \frac{a_3}{4} = \frac{1}{4 \cdot 3 \cdot 2}$, $\ldots$, $a_n = \frac{a_{n-1}}{n} = \frac{1}{n!}$

$\Rightarrow y = 0 + 0x + \frac{1}{2}x^2 + \frac{1}{3 \cdot 2}x^3 + \frac{1}{4 \cdot 3 \cdot 2}x^4 + + \ldots + \frac{1}{n!}x^n + \ldots$

$= \left(1 + 1x + \frac{1}{2}x^2 + \frac{1}{3 \cdot 2}x^3 + \frac{1}{4 \cdot 3 \cdot 2}x^4 + \ldots + \frac{1}{n!}x^n + \ldots\right) - 1 - x = \sum_{n=0}^{\infty} \frac{x^n}{n!} - 1 - x = e^x - x - 1$

21. $y' - xy = a_1 + (2a_2 - a_0)x + (3a_3 - a_1)x + \ldots + (na_n - a_{n-2})x^{n-1} + \ldots = 0 \Rightarrow a_1 = 0$, $2a_2 - a_0 = 0$, $3a_3 - a_1 = 0$, $4a_4 - a_2 = 0$ and in general $na_n - a_{n-2} = 0$. Since $y = 1$ when $x = 0$, we have $a_0 = 1$. Therefore $a_2 = \frac{a_0}{2} = \frac{1}{2}$, $a_3 = \frac{a_1}{3} = 0$, $a_4 = \frac{a_2}{4} = \frac{1}{2 \cdot 4}$, $a_5 = \frac{a_3}{5} = 0$, $\ldots$, $a_{2n} = \frac{1}{2 \cdot 4 \cdot 6 \cdots 2n}$ and $a_{2n+1} = 0$

$\Rightarrow y = 1 + \frac{1}{2}x^2 + \frac{1}{2 \cdot 4}x^4 + \frac{1}{2 \cdot 4 \cdot 6}x^6 + \ldots + \frac{1}{2 \cdot 4 \cdot 6 \cdots 2n}x^{2n} + \ldots = \sum_{n=0}^{\infty} \frac{x^{2n}}{2^n n!} = \sum_{n=0}^{\infty} \frac{\left(\frac{x^2}{2}\right)^n}{n!} = e^{x^2/2}$

23. $(1 - x)y' - y = (a_1 - a_0) + (2a_2 - a_1 - a_1)x + (3a_3 - 2a_2 - a_2)x^2 + (4a_4 - 3a_3 - a_3)x^3 + \ldots + (na_n - (n-1)a_{n-1} - a_{n-1})x^{n-1} + \ldots = 0 \Rightarrow a_1 - a_0 = 0$, $2a_2 - 2a_1 = 0$, $3a_3 - 3a_2 = 0$ and in general $(na_n - na_{n-1}) = 0$. Since $y = 2$ when $x = 0$, we have $a_0 = 2$. Therefore $a_1 = 2$, $a_2 = 2$, $\ldots$, $a_n = 2 \Rightarrow y = 2 + 2x + 2x^2 + \ldots = \sum_{n=0}^{\infty} 2x^n = \frac{2}{1 - x}$

25. $y = a_0 + a_1x + a_2x^2 + \ldots + a_nx^n + \ldots \Rightarrow y'' = 2a_2 + 3 \cdot 2a_3x + \ldots + n(n-1)a_nx^{n-2} + \ldots \Rightarrow y'' - y = (2a_2 - a_0) + (3 \cdot 2a_3 - a_1)x + (4 \cdot 3a_4 - a_2)x^2 + \ldots + (n(n-1)a_n - a_{n-2})x^{n-2} + \ldots = 0 \Rightarrow 2a_2 - a_0 = 0$, $3 \cdot 2a_3 - a_1 = 0$, $4 \cdot 3a_4 - a_2 = 0$ and in general $n(n-1)a_n - a_{n-2} = 0$. Since $y' = 1$ and $y = 0$ when $x = 0$, we have $a_0 = 0$ and $a_1 = 1$. Therefore $a_2 = 0$, $a_3 = \frac{1}{3 \cdot 2}$, $a_4 = 0$, $a_5 = \frac{1}{5 \cdot 4 \cdot 3 \cdot 2}$, $\ldots$, $a_{2n+1} = \frac{1}{(2n+1)!}$ and $a_{2n} = 0 \Rightarrow y = x + \frac{1}{3!}x^3 + \frac{1}{5!}x^5 + \ldots = \sum_{n=0}^{\infty} \frac{x^{2n+1}}{(2n+1)!} = \sinh x$

27. $y = a_0 + a_1x + a_2x^2 + \ldots + a_nx^n + \ldots \Rightarrow y'' = 2a_2 + 3 \cdot 2a_3x + \ldots + n(n-1)a_nx^{n-2} + \ldots \Rightarrow y'' + y = (2a_2 + a_0) + (3 \cdot 2a_3 + a_1)x + (4 \cdot 3a_4 + a_2)x^2 + \ldots + (n(n-1)a_n + a_{n-2})x^{n-2} + \ldots = x \Rightarrow 2a_2 + a_0 = 0$,

$3\cdot 2a_3 + a_1 = 1$, $4\cdot 3a_4 + a_2 = 0$ and in general $n(n-1)a_n + a_{n-2} = 0$. Since $y' = 1$ and $y = 2$ when $x = 0$, we have $a_0 = 2$ and $a_1 = 1$. Therefore $a_2 = -1$, $a_3 = 0$, $a_4 = \frac{1}{4\cdot 3}$, $a_5 = 0, \ldots, a_{2n} = -2\cdot\frac{(-1)^{n+1}}{(2n)!}$ and

$$a_{2n+1} = 0 \Rightarrow y = 2 + x - x^2 + 2\cdot\frac{x^4}{4!} + \ldots = 2 + x - 2\sum_{n=1}^{\infty}\frac{(-1)^{n+1}x^{2n}}{(2n)!}$$

29. $y = a_0 + a_1x + a_2x^2 + \ldots + a_nx^n + \ldots \Rightarrow y'' = 2a_2 + 3\cdot 2a_3x + \ldots + n(n-1)a_nx^{n-2} + \ldots \Rightarrow y'' - y$
$= (2a_2 - a_0) + (3\cdot 2a_3 - a_1)(x-2) + (4\cdot 3a_4 - a_2)(x-2)^2 + \ldots + (n(n-1)a_n - a_{n-2})(x-2)^{n-2} + \ldots = 0$
$\Rightarrow 2a_2 - a_0 = 0$, $3\cdot 2a_3 - a_1 = 0$, $4\cdot 3a_4 - a_2 = 0$ and in general $n(n-1)a_n - a_{n-2} = 0$. Since $y' = -2$ and $y = 0$ when $x = 2$, we have $a_0 = 0$ and $a_1 = -2$. Therefore $a_2 = 0$, $a_3 = \frac{-2}{3\cdot 2}$, $a_4 = 0$, $a_5 = \frac{-2}{5!}, \ldots, a_{2n} = 0$, and

$$a_{2n+1} = \frac{-2}{(2n+1)!} \Rightarrow y = -2(x-2) - \frac{2}{3!}(x-2)^3 - \ldots = \sum_{n=0}^{\infty}\frac{-2(x-2)^{2n+1}}{(2n+1)!}$$

31. $y'' + x^2y = 2a_2 + 6a_3x + (4\cdot 3a_4 + a_0)x^2 + \ldots + (n(n-1)a_n + a_{n-4})x^{n-2} + \ldots = x \Rightarrow 2a_2 = 0$, $6a_3 = 1$, $4\cdot 3a_4 + a_0 = 0$, $5\cdot 4a_5 + a_1 = 0$, and in general $n(n-1)a_n + a_{n-4} = 0$. Since $y' = b$ and $y = a$ when $x = 0$, we have $a_0 = a$ and $a_1 = b$. Therefore $a_2 = 0$, $a_3 = \frac{1}{2\cdot 3}$, $a_4 = -\frac{a}{3\cdot 4}$, $a_5 = -\frac{b}{4\cdot 5}$, $a_6 = 0$, $a_7 = \frac{1}{2\cdot 3\cdot 6\cdot 7}$

$$\Rightarrow y = a + bx + \frac{1}{2\cdot 3}x^3 - \frac{a}{3\cdot 4}x^4 - \frac{b}{4\cdot 5}x^5 - \frac{1}{2\cdot 3\cdot 6\cdot 7}x^7 + \frac{ax^8}{3\cdot 4\cdot 7\cdot 8} + \frac{bx^9}{4\cdot 5\cdot 8\cdot 9} + \ldots$$

33. $\int_0^{0.2} \sin x^2\,dx = \int_0^{0.2}\left(x^2 - \frac{x^6}{3!} + \frac{x^{10}}{5!} - \ldots\right)dx = \left[\frac{x^3}{3} - \frac{x^7}{7\cdot 3!} + \ldots\right]_0^{0.2} \approx \left[\frac{x^3}{3}\right]_0^{0.2} \approx 0.00267$ with error $|E| \le \frac{(.2)^7}{7\cdot 3!} \approx 0.0000003$

35. $\int_0^{0.1}\frac{1}{\sqrt{1+x^4}}\,dx = \int_0^{0.1}\left(1 - \frac{x^4}{2} + \frac{3x^8}{8} - \ldots\right)dx = \left[x - \frac{x^5}{10} + \ldots\right]_0^{0.1} \approx [x]_0^{0.1} \approx 0.1$ with error $|E| \le \frac{(0.1)^5}{10} = 0.000001$

37. $\int_0^{0.1}\frac{\sin x}{x}\,dx = \int_0^{0.1}\left(1 - \frac{x^2}{3!} + \frac{x^4}{5!} - \frac{x^6}{7!} + \ldots\right)dx = \left[x - \frac{x^3}{3\cdot 3!} + \frac{x^5}{5\cdot 5!} - \frac{x^7}{7\cdot 7!} + \ldots\right]_0^{0.1} \approx \left[x - \frac{x^3}{3\cdot 3!} + \frac{x^5}{5\cdot 5!}\right]_0^{0.1}$
≈ 0.0999444611

39. $(1+x^4)^{1/2} = (1)^{1/2} + \frac{\left(\frac{1}{2}\right)}{1}(1)^{-1/2}(x^4) + \frac{\left(\frac{1}{2}\right)\left(-\frac{1}{2}\right)}{2!}(1)^{-3/2}(x^4)^2 + \frac{\left(\frac{1}{2}\right)\left(-\frac{1}{2}\right)\left(-\frac{3}{2}\right)}{3!}(1)^{-5/2}(x^4)^3$
$+ \frac{\left(\frac{1}{2}\right)\left(-\frac{1}{2}\right)\left(-\frac{3}{2}\right)\left(-\frac{5}{2}\right)}{4!}(1)^{-7/2}(x^4)^4 + \ldots = 1 + \frac{x^4}{2} - \frac{x^8}{8} + \frac{x^{12}}{16} - \frac{5x^{16}}{128} + \ldots$

$$\Rightarrow \int_0^{0.1}\left(1 + \frac{x^4}{2} - \frac{x^8}{8} + \frac{x^{12}}{16} - \frac{5x^{16}}{128} + \ldots\right)dx = \left[x + \frac{x^5}{10} - \frac{x^9}{72} + \frac{x^{13}}{208} - \frac{5x^{17}}{2176} + \ldots\right]_0^{0.1} \approx 0.100001$$

41. $\int_0^1 \cos t^2\,dt = \int_0^1 \left(1 - \frac{t^4}{2} + \frac{t^8}{4!} - \frac{t^{12}}{6!} + \ldots\right) dt = \left[t - \frac{t^5}{10} + \frac{t^9}{9\cdot 4!} - \frac{t^{13}}{13\cdot 6!} + \ldots\right]_0^1 \Rightarrow |\text{error}| < \frac{1}{13\cdot 6!} \approx .00011$

43. $F(x) = \int_0^x \left(t^2 - \frac{t^6}{3!} + \frac{t^{10}}{5!} - \frac{t^{14}}{7!} + \ldots\right) dt = \left[\frac{t^3}{3} - \frac{t^7}{7\cdot 3!} + \frac{t^{11}}{11\cdot 5!} - \frac{t^{15}}{15\cdot 7!} + \ldots\right]_0^x \approx \frac{x^3}{3} - \frac{x^7}{7\cdot 3!} + \frac{x^{11}}{11\cdot 5!}$

$\Rightarrow |\text{error}| < \frac{1}{15\cdot 7!} \approx 0.00002$

45. (a) $F(x) = \int_0^x \left(t - \frac{t^3}{3} + \frac{t^5}{5} - \frac{t^7}{7} + \ldots\right) dt = \left[\frac{t^2}{2} - \frac{t^4}{12} + \frac{t^6}{30} - \ldots\right]_0^x \approx \frac{x^2}{2} - \frac{x^4}{12} \Rightarrow |\text{error}| < \frac{(0.5)^6}{30} \approx .00052$

(b) $|\text{error}| < \frac{1}{33\cdot 34} \approx .00089$ so $F(x) \approx \frac{x^2}{2} - \frac{x^4}{3\cdot 4} + \frac{x^6}{5\cdot 6} - \frac{x^8}{7\cdot 8} + \ldots + (-1)^{15}\frac{x^{32}}{31\cdot 32}$

47. $\frac{1}{x^2}\left(e^x - (1+x)\right) = \frac{1}{x^2}\left(\left(1 + x + \frac{x^2}{2} + \frac{x^3}{3!} + \ldots\right) - 1 - x\right) = \frac{1}{2} + \frac{x}{3!} + \frac{x^2}{4!} + \ldots \Rightarrow \lim_{x\to 0} \frac{e^x - (1+x)}{x^2}$

$= \lim_{x\to 0}\left(\frac{1}{2} + \frac{x}{3!} + \frac{x^2}{4!} + \ldots\right) = \frac{1}{2}$

49. $\frac{1}{t^4}\left(1 - \cos t - \frac{t^2}{2}\right) = \frac{1}{t^4}\left[1 - \frac{t^2}{2} - \left(1 - \frac{t^2}{2} + \frac{t^4}{4!} - \frac{t^6}{6!} + \ldots\right)\right] = -\frac{1}{4!} + \frac{t^2}{6!} - \frac{t^4}{8!} + \ldots \Rightarrow \lim_{t\to 0} \frac{1 - \cos t - \left(\frac{t^2}{2}\right)}{t^4}$

$= \lim_{t\to 0}\left(-\frac{1}{4!} + \frac{t^2}{6!} - \frac{t^4}{8!} + \ldots\right) = -\frac{1}{24}$

51. $\frac{1}{y^3}\left(y - \tan^{-1} y\right) = \frac{1}{y^3}\left[y - \left(y - \frac{y^3}{3} + \frac{y^5}{5} - \ldots\right)\right] = \frac{1}{3} - \frac{y^2}{5} + \frac{y^4}{7} - \ldots \Rightarrow \lim_{y\to 0} \frac{y - \tan^{-1} y}{y^3} = \lim_{y\to 0}\left(\frac{1}{3} - \frac{y^2}{5} + \frac{y^4}{7} - \ldots\right)$

$= \frac{1}{3}$

53. $x^2\left(-1 + e^{-1/x^2}\right) = x^2\left(-1 + 1 - \frac{1}{x^2} + \frac{1}{2x^4} - \frac{1}{6x^6} + \ldots\right) = -1 + \frac{1}{2x^2} - \frac{1}{6x^4} + \ldots \Rightarrow \lim_{x\to\infty} x^2\left(e^{-1/x^2} - 1\right)$

$= \lim_{x\to\infty}\left(-1 + \frac{1}{2x^2} - \frac{1}{6x^4} + \ldots\right) = -1$

55. $\frac{\ln(1+x^2)}{1 - \cos x} = \frac{\left(x^2 - \frac{x^4}{2} + \frac{x^6}{3} - \ldots\right)}{1 - \left(1 - \frac{x^2}{2!} + \frac{x^4}{4!} - \ldots\right)} = \frac{\left(1 - \frac{x^2}{2} + \frac{x^4}{3} - \ldots\right)}{\left(\frac{1}{2!} - \frac{x^2}{4!} + \ldots\right)} \Rightarrow \lim_{x\to 0} \frac{\ln(1+x^2)}{1 - \cos x} = \lim_{x\to 0} \frac{\left(1 - \frac{x^2}{2} + \frac{x^4}{3} - \ldots\right)}{\left(\frac{1}{2!} - \frac{x^2}{4!} + \ldots\right)} = 2!$

$= 2$

57. $\ln\left(\frac{1+x}{1-x}\right) = \ln(1+x) - \ln(1-x) = \left(x - \frac{x^2}{2} + \frac{x^3}{3} - \frac{x^4}{4} + \ldots\right) - \left(-x - \frac{x^2}{2} - \frac{x^3}{3} - \frac{x^4}{4} - \ldots\right) = 2\left(x + \frac{x^3}{3} + \frac{x^5}{5} + \ldots\right)$

59. $\tan^{-1} x = x - \frac{x^3}{3} + \frac{x^5}{5} - \frac{x^7}{7} + \frac{x^9}{9} - \ldots + \frac{(-1)^{n-1}x^{2n-1}}{2n-1} + \ldots \Rightarrow |\text{error}| = \left|\frac{(-1)^{n-1}x^{2n-1}}{2n-1}\right| = \frac{1}{2n-1}$ when $x = 1$;

$\frac{1}{2n-1} < \frac{1}{10^3} \Rightarrow n > \frac{1001}{2} = 500.5 \Rightarrow$ the first term not used is the $501^{\text{st}} \Rightarrow$ we must use 500 terms

61. $\tan^{-1} x = x - \frac{x^3}{3} + \frac{x^5}{5} - \frac{x^7}{7} + \frac{x^9}{9} - \ldots + \frac{(-1)^{n-1}x^{2n-1}}{2n-1} + \ldots$ and when the series representing $48 \tan^{-1}\left(\frac{1}{18}\right)$ has an error of magnitude less than 10^{-6}, then the series representing the sum
$48 \tan^{-1}\left(\frac{1}{18}\right) + 32 \tan^{-1}\left(\frac{1}{57}\right) - 20 \tan^{-1}\left(\frac{1}{239}\right)$ also has an error of magnitude less than 10^{-6}; thus
$|\text{error}| = \frac{\left(\frac{1}{18}\right)^{2n-1}}{2n-1} < \frac{1}{10^6} \Rightarrow n \geq 3$ using a calculator $\Rightarrow$ 3 terms

63. (a) $\left(1 - x^2\right)^{-1/2} \approx 1 + \frac{x^2}{2} + \frac{3x^4}{8} + \frac{5x^6}{16} \Rightarrow \sin^{-1} x \approx x + \frac{x^3}{6} + \frac{3x^5}{40} + \frac{5x^7}{112}$;

$$\lim_{n\to\infty} \left|\frac{1\cdot3\cdot5\cdots(2n-1)(2n+1)x^{2n+3}}{2\cdot4\cdot6\cdots(2n)(2n+2)(2n+3)} \cdot \frac{2\cdot4\cdot6\cdots(2n)(2n+1)}{1\cdot3\cdot5\cdots(2n-1)x^{2n+1}}\right| < 1 \Rightarrow x^2 \lim_{n\to\infty} \left|\frac{(2n+1)(2n+1)}{(2n+2)(2n+3)}\right| < 1$$

$\Rightarrow |x| < 1 \Rightarrow$ the radius of convergence is 1

(b) $\frac{d}{dx}\left(\cos^{-1} x\right) = -\left(1 - x^2\right)^{-1/2} \Rightarrow \cos^{-1} x = \frac{\pi}{2} - \sin^{-1} x \approx \frac{\pi}{2} - \left(x + \frac{x^3}{6} + \frac{3x^5}{40} + \frac{5x^7}{112}\right) \approx \frac{\pi}{2} - x - \frac{x^3}{6} - \frac{3x^5}{40} - \frac{5x^7}{112}$

65. $\frac{-1}{1+x} = -\frac{1}{1-(-x)} = -1 + x - x^2 + x^3 - \ldots \Rightarrow \frac{d}{dx}\left(\frac{-1}{1+x}\right) = \frac{1}{1+x^2} = \frac{d}{dx}\left(-1 + x - x^2 + x^3 - \ldots\right)$
$= 1 - 2x + 3x^2 - 4x^3 + \ldots$

67. Wallis' formula gives the approximation $\pi \approx 4\left[\frac{2\cdot4\cdot4\cdot6\cdot6\cdot8\cdots(2n-2)\cdot(2n)}{3\cdot3\cdot5\cdot5\cdot7\cdot7\cdots(2n-1)\cdot(2n-1)}\right]$ to produce the table

n	$\sim\pi$
10	3.221088998
20	3.181104886
30	3.167880758
80	3.151425420
90	3.150331383
93	3.150049112
94	3.149959030
95	3.149870848
100	3.149456425

At $n = 1929$ we obtain the first approximation accurate to 3 decimals: 3.141999845. At $n = 30{,}000$ we still do not obtain accuracy to 4 decimals: 3.141617732, so the convergence to π is very slow. Here is a Maple CAS procedure to produce these approximations:

```
pie :=
  proc(n)
  local i,j;
    a(2) := evalf(8/9);
    for i from 3 to n do a(i) := evalf(2*(2*i-2)*i/(2*i-1)^2*a(i-1)) od;
    [[j,4*a(j)] $ (j = n-5 .. n)]
  end
```

69. $(1-x^2)^{-1/2} = (1+(-x^2))^{-1/2} = (1)^{-1/2} + \left(-\frac{1}{2}\right)(1)^{-3/2}(-x^2) + \frac{\left(-\frac{1}{2}\right)\left(-\frac{3}{2}\right)(1)^{-5/2}(-x^2)^2}{2!}$

$+ \frac{\left(-\frac{1}{2}\right)\left(-\frac{3}{2}\right)\left(-\frac{5}{2}\right)(1)^{-7/2}(-x^2)^3}{3!} + \ldots = 1 + \frac{x^2}{2} + \frac{1\cdot 3x^4}{2^2\cdot 2!} + \frac{1\cdot 3\cdot 5x^6}{2^3\cdot 3!} + \ldots = 1 + \sum_{n=1}^{\infty} \frac{1\cdot 3\cdot 5\cdots(2n-1)x^{2n}}{2^n\cdot n!}$

$\Rightarrow \sin^{-1} x = \int_0^x (1-t^2)^{-1/2}\,dt = \int_0^x \left(1 + \sum_{n=1}^{\infty} \frac{1\cdot 3\cdot 5\cdots(2n-1)x^{2n}}{2^n\cdot n!}\right) dt = x + \sum_{n=1}^{\infty} \frac{1\cdot 3\cdot 5\cdots(2n-1)x^{2n+1}}{2\cdot 4\cdots(2n)(2n+1)}$,

where $|x| < 1$

71. (a) $\tan\left(\tan^{-1}(n+1) - \tan^{-1}(n-1)\right) = \frac{\tan\left(\tan^{-1}(n+1)\right) - \tan\left(\tan^{-1}(n-1)\right)}{1+\tan\left(\tan^{-1}(n+1)\right)\tan\left(\tan^{-1}(n-1)\right)} = \frac{(n+1)-(n-1)}{1+(n+1)(n-1)} = \frac{2}{n^2}$

(b) $\sum_{n=1}^{N} \tan^{-1}\left(\frac{2}{n^2}\right) = \sum_{n=1}^{N} \left[\tan^{-1}(n+1) - \tan^{-1}(n-1)\right] = \left(\tan^{-1}2 - \tan^{-1}0\right) + \left(\tan^{-1}3 - \tan^{-1}1\right)$

$+ \left(\tan^{-1}4 - \tan^{-1}2\right) + \ldots + \left(\tan^{-1}(N+1) - \tan^{-1}(N-1)\right) = \tan^{-1}(N+1) + \tan^{-1}N - \frac{\pi}{4}$

(c) $\sum_{n=1}^{\infty} \tan^{-1}\left(\frac{2}{n^2}\right) = \lim_{n\to\infty} \left[\tan^{-1}(N+1) + \tan^{-1}N - \frac{\pi}{4}\right] = \frac{\pi}{2} + \frac{\pi}{2} - \frac{\pi}{4} = \frac{3\pi}{4}$

CHAPTER 8 PRACTICE EXERCISES

1. converges to 1, since $\lim_{n\to\infty} a_n = \lim_{n\to\infty} \left(1 + \frac{(-1)^n}{n}\right) = 1$

3. converges to -1, since $\lim_{n\to\infty} a_n = \lim_{n\to\infty} \left(\frac{1-2^n}{2^n}\right) = \lim_{n\to\infty} \left(\frac{1}{2^n} - 1\right) = -1$

5. diverges, since $\left\{\sin \frac{n\pi}{2}\right\} = \{0, 1, 0, -1, 0, 1, \ldots\}$

7. converges to 0, since $\lim_{n\to\infty} a_n = \lim_{n\to\infty} \frac{\ln n^2}{n} = 2 \lim_{n\to\infty} \frac{\left(\frac{1}{n}\right)}{1} = 0$

9. converges to 1, since $\lim_{n\to\infty} a_n = \lim_{n\to\infty} \left(\frac{n+\ln n}{n}\right) = \lim_{n\to\infty} \frac{1+\left(\frac{1}{n}\right)}{1} = 1$

11. converges to e^{-5}, since $\lim_{n\to\infty} a_n = \lim_{n\to\infty} \left(\frac{n-5}{n}\right)^n = \lim_{n\to\infty} \left(1 + \frac{(-5)}{n}\right)^n = e^{-5}$ by Table 8.1

13. converges to 3, since $\lim_{n\to\infty} a_n = \lim_{n\to\infty} \left(\frac{3^n}{n}\right)^{1/n} = \lim_{n\to\infty} \frac{3}{n^{1/n}} = \frac{3}{1} = 3$ by Table 8.1

15. converges to ln 2, since $\lim_{n\to\infty} a_n = \lim_{n\to\infty} n\left(2^{1/n} - 1\right) = \lim_{n\to\infty} \frac{2^{1/n}-1}{\left(\frac{1}{n}\right)} = \lim_{n\to\infty} \frac{\left[\frac{\left(-2^{1/n}\ln 2\right)}{n^2}\right]}{\left(\frac{-1}{n^2}\right)} = \lim_{n\to\infty} 2^{1/n}\ln 2$
$= 2^0\cdot \ln 2 = \ln 2$

17. diverges, since $\lim_{n\to\infty} a_n = \lim_{n\to\infty} \frac{(n+1)!}{n!} = \lim_{n\to\infty} (n+1) = \infty$

19. $\frac{1}{(2n-3)(2n-1)} = \frac{\left(\frac{1}{2}\right)}{2n-3} - \frac{\left(\frac{1}{2}\right)}{2n-1} \Rightarrow s_n = \left[\frac{\left(\frac{1}{2}\right)}{3} - \frac{\left(\frac{1}{2}\right)}{5}\right] + \left[\frac{\left(\frac{1}{2}\right)}{5} - \frac{\left(\frac{1}{2}\right)}{7}\right] + \ldots + \left[\frac{\left(\frac{1}{2}\right)}{2n-3} - \frac{\left(\frac{1}{2}\right)}{2n-1}\right] = \frac{\left(\frac{1}{2}\right)}{3} - \frac{\left(\frac{1}{2}\right)}{2n-1}$

$\Rightarrow \lim_{n\to\infty} s_n = \lim_{n\to\infty} \left[\frac{1}{6} - \frac{\left(\frac{1}{2}\right)}{2n-1}\right] = \frac{1}{6}$

21. $\frac{9}{(3n-1)(3n+2)} = \frac{3}{3n-1} - \frac{3}{3n+2} \Rightarrow s_n = \left(\frac{3}{2} - \frac{3}{5}\right) + \left(\frac{3}{5} - \frac{3}{8}\right) + \left(\frac{3}{8} - \frac{3}{11}\right) + \ldots + \left(\frac{3}{3n-1} - \frac{3}{3n+2}\right)$

$= \frac{3}{2} - \frac{3}{3n+2} \Rightarrow \lim_{n\to\infty} s_n = \lim_{n\to\infty} \left(\frac{3}{2} - \frac{3}{3n+2}\right) = \frac{3}{2}$

23. $\sum_{n=0}^{\infty} e^{-n} = \sum_{n=0}^{\infty} \frac{1}{e^n}$, a convergent geometric series with $r = \frac{1}{e}$ and $a = 1 \Rightarrow$ the sum is $\frac{1}{1-\left(\frac{1}{e}\right)} = \frac{e}{e-1}$

25. diverges, a p-series with $p = \frac{1}{2}$

27. Since $f(x) = \frac{1}{x^{1/2}} \Rightarrow f'(x) = -\frac{1}{2x^{3/2}} < 0 \Rightarrow f(x)$ is decreasing $\Rightarrow a_{n+1} < a_n$, and $\lim_{n\to\infty} a_n = \lim_{n\to\infty} \frac{(-1)}{\sqrt{n}} = 0$, the series $\sum_{n=1}^{\infty} \frac{(-1)^n}{\sqrt{n}}$ converges by the Alternating Series Test. Since $\sum_{n=1}^{\infty} \frac{1}{\sqrt{n}}$ diverges, the given series converges conditionally.

29. The given series does not converge absolutely by the Direct Comparison Test since $\frac{1}{\ln(n+1)} > \frac{1}{n+1}$, which is the nth term of a divergent series. Since $f(x) = \frac{1}{\ln(x+1)} \Rightarrow f'(x) = -\frac{1}{(\ln(x+1))^2(x+1)} < 0 \Rightarrow f(x)$ is decreasing $\Rightarrow a_{n+1} < a_n$, and $\lim_{n\to\infty} a_n = \lim_{n\to\infty} \frac{1}{\ln(n+1)} = 0$, the given series converges conditionally by the Alternating Series Test.

31. converges absolutely by the Direct Comparison Test since $\frac{\ln n}{n^3} < \frac{n}{n^3} = \frac{1}{n^2}$, the nth term of a convergent p-series

33. $\lim_{n\to\infty} \frac{\left(\frac{1}{n\sqrt{n^2+1}}\right)}{\left(\frac{1}{n^2}\right)} = \sqrt{\lim_{n\to\infty} \frac{n^2}{n^2+1}} = \sqrt{1} = 1 \Rightarrow$ converges absolutely by the Limit Comparison Test

35. converges absolutely by the Ratio Test since $\lim_{n\to\infty} \left[\frac{n+2}{(n+1)!} \cdot \frac{n!}{n+1}\right] = \lim_{n\to\infty} \frac{n+2}{(n+1)^2} = 0 < 1$

37. converges absolutely by the Ratio Test since $\lim_{n\to\infty} \left[\frac{3^{n+1}}{(n+1)!} \cdot \frac{n!}{3^n}\right] = \lim_{n\to\infty} \frac{3}{n+1} = 0 < 1$

39. converges absolutely by the Limit Comparison Test since $\lim\limits_{n\to\infty} \dfrac{\left(\frac{1}{n^{3/2}}\right)}{\left(\frac{1}{\sqrt{n(n+1)(n+2)}}\right)} = \sqrt{\lim\limits_{n\to\infty} \dfrac{n(n+1)(n+2)}{n^3}}$ $= 1$

41. $\lim\limits_{n\to\infty} \left|\frac{u_{n+1}}{u_n}\right| < 1 \Rightarrow \lim\limits_{n\to\infty} \left|\frac{(x+4)^{n+1}}{(n+1)3^{n+1}} \cdot \frac{n3^n}{(x+4)^n}\right| < 1 \Rightarrow \frac{|x+4|}{3} \lim\limits_{n\to\infty} \left(\frac{n}{n+1}\right) < 1 \Rightarrow \frac{|x+4|}{3} < 1$

$\Rightarrow |x+4| < 3 \Rightarrow -3 < x+4 < 3 \Rightarrow -7 < x < -1$; at $x = -7$ we have $\sum\limits_{n=1}^{\infty} \frac{(-1)^n 3^n}{n3^n} = \sum\limits_{n=1}^{\infty} \frac{(-1)^n}{n}$, the alternating harmonic series, which converges conditionally; at $x = -1$ we have $\sum\limits_{n=1}^{\infty} \frac{3^n}{n3^n} = \sum\limits_{n=1}^{\infty} \frac{1}{n}$, the divergent harmonic series

(a) the radius is 3; the interval of convergence is $-7 \le x < -1$

(b) the interval of absolute convergence is $-7 < x < -1$

(c) the series converges conditionally at $x = -7$

43. $\lim\limits_{n\to\infty} \left|\frac{u_{n+1}}{u_n}\right| < 1 \Rightarrow \lim\limits_{n\to\infty} \left|\frac{(3x-1)^{n+1}}{(n+1)^2} \cdot \frac{n^2}{(3x-1)^n}\right| < 1 \Rightarrow |3x-1| \lim\limits_{n\to\infty} \frac{n^2}{(n+1)^2} < 1 \Rightarrow |3x-1| < 1$

$\Rightarrow -1 < 3x-1 < 1 \Rightarrow 0 < 3x < 2 \Rightarrow 0 < x < \frac{2}{3}$; at $x = 0$ we have $\sum\limits_{n=1}^{\infty} \frac{(-1)^{n-1}(-1)^n}{n^2} = \sum\limits_{n=1}^{\infty} \frac{(-1)^{2n-1}}{n^2}$ $= -\sum\limits_{n=1}^{\infty} \frac{1}{n^2}$, a nonzero constant multiple of a convergent p-series, which is absolutely convergent; at $x = \frac{2}{3}$ we have $\sum\limits_{n=1}^{\infty} \frac{(-1)^{n-1}(1)^n}{n^2} = \sum\limits_{n=1}^{\infty} \frac{(-1)^{n-1}}{n^2}$, which converges absolutely

(a) the radius is $\frac{1}{3}$; the interval of convergence is $0 \le x \le \frac{2}{3}$

(b) the interval of absolute convergence is $0 \le x \le \frac{2}{3}$

(c) there are no values for which the series converges conditionally

45. $\lim\limits_{n\to\infty} \left|\frac{u_{n+1}}{u_n}\right| < 1 \Rightarrow \lim\limits_{n\to\infty} \left|\frac{x^{n+1}}{(n+1)^{n+1}} \cdot \frac{n^n}{x^n}\right| < 1 \Rightarrow |x| \lim\limits_{n\to\infty} \left|\left(\frac{n}{n+1}\right)^n \left(\frac{1}{n+1}\right)\right| < 1 \Rightarrow \frac{|x|}{e} \lim\limits_{n\to\infty} \left(\frac{1}{n+1}\right) < 1$

$\Rightarrow \frac{|x|}{e} \cdot 0 < 1$, which holds for all x

(a) the radius is ∞; the series converges for all x

(b) the series converges absolutely for all x

(c) there are no values for which the series converges conditionally

47. $\lim\limits_{n\to\infty} \left|\frac{u_{n+1}}{u_n}\right| < 1 \Rightarrow \lim\limits_{n\to\infty} \left|\frac{(n+2)x^{2n+1}}{3^{n+1}} \cdot \frac{3^n}{(n+1)x^{2n-1}}\right| < 1 \Rightarrow \frac{x^2}{3} \lim\limits_{n\to\infty} \left(\frac{n+2}{n+1}\right) < 1 \Rightarrow -\sqrt{3} < x < \sqrt{3}$;

the series $\sum\limits_{n=1}^{\infty} -\frac{n+1}{\sqrt{3}}$ and $\sum\limits_{n=1}^{\infty} \frac{n+1}{\sqrt{3}}$, obtained with $x = \pm\sqrt{3}$, both diverge

(a) the radius is $\sqrt{3}$; the interval of convergence is $-\sqrt{3} < x < \sqrt{3}$

(b) the interval of absolute convergence is $-\sqrt{3} < x < \sqrt{3}$

(c) there are no values for which the series converges conditionally

49. $\lim\limits_{n\to\infty} \left|\frac{u_{n+1}}{u_n}\right| < 1 \Rightarrow \lim\limits_{n\to\infty} \left|\frac{\operatorname{csch}(n+1)x^{n+1}}{\operatorname{csch}(n)x^n}\right| < 1 \Rightarrow |x| \lim\limits_{n\to\infty} \left|\frac{\left(\frac{2}{e^{n+1}-e^{-n-1}}\right)}{\left(\frac{2}{e^n-e^{-n}}\right)}\right| < 1$

$\Rightarrow |x| \lim\limits_{n\to\infty} \left|\frac{e^{-1}-e^{-2n-1}}{1-e^{-2n-2}}\right| < 1 \Rightarrow \frac{|x|}{e} < 1 \Rightarrow -e < x < e$; the series $\sum\limits_{n=1}^{\infty} (\pm e)^n \operatorname{csch} n$, obtained with $x = \pm e$,

both diverge since $\lim\limits_{n\to\infty} (\pm e)^n \operatorname{csch} n \neq 0$

(a) the radius is e; the interval of convergence is $-e < x < e$

(b) the interval of absolute convergence is $-e < x < e$

(c) there are no values for which the series converges conditionally

51. The given series has the form $1 - x + x^2 - x^3 + \ldots + (-x)^n + \ldots = \frac{1}{1+x}$, where $x = \frac{1}{4}$; the sum is $\frac{1}{1+\left(\frac{1}{4}\right)} = \frac{4}{5}$

53. The given series has the form $x - \frac{x^3}{3!} + \frac{x^5}{5!} - \ldots + (-1)^n \frac{x^{2n+1}}{(2n+1)!} + \ldots = \sin x$, where $x = \pi$; the sum is $\sin \pi = 0$

55. The given series has the form $1 + x + \frac{x^2}{2!} + \frac{x^2}{3!} + \ldots + \frac{x^n}{n!} + \ldots = e^x$, where $x = \ln 2$; the sum is $e^{\ln(2)} = 2$

57. Consider $\frac{1}{1-2x}$ as the sum of a convergent geometric series with $a = 1$ and $r = 2x \Rightarrow \frac{1}{1-2x}$

$= 1 + (2x) + (2x)^2 + (2x)^3 + \ldots = \sum\limits_{n=0}^{\infty} (2x)^n = \sum\limits_{n=0}^{\infty} 2^n x^n$ where $|2x| < 1 \Rightarrow |x| < \frac{1}{2}$

59. $\sin x = \sum\limits_{n=0}^{\infty} \frac{(-1)^n x^{2n+1}}{(2n+1)!} \Rightarrow \sin \pi x = \sum\limits_{n=0}^{\infty} \frac{(-1)^n (\pi x)^{2n+1}}{(2n+1)!} = \sum\limits_{n=0}^{\infty} \frac{(-1)^n \pi^{2n+1} x^{2n+1}}{(2n+1)!}$

61. $\cos x = \sum\limits_{n=0}^{\infty} \frac{(-1)^n x^{2n}}{(2n)!} \Rightarrow \cos\left(x^{5/2}\right) = \sum\limits_{n=0}^{\infty} \frac{(-1)^n \left(x^{5/2}\right)^{2n}}{(2n)!} = \sum\limits_{n=0}^{\infty} \frac{(-1)^n x^{5n}}{(2n)!}$

63. $e^x = \sum\limits_{n=0}^{\infty} \frac{x^n}{n!} \Rightarrow e^{(\pi x/2)} = \sum\limits_{n=0}^{\infty} \frac{\left(\frac{\pi x}{2}\right)^n}{n!} = \sum\limits_{n=0}^{\infty} \frac{\pi^n x^n}{2^n n!}$

65. $f(x) = \sqrt{3+x^2} = \left(3+x^2\right)^{1/2} \Rightarrow f'(x) = x\left(3+x^2\right)^{-1/2} \Rightarrow f''(x) = -x^2\left(3+x^2\right)^{-3/2} + \left(3+x^2\right)^{-1/2}$

$\Rightarrow f'''(x) = 3x^3\left(3+x^2\right)^{-5/2} - 3x\left(3+x^2\right)^{-3/2}$; $f(-1) = 2$, $f'(-1) = -\frac{1}{2}$, $f''(-1) = -\frac{1}{8} + \frac{1}{2} = \frac{3}{8}$,

$f'''(-1) = -\frac{3}{32} + \frac{3}{8} = \frac{9}{32} \Rightarrow \sqrt{3+x^2} = 2 - \frac{(x+1)}{2 \cdot 1!} + \frac{3(x+1)^2}{2^3 \cdot 2!} + \frac{9(x+1)^3}{2^5 \cdot 3!} + \ldots$

67. $f(x) = \frac{1}{x+1} = (x+1)^{-1} \Rightarrow f'(x) = -(x+1)^{-2} \Rightarrow f''(x) = 2(x+1)^{-3} \Rightarrow f'''(x) = -6(x+1)^{-4};\ f(3) = \frac{1}{4},$
$f'(3) = -\frac{1}{4^2},\ f''(3) = \frac{2}{4^3},\ f'''(2) = \frac{-6}{4^4} \Rightarrow \frac{1}{x+1} = \frac{1}{4} - \frac{1}{4^2}(x-3) + \frac{1}{4^3}(x-3)^2 - \frac{1}{4^4}(x-3)^3 + \ldots$

69. Assume the solution has the form $y = a_0 + a_1x + a_2x^2 + \ldots + a_{n-1}x^{n-1} + a_nx^n + \ldots$
$\Rightarrow \frac{dy}{dx} = a_1 + 2a_2x + \ldots + na_nx^{n-1} + \ldots \Rightarrow \frac{dy}{dx} + y$
$= (a_1 + a_0) + (2a_2 + a_1)x + (3a_3 + a_2)x^2 + \ldots + (na_n + a_{n-1})x^{n-1} + \ldots = 0 \Rightarrow a_1 + a_0 = 0,\ 2a_2 + a_1 = 0,$
$3a_3 + a_2 = 0$ and in general $na_n + a_{n-1} = 0$. Since $y = -1$ when $x = 0$ we have $a_0 = -1$. Therefore $a_1 = 1$,
$a_2 = \frac{-a_1}{2 \cdot 1} = -\frac{1}{2},\ a_3 = \frac{-a_2}{3} = \frac{1}{3 \cdot 2},\ a_4 = \frac{-a_3}{4} = -\frac{1}{4 \cdot 3 \cdot 2}, \ldots,\ a_n = \frac{-a_{n-1}}{n} = \frac{-1}{n}\frac{(-1)^n}{(n-1)!} = \frac{(-1)^{n+1}}{n!}$
$\Rightarrow y = -1 + x - \frac{1}{2}x^2 + \frac{1}{3 \cdot 2}x^3 - \ldots + \frac{(-1)^{n+1}}{n!}x^n + \ldots = -\sum_{n=0}^{\infty} \frac{(-1)^n x^n}{n!} = -e^{-x}$

71. Assume the solution has the form $y = a_0 + a_1x + a_2x^2 + \ldots + a_{n-1}x^{n-1} + a_nx^n + \ldots$
$\Rightarrow \frac{dy}{dx} = a_1 + 2a_2x + \ldots + na_nx^{n-1} + \ldots \Rightarrow \frac{dy}{dx} + 2y$
$= (a_1 + 2a_0) + (2a_2 + 2a_1)x + (3a_3 + 2a_2)x^2 + \ldots + (na_n + 2a_{n-1})x^{n-1} + \ldots = 0$. Since $y = 3$ when $x = 0$ we have $a_0 = 3$. Therefore $a_1 = -2a_0 = -2(3) = -3(2),\ a_2 = -\frac{2}{2}a_1 = -\frac{2}{2}(-2 \cdot 3) = 3\left(\frac{2^2}{2}\right),\ a_3 = -\frac{2}{3}a_2$
$= -\frac{2}{3}\left[3\left(\frac{2^2}{2}\right)\right] = -3\left(\frac{2^3}{3 \cdot 2}\right), \ldots,\ a_n = \left(-\frac{2}{n}\right)a_{n-1} = \left(-\frac{2}{n}\right)\left(3\left(\frac{(-1)^{n-1}2^{n-1}}{(n-1)!}\right)\right) = 3\left(\frac{(-1)^n 2^n}{n!}\right)$
$\Rightarrow y = 3 - 3(2x) + 3\frac{(2)^2}{2}x^2 - 3\frac{(2)^3}{3 \cdot 2}x^3 + \ldots + 3\frac{(-1)^n 2^n}{n!}x^n + \ldots$
$= 3\left[1 - (2x) + \frac{(2x)^2}{2!} - \frac{(2x)^3}{3!} + \ldots + \frac{(-1)^n(2x)^n}{n!} + \ldots\right] = 3\sum_{n=0}^{\infty} \frac{(-1)^n(2x)^n}{n!} = 3e^{-2x}$

73. Assume the solution has the form $y = a_0 + a_1x + a_2x^2 + \ldots + a_{n-1}x^{n-1} + a_nx^n + \ldots$
$\Rightarrow \frac{dy}{dx} = a_1 + 2a_2x + \ldots + na_nx^{n-1} + \ldots \Rightarrow \frac{dy}{dx} - y$
$= (a_1 - a_0) + (2a_2 - a_1)x + (3a_3 - a_2)x^2 + \ldots + (na_n - a_{n-1})x^{n-1} + \ldots = 3x \Rightarrow a_1 - a_0 = 0,\ 2a_2 - a_1 = 3,$
$3a_3 - a_2 = 0$ and in general $na_n - a_{n-1} = 0$ for $n > 2$. Since $y = -1$ when $x = 0$ we have $a_0 = -1$. Therefore
$a_1 = -1,\ a_2 = \frac{3 + a_1}{2} = \frac{2}{2},\ a_3 = \frac{a_2}{3} = \frac{2}{3 \cdot 2},\ a_4 = \frac{a_3}{4} = \frac{2}{4 \cdot 3 \cdot 2}, \ldots,\ a_n = \frac{a_{n-1}}{n} = \frac{2}{n!}$
$\Rightarrow y = -1 - x + \left(\frac{2}{2}\right)x^2 + \frac{3}{3 \cdot 2}x^3 + \frac{2}{4 \cdot 3 \cdot 2}x^4 + \ldots + \frac{2}{n!}x^n + \ldots$
$= 2\left(1 + x + \frac{1}{2}x^2 + \frac{1}{3 \cdot 2}x^3 + \frac{1}{4 \cdot 3 \cdot 2}x^4 + \ldots + \frac{1}{n!}x^n + \ldots\right) - 3 - 3x = 2\sum_{n=0}^{\infty} \frac{x^n}{n!} - 3 - 3x = 2e^x - 3x - 3$

75. Assume the solution has the form $y = a_0 + a_1x + a_2x^2 + \ldots + a_{n-1}x^{n-1} + a_nx^n + \ldots$
$\Rightarrow \frac{dy}{dx} = a_1 + 2a_2x + \ldots + na_nx^{n-1} + \ldots \Rightarrow \frac{dy}{dx} - y$
$= (a_1 - a_0) + (2a_2 - a_1)x + (3a_3 - a_2)x^2 + \ldots + (na_n - a_{n-1})x^{n-1} + \ldots = x \Rightarrow a_1 - a_0 = 0,\ 2a_2 - a_1 = 1,$
$3a_3 - a_2 = 0$ and in general $na_n - a_{n-1} = 0$ for $n > 2$. Since $y = 1$ when $x = 0$ we have $a_0 = 1$. Therefore

$a_1 = 1,\ a_2 = \frac{1+a_1}{2} = \frac{2}{2},\ a_3 = \frac{a_2}{3} = \frac{2}{3\cdot 2},\ a_4 = \frac{a_3}{4} = \frac{2}{4\cdot 3\cdot 2}, \ldots,\ a_n = \frac{a_{n-1}}{n} = \frac{2}{n!}$

$\Rightarrow y = 1 + x + \left(\frac{2}{2}\right)x^2 + \frac{2}{3\cdot 2}x^3 + \frac{2}{4\cdot 2\cdot 2}x^4 + \ldots + \frac{2}{n!}x^n + \ldots$

$= 2\left(1 + x + \frac{1}{2}x^2 + \frac{1}{3\cdot 2}x^3 + \frac{1}{4\cdot 3\cdot 2}x^4 + \ldots + \frac{1}{n!}x^n + \ldots\right) - 1 - x = 2\sum_{n=0}^{\infty}\frac{x^n}{n!} - 1 - x = 2e^x - x - 1$

77. $\int_0^{1/2} \exp(-x^3)\,dx = \int_0^{1/2}\left(1 - x^3 + \frac{x^6}{2!} - \frac{x^9}{3!} + \frac{x^{12}}{4!} + \ldots\right)dx = \left[x - \frac{x^4}{4} + \frac{x^7}{7\cdot 2!} - \frac{x^{10}}{10\cdot 3!} + \frac{x^{13}}{13\cdot 4!} - \ldots\right]_0^{1/2}$

$\approx \frac{1}{2} - \frac{1}{2^4\cdot 4} + \frac{1}{2^7\cdot 7\cdot 2!} - \frac{1}{2^{10}\cdot 10\cdot 3!} + \frac{1}{2^{13}\cdot 13\cdot 4!} - \frac{1}{2^{16}\cdot 16\cdot 5!} \approx 0.484917143$

79. $\int_1^{1/2} \frac{\tan^{-1}x}{x}\,dx = \int_1^{1/2}\left(1 - \frac{x^2}{3} + \frac{x^4}{5} - \frac{x^6}{7} + \frac{x^8}{9} - \frac{x^{10}}{11} + \ldots\right)dx = \left[x - \frac{x^3}{9} + \frac{x^5}{25} - \frac{x^7}{49} + \frac{x^9}{81} - \frac{x^{11}}{121} + \ldots\right]_0^{1/2}$

$\approx \frac{1}{2} - \frac{1}{9\cdot 2^3} + \frac{1}{5^2\cdot 2^5} - \frac{1}{7^2\cdot 2^7} + \frac{1}{9^2\cdot 2^9} - \frac{1}{11^2\cdot 2^{11}} + \frac{1}{13^2\cdot 2^{13}} - \frac{1}{15^2\cdot 2^{15}} + \frac{1}{17^2\cdot 2^{17}} - \frac{1}{19^2\cdot 2^{19}} + \frac{1}{21^2\cdot 2^{21}}$

≈ 0.4872223583

81. $\lim_{x\to 0} \frac{7\sin x}{e^{2x}-1} = \lim_{x\to 0} \frac{7\left(x - \frac{x^3}{3!} + \frac{x^5}{5!} - \ldots\right)}{\left(2x + \frac{2^2x^2}{2!} + \frac{2^3x^3}{3!} + \ldots\right)} = \lim_{x\to 0} \frac{7\left(1 - \frac{x^2}{3!} + \frac{x^4}{5!} - \ldots\right)}{\left(2 + \frac{2^2x}{2!} + \frac{2^3x^2}{3!} + \ldots\right)} = \frac{7}{2}$

83. $\lim_{t\to 0}\left(\frac{1}{2-2\cos t} - \frac{1}{t^2}\right) = \lim_{t\to 0} \frac{t^2 - 2 + 2\cos t}{2t^2(1-\cos t)} = \lim_{t\to 0} \frac{t^2 - 2 + 2\left(1 - \frac{t^2}{2} + \frac{t^4}{4!} - \ldots\right)}{2t^2\left(1 - 1 + \frac{t^2}{2} - \frac{t^4}{4!} + \ldots\right)} = \lim_{t\to 0} \frac{2\left(\frac{t^4}{4!} - \frac{t^6}{6!} + \ldots\right)}{\left(t^4 - \frac{2t^6}{4!} + \ldots\right)}$

$= \lim_{t\to 0} \frac{2\left(\frac{1}{4!} - \frac{t^2}{6!} + \ldots\right)}{\left(1 - \frac{2t^2}{4!} + \ldots\right)} = \frac{1}{12}$

85. $\lim_{z\to 0} \frac{1-\cos^2 z}{\ln(1-z) + \sin z} = \lim_{z\to 0} \frac{1 - \left(1 - z^2 + \frac{z^4}{3} - \ldots\right)}{\left(-z - \frac{z^2}{2} - \frac{z^3}{3} - \ldots\right) + \left(z - \frac{z^3}{3!} + \frac{z^5}{5!} - \ldots\right)} = \lim_{z\to 0} \frac{\left(z^2 - \frac{z^4}{3} + \ldots\right)}{\left(-\frac{z^2}{2} - \frac{2z^3}{3} - \frac{z^4}{4} - \ldots\right)}$

$= \lim_{z\to 0} \frac{\left(1 - \frac{z^2}{3} + \ldots\right)}{\left(-\frac{1}{2} - \frac{2z}{3} - \frac{z^2}{4} - \ldots\right)} = -2$

87. $\lim_{x\to 0}\left(\frac{\sin 3x}{x^3}+\frac{r}{x^2}+s\right)=\lim_{x\to 0}\left[\frac{\left(3x-\frac{(3x)^3}{6}+\frac{(3x)^5}{120}-\ldots\right)}{x^3}+\frac{r}{x^2}+s\right]=\lim_{x\to 0}\left(\frac{3}{x^2}-\frac{9}{2}+\frac{81x^2}{40}+\ldots+\frac{r}{x^2}+s\right)=0$

$\Rightarrow \frac{r}{x^2}+\frac{3}{x^2}=0$ and $s-\frac{9}{2}=0 \Rightarrow r=-3$ and $s=\frac{9}{2}$

89. (a) $\sum_{n=1}^{\infty}\left(\sin\frac{1}{2n}-\sin\frac{1}{2n+1}\right)=\left(\sin\frac{1}{2}-\sin\frac{1}{3}\right)+\left(\sin\frac{1}{4}-\sin\frac{1}{5}\right)+\left(\sin\frac{1}{6}-\sin\frac{1}{7}\right)+\ldots+\left(\sin\frac{1}{2n}-\sin\frac{1}{2n+1}\right)$

$+\ldots=\sum_{n=2}^{\infty}(-1)^n\sin\frac{1}{n}$; $f(x)=\sin\frac{1}{x}\Rightarrow f'(x)=\frac{-\cos\left(\frac{1}{x}\right)}{x^2}<0$ if $x\ge 2\Rightarrow \sin\frac{1}{n+1}<\sin\frac{1}{n}$, and

$\lim_{n\to\infty}\sin\frac{1}{n}=0\Rightarrow\sum_{n=2}^{\infty}(-1)^n\sin\frac{1}{n}$ converges by the Alternating Series Test

(b) $|\text{error}|<\left|\sin\frac{1}{42}\right|\approx 0.02381$ and the sum is an underestimate because the remainder is positive

91. $\lim_{n\to\infty}\left|\frac{2\cdot5\cdot8\cdots(3n-1)(3n+2)x^{n+1}}{2\cdot4\cdot6\cdots(2n)(2n+2)}\cdot\frac{2\cdot4\cdot6\cdots(2n)}{2\cdot5\cdot8\cdots(3n-1)x^n}\right|<1\Rightarrow|x|\lim_{n\to\infty}\left|\frac{3n+2}{2n+2}\right|<1\Rightarrow|x|<\frac{2}{3}$

$\Rightarrow$ the radius of convergence is $\frac{2}{3}$

93. $\sum_{k=2}^{n}\ln\left(1-\frac{1}{k^2}\right)=\sum_{k=2}^{n}\left[\ln\left(1+\frac{1}{k}\right)+\ln\left(1-\frac{1}{k}\right)\right]=\sum_{k=2}^{n}\left[\ln(k+1)-\ln k+\ln(k-1)-\ln k\right]$

$=[\ln 3-\ln 2+\ln 1-\ln 2]+[\ln 4-\ln 3+\ln 2-\ln 3]+[\ln 5-\ln 4+\ln 3-\ln 4]+[\ln 6-\ln 5+\ln 4-\ln 5]$

$+\ldots+[\ln(n+1)-\ln n+\ln(n-1)-\ln n]=[\ln 1-\ln 2]+[\ln(n+1)-\ln n]$ after cancellation

$\Rightarrow\sum_{k=2}^{n}\ln\left(1-\frac{1}{k^2}\right)=\ln\left(\frac{n+1}{2n}\right)\Rightarrow\sum_{k=2}^{\infty}\ln\left(1-\frac{1}{k^2}\right)=\lim_{n\to\infty}\ln\left(\frac{n+1}{2n}\right)=\ln\frac{1}{2}$ is the sum

95. (a) $\lim_{n\to\infty}\left|\frac{1\cdot4\cdot7\cdots(3n-2)(3n+1)x^{3n+3}}{(3n+3)!}\cdot\frac{(3n)!}{1\cdot4\cdot7\cdots(3n-2)x^{3n}}\right|<1\Rightarrow|x^3|\lim_{n\to\infty}\frac{(3n+1)}{(3n+1)(3n+2)(3n+3)}$

$=|x^3|\cdot 0<1\Rightarrow$ the radius of convergence is ∞

(b) $y=1+\sum_{n=1}^{\infty}\frac{1\cdot4\cdot7\cdots(3n-2)}{(3n)!}x^{3n}\Rightarrow\frac{dy}{dx}=\sum_{n=1}^{\infty}\frac{1\cdot4\cdot7\cdots(3n-2)}{(3n-1)!}x^{3n-1}$

$\Rightarrow\frac{d^2y}{dx^2}=\sum_{n=1}^{\infty}\frac{1\cdot4\cdot7\cdots(3n-2)}{(3n-2)!}x^{3n-2}=x+\sum_{n=2}^{\infty}\frac{1\cdot4\cdot7\cdots(3n-5)}{(3n-3)!}x^{3n-2}$

$=x\left(1+\sum_{n=1}^{\infty}\frac{1\cdot4\cdot7\cdots(3n-2)}{(3n)!}x^{3n}\right)=xy+0\Rightarrow a=1$ and $b=0$

97. Yes, the series $\sum_{n=1}^{\infty}a_nb_n$ converges as we now show. Since $\sum_{n=1}^{\infty}a_n$ converges it follows that $a_n\to 0\Rightarrow a_n<1$ for $n>$ some index $N\Rightarrow a_nb_n<b_n$ for $n>N\Rightarrow\sum_{n=1}^{\infty}a_nb_n$ converges by the Direct Comparison Test with $\sum_{n=1}^{\infty}b_n$

99. $\sum_{n=1}^{\infty} (x_{n+1} - x_n) = \lim_{n\to\infty} \sum_{k=1}^{\infty} (x_{k+1} - x_k) = \lim_{n\to\infty} (x_{n+1} - x_1) = \lim_{n\to\infty} (x_{n+1}) - x_1 \Rightarrow$ both the series and sequence must either converge or diverge.

101. Newton's method gives $x_{n+1} = x_n - \frac{(x_n - 1)^{40}}{40(x_n - 1)^{39}} = \frac{39}{40}x_n + \frac{1}{40}$, and if the sequence $\{x_n\}$ has the limit L, then $L = \frac{39}{40}L + \frac{1}{40} \Rightarrow L = 1$

103. (a) $T = \frac{\left(\frac{1}{2}\right)}{2}\left(0 + 2\left(\frac{1}{2}\right)^2 e^{1/2} + e\right) = \frac{1}{8}e^{1/2} + \frac{1}{4}e \approx 0.885660616$

(b) $x^2e^x = x^2\left(1 + x + \frac{x^2}{2} + \ldots\right) = x^2 + x^3 + \frac{x^4}{2} + \ldots \Rightarrow \int_0^1 \left(x^2 + x^3 + \frac{x^4}{2}\right) dx = \left[\frac{x^3}{3} + \frac{x^4}{4} + \frac{x^5}{10}\right]_0^1 = \frac{41}{60} = 0.6833\overline{3}$

(c) If the second derivative is positive, the curve is concave upward and the polygonal line segments used in the trapezoidal rule lie above the curve. The trapezoidal approximation is therefore greater than the actual area under the graph.

(d) All terms in the Maclaurin series are positive. If we truncate the series, we are omitting positive terms and hence the estimate is too small.

(e) $\int_0^1 x^2e^x\, dx = \left[x^2e^x - 2xe^x + 2e^x\right]_0^1 = e - 2e + 2e - 2 = e - 2 \approx 0.7182818285$

CHAPTER 8 ADDITIONAL EXERCISES–THEORY, EXAMPLES, APPLICATIONS

1. converges since $\frac{1}{(3n-2)^{(2n+1)/2}} < \frac{1}{(3n-2)^{3/2}}$ and $\sum_{n=1}^{\infty} \frac{1}{(3n-2)^{3/2}}$ converges by the Limit Comparison Test: $\lim_{n\to\infty} \frac{\left(\frac{1}{n^{3/2}}\right)}{\left(\frac{1}{(3n-2)^{3/2}}\right)} = \lim_{n\to\infty} \left(\frac{3n-2}{n}\right)^{3/2} = 3^{3/2}$

3. diverges by the nth-Term Test since $\lim_{n\to\infty} a_n = \lim_{n\to\infty} (-1)^n \tanh n = \lim_{b\to\infty} (-1)^n\left(\frac{1-e^{-2n}}{1+e^{-2n}}\right) = \lim_{n\to\infty} (-1)^n$ does not exist

5. converges by the Direct Comparison Test: $a_1 = 1 = \frac{12}{(1)(3)(2)^2}$, $a_2 = \frac{1\cdot 2}{3\cdot 4} = \frac{12}{(2)(4)(3)^2}$, $a_3 = \left(\frac{2\cdot 3}{4\cdot 5}\right)\left(\frac{1\cdot 2}{3\cdot 4}\right) = \frac{12}{(3)(5)(4)^2}$, $a_4 = \left(\frac{3\cdot 4}{5\cdot 6}\right)\left(\frac{2\cdot 3}{4\cdot 5}\right)\left(\frac{1\cdot 2}{3\cdot 4}\right) = \frac{12}{(4)(6)(5)^2}$, $\ldots \Rightarrow 1 + \sum_{n=1}^{\infty} \frac{12}{(n+1)(n+3)(n+2)^2}$ represents the given series and $\frac{12}{(n+1)(n+3)(n+2)^2} < \frac{12}{n^4}$, which is the nth-term of a convergent p-series

7. diverges by the nth-Term Test since if $a_n \to L$ as $n \to \infty$, then $L = \frac{1}{1+L} \Rightarrow L^2 + L - 1 = 0 \Rightarrow L = \frac{-1 \pm \sqrt{5}}{2}$ $\neq 0$

9. $f(x) = \cos x$ with $a = \frac{\pi}{3} \Rightarrow f\left(\frac{\pi}{3}\right) = 0.5$, $f'\left(\frac{\pi}{3}\right) = -\frac{\sqrt{3}}{2}$, $f''\left(\frac{\pi}{3}\right) = -0.5$, $f'''\left(\frac{\pi}{3}\right) = \frac{\sqrt{3}}{2}$, $f^{(4)}\left(\frac{\pi}{3}\right) = 0.5$;

$$\cos x = \frac{1}{2} - \frac{\sqrt{3}}{2}\left(x - \frac{\pi}{3}\right) - \frac{1}{4}\left(x - \frac{\pi}{3}\right)^2 + \frac{\sqrt{3}}{12}\left(x - \frac{\pi}{3}\right)^3 + \ldots$$

11. $e^x = 1 + x + \frac{x^2}{2!} + \frac{x^3}{2!} + \ldots$ with $a = 0$

13. $f(x) = \cos x$ with $a = 22\pi \Rightarrow f(22\pi) = 1$, $f'(22\pi) = 0$, $f''(22\pi) = -1$, $f'''(22\pi) = 0$, $f^{(4)}(22\pi) = 1$, $f^{(5)}(22\pi) = 0$, $f^{(6)}(22\pi) = -1$; $\cos x = 1 - \frac{1}{2}(x - 22\pi)^2 + \frac{1}{4!}(x - 22\pi)^4 - \frac{1}{6!}(x - 22\pi)^6 + \ldots$

15. Yes, the sequence converges: $c_n = \left(a^n + b^n\right)^{1/n} \Rightarrow c_n = b\left(\left(\frac{a}{b}\right)^n + 1\right)^{1/n} \Rightarrow \lim\limits_{n\to\infty} c_n = \lim\limits_{n\to\infty} b\left(\left(\frac{a}{b}\right)^n + 1\right)^{1/n} = b$ since $0 < a < b$

17. $s_n = \sum\limits_{k=0}^{n-1} \int_k^{k+1} \frac{dx}{1+x^2} \Rightarrow s_n = \int_0^1 \frac{dx}{1+x^2} + \int_1^2 \frac{dx}{1+x^2} + \ldots + \int_{n-1}^{n} \frac{dx}{1+x^2} \Rightarrow s_n = \int_0^n \frac{dx}{1+x^2}$

$$\Rightarrow \lim_{n\to\infty} s_n = \lim_{n\to\infty}\left(\tan^{-1} n - \tan^{-1} 0\right) = \frac{\pi}{2}$$

19. (a) Each A_{n+1} fits into the corresponding upper triangular region, whose vertices are: $(n, f(n) - f(n+1))$, $(n+1, f(n+1))$ and $(n, f(n))$ along the line whose slope is $f(n+2) - f(n+1)$. All the A_n's fit into the first upper triangular region whose area is $\frac{f(1) - f(2)}{2} \Rightarrow \sum\limits_{n=1}^{\infty} A_n < \frac{f(1) - f(2)}{2}$

(b) If $A_k = \frac{f(k+1) + f(k)}{2} - \int_k^{k+1} f(x)\,dx$, then

$$\sum_{k=1}^{n-1} A_k = \frac{f(1) + f(2) + f(2) + f(3) + f(3) + \ldots + f(n-1) + f(n)}{2} - \int_1^2 f(x)\,dx - \int_2^3 f(x)\,dx - \ldots - \int_{n-1}^{n} f(x)\,dx$$

$$= \frac{f(1) + f(n)}{2} + \sum_{k=2}^{n-1} f(k) - \int_1^n f(x)\,dx \Rightarrow \sum_{k=1}^{n-1} A_k = \sum_{k=1}^{n} f(k) - \frac{f(1) + f(n)}{2} - \int_1^n f(x)\,dx < \frac{f(1) - f(2)}{2},$$ from

part (a). The sequence $\left\{\sum\limits_{k=1}^{n-1} A_k\right\}$ is bounded above and increasing, so it converges and the limit in question must exist.

(c) From part (b) we have $\sum\limits_{k=1}^{\infty} f(k) - \int_1^n f(x)\,dx < f(1) - \frac{f(2)}{2} + \frac{f(n)}{2}$

$$\Rightarrow \lim_{n\to\infty}\left[\sum_{k=1}^{n} f(k) - \int_1^n f(x)\,dx\right] < \lim_{n\to\infty}\left[f(1) - \frac{f(2)}{2} + \frac{f(n)}{2}\right] = f(1) - \frac{f(2)}{2}.$$ The sequence

$\left\{\sum\limits_{k=1}^{n} f(k) - \int_1^n f(x)\,dx\right\}$ is bounded and increasing, so it converges and the limit in question must exist.

31. $8x^2 - 2y^2 = 16 \Rightarrow \frac{x^2}{2} - \frac{y^2}{8} = 1 \Rightarrow c = \sqrt{a^2 + b^2}$
$= \sqrt{2+8} = \sqrt{10}$; asymptotes are $y = \pm 2x$

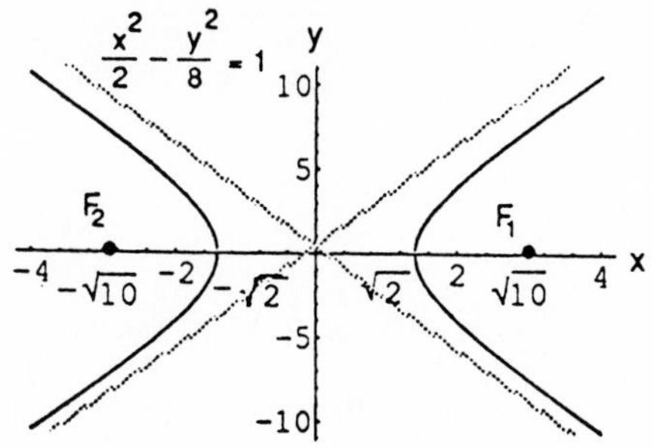

33. $8y^2 - 2x^2 = 16 \Rightarrow \frac{y^2}{2} - \frac{x^2}{8} = 1 \Rightarrow c = \sqrt{a^2 + b^2}$
$= \sqrt{2+8} = \sqrt{10}$; asymptotes are $y = \pm \frac{x}{2}$

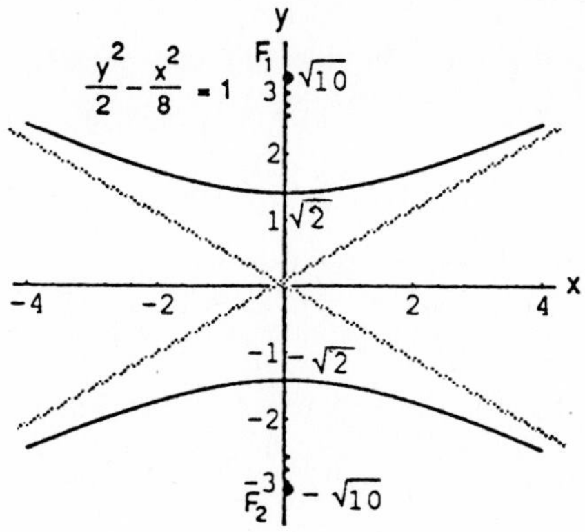

35. Foci: $(0, \pm\sqrt{2})$, Asymptotes: $y = \pm x \Rightarrow c = \sqrt{2}$ and $\frac{b}{a} = 1 \Rightarrow a = b \Rightarrow c^2 = a^2 + b^2 = 2a^2 \Rightarrow 2 = 2a^2$
$\Rightarrow a = 1 \Rightarrow b = 1 \Rightarrow y^2 - x^2 = 1$

37. Vertices: $(\pm 3, 0)$, Asymptotes: $y = \pm\frac{4}{3}x \Rightarrow a = 3$ and $\frac{b}{a} = \frac{4}{3} \Rightarrow b = \frac{4}{3}(3) = 4 \Rightarrow \frac{x^2}{9} - \frac{y^2}{16} = 1$

39. (a) $y^2 = 8x \Rightarrow 4p = 8 \Rightarrow p = 2 \Rightarrow$ directrix is $x = -2$, focus is $(2,0)$, and vertex is $(0,0)$; therefore the new directrix is $x = -1$, the new focus is $(3,-2)$, and the new vertex is $(1,-2)$

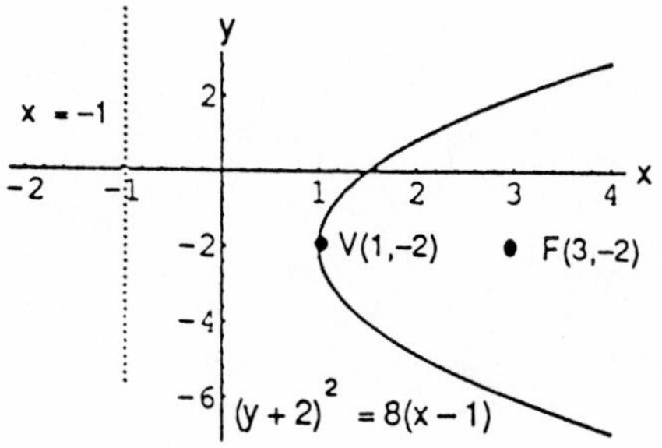

41. (a) $\frac{x^2}{16} + \frac{y^2}{9} = 1 \Rightarrow$ center is $(0,0)$, vertices are $(-4,0)$ and $(4,0)$; $c = \sqrt{a^2 - b^2} = \sqrt{7} \Rightarrow$ foci are $(\sqrt{7}, 0)$ and $(-\sqrt{7}, 0)$; therefore the new center is $(4,3)$, the new vertices are $(0,3)$ and $(8,3)$, and the new foci are $(4 \pm \sqrt{7}, 3)$

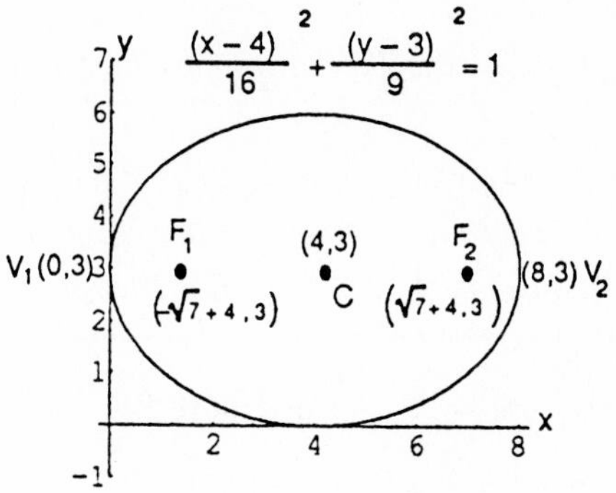

43. (a) $\frac{x^2}{16} - \frac{y^2}{9} = 1 \Rightarrow$ center is $(0,0)$, vertices are $(-4,0)$ and $(4,0)$, and the asymptotes are $\frac{x}{4} = \pm\frac{y}{3}$ or $y = \pm\frac{3x}{4}$; $c = \sqrt{a^2+b^2} = \sqrt{25} = 5 \Rightarrow$ foci are $(-5,0)$ and $(5,0)$; therefore the new center is $(2,0)$, the new vertices are $(-2,0)$ and $(6,0)$, the new foci are $(-3,0)$ and $(7,0)$, and the new asymptotes are $y = \pm\frac{3(x-2)}{4}$

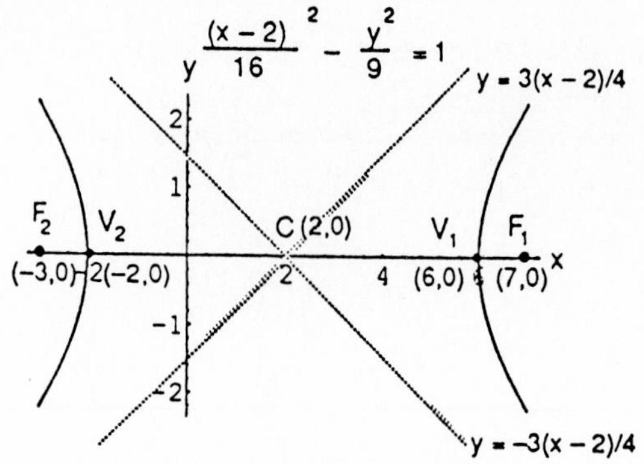

45. $y^2 = 4x \Rightarrow 4p = 4 \Rightarrow p = 1 \Rightarrow$ focus is $(1,0)$, directrix is $x = -1$, and vertex is $(0,0)$; therefore the new vertex is $(-2,-3)$, the new focus is $(-1,-3)$, and the new directrix is $x = -3$; the new equation is $(y+3)^2 = 4(x+2)$

47. $x^2 = 8y \Rightarrow 4p = 8 \Rightarrow p = 2 \Rightarrow$ focus is $(0,2)$, directrix is $y = -2$, and vertex is $(0,0)$; therefore the new vertex is $(1,-7)$, the new focus is $(1,-5)$, and the new directrix is $y = -9$; the new equation is $(x-1)^2 = 8(y+7)$

49. $\frac{x^2}{6} + \frac{y^2}{9} = 1 \Rightarrow$ center is $(0,0)$, vertices are $(0,3)$ and $(0,-3)$; $c = \sqrt{a^2-b^2} = \sqrt{9-6} = \sqrt{3} \Rightarrow$ foci are $(0,\sqrt{3})$ and $(0,-\sqrt{3})$; therefore the new center is $(-2,-1)$, the new vertices are $(-2,2)$ and $(-2,-4)$, and the new foci are $(-2,-1\pm\sqrt{3})$; the new equation is $\frac{(x+2)^2}{6} + \frac{(y+1)^2}{9} = 1$

51. $\frac{x^2}{3} + \frac{y^2}{2} = 1 \Rightarrow$ center is $(0,0)$, vertices are $(\sqrt{3},0)$ and $(-\sqrt{3},0)$; $c = \sqrt{a^2-b^2} = \sqrt{3-2} = 1 \Rightarrow$ foci are $(-1,0)$ and $(1,0)$; therefore the new center is $(2,3)$, the new vertices are $(2\pm\sqrt{3},3)$, and the new foci are $(1,3)$ and $(3,3)$; the new equation is $\frac{(x-2)^2}{3} + \frac{(y-3)^2}{2} = 1$

53. $\frac{x^2}{4} - \frac{y^2}{5} = 1 \Rightarrow$ center is $(0,0)$, vertices are $(2,0)$ and $(-2,0)$; $c = \sqrt{a^2+b^2} = \sqrt{4+5} = 3 \Rightarrow$ foci are $(3,0)$ and $(-3,0)$; the asymptotes are $\pm\frac{x}{2} = \frac{y}{\sqrt{5}} \Rightarrow y = \pm\frac{\sqrt{5}x}{2}$; therefore the new center is $(2,2)$, the new vertices are $(4,2)$ and $(0,2)$, and the new foci are $(5,2)$ and $(-1,2)$; the new asymptotes are $y - 2 = \pm\frac{\sqrt{5}(x-2)}{2}$; the new equation is $\frac{(x-2)^2}{4} - \frac{(y-2)^2}{5} = 1$

55. $y^2 - x^2 = 1 \Rightarrow$ center is $(0,0)$, vertices are $(0,1)$ and $(0,-1)$; $c = \sqrt{a^2+b^2} = \sqrt{1+1} = \sqrt{2} \Rightarrow$ foci are $(0,\pm\sqrt{2})$; the asymptotes are $y = \pm x$; therefore the new center is $(-1,-1)$, the new vertices are $(-1,0)$ and $(-1,-2)$, and the new foci are $(-1,-1\pm\sqrt{2})$; the new asymptotes are $y+1 = \pm(x+1)$; the new equation is $(y+1)^2 - (x+1)^2 = 1$

57. $x^2 + 4x + y^2 = 12 \Rightarrow x^2 + 4x + 4 + y^2 = 12 + 4 \Rightarrow (x+2)^2 + y^2 = 16$; this is a circle: center at $C(-2,0)$, $a = 4$

59. $x^2 + 2x + 4y - 3 = 0 \Rightarrow x^2 + 2x + 1 = -4y + 3 + 1 \Rightarrow (x+1)^2 = -4(y-1)$; this is a parabola: $V(-1,1)$, $F(-1,0)$

61. $x^2 + 5y^2 + 4x = 1 \Rightarrow x^2 + 4x + 4 + 5y^2 = 5 \Rightarrow (x+2)^2 + 5y^2 = 5 \Rightarrow \frac{(x+2)^2}{5} + y^2 = 1$; this is an ellipse: the center is $(-2,0)$, the vertices are $\left(-2 \pm \sqrt{5}, 0\right)$; $c = \sqrt{a^2 - b^2} = \sqrt{5-1} = 2 \Rightarrow$ the foci are $(-4,0)$ and $(0,0)$

63. $x^2 + 2y^2 - 2x - 4y = -1 \Rightarrow x^2 - 2x + 1 + 2\left(y^2 - 2y + 1\right) = 2 \Rightarrow (x-1)^2 + 2(y-1)^2 = 2$
$\Rightarrow \frac{(x-1)^2}{2} + (y-1)^2 = 1$; this is an ellipse: the center is $(1,1)$, the vertices are $\left(1 \pm \sqrt{2}, 1\right)$;
$c = \sqrt{a^2 - b^2} = \sqrt{2-1} = 1 \Rightarrow$ the foci are $(2,1)$ and $(0,1)$

65. $x^2 - y^2 - 2x + 4y = 4 \Rightarrow x^2 - 2x + 1 - \left(y^2 - 4y + 4\right) = 1 \Rightarrow (x-1)^2 - (y-2)^2 = 1$; this is a hyperbola: the center is $(1,2)$, the vertices are $(2,2)$ and $(0,2)$; $c = \sqrt{a^2 + b^2} = \sqrt{1+1} = \sqrt{2} \Rightarrow$ the foci are $\left(1 \pm \sqrt{2}, 2\right)$; the asymptotes are $y - 2 = \pm(x-1)$

67. $2x^2 - y^2 + 6y = 3 \Rightarrow 2x^2 - \left(y^2 - 6y + 9\right) = -6 \Rightarrow \frac{(y-3)^2}{6} - \frac{x^2}{3} = 1$; this is a hyperbola: the center is $(0,3)$, the vertices are $\left(0, 3 \pm \sqrt{6}\right)$; $c = \sqrt{a^2 + b^2} = \sqrt{6+3} = 3 \Rightarrow$ the foci are $(0,6)$ and $(0,0)$; the asymptotes are $\frac{y-3}{\sqrt{6}} = \pm \frac{x}{\sqrt{3}} \Rightarrow y = \pm\sqrt{2}x + 3$

69.

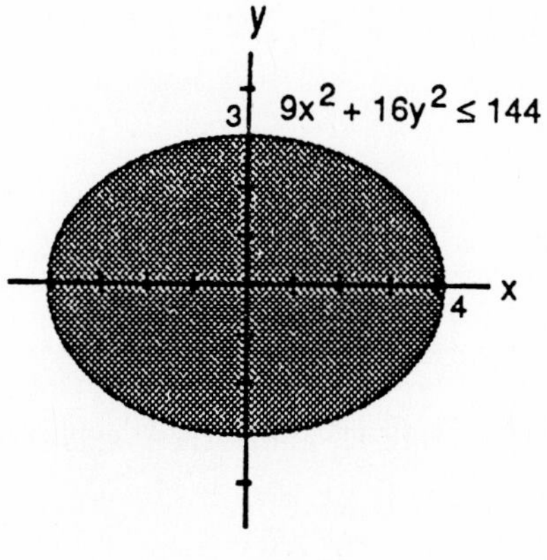

71.

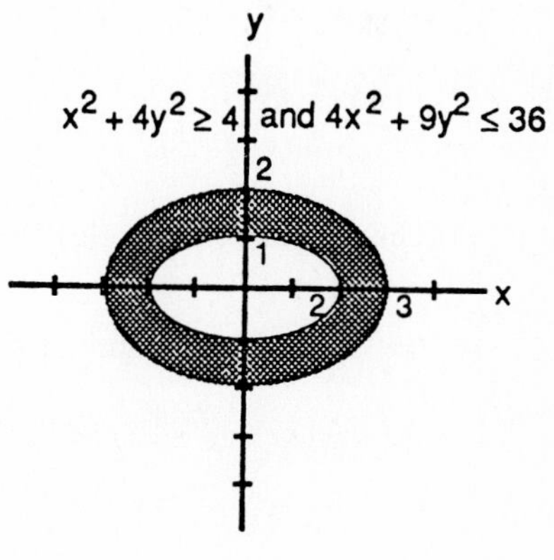

73.

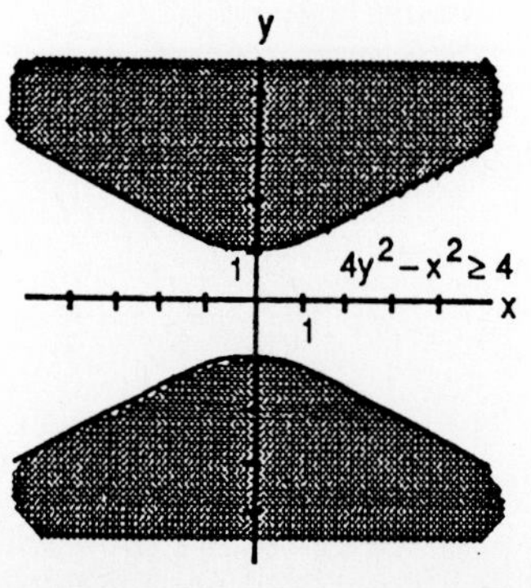

75. Volume of the Parabolic Solid: $V_1 = \int_0^{b/2} 2\pi x\left(h - \frac{4h}{b^2}x^2\right) dx = 2\pi h \int_0^{b/2} \left(x - \frac{4x^3}{b^2}\right) dx = 2\pi h\left[\frac{x^2}{2} - \frac{x^4}{b^2}\right]_0^{b/2}$

$= \frac{\pi h b^2}{8}$; Volume of the Cone: $V_2 = \frac{1}{3}\pi\left(\frac{b}{2}\right)^2 h = \frac{1}{3}\pi\left(\frac{b^2}{4}\right)h = \frac{\pi h b^2}{12}$; therefore $V_1 = \frac{3}{2}V_2$

77. A general equation of the circle is $x^2 + y^2 + ax + by + c = 0$, so we will substitute the three given points into this equation and solve the resulting system: $\left.\begin{aligned} a \quad + c &= -1 \\ b + c &= -1 \\ 2a + 2b + c &= -8 \end{aligned}\right\} \Rightarrow c = \frac{4}{3}$ and $a = b = -\frac{7}{3}$; therefore

$3x^2 + 3y^2 - 7x - 7y + 4 = 0$ represents the circle

79. $r^2 = (-2-1)^2 + (1-3)^2 = 13 \Rightarrow (x+2)^2 + (y-1)^2 = 13$ is an equation of the circle; the distance from the center to $(1.1, 2.8)$ is $\sqrt{(-2-1.1)^2 + (1-2.8)^2} = \sqrt{12.85} < \sqrt{13}$, the radius $\Rightarrow$ the point is inside the circle

81. (a) $y^2 = kx \Rightarrow x = \frac{y^2}{k}$; the volume of the solid formed by revolving R_1 about the y-axis is $V_1 = \int_0^{\sqrt{kx}} \pi\left(\frac{y^2}{k}\right)^2 dy$

$= \frac{\pi}{k^2}\int_0^{\sqrt{kx}} y^4\, dy = \frac{\pi x^2\sqrt{kx}}{5}$; the volume of the right circular cylinder formed by revolving PQ about the y-axis

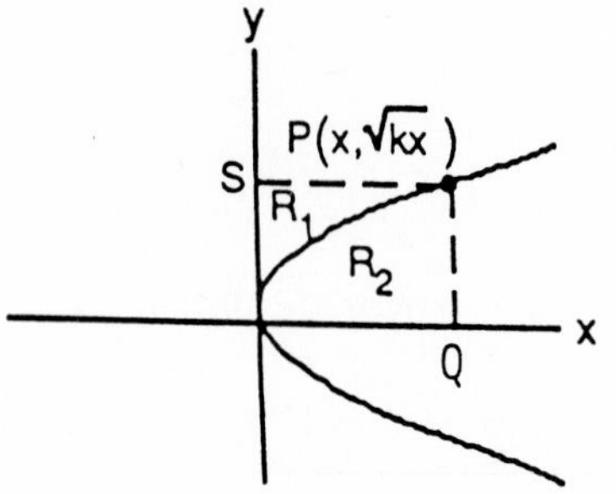

is $V_2 = \pi x^2\sqrt{kx} \Rightarrow$ the volume of the solid formed by revolving R_2 about the y-axis is $V_3 = V_2 - V_1$

$= \frac{4\pi x^2\sqrt{kx}}{5}$. Therefore we can see the ratio of V_3 to V_1 is 4:1.

(b) The volume of the solid formed by revolving R_2 about the x-axis is $V_1 = \int_0^x \pi\left(\sqrt{kt}\right)^2 dt = \pi k \int_0^x t\, dt$

$= \frac{\pi k x^2}{2}$. The volume of the right circular cylinder formed by revolving PS about the x-axis is

$V_2 = \pi\left(\sqrt{kx}\right)^2 x = \pi k x^2 \Rightarrow$ the volume of the solid formed by revolving R_1 about the x-axis is $V_3 = V_2 - V_1 = \pi k x^2 - \frac{\pi k x^2}{2} = \frac{\pi k x^2}{2}$. Therefore the ratio of V_3 to V_1 is 1:1.

83. Let $y = \sqrt{1 - \frac{x^2}{4}}$ on the interval $0 \le x \le 2$. The area of the inscribed rectangle is given by

$A(x) = 2x\left(2\sqrt{1 - \frac{x^2}{4}}\right) = 4x\sqrt{1 - \frac{x^2}{4}}$ (since the length is 2x and the height is 2y)

$\Rightarrow A'(x) = 4\sqrt{1 - \frac{x^2}{4}} - \frac{x^2}{\sqrt{1 - \frac{x^2}{4}}}$. Thus $A'(x) = 0 \Rightarrow 4\sqrt{1 - \frac{x^2}{4}} - \frac{x^2}{\sqrt{1 - \frac{x^2}{4}}} = 0 \Rightarrow 4\left(1 - \frac{x^2}{4}\right) - x^2 = 0 \Rightarrow x^2 = 2$

$\Rightarrow x = \sqrt{2}$ (only the positive square root lies in the interval). Since $A(0) = A(2) = 0$ we have that $A(\sqrt{2}) = 4$ is the maximum area when the length is $2\sqrt{2}$ and the height is $\sqrt{2}$.

85. $9x^2 - 4y^2 = 36 \Rightarrow y^2 = \frac{9x^2-36}{4} \Rightarrow y = \pm\frac{3}{2}\sqrt{x^2-4}$ on the interval $2 \le x \le 4 \Rightarrow V = \int_2^4 \pi\left(\frac{3}{2}\sqrt{x^2-4}\right)^2 dx$

$= \frac{9\pi}{4}\int_2^4 (x^2-4)\,dx = \frac{9\pi}{4}\left[\frac{x^3}{3} - 4x\right]_2^4 = \frac{9\pi}{4}\left[\left(\frac{64}{3} - 16\right) - \left(\frac{8}{3} - 8\right)\right] = \frac{9\pi}{4}\left(\frac{56}{3} - 8\right) = \frac{3\pi}{4}(56-24) = 24\pi$

87. Let $y = \sqrt{16 - \frac{16}{9}x^2}$ on the interval $-3 \le x \le 3$. Since the plate is symmetric about the y-axis, $\bar{x} = 0$. For a vertical strip: $(\tilde{x}, \tilde{y}) = \left(x, \frac{\sqrt{16-\frac{16}{9}x^2}}{2}\right)$, length $= \sqrt{16 - \frac{16}{9}x^2}$, width $= dx \Rightarrow$ area $= dA = \sqrt{16 - \frac{16}{9}x^2}\,dx$

$\Rightarrow$ mass $= dm = \delta\,dA = \delta\sqrt{16 - \frac{16}{9}x^2}\,dx$. Moment of the strip about the x-axis:

$\tilde{y}\,dm = \frac{\sqrt{16-\frac{16}{9}x^2}}{2}\left(\delta\sqrt{16 - \frac{16}{9}x^2}\right)dx = \delta\left(8 - \frac{8}{9}x^2\right)dx$ so the moment of the plate about the x-axis is

$M_x = \int \tilde{y}\,dm = \int_{-3}^{3} \delta\left(8 - \frac{8}{9}x^2\right)dx = \delta\left[8x - \frac{8}{27}x^3\right]_{-3}^{3} = 32\delta$; also the mass of the plate is

$M = \int_{-3}^{3} \delta\sqrt{16 - \frac{16}{9}x^2}\,dx = \int_{-3}^{3} 4\delta\sqrt{1-\left(\frac{1}{3}x\right)^2}\,dx = 4\delta\int_{-1}^{1} 3\sqrt{1-u^2}\,du$ where $u = \frac{x}{3} \Rightarrow 3\,du = dx$; $x = -3$

$\Rightarrow u = -1$ and $x = 3 \Rightarrow u = 1$. Hence, $4\delta\int_{-1}^{1} 3\sqrt{1-u^2}\,du = 12\delta\int_{-1}^{1}\sqrt{1-u^2}\,du$

$= 12\delta\left[\frac{1}{2}\left(u\sqrt{1-u^2} + \sin^{-1}u\right)\right]_{-1}^{1} = 6\pi\delta \Rightarrow \bar{y} = \frac{M_x}{M} = \frac{32\delta}{6\pi\delta} = \frac{16}{3\pi}$. Therefore the center of mass is $\left(0, \frac{16}{3\pi}\right)$.

89. $\frac{dr_A}{dt} = \frac{dr_B}{dt} \Rightarrow \frac{d}{dt}(r_A - r_B) = 0 \Rightarrow r_A - r_B = C$, a constant $\Rightarrow$ the points P(t) lie on a hyperbola with foci at A and B

91. PF will always equal PB because the string has constant length AB = FP + PA = AP + PB.

93. $x^2 = 4py$ and $y = p \Rightarrow x^2 = 4p^2 \Rightarrow x = \pm 2p$. Therefore the line $y = p$ cuts the parabola at points $(-2p, p)$ and $(2p, p)$, and these points are $\sqrt{[2p-(-2p)]^2 + (p-p)^2} = 4p$ units apart.

9.2 CLASSIFYING CONIC SECTIONS BY ECCENTRICITY

1. $16x^2 + 25y^2 = 400 \Rightarrow \frac{x^2}{25} + \frac{y^2}{16} = 1 \Rightarrow c = \sqrt{a^2 - b^2}$
$= \sqrt{25 - 16} = 3 \Rightarrow e = \frac{c}{a} = \frac{3}{5}$; $F(\pm 3, 0)$;
directrices are $x = 0 \pm \frac{a}{e} = \pm \frac{5}{\left(\frac{3}{5}\right)} = \pm \frac{25}{3}$

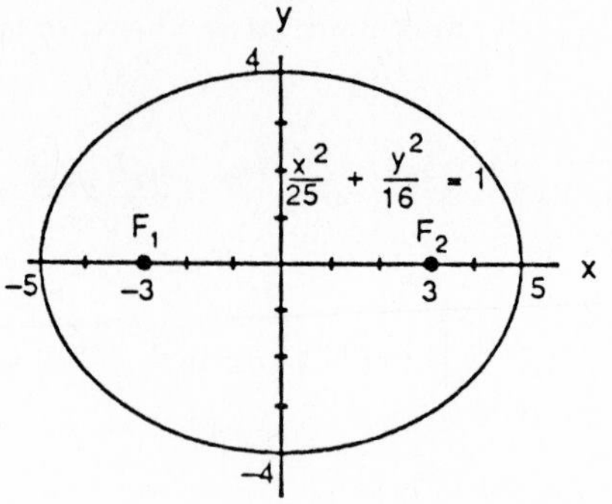

3. $2x^2 + y^2 = 2 \Rightarrow x^2 + \frac{y^2}{2} = 1 \Rightarrow c = \sqrt{a^2 - b^2}$
$= \sqrt{2 - 1} = 1 \Rightarrow e = \frac{c}{a} = \frac{1}{\sqrt{2}}$; $F(0, \pm 1)$;
directrices are $y = 0 \pm \frac{a}{e} = \pm \frac{\sqrt{2}}{\left(\frac{1}{\sqrt{2}}\right)} = \pm 2$

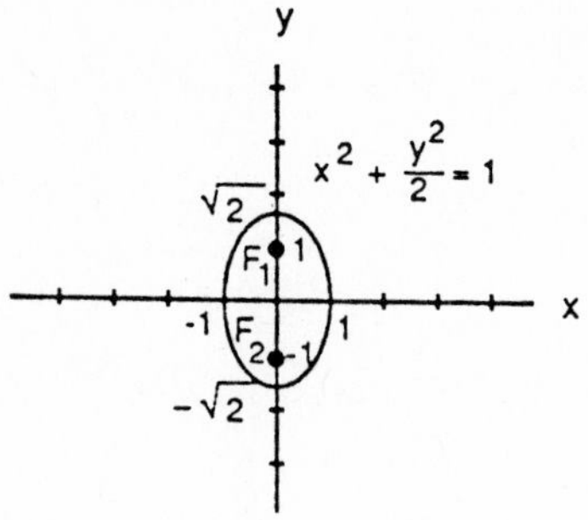

5. $3x^2 + 2y^2 = 6 \Rightarrow \frac{x^2}{2} + \frac{y^2}{3} = 1 \Rightarrow c = \sqrt{a^2 - b^2}$
$= \sqrt{3 - 2} = 1 \Rightarrow e = \frac{c}{a} = \frac{1}{\sqrt{3}}$; $F(0, \pm 1)$;
directrices are $y = 0 \pm \frac{a}{e} = \pm \frac{\sqrt{3}}{\left(\frac{1}{\sqrt{3}}\right)} = \pm 3$

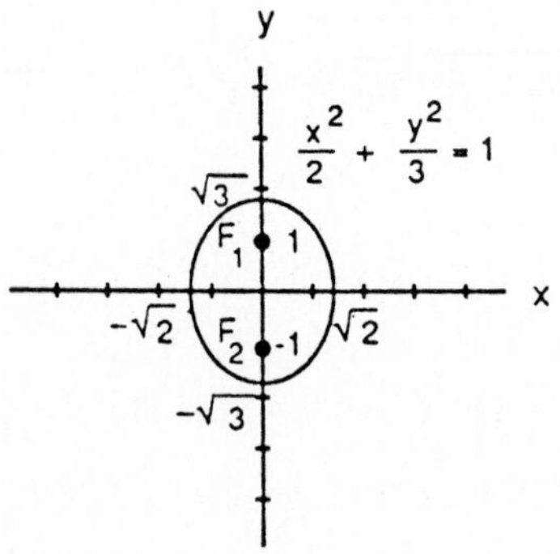

7. $6x^2 + 9y^2 = 54 \Rightarrow \frac{x^2}{9} + \frac{y^2}{6} = 1 \Rightarrow c = \sqrt{a^2 - b^2}$
$= \sqrt{9 - 6} = \sqrt{3} \Rightarrow e = \frac{c}{a} = \frac{\sqrt{3}}{3}$; $F(\pm\sqrt{3}, 0)$;
directrices are $x = 0 \pm \frac{a}{e} = \pm \frac{3}{\left(\frac{\sqrt{3}}{3}\right)} = \pm 3\sqrt{3}$

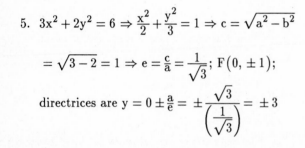

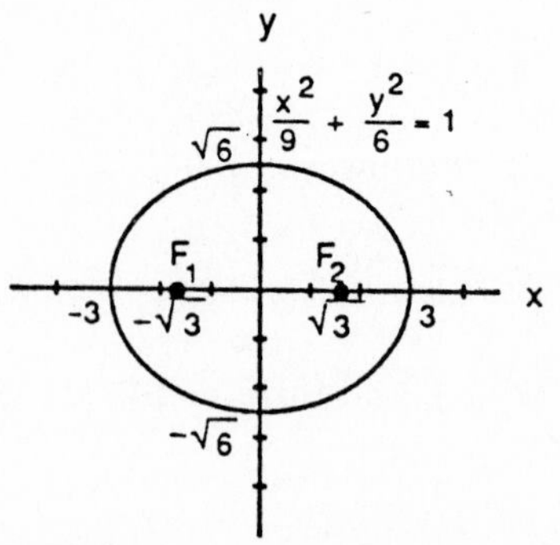

9. Foci: $(0, \pm 3)$, $e = 0.5 \Rightarrow c = 3$ and $a = \frac{c}{e} = \frac{3}{0.5} = 6 \Rightarrow b^2 = 36 - 9 = 27 \Rightarrow \frac{x^2}{27} + \frac{y^2}{36} = 1$

11. Vertices: $(0, \pm 70)$, $e = 0.1 \Rightarrow a = 70$ and $c = ae = 70(0.1) = 7 \Rightarrow b^2 = 4900 - 49 = 4851 \Rightarrow \frac{x^2}{4851} + \frac{y^2}{4900} = 1$

13. Focus: $(\sqrt{5},0)$, Directrix: $x = \frac{9}{\sqrt{5}} \Rightarrow c = ae = \sqrt{5}$ and $\frac{a}{e} = \frac{9}{\sqrt{5}} \Rightarrow \frac{ae}{e^2} = \frac{9}{\sqrt{5}} \Rightarrow \frac{\sqrt{5}}{e^2} = \frac{9}{\sqrt{5}} \Rightarrow e^2 = \frac{5}{9}$

$\Rightarrow e = \frac{\sqrt{5}}{3}$. Then $PF = \frac{\sqrt{5}}{3}PD \Rightarrow \sqrt{(x-\sqrt{5})^2 + (y-0)^2} = \frac{\sqrt{5}}{3}\left|x - \frac{9}{\sqrt{5}}\right| \Rightarrow (x-\sqrt{5})^2 + y^2 = \frac{5}{9}\left(x - \frac{9}{\sqrt{5}}\right)^2$

$\Rightarrow x^2 - 2\sqrt{5}x + 5 + y^2 = \frac{5}{9}\left(x^2 - \frac{18}{\sqrt{5}}x + \frac{81}{5}\right) \Rightarrow \frac{4}{9}x^2 + y^2 = 4 \Rightarrow \frac{x^2}{9} + \frac{y^2}{4} = 1$

15. Focus: $(-4,0)$, Directrix: $x = -16 \Rightarrow c = ae = 4$ and $\frac{a}{e} = 16 \Rightarrow \frac{ae}{e^2} = 16 \Rightarrow \frac{4}{e^2} = 16 \Rightarrow e^2 = \frac{1}{4} \Rightarrow e = \frac{1}{2}$. Then

$PF = \frac{1}{2}PD \Rightarrow \sqrt{(x+4)^2 + (y-0)^2} = \frac{1}{2}|x+16| \Rightarrow (x+4)^2 + y^2 = \frac{1}{4}(x+16)^2 \Rightarrow x^2 + 8x + 16 + y^2$

$= \frac{1}{4}(x^2 + 32x + 256) \Rightarrow \frac{3}{4}x^2 + y^2 = 48 \Rightarrow \frac{x^2}{64} + \frac{y^2}{48} = 1$

17. $e = \frac{4}{5} \Rightarrow$ take $c = 4$ and $a = 5$; $c^2 = a^2 - b^2$

$\Rightarrow 16 = 25 - b^2 \Rightarrow b^2 = 9 \Rightarrow b = 3$; therefore

$\frac{x^2}{25} + \frac{y^2}{9} = 1$

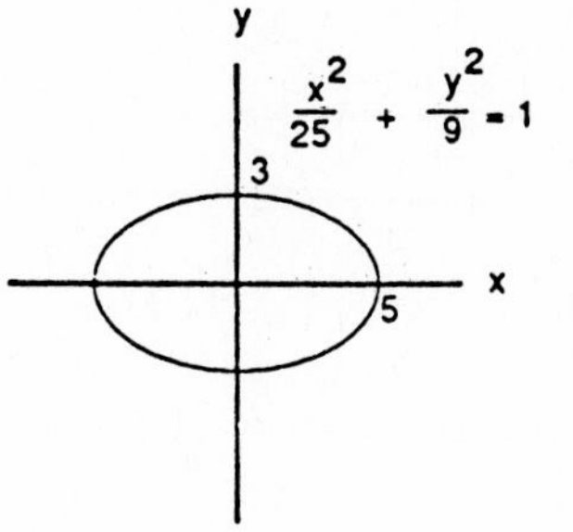

19. One axis is from $A(1,1)$ to $B(1,7)$ and is 6 units long; the other axis is from $C(3,4)$ to $D(-1,4)$ and is 4 units long. Therefore $a = 3$, $b = 2$ and the major axis is vertical. The center is the point $C(1,4)$ and the ellipse is given by $\frac{(x-1)^2}{4} + \frac{(y-4)^2}{9} = 1$; $c^2 = a^2 - b^2 = 3^2 - 2^2 = 5$ $\Rightarrow c = \sqrt{5}$; therefore the foci are $F(1, 4 \pm \sqrt{5})$, the eccentricity is $e = \frac{c}{a} = \frac{\sqrt{5}}{3}$, and the directrices are $y = 4 \pm \frac{a}{e} = 4 \pm \frac{3}{\left(\frac{\sqrt{5}}{3}\right)} = 4 \pm \frac{9\sqrt{5}}{5}$.

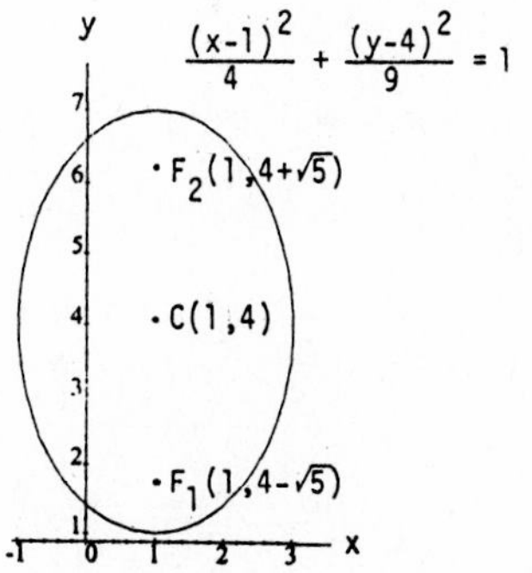

21. The ellipse must pass through $(0,0) \Rightarrow c = 0$; the point $(-1,2)$ lies on the ellipse $\Rightarrow -a + 2b = -8$. The ellipse is tangent to the x-axis $\Rightarrow$ its center is on the y-axis, so $a = 0$ and $b = -4 \Rightarrow$ the equation is $4x^2 + y^2 - 4y = 0$. Next, $4x^2 + y^2 - 4y + 16 = 16 \Rightarrow 4x^2 + (y-4)^2 = 16 \Rightarrow \frac{x^2}{4} + \frac{(y-4)^2}{16} = 1 \Rightarrow a = 4$ and $b = 2$ (now using the standard symbols) $\Rightarrow c^2 = a^2 - b^2 = 16 - 4 = 12 \Rightarrow c = \sqrt{12} \Rightarrow e = \frac{c}{a} = \frac{\sqrt{12}}{4} = \frac{\sqrt{3}}{2}$.

23. $x^2 - y^2 = 1 \Rightarrow c = \sqrt{a^2 + b^2} = \sqrt{1+1} = \sqrt{2} \Rightarrow e = \frac{c}{a}$ $= \frac{\sqrt{2}}{1} = \sqrt{2}$; asymptotes are $y = \pm x$; $F(\pm\sqrt{2}, 0)$; directrices are $x = 0 \pm \frac{a}{e} = \pm\frac{1}{\sqrt{2}}$

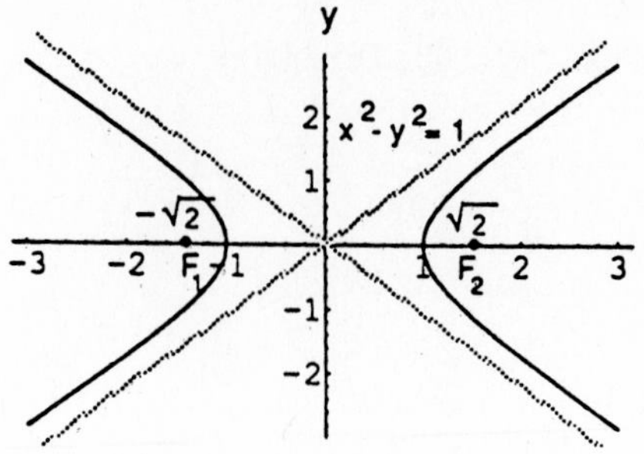

25. $y^2 - x^2 = 8 \Rightarrow \frac{y^2}{8} - \frac{x^2}{8} = 1 \Rightarrow c = \sqrt{a^2 + b^2}$ $= \sqrt{8+8} = 4 \Rightarrow e = \frac{c}{a} = \frac{4}{\sqrt{8}} = \sqrt{2}$; asymptotes are $y = \pm x$; $F(0, \pm 4)$; directrices are $y = 0 \pm \frac{a}{e}$ $= \pm\frac{\sqrt{8}}{\sqrt{2}} = \pm 2$

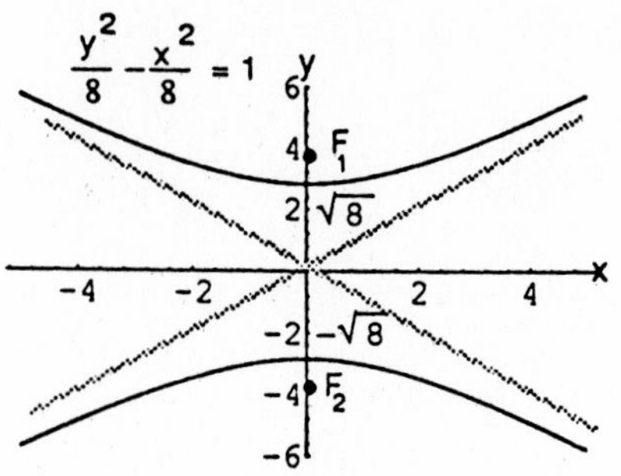

27. $8x^2 - 2y^2 = 16 \Rightarrow \frac{x^2}{2} - \frac{y^2}{8} = 1 \Rightarrow c = \sqrt{a^2 + b^2}$ $= \sqrt{2+8} = \sqrt{10} \Rightarrow e = \frac{c}{a} = \frac{\sqrt{10}}{\sqrt{2}} = \sqrt{5}$; asymptotes are $y = \pm 2x$; $F(\pm\sqrt{10}, 0)$; directrices are $x = 0 \pm \frac{a}{e}$ $= \pm\frac{\sqrt{2}}{\sqrt{5}} = \pm\frac{2}{\sqrt{10}}$

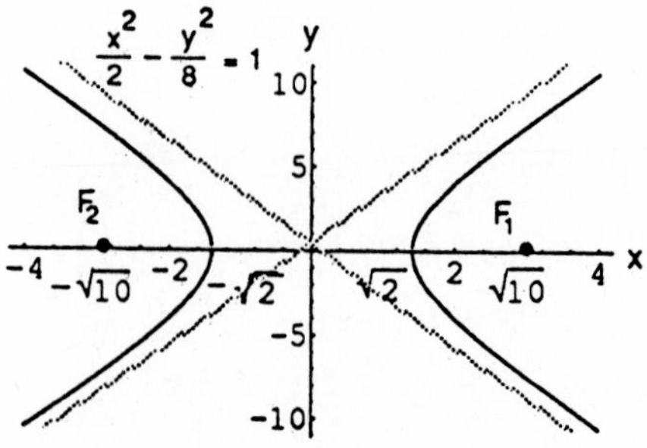

29. $8y^2 - 2x^2 = 16 \Rightarrow \frac{y^2}{2} - \frac{x^2}{8} = 1 \Rightarrow c = \sqrt{a^2 + b^2}$ $= \sqrt{2+8} = \sqrt{10} \Rightarrow e = \frac{c}{a} = \frac{\sqrt{10}}{\sqrt{2}} = \sqrt{5}$; asymptotes are $y = \pm\frac{x}{2}$; $F(0, \pm\sqrt{10})$; directrices are $y = 0 \pm \frac{a}{e}$ $= \pm\frac{\sqrt{2}}{\sqrt{5}} = \pm\frac{2}{\sqrt{10}}$

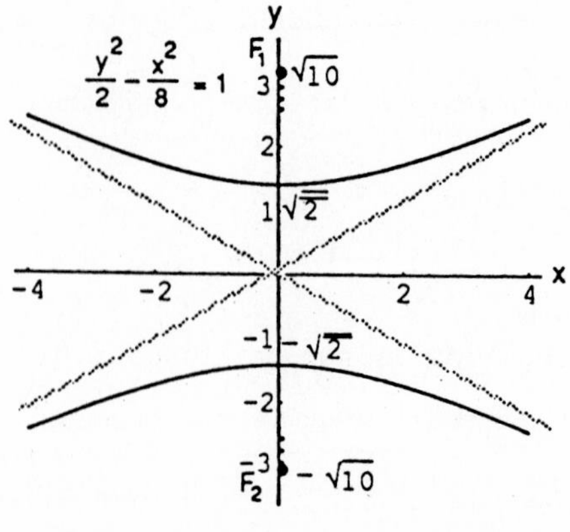

31. Vertices $(0, \pm 1)$ and $e = 3 \Rightarrow a = 1$ and $e = \frac{c}{a} = 3 \Rightarrow c = 3a = 3 \Rightarrow b^2 = c^2 - a^2 = 9 - 1 = 8 \Rightarrow y^2 - \frac{x^2}{8} = 1$

33. Foci $(\pm 3, 0)$ and $e = 3 \Rightarrow c = 3$ and $e = \frac{c}{a} = 3 \Rightarrow c = 3a \Rightarrow a = 1 \Rightarrow b^2 = c^2 - a^2 = 9 - 1 = 8 \Rightarrow x^2 - \frac{y^2}{8} = 1$

35. Focus $(4,0)$ and Directrix $x = 2 \Rightarrow c = ae = 4$ and $\frac{a}{e} = 2 \Rightarrow \frac{ae}{e^2} = 2 \Rightarrow \frac{4}{e^2} = 2 \Rightarrow e^2 = 2 \Rightarrow e = \sqrt{2}$. Then

$PF = \sqrt{2}\,PD \Rightarrow \sqrt{(x-4)^2 + (y-0)^2} = \sqrt{2}\,|x-2| \Rightarrow (x-4)^2 + y^2 = 2(x-2)^2 \Rightarrow x^2 - 8x + 16 + y^2$

$= 2(x^2 - 4x + 4) \Rightarrow -x^2 + y^2 = -8 \Rightarrow \frac{x^2}{8} - \frac{y^2}{8} = 1$

37. Focus $(-2,0)$ and Directrix $x = -\frac{1}{2} \Rightarrow c = ae = 2$ and $\frac{a}{e} = \frac{1}{2} \Rightarrow \frac{ae}{e^2} = \frac{1}{2} \Rightarrow \frac{2}{e^2} = \frac{1}{2} \Rightarrow e^2 = 4 \Rightarrow e = 2$. Then

$PF = 2PD \Rightarrow \sqrt{(x+2)^2 + (y-0)^2} = 2\left|x + \frac{1}{2}\right| \Rightarrow (x+2)^2 + y^2 = 4\left(x + \frac{1}{2}\right)^2 \Rightarrow x^2 + 4x + 4 + y^2 = 4\left(x^2 + x + \frac{1}{4}\right)$

$\Rightarrow -3x^2 + y^2 = -3 \Rightarrow x^2 - \frac{y^2}{3} = 1$

39. $\sqrt{(x-1)^2 + (y+3)^2} = \frac{3}{2}|y-2| \Rightarrow x^2 - 2x + 1 + y^2 + 6y + 9 = \frac{9}{4}(y^2 - 4y + 4) \Rightarrow 4x^2 - 5y^2 - 8x + 60y + 4 = 0$

$\Rightarrow 4(x^2 - 2x + 1) - 5(y^2 - 12y + 36) = -4 + 4 - 180 \Rightarrow \frac{(y-6)^2}{36} - \frac{(x-1)^2}{45} = 1$

41. To prove the reflective property for hyperbolas:

$\frac{x^2}{a^2} - \frac{y^2}{b^2} = 1 \Rightarrow a^2y^2 = b^2x^2 - a^2b^2$ and $\frac{dy}{dx} = \frac{xb^2}{ya^2}$.

Let $P(x_0, y_0)$ be a point of tangency (see the accompanying figure). The slope from P to $F(-c,0)$ is $\frac{y_0}{x_0 + c}$ and from P to $F_2(c,0)$ it is $\frac{y_0}{x_0 - c}$. Let the tangent through P meet the x-axis in point A, and define the angles $\angle F_1PA = \alpha$ and $\angle F_2PA = \beta$. We will show that $\tan\alpha = \tan\beta$. From the preliminary result in Exercise 22,

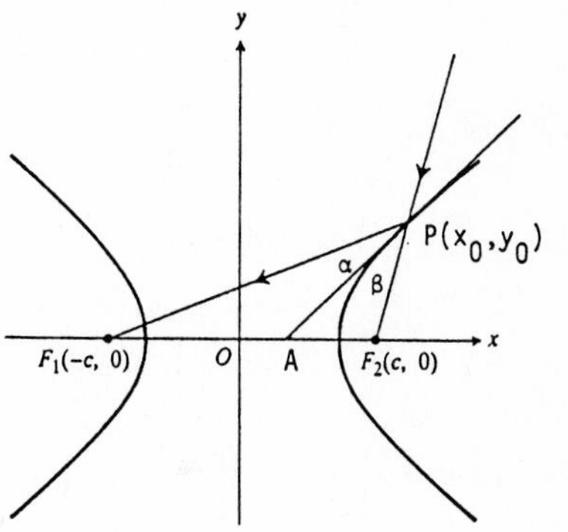

$$\tan\alpha = \frac{\left(\frac{x_0b^2}{y_0a^2} - \frac{y_0}{x_0+c}\right)}{1 + \left(\frac{x_0b^2}{y_0a^2}\right)\left(\frac{y_0}{x_0+c}\right)} = \frac{x_0^2b^2 + x_0b^2c - y_0^2a^2}{x_0y_0a^2 + y_0a^2c + x_0y_0b^2} = \frac{a^2b^2 + x_0b^2c}{x_0y_0c^2 + y_0a^2c} = \frac{b^2}{y_0c}.$$ In a similar manner,

$$\tan\beta = \frac{\left(\frac{y_0}{x_0-c} - \frac{x_0b^2}{y_0a^2}\right)}{1 + \left(\frac{y_0}{x_0-c}\right)\left(\frac{x_0b^2}{y_0a^2}\right)} = \frac{b^2}{y_0c}.$$ Since $\tan\alpha = \tan\beta$, and α and β are acute angles, we have $\alpha = \beta$.

9.3 QUADRATIC EQUATIONS AND ROTATIONS

1. $x^2 - 3xy + y^2 - x = 0 \Rightarrow B^2 - 4AC = (-3)^2 - 4(1)(1) = 5 > 0 \Rightarrow$ Hyperbola

3. $3x^2 - 7xy + \sqrt{17}y^2 = 1 \Rightarrow B^2 - 4AC = (-7)^2 - 4(3)\sqrt{17} \approx -0.477 < 0 \Rightarrow$ Ellipse

5. $x^2 + 2xy + y^2 + 2x - y + 2 = 0 \Rightarrow B^2 - 4AC = 2^2 - 4(1)(1) = 0 \Rightarrow$ Parabola

7. $x^2 + 4xy + 4y^2 - 3x = 6 \Rightarrow B^2 - 4AC = 4^2 - 4(1)(4) = 0 \Rightarrow$ Parabola

9. $xy + y^2 - 3x = 5 \Rightarrow B^2 - 4AC = 1^2 - 4(0)(1) = 1 > 0 \Rightarrow$ Hyperbola

11. $3x^2 - 5xy + 2y^2 - 7x - 14y = -1 \Rightarrow B^2 - 4AC = (-5)^2 - 4(3)(2) = 1 > 0 \Rightarrow$ Hyperbola

13. $x^2 - 3xy + 3y^2 + 6y = 7 \Rightarrow B^2 - 4AC = (-3)^2 - 4(1)(3) = -3 < 0 \Rightarrow$ Ellipse

15. $6x^2 + 3xy + 2y^2 + 17y + 2 = 0 \Rightarrow B^2 - 4AC = 3^2 - 4(6)(2) = -39 < 0 \Rightarrow$ Ellipse

17. $\cot 2\alpha = \frac{A-C}{B} = \frac{0}{1} = 0 \Rightarrow 2\alpha = \frac{\pi}{2} \Rightarrow \alpha = \frac{\pi}{4}$; therefore $x = x' \cos\alpha - y' \sin\alpha$,

$y = x' \sin\alpha + y' \cos\alpha \Rightarrow x = x'\frac{\sqrt{2}}{2} - y'\frac{\sqrt{2}}{2},\ y = x'\frac{\sqrt{2}}{2} + y'\frac{\sqrt{2}}{2}$

$\Rightarrow \left(\frac{\sqrt{2}}{2}x' - \frac{\sqrt{2}}{2}y'\right)\left(\frac{\sqrt{2}}{2}x' + \frac{\sqrt{2}}{2}y'\right) = 2 \Rightarrow \frac{1}{2}x'^2 - \frac{1}{2}y'^2 = 2 \Rightarrow x'^2 - y'^2 = 4 \Rightarrow$ Hyperbola

19. $\cot 2\alpha = \frac{A-C}{B} = \frac{3-1}{2\sqrt{3}} = \frac{1}{\sqrt{3}} \Rightarrow 2\alpha = \frac{\pi}{3} \Rightarrow \alpha = \frac{\pi}{6}$; therefore $x = x' \cos\alpha - y' \sin\alpha$,

$y = x' \sin\alpha + y' \cos\alpha \Rightarrow x = \frac{\sqrt{3}}{2}x' - \frac{1}{2}y',\ y = \frac{1}{2}x' + \frac{\sqrt{3}}{2}y'$

$\Rightarrow 3\left(\frac{\sqrt{3}}{2}x' - \frac{1}{2}y'\right)^2 + 2\sqrt{3}\left(\frac{\sqrt{3}}{2}x' + \frac{1}{2}y'\right)\left(\frac{1}{2}x' + \frac{\sqrt{3}}{2}y'\right) + \left(\frac{1}{2}x' + \frac{\sqrt{3}}{2}y'\right)^2 - 8\left(\frac{\sqrt{3}}{2}x' - \frac{1}{2}y'\right)$

$+ 8\sqrt{3}\left(\frac{1}{2}x' + \frac{\sqrt{3}}{2}y'\right) = 0 \Rightarrow 4x'^2 + 16y' = 0 \Rightarrow$ Parabola

21. $\cot 2\alpha = \frac{A-C}{B} = \frac{1-1}{-2} = 0 \Rightarrow 2\alpha = \frac{\pi}{2} \Rightarrow \alpha = \frac{\pi}{4}$; therefore $x = x' \cos\alpha - y' \sin\alpha$,

$y = x' \sin\alpha + y' \cos\alpha \Rightarrow x = \frac{\sqrt{2}}{2}x' - \frac{\sqrt{2}}{2}y',\ y = \frac{\sqrt{2}}{2}x' + \frac{\sqrt{2}}{2}y'$

$\Rightarrow \left(\frac{\sqrt{2}}{2}x' - \frac{\sqrt{2}}{2}y'\right)^2 - 2\left(\frac{\sqrt{2}}{2}x' - \frac{\sqrt{2}}{2}y'\right)\left(\frac{\sqrt{2}}{2}x' + \frac{\sqrt{2}}{2}y'\right) + \left(\frac{\sqrt{2}}{2}x' + \frac{\sqrt{2}}{2}y'\right)^2 = 2 \Rightarrow y'^2 = 1$

$\Rightarrow$ Parallel horizontal lines

23. $\cot 2\alpha = \frac{A-C}{B} = \frac{\sqrt{2}-\sqrt{2}}{2\sqrt{2}} = 0 \Rightarrow 2\alpha = \frac{\pi}{2} \Rightarrow \alpha = \frac{\pi}{4}$; therefore $x = x' \cos\alpha - y' \sin\alpha$,

$y = x' \sin\alpha + y' \cos\alpha \Rightarrow x = \frac{\sqrt{2}}{2}x' - \frac{\sqrt{2}}{2}y',\ y = \frac{\sqrt{2}}{2}x' + \frac{\sqrt{2}}{2}y'$

$\Rightarrow \sqrt{2}\left(\frac{\sqrt{2}}{2}x' - \frac{\sqrt{2}}{2}y'\right)^2 + 2\sqrt{2}\left(\frac{\sqrt{2}}{2}x' - \frac{\sqrt{2}}{2}y'\right)\left(\frac{\sqrt{2}}{2}x' + \frac{\sqrt{2}}{2}y'\right) + \sqrt{2}\left(\frac{\sqrt{2}}{2}x' + \frac{\sqrt{2}}{2}y'\right)^2$

$- 8\left(\frac{\sqrt{2}}{2}x' - \frac{\sqrt{2}}{2}y'\right) + 8\left(\frac{\sqrt{2}}{2}x' + \frac{\sqrt{2}}{2}y'\right) = 0 \Rightarrow 2\sqrt{2}x'^2 + 8\sqrt{2}y' = 0 \Rightarrow$ Parabola

25. $\cot 2\alpha = \frac{A-C}{B} = \frac{3-3}{2} = 0 \Rightarrow 2\alpha = \frac{\pi}{2} \Rightarrow \alpha = \frac{\pi}{4}$; therefore $x = x' \cos\alpha - y' \sin\alpha$,

$y = x' \sin\alpha + y' \cos\alpha \Rightarrow x = \frac{\sqrt{2}}{2}x' - \frac{\sqrt{2}}{2}y'$, $y = \frac{\sqrt{2}}{2}x' + \frac{\sqrt{2}}{2}y'$

$\Rightarrow 3\left(\frac{\sqrt{2}}{2}x' - \frac{\sqrt{2}}{2}y'\right)^2 + 2\left(\frac{\sqrt{2}}{2}x' - \frac{\sqrt{2}}{2}y'\right)\left(\frac{\sqrt{2}}{2}x' + \frac{\sqrt{2}}{2}y'\right) + 3\left(\frac{\sqrt{2}}{2}x' + \frac{\sqrt{2}}{2}y'\right)^2 = 19 \Rightarrow 4x'^2 + 2y'^2 = 19$

$\Rightarrow$ Ellipse

27. $\cot 2\alpha = \frac{14-2}{16} = \frac{3}{4} \Rightarrow \cos 2\alpha = \frac{3}{5}$ (if we choose 2α in Quadrant I); thus $\sin\alpha = \sqrt{\frac{1-\cos 2\alpha}{2}} = \sqrt{\frac{1-\left(\frac{3}{5}\right)}{2}} = \frac{1}{\sqrt{5}}$

and $\cos\alpha = \sqrt{\frac{1+\cos 2\alpha}{2}} = \sqrt{\frac{1+\left(\frac{3}{5}\right)}{2}} = \frac{2}{\sqrt{5}}$ (or $\sin\alpha = -\frac{2}{\sqrt{5}}$ and $\cos\alpha = \frac{1}{\sqrt{5}}$)

29. $\tan 2\alpha = \frac{-1}{1-3} = \frac{1}{2} \Rightarrow 2\alpha \approx 26.57^\circ \Rightarrow \alpha \approx 13.28^\circ \Rightarrow \sin\alpha \approx 0.23$, $\cos\alpha \approx 0.97$; then $A' \approx 0.88$, $B' \approx 0.00$, $C' \approx 3.10$, $D' \approx 0.74$, $E' \approx -1.20$, and $F' = -3 \Rightarrow 0.88x'^2 + 3.10y'^2 + 0.74x' - 1.20y' - 3 = 0$, an ellipse

31. $\tan 2\alpha = \frac{-4}{1-4} = \frac{4}{3} \Rightarrow 2\alpha \approx 53.13^\circ \Rightarrow \alpha \approx 26.56^\circ \Rightarrow \sin\alpha \approx 0.45$, $\cos\alpha \approx 0.89$; then $A' \approx 0.00$, $B' \approx 0.00$, $C' \approx 5.00$, $D' \approx 0$, $E' \approx 0$, and $F' = -5 \Rightarrow 5.00y'^2 - 5 = 0$ or $y' = \pm 1.00$, parallel lines

33. $\tan 2\alpha = \frac{5}{3-2} = 5 \Rightarrow 2\alpha \approx 78.69^\circ \Rightarrow \alpha \approx 39.34^\circ \Rightarrow \sin\alpha \approx 0.63$, $\cos\alpha \approx 0.77$; then $A' \approx 5.05$, $B' \approx 0.00$, $C' \approx -0.05$, $D' \approx -5.07$, $E' \approx -6.18$, and $F' = -1 \Rightarrow 5.05x'^2 - 0.05y'^2 - 5.07x' - 6.18y' - 1 = 0$, a hyperbola

35. $\alpha = 90^\circ \Rightarrow x = x'\cos 90^\circ - y'\sin 90^\circ = -y'$ and $y = x'\sin 90^\circ + y'\cos 90^\circ = x'$

(a) $\frac{x'^2}{b^2} + \frac{y'^2}{a^2} = 1$ (b) $\frac{y'^2}{a^2} - \frac{x'^2}{b^2} = 1$ (c) $x'^2 + y'^2 = a^2$

(d) $y = mx \Rightarrow y - mx = 0 \Rightarrow D = -m$ and $E = 1$; $\alpha = 90^\circ \Rightarrow D' = 1$ and $E' = m \Rightarrow my' + x' = 0 \Rightarrow y' = -\frac{1}{m}x'$

(e) $y = mx + b \Rightarrow y - mx - b = 0 \Rightarrow D = -m$ and $E = 1$; $\alpha = 90^\circ \Rightarrow D' = 1$, $E' = m$ and $F' = -b$ $\Rightarrow my' + x' - b = 0 \Rightarrow y' = -\frac{1}{m}x' + \frac{b}{m}$

37. (a) $A' = \cos 45^\circ \sin 45^\circ = \left(\frac{\sqrt{2}}{2}\right)\left(\frac{\sqrt{2}}{2}\right) = \frac{1}{2}$, $B' = 0$, $C' = -\cos 45^\circ \sin 45^\circ = -\frac{1}{2}$, $F' = -1$

$\Rightarrow \frac{1}{2}x'^2 - \frac{1}{2}y'^2 = 1 \Rightarrow x'^2 - y'^2 = 2$

(b) $A' = \frac{1}{2}$, $C' = -\frac{1}{2}$ (see part (a) above), $D' = E' = B' = 0$, $F' = -a \Rightarrow \frac{1}{2}x'^2 - \frac{1}{2}y'^2 = a \Rightarrow x'^2 - y'^2 = 2a$

39. Yes, the graph is a hyperbola: with $AC < 0$ we have $-4AC > 0$ and $B^2 - 4AC > 0$.

41. Let α be any angle. Then $A' = \cos^2\alpha + \sin^2\alpha = 1$, $B' = 0$, $C' = \sin^2\alpha + \cos^2\alpha = 1$, $D' = E' = 0$ and $F' = -a^2$ $\Rightarrow x'^2 + y'^2 = a^2$.

43. (a) $B^2 - 4AC = 4^2 - 4(1)(4) = 0$, so the discriminant indicates this conic is a parabola

(b) The left-hand side of $x^2 + 4xy + 4y^2 + 6x + 12y + 9 = 0$ factors as a perfect square: $(x + 2y + 3)^2 = 0$ $\Rightarrow x + 2y + 3 = 0 \Rightarrow 2y = -x - 3$; thus the curve is a degenerate parabola (i.e., a straight line).

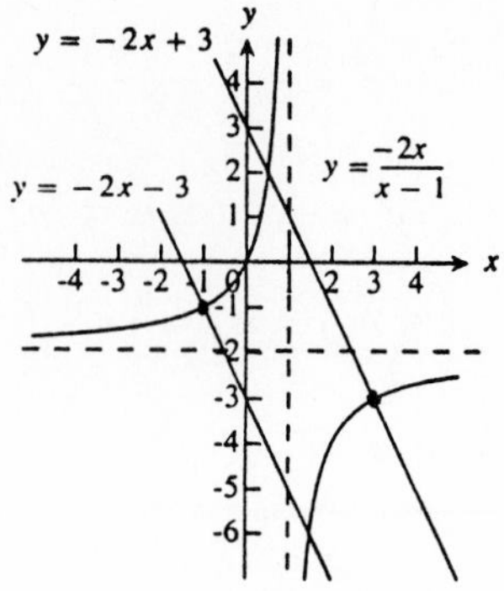

45. (a) $B^2 - 4AC = 1 - 4(0)(0) = 1 \Rightarrow$ hyperbola

(b) $xy + 2x - y = 0 \Rightarrow y(x-1) = -2x \Rightarrow y = \frac{-2x}{x-1}$

(c) $y = \frac{-2x}{x-1} \Rightarrow \frac{dy}{dx} = \frac{2}{(x-1)^2}$ and we want $\frac{-1}{\left(\frac{dy}{dx}\right)} = -2$, the slope of $y = -2x \Rightarrow -2 = -\frac{(x-1)^2}{2} \Rightarrow (x-1)^2 = 4$ $\Rightarrow x = 3$ or $x = -1$; $x = 3 \Rightarrow y = -3 \Rightarrow (3,-3)$ is a point on the hyperbola where the line with slope $m = -2$ is normal $\Rightarrow$ the line is $y + 3 = -2(x-3)$ or $y = -2x + 3$; $x = -1 \Rightarrow y = -1 \Rightarrow (-1,-1)$ is a point on the hyperbola where the line with slope $m = -2$ is normal $\Rightarrow$ the line is $y + 1 = -2(x+1)$ or $y = -2x - 3$

47. Assume the ellipse has been rotated to eliminate the xy-term $\Rightarrow$ the new equation is $A'x'^2 + C'y'^2 = 1 \Rightarrow$ the semi-axes are $\sqrt{\frac{1}{A'}}$ and $\sqrt{\frac{1}{C'}} \Rightarrow$ the area is $\pi\left(\sqrt{\frac{1}{A'}}\right)\left(\sqrt{\frac{1}{C'}}\right) = \frac{\pi}{\sqrt{A'C'}} = \frac{2\pi}{\sqrt{4A'C'}}$. Since $B^2 - 4AC$ $= B'^2 - 4A'C' = -4A'C'$ (because $B' = 0$) we find that the area is $\frac{2\pi}{\sqrt{4AC - B^2}}$ as claimed.

49. $B'^2 - 4A'C'$

$= \left(B \cos 2\alpha + (C - A) \sin 2\alpha\right)^2 - 4\left(A \cos^2\alpha + B \cos\alpha \sin\alpha + C \sin^2\alpha\right)\left(A \sin^2\alpha - B \cos\alpha \sin\alpha + C \cos^2\alpha\right)$

$= B^2 \cos^2 2\alpha + 2B(C-A) \sin 2\alpha \cos 2\alpha + (C-A)^2 \sin^2 2\alpha - 4A^2 \cos^2\alpha \sin^2\alpha + 4AB \cos^3\alpha \sin\alpha$
$- 4AC \cos^4\alpha - 4AB \cos\alpha \sin^3\alpha + 4B^2 \cos^2\alpha \sin^2\alpha - 4BC \cos^3\alpha \sin\alpha - 4AC \sin^4\alpha + 4BC \cos\alpha \sin^3\alpha$
$- 4C^2 \cos^2\alpha \sin^2\alpha$

$= B^2 \cos^2 2\alpha + 2BC \sin 2\alpha \cos 2\alpha - 2AB \sin 2\alpha \cos 2\alpha + C^2 \sin^2 2\alpha - 2AC \sin^2 2\alpha + A^2 \sin^2 2\alpha$
$- 4A^2 \cos^2\alpha \sin^2\alpha + 4AB \cos^3\alpha \sin\alpha - 4AC \cos^4\alpha - 4AB \cos\alpha \sin^3\alpha + B^2 \sin^2 2\alpha - 4BC \cos^3\alpha \sin\alpha$
$- 4AC \sin^4\alpha + 4BC \cos\alpha \sin^3\alpha - 4C^2 \cos^2\alpha \sin^2\alpha$

$= B^2 + 2BC(2 \sin\alpha \cos\alpha)\left(\cos^2\alpha - \sin^2\alpha\right) - 2AB(2 \sin\alpha \cos\alpha)\left(\cos^2\alpha - \sin^2\alpha\right) + C^2\left(4 \sin^2\alpha \cos^2\alpha\right)$
$- 2AC\left(4 \sin^2\alpha \cos^2\alpha\right) + A^2\left(4 \sin^2\alpha \cos^2\alpha\right) - 4A^2 \cos^2\alpha \sin^2\alpha + 4AB \cos^3\alpha \sin\alpha - 4AC \cos^4\alpha$
$- 4AB \cos\alpha \sin^3\alpha - 4BC \cos^3\alpha \sin\alpha - 4AC \sin^4\alpha + 4BC \cos\alpha \sin^3\alpha - 4C^2 \cos^2\alpha \sin^2\alpha$

$= B^2 - 8AC \sin^2\alpha \cos^2\alpha - 4AC \cos^4\alpha - 4AC \sin^4\alpha$

$= B^2 - 4AC\left(\cos^4\alpha + 2 \sin^2\alpha \cos^2\alpha + \sin^4\alpha\right)$

$= B^2 - 4AC\left(\cos^2\alpha + \sin^2\alpha\right)^2$

$= B^2 - 4AC$

9.4 PARAMETRIZATIONS OF CURVES

1. $x = \cos t,\ y = \sin t,\ 0 \le t \le \pi$
$\Rightarrow \cos^2 t + \sin^2 t = 1 \Rightarrow x^2 + y^2 = 1$

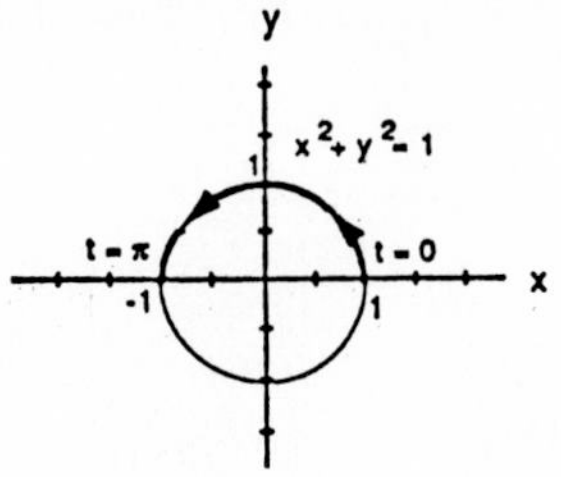

3. $x = \sin(2\pi(1 - t)),\ y = \cos(2\pi(1 - t)),\ 0 \le t \le 1$
$\Rightarrow \sin^2(2\pi(1 - t)) + \cos^2(2\pi(1 - t)) = 1$
$\Rightarrow x^2 + y^2 = 1$

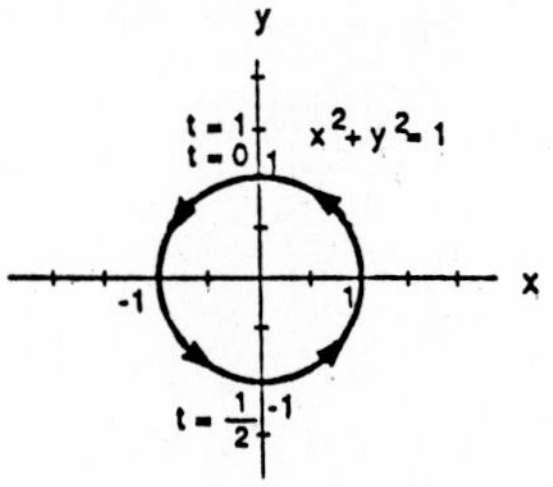

5. $x = 4 \cos t,\ y = 2 \sin t,\ 0 \le t \le 2\pi$
$\Rightarrow \frac{16 \cos^2 t}{16} + \frac{4 \sin^2 t}{4} = 1 \Rightarrow \frac{x^2}{16} + \frac{y^2}{4} = 1$

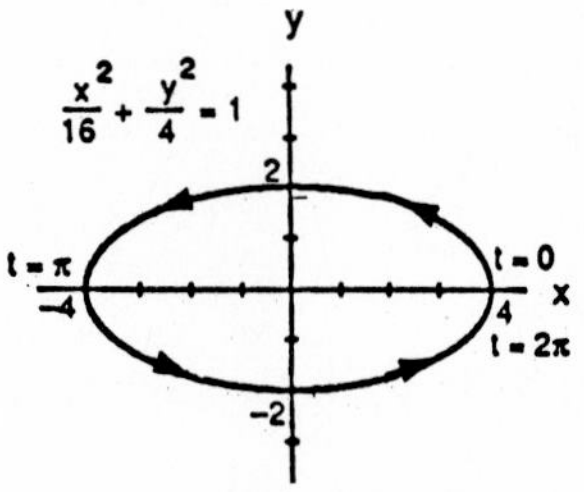

7. $x = 4 \cos t,\ y = 5 \sin t,\ 0 \le t \le \pi$
$\Rightarrow \frac{16 \cos^2 t}{16} + \frac{25 \sin^2 t}{25} = 1 \Rightarrow \frac{x^2}{16} + \frac{y^2}{25} = 1$

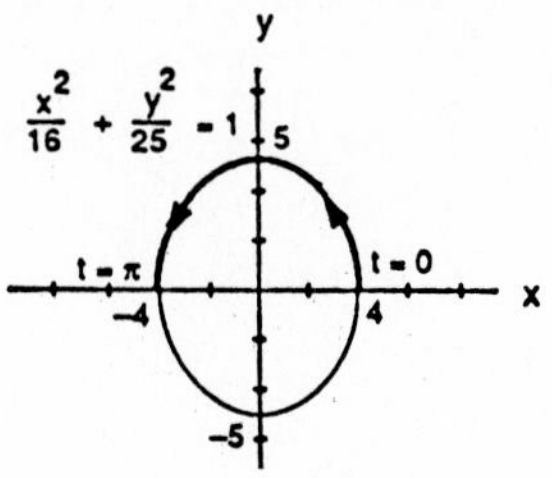

9. $x = 3t,\ y = 9t^2,\ -\infty < t < \infty \Rightarrow y = x^2$

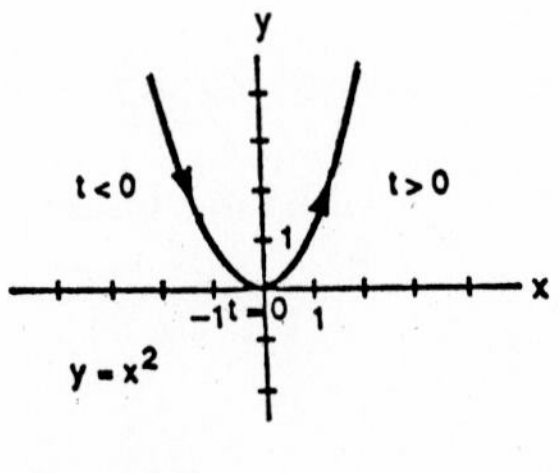

11. $x = t,\ y = \sqrt{t},\ t \ge 0 \Rightarrow y = \sqrt{x}$

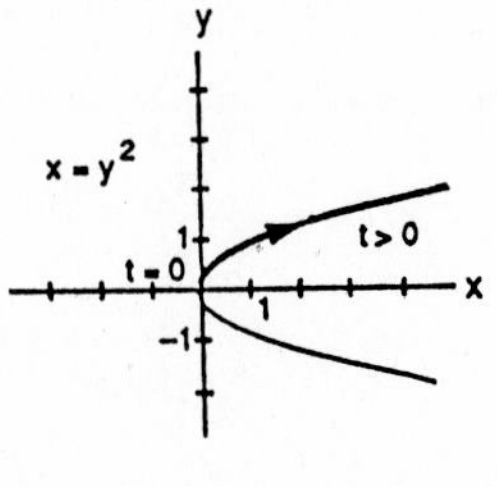

13. $x = -\sec t,\ y = \tan t,\ -\frac{\pi}{2} < t < \frac{\pi}{2}$
$\Rightarrow \sec^2 t - \tan^2 t = 1 \Rightarrow x^2 - y^2 = 1$

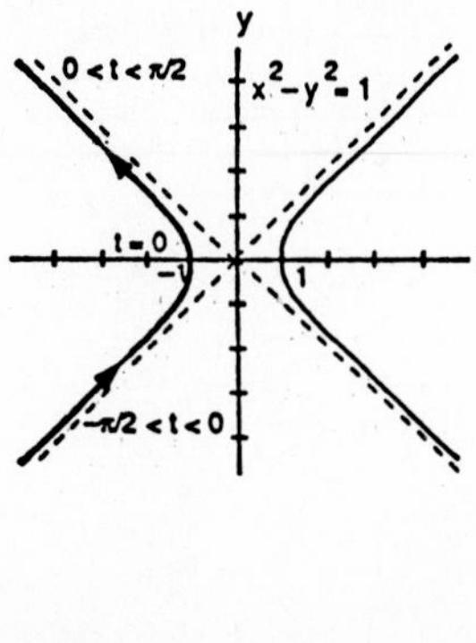

15. $x = 2t - 5,\ y = 4t - 7,\ -\infty < t < \infty$
$\Rightarrow x + 5 = 2t \Rightarrow 2(x + 5) = 4t$
$\Rightarrow y = 2(x + 5) - 7 \Rightarrow y = 2x + 3$

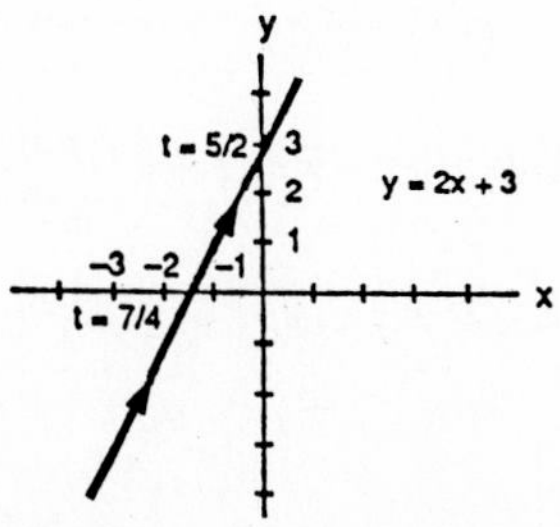

17. $x = t,\ y = 1 - t,\ 0 \le t \le 1$

$\Rightarrow y = 1 - x$

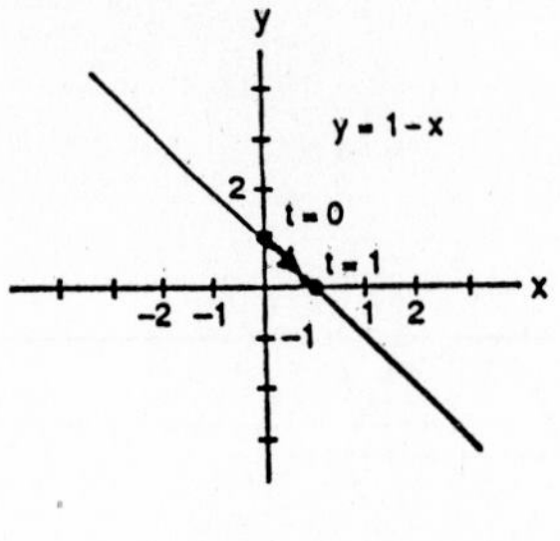

19. $x = t,\ y = \sqrt{1 - t^2},\ -1 \le t \le 0$

$\Rightarrow y = \sqrt{1 - x^2}$

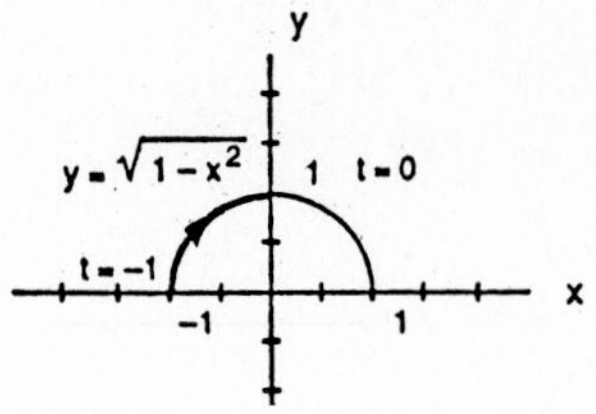

21. $x = t^2,\ y = \sqrt{t^4 + 1},\ t \ge 0$

$\Rightarrow y = \sqrt{x^2 + 1},\ x \ge 0$

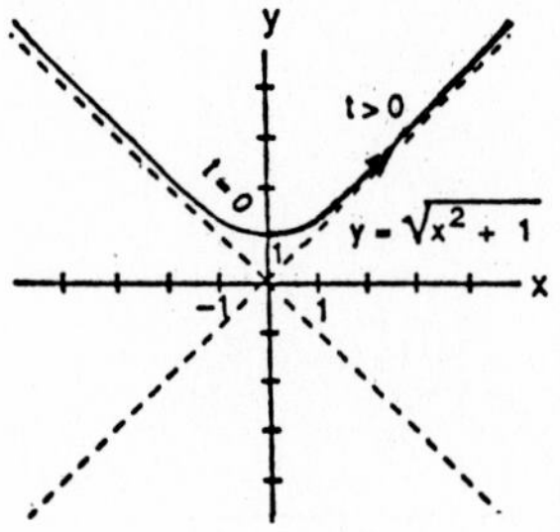

23. $x = -\cosh t,\ y = \sinh t,\ -\infty < 1 < \infty$

$\Rightarrow \cosh^2 t - \sinh^2 t = 1 \Rightarrow x^2 - y^2 = 1$

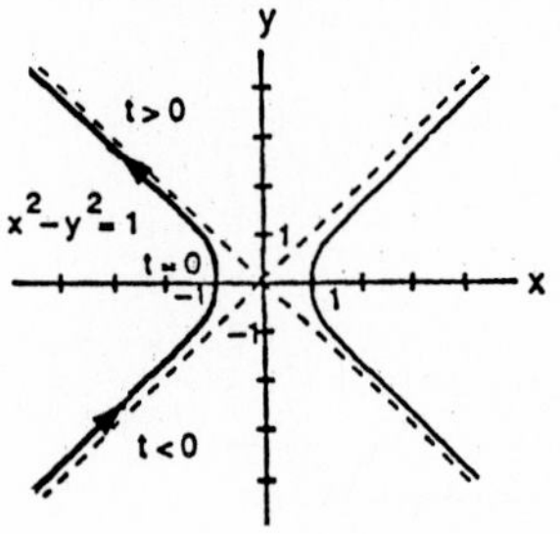

25. (a) $x = a \cos t,\ y = -a \sin t,\ 0 \le t \le 2\pi$

(b) $x = a \cos t,\ y = a \sin t,\ 0 \le t \le 2\pi$

(c) $x = a \cos t,\ y = -a \sin t,\ 0 \le t \le 4\pi$

(d) $x = a \cos t,\ y = a \sin t,\ 0 \le t \le 4\pi$

27. $x^2 + y^2 = a^2 \Rightarrow 2x + 2y \frac{dy}{dx} = 0 \Rightarrow \frac{dy}{dx} = -\frac{x}{y}$; let $t = \frac{dy}{dx} \Rightarrow -\frac{x}{y} = t \Rightarrow x = -yt$. Substitution yields $y^2t^2 + y^2 = a^2 \Rightarrow y = \frac{a}{\sqrt{1+t^2}}$ and $x = \frac{-at}{\sqrt{1+t}}$, $-\infty < t < \infty$

29. Extend the vertical line through A to the x-axis and let C be the point of intersection. Then $OC = AQ = x$ and $\tan t = \frac{2}{OC} = \frac{2}{x} \Rightarrow x = \frac{2}{\tan t} = 2 \cot t$; $\sin t = \frac{2}{OA}$ $\Rightarrow OA = \frac{2}{\sin t}$; and $(AB)(OA) = (AQ)^2 \Rightarrow AB\left(\frac{2}{\sin t}\right) = x^2$ $\Rightarrow AB\left(\frac{2}{\sin t}\right) = \left(\frac{2}{\tan t}\right)^2 \Rightarrow AB = \frac{2 \sin t}{\tan^2 t}$. Next

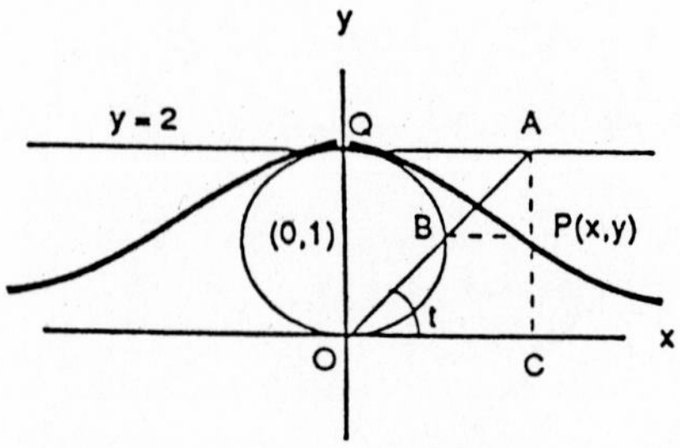

$y = 2 - AB \sin t \Rightarrow y = 2 - \left(\frac{2 \sin t}{\tan^2 t}\right) \sin t =$

$2 - \frac{2 \sin^2 t}{\tan^2 t} = 2 - 2 \cos^2 t = 2 \sin^2 t$. Therefore let $x = 2 \cot t$ and $y = 2 \sin^2 t$, $0 < t < \pi$.

31. (a) $x = x_0 + (x_1 - x_0)t$ and $y = y_0 + (y_1 - y_0)t \Rightarrow t = \frac{x - x_0}{x_1 - x_0} \Rightarrow y = y_0 + (y_1 - y_0)\left(\frac{x - x_0}{x_1 - x_0}\right)$

$\Rightarrow y - y_0 = \left(\frac{y_1 - y_0}{x_1 - x_0}\right)(x - x_0)$ which is an equation of the line through the points (x_0, y_0) and (x_1, y_1)

(b) Let $x_0 = y_0 = 0$ in (a) $\Rightarrow x = x_1 t$, $y = y_1 t$ (the answer is not unique)

(c) Let $(x_0, y_0) = (-1, 0)$ and $(x_1, y_1) = (0, 1)$ or let $(x_0, y_0) = (0, 1)$ and $(x_1, y_1) = (-1, 0)$ in part (a) $\Rightarrow x = -1 + t$, $y = t$ or $x = -t$, $y = 1 - t$ (the answer is not unique)

33. Arc PF = Arc AF since each is the distance rolled and $\frac{\text{Arc PF}}{b} = \angle FCP \Rightarrow \text{Arc PF} = b(\angle FCP)$; $\frac{\text{Arc AF}}{a} = \theta$ $\Rightarrow \text{Arc AF} = a\theta \Rightarrow a\theta = b(\angle FCP) \Rightarrow \angle FCP = \frac{a}{b}\theta$; $\angle OCG = \frac{\pi}{2} - \theta$; $\angle OCG = \angle OCP + \angle PCE$ $= \angle OCP + \left(\frac{\pi}{2} - \alpha\right)$. Now $\angle OCP = \pi - \angle FCP$ $= \pi - \frac{a}{b}\theta$. Thus $\angle OCG = \pi - \frac{a}{b}\theta + \frac{\pi}{2} - \alpha \Rightarrow \frac{\pi}{2} - \theta$ $= \pi - \frac{a}{b}\theta + \frac{\pi}{2} - \alpha \Rightarrow \alpha = \pi - \frac{a}{b}\theta + \theta = \pi - \left(\frac{a-b}{b}\theta\right)$.

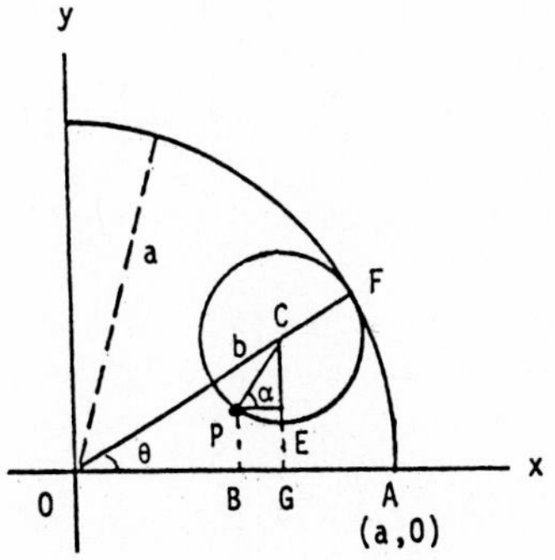

Then $x = OG - BG = OG - PE = (a - b)\cos\theta - b\cos\alpha = (a - b)\cos\theta - b\cos\left(\pi - \frac{a-b}{b}\theta\right)$

$= (a - b)\cos\theta + b\cos\left(\frac{a-b}{b}\theta\right)$. Also $y = EG = CG - CE = (a - b)\sin\theta - b\sin\alpha$

$= (a - b)\sin\theta - b\sin\left(\pi - \frac{a-b}{b}\theta\right) = (a - b)\sin\theta - b\sin\left(\frac{a-b}{b}\theta\right)$. Therefore

$x = (a - b)\cos\theta + b\cos\left(\frac{a-b}{b}\theta\right)$ and $y = (a - b)\sin\theta - b\sin\left(\frac{a-b}{b}\theta\right)$.

If $b = \frac{a}{4}$, then $x = \left(a - \frac{a}{4}\right)\cos\theta + \frac{a}{4}\cos\left(\frac{a - \left(\frac{a}{4}\right)}{\left(\frac{a}{4}\right)}\theta\right)$

$= \frac{3a}{4}\cos\theta + \frac{a}{4}\cos 3\theta = \frac{3a}{4}\cos\theta + \frac{a}{4}(\cos\theta\cos 2\theta - \sin\theta\sin 2\theta)$

$= \frac{3a}{4}\cos\theta + \frac{a}{4}\left((\cos\theta)(\cos^2\theta - \sin^2\theta) - (\sin\theta)(2\sin\theta\cos\theta)\right)$

$= \frac{3a}{4}\cos\theta + \frac{a}{4}\cos^3\theta - \frac{a}{4}\cos\theta\sin^2\theta - \frac{2a}{4}\sin^2\theta\cos\theta$

$= \frac{3a}{4}\cos\theta + \frac{a}{4}\cos^3\theta - \frac{3a}{4}(\cos\theta)(1 - \cos^2\theta) = a\cos^3\theta$;

$y = \left(a - \frac{a}{4}\right)\sin\theta - \frac{a}{4}\sin\left(\frac{a - \left(\frac{a}{4}\right)}{\left(\frac{a}{4}\right)}\theta\right) = \frac{3a}{4}\sin\theta - \frac{a}{4}\sin 3\theta = \frac{3a}{4}\sin\theta - \frac{a}{4}(\sin\theta\cos 2\theta + \cos\theta\sin 2\theta)$

$= \frac{3a}{4}\sin\theta - \frac{a}{4}\left((\sin\theta)(\cos^2\theta - \sin^2\theta) + (\cos\theta)(2\sin\theta\cos\theta)\right)$

$= \frac{3a}{4}\sin\theta - \frac{a}{4}\sin\theta\cos^2\theta + \frac{a}{4}\sin^3\theta - \frac{2a}{4}\cos^2\theta\sin\theta$

$= \frac{3a}{4}\sin\theta - \frac{3a}{4}\sin\theta\cos^2\theta + \frac{a}{4}\sin^3\theta$

$= \frac{3a}{4}\sin\theta - \frac{3a}{4}(\sin\theta)(1 - \sin^2\theta) + \frac{a}{4}\sin^3\theta = a\sin^3\theta$.

35. Draw line AM in the figure and note that $\angle AMO$ is a right angle since it is an inscribed angle which spans the diameter of a circle. Then $AN^2 = MN^2 + AM^2$. Now, $OA = a$, $\frac{AN}{a} = \tan t$, and $\frac{AM}{a} = \sin t$. Next $MN = OP \Rightarrow OP^2 = AN^2 - AM^2$

$= a^2 \tan^2 t - a^2 \sin^2 t \Rightarrow OP = \sqrt{a^2 \tan^2 t - a^2 \sin^2 t}$

$= (a \sin t)\sqrt{\sec^2 t - 1} = \frac{a \sin^2 t}{\cos t}$. In triangle BPO,

$x = OP \sin t = \frac{a \sin^3 t}{\cos t} = a \sin^2 t \tan t$ and

$y = OP \cos t = a \sin^2 t \Rightarrow x = a \sin^2 t \tan t$ and $y = a \sin^2 t$.

37. $D = \sqrt{(x-2)^2 + \left(y - \frac{1}{2}\right)^2} \Rightarrow D^2 = (x-2)^2 + \left(y - \frac{1}{2}\right)^2 = (t-2)^2 + \left(t^2 - \frac{1}{2}\right)^2 \Rightarrow D^2 = t^4 - 4t + \frac{17}{4}$

$\Rightarrow \frac{d(D^2)}{dt} = 4t^3 - 4 = 0 \Rightarrow t = 1$. The second derivative is always positive for $t \neq 0 \Rightarrow t = 1$ gives a local minimum for D^2 (and hence D) which is an absolute minimum since it is the only extremum $\Rightarrow$ the closest point on the parabola is $(1, 1)$.

39. (a)

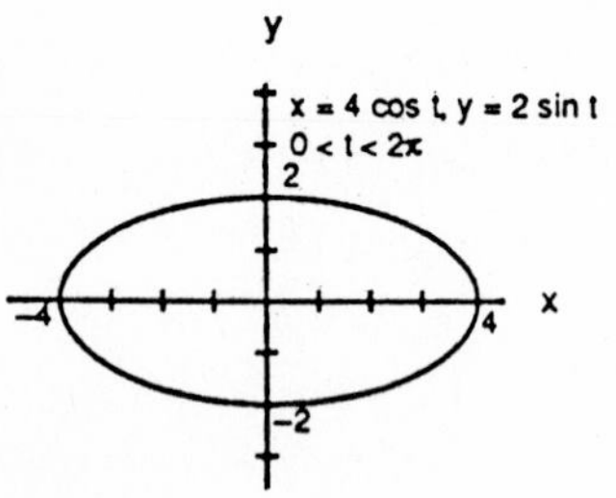

(b)

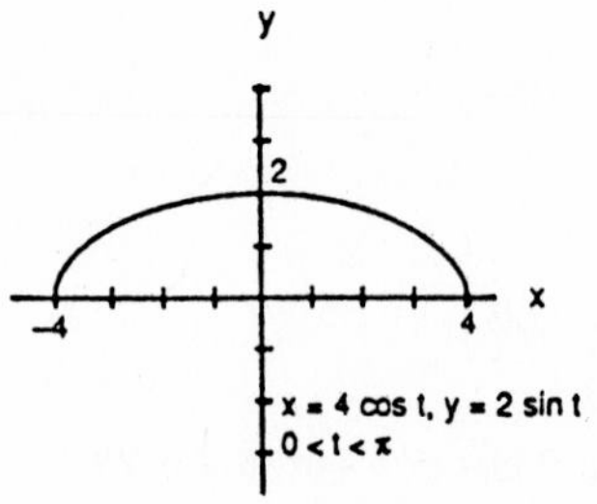

(c)

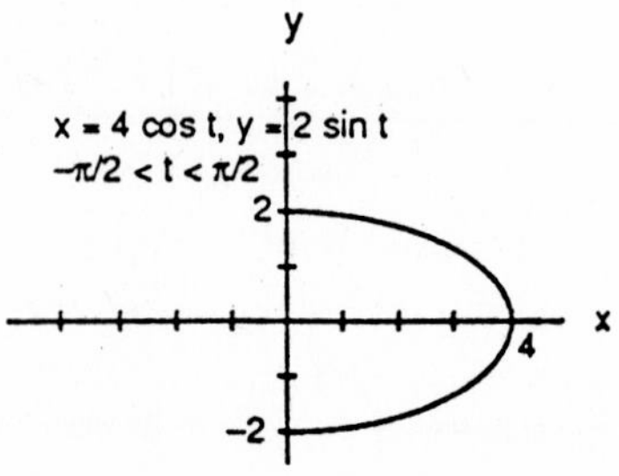

41.

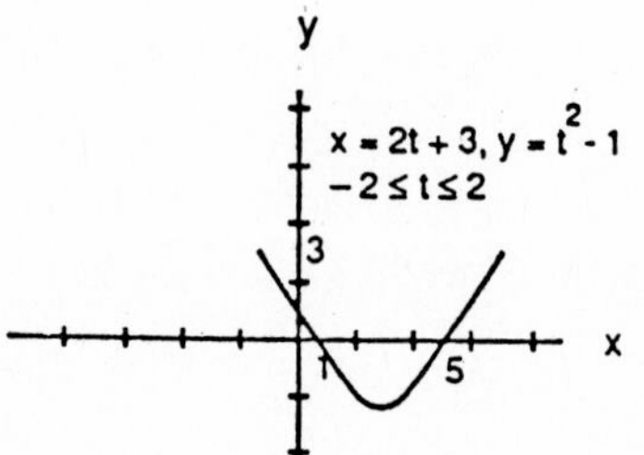

43. (a)

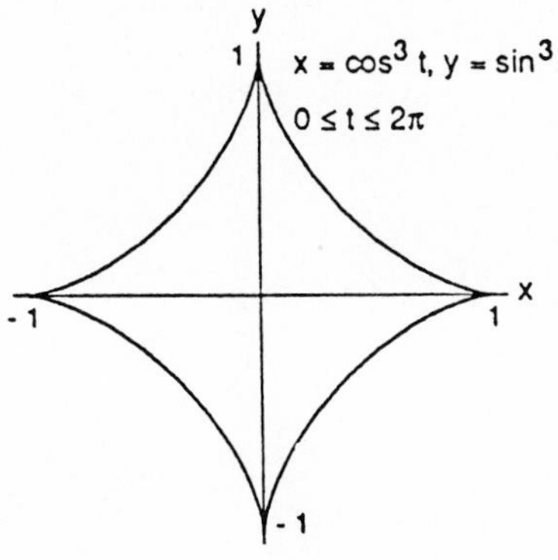

(b)

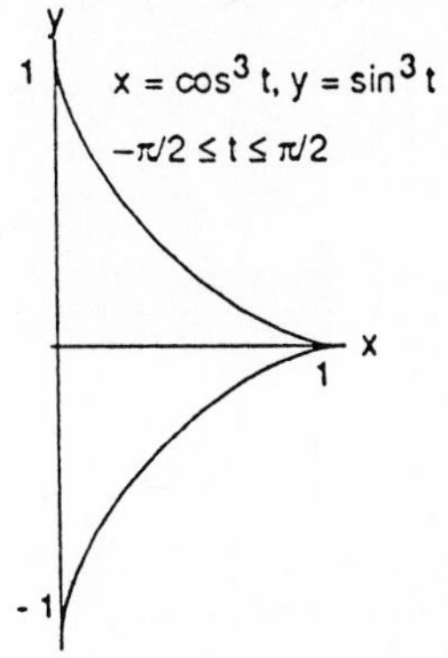

45. (a)

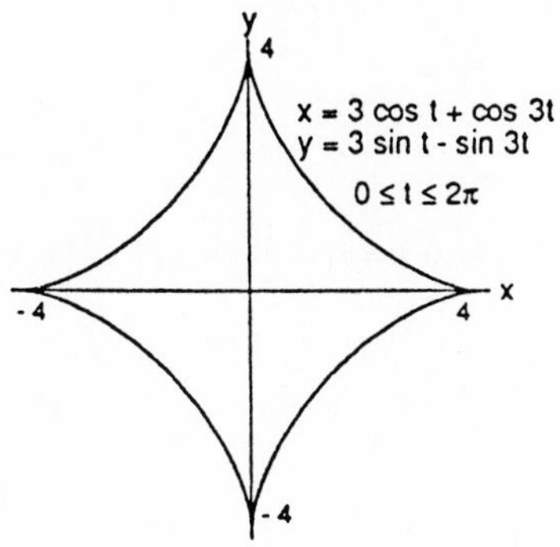

(b)

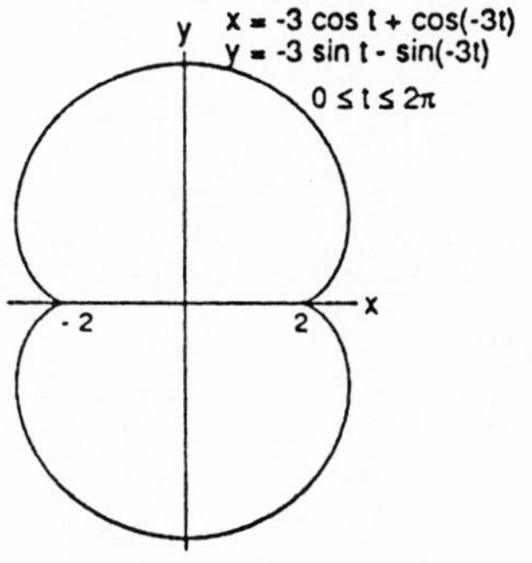

47. (a)

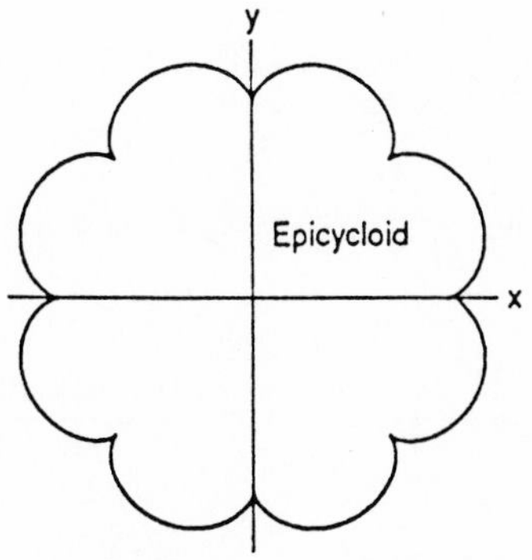

(b)

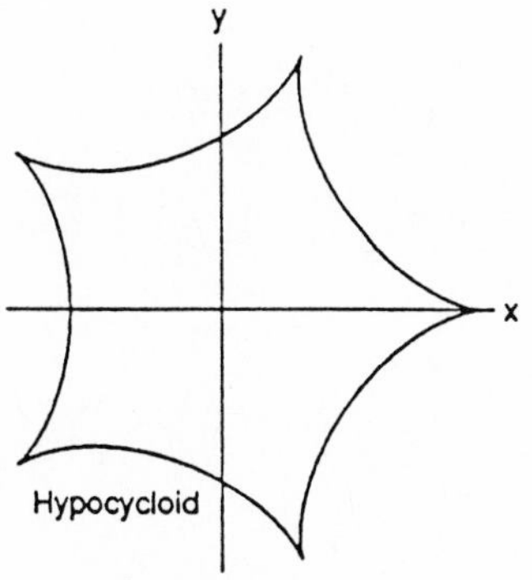

(c)

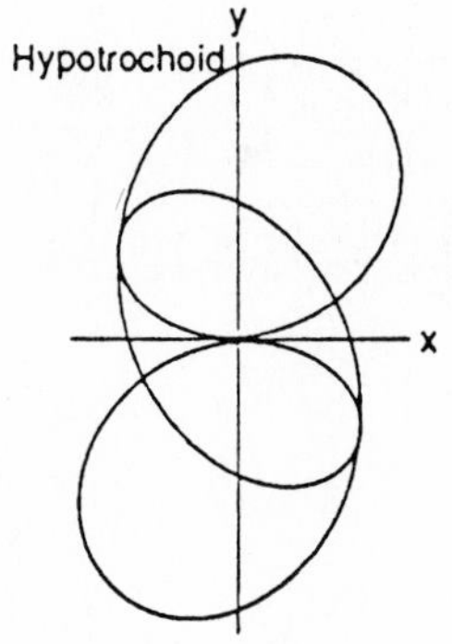

9.5 CALCULUS WITH PARAMETRIZED CURVES

1. $t = \frac{\pi}{4} \Rightarrow x = 2\cos\frac{\pi}{4} = \sqrt{2},\ y = 2\sin\frac{\pi}{4} = \sqrt{2};\ \frac{dx}{dt} = -2\sin t,\ \frac{dy}{dt} = 2\cos t \Rightarrow \frac{dy}{dx} = \frac{dy/dt}{dx/dt} = \frac{2\cos t}{-2\sin t} = -\cot t$

$\Rightarrow \left.\frac{dy}{dx}\right|_{t=\frac{\pi}{4}} = -\cot\frac{\pi}{4} = -1$; tangent line is $y - \sqrt{2} = -1\left(x - \sqrt{2}\right)$ or $y = -x + 2\sqrt{2}$; $\frac{dy'}{dt} = \csc^2 t$

$\Rightarrow \frac{d^2y}{dx^2} = \frac{dy'/dt}{dx/dt} = \frac{\csc^2 t}{-2\sin t} = -\frac{1}{2\sin^3 t} \Rightarrow \left.\frac{d^2y}{dx^2}\right|_{t=\frac{\pi}{4}} = -\sqrt{2}$

3. $t = \frac{\pi}{4} \Rightarrow x = 4 \sin \frac{\pi}{4} = 2\sqrt{2},\ y = 2 \cos \frac{\pi}{4} = \sqrt{2};\ \frac{dx}{dt} = 4 \cos t,\ \frac{dy}{dt} = -2 \sin t \Rightarrow \frac{dy}{dx} = \frac{dy/dt}{dx/dt} = \frac{-2 \sin t}{4 \cos t}$

$= -\frac{1}{2} \tan t \Rightarrow \left.\frac{dy}{dx}\right|_{t=\frac{\pi}{4}} = -\frac{1}{2} \tan \frac{\pi}{4} = -\frac{1}{2}$; tangent line is $y - \sqrt{2} = -\frac{1}{2}(x - 2\sqrt{2})$ or $y = -\frac{1}{2}x + 2\sqrt{2}$;

$\frac{dy'}{dt} = -\frac{1}{2} \sec^2 t \Rightarrow \frac{d^2y}{dx^2} = \frac{dy'/dt}{dx/dt} = \frac{-\frac{1}{2} \sec^2 t}{4 \cos t} = -\frac{1}{8 \cos^3 t} \Rightarrow \left.\frac{d^2y}{dx^2}\right|_{t=\frac{\pi}{4}} = -\frac{\sqrt{2}}{4}$

5. $t = \frac{1}{4} \Rightarrow x = \frac{1}{4},\ y = \frac{1}{2};\ \frac{dx}{dt} = 1,\ \frac{dy}{dt} = \frac{1}{2\sqrt{t}} \Rightarrow \frac{dy}{dx} = \frac{dy/dt}{dx/dt} = \frac{1}{2\sqrt{t}} \Rightarrow \left.\frac{dy}{dx}\right|_{t=\frac{1}{4}} = \frac{1}{2\sqrt{\frac{1}{4}}} = 1$; tangent line is

$y - \frac{1}{2} = 1 \cdot \left(x - \frac{1}{4}\right)$ or $y = x + \frac{1}{4};\ \frac{dy'}{dt} = -\frac{1}{4}t^{-3/2} \Rightarrow \frac{d^2y}{dx^2} = \frac{dy'/dt}{dx/dt} = -\frac{1}{4}t^{-3/2} \Rightarrow \left.\frac{d^2y}{dx^2}\right|_{t=\frac{1}{4}} = -2$

7. $t = \frac{\pi}{6} \Rightarrow x = \sec \frac{\pi}{6} = \frac{2}{\sqrt{3}},\ y = \tan \frac{\pi}{6} = \frac{1}{\sqrt{3}};\ \frac{dx}{dt} = \sec t \tan t,\ \frac{dy}{dt} = \sec^2 t \Rightarrow \frac{dy}{dx} = \frac{dy/dt}{dx/dt}$

$= \frac{\sec^2 t}{\sec t \tan t} = \csc t \Rightarrow \left.\frac{dy}{dx}\right|_{t=\frac{\pi}{6}} = \csc \frac{\pi}{6} = 2$; tangent line is $y - \frac{1}{\sqrt{3}} = 2\left(x - \frac{2}{\sqrt{3}}\right)$ or $y = 2x - \sqrt{3}$;

$\frac{dy'}{dt} = -\csc t \cot t \Rightarrow \frac{d^2y}{dx^2} = \frac{dy'/dt}{dx/dt} = \frac{-\csc t \cot t}{\sec t \tan t} = -\cot^3 t \Rightarrow \left.\frac{d^2y}{dx^2}\right|_{t=\frac{\pi}{6}} = -3\sqrt{3}$

9. $t = -1 \Rightarrow x = 5,\ y = 1;\ \frac{dx}{dt} = 4t,\ \frac{dy}{dt} = 4t^3 \Rightarrow \frac{dy}{dx} = \frac{dy/dt}{dx/dt} = \frac{4t^3}{4t} = t^2 \Rightarrow \left.\frac{dy}{dx}\right|_{t=-1} = (-1)^2 = 1$; tangent line is

$y - 1 = 1 \cdot (x - 5)$ or $y = x - 4;\ \frac{dy'}{dt} = 2t \Rightarrow \frac{d^2y}{dx^2} = \frac{dy'/dt}{dx/dt} = \frac{2t}{4t} = \frac{1}{2} \Rightarrow \left.\frac{d^2y}{dx^2}\right|_{t=-1} = \frac{1}{2}$

11. $t = \frac{\pi}{3} \Rightarrow x = \frac{\pi}{3} - \sin \frac{\pi}{3} = \frac{\pi}{3} - \frac{\sqrt{3}}{2},\ y = 1 - \cos \frac{\pi}{3} = 1 - \frac{1}{2} = \frac{1}{2};\ \frac{dx}{dt} = 1 - \cos t,\ \frac{dy}{dt} = \sin t \Rightarrow \frac{dy}{dx} = \frac{dy/dt}{dx/dt}$

$= \frac{\sin t}{1 - \cos t} \Rightarrow \left.\frac{dy}{dx}\right|_{t=\frac{\pi}{3}} = \frac{\sin\left(\frac{\pi}{3}\right)}{1 - \cos\left(\frac{\pi}{3}\right)} = \frac{\left(\frac{\sqrt{3}}{2}\right)}{\left(\frac{1}{2}\right)} = \sqrt{3}$; tangent line is $y - \frac{1}{2} = \sqrt{3}\left(x - \frac{\pi}{3} + \frac{\sqrt{3}}{2}\right)$

$\Rightarrow y = \sqrt{3}x - \frac{\pi\sqrt{3}}{3} + 2;\ \frac{dy'}{dt} = \frac{(1 - \cos t)(\cos t) - (\sin t)(\sin t)}{(1 - \cos t)^2} = \frac{-1}{1 - \cos t} \Rightarrow \frac{d^2y}{dx^2} = \frac{dy'/dt}{dx/dt} = \frac{\left(\frac{-1}{1 - \cos t}\right)}{1 - \cos t}$

$= \frac{-1}{(1 - \cos t)^2} \Rightarrow \left.\frac{d^2y}{dx^2}\right|_{t=\frac{\pi}{3}} = -4$

13. $x^2 - 2tx + 2t^2 = 4 \Rightarrow 2x \frac{dx}{dt} - 2x - 2t \frac{dx}{dt} + 4t = 0 \Rightarrow (2x - 2t) \frac{dx}{dt} = 2x - 4t \Rightarrow \frac{dx}{dt} = \frac{2x - 4t}{2x - 2t} = \frac{x - 2t}{x - t}$;

$2y^3 - 3t^2 = 4 \Rightarrow 6y^2 \frac{dy}{dt} - 6t = 0 \Rightarrow \frac{dy}{dt} = \frac{6t}{6y^2} = \frac{t}{y^2}$; thus $\frac{dy}{dx} = \frac{dy/dt}{dx/dt} = \frac{\left(\frac{t}{y^2}\right)}{\left(\frac{x - 2t}{x - t}\right)} = \frac{t(x - t)}{y^2(x - 2t)}$; $t = 2$

$\Rightarrow x^2 - 2(2)x + 2(2)^2 = 4 \Rightarrow x^2 - 4x + 4 = 0 \Rightarrow (x-2)^2 = 0 \Rightarrow x = 2;\ t = 2 \Rightarrow 2y^3 - 3(2)^2 = 4$

$\Rightarrow 2y^3 = 16 \Rightarrow y^3 = 8 \Rightarrow y = 2;$ therefore $\left.\frac{dy}{dx}\right|_{t=2} = \frac{2(2-2)}{(2)^2(2-2(2))} = 0$

15. $x + 2x^{3/2} = t^2 + t \Rightarrow \frac{dx}{dt} + 3x^{1/2}\frac{dx}{dt} = 2t + 1 \Rightarrow \left(1 + 3x^{1/2}\right)\frac{dx}{dt} = 2t + 1 \Rightarrow \frac{dx}{dt} = \frac{2t+1}{1+3x^{1/2}};\ y\sqrt{t+1} + 2t\sqrt{y} = 4$

$\Rightarrow \frac{dy}{dt}\sqrt{t+1} + y\left(\frac{1}{2}\right)(t+1)^{-1/2} + 2\sqrt{y} + 2t\left(\frac{1}{2}y^{-1/2}\right)\frac{dy}{dt} = 0 \Rightarrow \frac{dy}{dt}\sqrt{t+1} + \frac{y}{2\sqrt{t+1}} + 2\sqrt{y} + \left(\frac{t}{\sqrt{y}}\right)\frac{dy}{dt} = 0$

$\Rightarrow \left(\sqrt{t+1} + \frac{t}{\sqrt{y}}\right)\frac{dy}{dt} = \frac{-y}{2\sqrt{t+1}} - 2\sqrt{y} \Rightarrow \frac{dy}{dt} = \frac{\left(\frac{-y}{2\sqrt{t+1}} - 2\sqrt{y}\right)}{\left(\sqrt{t+1} + \frac{t}{\sqrt{y}}\right)} = \frac{-y\sqrt{y} - 4y\sqrt{t+1}}{2\sqrt{y}\,(t+1) + 2t\sqrt{t+1}};$ thus

$\frac{dy}{dx} = \frac{dy/dt}{dx/dt} = \frac{\left(\frac{-y\sqrt{y} - 4y\sqrt{t+1}}{2\sqrt{y}\,(t+1) + 2t\sqrt{t+1}}\right)}{\left(\frac{2t+1}{1+3x^{1/2}}\right)};\ t = 0 \Rightarrow x + 2x^{3/2} = 0 \Rightarrow x\left(1 + 2x^{1/2}\right) = 0 \Rightarrow x = 0;\ t = 0$

$\Rightarrow y\sqrt{0+1} + 2(0)\sqrt{y} = 4 \Rightarrow y = 4;$ therefore $\left.\frac{dy}{dx}\right|_{t=0} = \frac{\left(\frac{-4\sqrt{4} - 4(4)\sqrt{0+1}}{2\sqrt{4}(0+1) + 2(0)\sqrt{0+1}}\right)}{\left(\frac{2(0)+1}{1+3(0)^{1/2}}\right)} = -6$

17. $\frac{dx}{dt} = -\sin t$ and $\frac{dy}{dt} = 1 + \cos t \Rightarrow \sqrt{\left(\frac{dx}{dt}\right)^2 + \left(\frac{dy}{dt}\right)^2} = \sqrt{(-\sin t)^2 + (1+\cos t)^2} = \sqrt{2 + 2\cos t}$

$\Rightarrow \text{Length} = \int_0^{\pi} \sqrt{2 + 2\cos t}\, dt = \sqrt{2}\int_0^{\pi} \sqrt{\left(\frac{1-\cos t}{1-\cos t}\right)(1+\cos t)}\, dt = \sqrt{2}\int_0^{\pi} \sqrt{\frac{\sin^2 t}{1-\cos t}}\, dt$

$= \sqrt{2}\int_0^{\pi} \frac{\sin t}{\sqrt{1-\cos t}}\, dt$ (since $\sin t \ge 0$ on $[0, \pi]$); $[u = 1 - \cos t \Rightarrow du = \sin t\, dt;\ t = 0 \Rightarrow u = 0,$

$t = \pi \Rightarrow u = 2] \rightarrow \sqrt{2}\int_0^2 u^{-1/2}\, du = \sqrt{2}\left[2u^{1/2}\right]_0^2 = 4$

$u^{-1/2}\,du = 2u^{1/2}$ $= 2\sqrt{2}\sqrt{1-\cos t}\,\Big|_0^{\pi}$

19. $\frac{dx}{dt} = t$ and $\frac{dy}{dt} = (2t+1)^{1/2} \Rightarrow \sqrt{\left(\frac{dx}{dt}\right)^2 + \left(\frac{dy}{dt}\right)^2} = \sqrt{t^2 + (2t+1)} = \sqrt{(t+1)^2} = |t+1| = t + 1$ since $0 \le t \le 4$

$\Rightarrow \text{Length} = \int_0^4 (t+1)\, dt = \left[\frac{t^2}{2} + t\right]_0^4 = (8 + 4) = 12$

21. $\frac{dx}{dt} = 8t \cos t$ and $\frac{dy}{dt} = 8t \sin t \Rightarrow \sqrt{\left(\frac{dx}{dt}\right)^2 + \left(\frac{dy}{dt}\right)^2} = \sqrt{(8t \cos t)^2 + (8t \sin t)^2} = \sqrt{64t^2 \cos^2 t + 64t^2 \sin^2 t}$

$= |8t| = 8t$ since $0 \le t \le \frac{\pi}{2} \Rightarrow \text{Length} = \int_0^{\pi/2} 8t\, dt = \left[4t^2\right]_0^{\pi/2} = \pi^2$

23. $\frac{dx}{dt} = -\sin t$ and $\frac{dy}{dt} = \cos t \Rightarrow \sqrt{\left(\frac{dx}{dt}\right)^2 + \left(\frac{dy}{dt}\right)^2} = \sqrt{(-\sin t)^2 + (\cos t)^2} = 1 \Rightarrow \text{Area} = \int 2\pi y\, ds$

$= \int_0^{2\pi} 2\pi(2 + \sin t)(1)\, dt = 2\pi[2t - \cos t]_0^{2\pi} = 2\pi[(4\pi - 1) - (0 - 1)] = 8\pi^2$

25. $\frac{dx}{dt} = 1$ and $\frac{dy}{dt} = t + \sqrt{2} \Rightarrow \sqrt{\left(\frac{dx}{dt}\right)^2 + \left(\frac{dy}{dt}\right)^2} = \sqrt{1^2 + \left(t + \sqrt{2}\right)^2} = \sqrt{t^2 + 2\sqrt{2}\,t + 3} \Rightarrow \text{Area} = \int 2\pi x\, ds$

$= \int_{-\sqrt{2}}^{\sqrt{2}} 2\pi\left(t + \sqrt{2}\right)\sqrt{t^2 + 2\sqrt{2}\,t + 3}\, dt$; $\left[u = t^2 + 2\sqrt{2}\,t + 3 \Rightarrow du = \left(2t + 2\sqrt{2}\right) dt;\ t = -\sqrt{2} \Rightarrow u = 1,\right.$

$\left. t = \sqrt{2} \Rightarrow u = 9\right] \to \int_1^9 \pi\sqrt{u}\, du = \left[\frac{2}{3}\pi u^{3/2}\right]_1^9 = \frac{2\pi}{3}(27 - 1) = \frac{52\pi}{3}$

27. $\frac{dx}{dt} = 2$ and $\frac{dy}{dt} = 1 \Rightarrow \sqrt{\left(\frac{dx}{dt}\right)^2 + \left(\frac{dy}{dt}\right)^2} = \sqrt{2^2 + 1^2} = \sqrt{5} \Rightarrow \text{Area} = \int 2\pi y\, ds = \int_0^1 2\pi(t + 1)\sqrt{5}\, dt$

$= 2\pi\sqrt{5}\left[\frac{t^2}{2} + t\right]_0^1 = 3\pi\sqrt{5}$. Check: slant height is $\sqrt{5} \Rightarrow$ Area is $\pi(1 + 2)\sqrt{5} = 3\pi\sqrt{5}$.

29. (a) Let the density be $\delta = 1$. Then $x = \cos t + t \sin t \Rightarrow \frac{dx}{dt} = t \cos t$, and $y = \sin t - t \cos t \Rightarrow \frac{dy}{dt} = t \sin t$

$\Rightarrow dm = 1 \cdot ds = \sqrt{\left(\frac{dx}{dt}\right)^2 + \left(\frac{dy}{dt}\right)^2}\, dt = \sqrt{(t \cos t)^2 + (t \sin t)^2} = |t|\, dt = t\, dt$ since $0 \le t \le \frac{\pi}{2}$. The curve's

mass is $M = \int dm = \int_0^{\pi/2} t\, dt = \frac{\pi^2}{8}$. Also $M_x = \int \tilde{y}\, dm = \int_0^{\pi/2} (\sin t - t \cos t)\, t\, dt$

$= \int_0^{\pi/2} t \sin t\, dt - \int_0^{\pi/2} t^2 \cos t\, dt = [\sin t - t \cos t]_0^{\pi/2} - \left[t^2 \sin t - 2 \sin t + 2t \cos t\right]_0^{\pi/2} = 3 - \frac{\pi^2}{4}$, where

we integrated by parts. Therefore, $\overline{y} = \frac{M_x}{M} = \frac{\left(3 - \frac{\pi^2}{4}\right)}{\left(\frac{\pi^2}{8}\right)} = \frac{24}{\pi^2} - 2$. Next, $M_y = \int \tilde{x}\, dm$

$= \int_0^{\pi/2} (\cos t + t \sin t)\, t\, dt = \int_0^{\pi/2} t \cos t\, dt + \int_0^{\pi/2} t^2 \sin t\, dt$

$= \left[\cos t + t \sin t\right]_0^{\pi/2} + \left[-t^2 \cos t + 2 \cos t + 2t \sin t\right]_0^{\pi/2} = \frac{3\pi}{2} - 3$, again integrating by parts.

Hence $\bar{x} = \frac{M_y}{M} = \frac{\left(\frac{3\pi}{2} - 3\right)}{\left(\frac{\pi^2}{8}\right)} = \frac{12}{\pi} - \frac{24}{\pi^2}$. Therefore $(\bar{x}, \bar{y}) = \left(\frac{12}{\pi} - \frac{24}{\pi^2}, \frac{24}{\pi^2} - 2\right)$

(b) $(\bar{x}, \bar{y}) \approx (1.4, 0.4)$

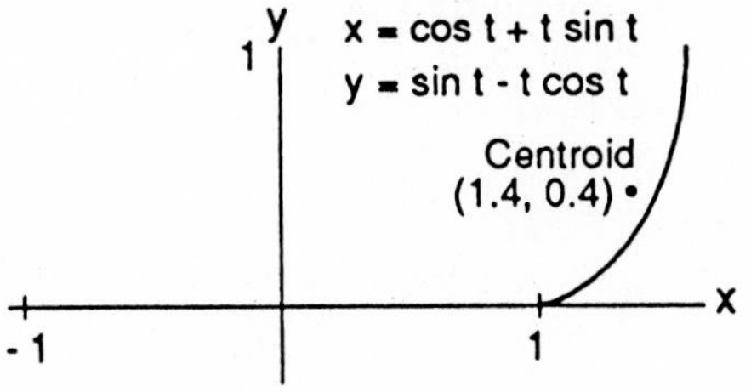

31. (a) Let the density be $\delta = 1$. Then $x = \cos t \Rightarrow \frac{dx}{dt} = -\sin t$, and $y = t + \sin t \Rightarrow \frac{dy}{dt} = 1 + \cos t$

$\Rightarrow dm = 1 \cdot ds = \sqrt{\left(\frac{dx}{dt}\right)^2 + \left(\frac{dy}{dt}\right)^2}\, dt = \sqrt{(-\sin t)^2 + (1 + \cos t)^2}\, dt = \sqrt{2 + 2\cos t}\, dt$. The curve's mass

is $M = \int dm = \int_0^\pi \sqrt{2 + 2\cos t}\, dt = \sqrt{2} \int_0^\pi \sqrt{1 + \cos t}\, dt = \sqrt{2} \int_0^\pi \sqrt{2\cos^2\left(\frac{t}{2}\right)}\, dt = 2 \int_0^\pi \left|\cos\left(\frac{t}{2}\right)\right| dt$

$= 2 \int_0^\pi \cos\left(\frac{t}{2}\right) dt \left(\text{since } 0 \le t \le \pi \Rightarrow 0 \le \frac{t}{2} \le \frac{\pi}{2}\right) = 2\left[2 \sin\left(\frac{t}{2}\right)\right]_0^\pi = 4$. Also $M_x = \int \tilde{y}\, dm$

$= \int_0^\pi (t + \sin t)\left(2 \cos \frac{t}{2}\right) dt = \int_0^\pi 2t \cos\left(\frac{t}{2}\right) dt + \int_0^\pi 2 \sin t \cos\left(\frac{t}{2}\right) dt$

$= 2\left[4 \cos\left(\frac{t}{2}\right) + 2t \sin\left(\frac{t}{2}\right)\right]_0^\pi + 2\left[-\frac{1}{3}\cos\left(\frac{3}{2}t\right) - \cos\left(\frac{1}{2}t\right)\right]_0^\pi = 4\pi - \frac{16}{3} \Rightarrow \bar{y} = \frac{M_x}{M} = \frac{\left(4\pi - \frac{16}{3}\right)}{4} = \pi - \frac{4}{3}$.

Next $M_y = \int \tilde{x}\, dm = \int_0^\pi (\cos t)\left(2 \cos \frac{t}{2}\right) dt = 2 \int_0^\pi \cos t \cos\left(\frac{t}{2}\right) dt = 2\left[\sin\left(\frac{t}{2}\right) + \frac{\sin\left(\frac{3}{2}t\right)}{3}\right]_0^\pi = 2 - \frac{2}{3}$

$= \frac{4}{3} \Rightarrow \bar{x} = \frac{M_y}{M} = \frac{\left(\frac{4}{3}\right)}{4} = \frac{1}{3}$. Therefore $(\bar{x}, \bar{y}) = \left(\frac{1}{3}, \pi - \frac{4}{3}\right)$

(b) $(\bar{x}, \bar{y}) \approx (0.33, 1.81)$

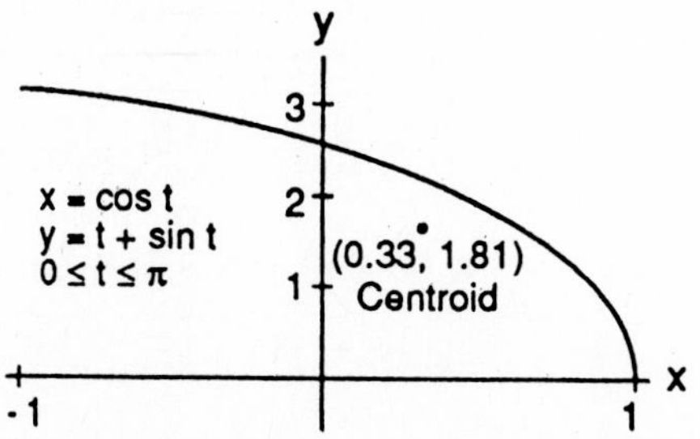

33. (a) $\frac{dx}{dt} = -2 \sin 2t$ and $\frac{dy}{dt} = 2 \cos 2t \Rightarrow \sqrt{\left(\frac{dx}{dt}\right)^2 + \left(\frac{dy}{dt}\right)^2} = \sqrt{(-2 \sin 2t)^2 + (2 \cos 2t)^2} = 2$

$\Rightarrow \text{Length} = \int_0^{\pi/2} 2\, dt = [2t]_0^{\pi/2} = \pi$

(b) $\frac{dx}{dt} = \pi \cos \pi t$ and $\frac{dy}{dt} = -\pi \sin \pi t = \sqrt{\left(\frac{dx}{dt}\right)^2 + \left(\frac{dy}{dt}\right)^2} = \sqrt{(\pi \cos \pi t)^2 + (-\pi \sin \pi t)^2} = \pi$

$\Rightarrow \text{Length} = \int_{-1/2}^{1/2} \pi\, dt = [\pi t]_{-1/2}^{1/2} = \pi$

35. $x = x \Rightarrow \frac{dy}{dx} = 1$, and $y = f(x) \Rightarrow \frac{dy}{dx} = f'(x)$; then $\text{Length} = \int_a^b \sqrt{\left(\frac{dx}{dt}\right)^2 + \left(\frac{dy}{dt}\right)^2}\, dt = \int_a^b \sqrt{\left(\frac{dx}{dx}\right)^2 + \left(\frac{dy}{dx}\right)^2}\, dx$

$= \int_a^b \sqrt{1 + \left(\frac{dy}{dx}\right)^2}\, dx = \int_a^b \sqrt{1 + [f'(x)]^2}\, dx$

37. For one arch of the cycloid we use the interval $0 \le \theta \le 2\pi$. Then, $A = \int_0^{2\pi} y(\theta)\, dx = \int_0^{2\pi} a(1 - \cos \theta)\left(\frac{dx}{d\theta}\right) d\theta$

and $\frac{dx}{d\theta} = a(1 - \cos \theta) \Rightarrow A = \int_0^{2\pi} a^2(1 - \cos \theta)^2\, d\theta = a^2 \int_0^{2\pi} (1 - 2 \cos \theta + \cos^2 \theta)\, d\theta$

$= a^2\left[\int_0^{2\pi} d\theta - 2\int_0^{2\pi} \cos \theta\, d\theta + \int_0^{2\pi} \frac{1}{2}(1 + \cos 2\theta)\, d\theta\right] = a^2\left([\theta]_0^{2\pi} - 2[\sin \theta]_0^{2\pi} + \frac{1}{2}\left[\theta + \frac{1}{2} \sin 2\theta\right]_0^{2\pi}\right)$

$= a^2\left[(2\pi - 0) - 2(0 - 0) + \frac{1}{2}(2\pi - 0)\right] = 3\pi a^2$

39. $x = \theta - \sin \theta$ and $y = 1 - \cos \theta$, $0 \le \theta \le 2\pi \Rightarrow ds = \sqrt{(1 - \cos \theta)^2 + \sin^2 \theta}\, d\theta = \sqrt{2 - 2 \cos \theta}\, d\theta \Rightarrow S = \int 2\pi y\, ds$

$= 2\sqrt{2}\int_0^{2\pi} \pi(1 - \cos \theta)^{3/2}\, d\theta = 2\sqrt{2}\int_0^{2\pi} \pi\left(\sqrt{2} \sin \frac{\theta}{2}\right)^3 d\theta = 8\pi \int_0^{2\pi} \left(1 - \cos^2 \frac{\theta}{2}\right)\left(\sin \frac{\theta}{2}\right) d\theta$

$= 8\pi\left[-2 \cos \frac{\theta}{2} + \frac{2}{3} \cos^3 \frac{\theta}{2}\right]_0^{2\pi} = \frac{64\pi}{3}$

41. $\frac{dx}{dt} = \cos t$ and $\frac{dy}{dt} = 2 \cos 2t \Rightarrow \frac{dy}{dx} = \frac{dy/dt}{dx/dt} = \frac{2 \cos 2t}{\cos t} = \frac{2(2 \cos^2 t - 1)}{\cos t}$; then $\frac{dy}{dx} = 0 \Rightarrow \frac{2(2 \cos^2 t - 1)}{\cos t} = 0$

$\Rightarrow 2 \cos^2 t - 1 = 0 \Rightarrow \cos t = \pm\frac{1}{\sqrt{2}} \Rightarrow t = \frac{\pi}{4}, \frac{3\pi}{4}, \frac{5\pi}{4}, \frac{7\pi}{4}$. In the 1st quadrant: $t = \frac{\pi}{4} \Rightarrow x = \sin \frac{\pi}{4} = \frac{\sqrt{2}}{2}$ and

$y = \sin 2\left(\frac{\pi}{4}\right) = 1 \Rightarrow \left(\frac{\sqrt{2}}{2}, 1\right)$ is the point where the tangent line is horizontal. At the origin: $x = 0$ and $y = 0$

$\Rightarrow \sin t = 0 \Rightarrow t = 0$ or $t = \pi$ and $\sin 2t = 0 \Rightarrow t = 0, \frac{\pi}{2}, \pi, \frac{3\pi}{2}$; thus $t = 0$ and $t = \pi$ give the tangent lines at

the origin. Tangents at origin: $\left.\frac{dy}{dx}\right|_{t=0} = 2 \Rightarrow y = 2x$ and $\left.\frac{dy}{dx}\right|_{t=\pi} = -2 \Rightarrow y = -2x$

43.

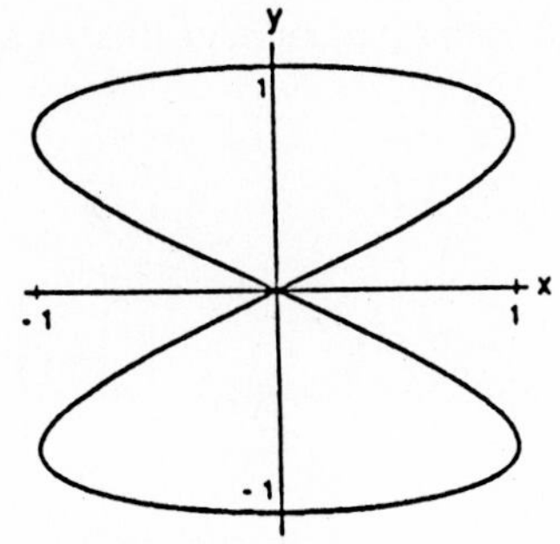

45.

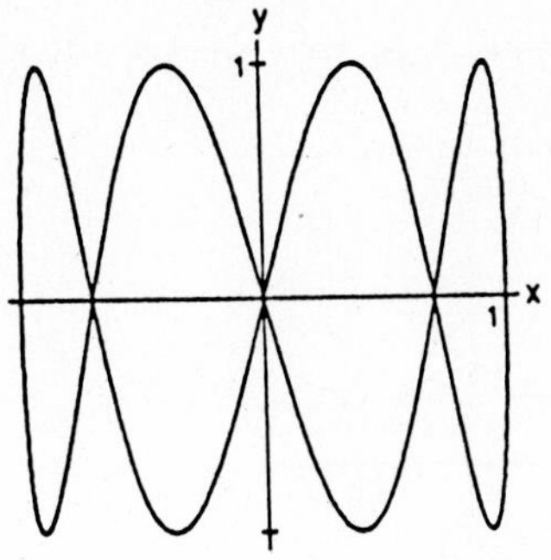

47.

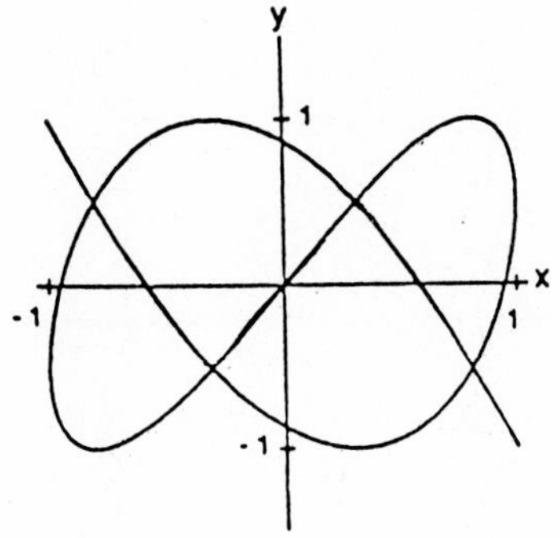

49.

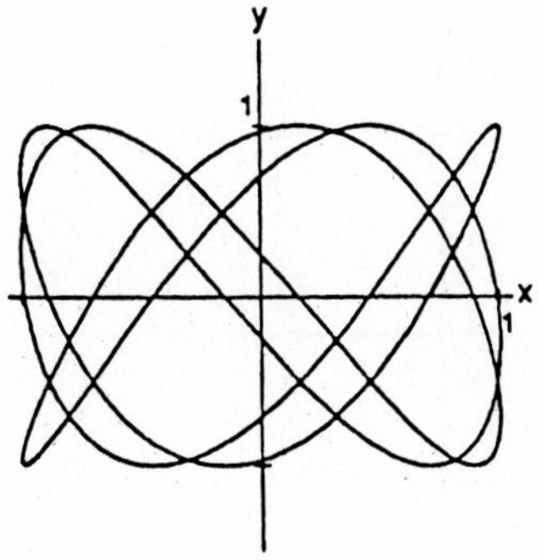

9.6 POLAR COORDINATES

1. a, e; b, g; c, h; d, f

3. (a) $\left(2, \frac{\pi}{2} + 2n\pi\right)$ and $\left(-2, \frac{\pi}{2} + (2n+1)\pi\right)$, n an integer

   (b) $(2, 2n\pi)$ and $(-2, (2n+1)\pi)$, n an integer

   (c) $\left(2, \frac{3\pi}{2} + 2n\pi\right)$ and $\left(-2, \frac{3\pi}{2} + (2n+1)\pi\right)$, n an integer

   (d) $(2, (2n+1)\pi)$ and $(-2, 2n\pi)$, n an integer

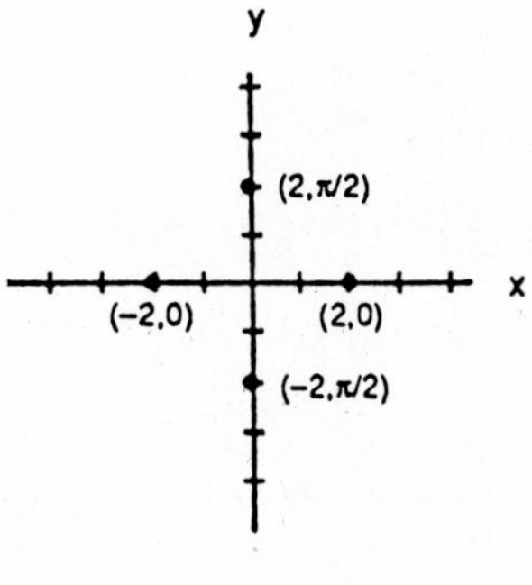

5. (a) $x = r\cos\theta = 3\cos 0 = 3$, $y = r\sin\theta = 3\sin 0 = 0 \Rightarrow$ Cartesian coordinates are $(3, 0)$

   (b) $x = r\cos\theta = -3\cos 0 = -3$, $y = r\sin\theta = -3\sin 0 = 0 \Rightarrow$ Cartesian coordinates are $(-3, 0)$

   (c) $x = r\cos\theta = 2\cos\frac{2\pi}{3} = -1$, $y = r\sin\theta = 2\sin\frac{2\pi}{3} = \sqrt{3} \Rightarrow$ Cartesian coordinates are $\left(-1, \sqrt{3}\right)$

   (d) $x = r\cos\theta = 2\cos\frac{7\pi}{3} = 1$, $y = r\sin\theta = 2\sin\frac{7\pi}{3} = \sqrt{3} \Rightarrow$ Cartesian coordinates are $\left(1, \sqrt{3}\right)$

   (e) $x = r\cos\theta = -3\cos\pi = 3$, $y = r\sin\theta = -3\sin\pi = 0 \Rightarrow$ Cartesian coordinates are $(3, 0)$

   (f) $x = r\cos\theta = 2\cos\frac{\pi}{3} = 1$, $y = r\sin\theta = 2\sin\frac{\pi}{3} = \sqrt{3} \Rightarrow$ Cartesian coordinates are $\left(1, \sqrt{3}\right)$

   (g) $x = r\cos\theta = -3\cos 2\pi = -3$, $y = r\sin\theta = -3\sin 2\pi = 0 \Rightarrow$ Cartesian coordinates are $(-3, 0)$

(h) $x = r\cos\theta = -2\cos\left(-\frac{\pi}{3}\right) = -1$, $y = r\sin\theta = -2\sin\left(-\frac{\pi}{3}\right) = \sqrt{3} \Rightarrow$ Cartesian coordinates are $(-1, \sqrt{3})$

7.

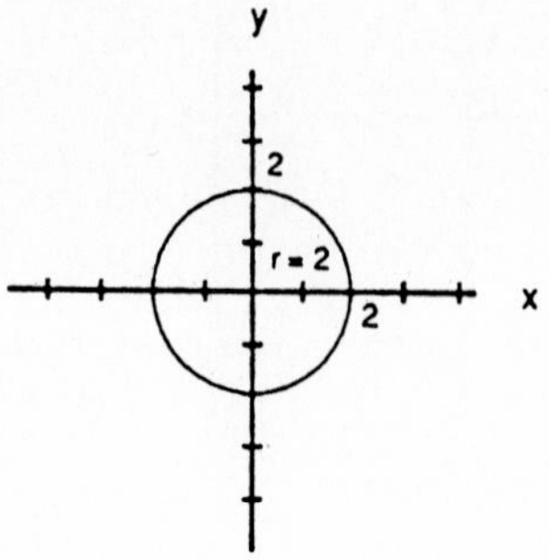

9.

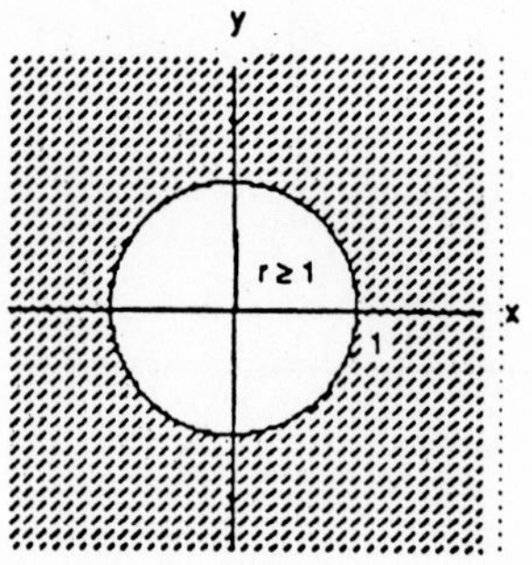

11.

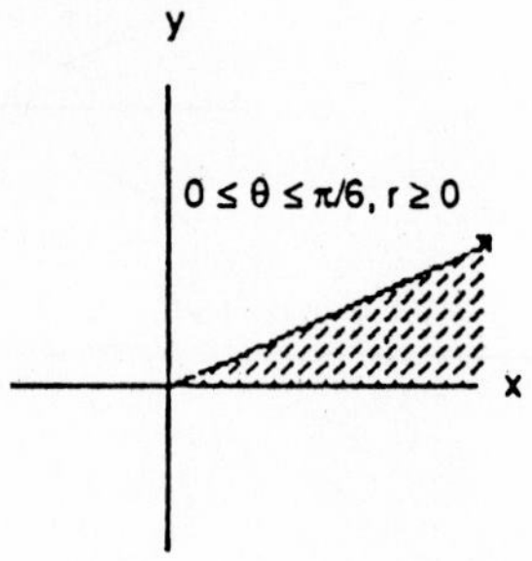

13.

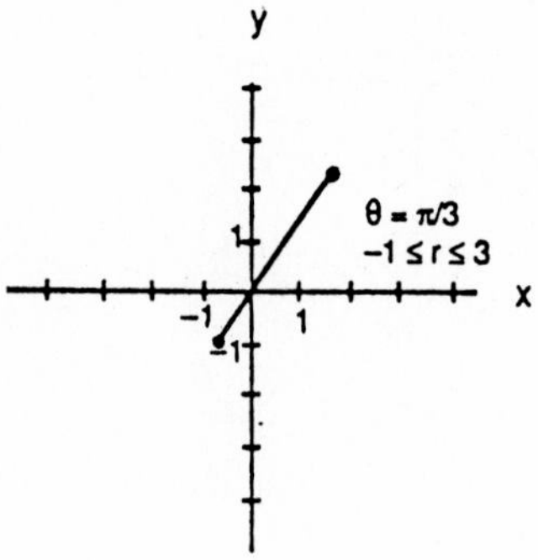

15.

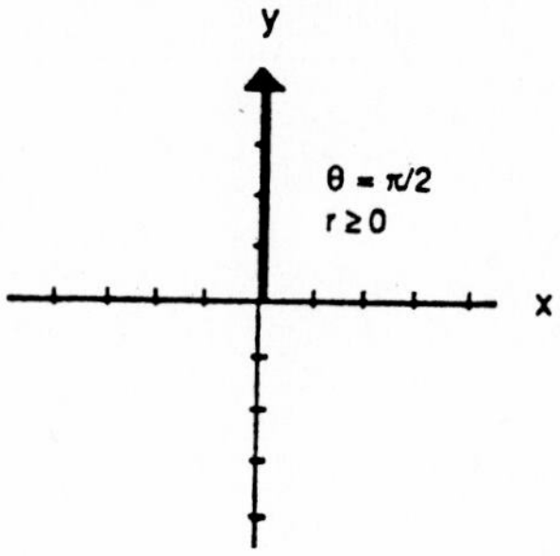

17.

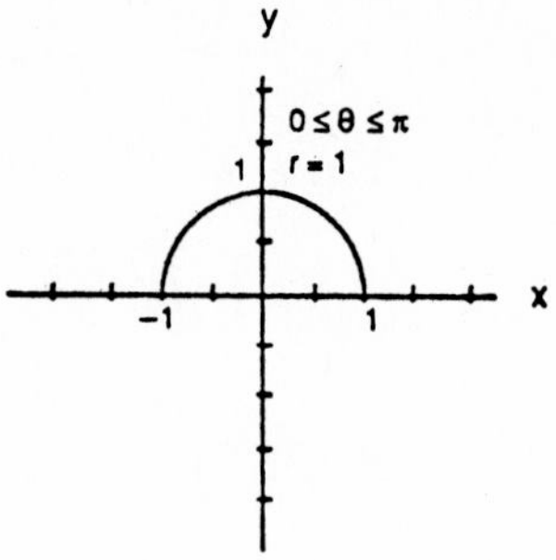

19.

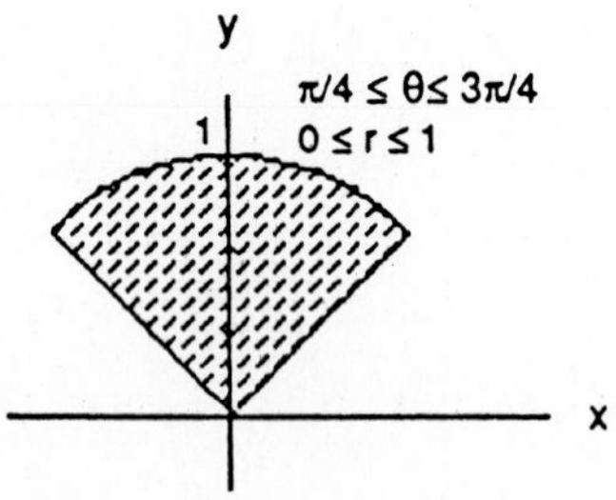

21.

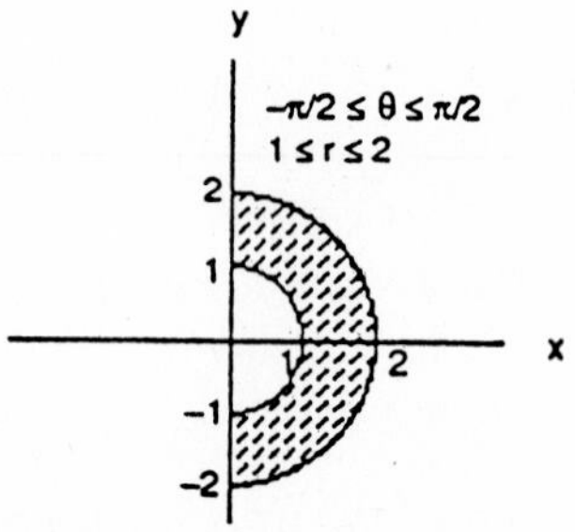

23. $r\cos\theta = 2 \Rightarrow x = 2$, vertical line through $(2,0)$

25. $r\sin\theta = 0 \Rightarrow y = 0$, the x-axis

27. $r = 4\csc\theta \Rightarrow r = \frac{4}{\sin\theta} \Rightarrow r\sin\theta = 4 \Rightarrow y = 4$, a horizontal line through $(0,4)$

29. $r\cos\theta + r\sin\theta = 1 \Rightarrow x + y = 1$, line with slope $m = -1$ and intercept $b = 1$

31. $r^2 = 1 \Rightarrow x^2 + y^2 = 1$, circle with center $C = (0,0)$ and radius 1

33. $r = \frac{5}{\sin\theta - 2\cos\theta} \Rightarrow r\sin\theta - 2r\cos\theta = 5 \Rightarrow y - 2x = 5$, line with slope $m = 2$ and intercept $b = 5$

35. $r = \cot\theta\csc\theta = \left(\frac{\cos\theta}{\sin\theta}\right)\left(\frac{1}{\sin\theta}\right) \Rightarrow r\sin^2\theta = \cos\theta \Rightarrow r^2\sin^2\theta = r\cos\theta \Rightarrow y^2 = x$, parabola with vertex $(0,0)$ which opens to the right

37. $r = (\csc\theta)\, e^{r\cos\theta} \Rightarrow r\sin\theta = e^{r\cos\theta} \Rightarrow y = e^x$, graph of the natural exponential function

39. $r^2 + 2r^2 \cos\theta \sin\theta = 1 \Rightarrow x^2 + y^2 + 2xy = 1 \Rightarrow x^2 + 2xy + y^2 = 1 \Rightarrow (x+y)^2 = 1 \Rightarrow x + y = \pm 1$, two parallel straight lines of slope -1 and y-intercepts $b = \pm 1$

41. $r^2 = -4r \cos\theta \Rightarrow x^2 + y^2 = -4x \Rightarrow x^2 + 4x + y^2 = 0 \Rightarrow x^2 + 4x + 4 + y^2 = 4 \Rightarrow (x+2)^2 + y^2 = 4$, a circle with center $C(-2, 0)$ and radius 2

43. $r = 8 \sin\theta \Rightarrow r^2 = 8r \sin\theta \Rightarrow x^2 + y^2 = 8y \Rightarrow x^2 + y^2 - 8y = 0 \Rightarrow x^2 + y^2 - 8y + 16 = 16$
$\Rightarrow x^2 + (y-4)^2 = 16$, a circle with center $C(0, 4)$ and radius 4

45. $r = 2\cos\theta + 2\sin\theta \Rightarrow r^2 = 2r\cos\theta + 2r\sin\theta \Rightarrow x^2 + y^2 = 2x + 2y \Rightarrow x^2 - 2x + y^2 - 2y = 0$
$\Rightarrow (x-1)^2 + (y-1)^2 = 2$, a circle with center $C(1, 1)$ and radius $\sqrt{2}$

47. $r \sin\left(\theta + \frac{\pi}{6}\right) = 2 \Rightarrow r\left(\sin\theta \cos\frac{\pi}{6} + \cos\theta \sin\frac{\pi}{6}\right) = 2 \Rightarrow \frac{\sqrt{3}}{2} r \sin\theta + \frac{1}{2} r \cos\theta = 2 \Rightarrow \frac{\sqrt{3}}{2} y + \frac{1}{2} x = 2$
$\Rightarrow \sqrt{3}\,y + x = 4$, line with slope $m = -\frac{1}{\sqrt{3}}$ and intercept $b = \frac{4}{\sqrt{3}}$

49. $x = 7 \Rightarrow r\cos\theta = 7$

51. $x = y \Rightarrow r\cos\theta = r\sin\theta \Rightarrow \theta = \frac{\pi}{4}$

53. $x^2 + y^2 = 4 \Rightarrow r^2 = 4 \Rightarrow r = 2$ or $r = -2$

55. $\frac{x^2}{9} + \frac{y^2}{4} = 1 \Rightarrow 4x^2 + 9y^2 = 36 \Rightarrow 4r^2\cos^2\theta + 9r^2\sin^2\theta = 36$

57. $y^2 = 4x \Rightarrow r^2\sin^2\theta = 4r\cos\theta \Rightarrow r\sin^2\theta = 4\cos\theta$

59. $x^2 + (y-2)^2 = 4 \Rightarrow x^2 + y^2 - 4y + 4 = 4 \Rightarrow x^2 + y^2 = 4y \Rightarrow r^2 = 4r\sin\theta \Rightarrow r = 4\sin\theta$

61. $(x-3)^2 + (y+1)^2 = 4 \Rightarrow x^2 - 6x + 9 + y^2 + 2y + 1 = 4 \Rightarrow x^2 + y^2 = 6x - 2y - 6 \Rightarrow r^2 = 6r\cos\theta - 2r\sin\theta - 6$

63. $(0, \theta)$ where θ is any angle

9.7 GRAPHING IN POLAR COORDINATES

1. $1 + \cos(-\theta) = 1 + \cos\theta = r \Rightarrow$ symmetric about the x-axis; $1 + \cos(-\theta) \neq -r$ and $1 + \cos(\pi - \theta) = 1 - \cos\theta \neq r$ $\Rightarrow$ not symmetric about the y-axis; therefore not symmetric about the origin

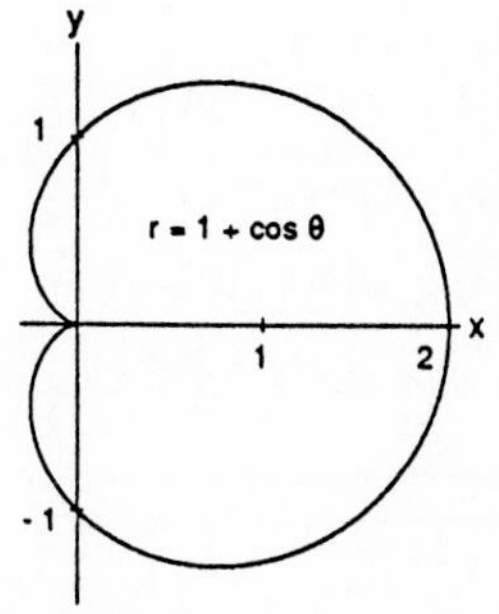

3. $1-\sin(-\theta)=1+\sin\theta\neq r$ and $1-\sin(\pi-\theta)$
$=1-\sin\theta\neq -r \Rightarrow$ not symmetric about the x-axis;
$1-\sin(\pi-\theta)=1-\sin\theta=r\Rightarrow$ symmetric about the y-axis; therefore not symmetric about the origin

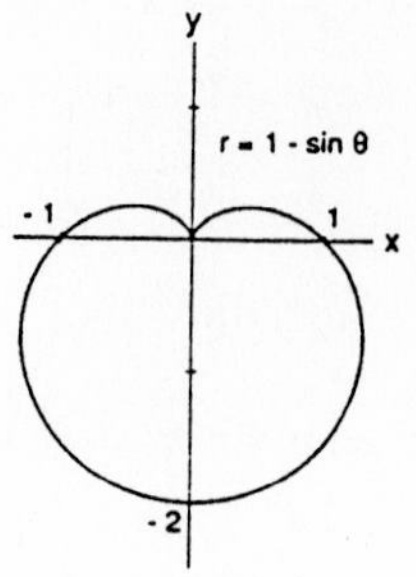

5. $2+\sin(-\theta)=2-\sin\theta\neq r$ and $2+\sin(\pi-\theta)$
$=2+\sin\theta\neq r\Rightarrow$ not symmetric about the x-axis;
$2+\sin(\pi-\theta)=2+\sin\theta=r\Rightarrow$ symmetric about the y-axis; therefore not symmetric about the origin

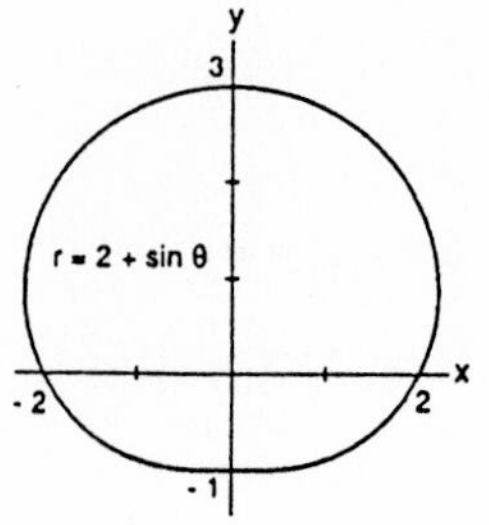

7. $\sin\left(-\frac{\theta}{2}\right)=-\sin\left(\frac{\theta}{2}\right)=-r\Rightarrow$ symmetric about the y-axis;
$\sin\left(-\frac{\theta}{2}\right)=-\sin\left(\frac{\theta}{2}\right)=-r\ \neq r$ and $\sin\left(\frac{\pi-\theta}{2}\right)=\sin\left(\frac{\pi}{2}-\frac{\theta}{2}\right)$
$=\cos\left(\frac{\theta}{2}\right)\neq -r$, but clearly the graph is symmetric about the x-axis and the origin. The symmetry tests as stated do not necessarily tell when a graph is not symmetric. Note that $\sin\left(\frac{2\pi-\theta}{2}\right)=\sin\left(\frac{\theta}{2}\right)$, so the graph is symmetric about the x-axis, and hence the origin.

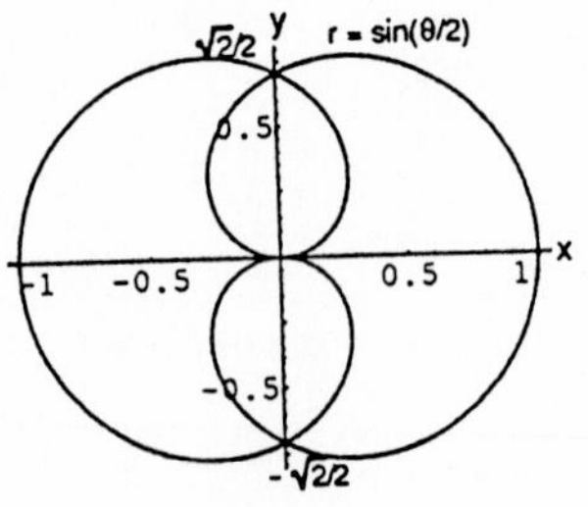

9. $\cos(-\theta)=\cos\theta=r^2\Rightarrow(r,-\theta)$ and $(-r,-\theta)$ are on the graph when (r,θ) is on the graph $\Rightarrow$ symmetric about the x-axis and the y-axis; therefore symmetric about the origin

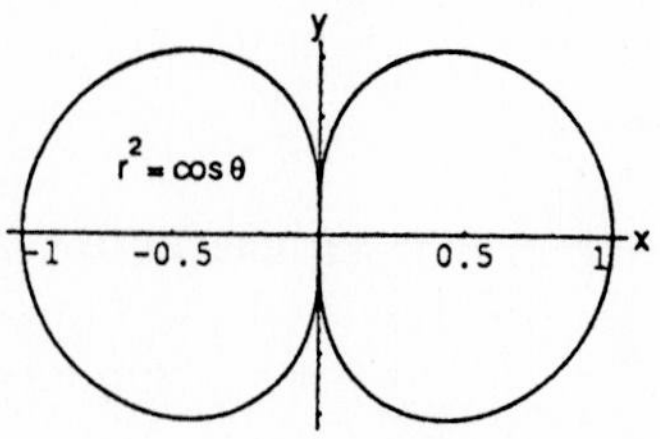

11. $-\sin(\pi-\theta)=-\sin\theta=r^2\Rightarrow(r,\pi-\theta)$ and $(-r,\pi-\theta)$ are on the graph when (r,θ) is on the graph $\Rightarrow$ symmetric about the y-axis and the x-axis; therefore symmetric about the origin

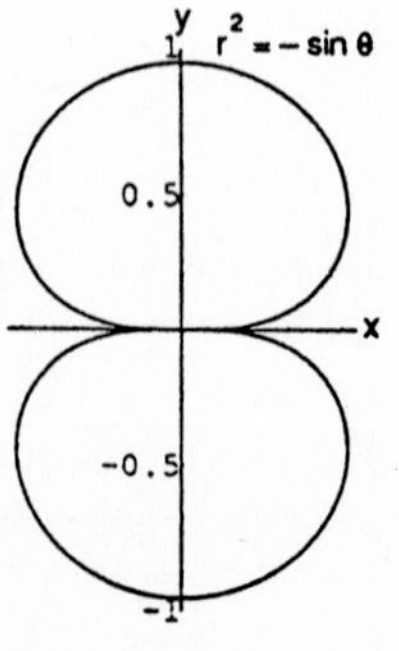

13. Since $(\pm r, -\theta)$ are on the graph when (r, θ) is on the graph $\left((\pm r)^2 = 4\cos 2(-\theta) \Rightarrow r^2 = 4\cos 2\theta\right)$, the graph is symmetric about the x-axis and the y-axis $\Rightarrow$ the graph is symmetric about the origin

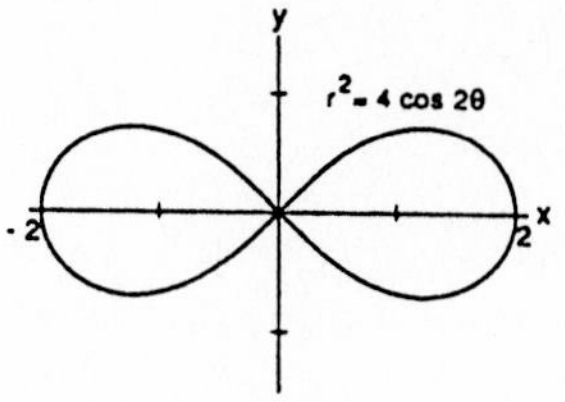

15. Since (r, θ) on the graph $\Rightarrow (-r, \theta)$ is on the graph $\left((\pm r)^2 = -\sin 2\theta \Rightarrow r^2 = -\sin 2\theta\right)$, the graph is symmetric about the origin. But $-\sin 2(-\theta) = -(-\sin 2\theta)$ $\sin 2\theta \neq r^2$ and $-\sin 2(\pi - \theta) = -\sin(2\pi - 2\theta)$ $= -\sin(-2\theta) = -(-\sin 2\theta) = \sin 2\theta \neq r^2 \Rightarrow$ the graph is not symmetric about the x-axis; therefore the graph is not symmetric about the y-axis

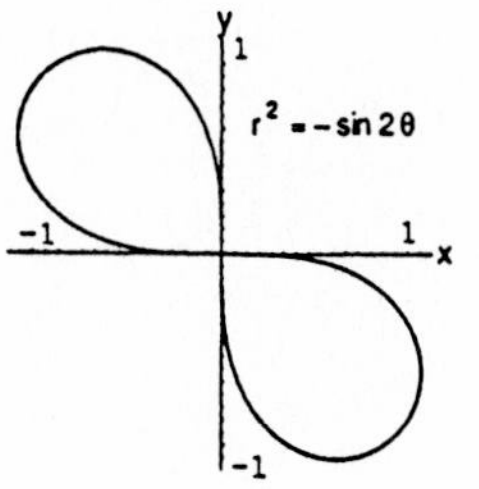

17. $\theta = \frac{\pi}{2} \Rightarrow r = -1 \Rightarrow \left(-1, \frac{\pi}{2}\right)$, and $\theta = -\frac{\pi}{2} \Rightarrow r = -1$ $\Rightarrow \left(-1, -\frac{\pi}{2}\right)$; $r' = \frac{dr}{d\theta} = -\sin\theta$; Slope $= \frac{r'\sin\theta + r\cos\theta}{r'\cos\theta - r\sin\theta}$

$= \frac{-\sin^2\theta + r\cos\theta}{-\sin\theta\cos\theta - r\sin\theta} \Rightarrow$ Slope at $\left(-1, \frac{\pi}{2}\right)$ is

$\frac{-\sin^2\left(\frac{\pi}{2}\right) + (-1)\cos\frac{\pi}{2}}{-\sin\frac{\pi}{2}\cos\frac{\pi}{2} - (-1)\sin\frac{\pi}{2}} = -1$; Slope at $\left(-1, -\frac{\pi}{2}\right)$ is

$$\frac{-\sin^2\left(-\frac{\pi}{2}\right) + (-1)\cos\left(-\frac{\pi}{2}\right)}{-\sin\left(-\frac{\pi}{2}\right)\cos\left(-\frac{\pi}{2}\right) - (-1)\sin\left(-\frac{\pi}{2}\right)} = 1$$

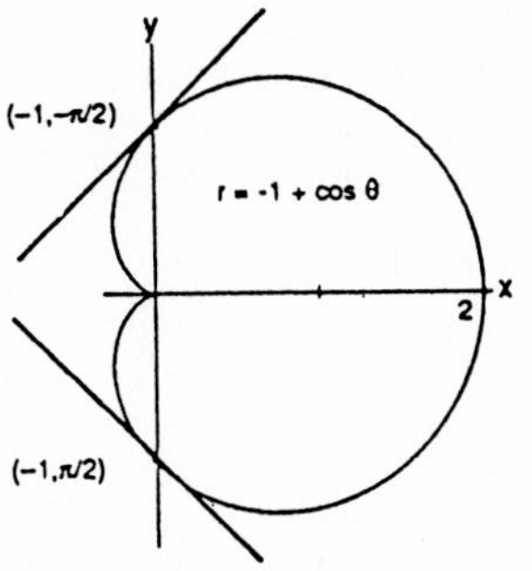

19. $\theta = \frac{\pi}{4} \Rightarrow r = 1 \Rightarrow \left(1, \frac{\pi}{4}\right)$; $\theta = -\frac{\pi}{4} \Rightarrow r = -1 \Rightarrow \left(-1, -\frac{\pi}{4}\right)$;

$\theta = \frac{3\pi}{4} \Rightarrow r = -1 \Rightarrow \left(-1, \frac{3\pi}{4}\right)$; $\theta = -\frac{3\pi}{4} \Rightarrow r = 1 \Rightarrow \left(1, -\frac{3\pi}{4}\right)$;

$r' = \frac{dr}{d\theta} = 2\cos 2\theta$;

Slope $= \frac{r'\sin\theta + r\cos\theta}{r'\cos\theta - r\sin\theta} = \frac{2\cos 2\theta\sin\theta + r\cos\theta}{2\cos 2\theta\cos\theta - r\sin\theta}$

$\Rightarrow$ Slope at $\left(1, \frac{\pi}{4}\right)$ is $\frac{2\cos\left(\frac{\pi}{2}\right)\sin\left(\frac{\pi}{4}\right) + (1)\cos\left(\frac{\pi}{4}\right)}{2\cos\left(\frac{\pi}{2}\right)\cos\left(\frac{\pi}{4}\right) - (1)\sin\left(\frac{\pi}{4}\right)} = 1$;

Slope at $\left(-1, -\frac{\pi}{4}\right)$ is $\frac{2\cos\left(-\frac{\pi}{2}\right)\sin\left(-\frac{\pi}{4}\right) + (-1)\cos\left(-\frac{\pi}{4}\right)}{2\cos\left(-\frac{\pi}{2}\right)\cos\left(-\frac{\pi}{4}\right) - (-1)\sin\left(-\frac{\pi}{4}\right)} = 1$;

Slope at $\left(-1, \frac{3\pi}{4}\right)$ is $\frac{2\cos\left(\frac{3\pi}{2}\right)\sin\left(\frac{3\pi}{4}\right) + (-1)\cos\left(\frac{3\pi}{4}\right)}{2\cos\left(\frac{3\pi}{2}\right)\cos\left(\frac{3\pi}{4}\right) - (-1)\sin\left(\frac{3\pi}{4}\right)} = 1$;

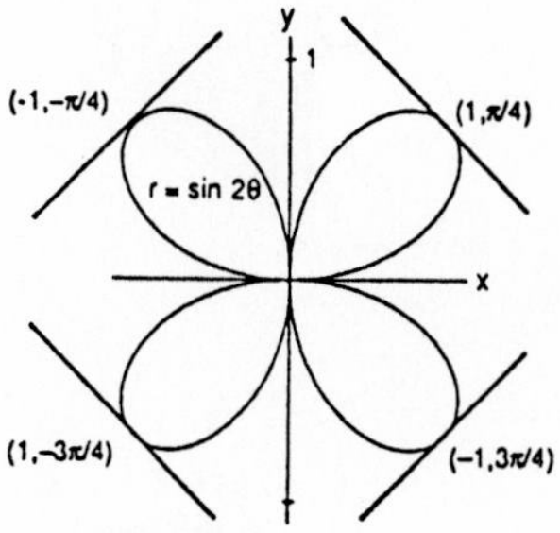

$$\text{Slope at } \left(1, -\frac{3\pi}{4}\right) \text{ is } \frac{2\cos\left(-\frac{3\pi}{2}\right)\sin\left(-\frac{3\pi}{4}\right)+(1)\cos\left(-\frac{3\pi}{4}\right)}{2\cos\left(-\frac{3\pi}{2}\right)\cos\left(-\frac{3\pi}{4}\right)-(1)\sin\left(-\frac{3\pi}{4}\right)} = -1$$

21. (a)

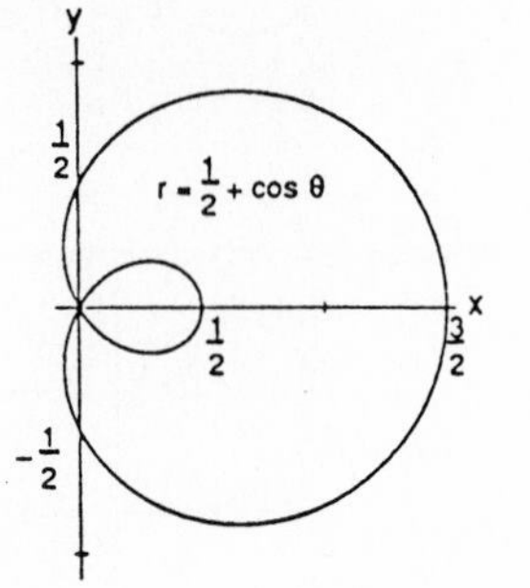

(b)

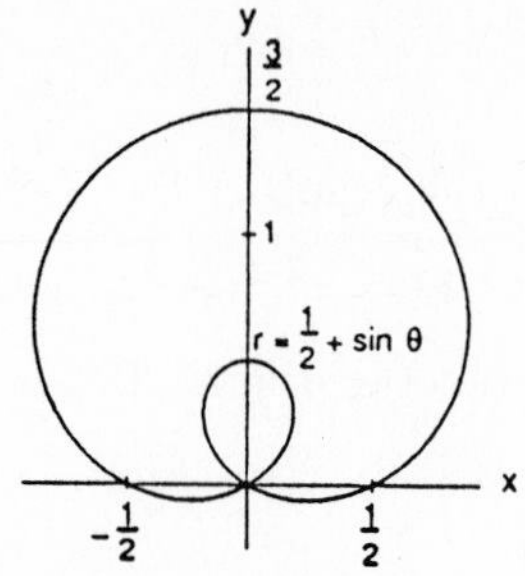

23. (a)

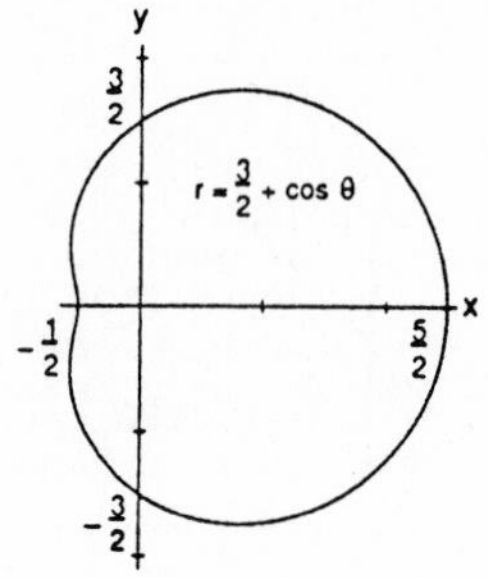

(b)

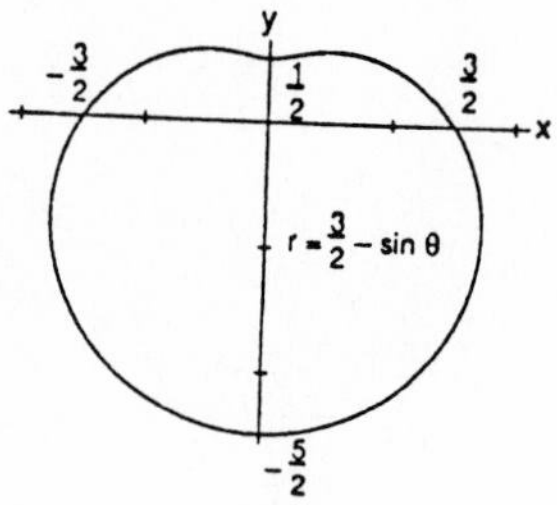

25.

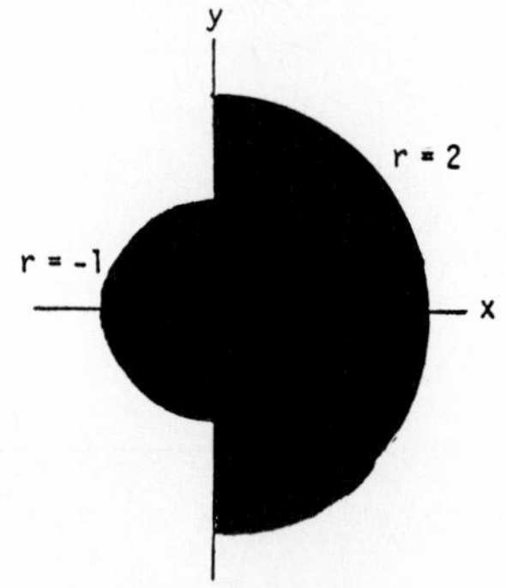

27.

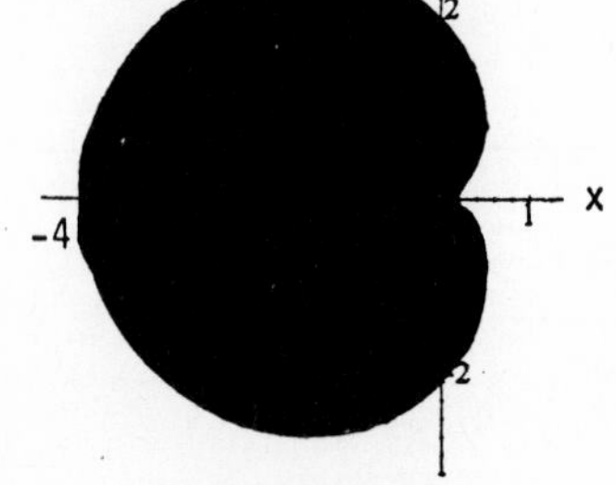

23. $r = \frac{6}{1+\cos\theta},\ 0 \le \theta \le \frac{\pi}{2} \Rightarrow \frac{dr}{d\theta} = \frac{6\sin\theta}{(1+\cos\theta)^2}$; therefore Length $= \int_0^{\pi/2} \sqrt{\left(\frac{6}{1+\cos\theta}\right)^2 + \left(\frac{6\sin\theta}{(1+\cos\theta)^2}\right)^2}\, d\theta$

$$= \int_0^{\pi/2} \sqrt{\frac{36}{(1+\cos\theta)^2} + \frac{36\sin^2\theta}{(1+\cos^2\theta)^4}}\, d\theta = 6\int_0^{\pi/2} \left|\frac{1}{1+\cos\theta}\right| \sqrt{1+\frac{\sin^2\theta}{(1+\cos\theta)^2}}\, d\theta$$

$$= \left(\text{since } \frac{1}{1+\cos\theta} > 0 \text{ on } 0 \le \theta \le \frac{\pi}{2}\right)\ 6\int_0^{\pi/2} \left(\frac{1}{1+\cos\theta}\right) \sqrt{\frac{1+2\cos\theta+\cos^2\theta+\sin^2\theta}{(1+\cos\theta)^2}}\, d\theta$$

$$= 6\int_0^{\pi/2} \left(\frac{1}{1+\cos\theta}\right)\sqrt{\frac{2+2\cos\theta}{(1+\cos\theta)^2}}\, d\theta = 6\sqrt{2}\int_0^{\pi/2} \frac{d\theta}{(1+\cos\theta)^{3/2}} = 6\sqrt{2}\int_0^{\pi/2} \frac{d\theta}{\left(2\cos^2\frac{\theta}{2}\right)^{3/2}} = 6\int_0^{\pi/2} \left|\sec^3\frac{\theta}{2}\right| d\theta$$

$$= 6\int_0^{\pi/2} \sec^3\frac{\theta}{2}\, d\theta = 12\int_0^{\pi/4} \sec^3 u\, du = (\text{use tables})\ 6\left(\left[\frac{\sec u \tan u}{2}\right]_0^{\pi/4} + \frac{1}{2}\int_0^{\pi/4} \sec u\, du\right)$$

$$= 6\left(\frac{1}{\sqrt{2}} + \left[\frac{1}{2}\ln|\sec u + \tan u|\right]_0^{\pi/4}\right) = 3\left[\sqrt{2} + \ln\left(1+\sqrt{2}\right)\right]$$

25. $r = \cos^3\frac{\theta}{3} \Rightarrow \frac{dr}{d\theta} = -\sin\frac{\theta}{3}\cos^2\frac{\theta}{3}$; therefore Length $= \int_0^{\pi/4} \sqrt{\left(\cos^3\frac{\theta}{3}\right)^2 + \left(-\sin\frac{\theta}{3}\cos^2\frac{\theta}{3}\right)^2}\, d\theta$

$$= \int_0^{\pi/4} \sqrt{\cos^6\left(\frac{\theta}{3}\right) + \sin^2\left(\frac{\theta}{3}\right)\cos^4\left(\frac{\theta}{3}\right)}\, d\theta = \int_0^{\pi/4} \left(\cos^2\frac{\theta}{3}\right)\sqrt{\cos^2\left(\frac{\theta}{3}\right)+\sin^2\left(\frac{\theta}{3}\right)}\, d\theta = \int_0^{\pi/4} \cos^2\left(\frac{\theta}{3}\right) d\theta$$

$$= \int_0^{\pi/4} \frac{1+\cos\left(\frac{2\theta}{3}\right)}{2}\, d\theta = \frac{1}{2}\left[\theta + \frac{3}{2}\sin\frac{2\theta}{3}\right]_0^{\pi/4} = \frac{\pi}{8} + \frac{3}{8}$$

27. $r = \sqrt{1+\cos 2\theta} \Rightarrow \frac{dr}{d\theta} = \frac{1}{2}(1+\cos 2\theta)^{-1/2}(-2\sin 2\theta)$; therefore Length $= \int_0^{\pi\sqrt{2}} \sqrt{(1+\cos 2\theta) + \frac{\sin^2 2\theta}{(1+\cos 2\theta)}}\, d\theta$

$$= \int_0^{\pi\sqrt{2}} \sqrt{\frac{1+2\cos 2\theta+\cos^2 2\theta+\sin^2 2\theta}{1+\cos 2\theta}}\, d\theta = \int_0^{\pi\sqrt{2}} \sqrt{\frac{2+2\cos 2\theta}{1+\cos 2\theta}}\, d\theta = \int_0^{\pi\sqrt{2}} \sqrt{2}\, d\theta = \left[\sqrt{2}\,\theta\right]_0^{\pi\sqrt{2}} = 2\pi$$

29. $r = \sqrt{\cos 2\theta},\ 0 \le \theta \le \frac{\pi}{4} \Rightarrow \frac{dr}{d\theta} = \frac{1}{2}(\cos 2\theta)^{-1/2}(-\sin 2\theta)(2) = \frac{-\sin 2\theta}{\sqrt{\cos 2\theta}}$; therefore Surface Area

$$= \int_0^{\pi/4} (2\pi r\cos\theta)\sqrt{\left(\sqrt{\cos 2\theta}\right)^2 + \left(\frac{-\sin 2\theta}{\sqrt{\cos 2\theta}}\right)^2}\, d\theta = \int_0^{\pi/4} \left(2\pi\sqrt{\cos 2\theta}\right)(\cos\theta)\sqrt{\cos 2\theta + \frac{\sin^2 2\theta}{\cos 2\theta}}\, d\theta$$

$$= \int_0^{\pi/4} \left(2\pi\sqrt{\cos 2\theta}\right)(\cos\theta)\sqrt{\frac{1}{\cos 2\theta}}\, d\theta = \int_0^{\pi/4} 2\pi\cos\theta\, d\theta = [2\pi\sin\theta]_0^{\pi/4} = \pi\sqrt{2}$$

31. $r^2 = \cos 2\theta \Rightarrow r = \pm\sqrt{\cos 2\theta}$; use $r = \sqrt{\cos 2\theta}$ on $\left[0, \frac{\pi}{4}\right] \Rightarrow \frac{dr}{d\theta} = \frac{1}{2}(\cos 2\theta)^{-1/2}(-\sin 2\theta)(2) = \frac{-\sin 2\theta}{\sqrt{\cos 2\theta}}$;

therefore Surface Area $= 2\int_0^{\pi/4} \left(2\pi\sqrt{\cos 2\theta}\right)(\sin\theta)\sqrt{\cos 2\theta + \frac{\sin^2 2\theta}{\cos 2\theta}}\, d\theta = 4\pi\int_0^{\pi/4} \sqrt{\cos 2\theta}\,(\sin\theta)\sqrt{\frac{1}{\cos 2\theta}}\, d\theta$

$= 4\pi\int_0^{\pi/4} \sin\theta\, d\theta = 4\pi[-\cos\theta]_0^{\pi/4} = 4\pi\left[-\frac{\sqrt{2}}{2} - (-1)\right] = 2\pi\left(2 - \sqrt{2}\right)$

33. Let $r = f(\theta)$. Then $x = f(\theta)\cos\theta \Rightarrow \frac{dx}{d\theta} = f'(\theta)\cos\theta - f(\theta)\sin\theta \Rightarrow \left(\frac{dx}{d\theta}\right)^2 = \left[f'(\theta)\cos\theta - f(\theta)\sin\theta\right]^2$

$= \left[f'(\theta)\right]^2\cos^2\theta - 2f'(\theta)\,f(\theta)\sin\theta\cos\theta + [f(\theta)]^2\sin^2\theta$; $y = f(\theta)\sin\theta \Rightarrow \frac{dy}{d\theta} = f'(\theta)\sin\theta + f(\theta)\cos\theta$

$\Rightarrow \left(\frac{dy}{d\theta}\right)^2 = \left[f'(\theta)\sin\theta + f(\theta)\cos\theta\right]^2 = \left[f'(\theta)\right]^2\sin^2\theta + 2f'(\theta)f(\theta)\sin\theta\cos\theta + [f(\theta)]^2\cos^2\theta$. Therefore

$\left(\frac{dx}{d\theta}\right)^2 + \left(\frac{dy}{d\theta}\right)^2 = \left[f'(\theta)\right]^2\left(\cos^2\theta + \sin^2\theta\right) + [f(\theta)]^2\left(\cos^2\theta + \sin^2\theta\right) = \left[f'(\theta)\right]^2 + [f(\theta)]^2 = r^2 + \left(\frac{dr}{d\theta}\right)^2$.

Thus, $L = \int_\alpha^\beta \sqrt{\left(\frac{dx}{d\theta}\right)^2 + \left(\frac{dy}{d\theta}\right)^2}\, d\theta = \int_\alpha^\beta \sqrt{r^2 + \left(\frac{dr}{d\theta}\right)^2}\, d\theta$.

35. $r = 2f(\theta)$, $\alpha \le \theta \le \beta \Rightarrow \frac{dr}{d\theta} = 2f'(\theta) \Rightarrow r^2 + \left(\frac{dr}{d\theta}\right)^2 = [2f(\theta)]^2 + \left[2f'(\theta)\right]^2 \Rightarrow$ Length $= \int_\alpha^\beta \sqrt{4[f(\theta)]^2 + 4\left[f'(\theta)\right]^2}\, d\theta$

$= 2\int_\alpha^\beta \sqrt{[f(\theta)]^2 + \left[f'(\theta)\right]^2}\, d\theta$ which is twice the length of the curve $r = f(\theta)$ for $\alpha \le \theta \le \beta$.

37. $$\bar{x} = \frac{\frac{2}{3}\int_0^{2\pi} r^3\cos\theta\, d\theta}{\int_0^{2\pi} r^2\, d\theta} = \frac{\frac{2}{3}\int_0^{2\pi} [a(1+\cos\theta)]^3(\cos\theta)\, d\theta}{\int_0^{2\pi} [a(1+\cos\theta)]^2\, d\theta} = \frac{\frac{2}{3}a^3\int_0^{2\pi}\left(1 + 3\cos\theta + 3\cos^2\theta + \cos^3\theta\right)(\cos\theta)\, d\theta}{a^2\int_0^{2\pi}\left(1 + 2\cos\theta + \cos^2\theta\right) d\theta}$$

$$= \frac{\frac{2}{3}a\int_0^{2\pi}\left[\cos\theta + 3\left(\frac{1+\cos 2\theta}{2}\right) + 3\left(1 - \sin^2\theta\right)(\cos\theta) + \left(\frac{1+\cos 2\theta}{2}\right)^2\right] d\theta}{\int_0^{2\pi}\left[1 + 2\cos\theta + \left(\frac{1+\cos 2\theta}{2}\right)\right] d\theta}$$ = (After considerable algebra using

the identity $\cos^2 A = \frac{1+\cos 2A}{2}$) $$\frac{a\int_0^{2\pi}\left(\frac{15}{12} + \frac{8}{3}\cos\theta + \frac{4}{3}\cos 2\theta - 2\cos\theta\sin^2\theta + \frac{1}{12}\cos 4\theta\right) d\theta}{\int_0^{2\pi}\left(\frac{3}{2} + 2\cos\theta + \frac{1}{2}\cos 2\theta\right) d\theta}$$

$$= \frac{a\left[\frac{15}{12}\theta + \frac{8}{3}\sin\theta + \frac{2}{3}\sin 2\theta - \frac{2}{3}\sin^3\theta + \frac{1}{48}\sin 4\theta\right]_0^{2\pi}}{\left[\frac{3}{2}\theta + 2\sin\theta + \frac{1}{4}\sin 2\theta\right]_0^{2\pi}} = \frac{a\left(\frac{15}{6}\pi\right)}{3\pi} = \frac{5}{6}a;$$

$$\overline{y} = \frac{\frac{2}{3}\int_0^{2\pi} r^3 \sin\theta\, d\theta}{\int_0^{2\pi} r^2\, d\theta} = \frac{\frac{2}{3}\int_0^{2\pi} [a(1+\cos\theta)]^3(\sin\theta)\, d\theta}{3\pi}; \left[u = a(1+\cos\theta) \Rightarrow -\frac{1}{a}\,du = \sin\theta\, d\theta;\ \theta = 0 \Rightarrow u = 2a;\right.$$

$$\left.\theta = 2\pi \Rightarrow u = 2a\right] \to \frac{\frac{2}{3}\int_{2a}^{2a} -\frac{1}{a}u^3\, du}{3\pi} = \frac{0}{3\pi} = 0.$$ Therefore the centroid is $(\overline{x}, \overline{y}) = \left(\frac{5}{6}a, 0\right)$

CHAPTER 9 PRACTICE EXERCISES

1. $x^2 = -4y \Rightarrow y = -\frac{x^2}{4} \Rightarrow 4p = 4 \Rightarrow p = 1;$

therefore Focus is $(0, -1)$, Directrix is $y = 1$

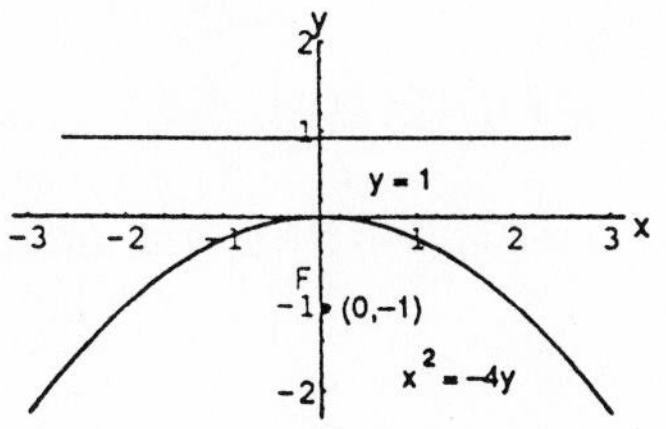

3. $y^2 = 3x \Rightarrow x = \frac{y^2}{3} \Rightarrow 4p = 3 \Rightarrow p = \frac{3}{4};$

therefore Focus is $\left(\frac{3}{4}, 0\right)$, Directrix is $x = -\frac{3}{4}$

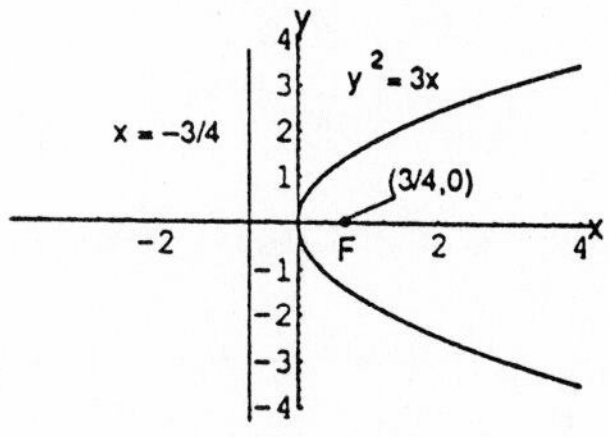

5. $16x^2 + 7y^2 = 112 \Rightarrow \frac{x^2}{7} + \frac{y^2}{16} = 1$

$\Rightarrow c^2 = 16 - 7 = 9 \Rightarrow c = 3;\ e = \frac{c}{a} = \frac{3}{4}$

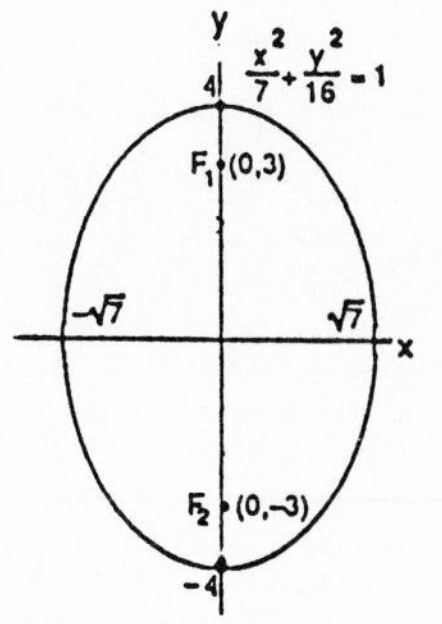

7. $3x^2 - y^2 = 3 \Rightarrow x^2 - \frac{y^2}{3} = 1 \Rightarrow c^2 = 1 + 3 = 4$

$\Rightarrow c = 2;\ e = \frac{c}{a} = \frac{2}{1} = 2$; the asymptotes are

$y = \pm\sqrt{3}\,x$

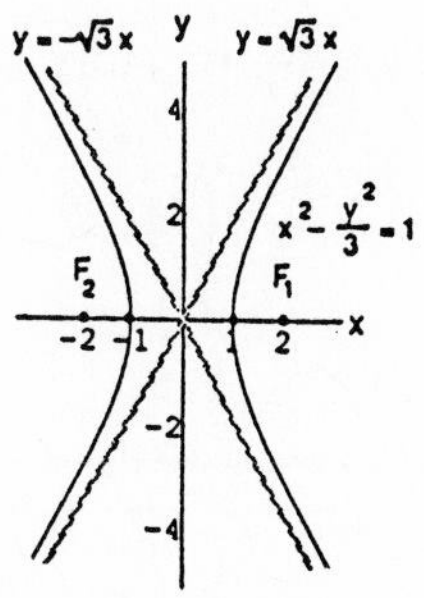

9. $x^2 = -12y \Rightarrow -\frac{x^2}{12} = y \Rightarrow 4p = 12 \Rightarrow p = 3 \Rightarrow$ focus is $(0, -3)$, directrix is $y = 3$, vertex is $(0, 0)$; therefore new vertex is $(2, 3)$, new focus is $(2, 0)$, new directrix is $y = 6$, and the new equation is $(x - 2)^2 = -12(y - 3)$

11. $\frac{x^2}{9}+\frac{y^2}{25}=1 \Rightarrow a=5$ and $b=3 \Rightarrow c=\sqrt{25-9}=4 \Rightarrow$ foci are $(0,\pm 4)$, vertices are $(0,\pm 5)$, center is $(0,0)$; therefore the new center is $(-3,-5)$, new foci are $(-3,-1)$ and $(-3,-9)$, new vertices are $(-3,-10)$ and $(-3,0)$, and the new equation is $\frac{(x+3)^2}{9}+\frac{(y+5)^2}{25}=1$

13. $\frac{y^2}{8}-\frac{x^2}{2}=1 \Rightarrow a=2\sqrt{2}$ and $b=\sqrt{2} \Rightarrow c=\sqrt{8+2}=\sqrt{10} \Rightarrow$ foci are $(0,\pm\sqrt{10})$, vertices are $(0,\pm 2\sqrt{2})$, center is $(0,0)$, and the asymptotes are $y=\pm 2x$; therefore the new center is $(2,2\sqrt{2})$, new foci are $(2,2\sqrt{2}\pm\sqrt{10})$, new vertices are $(2,4\sqrt{2})$ and $(2,0)$, the new asymptotes are $y=2x-4+2\sqrt{2}$ and $y=-2x+4+2\sqrt{2}$; the new equation is $\frac{(y-2\sqrt{2})^2}{8}-\frac{(x-2)^2}{2}=1$

15. $x^2-4x-4y^2=0 \Rightarrow x^2-4x+4-4y^2=4 \Rightarrow (x-2)^2-4y^2=4 \Rightarrow \frac{(x-2)^2}{4}-y^2=1$, a hyperbola; $a=2$ and $b=1 \Rightarrow c=\sqrt{1+4}=\sqrt{5}$; the center is $(2,0)$, the vertices are $(0,0)$ and $(4,0)$; the foci are $(2\pm\sqrt{5},0)$ and the asymptotes are $y=\pm\frac{x-2}{2}$

17. $y^2-2y+16x=-49 \Rightarrow y^2-2y+1=-16x-48 \Rightarrow (y-1)^2=-16(x+3)$, a parabola; the vertex is $(-3,1)$; $4p=16 \Rightarrow p=4 \Rightarrow$ the focus is $(-7,1)$ and the directrix is $x=1$

19. $9x^2+16y^2+54x-64y=-1 \Rightarrow 9(x^2+6x)+16(y^2-4y)=-1 \Rightarrow 9(x^2+6x+9)+16(y^2-4y+4)=144$ $\Rightarrow 9(x+3)^2+16(y-2)^2=144 \Rightarrow \frac{(x+3)^2}{16}+\frac{(y-2)^2}{9}=1$, an ellipse; the center is $(-3,2)$; $a=4$ and $b=3$ $\Rightarrow c=\sqrt{16-9}=\sqrt{7}$; the foci are $(-3\pm\sqrt{7},2)$; the vertices are $(1,2)$ and $(-7,2)$

21. $x^2+y^2-2x-2y=0 \Rightarrow x^2-2x+1+y^2-2y+1=2 \Rightarrow (x-1)^2+(y-1)^2=2$, a circle with center $(1,1)$ and radius $=\sqrt{2}$

23. $B^2-4AC=1-4(1)(1)=-3<0 \Rightarrow$ ellipse

25. $B^2-4AC=3^2-4(1)(2)=1>0 \Rightarrow$ hyperbola

27. $x^2-2xy+y^2=0 \Rightarrow (x-y)^2=0 \Rightarrow x-y=0$ or $y=x$, a straight line

29. $B^2-4AC=1^2-4(2)(2)=-15<0 \Rightarrow$ ellipse; $\cot 2\alpha=\frac{A-C}{B}=0 \Rightarrow 2\alpha=\frac{\pi}{2} \Rightarrow \alpha=\frac{\pi}{4}$; $x=\frac{\sqrt{2}}{2}x'-\frac{\sqrt{2}}{2}y'$ and $y=\frac{\sqrt{2}}{2}x'+\frac{\sqrt{2}}{2}y' \Rightarrow 2\left(\frac{\sqrt{2}}{2}x'-\frac{\sqrt{2}}{2}y'\right)^2+\left(\frac{\sqrt{2}}{2}x'-\frac{\sqrt{2}}{2}y'\right)\left(\frac{\sqrt{2}}{2}x'+\frac{\sqrt{2}}{2}y'\right)+2\left(\frac{\sqrt{2}}{2}x'+\frac{\sqrt{2}}{2}y'\right)^2-15=0$ $\Rightarrow 5x'^2+3y'^2=30$

31. $B^2-4AC=(2\sqrt{3})^2-4(1)(-1)=16 \Rightarrow$ hyperbola; $\cot 2\alpha=\frac{A-C}{B}=\frac{1}{\sqrt{3}} \Rightarrow 2\alpha=\frac{\pi}{3} \Rightarrow \alpha=\frac{\pi}{6}$; $x=\frac{\sqrt{3}}{2}x'-\frac{1}{2}y'$ and $y=\frac{1}{2}x'+\frac{\sqrt{3}}{2}y' \Rightarrow \left(\frac{\sqrt{3}}{2}x'-\frac{1}{2}y'\right)^2+2\sqrt{3}\left(\frac{\sqrt{3}}{2}x'-\frac{1}{2}y'\right)\left(\frac{1}{2}x'+\frac{\sqrt{3}}{2}y'\right)-\left(\frac{1}{2}x'+\frac{\sqrt{3}}{2}y'\right)^2=4$ $\Rightarrow 2x'^2-2y'^2=4 \Rightarrow x'^2-y'^2=2$

33. $x = \frac{t}{2}$ and $y = t + 1 \Rightarrow 2x = t \Rightarrow y = 2x + 1$

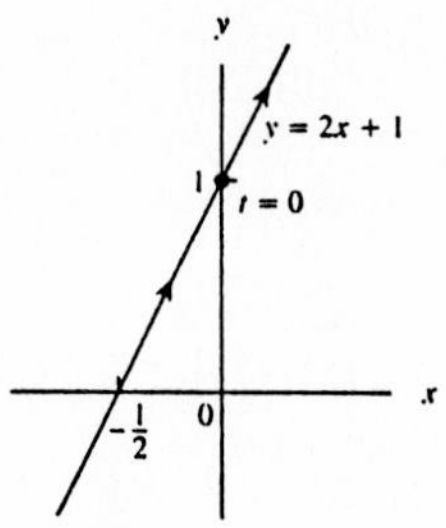

35. $x = \frac{1}{2}\tan t$ and $y = \frac{1}{2}\sec t \Rightarrow x^2 = \frac{1}{4}\tan^2 t$ and $y^2 = \frac{1}{4}\sec^2 t \Rightarrow 4x^2 = \tan^2 t$ and $4y^2 = \sec^2 t \Rightarrow 4x^2 + 1 = 4y^2 \Rightarrow 4y^2 - 4x^2 = 1$

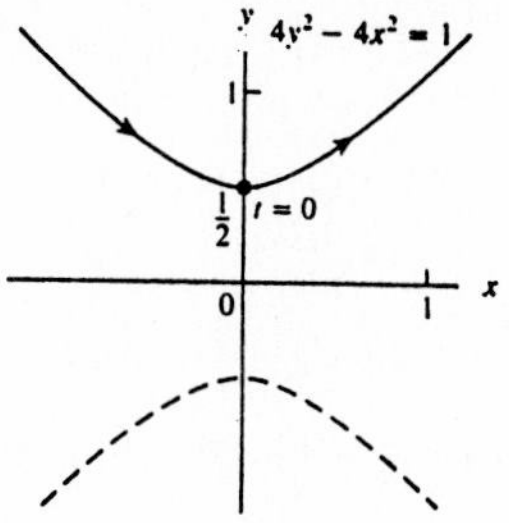

37. $x = -\cos t$ and $y = \cos^2 t \Rightarrow y = (-x)^2 = x^2$

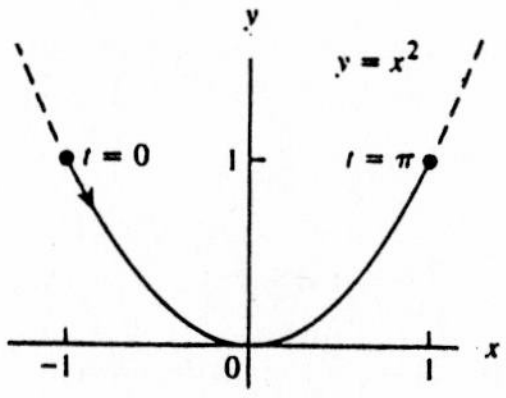

39. $16x^2 + 9y^2 = 144 \Rightarrow \frac{x^2}{9} + \frac{y^2}{16} = 1 \Rightarrow a = 3$ and $b = 4 \Rightarrow x = 3\cos t$ and $y = 4\sin t,\ 0 \le t \le 2\pi$

41. $x = \frac{1}{2}\tan t,\ y = \frac{1}{2}\sec t \Rightarrow \frac{dy}{dx} = \frac{dy/dt}{dx/dt} = \frac{\frac{1}{2}\sec t\tan t}{\frac{1}{2}\sec^2 t} = \frac{\tan t}{\sec t} = \sin t \Rightarrow \left.\frac{dy}{dx}\right|_{t=\pi/3} = \sin\frac{\pi}{3} = \frac{\sqrt{3}}{2};\ t = \frac{\pi}{3}$

$\Rightarrow x = \frac{1}{2}\tan\frac{\pi}{3} = \frac{\sqrt{3}}{2}$ and $y = \frac{1}{2}\sec\frac{\pi}{3} = 1 \Rightarrow y = \frac{\sqrt{3}}{2}x + \frac{1}{4};\ \frac{d^2y}{dx^2} = \frac{dy'/dt}{dx/dt} = \frac{\cos t}{\frac{1}{2}\sec^2 t} = 2\cos^3 t \Rightarrow \left.\frac{d^2y}{dx^2}\right|_{t=\pi/3}$

$= 2\cos^3\left(\frac{\pi}{3}\right) = \frac{1}{4}$

43. $x = e^{2t} - \frac{t}{8}$ and $y = e^t,\ 0 \le t \le \ln 2 \Rightarrow \frac{dx}{dt} = 2e^{2t} - \frac{1}{8}$ and $\frac{dy}{dt} = e^t \Rightarrow \text{Length} = \int_0^{\ln 2} \sqrt{\left(2e^{2t} - \frac{1}{8}\right)^2 + \left(e^t\right)^2}\, dt$

$= \int_0^{\ln 2} \sqrt{4e^{4t} + \frac{1}{2}e^{2t} + \frac{1}{64}}\, dt = \int_0^{\ln 2} \sqrt{\left(2e^{2t} + \frac{1}{8}\right)^2}\, dt = \int_0^{\ln 2} \left(2e^{2t} + \frac{1}{8}\right) dt = \left[e^{2t} + \frac{t}{8}\right]_0^{\ln 2} = 3 + \frac{\ln 2}{8}$

45. $x = \frac{t^2}{2}$ and $y = 2t,\ 0 \le t \le \sqrt{5} \Rightarrow \frac{dx}{dt} = t$ and $\frac{dy}{dt} = 2 \Rightarrow \text{Surface Area} = \int_0^{\sqrt{5}} 2\pi(2t)\sqrt{t^2 + 4}\, dt = \int_4^9 2\pi u^{1/2}\, du$

$= 2\pi\left[\frac{2}{3}u^{3/2}\right]_4^9 = \frac{76\pi}{3}$, where $u = t^2 + 4 \Rightarrow du = 2t\, dt;\ t = 0 \Rightarrow u = 4,\ t = \sqrt{5} \Rightarrow u = 9$

47.

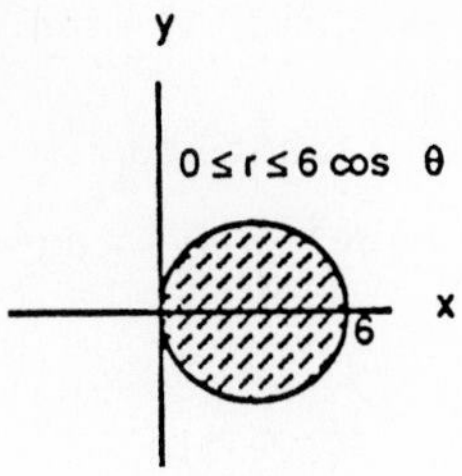

49. d

51. l

53. k

55. i

57. r = sin θ and r = 1 + sin $\theta \Rightarrow \sin\theta = 1 + \sin\theta \Rightarrow 0 = 1$ so no solutions exist. There are no points of intersection found by solving the system. The point of intersection (0, 0) is found by graphing.

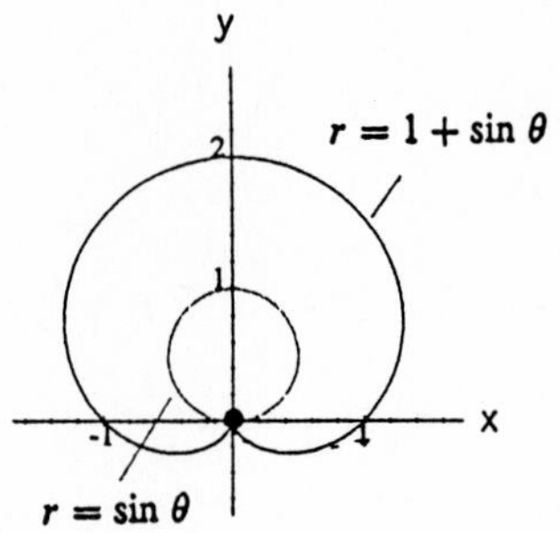

59. r = 1 + cos θ and r = 1 − cos $\theta \Rightarrow 1 + \cos\theta = 1 - \cos\theta$
$\Rightarrow 2\cos\theta = 0 \Rightarrow \cos\theta = 0 \Rightarrow \theta = \frac{\pi}{2}, \frac{3\pi}{2}; \theta = \frac{\pi}{2}$ or $\frac{3\pi}{2}$
$\Rightarrow$ r = 1. The points of intersection are $\left(1, \frac{\pi}{2}\right)$ and $\left(1, \frac{3\pi}{2}\right)$.
The point of intersection (0, 0) is found by graphing.

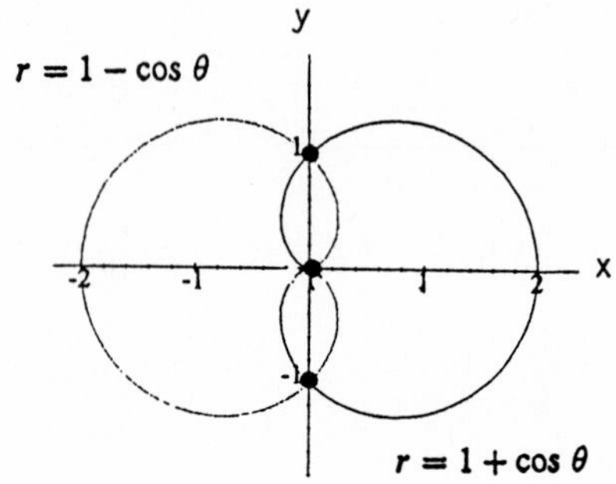

61. r = 1 + sin θ and r = −1 + sin θ intersect at all points of r = 1 + sin θ because the graphs coincide. This can be seen by graphing them.

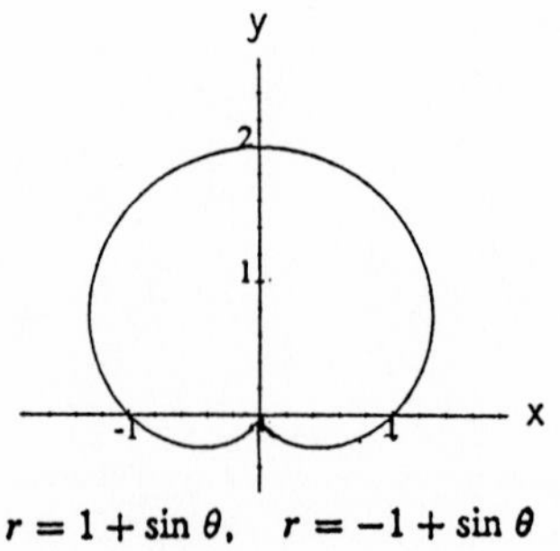

63. $r = \sec\theta$ and $r = 2\sin\theta \Rightarrow \sec\theta = 2\sin\theta$

$\Rightarrow 1 = 2\sin\theta\cos\theta \Rightarrow 1 = \sin 2\theta \Rightarrow 2\theta = \frac{\pi}{2} \Rightarrow \theta = \frac{\pi}{4}$

$\Rightarrow r = 2\sin\frac{\pi}{4} = \sqrt{2} \Rightarrow$ the point of intersection is

$\left(\sqrt{2}, \frac{\pi}{4}\right)$. No other points of intersection exist.

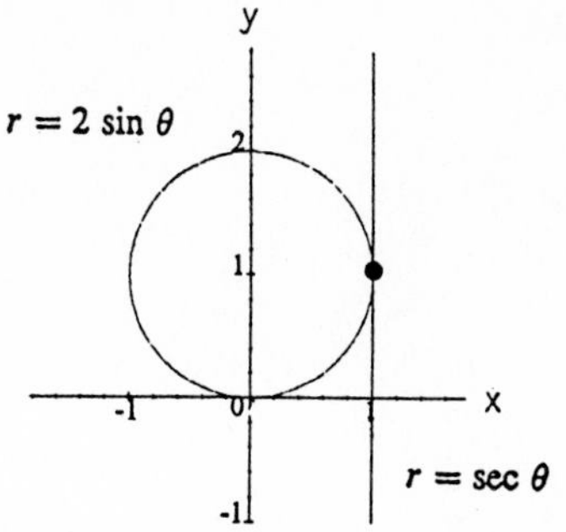

65. $r^2 = \cos 2\theta \Rightarrow r = 0$ when $\cos 2\theta = 0 \Rightarrow 2\theta = \frac{\pi}{2}, \frac{3\pi}{2} \Rightarrow \theta = \frac{\pi}{4}, \frac{3\pi}{4}$; $\theta_1 = \frac{\pi}{4} \Rightarrow m_1 = \tan\frac{\pi}{4} = 1 \Rightarrow y = x$ is one tangent line; $\theta_2 = \frac{3\pi}{4} \Rightarrow m_2 = \tan\frac{3\pi}{4} = -1 \Rightarrow y = -x$ is the other tangent line

67. The tips of the petals are at $\theta = \frac{\pi}{4}, \frac{3\pi}{4}, \frac{5\pi}{4}, \frac{7\pi}{4}$ and $r = 1$ at those values of θ. Then for $\theta = \frac{\pi}{4}$, the tangent line is $r\cos\left(\theta - \frac{\pi}{4}\right) = 1$; for $\theta = \frac{3\pi}{4}$, $r\cos\left(\theta - \frac{3\pi}{4}\right) = 1$; for $\theta = \frac{5\pi}{4}$, $r\cos\left(\theta - \frac{5\pi}{4}\right) = 1$; and for $\theta = \frac{7\pi}{4}$, $r\cos\left(\theta - \frac{7\pi}{4}\right) = 1$.

69. $r\cos\left(\theta + \frac{\pi}{3}\right) = 2\sqrt{3} \Rightarrow r\left(\cos\theta\cos\frac{\pi}{3} - \sin\theta\sin\frac{\pi}{3}\right)$

$= 2\sqrt{3} \Rightarrow \frac{1}{2}r\cos\theta - \frac{\sqrt{3}}{2}r\sin\theta = 2\sqrt{3}$

$\Rightarrow r\cos\theta - \sqrt{3}r\sin\theta = 4\sqrt{3} \Rightarrow x - \sqrt{3}y = 4\sqrt{3}$

$\Rightarrow y = \frac{\sqrt{3}}{3}x - 4$

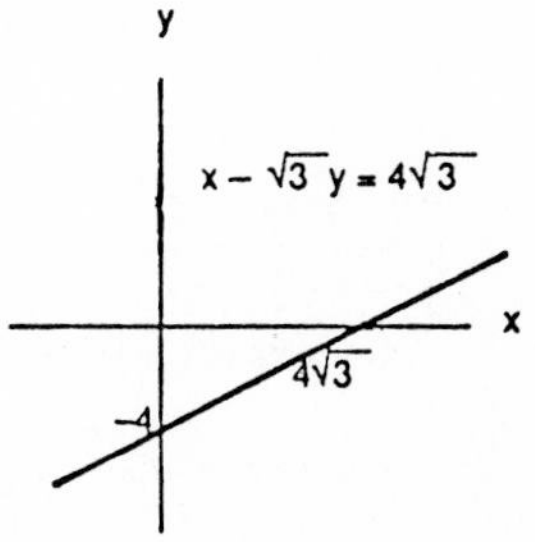

71. $r = 2\sec\theta \Rightarrow r = \frac{2}{\cos\theta} \Rightarrow r\cos\theta = 2 \Rightarrow x = 2$

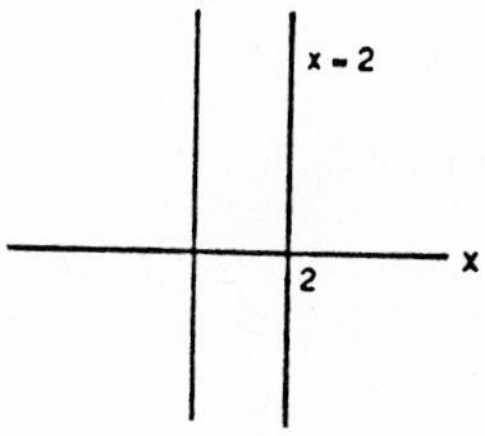

73. $r = -\frac{3}{2}\csc\theta \Rightarrow r\sin\theta = -\frac{3}{2} \Rightarrow y = -\frac{3}{2}$

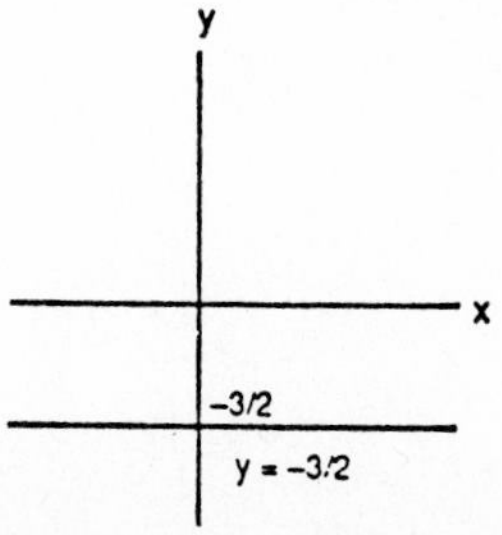

75. $r = -4 \sin \theta \Rightarrow r^2 = -4r \sin \theta \Rightarrow x^2 + y^2 + 4y = 0$
$\Rightarrow x^2 + (y+2)^2 = 4$; circle with center $(0, 2)$ and radius 2.

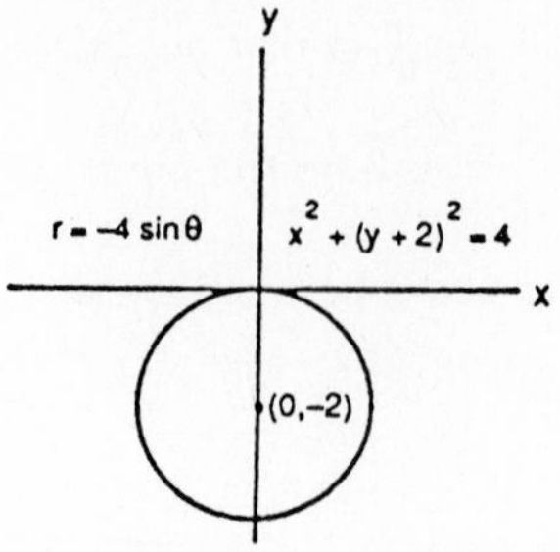

77. $r = 2\sqrt{2} \cos \theta \Rightarrow r^2 = 2\sqrt{2}\, r \cos \theta \Rightarrow x^2 + y^2 - 2\sqrt{2}\, x = 0$
$\Rightarrow \left(x - \sqrt{2}\right)^2 + y^2 = 2$; circle with center $\left(\sqrt{2}, 0\right)$ and radius $\sqrt{2}$

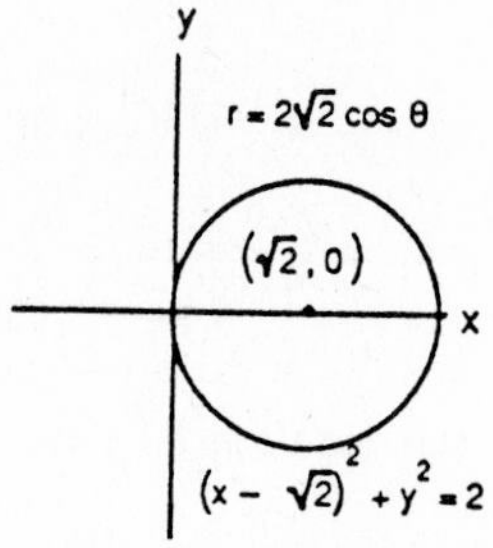

79. $x^2 + y^2 + 5y = 0 \Rightarrow x^2 + \left(y + \frac{5}{2}\right)^2 = \frac{25}{4} \Rightarrow C = \left(0, -\frac{5}{2}\right)$
and $a = \frac{5}{2}$; $r^2 + 5r \sin \theta = 0 \Rightarrow r = -5 \sin \theta$

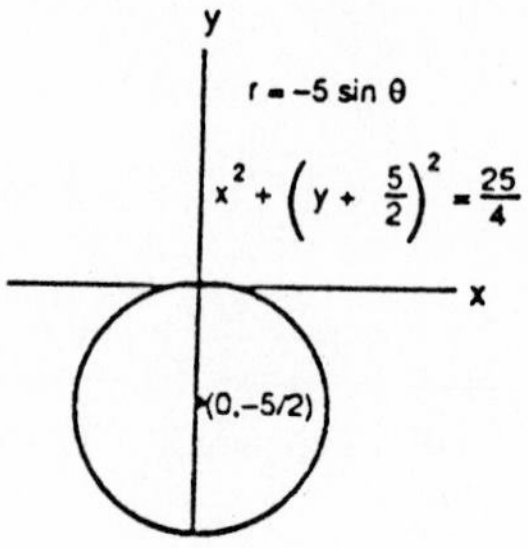

81. $x^2 + y^2 - 3x = 0 \Rightarrow \left(x - \frac{3}{2}\right)^2 + y^2 = \frac{9}{4} \Rightarrow C = \left(\frac{3}{2}, 0\right)$ and
$a = \frac{3}{2}$; $r^2 - 3r \cos \theta = 0 \Rightarrow r = 3 \cos \theta$

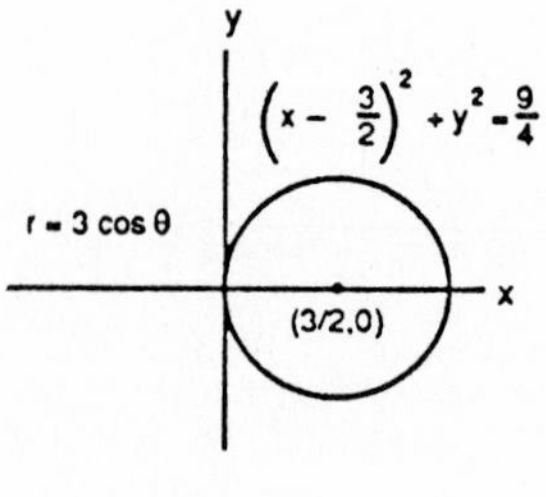

83. $r = \frac{2}{1 + \cos \theta} \Rightarrow e = 1 \Rightarrow$ parabola with vertex at $(1, 0)$

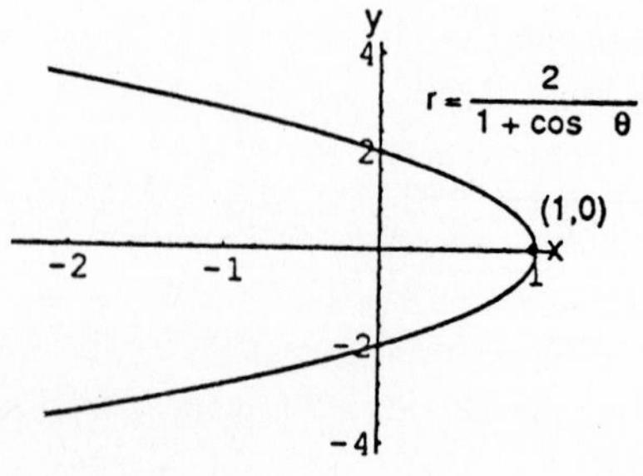

85. $r = \frac{6}{1-2\cos\theta} \Rightarrow e = 2 \Rightarrow$ hyperbola; $ke = 6 \Rightarrow 2k = 6$
$\Rightarrow k = 3 \Rightarrow$ vertices are $(2, \pi)$ and $(6, \pi)$

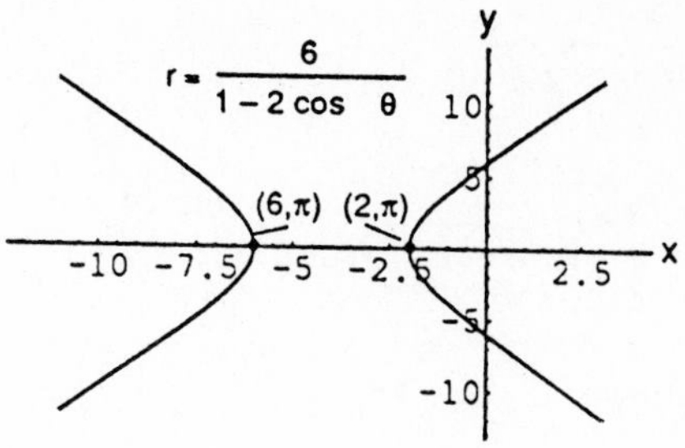

87. $e = 2$ and $r\cos\theta = 2 \Rightarrow x = 2$ is directrix $\Rightarrow k = 2$; the conic is a hyperbola; $r = \frac{ke}{1+e\cos\theta} \Rightarrow r = \frac{(2)(2)}{1+2\cos\theta}$
$\Rightarrow r = \frac{4}{1+2\cos\theta}$

89. $e = \frac{1}{2}$ and $r\sin\theta = 2 \Rightarrow y = 2$ is directrix $\Rightarrow k = 2$; the conic is an ellipse; $r = \frac{ke}{1+e\sin\theta} \Rightarrow r = \frac{(2)\left(\frac{1}{2}\right)}{1+\left(\frac{1}{2}\right)\sin\theta}$
$\Rightarrow r = \frac{2}{2+\sin\theta}$

91. $A = 2\int_0^{\pi} \frac{1}{2}r^2\,d\theta = \int_0^{\pi} (2-\cos\theta)^2\,d\theta = \int_0^{\pi} \left(4 - 2\cos\theta + \cos^2\theta\right)d\theta = \int_0^{\pi} \left(4 - 2\cos\theta + \frac{1+\cos 2\theta}{2}\right)d\theta$

$= \int_0^{\pi} \left(\frac{9}{2} - 2\cos\theta + \frac{\cos 2\theta}{2}\right)d\theta = \left[\frac{9}{2}\theta - 2\sin\theta + \frac{\sin 2\theta}{4}\right]_0^{\pi} = \frac{9}{2}\pi$

93. $r = 1 + \cos 2\theta$ and $r = 1 \Rightarrow 1 = 1 + \cos 2\theta \Rightarrow 0 = \cos 2\theta \Rightarrow 2\theta = \frac{\pi}{2} \Rightarrow \theta = \frac{\pi}{4}$; therefore

$A = 4\int_0^{\pi/4} \frac{1}{2}\left[(1+\cos 2\theta)^2 - 1^2\right]d\theta = 2\int_0^{\pi/4} \left(1 + 2\cos 2\theta + \cos^2 2\theta - 1\right)d\theta$

$= 2\int_0^{\pi/4} \left(2\cos 2\theta + \frac{1}{2} + \frac{\cos 4\theta}{2}\right)d\theta = 2\left[\sin 2\theta + \frac{1}{2}\theta + \frac{\sin 4\theta}{8}\right]_0^{\pi/4} = 2\left(1 + \frac{\pi}{8} + 0\right) = 2 + \frac{\pi}{4}$

95. $r = -1 + \cos\theta \Rightarrow \frac{dr}{d\theta} = -\sin\theta$; Length $= \int_0^{2\pi} \sqrt{(-1+\cos\theta)^2 + (-\sin\theta)^2}\,d\theta = \int_0^{2\pi} \sqrt{2 - 2\cos\theta}\,d\theta$

$= \int_0^{2\pi} \sqrt{\frac{4(1-\cos\theta)}{2}}\,d\theta = \int_0^{2\pi} 2\sin\frac{\theta}{2}\,d\theta = \left[-4\cos\frac{\theta}{2}\right]_0^{2\pi} = (-4)(-1) - (-4)(1) = 8$

97. $r = 8\sin^3\left(\frac{\theta}{3}\right),\ 0 \le \theta \le \frac{\pi}{4} \Rightarrow \frac{dr}{d\theta} = 8\sin^2\left(\frac{\theta}{3}\right)\cos\left(\frac{\theta}{3}\right);\ r^2 + \left(\frac{dr}{d\theta}\right)^2 = \left[8\sin^3\left(\frac{\theta}{3}\right)\right]^2 + \left[8\sin^2\left(\frac{\theta}{3}\right)\cos\left(\frac{\theta}{3}\right)\right]^2$

$= 64\sin^4\left(\frac{\theta}{3}\right) \Rightarrow L = \int_0^{\pi/4} \sqrt{64\sin^4\left(\frac{\theta}{3}\right)}\,d\theta = \int_0^{\pi/4} 8\sin^2\left(\frac{\theta}{3}\right)d\theta = \int_0^{\pi/4} 8\left[\frac{1-\cos\left(\frac{2\theta}{3}\right)}{2}\right]d\theta$

$= \int_0^{\pi/4} \left[4 - 4\cos\left(\frac{2\theta}{3}\right)\right]d\theta = \left[4\theta - 6\sin\left(\frac{2\theta}{3}\right)\right]_0^{\pi/4} = 4\left(\frac{\pi}{4}\right) - 6\sin\left(\frac{\pi}{6}\right) - 0 = \pi - 3$

99. $r = \sqrt{\cos 2\theta} \Rightarrow \frac{dr}{d\theta} = \frac{-\sin 2\theta}{\sqrt{\cos 2\theta}}$; Surface Area $= \int_0^{\pi/4} 2\pi(r \sin \theta)\sqrt{r^2 + \left(\frac{dr}{d\theta}\right)^2}\, d\theta$

$$= \int_0^{\pi/4} 2\pi\sqrt{\cos 2\theta}\,(\sin \theta)\sqrt{\cos 2\theta + \frac{\sin^2 2\theta}{\cos 2\theta}}\, d\theta = \int_0^{\pi/4} 2\pi\sqrt{\cos 2\theta}\,(\sin \theta)\sqrt{\frac{1}{\cos 2\theta}}\, d\theta = \int_0^{\pi/4} 2\pi \sin \theta\, d\theta$$

$$= \left[2\pi(-\cos \theta)\right]_0^{\pi/4} = 2\pi\left(1 - \frac{\sqrt{2}}{2}\right) = \left(2 - \sqrt{2}\right)\pi$$

101. (a) Around the x-axis: $9x^2 + 4y^2 = 36 \Rightarrow y^2 = 9 - \frac{9}{4}x^2 \Rightarrow y = \pm\sqrt{9 - \frac{9}{4}x^2}$ and we use the positive root:

$$V = 2\int_0^2 \pi\left(\sqrt{9 - \frac{9}{4}x^2}\right)^2 dx = 2\int_0^2 \pi\left(9 - \frac{9}{4}x^2\right) dx = 2\pi\left[9x - \frac{3}{4}x^3\right]_0^2 = 24\pi$$

(b) Around the y-axis: $9x^2 + 4y^2 = 36 \Rightarrow x^2 = 4 - \frac{4}{9}y^2 \Rightarrow x = \pm\sqrt{4 - \frac{4}{9}y^2}$ and we use the positive root:

$$V = 2\int_0^3 \pi\left(\sqrt{4 - \frac{4}{9}y^2}\right)^2 dy = 2\int_0^3 \pi\left(4 - \frac{4}{9}y^2\right) dy = 2\pi\left[4y - \frac{4}{27}y^3\right]_0^3 = 16\pi$$

103. Each portion of the wave front reflects to the other focus, and since the wave front travels at a constant speed as it expands, the different portions of the wave arrive at the second focus simultaneously, from all directions, causing a spurt at the second focus.

105. The time for the bullet to hit the target remains constant, say $t = t_0$. Let the time it takes for sound to travel from the target to the listener be t_2. Since the listener hears the sounds simultaneously, $t_1 = t_0 + t_2$ where t_1 is the time for the sound to travel from the rifle to the listener. If v is the velocity of sound, then $vt_1 = vt_0 + vt_2$ or $vt_1 - vt_2 = vt_0$. Now vt_1 is the distance from the rifle to the listener and vt_2 is the distance from the target to the listener. Therefore the difference of the distances is constant since vt_0 is constant so the listener is on a branch of a hyperbola with foci at the rifle and the target. The branch is the one with the target as focus.

107. (a) $r = \frac{k}{1 + e \cos \theta} \Rightarrow r + er \cos \theta = k \Rightarrow \sqrt{x^2 + y^2} + ex = k \Rightarrow \sqrt{x^2 + y^2} = k - ex \Rightarrow x^2 + y^2$

$= k^2 - 2kex + e^2x^2 \Rightarrow x^2 - e^2x^2 + y^2 + 2kex - k^2 = 0 \Rightarrow \left(1 - e^2\right)x^2 + y^2 + 2kex - k^2 = 0$

(b) $c = 0 \Rightarrow x^2 + y^2 - k^2 = 0 \Rightarrow x^2 + y^2 = k^2 \Rightarrow$ circle;

$0 < e < 1 \Rightarrow e^2 < 1 \Rightarrow e^2 - 1 < 0 \Rightarrow B^2 - 4AC = 0^2 - 4\left(1 - e^2\right)(1) = 4\left(e^2 - 1\right) < 0 \Rightarrow$ ellipse;

$e = 1 \Rightarrow B^2 - 4AC = 0^2 - 4(0)(1) = 0 \Rightarrow$ parabola;

$e > 1 \Rightarrow e^2 > 1 \Rightarrow B^2 - 4AC = 0^2 - 4\left(1 - e^2\right)(1) = 4e^2 - 4 > 0 \Rightarrow$ hyperbola

109. $\beta = \psi_2 - \psi_1 \Rightarrow \tan \beta = \tan(\psi_2 - \psi_1) = \frac{\tan \psi_2 - \tan \psi_1}{1 + \tan \psi_2 \tan \psi_1}$;

the curves will be orthogonal when $\tan \beta$ is undefined, or

when $\tan \psi_2 = \frac{-1}{\tan \psi_1} \Rightarrow \frac{r}{g'(\theta)} = \frac{-1}{\left[\frac{r}{f'(\theta)}\right]} \Rightarrow r^2 = -f'(\theta)\, g'(\theta)$

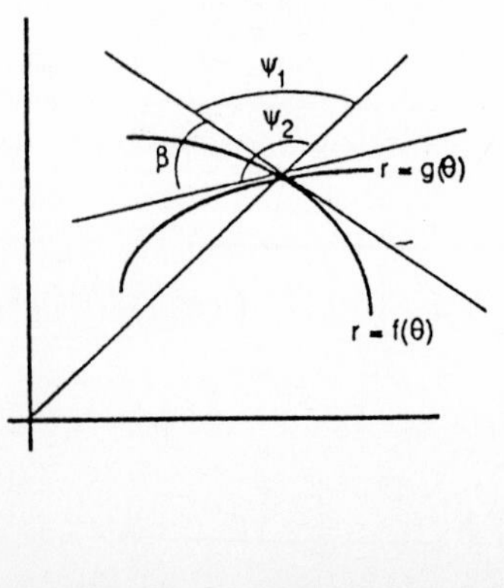

111. $r = 2a \sin 3\theta \Rightarrow \frac{dr}{d\theta} = 6a \cos 3\theta \Rightarrow \tan \psi = \frac{r}{\left(\frac{dr}{d\theta}\right)} = \frac{2a \sin 3\theta}{6a \cos 3\theta} = \frac{1}{3} \tan 3\theta$; when $\theta = \frac{\pi}{6}$, $\tan \psi = \frac{1}{3} \tan \frac{\pi}{2}$ $\Rightarrow \psi = \frac{\pi}{2}$

113. $\tan \psi_1 = \frac{\sqrt{3} \cos \theta}{-\sqrt{3} \sin \theta} = -\cot \theta$ is $-\frac{1}{\sqrt{3}}$ at $\theta = \frac{\pi}{3}$; $\tan \psi_2 = \frac{\sin \theta}{\cos \theta} = \tan \theta$ is $\sqrt{3}$ at $\theta = \frac{\pi}{3}$; since the product of these slopes is -1, the tangents are perpendicular

115. $r_1 = \frac{1}{1 - \cos \theta} \Rightarrow \frac{dr_1}{d\theta} = -\frac{\sin \theta}{(1 - \cos \theta)^2}$; $r_2 = \frac{3}{1 + \cos \theta} \Rightarrow \frac{dr_2}{d\theta} = \frac{3 \sin \theta}{(1 + \cos \theta)^2}$; $\frac{1}{1 - \cos \theta} = \frac{3}{1 + \cos \theta}$ $\Rightarrow 1 + \cos \theta = 3 - 3 \cos \theta \Rightarrow 4 \cos \theta = 2 \Rightarrow \cos \theta = \frac{1}{2} \Rightarrow \theta = \pm\frac{\pi}{3} \Rightarrow r_1 = r_2 = 2 \Rightarrow$ the curves intersect at the points $\left(2, \pm\frac{\pi}{3}\right)$; $\tan \psi_1 = \frac{\left(\frac{1}{1 - \cos \theta}\right)}{\left[\frac{-\sin \theta}{(1 - \cos \theta)^2}\right]} = -\frac{1 - \cos \theta}{\sin \theta}$ is $-\frac{1}{\sqrt{3}}$ at $\theta = \frac{\pi}{3}$; $\tan \psi_2 = \frac{\left(\frac{3}{1 + \cos \theta}\right)}{\left[\frac{3 \sin \theta}{(1 + \cos \theta)^2}\right]} = \frac{1 + \cos \theta}{\sin \theta}$ is $\sqrt{3}$ at $\theta = \frac{\pi}{3}$; therefore $\tan \beta$ is undefined at $\theta = \frac{\pi}{3}$ since $1 + \tan \psi_1 \tan \psi_2 = 1 + \left(-\frac{1}{\sqrt{3}}\right)\left(\sqrt{3}\right) = 0 \Rightarrow \beta = \frac{\pi}{2}$; $\tan \psi_1\big|_{\theta = -\pi/3} = -\frac{1 - \cos\left(-\frac{\pi}{3}\right)}{\sin\left(-\frac{\pi}{3}\right)} = \frac{1}{\sqrt{3}}$ and $\tan \psi_2\big|_{\theta = -\pi/3} = \frac{1 + \cos\left(-\frac{\pi}{3}\right)}{\sin\left(-\frac{\pi}{3}\right)} = -\sqrt{3} \Rightarrow \tan \beta$ is also undefined at $\theta = -\frac{\pi}{3} \Rightarrow \beta = \frac{\pi}{2}$

117. $r_1 = \frac{a}{1 + \cos \theta} \Rightarrow \frac{dr_1}{d\theta} = \frac{a \sin \theta}{(1 + \cos \theta)^2}$ and $r_2 = \frac{b}{1 - \cos \theta} \Rightarrow \frac{dr_2}{d\theta} = -\frac{b \sin \theta}{(1 - \cos \theta)^2}$; then $\tan \psi_1 = \frac{\left(\frac{a}{1 + \cos \theta}\right)}{\left[\frac{a \sin \theta}{(1 + \cos \theta)^2}\right]} = \frac{1 + \cos \theta}{\sin \theta}$ and $\tan \psi_2 = \frac{\left(\frac{b}{1 - \cos \theta}\right)}{\left[\frac{-b \sin \theta}{(1 - \cos \theta)^2}\right]} = \frac{1 - \cos \theta}{-\sin \theta} \Rightarrow 1 + \tan \psi_1 \tan \psi_2$ $= 1 + \left(\frac{1 + \cos \theta}{\sin \theta}\right)\left(\frac{1 - \cos \theta}{-\sin \theta}\right) = 1 - \frac{1 - \cos^2 \theta}{\sin^2 \theta} = 0 \Rightarrow \beta$ is undefined $\Rightarrow$ the parabolas are orthogonal at each point of intersection

119. $r = 3 \sec \theta \Rightarrow r = \frac{3}{\cos \theta}$; $\frac{3}{\cos \theta} = 4 + 4 \cos \theta \Rightarrow 3 = 4 \cos \theta + 4 \cos^2 \theta \Rightarrow (2 \cos \theta + 3)(2 \cos \theta - 1) = 0$ $\Rightarrow \cos \theta = \frac{1}{2}$ or $\cos \theta = -\frac{3}{2} \Rightarrow \theta = \frac{\pi}{3}$ or $\frac{5\pi}{3}$ (the second equation has no solutions); $\tan \psi_2 = \frac{4(1 + \cos \theta)}{-4 \sin \theta}$ $= -\frac{1 + \cos \theta}{\sin \theta}$ is $-\sqrt{3}$ at $\frac{\pi}{3}$ and $\tan \psi_1 = \frac{3 \sec \theta}{3 \sec \theta \tan \theta} = \cot \theta$ is $\frac{1}{\sqrt{3}}$ at $\frac{\pi}{3}$. Then $\tan \beta$ is undefined since $1 + \tan \psi_1 \tan \psi_2 = 1 + \left(\frac{1}{\sqrt{3}}\right)\left(-\sqrt{3}\right) = 0 \Rightarrow \beta = \frac{\pi}{2}$. Also, $\tan \psi_2\big|_{5\pi/3} = \sqrt{3}$ and $\tan \psi_1\big|_{5\pi/3} = -\frac{1}{\sqrt{3}}$ $\Rightarrow 1 + \tan \psi_1 \tan \psi_2 = 1 + \left(-\frac{1}{\sqrt{3}}\right)\left(\sqrt{3}\right) = 0 \Rightarrow \tan \beta$ is also undefined $\Rightarrow \beta = \frac{\pi}{2}$.

121. $\frac{1}{1-\cos\theta} = \frac{1}{1-\sin\theta} \Rightarrow 1-\cos\theta = 1-\sin\theta \Rightarrow \cos\theta = \sin\theta \Rightarrow \theta = \frac{\pi}{4}$; $\tan\psi_1 = \frac{\left(\frac{1}{1-\cos\theta}\right)}{\left[\frac{-\sin\theta}{(1-\cos\theta)^2}\right]} = \frac{1-\cos\theta}{-\sin\theta}$;

$\tan\psi_2 = \frac{\left(\frac{1}{1-\sin\theta}\right)}{\left[\frac{\cos\theta}{(1-\sin\theta)^2}\right]} = \frac{1-\sin\theta}{\cos\theta}$. Thus at $\theta = \frac{\pi}{4}$, $\tan\psi_1 = \frac{1-\cos\left(\frac{\pi}{4}\right)}{-\sin\left(\frac{\pi}{4}\right)} = 1-\sqrt{2}$ and

$\tan\psi_2 = \frac{1-\sin\left(\frac{\pi}{4}\right)}{\cos\left(\frac{\pi}{4}\right)} = \sqrt{2}-1$. Then $\tan\beta = \frac{(\sqrt{2}-1)-(1-\sqrt{2})}{1+(\sqrt{2}-1)(1-\sqrt{2})} = \frac{2\sqrt{2}-2}{2\sqrt{2}-2} = 1 \Rightarrow \beta = \frac{\pi}{4}$

123. (a) $\tan\alpha = \frac{r}{\left(\frac{dr}{d\theta}\right)} \Rightarrow \frac{dr}{r} = \frac{d\theta}{\tan\alpha} \Rightarrow \ln r = \frac{\theta}{\tan\alpha} + C$ (by integration) $\Rightarrow r = Be^{\theta/(\tan\alpha)}$ for some constant B;

$A = \frac{1}{2}\int_{\theta_1}^{\theta_2} B^2 e^{2\theta/(\tan\alpha)}\, d\theta = \left[\frac{B^2(\tan\alpha)\, e^{2\theta/(\tan\alpha)}}{4}\right]_{\theta_1}^{\theta_2} = \frac{\tan\alpha}{4}\left[B^2 e^{2\theta_2/(\tan\alpha)} - B^2 e^{2\theta_1/(\tan\alpha)}\right]$

$= \frac{\tan\alpha}{4}\left(r_2^2 - r_1^2\right)$ since $r_2^2 = B^2 e^{2\theta_2/(\tan\alpha)}$ and $r_1^2 = B^2 e^{2\theta_1/(\tan\alpha)}$; constant of proportionality $K = \frac{\tan\alpha}{4}$

(b) $\tan\alpha = \frac{r}{\left(\frac{dr}{d\theta}\right)} \Rightarrow \frac{dr}{d\theta} = \frac{r}{\tan\alpha} \Rightarrow \left(\frac{dr}{d\theta}\right)^2 = \frac{r^2}{\tan^2\alpha} \Rightarrow r^2 + \left(\frac{dr}{d\theta}\right)^2 = r^2 + \frac{r^2}{\tan^2\alpha} = r^2\left(\frac{\tan^2\alpha+1}{\tan^2\alpha}\right)$

$= r^2\left(\frac{\sec^2\alpha}{\tan^2\alpha}\right) \Rightarrow \text{Length} = \int_{\theta_1}^{\theta_2} r\left(\frac{\sec\alpha}{\tan\alpha}\right) d\theta = \int_{\theta_1}^{\theta_2} Be^{\theta/(\tan\alpha)} \cdot \frac{\sec\alpha}{\tan\alpha}\, d\theta = \left[B(\sec\alpha)\, e^{\theta/(\tan\alpha)}\right]_{\theta_1}^{\theta_2}$

$= (\sec\alpha)\left[Be^{\theta_2/(\tan\alpha)} - Be^{\theta_1(\tan\alpha)}\right] = K(r_2 - r_1)$ where $K = \sec\alpha$ is the constant of proportionality

CHAPTER 9 ADDITIONAL EXERCISES–THEORY, EXAMPLES, APPLICATIONS

1. Directrix $x = 3$ and focus $(4,0) \Rightarrow$ vertex is $\left(\frac{7}{2}, 0\right)$

$\Rightarrow p = \frac{1}{2} \Rightarrow$ the equation is $x - \frac{7}{2} = \frac{y^2}{2}$

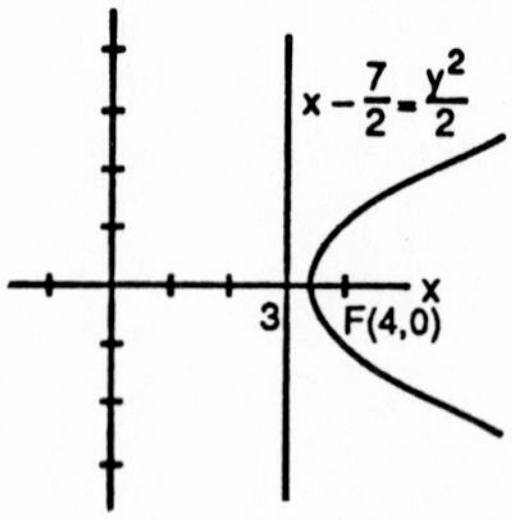

3. $x^2 = 4y \Rightarrow$ vertex is $(0,0)$ and $p = 1 \Rightarrow$ focus is $(0,1)$; thus the distance from $P(x,y)$ to the vertex is $\sqrt{x^2+y^2}$

and the distance from P to the focus is $\sqrt{x^2+(y-1)^2} \Rightarrow \sqrt{x^2+y^2} = 2\sqrt{x^2+(y-1)^2}$

$\Rightarrow x^2+y^2 = 4\left[x^2+(y-1)^2\right] \Rightarrow x^2+y^2 = 4x^2+4y^2-8y+4 \Rightarrow 3x^2+3y^2-8y+4 = 0$, which is a circle

5. Vertices are $(0, \pm 2) \Rightarrow a = 2$; $e = \frac{c}{a} \Rightarrow 0.5 = \frac{c}{2} \Rightarrow c = 1 \Rightarrow$ foci are $(0, \pm 1)$

7. Let the center of the hyperbola be $(0, y)$.

(a) Directrix $y = -1$, focus $(0, -7)$ and $e = 2 \Rightarrow c - \frac{a}{e} = 6 \Rightarrow \frac{a}{e} = c - 6 \Rightarrow a = 2c - 12$. Also $c = ae = 2a$ $\Rightarrow a = 2(2a) - 12 \Rightarrow a = 4 \Rightarrow c = 8$; $y - (-1) = \frac{a}{e} = \frac{4}{2} = 2 \Rightarrow y = 1 \Rightarrow$ the center is $(0, 1)$; $c^2 = a^2 + b^2$ $\Rightarrow b^2 = c^2 - a^2 = 64 - 16 = 48$; therefore the equation is $\frac{(y-1)^2}{16} - \frac{x^2}{48} = 1$

(b) $e = 5 \Rightarrow c - \frac{a}{e} = 6 \Rightarrow \frac{a}{e} = c - 6 \Rightarrow a = 5c - 30$. Also, $c = ae = 5a \Rightarrow a = 5(5a) - 30 \Rightarrow 24a = 30 \Rightarrow a = \frac{5}{4}$ $\Rightarrow c = \frac{25}{4}$; $y - (-1) = \frac{a}{e} = \frac{\left(\frac{5}{4}\right)}{5} = \frac{1}{4} \Rightarrow y = -\frac{3}{4} \Rightarrow$ the center is $\left(0, -\frac{3}{4}\right)$; $c^2 = a^2 + b^2 \Rightarrow b^2 = c^2 - a^2$ $= \frac{625}{16} - \frac{24}{16} = \frac{75}{2}$; therefore the equation is $\frac{\left(y + \frac{3}{4}\right)^2}{\left(\frac{25}{16}\right)} - \frac{x^2}{\left(\frac{75}{2}\right)} = 1$ or $\frac{16\left(y + \frac{3}{4}\right)^2}{25} - \frac{2x^2}{75} = 1$

9. $xy = 2 \Rightarrow x\frac{dy}{dx} + y = 0 \Rightarrow \frac{dy}{dx} = -\frac{y}{x}$; $x^2 - y^2 = 3$ $\Rightarrow 2x - 2y\frac{dy}{dx} = 0 \Rightarrow \frac{dy}{dx} = \frac{x}{y}$. If (x_0, y_0) is a point of intersection, then the product of the slopes is $\left(-\frac{y_0}{x_0}\right)\left(\frac{x_0}{y_0}\right) = -1 \Rightarrow$ the curves are orthogonal.

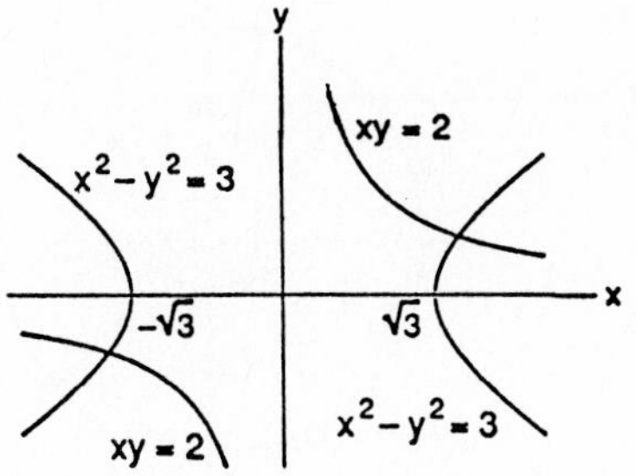

11. $2x^2 + 3y^2 = a^2 \Rightarrow 4x + 6y\frac{dy}{dx} = 0 \Rightarrow \frac{dy}{dx} = -\frac{2x}{3y}$; $ky^2 = x^3$ where k is a constant $\Rightarrow 2ky\frac{dy}{dx} = 3x^2 \Rightarrow \frac{dy}{dx} = \frac{3x^2}{2ky} = \frac{3x^2y}{2ky^2}$ $= \frac{3x^2y}{2x^3}$ (since $ky^2 = x^3$). If (x_0, y_0) is a point of intersection then the product of the slopes is $\left(-\frac{2x_0}{3y_0}\right)\left(\frac{3x_0^2y_0}{2x_0^3}\right) = -1$ $\Rightarrow$ the curves are orthogonal at their points of intersection.

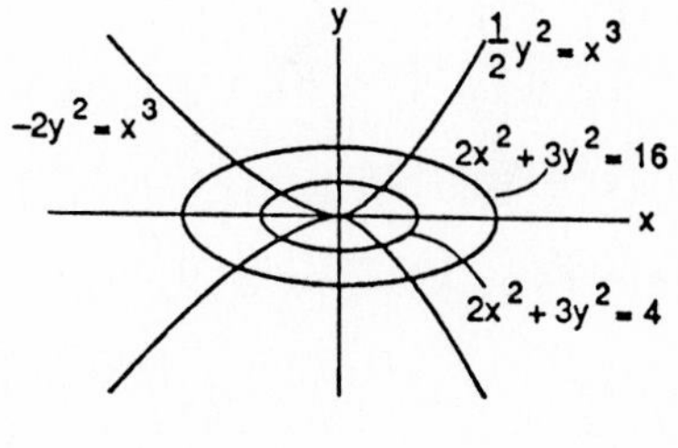

13. $y^2 = 4px \Rightarrow 2y\frac{dy}{dx} = 4p \Rightarrow \frac{dy}{dx} = \frac{2p}{y} \Rightarrow m_{tan} = \frac{2p}{y_1}$ at $P(x_1, y_1) \Rightarrow$ the tangent line is $y - y_1 = \left(\frac{2p}{y_1}\right)(x - x_1)$ and meets the axis of symmetry when $y = 0$ $\Rightarrow -y_1 = \left(\frac{2p}{y_1}\right)(x - x_1) \Rightarrow -\frac{y_1^2}{2p} + x_1 = x \Rightarrow -\frac{4px_1}{2p} + x_1$ $= x \Rightarrow x = -x_1$; that is, the tangent line meets the axis of symmetry x_1 units to the left of the vertex.

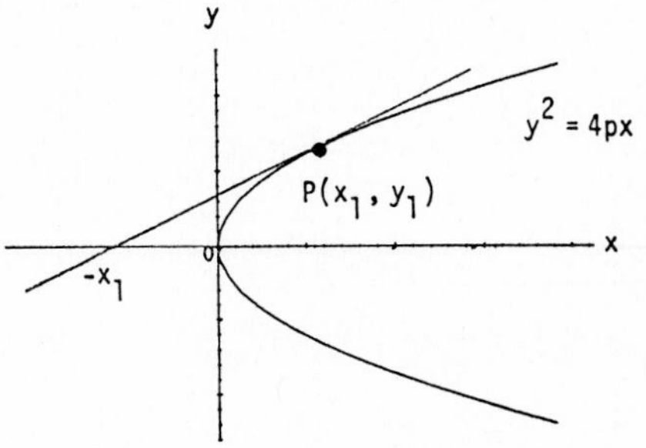

15. $xy = a^2$ is a hyperbola whose asymptotes are the x and y axes; $xy = a^2 \Rightarrow y + x\frac{dy}{dx} = 0 \Rightarrow \frac{dy}{dx} = -\frac{y}{x}$. Let $P(x_1, y_1)$ be a point on the hyperbola $\Rightarrow m_{tan} = -\frac{y_1}{x_1} \Rightarrow$ the equation of the tangent line is $y - y_1 = \left(-\frac{y_1}{x_1}\right)(x - x_1)$. The tangent line intersects the coordinate axes to form the triangle (see figure). If $x = 0$, then $y - y_1 = \left(-\frac{y_1}{x_1}\right)(0 - x_1)$ $\Rightarrow y - y_1 = y_1 \Rightarrow y = 2y_1$; if $y = 0$, then $-y_1 = \left(-\frac{y_1}{x_1}\right)(x - x_1)$ $\Rightarrow x_1 = x - x_1 \Rightarrow x = 2x_1$. Therefore the area is $A = \frac{1}{2}(2x_1)(2y_1) = 2x_1y_1 = 2a^2$.

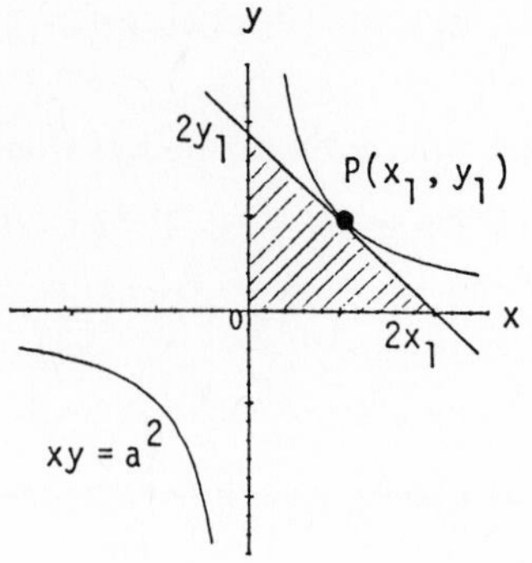

17.

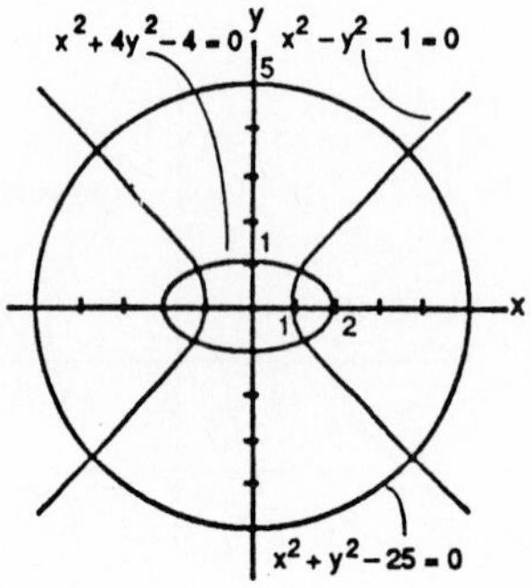

19.

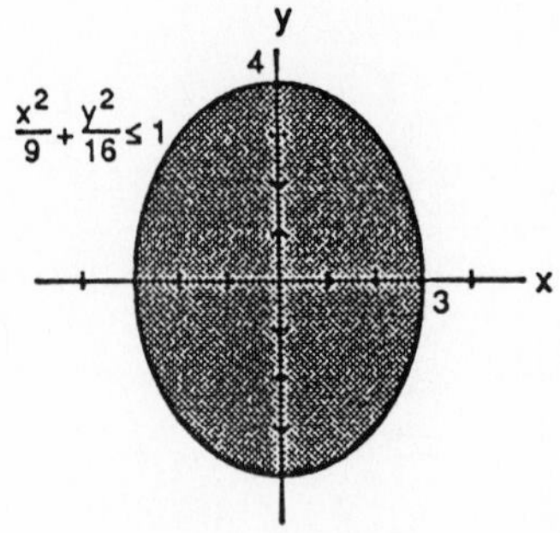

21. $(9x^2 + 4y^2 - 36)(4x^2 + 9y^2 - 16) \le 0 \Rightarrow 9x^2 + 4y^2 - 36 \le 0$ and $4x^2 + 9y^2 - 16 \ge 0$ or $9x^2 + 4y^2 - 36 \ge 0$ and $4x^2 + 9y^2 - 16 \le 0$

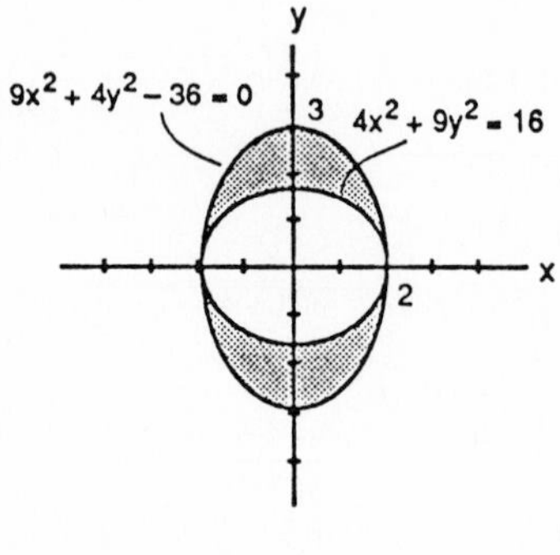

23. $x^4 - (y^2 - 9)^2 = 0 \Rightarrow x^2 - (y^2 - 9) = 0$ or $x^2 + (y^2 - 9) = 0 \Rightarrow y^2 - x^2 = 9$ or $x^2 + y^2 = 9$

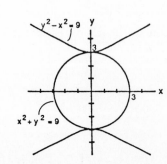

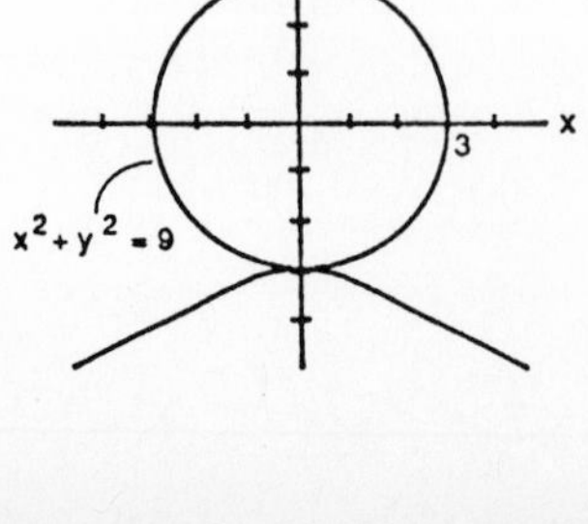

25. Arc PF = Arc AF since each is the distance rolled;

$\angle PCF = \frac{\text{Arc PF}}{b} \Rightarrow \text{Arc PF} = b(\angle PCF)$; $\theta = \frac{\text{Arc AF}}{a}$

$\Rightarrow \text{Arc AF} = a\theta \Rightarrow a\theta = b(\angle PCF) \Rightarrow \angle PCF = \left(\frac{a}{b}\right)\theta$;

$\angle OCB = \frac{\pi}{2} - \theta$ and $\angle OCB = \angle PCF - \angle PCE$

$= \angle PCF - \left(\frac{\pi}{2} - \alpha\right) = \left(\frac{a}{b}\right)\theta - \left(\frac{\pi}{2} - \alpha\right) \Rightarrow \frac{\pi}{2} - \theta$

$= \left(\frac{a}{b}\right)\theta - \left(\frac{\pi}{2} - \alpha\right) \Rightarrow \frac{\pi}{2} - \theta = \left(\frac{a}{b}\right)\theta - \frac{\pi}{2} + \alpha$

$\Rightarrow \alpha = \pi - \theta - \left(\frac{a}{b}\right)\theta \Rightarrow \alpha = \pi - \left(\frac{a+b}{b}\right)\theta.$

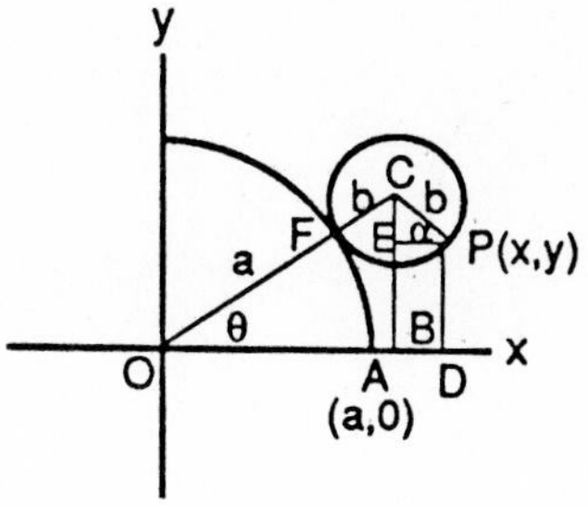

Now $x = OB + BD = OB + EP = (a+b)\cos\theta + b\cos\alpha = (a+b)\cos\theta + b\cos\left(\pi - \left(\frac{a+b}{b}\right)\theta\right)$

$= (a+b)\cos\theta + b\cos\pi\cos\left(\left(\frac{a+b}{b}\right)\theta\right) + b\sin\pi\sin\left(\left(\frac{a+b}{b}\right)\theta\right) = (a+b)\cos\theta - b\cos\left(\left(\frac{a+b}{b}\right)\theta\right)$ and

$y = PD = CB - CE = (a+b)\sin\theta - b\sin\alpha = (a+b)\sin\theta - b\sin\left(\pi - \left(\frac{a+b}{b}\right)\theta\right)$

$= (a+b)\sin\theta - b\sin\pi\cos\left(\left(\frac{a+b}{b}\right)\theta\right) + b\cos\pi\sin\left(\left(\frac{a+b}{b}\right)\theta\right) = (a+b)\sin\theta - b\sin\left(\left(\frac{a+b}{b}\right)\theta\right)$;

therefore $x = (a+b)\cos\theta - b\cos\left(\left(\frac{a+b}{b}\right)\theta\right)$ and $y = (a+b)\sin\theta - b\sin\left(\left(\frac{a+b}{b}\right)\theta\right)$

27. $x = \frac{1-t^2}{1+t^2} \Rightarrow x^2 = \frac{(1-t^2)^2}{(1+t^2)^2}$ and $y = \frac{2t}{1+t^2} \Rightarrow y^2 = \frac{4t^2}{(1+t^2)^2}$

$\Rightarrow x^2 + y^2 = \frac{(1-t^2)^2 + 4t^2}{(1+t^2)^2} = \frac{t^4 + 2t^2 + 1}{(1+t^2)^2} = \frac{(t^2+1)^2}{(1+t^2)^2} = 1$;

$y = 0 \Rightarrow \frac{2t}{1+t^2} = 0 \Rightarrow t = 0 \Rightarrow x = 1 \Rightarrow (-1,0)$ is not covered; $t = -1$ gives $(0,-1)$, $t = 0$ gives $(1,0)$, and $t = 1$ gives $(0,1)$. Note that as $t \to \pm\infty$, $x \to -1$ and $y \to 0$.

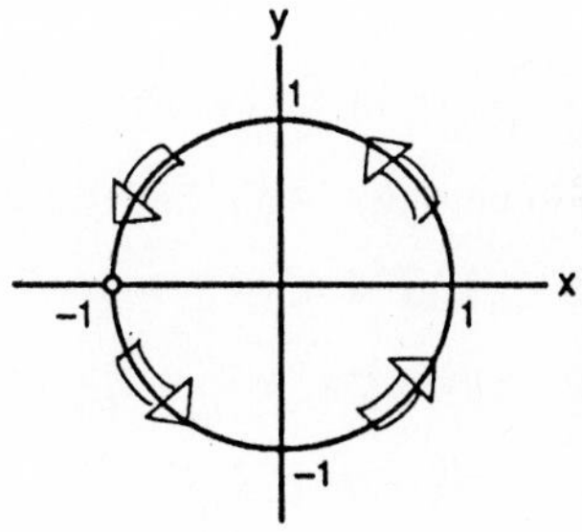

29. (a) $x = e^{2t}\cos t$ and $y = e^{2t}\sin t \Rightarrow x^2 + y^2 = e^{4t}\cos^2 t + e^{4t}\sin^2 t = e^{4t}$. Also $\frac{y}{x} = \frac{e^{2t}\sin t}{e^{2t}\cos t} = \tan t$

$\Rightarrow t = \tan^{-1}\left(\frac{y}{x}\right) \Rightarrow x^2 + y^2 = e^{4\tan^{-1}(y/x)}$ is the Cartesian equation. Since $r^2 = x^2 + y^2$ and $\theta = \tan^{-1}\left(\frac{y}{x}\right)$, the polar equation is $r^2 = e^{4\theta}$ or $r = e^{2\theta}$ for $r > 0$

(b) $ds^2 = r^2\,d\theta^2 + dr^2$; $r = e^{2\theta} \Rightarrow dr = 2e^{2\theta}\,d\theta$

$\Rightarrow ds^2 = r^2\,d\theta^2 + \left(2e^{2\theta}\,d\theta\right)^2 = \left(e^{2\theta}\right)^2 d\theta^2 + 4e^{4\theta}\,d\theta^2$

$= 5e^{4\theta}\,d\theta^2 \Rightarrow ds = \sqrt{5}\,e^{2\theta}\,d\theta \Rightarrow L = \int_0^{2\pi} \sqrt{5}\,e^{2\theta}\,d\theta$

$= \left[\frac{\sqrt{5}\,e^{2\theta}}{2}\right]_0^{2\pi} = \frac{\sqrt{5}}{2}\left(e^{4\pi} - 1\right)$

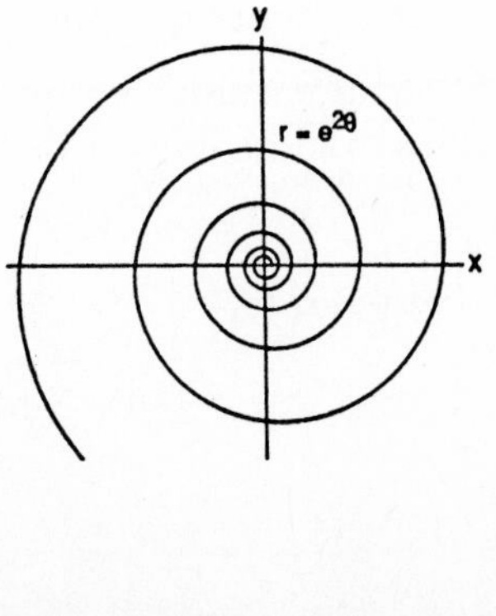

31. $r = 1 + \cos\theta$ and $S = \int 2\pi\rho\, ds$, where $\rho = y = r\sin\theta$; $ds = \sqrt{r^2\,d\theta^2 + dr^2}$

$= \sqrt{(1+\cos\theta)^2\,d\theta^2 + \sin^2\theta\,d\theta^2}\ \sqrt{1 + 2\cos\theta + \cos^2\theta + \sin^2\theta}\,d\theta = \sqrt{2 + 2\cos\theta}\,d\theta = \sqrt{4\cos^2\left(\frac{\theta}{2}\right)}\,d\theta$

$= 2\cos\left(\frac{\theta}{2}\right)d\theta$ since $0 \le \theta \le \frac{\pi}{2}$. Then $S = \int_0^{\pi/2} 2\pi(r\sin\theta)\cdot 2\cos\left(\frac{\theta}{2}\right)d\theta = \int_0^{\pi/2} 4\pi(1+\cos\theta)\cdot\sin\theta\cos\left(\frac{\theta}{2}\right)d\theta$

$= \int_0^{\pi/2} 4\pi\left[2\cos^2\left(\frac{\theta}{2}\right)\right]\left[2\sin\left(\frac{\theta}{2}\right)\cos\left(\frac{\theta}{2}\right)\cos\left(\frac{\theta}{2}\right)\right]d\theta = \int_0^{\pi/2} 16\pi\cos^4\left(\frac{\theta}{2}\right)\sin\left(\frac{\theta}{2}\right)d\theta = \left[\frac{-32\pi\cos^5\left(\frac{\theta}{2}\right)}{5}\right]_0^{\pi/2}$

$= \frac{(-32\pi)\left(\frac{\sqrt{2}}{2}\right)^5}{5} - \left(-\frac{32\pi}{5}\right) = \frac{32\pi - 4\pi\sqrt{2}}{5}$

33. $e = 2$ and $r\cos\theta = 2 \Rightarrow x = 2$ is the directrix $\Rightarrow k = 2$; the conic is a hyperbola with $r = \frac{ke}{1 + e\cos\theta}$

$\Rightarrow r = \frac{(2)(2)}{1 + 2\cos\theta} = \frac{4}{1 + 2\cos\theta}$

35. $e = \frac{1}{2}$ and $r\sin\theta = 2 \Rightarrow y = 2$ is the directrix $\Rightarrow k = 2$; the conic is an ellipse with $r = \frac{ke}{1 + e\sin\theta}$

$\Rightarrow r = \frac{2\left(\frac{1}{2}\right)}{1 + \left(\frac{1}{2}\right)\sin\theta} = \frac{2}{2 + \sin\theta}$

37. The length of the rope is $L = 2x + 2c + y \ge 8C$.

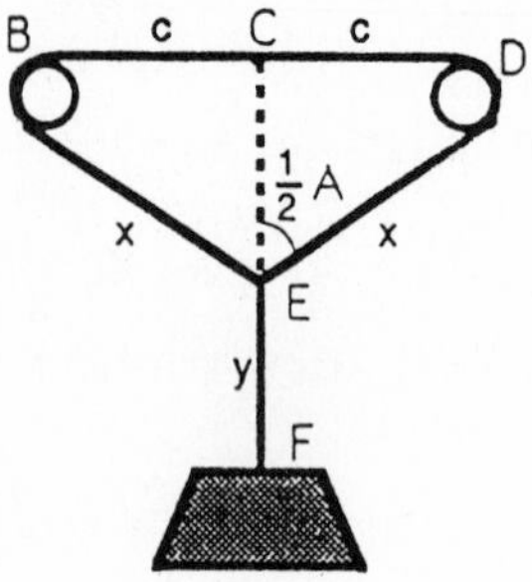

(a) The angle A ($\angle BED$) occurs when the distance $CF = \ell$ is maximized. Now $\ell = \sqrt{x^2 - c^2} + y$

$\Rightarrow \ell = \sqrt{x^2 - c^2} + L - 2x - 2c$

$\Rightarrow \frac{d\ell}{dx} = \frac{1}{2}(x^2 - c^2)^{-1/2}(2x) - 2 = \frac{x}{\sqrt{x^2 - c^2}} - 2.$

Thus $\frac{d\ell}{dx} = 0 \Rightarrow \frac{x}{\sqrt{x^2 - c^2}} - 2 = 0 \Rightarrow x = 2\sqrt{x^2 - c^2}$

$\Rightarrow x^2 = 4x^2 - 4c^2 \Rightarrow 3x^2 = 4c^2 \Rightarrow \frac{c^2}{x^2} = \frac{3}{4} \Rightarrow \frac{c}{x} = \frac{\sqrt{3}}{2}.$

Since $\frac{c}{x} = \sin\frac{A}{2}$ we have $\sin\frac{A}{2} = \frac{\sqrt{3}}{2} \Rightarrow \frac{A}{2} = 60^\circ \Rightarrow A = 120^\circ$

(b) If the ring is fixed at E (i.e., y is held constant) and E is moved to the right, for example, the rope will slip around the pegs so that BE lengthens and DE becomes shorter $\Rightarrow$ BE + ED is always $2x = L - y - 2c$, which is constant $\Rightarrow$ the point E lies on an ellipse with the pegs as foci.

(c) Minimal potential energy occurs when the weight is at its lowest point $\Rightarrow$ E is at the intersection of the ellipse and its minor axis.

39. If the vertex is $(0,0)$, then the focus is $(p,0)$. Let $P(x,y)$ be the present position of the comet. Then $\sqrt{(x-p)^2 + y^2} = 4\times 10^7$. Since $y^2 = 4px$ we have $\sqrt{(x-p)^2 + 4px} = 4\times 10^7 \Rightarrow (x-p)^2 + 4px = 16\times 10^{14}$.

Also, $x - p = 4 \times 10^7 \cos 60^\circ = 2 \times 10^7 \Rightarrow x = p + 2 \times 10^7$. Therefore $\left(2 \times 10^7\right)^2 + 4p\left(p + 2 \times 10^7\right) = 16 \times 10^{14}$ $\Rightarrow 4 \times 10^{14} + 4p^2 + 8p \times 10^7 = 16 \times 10^{14} \Rightarrow 4p^2 + 8p \times 10^7 - 12 \times 10^{14} = 0 \Rightarrow p^2 + 2p \times 10^7 - 3 \times 10^{14} = 0$ $\Rightarrow \left(p + 3 \times 10^7\right)\left(p - 10^7\right) = 0 \Rightarrow p = -3 \times 10^7$ or $p = 10^7$. Since p is positive we obtain $p = 10^7$ miles.

41. $\cot \alpha = \frac{A - C}{B} = 0 \Rightarrow \alpha = 45^\circ$ is the angle of rotation $\Rightarrow A' = \cos^2 45^\circ + \cos 45^\circ \sin 45^\circ + \sin^2 45^\circ = \frac{3}{2}$, $B' = 0$, and $C' = \sin^2 45^\circ - \sin 45^\circ \cos 45^\circ + \cos^2 45^\circ = \frac{1}{2} \Rightarrow \frac{3}{2}x'^2 + \frac{1}{2}y'^2 = 1 \Rightarrow b = \sqrt{\frac{2}{3}}$ and $a = \sqrt{2} \Rightarrow c^2 = a^2 - b^2$ $= 2 - \frac{2}{3} = \frac{4}{3} \Rightarrow c = \frac{2}{\sqrt{3}}$. Therefore the eccentricity is $e = \frac{c}{a} = \frac{\left(\frac{2}{\sqrt{3}}\right)}{\sqrt{2}} = \sqrt{\frac{2}{3}} \approx 0.82$.

43. $\sqrt{x} + \sqrt{y} = 1 \Rightarrow x + 2\sqrt{xy} + y = 1 \Rightarrow 2\sqrt{xy} = 1 - (x + y) \Rightarrow 4xy = 1 - 2(x + y) + (x + y)^2$ $\Rightarrow 4xy = x^2 + 2xy + y^2 - 2x - 2y + 1 \Rightarrow x^2 - 2xy + y^2 - 2x - 2y + 1 = 0 \Rightarrow B^2 - 4AC = (-2)^2 - 4(1)(1) = 0$ $\Rightarrow$ the curve is part of a parabola

45. (a) The equation of a parabola with focus $(0, 0)$ and vertex $(a, 0)$ is $r = \frac{2a}{1 + \cos \theta}$ and rotating this parabola through $\alpha = 45^\circ$ gives $r = \frac{2a}{1 + \cos\left(\theta - \frac{\pi}{4}\right)}$.

(b) Foci at $(0, 0)$ and $(2, 0) \Rightarrow$ the center is $(1, 0) \Rightarrow a = 3$ and $c = 1$ since one vertex is at $(4, 0)$. Then $e = \frac{c}{a}$ $= \frac{1}{3}$. For ellipses with one focus at the origin and major axis along the x-axis we have $r = \frac{a\left(1 - e^2\right)}{1 - e \cos \theta}$ $= \frac{3\left(1 - \frac{1}{9}\right)}{1 - \left(\frac{1}{3}\right) \cos \theta} = \frac{8}{3 - \cos \theta}$.

(c) Center at $\left(2, \frac{\pi}{2}\right)$ and focus at $(0, 0) \Rightarrow c = 2$; center at $\left(2, \frac{\pi}{2}\right)$ and vertex at $\left(1, \frac{\pi}{2}\right) \Rightarrow a = 1$. Then $e = \frac{c}{a}$ $= \frac{2}{1} = 2$. Also $k = ae - \frac{a}{e} = (1)(2) - \frac{1}{2} = \frac{3}{2}$. Therefore $r = \frac{ke}{1 + e \sin \theta} = \frac{\left(\frac{3}{2}\right)(2)}{1 + 2 \sin \theta} = \frac{3}{1 + 2 \sin \theta}$.

47. Arc PT = Arc TO since each is the same distance rolled. Now Arc PT $= a(\angle TAP)$ and Arc TO $= a(\angle TBO)$ $\Rightarrow \angle TAP = \angle TBO$. Since $AP = a = BO$ we have that ΔADP is congruent to $\Delta BCO \Rightarrow CO = DP \Rightarrow OP$ is parallel to $AB \Rightarrow \angle TBO = \angle TAP = \theta$. Then OPDC is a square $\Rightarrow r = CD = AB - AD - CB = AB - 2CB$ $\Rightarrow r = 2a - 2a \cos \theta = 2a(1 - \cos \theta)$, which is the polar equation of a cardiod.

49.

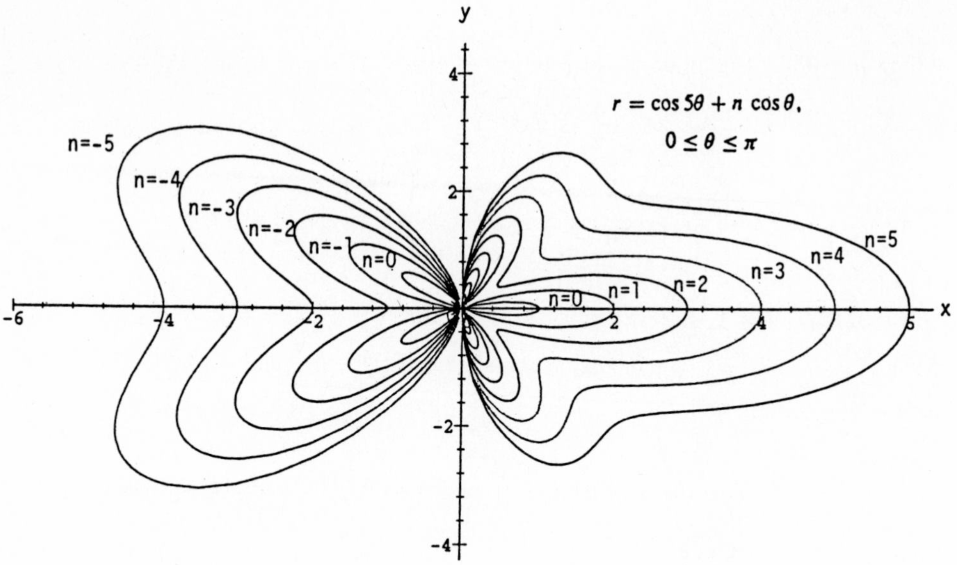
y
x
$r = \cos 5\theta + n\cos\theta,$
$0 \leq \theta \leq \pi$
n=-5
n=-4
n=-3
n=-2
n=-1
n=0
n=0
n=1
n=2
n=3
n=4
n=5
4
2
-2
-4
-6
-4
-2
2
4
6

NOTES:

CHAPTER 10 VECTORS AND ANALYTIC GEOMETRY IN SPACE

10.1 VECTORS IN THE PLANE

1. (a)

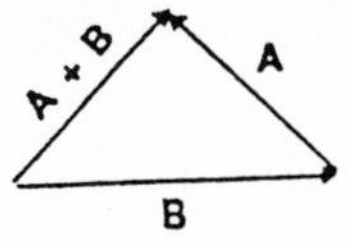

(b)

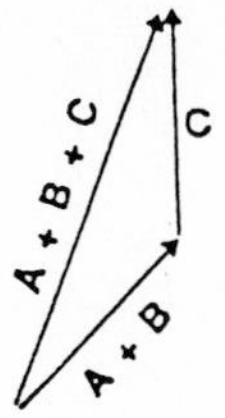

(c)

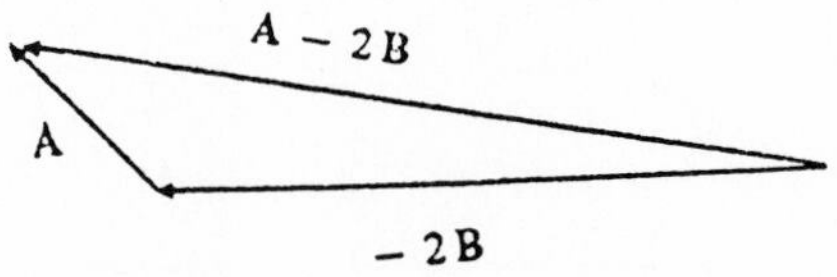

(d)

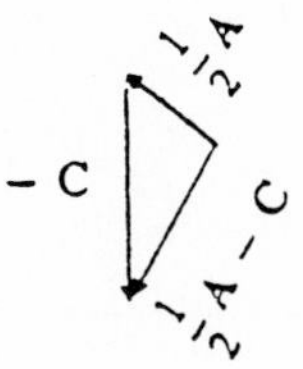

3. $\mathbf{A} + 2\mathbf{B} = (2\mathbf{i} - 7\mathbf{j}) + 2(\mathbf{i} + 6\mathbf{j}) = (2\mathbf{i} - 7\mathbf{j}) + (2\mathbf{i} + 12\mathbf{j}) = 4\mathbf{i} + 5\mathbf{j}$

5. $3\mathbf{A} - \frac{1}{\pi}\mathbf{C} = 3(2\mathbf{i} - 7\mathbf{j}) - \frac{1}{\pi}\left(\sqrt{3}\mathbf{i} - \pi\mathbf{j}\right) = (6\mathbf{i} - 21\mathbf{j}) - \left(\frac{\sqrt{3}}{\pi}\mathbf{i} - \mathbf{j}\right) = \left(6 - \frac{\sqrt{3}}{\pi}\right)\mathbf{i} - 20\mathbf{j}$

7. (a) $\mathbf{w} = \mathbf{u} + \mathbf{v}$ (b) $\mathbf{v} = \mathbf{w} + (-\mathbf{u}) = \mathbf{w} - \mathbf{u}$

9. $\overrightarrow{P_1P_2} = (2 - 5)\mathbf{i} + (9 - 7)\mathbf{j} = -3\mathbf{i} + 2\mathbf{j}$

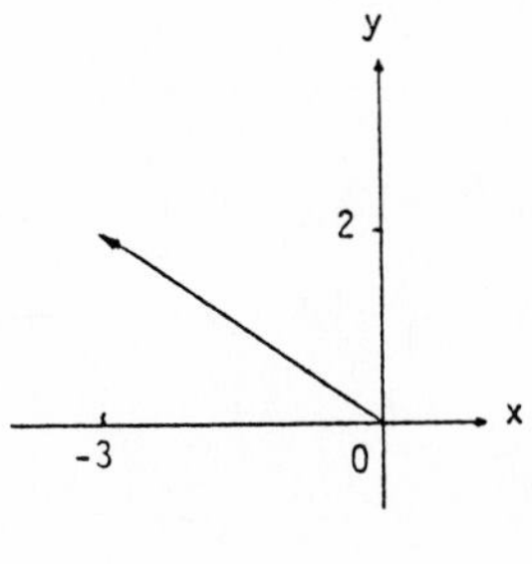

11. $\overrightarrow{AB} = (-10 - (-5))\mathbf{i} + (8 - 3)\mathbf{j} = -5\mathbf{i} + 5\mathbf{j}$

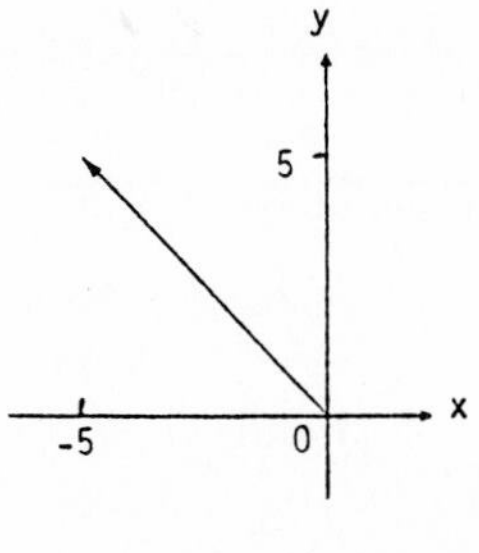

13. $\overrightarrow{P_1P_2} = (2-1)\mathbf{i} + (-1-3)\mathbf{j} = \mathbf{i} - 4\mathbf{j}$

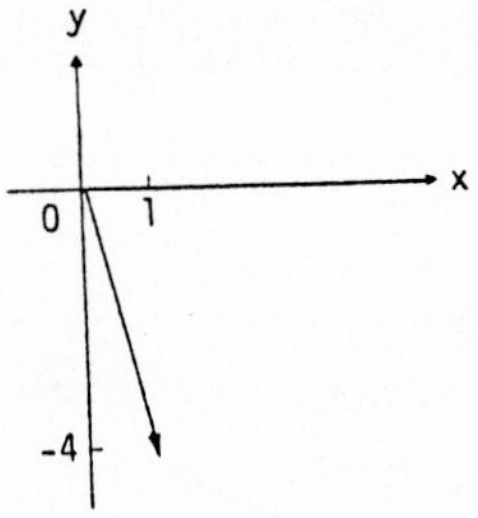

15. $\overrightarrow{AB} = (2-1)\mathbf{i} + (0-(-1))\mathbf{j} = \mathbf{i} + \mathbf{j}$ and

$\overrightarrow{CD} = (-2-(-1))\mathbf{i} + (2-3)\mathbf{j} = -\mathbf{i} - \mathbf{j}$

$\Rightarrow \overrightarrow{AB} + \overrightarrow{CD} = \mathbf{0}$

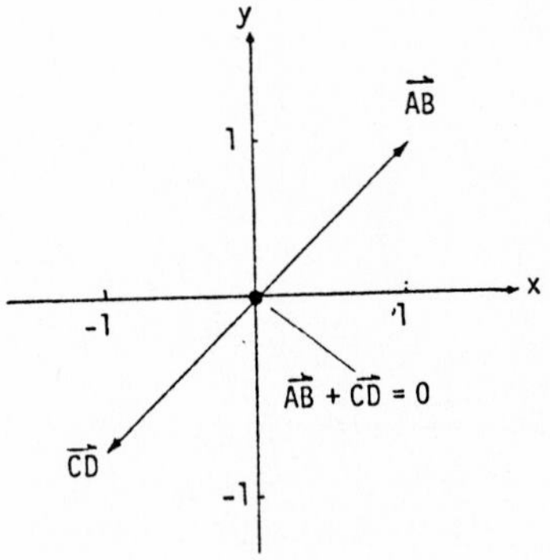

17. $\overrightarrow{AB} = (a-2)\mathbf{i} + (b-9)\mathbf{j} = 3\mathbf{i} - \mathbf{j} \Rightarrow a-2 = 3$ and $b-9 = -1 \Rightarrow a = 5$ and $b = 8 \Rightarrow$ B is the point $(5, 8)$

19. $\mathbf{u} = \left(\cos\frac{\pi}{6}\right)\mathbf{i} + \left(\sin\frac{\pi}{6}\right)\mathbf{j} = \frac{\sqrt{3}}{2}\mathbf{i} + \frac{1}{2}\mathbf{j}$;

$\mathbf{u} = \left(\cos\frac{2\pi}{3}\right)\mathbf{i} + \left(\sin\frac{2\pi}{3}\right)\mathbf{j} = -\frac{1}{2}\mathbf{i} + \frac{\sqrt{3}}{2}\mathbf{j}$

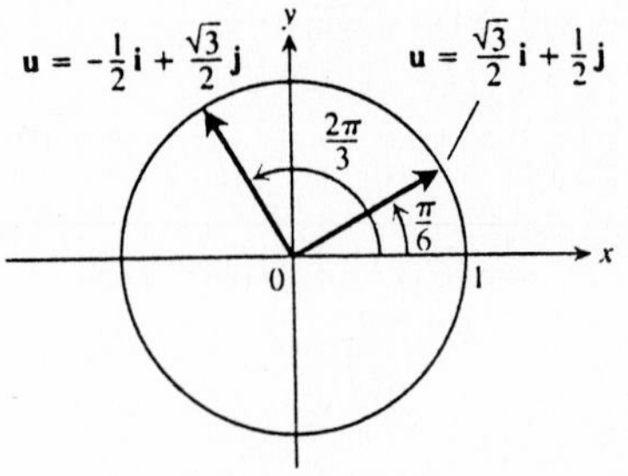

21. $\mathbf{u} = \left(\cos\left(\frac{\pi}{2} + \frac{3\pi}{4}\right)\right)\mathbf{i} + \left(\sin\left(\frac{\pi}{2} + \frac{3\pi}{4}\right)\right)\mathbf{j}$

$= \left(\cos\left(\frac{5\pi}{4}\right)\right)\mathbf{i} + \left(\sin\left(\frac{5\pi}{4}\right)\right)\mathbf{j}$

$= -\frac{\sqrt{2}}{2}\mathbf{i} - \frac{\sqrt{2}}{2}\mathbf{j}$

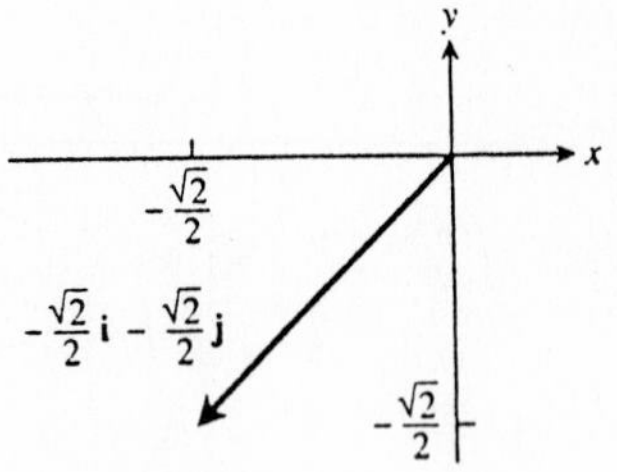

23. $|6\mathbf{i} - 8\mathbf{j}| = \sqrt{36 + 64} = 10 \Rightarrow \frac{\mathbf{v}}{|\mathbf{v}|} = \frac{6}{10}\mathbf{i} - \frac{8}{10}\mathbf{j} = \frac{3}{5}\mathbf{i} - \frac{4}{5}\mathbf{j}$

25. $\frac{dy}{dx} = 2x\,|_{x=2} = 4 \Rightarrow \mathbf{i} + 4\mathbf{j}$ is tangent to the curve at $(2, 4)$

$\Rightarrow \mathbf{u} = \frac{1}{\sqrt{17}}\mathbf{i} + \frac{4}{\sqrt{17}}\mathbf{j}$ and $-\mathbf{u} = -\frac{1}{\sqrt{17}}\mathbf{i} - \frac{4}{\sqrt{17}}\mathbf{j}$ are unit

tangent vectors; $\mathbf{n} = \frac{4}{\sqrt{17}}\mathbf{i} - \frac{1}{\sqrt{17}}\mathbf{j}$ and $-\mathbf{n} = -\frac{4}{\sqrt{17}}\mathbf{i} + \frac{1}{\sqrt{17}}\mathbf{j}$

are unit normal vectors

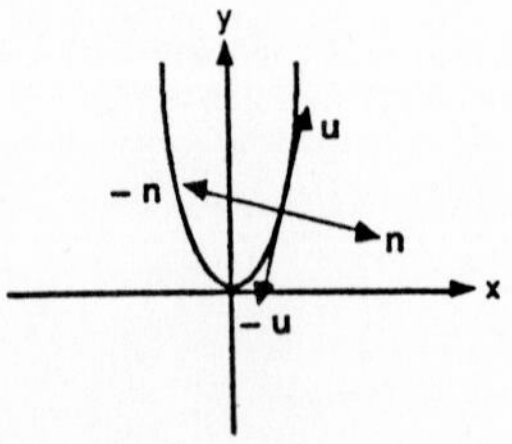

27. $\frac{dy}{dx} = \frac{1}{1+x^2}\Big|_{x=1} = \frac{1}{2} \Rightarrow \mathbf{i} + \frac{1}{2}\mathbf{j}$ is tangent to the curve at $(1,1) \Rightarrow 2\mathbf{i}+\mathbf{j}$ is tangent $\Rightarrow \mathbf{u} = \frac{2}{\sqrt{5}}\mathbf{i} + \frac{1}{\sqrt{5}}\mathbf{j}$ and $-\mathbf{u} = -\frac{2}{\sqrt{5}}\mathbf{i} - \frac{1}{\sqrt{5}}\mathbf{j}$ are unit tangent vectors; $\mathbf{n} = -\frac{1}{\sqrt{5}}\mathbf{i} + \frac{2}{\sqrt{5}}\mathbf{j}$ and $-\mathbf{n} = \frac{1}{\sqrt{5}}\mathbf{i} - \frac{2}{\sqrt{5}}\mathbf{j}$ are unit normal vectors

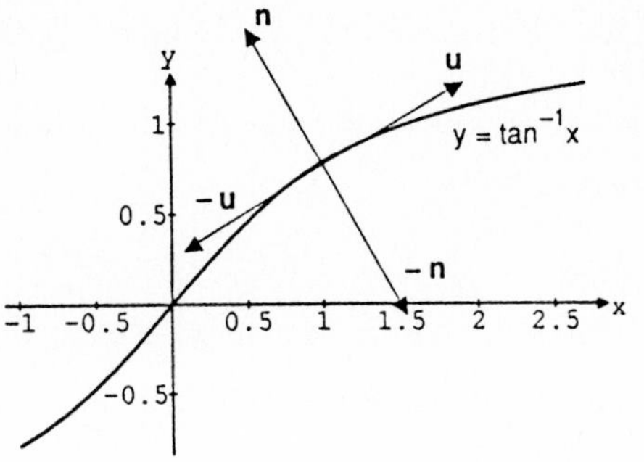

29. $6x + 8y + 8x\frac{dy}{dx} + 4y\frac{dy}{dx} = 0 \Rightarrow \frac{dy}{dx} = -\frac{3x+4y}{4x+2y}\Big|_{(1,0)} = -\frac{3}{4} \Rightarrow 4\mathbf{i} - 3\mathbf{j}$ is tangent to the curve at $(1,0)$ $\Rightarrow \mathbf{u} = \pm\frac{1}{5}(4\mathbf{i} - 3\mathbf{j})$ are unit tangent vectors and $\mathbf{v} = \pm\frac{1}{5}(3\mathbf{i} + 4\mathbf{j})$ are unit normal vectors

31. $\frac{dy}{dx} = \sqrt{3+x^4}\Big|_{(0,0)} = \sqrt{3} \Rightarrow \mathbf{i} + \sqrt{3}\mathbf{j}$ is tangent to the curve at $(0,0) \Rightarrow \mathbf{u} = \pm\frac{1}{2}(\mathbf{i} + \sqrt{3}\mathbf{j})$ are unit tangent vectors and $\mathbf{v} = \pm\frac{1}{2}(-\sqrt{3}\mathbf{i} + \mathbf{j})$ are unit normal vectors

33. $\mathbf{v} = 5\mathbf{i} + 12\mathbf{j} \Rightarrow |\mathbf{v}| = \sqrt{25+144} = 13 \Rightarrow \mathbf{v} = |\mathbf{v}|\left(\frac{\mathbf{v}}{|\mathbf{v}|}\right) = 13\left(\frac{5}{13}\mathbf{i} + \frac{12}{13}\mathbf{j}\right)$

35. $\mathbf{v} = 3\mathbf{i} - 4\mathbf{j} \Rightarrow |\mathbf{v}| = \sqrt{9+16} = 5 \Rightarrow \mathbf{u} = \pm\left(\frac{\mathbf{v}}{|\mathbf{v}|}\right) = \pm\frac{1}{5}(3\mathbf{i} - 4\mathbf{j})$

37. $\mathbf{A} = -3\mathbf{B} \Rightarrow \mathbf{A}$ and $\mathbf{B}$ have opposite directions

39. If $|x|$ is the magnitude of the x-component, then $\cos 30° = \frac{|x|}{|F|} \Rightarrow |x| = |F|\cos 30° = (10)\left(\frac{\sqrt{3}}{2}\right) = 5\sqrt{3}$ lb $\Rightarrow x = 5\sqrt{3}\,\mathbf{i}$;
if $|y|$ is the magnitude of the y-component, then $\sin 30° = \frac{|y|}{|F|} \Rightarrow |y| = |F|\sin 30° = (10)\left(\frac{1}{2}\right) = 5$ lb $\Rightarrow y = 5\,\mathbf{j}$.

41. $2\mathbf{i} + \mathbf{j} = \alpha(\mathbf{i}+\mathbf{j}) + \beta(\mathbf{i}-\mathbf{j}) = (\alpha+\beta)\mathbf{i} + (\alpha-\beta)\mathbf{j} \Rightarrow \alpha + \beta = 2$ and $\alpha - \beta = 1 \Rightarrow 2\alpha = 3 \Rightarrow \alpha = \frac{3}{2}$ and $\beta = \alpha - 1 = \frac{1}{2}$

43. (a) The tree is located at the tip of the vector $\overrightarrow{OP} = (5\cos 60°)\mathbf{i} + (5\sin 60°)\mathbf{j} = \frac{5}{2}\mathbf{i} + \frac{5\sqrt{3}}{2}\mathbf{j} \Rightarrow P = \left(\frac{5}{2}, \frac{5\sqrt{3}}{2}\right)$

(b) The telephone pole is located at the point Q, which is the tip of the vector $\overrightarrow{OP} + \overrightarrow{PQ}$
$= \left(\frac{5}{2}\mathbf{i} + \frac{5\sqrt{3}}{2}\mathbf{j}\right) + (10\cos 315°)\mathbf{i} + (10\sin 315°)\mathbf{j} = \left(\frac{5}{2} + \frac{\sqrt{2}}{2}\right)\mathbf{i} + \left(\frac{5\sqrt{3}}{2} - \frac{10\sqrt{2}}{2}\right)\mathbf{j}$
$\Rightarrow Q = \left(\frac{5+\sqrt{2}}{2}, \frac{5\sqrt{3} - 10\sqrt{2}}{2}\right)$

45. The slope of $-\mathbf{v} = -a\mathbf{i} - b\mathbf{j}$ is $\frac{-b}{-a} = \frac{b}{a}$, which is the same as the slope of $\mathbf{v}$.

10.2 CARTESIAN (RECTANGULAR) COORDINATES AND VECTORS IN SPACE

1. The line through the point $(2,3,0)$ parallel to the z-axis

3. The x-axis

5. The circle $x^2+y^2=4$ in the xy-plane

7. The circle $x^2+z^2=4$ in the xz-plane

9. The circle $y^2+z^2=1$ in the yz-plane

11. The circle $x^2+y^2=16$ in the xy-plane

13. (a) The first quadrant of the xy-plane (b) The fourth quadrant of the xy-plane

15. (a) The ball of radius 1 centered at the origin
(b) All points at distance greater than 1 unit from the origin

17. (a) The upper hemisphere of radius 1 centered at the origin
(b) The solid upper hemisphere of radius 1 centered at the origin

19. (a) $x=3$ (b) $y=-1$ (c) $z=-2$

21. (a) $z=1$ (b) $x=3$ (c) $y=-1$

23. (a) $x^2+(y-2)^2=4,\ z=0$ (b) $(y-2)^2+z^2=4,\ x=0$ (c) $x^2+z^2=4,\ y=2$

25. (a) $y=3,\ z=-1$ (b) $x=1,\ z=-1$ (c) $x=1,\ y=3$

27. $x^2+y^2+z^2=25,\ z=3$

29. $0\le z\le 1$

31. $z\le 0$

33. (a) $(x-1)^2+(y-1)^2+(z-1)^2<1$ (b) $(x-1)^2+(y-1)^2+(z-1)^2>1$

35. length $=|2\mathbf{i}+\mathbf{j}-2\mathbf{k}|=\sqrt{2^2+1^2+(-2)^2}=3$, the direction is $\frac{2}{3}\mathbf{i}+\frac{1}{3}\mathbf{j}-\frac{2}{3}\mathbf{k}\Rightarrow 2\mathbf{i}+\mathbf{j}-2\mathbf{k}=3\left(\frac{2}{3}\mathbf{i}+\frac{1}{3}\mathbf{j}-\frac{2}{3}\mathbf{k}\right)$

37. length $=|\mathbf{i}+4\mathbf{j}-8\mathbf{k}|=\sqrt{1+16+64}=9$, the direction is $\frac{1}{9}\mathbf{i}+\frac{4}{9}\mathbf{j}-\frac{8}{9}\mathbf{k}\Rightarrow \mathbf{i}+4\mathbf{j}-8\mathbf{k}=9\left(\frac{1}{9}\mathbf{i}+\frac{4}{9}\mathbf{j}-\frac{8}{9}\mathbf{k}\right)$

39. length $=|5\mathbf{k}|=\sqrt{25}=5$, the direction is $\mathbf{k}\Rightarrow 5\mathbf{k}=5(\mathbf{k})$

41. length $=\left|\frac{3}{5}\mathbf{i}+\frac{4}{5}\mathbf{k}\right|=\sqrt{\frac{9}{25}+\frac{16}{25}}=1$, the direction is $\frac{3}{5}\mathbf{i}+\frac{4}{5}\mathbf{k}\Rightarrow \frac{3}{5}\mathbf{i}+\frac{4}{5}\mathbf{k}=1\left(\frac{3}{5}\mathbf{i}+\frac{4}{5}\mathbf{k}\right)$

43. length $=\left|\frac{1}{\sqrt{6}}\mathbf{i}-\frac{1}{\sqrt{6}}\mathbf{j}-\frac{1}{\sqrt{6}}\mathbf{k}\right|=\sqrt{3\left(\frac{1}{\sqrt{6}}\right)^2}=\sqrt{\frac{1}{2}}$, the direction is $\frac{1}{\sqrt{3}}\mathbf{i}-\frac{1}{\sqrt{3}}\mathbf{j}-\frac{1}{\sqrt{3}}\mathbf{k}$
$\Rightarrow \frac{1}{\sqrt{6}}\mathbf{i}-\frac{1}{\sqrt{6}}\mathbf{j}-\frac{1}{\sqrt{6}}\mathbf{k}=\sqrt{\frac{1}{2}}\left(\frac{1}{\sqrt{3}}\mathbf{i}-\frac{1}{\sqrt{3}}\mathbf{j}-\frac{1}{\sqrt{3}}\mathbf{k}\right)$

45. (a) $2\mathbf{i}$ (b) $-\sqrt{3}\mathbf{k}$ (c) $\frac{3}{10}\mathbf{j}+\frac{2}{5}\mathbf{k}$ (d) $6\mathbf{i}-2\mathbf{j}+3\mathbf{k}$

47. $|\mathbf{A}|=\sqrt{12^2+5^2}=\sqrt{169}=13$; $\frac{\mathbf{A}}{|\mathbf{A}|}=\frac{1}{13}\mathbf{A}=\frac{1}{13}(12\mathbf{i}-5\mathbf{k})\Rightarrow$ the desired vector is $\frac{7}{13}(12\mathbf{i}-5\mathbf{k})$

49. $|\mathbf{A}|=|2\mathbf{i}-3\mathbf{j}+6\mathbf{k}|=\sqrt{2^2+(-3)^2+6^2}=\sqrt{49}=7$; $\frac{\mathbf{A}}{|\mathbf{A}|}=\frac{2}{7}\mathbf{i}-\frac{3}{7}\mathbf{j}+\frac{6}{7}\mathbf{k}\Rightarrow$ the desired vector is
$-5\left(\frac{2}{7}\mathbf{i}-\frac{3}{7}\mathbf{j}+\frac{6}{7}\mathbf{k}\right)=-\frac{10}{7}\mathbf{i}+\frac{15}{7}\mathbf{j}-\frac{30}{7}\mathbf{k}$

51. (a) the distance = the length $= \left|\overrightarrow{P_1P_2}\right| = |2\mathbf{i} + 2\mathbf{j} - \mathbf{k}| = \sqrt{2^2 + 2^2 + (-1)^2} = 3$

(b) $2\mathbf{i} + 2\mathbf{j} - \mathbf{k} = 3\left(\frac{2}{3}\mathbf{i} + \frac{2}{3}\mathbf{j} - \frac{1}{3}\mathbf{k}\right) \Rightarrow$ the direction is $\frac{2}{3}\mathbf{i} + \frac{2}{3}\mathbf{j} - \frac{1}{3}\mathbf{k}$

(c) the midpoint is $\left(2, 2, \frac{1}{2}\right)$

53. (a) the distance = the length $= \left|\overrightarrow{P_1P_2}\right| = |3\mathbf{i} - 6\mathbf{j} + 2\mathbf{k}| = \sqrt{9 + 36 + 4} = 7$

(b) $3\mathbf{i} - 6\mathbf{j} + 2\mathbf{k} = 7\left(\frac{3}{7}\mathbf{i} - \frac{6}{7}\mathbf{j} + \frac{2}{7}\mathbf{k}\right) \Rightarrow$ the direction is $\frac{3}{7}\mathbf{i} - \frac{6}{7}\mathbf{j} + \frac{2}{7}\mathbf{k}$

(c) the midpoint is $\left(\frac{5}{2}, 1, 6\right)$

55. (a) the distance = the length $= \left|\overrightarrow{P_1P_2}\right| = |2\mathbf{i} - 2\mathbf{j} - 2\mathbf{k}| = \sqrt{3 \cdot 2^2} = 2\sqrt{3}$

(b) $2\mathbf{i} - 2\mathbf{j} - 2\mathbf{k} = 2\sqrt{3}\left(\frac{1}{\sqrt{3}}\mathbf{i} - \frac{1}{\sqrt{3}}\mathbf{j} - \frac{1}{\sqrt{3}}\mathbf{k}\right) \Rightarrow$ the direction is $\frac{1}{\sqrt{3}}\mathbf{i} - \frac{1}{\sqrt{3}}\mathbf{j} - \frac{1}{\sqrt{3}}\mathbf{k}$

(c) the midpoint is $(1, -1, -1)$

57. $\overrightarrow{AB} = (5 - a)\mathbf{i} + (1 - b)\mathbf{j} + (3 - c)\mathbf{k} = \mathbf{i} + 4\mathbf{j} - 2\mathbf{k} \Rightarrow 5 - a = 1,\ 1 - b = 4$, and $3 - c = -2 \Rightarrow a = 4,\ b = -3$, and $c = 5 \Rightarrow$ A is the point $(4, -3, 5)$

59. center $(-2, 0, 2)$, radius $2\sqrt{2}$

61. center $\left(\sqrt{2}, \sqrt{2}, -\sqrt{2}\right)$, radius $\sqrt{2}$

63. $(x - 1)^2 + (y - 2)^2 + (z - 3)^2 = 14$

65. $(x + 2)^2 + y^2 + z^2 = 3$

67. $x^2 + y^2 + z^2 + 4x - 4z = 0 \Rightarrow \left(x^2 + 4x + 4\right) + y^2 + \left(z^2 - 4z + 4\right) = 4 + 4 \Rightarrow (x + 2)^2 + (y - 0)^2 + (z - 2)^2 = \left(\sqrt{8}\right)^2$
$\Rightarrow$ the center is at $(-2, 0, 2)$ and the radius is $\sqrt{8}$

69. $2x^2 + 2y^2 + 2z^2 + x + y + z = 9 \Rightarrow x^2 + \frac{1}{2}x + y^2 + \frac{1}{2}y + z^2 + \frac{1}{2}z = \frac{9}{2}$

$\Rightarrow \left(x^2 + \frac{1}{2}x + \frac{1}{16}\right) + \left(y^2 + \frac{1}{2}y + \frac{1}{16}\right) + \left(z^2 + \frac{1}{2}z + \frac{1}{16}\right) = \frac{9}{2} + \frac{3}{16} = \frac{75}{16} \Rightarrow \left(x + \frac{1}{4}\right)^2 + \left(y + \frac{1}{4}\right)^2 + \left(z + \frac{1}{4}\right)^2 = \left(\frac{5\sqrt{3}}{4}\right)^2$

$\Rightarrow$ the center is at $\left(-\frac{1}{4}, -\frac{1}{4}, -\frac{1}{4}\right)$ and the radius is $\frac{5\sqrt{3}}{4}$

71. (a) the distance between (x, y, z) and $(x, 0, 0)$ is $\sqrt{y^2 + z^2}$

(b) the distance between (x, y, z) and $(0, y, 0)$ is $\sqrt{x^2 + z^2}$

(c) the distance between (x, y, z) and $(0, 0, z)$ is $\sqrt{x^2 + y^2}$

73. (a) the midpoint of AB is $M\left(\frac{5}{2}, \frac{5}{2}, 0\right)$ and $\overrightarrow{CM} = \left(\frac{5}{2} - 1\right)\mathbf{i} + \left(\frac{5}{2} - 1\right)\mathbf{j} + (0 - 3)\mathbf{k} = \frac{3}{2}\mathbf{i} + \frac{3}{2}\mathbf{j} - 3\mathbf{k}$

(b) the desired vector is $\left(\frac{2}{3}\right)\overrightarrow{CM} = \frac{2}{3}\left(\frac{3}{2}\mathbf{i} + \frac{3}{2}\mathbf{j} - 3\mathbf{k}\right) = \mathbf{i} + \mathbf{j} - 2\mathbf{k}$

(c) the vector whose sum is the vector from the origin to C and the result of part (b) will terminate at the center of mass $\Rightarrow$ the terminal point of $(\mathbf{i} + \mathbf{j} + 3\mathbf{k}) + (\mathbf{i} + \mathbf{j} - 2\mathbf{k}) = 2\mathbf{i} + 2\mathbf{j} + \mathbf{k}$ is the point $(2, 2, 1)$, which is the location of the center of mass

75. Without loss of generality we identify the vertices of the quadrilateral such that $A(0, 0, 0)$, $B(x_b, 0, 0)$, $C(x_c, y_c, 0)$ and $D(x_d, y_d, z_d) \Rightarrow$ the midpoint of AB is $M_{AB}\left(\frac{x_b}{2}, 0, 0\right)$, the midpoint of BC is

$M_{BC}\left(\frac{x_b+x_c}{2},\frac{y_c}{2},0\right)$, the midpoint of CD is $M_{CD}\left(\frac{x_c+x_d}{2},\frac{y_c+y_d}{2},\frac{z_d}{2}\right)$ and the midpoint of AD is

$M_{AD}\left(\frac{x_d}{2},\frac{y_d}{2},\frac{z_d}{2}\right) \Rightarrow$ the midpoint of $M_{AB}M_{CD}$ is $\left(\frac{\frac{x_b}{2}+\frac{x_c+x_d}{2}}{2},\frac{y_c+y_d}{4},\frac{z_d}{4}\right)$ which is the same as the midpoint

of $M_{AD}M_{BC} = \left(\frac{\frac{x_b+x_c}{2}+\frac{x_d}{2}}{2},\frac{y_c+y_d}{4},\frac{z_d}{4}\right)$.

77. Without loss of generality we can coordinatize the vertices of the triangle such that A(0,0), B(b,0) and $C(x_c,y_c) \Rightarrow$ a is located at $\left(\frac{b+x_c}{2},\frac{y_c}{2}\right)$, b is at $\left(\frac{x_c}{2},\frac{y_c}{2}\right)$ and c is at $\left(\frac{b}{2},0\right)$. Therefore, $\vec{Aa} = \left(\frac{b}{2}+\frac{x_c}{2}\right)\mathbf{i}+\left(\frac{y_c}{2}\right)\mathbf{j}$, $\vec{Bb} = \left(\frac{x_c}{2}-b\right)\mathbf{i}+\left(\frac{y_c}{2}\right)\mathbf{j}$, and $\vec{Cc} = \left(\frac{b}{2}-x_c\right)\mathbf{i}+(-y_c)\mathbf{j} \Rightarrow \vec{Aa}+\vec{Bb}+\vec{Cc} = \mathbf{0}$.

10.3 DOT PRODUCTS

NOTE: In Exercises 1-9 below we calculate $\text{proj}_{\mathbf{A}}\,\mathbf{B}$ as the vector $\left(\frac{|\mathbf{B}|\cos\theta}{|\mathbf{A}|}\right)\mathbf{A}$, so the scalar multiplier of $\mathbf{A}$ is the number in column 5 divided by the number in column 2.

	$\mathbf{A}\cdot\mathbf{B}$	$\lvert\mathbf{A}\rvert$	$\lvert\mathbf{B}\rvert$	$\cos\theta$	$\lvert\mathbf{B}\rvert\cos\theta$	$\text{proj}_{\mathbf{A}}\,\mathbf{B}$
1.	-25	5	5	-1	-5	$-2\mathbf{i}+4\mathbf{j}-\sqrt{5}\mathbf{k}$
3.	25	15	5	$\frac{1}{3}$	$\frac{5}{3}$	$\frac{1}{9}(10\mathbf{i}+11\mathbf{j}-2\mathbf{k})$
5.	0	$\sqrt{53}$	1	0	0	$\mathbf{0}$
7.	2	$\sqrt{34}$	$\sqrt{3}$	$\frac{2}{\sqrt{3}\sqrt{34}}$	$\frac{2}{\sqrt{34}}$	$\frac{1}{17}(5\mathbf{j}-3\mathbf{k})$
9.	$\sqrt{3}-\sqrt{2}$	$\sqrt{2}$	3	$\frac{\sqrt{3}-\sqrt{2}}{3\sqrt{2}}$	$\frac{\sqrt{3}-\sqrt{2}}{\sqrt{2}}$	$\frac{\sqrt{3}-\sqrt{2}}{2}(-\mathbf{i}+\mathbf{j})$

11. $\mathbf{B} = \left(\frac{\mathbf{A}\cdot\mathbf{B}}{\mathbf{A}\cdot\mathbf{A}}\,\mathbf{A}\right)+\left(\mathbf{B}-\frac{\mathbf{A}\cdot\mathbf{B}}{\mathbf{A}\cdot\mathbf{A}}\,\mathbf{A}\right) = \frac{3}{2}(\mathbf{i}+\mathbf{j})+\left[(3\mathbf{j}+4\mathbf{k})-\frac{3}{2}(\mathbf{i}+\mathbf{j})\right] = \left(\frac{3}{2}\mathbf{i}+\frac{3}{2}\mathbf{j}\right)+\left(-\frac{3}{2}\mathbf{i}+\frac{3}{2}\mathbf{j}+4\mathbf{k}\right)$, where $\mathbf{A}\cdot\mathbf{B} = 3$ and $\mathbf{A}\cdot\mathbf{A} = 2$

13. $\mathbf{B} = \left(\frac{\mathbf{A}\cdot\mathbf{B}}{\mathbf{A}\cdot\mathbf{A}}\,\mathbf{A}\right)+\left(\mathbf{B}-\frac{\mathbf{A}\cdot\mathbf{B}}{\mathbf{A}\cdot\mathbf{A}}\,\mathbf{A}\right) = \frac{14}{3}(\mathbf{i}+2\mathbf{j}-\mathbf{k})+\left[(8\mathbf{i}+4\mathbf{j}-12\mathbf{k})-\left(\frac{14}{3}\mathbf{i}+\frac{28}{3}\mathbf{j}-\frac{14}{3}\mathbf{k}\right)\right]$
$= \left(\frac{14}{3}\mathbf{i}+\frac{28}{3}\mathbf{j}-\frac{14}{3}\mathbf{k}\right)+\left(\frac{10}{3}\mathbf{i}-\frac{16}{3}\mathbf{j}-\frac{22}{3}\mathbf{k}\right)$, where $\mathbf{A}\cdot\mathbf{B} = 28$ and $\mathbf{A}\cdot\mathbf{A} = 6$

15. The sum of two vectors of equal length is *always* orthogonal to their difference, as we can see from the equation $(\mathbf{v}_1+\mathbf{v}_2)\cdot(\mathbf{v}_1-\mathbf{v}_2) = \mathbf{v}_1\cdot\mathbf{v}_1+\mathbf{v}_2\cdot\mathbf{v}_1-\mathbf{v}_1\cdot\mathbf{v}_2-\mathbf{v}_2\cdot\mathbf{v}_2 = |\mathbf{v}_1|^2-|\mathbf{v}_2|^2 = 0$

17. Let $\mathbf{u}$ and $\mathbf{v}$ be the sides of a rhombus $\Rightarrow$ the diagonals are $\mathbf{d}_1 = \mathbf{u}+\mathbf{v}$ and $\mathbf{d}_2 = -\mathbf{u}+\mathbf{v}$ $\Rightarrow \mathbf{d}_1\cdot\mathbf{d}_2 = (\mathbf{u}+\mathbf{v})\cdot(-\mathbf{u}+\mathbf{v}) = -\mathbf{u}\cdot\mathbf{u}+\mathbf{u}\cdot\mathbf{v}-\mathbf{v}\cdot\mathbf{u}+\mathbf{v}\cdot\mathbf{v} = |\mathbf{v}|^2-|\mathbf{u}|^2 = 0$ because $|\mathbf{u}| = |\mathbf{v}|$, since a rhombus has equal sides.

19. Clearly the diagonals of a rectangle are equal in length. What is not as obvious is the statement that equal diagonals happen only in a rectangle. We show this is true by letting the opposite sides of a parallelogram be the vectors $(v_1\mathbf{i}+v_2\mathbf{j})$ and $(u_1\mathbf{i}+u_2\mathbf{j})$. The equal diagonals of the parallelogram are

$d_1=(v_1\mathbf{i}+v_2\mathbf{j})+(u_1\mathbf{i}+u_2\mathbf{j})$ and $d_2=(v_1\mathbf{i}+v_2\mathbf{j})-(u_1\mathbf{i}+u_2\mathbf{j})$. Hence $|d_1|=|d_2|=|(v_1\mathbf{i}+v_2\mathbf{j})+(u_1\mathbf{i}+u_2\mathbf{j})|$ $=|(v_1\mathbf{i}+v_2\mathbf{j})-(u_1\mathbf{i}+u_2\mathbf{j})|\Rightarrow|(v_1+u_1)\mathbf{i}+(v_2+u_2)\mathbf{j}|=|(v_1-u_1)\mathbf{i}+(v_2-u_2)\mathbf{j}|$

$\Rightarrow\sqrt{(v_1+u_1)^2+(v_2+u_2)^2}=\sqrt{(v_1-u_1)^2+(v_2-u_2)^2}\Rightarrow v_1^2+2v_1u_1+u_1^2+v_2^2+2v_2u_2+u_2^2$ $=v_1^2-2v_1u_1+u_1^2+v_2^2-2v_2u_2+u_2^2\Rightarrow 2(v_1u_1+v_2u_2)=-2(v_1u_1+v_2u_2)\Rightarrow v_1u_1+v_2u_2=0$

$\Rightarrow(v_1\mathbf{i}+v_2\mathbf{j})\cdot(u_1\mathbf{i}+u_2\mathbf{j})=0\Rightarrow$ the vectors $(v_1\mathbf{i}+v_2\mathbf{j})$ and $(u_1\mathbf{i}+u_2\mathbf{j})$ are perpendicular and the parallelogram must be a rectangle.

21. Let M be the midpoint of OB. By the Pythagorean Theorem $OB=\sqrt{1^2+1^2}=\sqrt{2}$ and $OM=\frac{\sqrt{2}}{2}$. Hence the angle θ between $\overrightarrow{OB}$ and $\overrightarrow{OD}$ has a tangent of $\frac{DM}{OB}=\frac{1}{\left(\frac{\sqrt{2}}{2}\right)}=\frac{2}{\sqrt{2}}=\sqrt{2}$. Therefore, $\tan\theta=\sqrt{2}$ $\Rightarrow\theta=\tan^{-1}\sqrt{2}\approx 54.7^\circ$.

23. $\theta=\cos^{-1}\left(\frac{\mathbf{A}\cdot\mathbf{B}}{|\mathbf{A}||\mathbf{B}|}\right)=\cos^{-1}\left(\frac{(2)(1)+(1)(2)+(0)(-1)}{\sqrt{2^2+1^2+0^2}\sqrt{1^2+2^2+(-1)^2}}\right)=\cos^{-1}\left(\frac{4}{\sqrt{5}\sqrt{6}}\right)=\cos^{-1}\left(\frac{4}{\sqrt{30}}\right)\approx 0.75$ rad

25. $\theta=\cos^{-1}\left(\frac{\mathbf{A}\cdot\mathbf{B}}{|\mathbf{A}||\mathbf{B}|}\right)=\cos^{-1}\left(\frac{(\sqrt{3})(\sqrt{3})+(-7)(1)+(0)(-2)}{\sqrt{(\sqrt{3})^2+(-7)^2+0^2}\sqrt{(\sqrt{3})^2+(1)^2+(-2)^2}}\right)=\cos^{-1}\left(\frac{3-7}{\sqrt{52}\sqrt{8}}\right)$ $=\cos^{-1}\left(\frac{-1}{\sqrt{26}}\right)\approx 1.77$ rad

27. $\overrightarrow{AB}=3\mathbf{i}+\mathbf{j}-3\mathbf{k}$, $\overrightarrow{AC}=2\mathbf{i}-2\mathbf{j}$, $\overrightarrow{BA}=-3\mathbf{i}-\mathbf{j}+3\mathbf{k}$, $\overrightarrow{CA}=-2\mathbf{i}+2\mathbf{j}$, $\overrightarrow{CB}=\mathbf{i}+3\mathbf{j}-3\mathbf{k}$, $\overrightarrow{BC}=-\mathbf{i}-3\mathbf{j}+3\mathbf{k}$; thus

$\angle A=\cos^{-1}\left(\frac{\overrightarrow{AB}\cdot\overrightarrow{AC}}{|\overrightarrow{AB}||\overrightarrow{AC}|}\right)=\cos^{-1}\left(\frac{4}{\sqrt{152}}\right)\approx 1.24\text{ rad}\approx 71.07^\circ$; $\angle B=\cos^{-1}\left(\frac{\overrightarrow{BA}\cdot\overrightarrow{BC}}{|\overrightarrow{BA}||\overrightarrow{BC}|}\right)=\cos^{-1}\left(\frac{15}{19}\right)$ $\approx 0.66\text{ rad}\approx 37.86^\circ$; $\angle C=\cos^{-1}\left(\frac{\overrightarrow{CA}\cdot\overrightarrow{CB}}{|\overrightarrow{CA}||\overrightarrow{CB}|}\right)=\cos^{-1}\left(\frac{4}{\sqrt{152}}\right)\approx 1.24\text{ rad}\approx 71.07^\circ$

29. Let $\mathbf{A}=\mathbf{i}+\mathbf{k}$ and $\mathbf{B}=\mathbf{i}+\mathbf{j}+\mathbf{k}\Rightarrow\theta=\cos^{-1}\left(\frac{\mathbf{A}\cdot\mathbf{B}}{|\mathbf{A}||\mathbf{B}|}\right)=\cos^{-1}\left(\frac{2}{\sqrt{2}\sqrt{3}}\right)\approx 0.62\text{ rad}\approx 35.26^\circ$

31. (a) Since $|\cos\theta|\le 1$, we have $|\mathbf{u}\cdot\mathbf{v}|=|\mathbf{u}||\mathbf{v}||\cos\theta|\le|\mathbf{u}||\mathbf{v}|(1)=|\mathbf{u}||\mathbf{v}|$.
(b) We have equality precisely when $|\cos\theta|=1$ or when one or both of $\mathbf{u}$ and $\mathbf{v}$ is $\mathbf{0}$. In the case of nonzero vectors, we have equality when $\theta=0$ or π, i.e., when the vectors are parallel.

33. $\mathbf{v}\cdot\mathbf{u}_1=(a\mathbf{u}_1+b\mathbf{u}_2)\cdot\mathbf{u}_1=a\mathbf{u}_1\cdot\mathbf{u}_1+b\mathbf{u}_2\cdot\mathbf{u}_1=a|\mathbf{u}_1|^2+b(\mathbf{u}_2\cdot\mathbf{u}_1)=a(1)^2+b(0)=a$

35. (a) $|\mathbf{D}|^2=\mathbf{D}\cdot\mathbf{D}=25\mathbf{A}\cdot\mathbf{A}+36\mathbf{B}\cdot\mathbf{B}+9\mathbf{C}\cdot\mathbf{C}=25+36+9=70\Rightarrow|\mathbf{D}|=\sqrt{70}$
(b) $|\mathbf{D}|^2=25|\mathbf{A}|^2+36|\mathbf{B}|^2+9|\mathbf{C}|^2=(25)(4)+(36)(9)+(9)(16)=568\Rightarrow|\mathbf{D}|=\sqrt{568}$

37. $P(0,0,0)$, $Q(1,1,1)$ and $\mathbf{F} = 5\mathbf{k} \Rightarrow \overrightarrow{PQ} = \mathbf{i}+\mathbf{j}+\mathbf{k}$ and $\mathbf{W} = \mathbf{F}\cdot\overrightarrow{PQ} = (5\mathbf{k})\cdot(\mathbf{i}+\mathbf{j}+\mathbf{k}) = 5\ \mathrm{N}\cdot\mathrm{m} = 5\ \mathrm{J}$

39. $\mathbf{W} = |\mathbf{F}|\left|\overrightarrow{PQ}\right| \cos\theta = (200)(20)(\cos 30^\circ) = 2000\sqrt{3} = 3464.10\ \mathrm{N}\cdot\mathrm{m} = 3464.10\ \mathrm{J}$

41. $P(x_1,y_1) = P\left(x_1, \frac{c}{b} - \frac{a}{b}x_1\right)$ and $Q(x_2,y_2) = Q\left(x_2, \frac{c}{b} - \frac{a}{b}x_2\right)$ are any two points P and Q on the line with $b \neq 0$ $\Rightarrow \overrightarrow{PQ} = (x_2-x_1)\mathbf{i} + \frac{a}{b}(x_2-x_1)\mathbf{j} \Rightarrow \overrightarrow{PQ}\cdot\mathbf{v} = \left[(x_2-x_1)\mathbf{i} + \frac{a}{b}(x_2-x_1)\mathbf{j}\right]\cdot(a\mathbf{i}+b\mathbf{j}) = a(x_2-x_1) + b\left(\frac{a}{b}\right)(x_2-x_1)$ $= 0 \Rightarrow \mathbf{v}$ is perpendicular to $\overrightarrow{PQ}$ for $b \neq 0$. If $b = 0$, then $\mathbf{v} = a\mathbf{i}$ is perpendicular to the vertical line $ax = c$.

Alternatively, the slope of $\mathbf{v}$ is $\frac{b}{a}$ and the slope of the line $ax + by = c$ is $-\frac{a}{b}$, so the slopes are negative reciprocals $\Rightarrow$ the vector $\mathbf{v}$ and the line are perpendicular.

43. $\mathbf{v} = \mathbf{i} + 2\mathbf{j}$ is perpendicular to the line $x + 2y = c$;
$P(2,1)$ on the line $\Rightarrow 2 + 2 = c \Rightarrow x + 2y = 4$

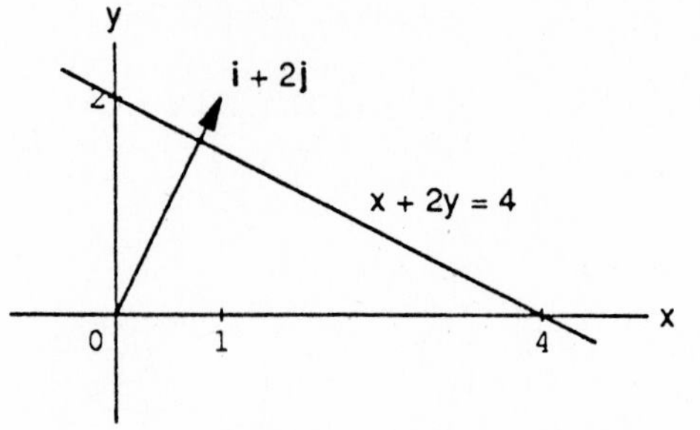

45. $\mathbf{v} = -2\mathbf{i} + \mathbf{j}$ is perpendicular to the line $-2x + y = c$;
$P(-2,-7)$ on the line $\Rightarrow (-2)(-2) - 7 = c \Rightarrow -2x + y = -3$

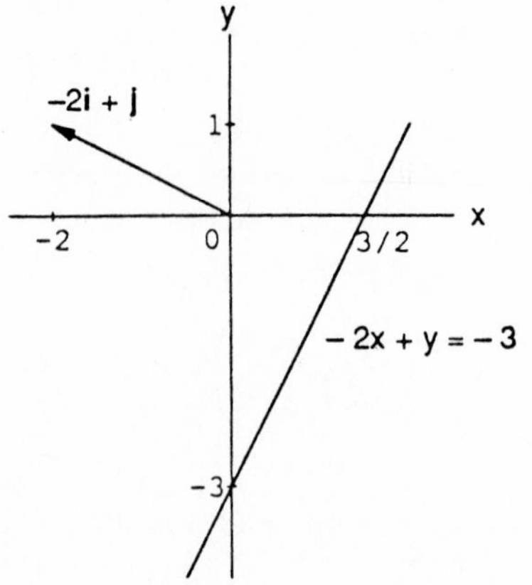

47. $\mathbf{v} = \mathbf{i} - \mathbf{j}$ is parallel to the line $x + y = c$;
$P(-2,1)$ on the line $\Rightarrow -2 + 1 = c \Rightarrow x + y = -1$

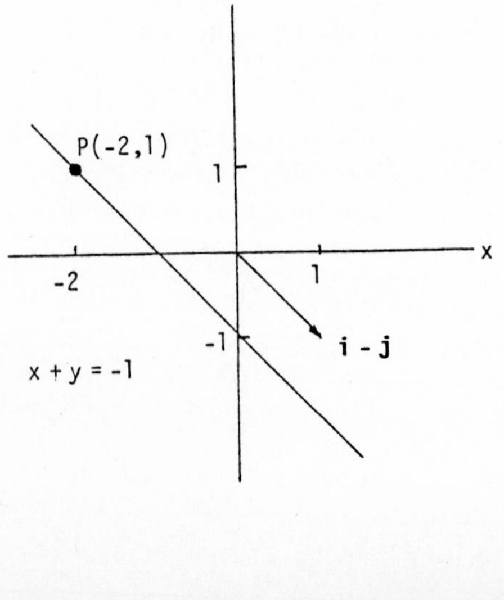

49. $\mathbf{v} = -\mathbf{i} - 2\mathbf{j}$ is parallel to the line $2x - y = c$;

$P(1,2)$ on the line $\Rightarrow (2)(1) - 2 = c \Rightarrow 2x - y = 0$

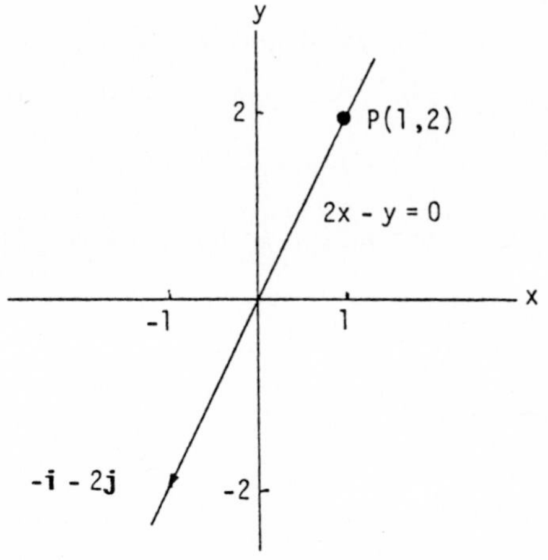

In Exercises 51-55 we use the fact that $\mathbf{n} = a\mathbf{i} + b\mathbf{j}$ is normal to the line $ax + by = c$.

51. $\mathbf{n}_1 = 3\mathbf{i} + \mathbf{j}$ and $\mathbf{n}_2 = 2\mathbf{i} - \mathbf{j} \Rightarrow \theta = \cos^{-1}\left(\frac{\mathbf{n}_1 \cdot \mathbf{n}_2}{|\mathbf{n}_1||\mathbf{n}_2|}\right) = \cos^{-1}\left(\frac{6-1}{\sqrt{10}\,\sqrt{5}}\right) = \cos^{-1}\left(\frac{1}{\sqrt{2}}\right) = \frac{\pi}{4}$

53. $\mathbf{n}_1 = \sqrt{3}\mathbf{i} - \mathbf{j}$ and $\mathbf{n}_2 = \mathbf{i} - \sqrt{3}\mathbf{j} \Rightarrow \theta = \cos^{-1}\left(\frac{\mathbf{n}_1 \cdot \mathbf{n}_2}{|\mathbf{n}_1||\mathbf{n}_2|}\right) = \cos^{-1}\left(\frac{\sqrt{3}+\sqrt{3}}{\sqrt{4}\,\sqrt{4}}\right) = \cos^{-1}\left(\frac{\sqrt{3}}{2}\right) = \frac{\pi}{6}$

55. $\mathbf{n}_1 = 3\mathbf{i} - 4\mathbf{j}$ and $\mathbf{n}_2 = \mathbf{i} - \mathbf{j} \Rightarrow \theta = \cos^{-1}\left(\frac{\mathbf{n}_1 \cdot \mathbf{n}_2}{|\mathbf{n}_1||\mathbf{n}_2|}\right) = \cos^{-1}\left(\frac{3+4}{\sqrt{25}\,\sqrt{2}}\right) = \cos^{-1}\left(\frac{7}{5\sqrt{2}}\right) \approx 0.14$ rad

57. The angle between the corresponding normals is equal to the angle between the corresponding tangents. The points of intersection are $\left(-\frac{\sqrt{3}}{2}, \frac{3}{4}\right)$ and $\left(\frac{\sqrt{3}}{2}, \frac{3}{4}\right)$. At $\left(-\frac{\sqrt{3}}{2}, \frac{3}{4}\right)$ the tangent line for $f(x) = x^2$ is $y - \frac{3}{4} = f'\left(-\frac{\sqrt{3}}{2}\right)\left(x - \left(-\frac{\sqrt{3}}{2}\right)\right) \Rightarrow y = -\sqrt{3}\left(x + \frac{\sqrt{3}}{2}\right) + \frac{3}{4} \Rightarrow y = -\sqrt{3}x - \frac{3}{4}$, and the tangent line for $f(x) = \left(\frac{3}{2}\right) - x^2$ is $y - \frac{3}{4} = f'\left(-\frac{\sqrt{3}}{2}\right)\left(x - \left(-\frac{\sqrt{3}}{2}\right)\right) \Rightarrow y = \sqrt{3}\left(x + \frac{\sqrt{3}}{2}\right) + \frac{3}{4} = \sqrt{3}x + \frac{9}{4}$. The corresponding normals are $\mathbf{n}_1 = \sqrt{3}\mathbf{i} + \mathbf{j}$ and $\mathbf{n}_2 = -\sqrt{3}\mathbf{i} + \mathbf{j}$. The angle at $\left(-\frac{\sqrt{3}}{2}, \frac{3}{4}\right)$ is $\theta = \cos^{-1}\left(\frac{\mathbf{n}_1 \cdot \mathbf{n}_2}{|\mathbf{n}_1||\mathbf{n}_2|}\right)$ $= \cos^{-1}\left(\frac{-3+1}{\sqrt{4}\,\sqrt{4}}\right) = \cos^{-1}\left(-\frac{1}{2}\right) = \frac{2\pi}{3}$, the angle is $\frac{\pi}{3}$ and $\frac{2\pi}{3}$. At $\left(\frac{\sqrt{3}}{2}, \frac{3}{4}\right)$ the tangent line for $f(x) = x^2$ is $y = \sqrt{3}\left(x + \frac{\sqrt{3}}{2}\right) + \frac{3}{4} = \sqrt{3}x + \frac{9}{4}$ and the tangent line for $f(x) = \frac{3}{2} - x^2$ is $y = -\sqrt{3}\left(x + \frac{\sqrt{3}}{2}\right) + \frac{3}{4}$ $= -\sqrt{3}x - \frac{3}{4}$. The corresponding normals are $\mathbf{n}_1 = -\sqrt{3}\mathbf{i} + \mathbf{j}$ and $\mathbf{n}_2 = \sqrt{3}\mathbf{i} + \mathbf{j}$. The angle at $\left(\frac{\sqrt{3}}{2}, \frac{3}{4}\right)$ is $\theta = \cos^{-1}\left(\frac{\mathbf{n}_1 \cdot \mathbf{n}_2}{|\mathbf{n}_1||\mathbf{n}_2|}\right) = \cos^{-1}\left(\frac{-3+1}{\sqrt{4}\,\sqrt{4}}\right) = \cos^{-1}\left(-\frac{1}{2}\right) = \frac{2\pi}{3}$, the angle is $\frac{\pi}{3}$ and $\frac{2\pi}{3}$.

59. The curves intersect when $y = x^3 = (y^2)^3 = y^6 \Rightarrow y = 0$ or $y = 1$. The points of intersection are $(0,0)$ and $(1,1)$. Note that $y \geq 0$ since $y = y^6$. At $(0,0)$ the tangent line for $y = x^3$ is $y = 0$ and the tangent line for $y = \sqrt{x}$ is $x = 0$. Therefore, the angle of intersection at $(0,0)$ is $\frac{\pi}{2}$. At $(1,1)$ the tangent line for $y = x^3$ is $y = 3x - 2$ and the tangent line for $y = \sqrt{x}$ is $y = \frac{1}{2}x + \frac{1}{2}$. The corresponding normal vectors are $\mathbf{n}_1 = -3\mathbf{i} + \mathbf{j}$ and $\mathbf{n}_2 = -\frac{1}{2}\mathbf{i} + \mathbf{j} \Rightarrow \theta = \cos^{-1}\left(\frac{\mathbf{n}_1 \cdot \mathbf{n}_2}{|\mathbf{n}_1||\mathbf{n}_2|}\right) = \cos^{-1}\left(\frac{1}{\sqrt{2}}\right) = \frac{\pi}{4}$, the angle is $\frac{\pi}{4}$ and $\frac{3\pi}{4}$.

10.4 CROSS PRODUCTS

1. $\mathbf{A} \times \mathbf{B} = \begin{vmatrix} \mathbf{i} & \mathbf{j} & \mathbf{k} \\ 2 & -2 & -1 \\ 1 & 0 & -1 \end{vmatrix} = 3\left(\frac{2}{3}\mathbf{i} + \frac{1}{3}\mathbf{j} + \frac{2}{3}\mathbf{k}\right) \Rightarrow$ length $= 3$ and the direction is $\frac{2}{3}\mathbf{i} + \frac{1}{3}\mathbf{j} + \frac{2}{3}\mathbf{k}$;

   $\mathbf{B} \times \mathbf{A} = -(\mathbf{A} \times \mathbf{B}) = -3\left(\frac{2}{3}\mathbf{i} + \frac{1}{3}\mathbf{j} + \frac{2}{3}\mathbf{k}\right) \Rightarrow$ length $= 3$ and the direction is $-\frac{2}{3}\mathbf{i} - \frac{1}{3}\mathbf{j} - \frac{2}{3}\mathbf{k}$

3. $\mathbf{A} \times \mathbf{B} = \begin{vmatrix} \mathbf{i} & \mathbf{j} & \mathbf{k} \\ 2 & -2 & 4 \\ -1 & 1 & -2 \end{vmatrix} = \mathbf{0} \Rightarrow$ length $= 0$ and has no direction

   $\mathbf{B} \times \mathbf{A} = -(\mathbf{A} \times \mathbf{B}) = \mathbf{0} \Rightarrow$ length $= 0$ and has no direction

5. $\mathbf{A} \times \mathbf{B} = \begin{vmatrix} \mathbf{i} & \mathbf{j} & \mathbf{k} \\ 2 & 0 & 0 \\ 0 & -3 & 0 \end{vmatrix} = -6(\mathbf{k}) \Rightarrow$ length $= 6$ and the direction is $-\mathbf{k}$

   $\mathbf{B} \times \mathbf{A} = -(\mathbf{A} \times \mathbf{B}) = 6(\mathbf{k}) \Rightarrow$ length $= 6$ and the direction is $\mathbf{k}$

7. $\mathbf{A} \times \mathbf{B} = \begin{vmatrix} \mathbf{i} & \mathbf{j} & \mathbf{k} \\ -8 & -2 & -4 \\ 2 & 2 & 1 \end{vmatrix} = 6\mathbf{i} - 12\mathbf{k} \Rightarrow$ length $= 6\sqrt{5}$ and the direction is $\frac{1}{\sqrt{5}}\mathbf{i} - \frac{2}{\sqrt{5}}\mathbf{k}$

   $\mathbf{B} \times \mathbf{A} = -(\mathbf{A} \times \mathbf{B}) = -(6\mathbf{i} - 12\mathbf{k}) \Rightarrow$ length $= 6\sqrt{5}$ and the direction is $-\frac{1}{\sqrt{5}}\mathbf{i} + \frac{2}{\sqrt{5}}\mathbf{k}$

9. $\mathbf{A}\times\mathbf{B}=\begin{vmatrix}\mathbf{i} & \mathbf{j} & \mathbf{k}\\ 1 & 0 & 0\\ 0 & 1 & 0\end{vmatrix}=\mathbf{k}$

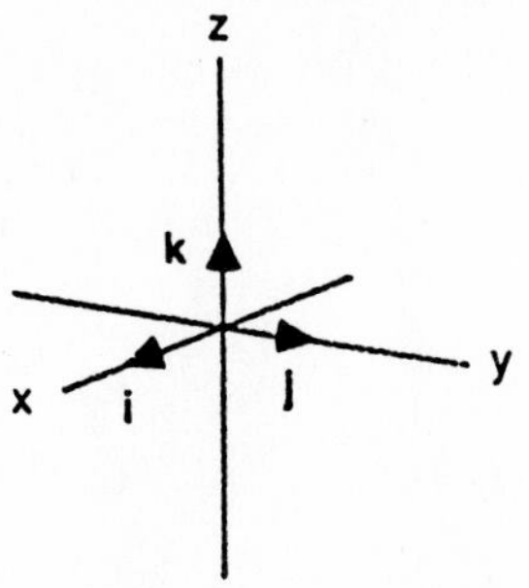

11. $\mathbf{A}\times\mathbf{B}=\begin{vmatrix}\mathbf{i} & \mathbf{j} & \mathbf{k}\\ 1 & 0 & -1\\ 0 & 1 & 1\end{vmatrix}=\mathbf{i}-\mathbf{j}+\mathbf{k}$

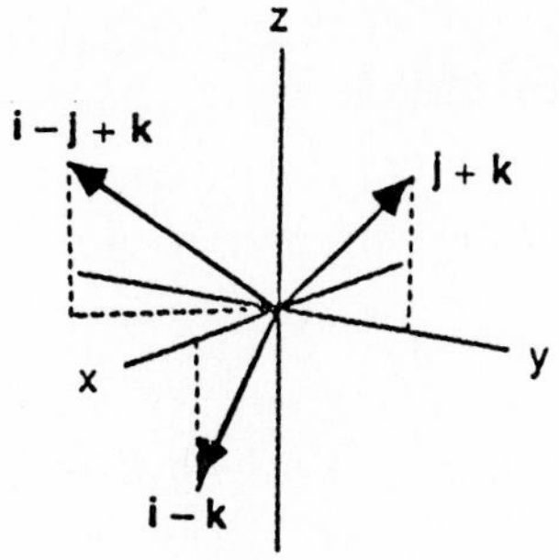

13. $\mathbf{A}\times\mathbf{B}=\begin{vmatrix}\mathbf{i} & \mathbf{j} & \mathbf{k}\\ 1 & 1 & 0\\ 1 & -1 & 0\end{vmatrix}=-2\mathbf{k}$

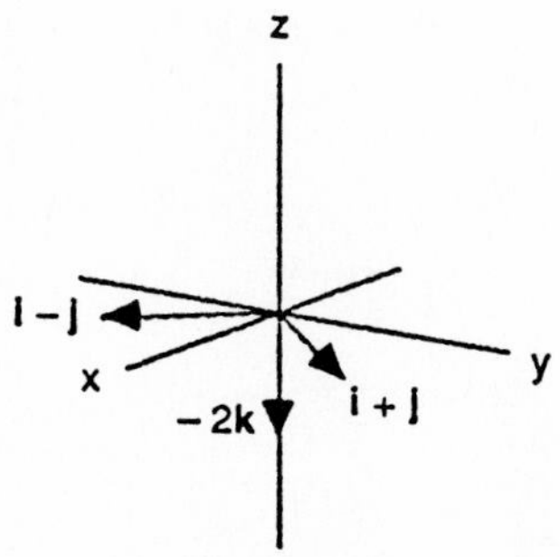

15. (a) $\overrightarrow{PQ}\times\overrightarrow{PR}=\begin{vmatrix}\mathbf{i} & \mathbf{j} & \mathbf{k}\\ 1 & 1 & -3\\ -1 & 3 & -1\end{vmatrix}=8\mathbf{i}+4\mathbf{j}+4\mathbf{k}\Rightarrow \text{Area}=\frac{1}{2}\left|\overrightarrow{PQ}\times\overrightarrow{PR}\right|=\frac{1}{2}\sqrt{64+16+16}=2\sqrt{6}$

(b) $\mathbf{u}=\pm\frac{\overrightarrow{PQ}\times\overrightarrow{PR}}{\left|\overrightarrow{PQ}\times\overrightarrow{PR}\right|}=\pm\frac{1}{\sqrt{6}}(2\mathbf{i}+\mathbf{j}+\mathbf{k})$

17. (a) $\overrightarrow{PQ}\times\overrightarrow{PR}=\begin{vmatrix}\mathbf{i} & \mathbf{j} & \mathbf{k}\\ 1 & 1 & 1\\ 1 & 1 & 0\end{vmatrix}=-\mathbf{i}+\mathbf{j}\Rightarrow \text{Area}=\frac{1}{2}\left|\overrightarrow{PQ}\times\overrightarrow{PR}\right|=\frac{1}{2}\sqrt{1+1}=\frac{\sqrt{2}}{2}$

(b) $\mathbf{u}=\pm\frac{\overrightarrow{PQ}\times\overrightarrow{PR}}{\left|\overrightarrow{PQ}\times\overrightarrow{PR}\right|}=\pm\frac{1}{\sqrt{2}}(-\mathbf{i}+\mathbf{j})=\pm\frac{1}{\sqrt{2}}(\mathbf{i}-\mathbf{j})$

19. (a) $\mathbf{A}\cdot\mathbf{B} = -6$, $\mathbf{A}\cdot\mathbf{C} = -81$, $\mathbf{B}\cdot\mathbf{C} = 18 \Rightarrow$ none

(b) $\mathbf{A}\times\mathbf{B} = \begin{vmatrix} \mathbf{i} & \mathbf{j} & \mathbf{k} \\ 5 & -1 & 1 \\ 0 & 1 & -5 \end{vmatrix} \neq \mathbf{0}$, $\mathbf{A}\times\mathbf{C} = \begin{vmatrix} \mathbf{i} & \mathbf{j} & \mathbf{k} \\ 5 & -1 & 1 \\ -15 & 3 & -3 \end{vmatrix} = \mathbf{0}$, $\mathbf{B}\times\mathbf{C} = \begin{vmatrix} \mathbf{i} & \mathbf{j} & \mathbf{k} \\ 0 & 1 & -5 \\ -15 & 3 & -3 \end{vmatrix} \neq \mathbf{0}$

$\Rightarrow$ **A** and **C** are parallel

21. $\left|\overrightarrow{PQ}\times\mathbf{F}\right| = \left|\overrightarrow{PQ}\right|\left|\mathbf{F}\right|\sin(60^\circ) = \frac{2}{3}\cdot 30\cdot\frac{\sqrt{3}}{2}$ ft · lb $= 10\sqrt{3}$ ft · lb

23. If $\mathbf{A} = a_1\mathbf{i} + a_2\mathbf{j} + a_3\mathbf{k}$, $\mathbf{B} = b_1\mathbf{i} + b_2\mathbf{j} + b_3\mathbf{k}$, and $\mathbf{C} = c_1\mathbf{i} + c_2\mathbf{j} + c_3\mathbf{k}$, then $\mathbf{A}\cdot(\mathbf{B}\times\mathbf{C}) = \begin{vmatrix} a_1 & a_2 & a_3 \\ b_1 & b_2 & b_3 \\ c_1 & c_2 & c_3 \end{vmatrix}$,

$\mathbf{B}\cdot(\mathbf{C}\times\mathbf{A}) = \begin{vmatrix} b_1 & b_2 & b_3 \\ c_1 & c_2 & c_3 \\ a_1 & a_2 & a_3 \end{vmatrix}$ and $\mathbf{C}\cdot(\mathbf{A}\times\mathbf{B}) = \begin{vmatrix} c_1 & c_2 & c_3 \\ a_1 & a_2 & a_3 \\ b_1 & b_2 & b_3 \end{vmatrix}$ which all have the same value, since the interchanging of two pair of rows in a determinant does not change its value $\Rightarrow$ the volume is

$\left|(\mathbf{A}\times\mathbf{B})\cdot\mathbf{C}\right| = \text{abs}\begin{vmatrix} 2 & 0 & 0 \\ 0 & 2 & 0 \\ 0 & 0 & 2 \end{vmatrix} = 8$

25. $\left|(\mathbf{A}\times\mathbf{B})\cdot\mathbf{C}\right| = \text{abs}\begin{vmatrix} 2 & 1 & 0 \\ 2 & -1 & 1 \\ 1 & 0 & 2 \end{vmatrix} = |-7| = 7$ (for details about verification, see Exercise 23)

27. (a) true, $|\mathbf{A}| = \sqrt{a_1^2 + a_2^2 + a_3^2} = \sqrt{\mathbf{A}\cdot\mathbf{A}}$

(b) not always true, $\mathbf{A}\cdot\mathbf{A} = |\mathbf{A}|^2$

(c) true, $\mathbf{A}\times\mathbf{0} = \begin{vmatrix} \mathbf{i} & \mathbf{j} & \mathbf{k} \\ a_1 & a_2 & a_3 \\ 0 & 0 & 0 \end{vmatrix} = 0\mathbf{i} + 0\mathbf{j} + 0\mathbf{k} = \mathbf{0}$

(d) true, $\mathbf{A}\times(-\mathbf{A}) = \begin{vmatrix} \mathbf{i} & \mathbf{j} & \mathbf{k} \\ a_1 & a_2 & a_3 \\ -a_1 & -a_2 & -a_3 \end{vmatrix} = (-a_2a_3 + a_2a_3)\mathbf{i} + (-a_1a_3 + a_1a_3)\mathbf{j} + (-a_1a_2 + a_1a_2)\mathbf{k} = \mathbf{0}$

(e) not always true, $\mathbf{i}\times\mathbf{j} = \mathbf{k} \neq -\mathbf{k} = \mathbf{j}\times\mathbf{i}$ for example

(f) true, Eqn. (6)

(g) true, $(\mathbf{A}\times\mathbf{B})\cdot\mathbf{B}=\mathbf{A}\cdot(\mathbf{B}\times\mathbf{B})=\mathbf{A}\cdot\mathbf{0}=0$

(h) true, Eqn. (13)

29. (a) $\text{proj}_{\mathbf{B}}\,\mathbf{A}=\left(\frac{\mathbf{A}\cdot\mathbf{B}}{\mathbf{B}\cdot\mathbf{B}}\right)\mathbf{B}$ (b) $\pm(\mathbf{A}\times\mathbf{B})$ (c) $\pm(\mathbf{A}\times\mathbf{B})\times\mathbf{C}$ (d) $|(\mathbf{A}\times\mathbf{B})\cdot\mathbf{C}|$

31. (a) yes, $\mathbf{A}\times\mathbf{B}$ and $\mathbf{C}$ are both vectors (b) no, $\mathbf{A}$ is a vector but $\mathbf{B}\cdot\mathbf{C}$ is a scalar
(c) yes, $\mathbf{A}$ and $\mathbf{A}\times\mathbf{C}$ are both vectors (d) no, $\mathbf{A}$ is a vector but $\mathbf{B}\cdot\mathbf{C}$ is a scalar

33. No, $\mathbf{B}$ need not equal $\mathbf{C}$. For example, $\mathbf{i}+\mathbf{j}\neq-\mathbf{i}+\mathbf{j}$, but $\mathbf{i}\times(\mathbf{i}+\mathbf{j})=\mathbf{i}\times\mathbf{i}+\mathbf{i}\times\mathbf{j}=\mathbf{0}+\mathbf{k}=\mathbf{k}$ and $\mathbf{i}\times(-\mathbf{i}+\mathbf{j})=-\mathbf{i}\times\mathbf{i}+\mathbf{i}\times\mathbf{j}=\mathbf{0}+\mathbf{k}=\mathbf{k}$.

35. $\overrightarrow{AB}=-\mathbf{i}+\mathbf{j}$ and $\overrightarrow{AD}=-\mathbf{i}-\mathbf{j}\Rightarrow\overrightarrow{AB}\times\overrightarrow{AD}=\begin{vmatrix}\mathbf{i}&\mathbf{j}&\mathbf{k}\\-1&1&0\\-1&-1&0\end{vmatrix}=2\mathbf{k}\Rightarrow\text{area}=\left|\overrightarrow{AB}\times\overrightarrow{AD}\right|=2$

37. $\overrightarrow{AB}=3\mathbf{i}-2\mathbf{j}$ and $\overrightarrow{AD}=5\mathbf{i}+\mathbf{j}\Rightarrow\overrightarrow{AB}\times\overrightarrow{AD}=\begin{vmatrix}\mathbf{i}&\mathbf{j}&\mathbf{k}\\3&-2&0\\5&1&0\end{vmatrix}=13\mathbf{k}\Rightarrow\text{area}=\left|\overrightarrow{AB}\times\overrightarrow{AD}\right|=13$

39. $\overrightarrow{AB}=-2\mathbf{i}+3\mathbf{j}$ and $\overrightarrow{AC}=3\mathbf{i}+\mathbf{j}\Rightarrow\overrightarrow{AB}\times\overrightarrow{AC}=\begin{vmatrix}\mathbf{i}&\mathbf{j}&\mathbf{k}\\-2&3&0\\3&1&0\end{vmatrix}=-11\mathbf{k}\Rightarrow\text{area}=\frac{1}{2}\left|\overrightarrow{AB}\times\overrightarrow{AC}\right|=\frac{11}{2}$

41. $\overrightarrow{AB}=6\mathbf{i}-5\mathbf{j}$ and $\overrightarrow{AC}=11\mathbf{i}-5\mathbf{j}\Rightarrow\overrightarrow{AB}\times\overrightarrow{AC}=\begin{vmatrix}\mathbf{i}&\mathbf{j}&\mathbf{k}\\6&-5&0\\11&-5&0\end{vmatrix}=25\mathbf{k}\Rightarrow\text{area}=\frac{1}{2}\left|\overrightarrow{AB}\times\overrightarrow{AC}\right|=\frac{25}{2}$

43. If $\mathbf{A}=a_1\mathbf{i}+a_2\mathbf{j}$ and $\mathbf{B}=b_1\mathbf{i}+b_2\mathbf{j}$, then $\mathbf{A}\times\mathbf{B}=\begin{vmatrix}\mathbf{i}&\mathbf{j}&\mathbf{k}\\a_1&a_2&0\\b_1&b_2&0\end{vmatrix}=\begin{vmatrix}a_1&a_2\\b_1&b_2\end{vmatrix}\mathbf{k}$ and the triangle's area is $\frac{1}{2}|\mathbf{A}\times\mathbf{B}|=\pm\frac{1}{2}\begin{vmatrix}a_1&a_2\\b_1&b_2\end{vmatrix}$. The applicable sign is (+) if the acute angle from $\mathbf{A}$ to $\mathbf{B}$ runs counterclockwise in the xy-plane, and (−) if it runs clockwise, because the area must be a nonnegative number.

10.5 LINES AND PLANES IN SPACE

1. The direction $\mathbf{i}+\mathbf{j}+\mathbf{k}$ and $P(3,-4,-1)\Rightarrow x=3+t,\ y=-4+t,\ z=-1+t$

3. The direction $\overrightarrow{PQ}=5\mathbf{i}+5\mathbf{j}-5\mathbf{k}$ and $P(-2,0,3)\Rightarrow x=-2+5t,\ y=5t,\ z=3-5t$

5. The direction $2\mathbf{j}+\mathbf{k}$ and $P(0,0,0) \Rightarrow x=0,\ y=2t,\ z=t$

7. The direction $\mathbf{k}$ and $P(1,1,1) \Rightarrow x=1,\ y=1,\ z=1+t$

9. The direction $\mathbf{i}+2\mathbf{j}+2\mathbf{k}$ and $P(0,-7,0) \Rightarrow x=t,\ y=-7+2t,\ z=2t$

11. The direction $\mathbf{i}$ and $P(0,0,0) \Rightarrow x=t,\ y=0,\ z=0$

13. The direction $\overrightarrow{PQ}=\mathbf{i}+\mathbf{j}+\frac{3}{2}\mathbf{k}$ and $P(0,0,0) \Rightarrow x=t,\ y=t,\ z=\frac{3}{2}t$, where $0 \le t \le 1$

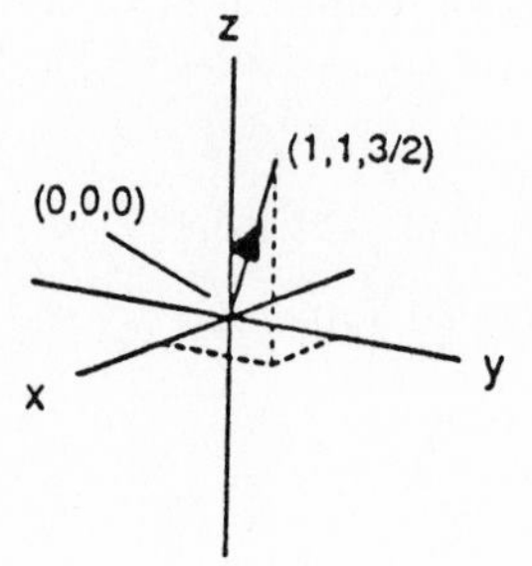

15. The direction $\overrightarrow{PQ}=\mathbf{j}$ and $P(1,1,0) \Rightarrow x=1,\ y=1+t,\ z=0$, where $-1 \le t \le 0$

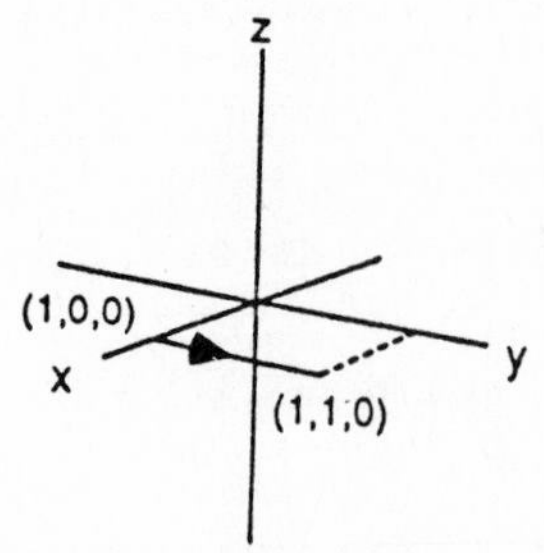

17. The direction $\overrightarrow{PQ}=-2\mathbf{j}$ and $P(0,1,1) \Rightarrow x=0,\ y=1-2t,\ z=1$, where $0 \le t \le 1$

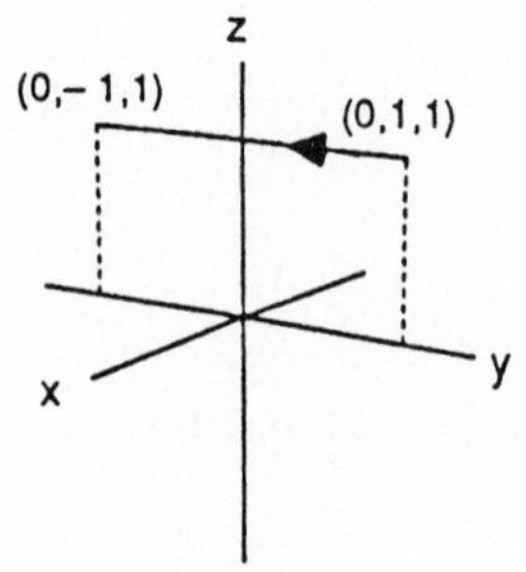

19. The direction $\overrightarrow{PQ}=-2\mathbf{i}+2\mathbf{j}-2\mathbf{k}$ and $P(2,0,2) \Rightarrow x=2-2t$, $y=2t,\ z=2-2t$, where $0 \le t \le 1$

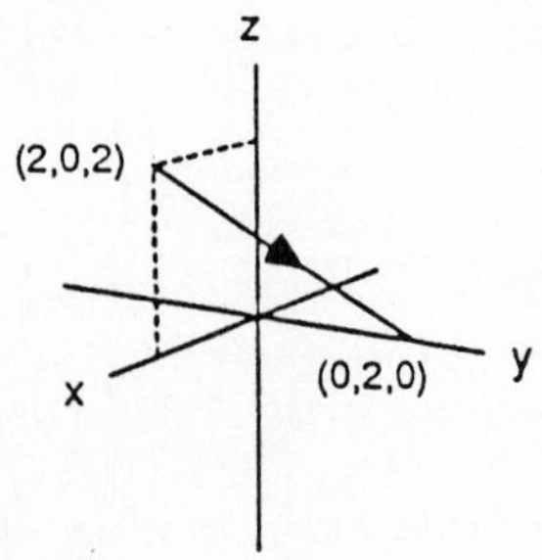

21. $3(x-0)+(-2)(y-2)+(-1)(z+1)=0 \Rightarrow 3x-2y-z=-3$

23. $\overrightarrow{PQ}=\mathbf{i}-\mathbf{j}+3\mathbf{k}$, $\overrightarrow{PS}=-\mathbf{i}-3\mathbf{j}+2\mathbf{k} \Rightarrow \overrightarrow{PQ}\times\overrightarrow{PS}=\begin{vmatrix} \mathbf{i} & \mathbf{j} & \mathbf{k} \\ 1 & -1 & 3 \\ -1 & -3 & 2 \end{vmatrix}=7\mathbf{i}-5\mathbf{j}-4\mathbf{k}$ is normal to the plane

$\Rightarrow 7(x-2)+(-5)(y-0)+(-4)(z-2)=0 \Rightarrow 7x-5y-4z=6$

25. $\mathbf{n}=\mathbf{i}+3\mathbf{j}+4\mathbf{k}$, $P(2,4,5)=(1)(x-2)+(3)(y-4)+(4)(z-5)=0 \Rightarrow x+3y+4z=34$

27. $\begin{cases} x=2t+1=\ \ s+2 \\ y=3t+2=2s+4 \end{cases} \Rightarrow \begin{cases} 2t-\ \ s=1 \\ 3t-2s=2 \end{cases} \Rightarrow \begin{cases} 4t-2s=2 \\ 3t-2s=2 \end{cases} \Rightarrow t=0$ and $s=-1$; then $z=4t+3=-4s-1$

$\Rightarrow 4(0)+3=(-4)(-1)-1$ is satisfied $\Rightarrow$ the lines do intersect when $t=0$ and $s=-1 \Rightarrow$ the point of intersection is $x=1$, $y=2$, and $z=3$ or $P(1,2,3)$. A vector normal to the plane determined by these lines is

$\mathbf{n}_1\times\mathbf{n}_2=\begin{vmatrix} \mathbf{i} & \mathbf{j} & \mathbf{k} \\ 2 & 3 & 4 \\ 1 & 2 & -4 \end{vmatrix}=-20\mathbf{i}+12\mathbf{j}+\mathbf{k}$, where $\mathbf{n}_1$ and $\mathbf{n}_2$ are directions of the lines $\Rightarrow$ the plane

containing the lines is represented by $(-20)(x-1)+(12)(y-2)+(1)(z-3)=0 \Rightarrow -20x+12y+z=7$.

29. The cross product of $\mathbf{i}+\mathbf{j}-\mathbf{k}$ and $-4\mathbf{i}+2\mathbf{j}-2\mathbf{k}$ has the same direction as the normal to the plane

$\Rightarrow \mathbf{n}=\begin{vmatrix} \mathbf{i} & \mathbf{j} & \mathbf{k} \\ 1 & 1 & -1 \\ -4 & 2 & -2 \end{vmatrix}=6\mathbf{j}+6\mathbf{k}$. Select a point on either line, such as $P(-1,2,1)$. Since the lines are given

to intersect, the desired plane is $0(x+1)+6(y-2)+6(z-1)=0 \Rightarrow 6y+6z=18 \Rightarrow y+z=3$.

31. $\mathbf{n}_1\times\mathbf{n}_2=\begin{vmatrix} \mathbf{i} & \mathbf{j} & \mathbf{k} \\ 2 & 1 & -1 \\ 1 & 2 & 1 \end{vmatrix}=3\mathbf{i}-3\mathbf{j}+3\mathbf{k}$ is a vector in the direction of the line of intersection of the planes

$\Rightarrow 3(x-2)+(-3)(y-1)+3(z+1)=0 \Rightarrow 3x-3y+3z=0 \Rightarrow x-y+z=0$ is the desired plane containing $P_0(2,1,-1)$

33. $S(0,0,12)$, $P(0,0,0)$ and $\mathbf{v}=4\mathbf{i}-2\mathbf{j}+2\mathbf{k} \Rightarrow \overrightarrow{PS}\times\mathbf{v}=\begin{vmatrix} \mathbf{i} & \mathbf{j} & \mathbf{k} \\ 0 & 0 & 12 \\ 4 & -2 & 2 \end{vmatrix}=24\mathbf{i}+48\mathbf{j}=24(\mathbf{i}+2\mathbf{j})$

$\Rightarrow d=\frac{|\overrightarrow{PS}\times\mathbf{v}|}{|\mathbf{v}|}=\frac{24\sqrt{1+4}}{\sqrt{16+4+4}}=\frac{24\sqrt{5}}{\sqrt{24}}=\sqrt{5\cdot 24}=2\sqrt{30}$ is the distance from S to the line

35. S(2, 1, 3), P(2, 1, 3) and $\mathbf{v} = 2\mathbf{i} + 6\mathbf{j} \Rightarrow \overrightarrow{PS} \times \mathbf{v} = \mathbf{0} \Rightarrow d = \frac{|\overrightarrow{PS} \times \mathbf{v}|}{|\mathbf{v}|} = \frac{0}{\sqrt{40}} = 0$ is the distance from S to the line (i.e., the point S lies on the line)

37. S(3, −1, 4), P(4, 3, −5) and $\mathbf{v} = -\mathbf{i} + 2\mathbf{j} + 3\mathbf{k} \Rightarrow \overrightarrow{PS} \times \mathbf{v} = \begin{vmatrix} \mathbf{i} & \mathbf{j} & \mathbf{k} \\ -1 & -4 & 9 \\ -1 & 2 & 3 \end{vmatrix} = -30\mathbf{i} - 6\mathbf{j} - 6\mathbf{k}$

$\Rightarrow d = \frac{|\overrightarrow{PS} \times \mathbf{v}|}{|\mathbf{v}|} = \frac{\sqrt{900+36+36}}{\sqrt{1+4+9}} = \frac{\sqrt{972}}{\sqrt{14}} = \frac{\sqrt{486}}{\sqrt{7}} = \frac{\sqrt{81 \cdot 6}}{\sqrt{7}} = \frac{9\sqrt{42}}{7}$ is the distance from S to the line

39. S(2, −3, 4), $x + 2y + 2z = 13$ and P(13, 0, 0) is on the plane $\Rightarrow \overrightarrow{PS} = -11\mathbf{i} - 3\mathbf{j} + 4\mathbf{k}$ and $\mathbf{n} = \mathbf{i} + 2\mathbf{j} + 2\mathbf{k}$

$\Rightarrow d = \left|\overrightarrow{PS} \cdot \frac{\mathbf{n}}{|\mathbf{n}|}\right| = \left|\frac{-11-6+8}{\sqrt{1+4+4}}\right| = \left|\frac{-9}{\sqrt{9}}\right| = 3$

41. S(0, 1, 1), $4y + 3z = -12$ and P(0, −3, 0) is on the plane $\Rightarrow \overrightarrow{PS} = 4\mathbf{j} + \mathbf{k}$ and $\mathbf{n} = 4\mathbf{j} + 3\mathbf{k}$

$\Rightarrow d = \left|\overrightarrow{PS} \cdot \frac{\mathbf{n}}{|\mathbf{n}|}\right| = \left|\frac{16+3}{\sqrt{16+9}}\right| = \frac{19}{5}$

43. S(0, −1, 0), $2x + y + 2z = 4$ and P(2, 0, 0) is on the plane $\Rightarrow \overrightarrow{PS} = -2\mathbf{i} - \mathbf{j}$ and $\mathbf{n} = 2\mathbf{i} + \mathbf{j} + 2\mathbf{k}$

$\Rightarrow d = \left|\overrightarrow{PS} \cdot \frac{\mathbf{n}}{|\mathbf{n}|}\right| = \left|\frac{-4-1+0}{\sqrt{4+1+4}}\right| = \frac{5}{3}$

45. The point P(1, 0, 0) is on the first plane and S(10, 0, 0) is a point on the second plane $\Rightarrow \overrightarrow{PS} = 9\mathbf{i}$, and $\mathbf{n} = \mathbf{i} + 2\mathbf{j} + 6\mathbf{k}$ is normal to the first plane ⇒ the distance from S to the first plane is $d = \left|\overrightarrow{PS} \cdot \frac{\mathbf{n}}{|\mathbf{n}|}\right|$ $= \left|\frac{9}{\sqrt{1+4+36}}\right| = \frac{9}{\sqrt{41}}$, which is also the distance between the planes.

47. $\mathbf{n}_1 = \mathbf{i} + \mathbf{j}$ and $\mathbf{n}_2 = 2\mathbf{i} + \mathbf{j} - 2\mathbf{k} \Rightarrow \theta = \cos^{-1}\left(\frac{\mathbf{n}_1 \cdot \mathbf{n}_2}{|\mathbf{n}_1||\mathbf{n}_2|}\right) = \cos^{-1}\left(\frac{2+1}{\sqrt{2}\sqrt{9}}\right) = \cos^{-1}\left(\frac{1}{\sqrt{2}}\right) = \frac{\pi}{4}$

49. $\mathbf{n}_1 = 2\mathbf{i} + 2\mathbf{j} + 2\mathbf{k}$ and $\mathbf{n}_2 = 2\mathbf{i} - 2\mathbf{j} - \mathbf{k} \Rightarrow \theta = \cos^{-1}\left(\frac{\mathbf{n}_1 \cdot \mathbf{n}_2}{|\mathbf{n}_1||\mathbf{n}_2|}\right) = \cos^{-1}\left(\frac{4-4-2}{\sqrt{12}\sqrt{9}}\right) = \cos^{-1}\left(\frac{-1}{3\sqrt{3}}\right) \approx 1.76$ rad

51. $\mathbf{n}_1 = 2\mathbf{i} + 2\mathbf{j} - \mathbf{k}$ and $\mathbf{n}_2 = \mathbf{i} + 2\mathbf{j} + \mathbf{k} \Rightarrow \theta = \cos^{-1}\left(\frac{\mathbf{n}_1 \cdot \mathbf{n}_2}{|\mathbf{n}_1||\mathbf{n}_2|}\right) = \cos^{-1}\left(\frac{2+4-1}{\sqrt{9}\sqrt{6}}\right) = \cos^{-1}\left(\frac{5}{3\sqrt{6}}\right) \approx 0.82$ rad

53. $2x - y + 3z = 6 \Rightarrow 2(1-t) - (3t) + 3(1+t) = 6 \Rightarrow -2t + 5 = 6 \Rightarrow t = -\frac{1}{2} \Rightarrow x = \frac{3}{2}, y = -\frac{3}{2}$ and $z = \frac{1}{2}$

$\Rightarrow \left(\frac{3}{2}, -\frac{3}{2}, \frac{1}{2}\right)$ is the point

55. $x + y + z = 2 \Rightarrow (1+2t) + (1+5t) + (3t) = 2 \Rightarrow 10t + 2 = 2 \Rightarrow t = 0 \Rightarrow x = 1, y = 1$ and $z = 0$
$\Rightarrow (1, 1, 0)$ is the point

57. $\mathbf{n}_1 = \mathbf{i}+\mathbf{j}+\mathbf{k}$ and $\mathbf{n}_2 = \mathbf{i}+\mathbf{j} \Rightarrow \mathbf{n}_1 \times \mathbf{n}_2 = \begin{vmatrix} \mathbf{i} & \mathbf{j} & \mathbf{k} \\ 1 & 1 & 1 \\ 1 & 1 & 0 \end{vmatrix} = -\mathbf{i}+\mathbf{j}$, the direction of the desired line; $(1,1,-1)$ is on both planes $\Rightarrow$ the desired line is $x = 1-t$, $y = 1+t$, $z = -1$

59. $\mathbf{n}_1 = \mathbf{i}-2\mathbf{j}+4\mathbf{k}$ and $\mathbf{n}_2 = \mathbf{i}+\mathbf{j}-2\mathbf{k} \Rightarrow \mathbf{n}_1 \times \mathbf{n}_2 = \begin{vmatrix} \mathbf{i} & \mathbf{j} & \mathbf{k} \\ 1 & -2 & 4 \\ 1 & 1 & -2 \end{vmatrix} = 6\mathbf{j}+3\mathbf{k}$, the direction of the desired line; $(4,3,1)$ is on both planes $\Rightarrow$ the desired line is $x = 4$, $y = 3+6t$, $z = 1+3t$

61. L1 & L2: $x = 3+2t = 1+4s$ and $y = -1+4t = 1+2s \Rightarrow \begin{cases} 2t-4s = -2 \\ 4t-2s = 2 \end{cases} \Rightarrow \begin{cases} 2t-4s = -2 \\ 2t-s = 1 \end{cases}$

$\Rightarrow -3s = -3 \Rightarrow s = 1$ and $t = 1 \Rightarrow$ on L1, $z = 1$ and on L2, $z = 1 \Rightarrow$ L1 and L2 intersect at $(5,3,1)$.

L2 & L3: The direction of L2 is $\frac{1}{6}(4\mathbf{i}+2\mathbf{j}+4\mathbf{k}) = \frac{1}{3}(2\mathbf{i}+\mathbf{j}+2\mathbf{k})$ which is the same as the direction $\frac{1}{3}(2\mathbf{i}+\mathbf{j}+2\mathbf{k})$ of L3; hence L2 and L3 are parallel.

L1 & L3: $x = 3+2t = 3+2r$ and $y = -1+4t = 2+r \Rightarrow \begin{cases} 2t-2r = 0 \\ 4t-r = 3 \end{cases} \Rightarrow \begin{cases} t-r = 0 \\ 4t-r = 3 \end{cases} \Rightarrow 3t = 3$

$\Rightarrow t = 1$ and $r = 1 \Rightarrow$ on L1, $z = 2$ while on L3, $z = 0 \Rightarrow$ L1 and L2 do not intersect. The direction of L1 is $\frac{1}{\sqrt{21}}(2\mathbf{i}+4\mathbf{j}-\mathbf{k})$ while the direction of L3 is $\frac{1}{3}(2\mathbf{i}+\mathbf{j}+2\mathbf{k})$ and neither is a multiple of the other; hence L1 and L3 are skew.

63. $x = 2+2t$, $y = -4-t$, $z = 7+3t$; $x = -2-t$, $y = -2+\frac{1}{2}t$, $z = 1-\frac{3}{2}t$

65. $x = 0 \Rightarrow t = -\frac{1}{2}$, $y = -\frac{1}{2}$, $z = -\frac{3}{2} \Rightarrow \left(0,-\frac{1}{2},-\frac{3}{2}\right)$; $y = 0 \Rightarrow t = -1$, $x = -1$, $z = -3 \Rightarrow (-1,0,-3)$; $z = 0$ $\Rightarrow t = 0$, $x = 1$, $y = -1 \Rightarrow (1,-1,0)$

67. With substitution of the line into the plane we have $2(1-2t)+(2+5t)-(-3t) = 8 \Rightarrow 2-4t+2+5t+3t = 8$ $\Rightarrow 4t+4 = 8 \Rightarrow t = 1 \Rightarrow$ the point $(-1,7,-3)$ is contained in both the line and plane, so they are not parallel.

69. There are many possible answers. One is found as follows: eliminate t to get $t = x-1 = 2-y = \frac{z-3}{2}$ $\Rightarrow x-1 = 2-y$ and $2-y = \frac{z-3}{2} \Rightarrow x+y = 3$ and $2y+z = 7$ are two such planes.

71. The points $(a,0,0)$, $(0,b,0)$ and $(0,0,c)$ are the x, y, and z intercepts of the plane. Since a, b, and c are all nonzero, the plane must intersect all three coordinate axes and cannot pass through the origin. Thus, $\frac{x}{a}+\frac{y}{b}+\frac{z}{c} = 1$ describes all planes <u>except</u> those through the origin or parallel to a coordinate axis.

73. (a) $\overrightarrow{EP} = c\overrightarrow{EP_1} \Rightarrow -x_0\mathbf{i}+y\mathbf{j}+z\mathbf{k} = c\left[(x_1-x_0)\mathbf{i}+y_1\mathbf{j}+z_1\mathbf{k}\right] \Rightarrow -x_0 = c(x_1-x_0)$, $y = cy_1$ and $z = cz_1$, where c is a positive real number

(b) At $x_1 = 0 \Rightarrow c = 1 \Rightarrow y = y_1$ and $z = z_1$; at $x_1 = x_0 \Rightarrow x_0 = 0$, $y = 0$, $z = 0$; $\lim\limits_{x_0 \to \infty} c = \lim\limits_{x_0 \to \infty} \frac{-x_0}{x_1-x_0}$ $= \lim\limits_{x_0 \to \infty} \frac{-1}{-1} = 1 \Rightarrow c \to 1$ so that $y \to y_1$ and $z \to z_1$

10.6 CYLINDERS AND QUADRIC SURFACES

1. d, ellipsoid
3. a, cylinder
5. l, hyperbolic paraboloid
7. b, cylinder
9. k, hyperbolic paraboloid
11. h, cone

13. $x^2 + y^2 = 4$

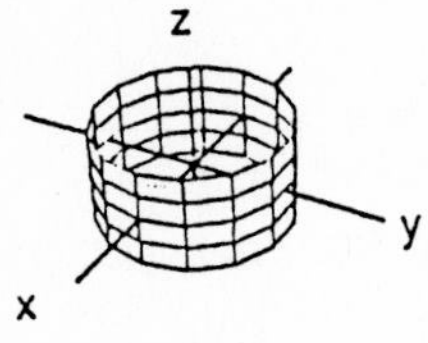

15. $z = y^2 - 1$

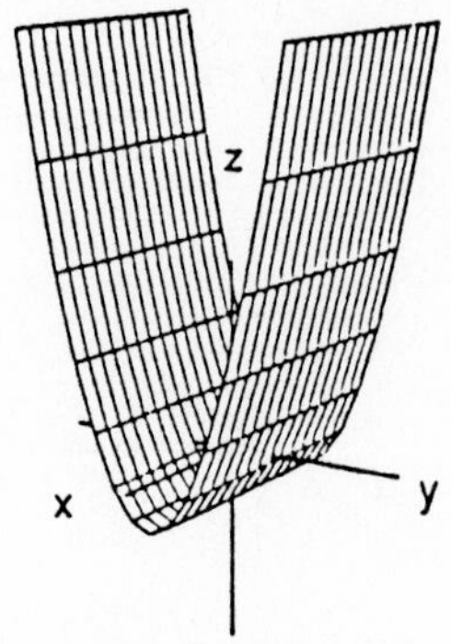

17. $x^2 + 4z^2 = 16$

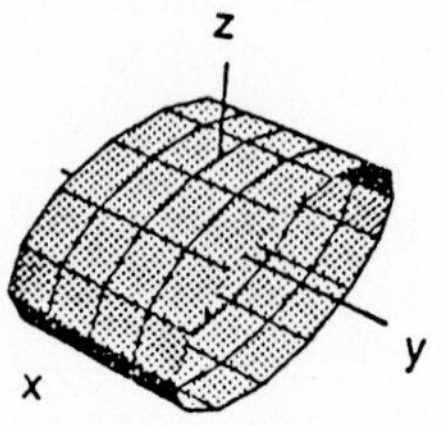

19. $z^2 - y^2 = 1$

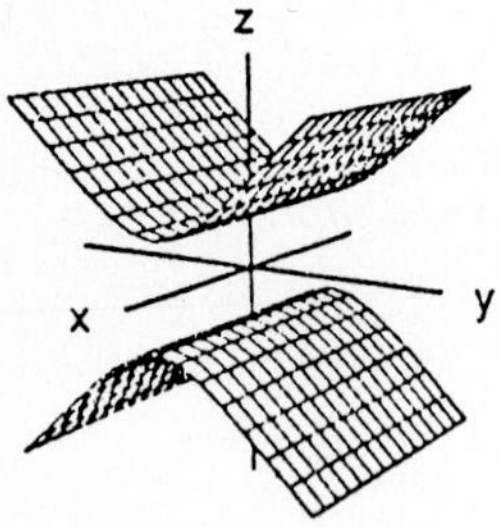

21. $9x^2 + y^2 + z^2 = 9$

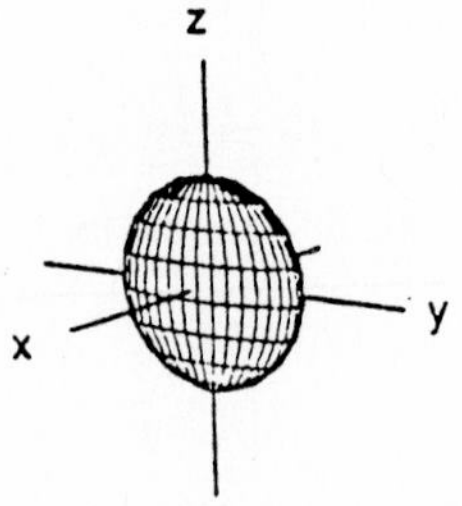

23. $4x^2 + 9y^2 + 4z^2 = 36$

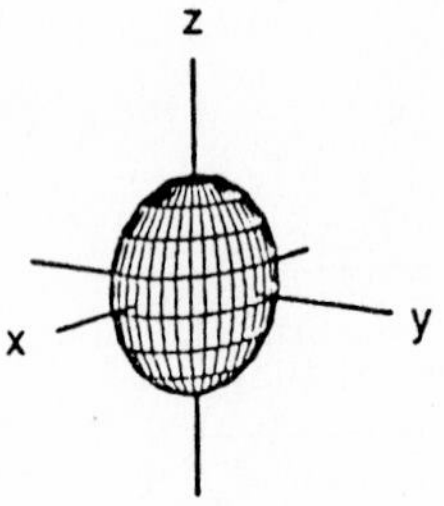

25. $x^2 + 4y^2 = z$

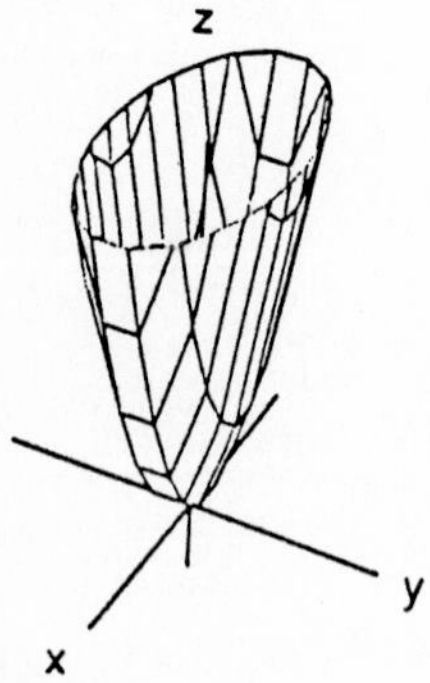

27. $z = 8 - x^2 - y^2$

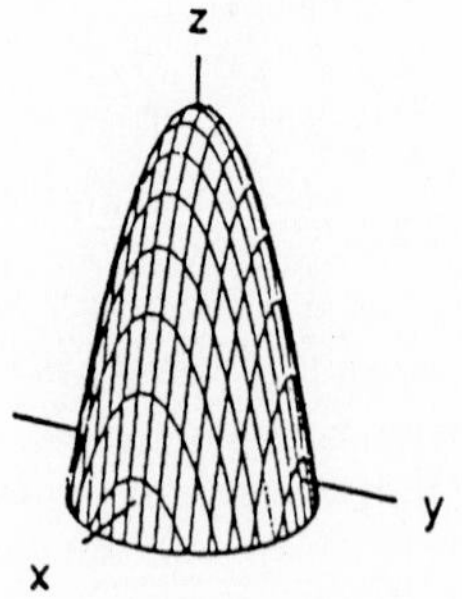

29. $x = 4 - 4y^2 - z^2$

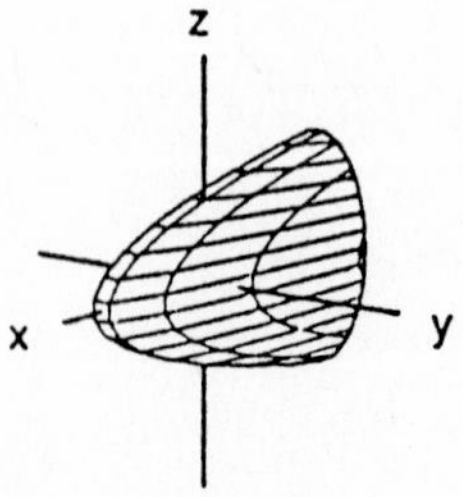

31. $x^2 + y^2 = z^2$

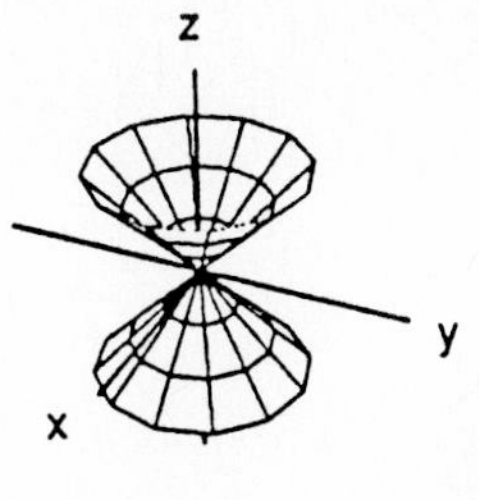

33. $4x^2 + 9z^2 = 9y^2$

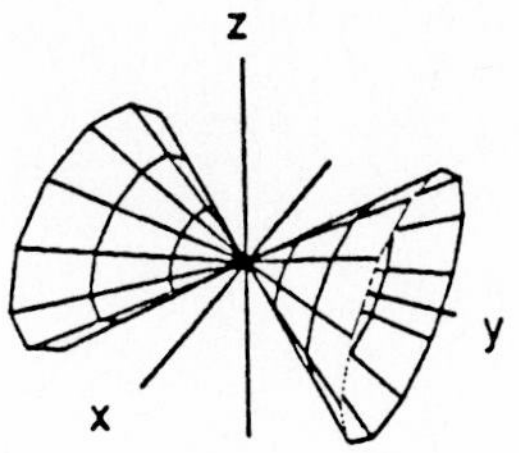

35. $x^2 + y^2 - z^2 = 1$

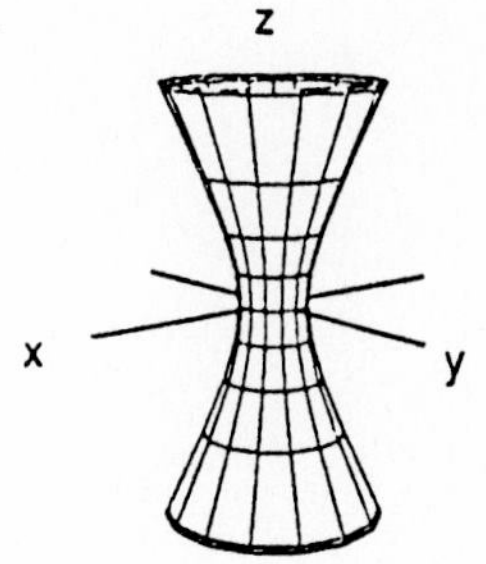

37. $\frac{y^2}{4} + \frac{z^2}{9} - \frac{x^2}{4} = 1$

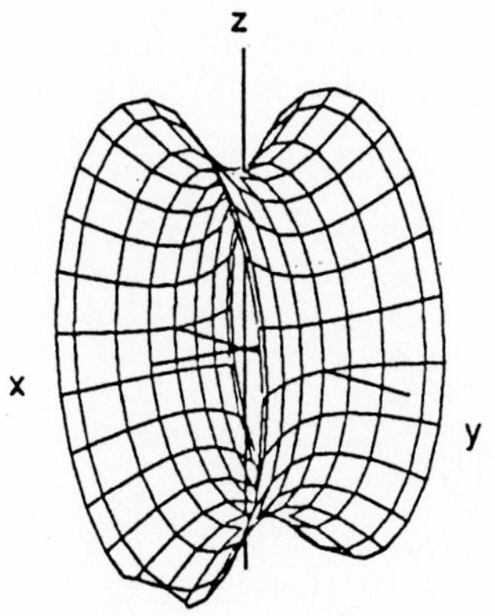

39. $z^2 - x^2 - y^2 = 1$

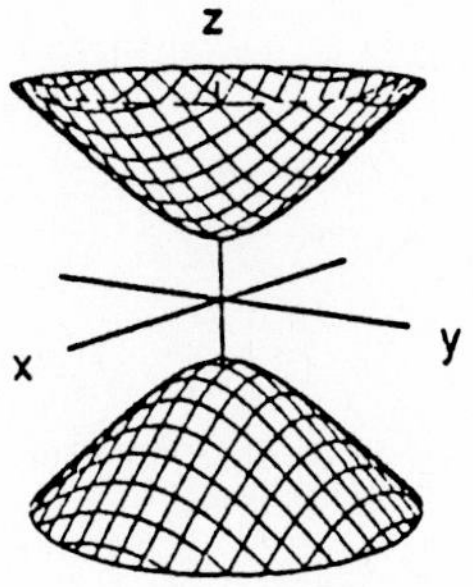

41. $\frac{x^2}{4} - y^2 - \frac{z^2}{4} = 1$

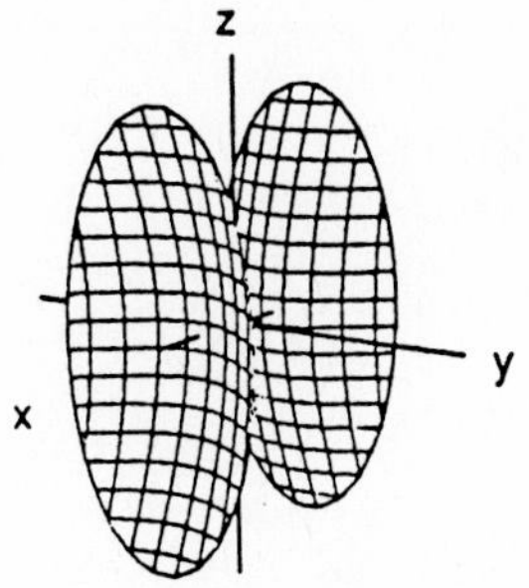

43. $y^2 - x^2 = z$

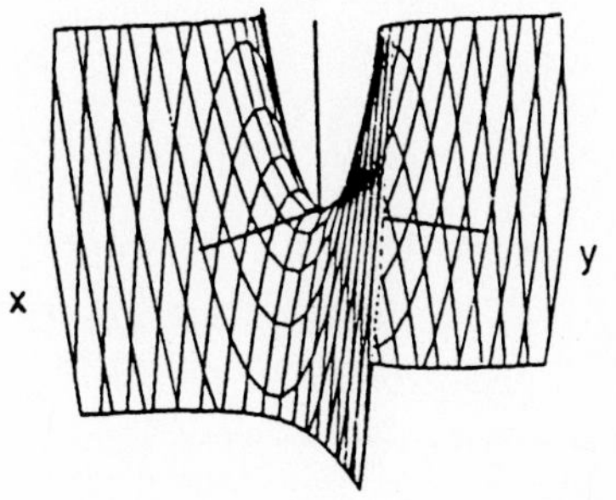

45. $x^2 + y^2 + z^2 = 4$

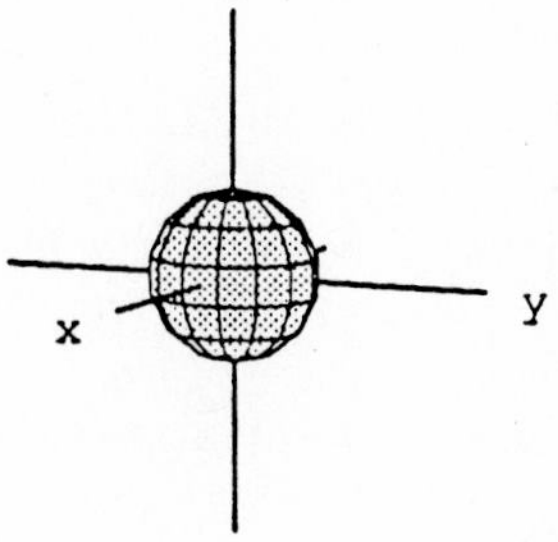

47. $z = 1 + y^2 - x^2$

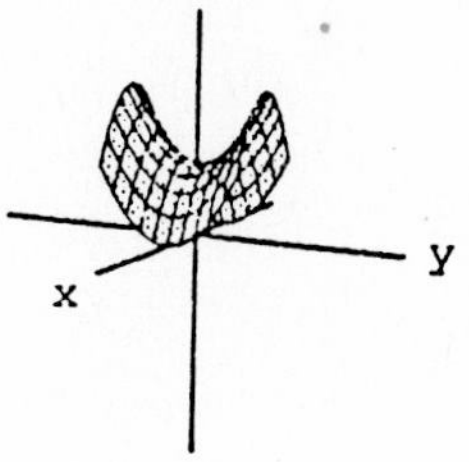

49. $y = -(x^2 + z^2)$

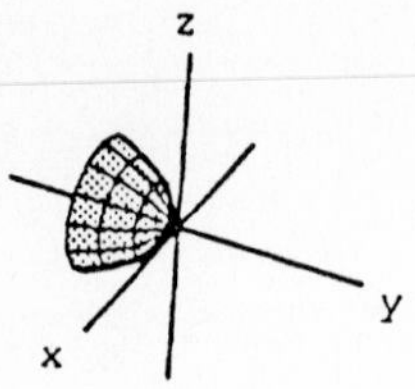

51. $16x^2 + 4y^2 = 1$

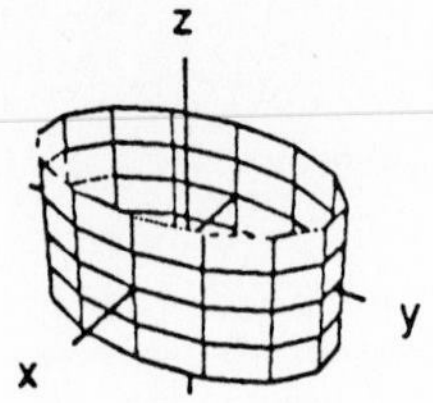

53. $x^2 + y^2 - z^2 = 4$

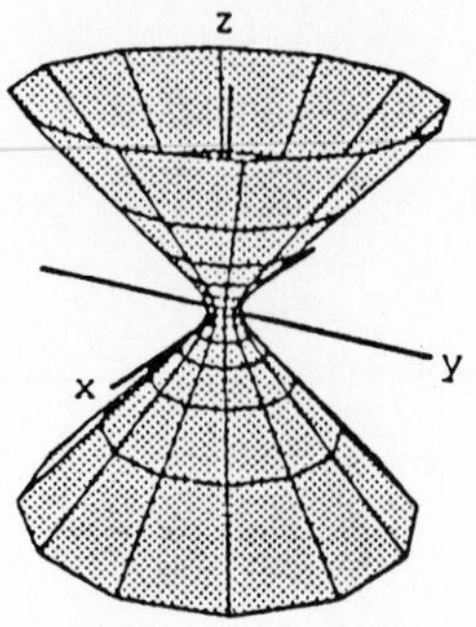

55. $x^2 + z^2 = y$

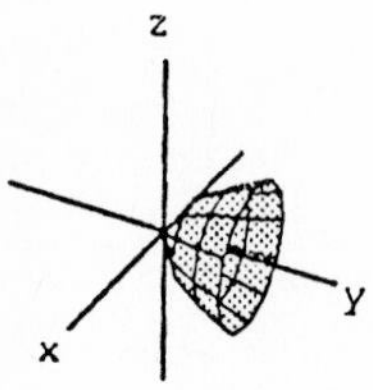

57. $x^2 + z^2 = 1$

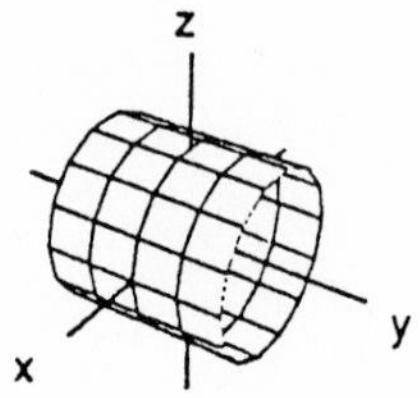

59. $16y^2 + 9z^2 = 4x^2$

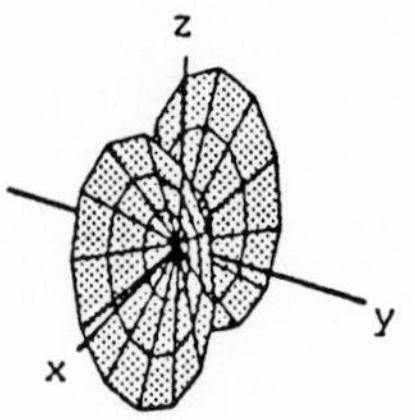

61. $9x^2 + 4y^2 + z^2 = 36$

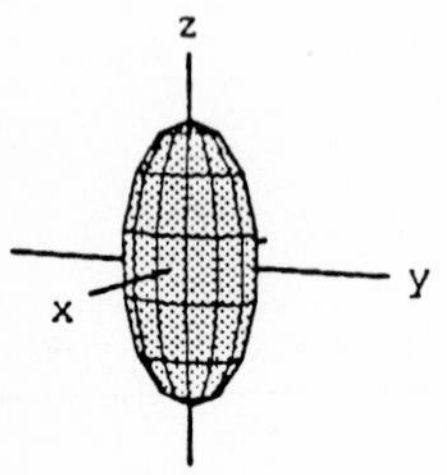

63. $x^2 + y^2 - 16z^2 = 16$

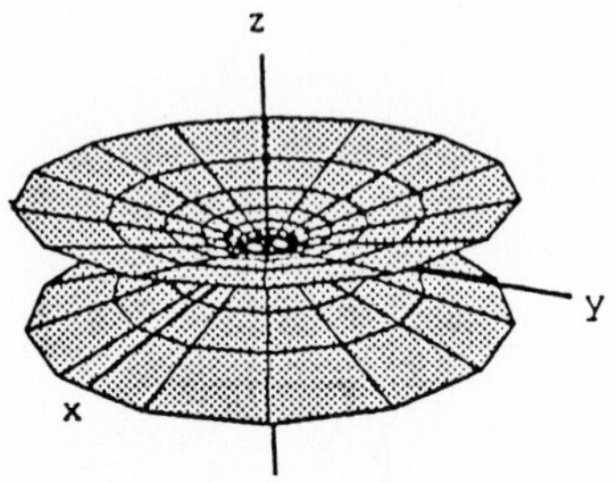

65. $z = -(x^2 + y^2)$

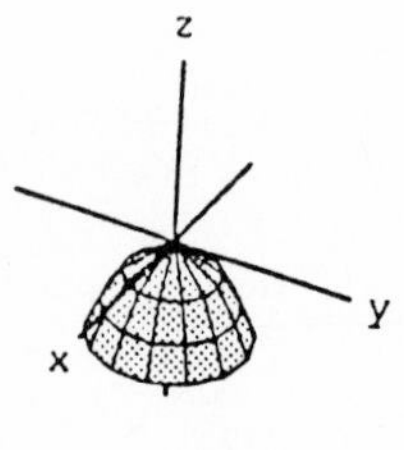

67. $x^2 - 4y^2 = 1$

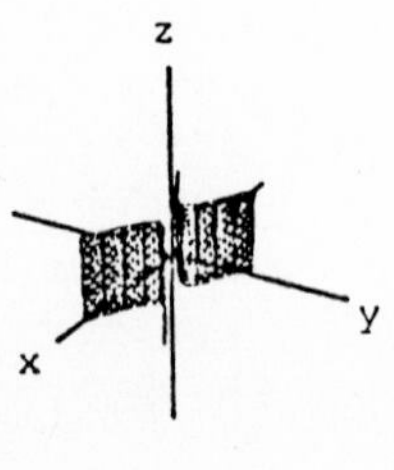

69. $4y^2 + z^2 - 4x^2 = 4$

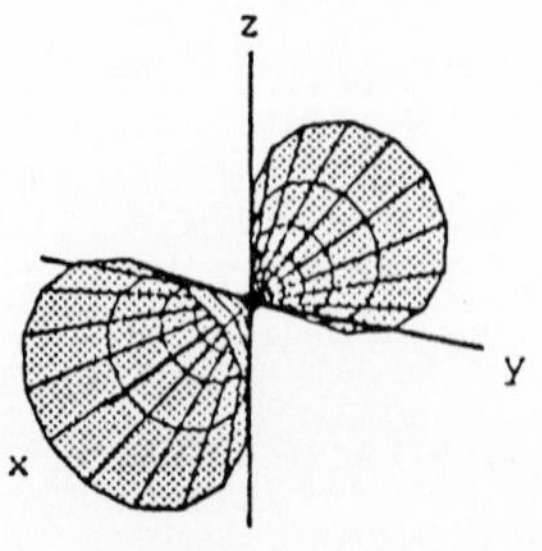

71. $x^2 + y^2 = z$

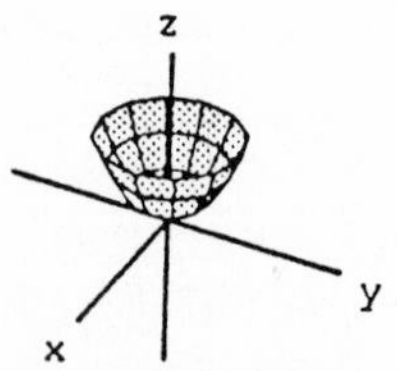

73. $yz = 1$

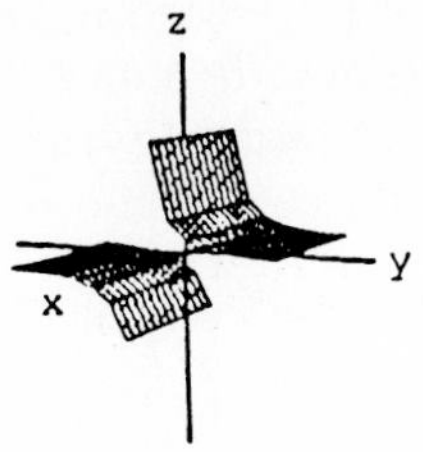

75. $9x^2 + 16y^2 = 4z^2$

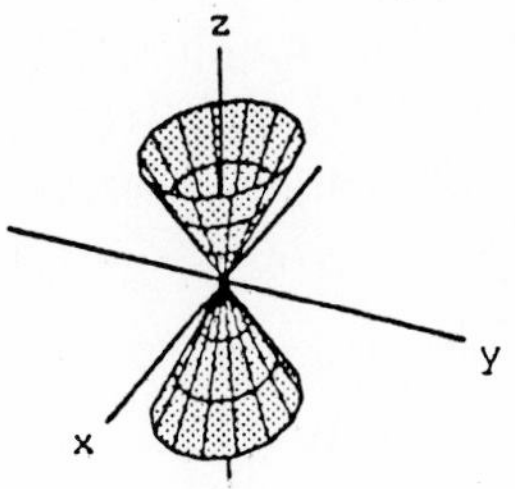

77. (a) If $x^2 + \frac{y^2}{4} + \frac{z^2}{9} = 1$ and $z = c$, then $x^2 + \frac{y^2}{4} = \frac{9 - c^2}{9} \Rightarrow \frac{x^2}{\left(\frac{9-c^2}{9}\right)} + \frac{y^2}{\left[\frac{4(9-c^2)}{9}\right]} = 1 \Rightarrow A = ab\pi$

$= \pi\left(\frac{\sqrt{9-c^2}}{3}\right)\left(\frac{2\sqrt{9-c^2}}{3}\right) = \frac{2\pi(9-c^2)}{9}$

(b) From part (a), each slice has the area $\frac{2\pi(9-z^2)}{9}$, where $-3 \le z \le 3$. Thus $V = 2\int_0^3 \frac{2\pi}{9}(9 - z^2)\,dz$

$= \frac{4\pi}{9}\int_0^3 (9 - z^2)\,dz = \frac{4\pi}{9}\left[9z - \frac{z^3}{3}\right]_0^3 = \frac{4\pi}{9}(27 - 9) = 8\pi$

(c) $\frac{x^2}{a^2} + \frac{y^2}{b^2} + \frac{z^2}{c^2} = 1 \Rightarrow \frac{x^2}{\left[\frac{a^2(c^2-z^2)}{c^2}\right]} + \frac{y^2}{\left[\frac{b^2(c^2-z^2)}{c^2}\right]} = 1 \Rightarrow A = \pi\left(\frac{a\sqrt{c^2-z^2}}{c}\right)\left(\frac{b\sqrt{c^2-z^2}}{c}\right)$

$\Rightarrow V = 2\int_0^c \frac{\pi ab}{c^2}(c^2 - z^2)\,dz = \frac{2\pi ab}{c^2}\left[c^2 z - \frac{z^3}{3}\right]_0^c = \frac{2\pi ab}{c^2}\left(\frac{2}{3}c^3\right) = \frac{4\pi abc}{3}$. Note that if $r = a = b = c$,

then $V = \frac{4\pi r^3}{3}$, which is the volume of a sphere.

79. We calculate the volume by the slicing method, taking slices parallel to the xy-plane. For fixed z, $\frac{x^2}{a^2} + \frac{y^2}{b^2} = \frac{z}{c}$ gives the ellipse $\frac{x^2}{\left(\frac{za^2}{c}\right)} + \frac{y^2}{\left(\frac{zb^2}{c}\right)} = 1$. The area of this ellipse is $\pi\left(a\sqrt{\frac{z}{c}}\right)\left(b\sqrt{\frac{z}{c}}\right) = \frac{\pi abz}{c}$ (see Exercise 77a). Hence the volume is given by $V = \int_0^h \frac{\pi abz}{c}\,dz = \left[\frac{\pi abz^2}{2c}\right]_0^h = \frac{\pi abh^2}{c}$. Now the area of the elliptic base when $z = h$ is $A = \frac{\pi abh}{c}$, as determined previously. Thus, $V = \frac{\pi abh^2}{c} = \frac{1}{2}\left(\frac{\pi abh}{c}\right)h = \frac{1}{2}(\text{base})(\text{altitude})$, as claimed.

81. $y = y_1 \Rightarrow \frac{z}{c} = \frac{y_1^2}{b^2} - \frac{x^2}{a^2}$, a parabola in the plane $y = y_1 \Rightarrow$ vertex when $\frac{dz}{dx} = 0$ or $c\,\frac{dz}{dx} = -\frac{2x}{a^2} = 0 \Rightarrow x = 0$

$\Rightarrow \text{Vertex}\left(0, y_1, \frac{cy_1^2}{b^2}\right)$; writing the parabola as $x^2 = -\frac{a^2}{c}z + \frac{cy_1^2}{b^2}$ we see that $4p = -\frac{a^2}{c} \Rightarrow p = -\frac{a^2}{4c}$

$\Rightarrow \text{Focus}\left(0, y_1, \frac{cy_1^2}{b^2} - \frac{a^2}{4c}\right)$

83. No, it is not mere coincidence. A plane parallel to one of the coordinate planes will set one of the variables x, y, or z equal to a constant in the general equation $Ax^2 + By^2 + Cz^2 + Dxy + Eyz + Fxz + Gx + Hy + Jz + K = 0$ for a quadric surface. The resulting equation then has the general form for a conic in that parallel plane. For example, setting $y = y_1$ results in the equation $Ax^2 + Cz^2 + D'x + E'z + Fxz + Gx + Jz + K' = 0$ where $D' = Dy_1$, $E' = Ey_1$, and $K' = K + By_1^2 + Hy_1$, which is the general form of a conic section in the plane $y = y_1$ by Section 9.3.

85. $z = y^2$

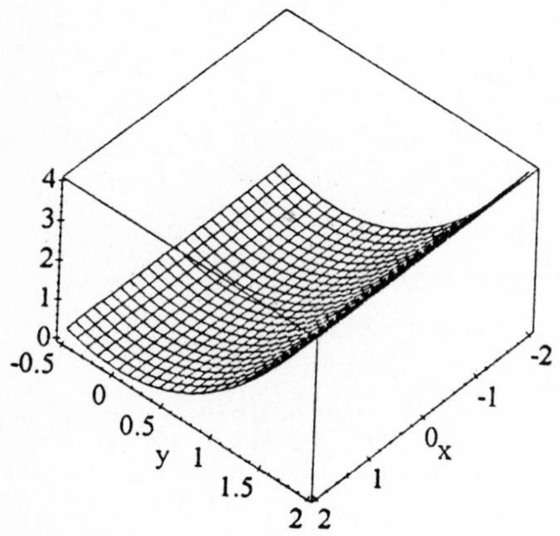

87. $z = x^2 + y^2$

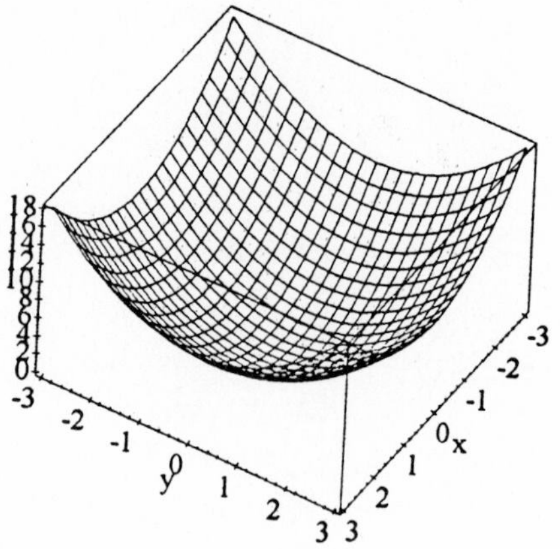

10.7 CYLINDRICAL AND SPHERICAL COORDINATES

	Rectangular	Cylindrical	Spherical
1.	$(0,0,0)$	$(0,0,0)$	$(0,0,0)$
3.	$(0,1,0)$	$\left(1,\frac{\pi}{2},0\right)$	$\left(1,\frac{\pi}{2},\frac{\pi}{2}\right)$
5.	$(1,0,0)$	$(1,0,0)$	$\left(1,\frac{\pi}{2},0\right)$
7.	$(0,1,1)$	$\left(1,\frac{\pi}{2},1\right)$	$\left(\sqrt{2},\frac{\pi}{4},\frac{\pi}{2}\right)$
9.	$\left(0,-2\sqrt{2},0\right)$	$\left(2\sqrt{2},\frac{3\pi}{2},0\right)$	$\left(2\sqrt{2},\frac{\pi}{2},\frac{3\pi}{2}\right)$

11. $r = 0 \Rightarrow$ rectangular, $x^2 + y^2 = 0$; spherical, $\phi = 0$ or $\phi = \pi$; the z-axis

13. $z = 0 \Rightarrow$ cylindrical, $z = 0$; spherical, $\phi = \frac{\pi}{2}$; the xy-plane

15. $z = \sqrt{x^2 + y^2}$, $z \le 1 \Rightarrow$ cylindrical, $z = r$, $0 \le r \le 1$; spherical, $\phi = \tan^{-1} \frac{\sqrt{x^2+y^2}}{z} = \tan^{-1} 1 = \frac{\pi}{4}$,
$\rho = \sqrt{x^2+y^2+z^2} = \sqrt{x^2+y^2+x^2+y^2} = \sqrt{2(x^2+y^2)} = \sqrt{2z^2} = \sqrt{2}\,|z| \le \sqrt{2} \Rightarrow 0 \le \rho \le \sqrt{2}$; a (finite) cone

17. $\rho \sin\phi \cos\theta = 0 \Rightarrow$ rectangular, $x = 0$; cylindrical $\theta = \frac{\pi}{2}$; the yz-plane

19. $x^2 + y^2 + z^2 = 4 \Rightarrow$ cylindrical, $r^2 + z^2 = 4$; spherical, $\rho = 2$; a sphere of radius 2 centered at the origin

21. $\rho = 5\cos\phi \Rightarrow$ rectangular, $\sqrt{x^2+y^2+z^2} = 5\cos\left(\cos^{-1}\left(\frac{z}{\sqrt{x^2+y^2+z^2}}\right)\right) = \sqrt{x^2+y^2+z^2} = \frac{5z}{\sqrt{x^2+y^2+z^2}}$

$\Rightarrow x^2+y^2+z^2 = 5z \Rightarrow x^2+y^2+z^2 - 5z + \frac{25}{4} = \frac{25}{4} \Rightarrow x^2+y^2+\left(z-\frac{5}{2}\right)^2 = \frac{25}{4}$; cylindrical,
$r^2 + \left(z - \frac{5}{2}\right)^2 = \frac{25}{4} \Rightarrow r^2 + z^2 = 5z$, a sphere of radius $\frac{5}{2}$ centered at $\left(0, 0, \frac{5}{2}\right)$ (rectangular)

23. $r = \csc\theta \Rightarrow$ rectangular, $r = \frac{r}{y} \Rightarrow y = 1$ since $r \ne 0$; spherical, $\rho\sin\phi = \csc\theta \Rightarrow \rho\sin\phi\sin\theta = 1$, the plane $y = 1$

25. $\rho = \sqrt{2}\sec\phi \Rightarrow \rho = \frac{\sqrt{2}}{\cos\phi} \Rightarrow$ rectangular, $\sqrt{x^2+y^2+z^2} = \frac{\sqrt{2}}{\cos\left(\cos^{-1}\left(\frac{z}{\sqrt{x^2+y^2+z^2}}\right)\right)}$

$\Rightarrow \sqrt{x^2+y^2+z^2} = \frac{\sqrt{2}}{\left(\frac{z}{\sqrt{x^2+y^2+z^2}}\right)} \Rightarrow z\sqrt{x^2+y^2+z^2} = \sqrt{2}\sqrt{x^2+y^2+z^2} \Rightarrow z = \sqrt{2}$ since $x^2+y^2+z^2 \ne 0$;

cylindrical, $z = \sqrt{2}$, the plane $z = \sqrt{2}$

27. $x^2+y^2+(z-1)^2 = 1$, $z \le 1 \Rightarrow$ cylindrical, $r^2 + (z-1)^2 = 1 \Rightarrow r^2 + z^2 - 2z + 1 = 1 \Rightarrow r^2 + z^2 = 2z$, $z \le 1$;
spherical, $x^2+y^2+z^2-2z = 0 \Rightarrow \rho^2 - 2\rho\cos\phi = 0 \Rightarrow \rho(\rho - 2\cos\phi) = 0 \Rightarrow \rho = 2\cos\phi$, $\frac{\pi}{4} \le \phi \le \frac{\pi}{2}$ since $\rho \ne 0$, the lower half (hemisphere) of the sphere of radius 1 centered at $(0, 0, 1)$ (rectangular)

29. $\rho = 3$, $\frac{\pi}{3} \le \phi \le \frac{2\pi}{3} \Rightarrow$ rectangular, $\sqrt{x^2+y^2+z^2} = 3$ and $3\cos\left(\frac{\pi}{3}\right) \ge z \ge 3\cos\left(\frac{2\pi}{3}\right) \Rightarrow x^2+y^2+z^2 = 9$ and $-\frac{3}{2} \le z \le \frac{3}{2}$; cylindrical, $r^2 + z^2 = 9$ and $-\frac{3}{2} \le z \le \frac{3}{2}$, the portion of the sphere of radius 3 centered at the origin between the planes $z = -\frac{3}{2}$ and $z = \frac{3}{2}$

31. $z = 4 - 4r^2$, $0 \le r \le 1 \Rightarrow$ spherical, $\rho\cos\phi = 4 - 4\rho^2\sin^2\phi$ and $0 \le \phi \le \frac{\pi}{2}$; rectangular $z = 4 - 4(x^2+y^2)$ and $0 \le z \le 4$, the upper portion cut from the paraboloid $z = 4 - 4(x^2+y^2)$ by the xy-plane

33. $\phi = \frac{3\pi}{4}$, $0 \le \rho \le \sqrt{2} \Rightarrow$ rectangular, $\cos\frac{3\pi}{4} = \cos\phi = \frac{z}{\sqrt{x^2+y^2+z^2}} \Rightarrow -\frac{1}{\sqrt{2}} = \frac{z}{\sqrt{x^2+y^2+z^2}}$

$\Rightarrow \sqrt{x^2+y^2+z^2} = -\sqrt{2}z \Rightarrow x^2+y^2+z^2 = 2z^2 \Rightarrow x^2+y^2-z^2 = 0$ with $z \le 0 \Rightarrow z = -\sqrt{x^2+y^2}$ and

$0 \ge z \ge \sqrt{2}\cos\frac{3\pi}{4} \Rightarrow z = -\sqrt{x^2+y^2}$ and $-1 \le z \le 0$; cylindrical $x^2+y^2-z^2 = 0 \Rightarrow r^2 - z^2 = 0$

$\Rightarrow r = -z$ or $r = z$, but $r \geq 0$ and $z \leq 0 \Rightarrow r = -z$, a cone with vertex at the origin and base the circle $x^2 + y^2 = 1$ in the plane $z = -1$

35. $z + r^2 \cos 2\theta = 0 \Rightarrow z + r^2(\cos^2\theta - \sin^2\theta) = 0 \Rightarrow z + (r \cos \theta)^2 - (r \sin \theta)^2 = 0 \Rightarrow$ rectangular, $z + x^2 - y^2 = 0$ or $z = y^2 - x^2$; spherical, $z + r^2 \cos 2\theta = 0 \Rightarrow \rho \cos \phi + (\rho \sin \phi)^2 (\cos 2\theta) = 0 \Rightarrow \rho(\cos \phi + \rho \sin^2\phi \cos 2\theta) = 0$ $\Rightarrow \cos \phi + \rho \sin^2\phi \cos 2\theta = 0$ since $\rho \neq 0$, a hyperbolic paraboloid

37. $r^2 + z^2 = 4r \cos \theta + 6r \sin \theta + 2z \Rightarrow x^2 + y^2 + z^2 = 4x + 6y + 2z \Rightarrow (x^2 - 4x + 4) + (y^2 - 6y + 9) + (z^2 - 2z + 1)$ $= 14 \Rightarrow (x-2)^2 + (y-3)^2 + (z-1)^2 = 14 \Rightarrow$ the center is located at $(2, 3, 1)$ in rectangular coordinates

39. Right circular cylinder parallel to the z-axis generated by the circle $r = -2 \sin \theta$ in the $r\theta$-plane

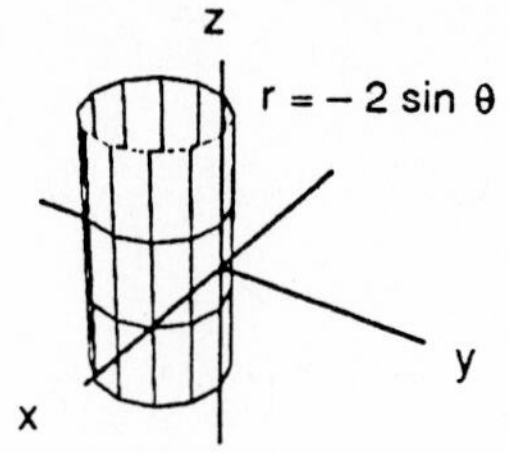

41. Cylinder of lines parallel to the z-axis generated by the cardioid $r = 1 - \cos \theta$ in the $r\theta$-plane

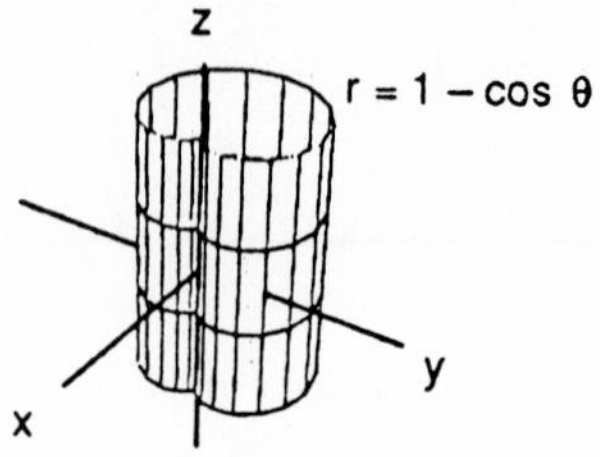

43. Cardioid of revolution symmetric about the y-axis, cusp at the origin pointing down

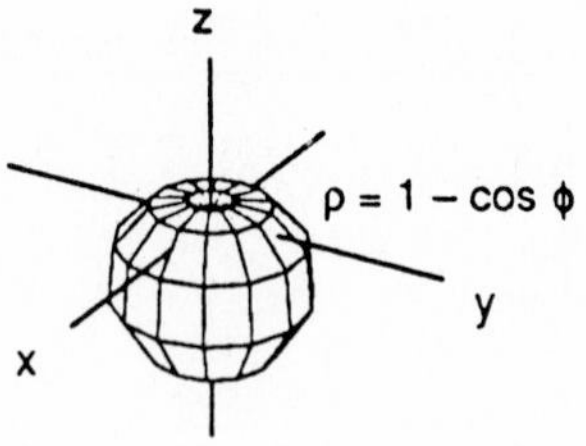

45. (a) $z = c \Rightarrow \rho \cos \phi = c \Rightarrow \rho = \frac{c}{\csc \phi} \Rightarrow \rho = c \sec \phi$

(b) The xy-plane is perpendicular to the z-axis $\Rightarrow \phi = \frac{\pi}{2}$

47. (a) A plane perpendicular to the x-axis has the form $x = a$ in rectangular coordinates $\Rightarrow r \cos \theta = a$ $\Rightarrow r = \frac{a}{\cos \theta} \Rightarrow r = a \sec \theta$, in cylindrical coordinates

(b) A plane perpendicular to the y-axis has the form y = b in rectangular coordinates $\Rightarrow$ r sin θ = b $\Rightarrow r = \frac{b}{\sin\theta} \Rightarrow r = b\csc\theta$, in cylindrical coordinates

49. The equation r = f(z) implies that the point $(r,\theta,z) = (f(z),\theta,z)$ will lie on the surface for all θ. In particular $(f(z),\theta+\pi,z)$ lies on the surface whenever $(f(z),\theta,z)$ does $\Rightarrow$ the surface is symmetric with respect to the z-axis.

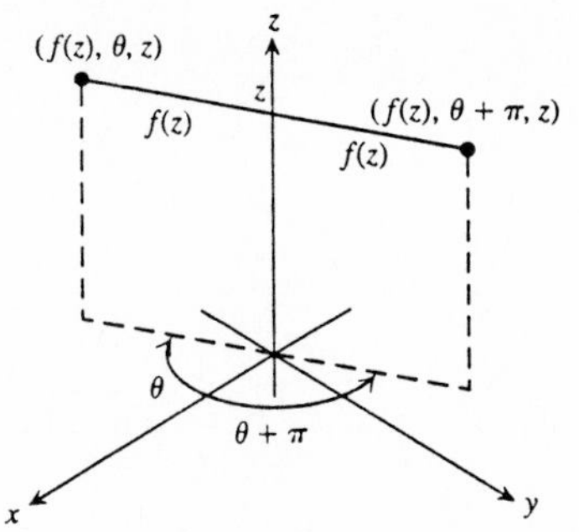

CHAPTER 10 PRACTICE EXERCISES

1. $\theta = 0 \Rightarrow \mathbf{u} = \mathbf{i}$; $\theta = \frac{\pi}{2} \Rightarrow \mathbf{u} = \mathbf{j}$; $\theta = \frac{2\pi}{3} \Rightarrow \mathbf{u} = -\frac{1}{2}\mathbf{i} + \frac{\sqrt{3}}{2}\mathbf{j}$; $\theta = \frac{5\pi}{4} \Rightarrow \mathbf{u} = -\frac{\sqrt{2}}{2}\mathbf{i} - \frac{\sqrt{2}}{2}\mathbf{j}$; $\theta = \frac{5\pi}{3} \Rightarrow \mathbf{u} = \frac{1}{2}\mathbf{i} - \frac{\sqrt{3}}{2}\mathbf{j}$

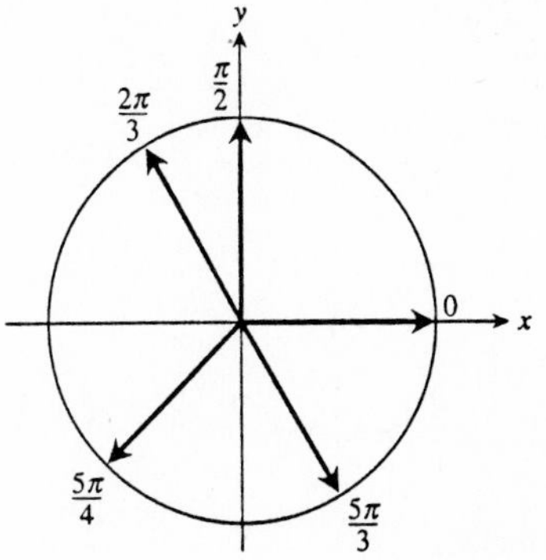

3. length $= \left|\sqrt{2}\mathbf{i} + \sqrt{2}\mathbf{j}\right| = \sqrt{2+2} = 2$, $\sqrt{2}\mathbf{i} + \sqrt{2}\mathbf{j} = 2\left(\frac{1}{\sqrt{2}}\mathbf{i} + \frac{1}{\sqrt{2}}\mathbf{j}\right) \Rightarrow$ the direction is $\frac{1}{\sqrt{2}}\mathbf{i} + \frac{1}{\sqrt{2}}\mathbf{j}$

5. length $= |2\mathbf{i} - 3\mathbf{j} + 6\mathbf{k}| = \sqrt{4+9+36} = 7$, $2\mathbf{i} - 3\mathbf{j} + 6\mathbf{k} = 7\left(\frac{2}{7}\mathbf{i} - \frac{3}{7}\mathbf{j} + \frac{6}{7}\mathbf{k}\right) \Rightarrow$ the direction is $\frac{2}{7}\mathbf{i} - \frac{3}{7}\mathbf{j} + \frac{6}{7}\mathbf{k}$

7. $2\frac{\mathbf{A}}{|\mathbf{A}|} = 2 \cdot \frac{4\mathbf{i} - \mathbf{j} + 4\mathbf{k}}{\sqrt{4^2 + (-1)^2 + 4^2}} = 2 \cdot \frac{4\mathbf{i} - \mathbf{j} + 4\mathbf{k}}{\sqrt{33}} = \frac{8}{\sqrt{33}}\mathbf{i} - \frac{2}{\sqrt{33}}\mathbf{j} + \frac{8}{\sqrt{33}}\mathbf{k}$

9. (a) $\overrightarrow{BD} = \overrightarrow{AD} - \overrightarrow{AB}$

(b) $\overrightarrow{AP} = \overrightarrow{AB} + \frac{1}{2}\overrightarrow{BD} = \overrightarrow{AB} + \frac{1}{2}\left(\overrightarrow{AD} - \overrightarrow{AB}\right) = \overrightarrow{AB} + \frac{1}{2}\overrightarrow{AD} - \frac{1}{2}\overrightarrow{AB} = \frac{1}{2}\overrightarrow{AB} + \frac{1}{2}\overrightarrow{AD} = \frac{1}{2}\left(\overrightarrow{AB} + \overrightarrow{AD}\right)$

(c) $\overrightarrow{PC} = \overrightarrow{AC} - \overrightarrow{AP} = \left(\overrightarrow{AB} + \overrightarrow{AD}\right) - \frac{1}{2}\left(\overrightarrow{AB} + \overrightarrow{AD}\right) = \frac{1}{2}\left(\overrightarrow{AB} + \overrightarrow{AD}\right) = \overrightarrow{AP} \Rightarrow$ P is the midpoint of AC

11.

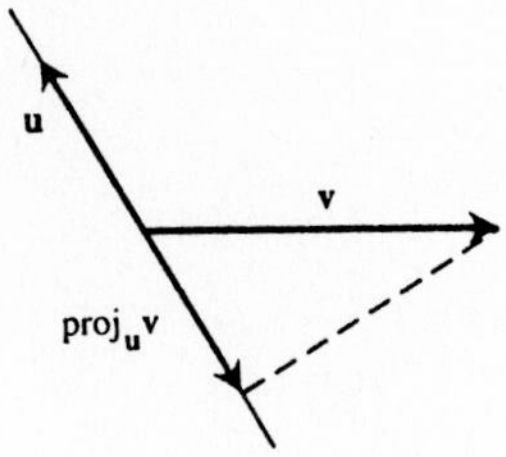

13. $|\mathbf{A}| = \sqrt{1+1} = \sqrt{2}$, $|\mathbf{B}| = \sqrt{4+1+4} = 3$, $\mathbf{A}\cdot\mathbf{B} = 3$, $\mathbf{B}\cdot\mathbf{A} = 3$, $\mathbf{A}\times\mathbf{B} = \begin{vmatrix} \mathbf{i} & \mathbf{j} & \mathbf{k} \\ 1 & 1 & 0 \\ 2 & 1 & -2 \end{vmatrix} = -2\mathbf{i} + 2\mathbf{j} - \mathbf{k}$,

$\mathbf{B}\times\mathbf{A} = -(\mathbf{A}\times\mathbf{B}) = 2\mathbf{i} - 2\mathbf{j} + \mathbf{k}$, $|\mathbf{A}\times\mathbf{B}| = \sqrt{4+4+1} = 3$, $\theta = \cos^{-1}\left(\frac{\mathbf{A}\cdot\mathbf{B}}{|\mathbf{A}||\mathbf{B}|}\right) = \cos^{-1}\left(\frac{1}{\sqrt{2}}\right) = \frac{\pi}{4}$,

$|\mathbf{B}|\cos\theta = \frac{3}{\sqrt{2}}$, $\text{proj}_{\mathbf{A}}\,\mathbf{B} = \left(\frac{\mathbf{A}\cdot\mathbf{B}}{\mathbf{A}\cdot\mathbf{A}}\right)\mathbf{A} = \frac{3}{2}(\mathbf{i}+\mathbf{j})$

15. $\mathbf{B} = \left(\frac{\mathbf{A}\cdot\mathbf{B}}{\mathbf{A}\cdot\mathbf{A}}\right)\mathbf{A} + \left[\mathbf{B} - \left(\frac{\mathbf{A}\cdot\mathbf{B}}{\mathbf{A}\cdot\mathbf{A}}\right)\mathbf{A}\right] = \frac{4}{3}(2\mathbf{i}+\mathbf{j}-\mathbf{k}) + \left[(\mathbf{i}+\mathbf{j}-5\mathbf{k}) - \frac{4}{3}(2\mathbf{i}+\mathbf{j}-\mathbf{k})\right] = \frac{4}{3}(2\mathbf{i}+\mathbf{j}-\mathbf{k}) - \frac{1}{3}(5\mathbf{i}+\mathbf{j}+11\mathbf{k})$, where $\mathbf{A}\cdot\mathbf{B} = 8$ and $\mathbf{A}\cdot\mathbf{A} = 6$

17. $\mathbf{A}\times\mathbf{B} = \begin{vmatrix} \mathbf{i} & \mathbf{j} & \mathbf{k} \\ 1 & 0 & 0 \\ 1 & 1 & 0 \end{vmatrix} = \mathbf{k}$

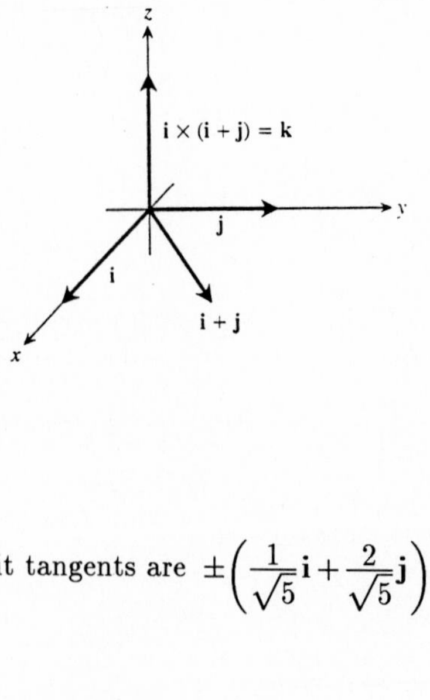

19. $y = \tan x \Rightarrow [y']_{\pi/4} = [\sec^2 x]_{\pi/4} = 2 = \frac{2}{1} \Rightarrow \mathbf{T} = \mathbf{i} + 2\mathbf{j} \Rightarrow$ the unit tangents are $\pm\left(\frac{1}{\sqrt{5}}\mathbf{i} + \frac{2}{\sqrt{5}}\mathbf{j}\right)$ and the unit normals are $\pm\left(-\frac{2}{\sqrt{5}}\mathbf{i} + \frac{1}{\sqrt{5}}\mathbf{j}\right)$

21. Let $\mathbf{A} = a_1\mathbf{i} + a_2\mathbf{j} + a_3\mathbf{k}$ and $\mathbf{B} = b_1\mathbf{i} + b_2\mathbf{j} + b_3\mathbf{k}$. Then $|\mathbf{A}+\mathbf{B}|^2 + |\mathbf{A}-\mathbf{B}|^2$

$= \left[(a_1+b_1)^2 + (a_2+b_2)^2 + (a_3+b_3)^2\right] + \left[(a_1-b_1)^2 + (a_2-b_2)^2 + (a_3-b_3)^2\right]$

$= \left(a_1^2 + 2a_1b_1 + b_1^2 + a_2^2 + 2a_2b_2 + b_2^2 + a_3^2 + 2a_3b_3 + b_3^2\right)$

$+ \left(a_1^2 - 2a_1b_1 + b_1^2 + a_2^2 - 2a_2b_2 + b_2^2 + a_3^2 - 2a_3b_3 + b_3^2\right)$

$= 2\left(a_1^2 + a_2^2 + a_3^2\right) + 2\left(b_1^2 + b_2^2 + b_3^2\right) = 2\,|\mathbf{A}|^2 + 2\,|\mathbf{B}|^2.$

23. Let $\mathbf{v} = v_1\mathbf{i} + v_2\mathbf{j} + v_3\mathbf{k}$ and $\mathbf{w} = w_1\mathbf{i} + w_2\mathbf{j} + w_3\mathbf{k}$. Then $|\mathbf{v} - 2\mathbf{w}|^2 = |(v_1\mathbf{i} + v_2\mathbf{j} + v_3\mathbf{k}) - 2(w_1\mathbf{i} + w_2\mathbf{j} + w_3\mathbf{k})|^2$

$= |(v_1 - 2w_1)\mathbf{i} + (v_2 - 2w_2)\mathbf{j} + (v_3 - 2w_3)\mathbf{k}|^2 = \left(\sqrt{(v_1 - 2w_1)^2 + (v_2 - 2w_2)^2 + (v_3 - 2w_3)^2}\right)^2$

$= \left(v_1^2 + v_2^2 + v_3^2\right) - 4(v_1w_1 + v_2w_2 + v_3w_3) + 4\left(w_1^2 + w_2^2 + w_3^2\right) = |\mathbf{v}|^2 - 4\mathbf{v}\cdot\mathbf{w} + 4\,|\mathbf{w}|^2$

$= |\mathbf{v}|^2 - 4\,|\mathbf{u}|\,|\mathbf{w}|\cos\theta + 4\,|\mathbf{w}|^2 = 4 - 4(2)(3)\left(\cos\frac{\pi}{3}\right) + 36 = 40 - 24\left(\frac{1}{2}\right) = 40 - 12 = 28 \Rightarrow |\mathbf{v} - 2\mathbf{w}| = \sqrt{28}$

$= 2\sqrt{7}$

25. (a) $\text{area} = |\mathbf{A}\times\mathbf{B}| = \text{abs}\begin{vmatrix} \mathbf{i} & \mathbf{j} & \mathbf{k} \\ 1 & 1 & -1 \\ 2 & 1 & 1 \end{vmatrix} = |2\mathbf{i} - 3\mathbf{j} - \mathbf{k}| = \sqrt{4 + 9 + 1} = \sqrt{14}$

(b) $\text{volume} = \mathbf{A}\cdot(\mathbf{B}\times\mathbf{C}) = \begin{vmatrix} 1 & 1 & -1 \\ 2 & 1 & 1 \\ -1 & -2 & 3 \end{vmatrix} = 1(3 + 2) + 1(-1 - 6) - 1(-4 + 1) = 1$

27. The desired vector is $\mathbf{n}\times\mathbf{v}$ or $\mathbf{v}\times\mathbf{n}$ since $\mathbf{n}\times\mathbf{v}$ is perpendicular to both $\mathbf{n}$ and $\mathbf{v}$ and, therefore, also parallel to the plane.

29. The line L passes through the point $P(0, 0, -1)$ parallel to $\mathbf{v} = -\mathbf{i} + \mathbf{j} + \mathbf{k}$. With $\overrightarrow{PS} = 2\mathbf{i} + 2\mathbf{j} + \mathbf{k}$ and

$\overrightarrow{PS}\times\mathbf{v} = \begin{vmatrix} \mathbf{i} & \mathbf{j} & \mathbf{k} \\ 2 & 2 & 1 \\ -1 & 1 & 1 \end{vmatrix} = (2 - 1)\mathbf{i} + (-1 - 2)\mathbf{j} + (2 + 2)\mathbf{k} = \mathbf{i} - 3\mathbf{j} + 4\mathbf{k}$, we find the distance

$d = \frac{|\overrightarrow{PS}\times\mathbf{v}|}{|\mathbf{v}|} = \frac{\sqrt{1 + 9 + 16}}{\sqrt{1 + 1 + 1}} = \frac{\sqrt{26}}{\sqrt{3}} = \frac{\sqrt{78}}{3}.$

31. Parametric equations for the line are $x = 1 - 3t$, $y = 2$, $z = 3 + 7t$.

33. The point $P(4, 0, 0)$ lies on the plane $x - y = 4$, and $\overrightarrow{PS} = (6 - 4)\mathbf{i} + 0\mathbf{j} + (-6 + 0)\mathbf{k} = 2\mathbf{i} - 6\mathbf{k}$ with $\mathbf{n} = \mathbf{i} - \mathbf{j}$

$\Rightarrow d = \frac{|\mathbf{n}\cdot\overrightarrow{PS}|}{|\mathbf{n}|} = \left|\frac{2 + 0 + 0}{\sqrt{1 + 1 + 0}}\right| = \frac{2}{\sqrt{2}} = \sqrt{2}.$

35. $P(3, -2, 1)$ and $\mathbf{n} = 2\mathbf{i} + \mathbf{j} - \mathbf{k} \Rightarrow (2)(x - 3) + (1)(y - (-2)) + (-1)(z - 1) = 0 \Rightarrow 2x + y - z = 3$

37. $P(1,-1,2)$, $Q(2,1,3)$ and $R(-1,2,-1) \Rightarrow \overrightarrow{PQ} = \mathbf{i} + 2\mathbf{j} + \mathbf{k}$, $\overrightarrow{PR} = -2\mathbf{i} + 3\mathbf{j} - 3\mathbf{k}$ and $\overrightarrow{PQ} \times \overrightarrow{PR}$

$$= \begin{vmatrix} \mathbf{i} & \mathbf{j} & \mathbf{k} \\ 1 & 2 & 1 \\ -2 & 3 & -3 \end{vmatrix} = -9\mathbf{i} + \mathbf{j} + 7\mathbf{k} \text{ is normal to the plane} \Rightarrow (-9)(x-1) + (1)(y+1) + (7)(z-2) = 0$$

$\Rightarrow -9x + y + 7z = 4$

39. $\left(0, -\frac{1}{2}, -\frac{3}{2}\right)$, since $t = -\frac{1}{2}$, $y = -\frac{1}{2}$ and $z = -\frac{3}{2}$ when $x = 0$; $(-1, 0, -3)$, since $t = -1$, $x = -1$ and $z = -3$ when $y = 0$; $(1, -1, 0)$, since $t = 0$, $x = 1$ and $y = -1$ when $z = 0$

41. $\mathbf{n}_1 = \mathbf{i}$ and $\mathbf{n}_2 = \mathbf{i} + \mathbf{j} + \sqrt{2}\mathbf{k} \Rightarrow$ the desired angle is $\cos^{-1}\left(\frac{\mathbf{n}_1 \cdot \mathbf{n}_2}{|\mathbf{n}_1||\mathbf{n}_2|}\right) = \cos^{-1}\left(\frac{1}{2}\right) = \frac{\pi}{3}$

43. The direction of the line is $\mathbf{n}_1 \times \mathbf{n}_2 = \begin{vmatrix} \mathbf{i} & \mathbf{j} & \mathbf{k} \\ 1 & 2 & 1 \\ 1 & -1 & 2 \end{vmatrix} = 5\mathbf{i} - \mathbf{j} - 3\mathbf{k}$. Since the point $(-5, 3, 0)$ is on both planes, the desired line is $x = -5 + 5t$, $y = 3 - t$, $z = -3t$.

45. (a) The corresponding normals are $\mathbf{n}_1 = 3\mathbf{i} + 6\mathbf{k}$ and $\mathbf{n}_2 = 2\mathbf{i} + 2\mathbf{j} - \mathbf{k}$ and since $\mathbf{n}_1 \cdot \mathbf{n}_2 = (3)(2) + (0)(2) + (6)(-1) = 6 + 0 - 6 = 0$, we have that the planes are orthogonal

(b) The line of intersection is parallel to $\mathbf{n}_1 \times \mathbf{n}_2 = \begin{vmatrix} \mathbf{i} & \mathbf{j} & \mathbf{k} \\ 3 & 0 & 6 \\ 2 & 2 & -1 \end{vmatrix} = -12\mathbf{i} + 15\mathbf{j} + 6\mathbf{k}$. Now to find a point in the intersection, solve $\begin{cases} 3x + 6z = 1 \\ 2x + 2y - z = 3 \end{cases} \Rightarrow \begin{cases} 3x + 6z = 1 \\ 12x + 12y - 6z = 18 \end{cases} \Rightarrow 15x + 12y = 19 \Rightarrow x = 0$ and $y = \frac{19}{12}$ $\Rightarrow \left(0, \frac{19}{12}, \frac{1}{6}\right)$ is a point on the line we seek. Therefore, the line is $x = -12t$, $y = \frac{19}{12} + 15t$ and $z = \frac{1}{6} + 6t$.

47. Yes; $\mathbf{v} \cdot \mathbf{n} = (2\mathbf{i} - 4\mathbf{j} + \mathbf{k}) \cdot (2\mathbf{i} + \mathbf{j} + 0\mathbf{k}) = 2 \cdot 2 - 4 \cdot 1 + 1 \cdot 0 = 0 \Rightarrow$ the vector is orthogonal to the plane's normal $\Rightarrow \mathbf{v}$ is parallel to the plane

49. A normal to the plane is $\mathbf{n} = \overrightarrow{AB} \times \overrightarrow{AC} = \begin{vmatrix} \mathbf{i} & \mathbf{j} & \mathbf{k} \\ 2 & 0 & -1 \\ 2 & -1 & 0 \end{vmatrix} = -\mathbf{i} - 2\mathbf{j} - 2\mathbf{k} \Rightarrow$ the distance is $d = \left|\frac{\overrightarrow{AP} \cdot \mathbf{n}}{\mathbf{n}}\right|$

$$= \left|\frac{(\mathbf{i} + 4\mathbf{j}) \cdot (-\mathbf{i} - 2\mathbf{j} - 2\mathbf{k})}{\sqrt{1 + 4 + 4}}\right| = \left|\frac{-1 - 8 + 0}{3}\right| = 3$$

51. $\mathbf{n} = 2\mathbf{i} - \mathbf{j} - \mathbf{k}$ is normal to the plane $\Rightarrow \mathbf{n} \times \mathbf{v} = \begin{vmatrix} \mathbf{i} & \mathbf{j} & \mathbf{k} \\ 2 & -1 & -1 \\ 1 & 1 & 1 \end{vmatrix} = 0\mathbf{i} - 3\mathbf{j} + 3\mathbf{k} = -3\mathbf{j} + 3\mathbf{k}$ is orthogonal to $\mathbf{v}$ and parallel to the plane

53. A vector parallel to the line of intersection is $\mathbf{v} = \mathbf{n}_1 \times \mathbf{n}_2 = \begin{vmatrix} \mathbf{i} & \mathbf{j} & \mathbf{k} \\ 1 & 2 & 1 \\ 1 & -1 & 2 \end{vmatrix} = 5\mathbf{i} - \mathbf{j} - 3\mathbf{k}$

$\Rightarrow |\mathbf{v}| = \sqrt{25 + 1 + 9} = \sqrt{35} \Rightarrow 2\left(\frac{\mathbf{v}}{|\mathbf{v}|}\right) = \frac{2}{\sqrt{35}}(5\mathbf{i} - \mathbf{j} - 3\mathbf{k})$ is the desired vector.

55. The line is represented by $x = 3 + 2t$, $y = 2 - t$, and $z = 1 + 2t$. It meets the plane $2x - y + 2z = -2$ when $2(3 + 2t) - (2 - t) + 2(1 + 2t) = -2 \Rightarrow t = -\frac{8}{9} \Rightarrow$ the point is $\left(\frac{11}{9}, \frac{26}{9}, -\frac{7}{9}\right)$.

57. The intersection occurs when $(3 + 2t) + 3(2t) - t = -4 \Rightarrow t = -1 \Rightarrow$ the point is $(1, -2, -1)$. The required line must be perpendicular to both the given line and to the normal, and hence is parallel to $\begin{vmatrix} \mathbf{i} & \mathbf{j} & \mathbf{k} \\ 2 & 2 & 1 \\ 1 & 3 & -1 \end{vmatrix}$

$= -5\mathbf{i} + 3\mathbf{j} + 4\mathbf{k} \Rightarrow$ the line is represented by $x = 1 - 5t$, $y = -2 + 3t$, and $z = -1 + 4t$.

59. The vector $\overrightarrow{AB} \times \overrightarrow{CD} = \begin{vmatrix} \mathbf{i} & \mathbf{j} & \mathbf{k} \\ 3 & -2 & 4 \\ \frac{26}{5} & 0 & -\frac{26}{5} \end{vmatrix} = \frac{26}{5}(2\mathbf{i} + 7\mathbf{j} + 2\mathbf{k})$ is normal to the plane and $A(-2, 0, -3)$ lies on the plane $\Rightarrow 2(x + 2) + 7(y - 0) + 2(z - (-3)) = 0 \Rightarrow 2x + 7y + 2z + 10 = 0$ is an equation of the plane.

61. The vector $\overrightarrow{PQ} \times \overrightarrow{PR} = \begin{vmatrix} \mathbf{i} & \mathbf{j} & \mathbf{k} \\ 2 & -1 & 3 \\ -3 & 0 & 1 \end{vmatrix} = -\mathbf{i} - 11\mathbf{j} - 3\mathbf{k}$ is normal to the plane.

(a) No, the plane is not orthogonal to $\overrightarrow{PQ} \times \overrightarrow{PR}$.
(b) No, these equations represent a line, not a plane.
(c) No, the plane $(x + 2) + 11(y - 1) - 3z = 0$ has normal $\mathbf{i} + 11\mathbf{j} - 3\mathbf{k}$ which is not parallel to $\overrightarrow{PQ} \times \overrightarrow{PR}$.
(d) No, this vector equation is equivalent to the equations $3y + 3z = 3$, $3x - 2z = -6$, and $3x + 2y = -4$ $\Rightarrow x = -\frac{4}{3} - \frac{2}{3}t$, $y = t$, $z = 1 - t$, which represents a line, not a plane.
(e) Yes, this is a plane containing the point $R(-2, 1, 0)$ with normal $\overrightarrow{PQ} \times \overrightarrow{PR}$.

63. $\vec{AB} = -2\mathbf{i}+\mathbf{j}+\mathbf{k}$, $\vec{CD} = \mathbf{i}+4\mathbf{j}-\mathbf{k}$, and $\vec{AC} = 2\mathbf{i}+\mathbf{j} \Rightarrow \mathbf{n} = \begin{vmatrix} \mathbf{i} & \mathbf{j} & \mathbf{k} \\ -2 & 1 & 1 \\ 1 & 4 & -1 \end{vmatrix} = -5\mathbf{i}-\mathbf{j}-9\mathbf{k} \Rightarrow$ the distance is

$$d = \left| \frac{(2\mathbf{i}+\mathbf{j})\cdot(-5\mathbf{i}-\mathbf{j}-9\mathbf{k})}{\sqrt{25+1+81}} \right| = \frac{11}{\sqrt{107}}$$

65. $x^2+y^2+z^2=4$

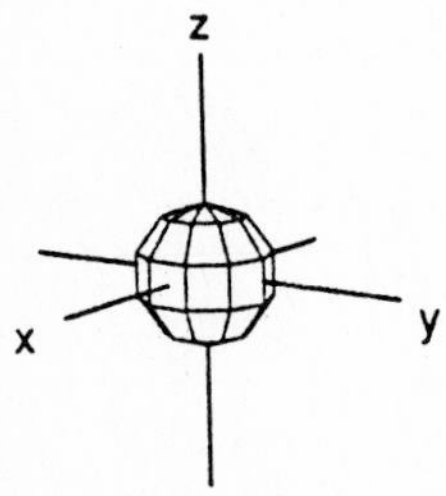

67. $4x^2+4y^2+z^2=4$

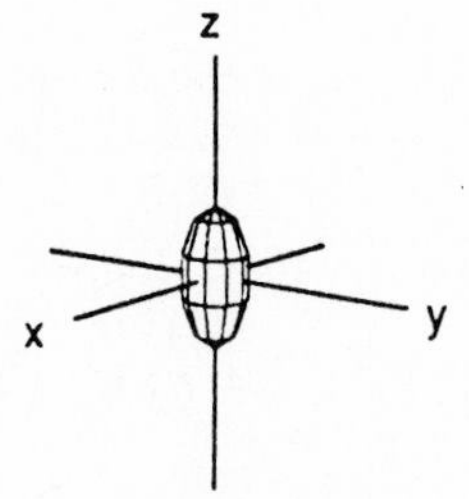

69. $z=-\left(x^2+y^2\right)$

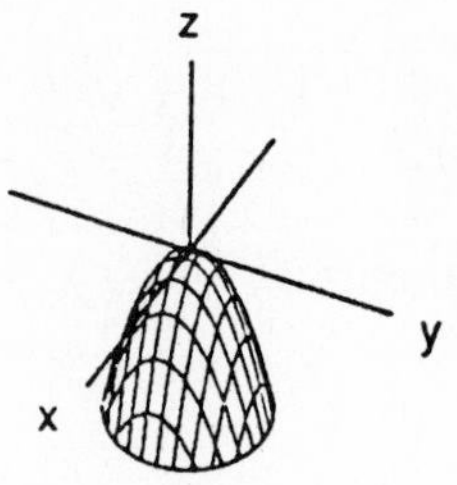

71. $x^2+y^2=z^2$

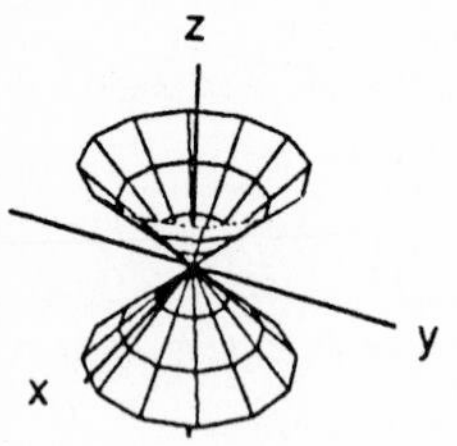

73. $x^2+y^2-z^2=4$

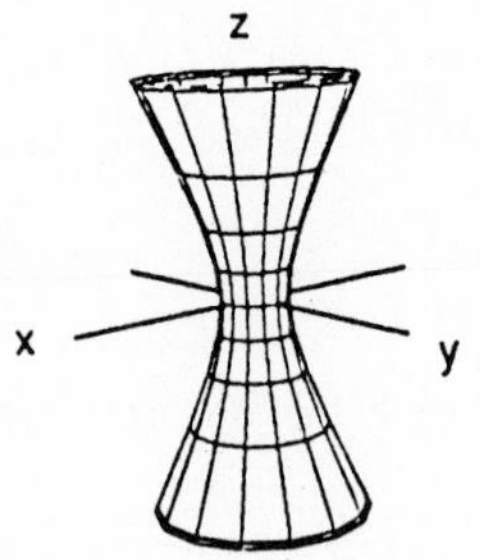

75. $y^2-x^2-z^2=1$

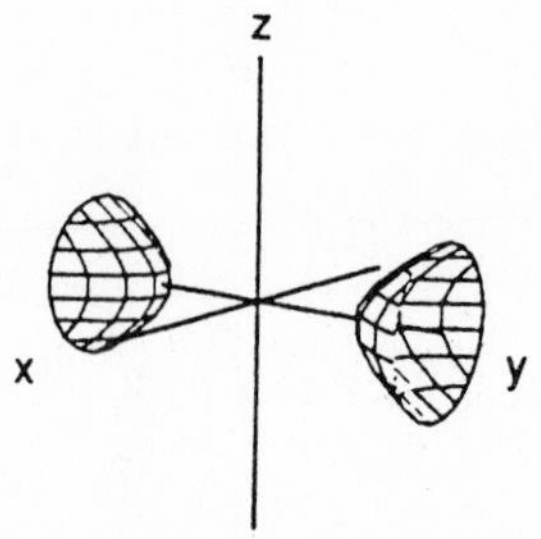

77. The y-axis in the xy-plane; the yz-plane in three dimensional space

79. The circle centered at $(0,0)$ with radius 2 in the xy-plane; the cylinder parallel to the z-axis in three dimensional space with the circle as a generating curve

81. The parabola $x=y^2$ in the xy-plane; the cylinder parallel to the z-axis in three dimensional space with the parabola as a generating curve

83. A cardioid in the $r\theta$-plane; a cylinder parallel to the z-axis in three dimensional space with the cardioid as a generating curve

85. A horizontal lemniscate of length $2\sqrt{2}$ in the $r\theta$-plane; the cylinder parallel to the z-axis in three dimensional space with the lemniscate as a generating curve

87. The sphere of radius 2 centered at the origin

89. The upper nappe of a cone having its vertex at the origin and making an angle of $\frac{\pi}{6}$ rad with the z-axis

91. The upper hemisphere of the sphere of radius 1 centered at the origin

	Rectangular	Cylindrical	Spherical
93.	$(1,0,0)$	$(1,0,0)$	$\left(1,\frac{\pi}{2},0\right)$
95.	$(0,1,1)$	$\left(1,\frac{\pi}{2},1\right)$	$\left(\sqrt{2},\frac{\pi}{4},\frac{\pi}{2}\right)$
97.	$(-1,0,-1)$	$(1,\pi,-1)$	$\left(\sqrt{2},\frac{3\pi}{4},\pi\right)$

99. $z = 2 \Rightarrow$ cylindrical, $z = 2$; spherical, $\rho \cos \phi = 2$; a plane parallel to the xy-plane

101. $x^2 + y^2 + (z+1)^2 = 1 \Rightarrow$ cylindrical, $r^2 + (z+1)^2 = 1 \Rightarrow r^2 + z^2 + 2z + 1 = 1 \Rightarrow r^2 + z^2 = -2z$; spherical, $x^2 + y^2 + z^2 + 2z = 0 \Rightarrow \rho^2 + 2\rho \cos \phi = 0 \Rightarrow \rho(\rho + 2 \cos \phi) = 0 \Rightarrow \rho = -2 \cos \phi$ (since $\rho \neq 0$), a sphere of radius 1 centered at $(0,0,-1)$ (rectangular)

103. $z = r^2 \Rightarrow$ rectangular, $z = x^2 + y^2$; spherical, $\rho \cos \phi = \rho^2 \sin^2 \phi \Rightarrow \rho^2 \sin^2 \phi - \rho \cos \phi = 0$ $\Rightarrow \rho\left(\rho \sin^2 \phi - \cos \phi\right) = 0 \Rightarrow \rho = \frac{\cos \phi}{\sin^2 \phi}$, $0 < \phi \leq \frac{\pi}{2}$ (since $\rho \neq 0$ unless $\phi = \pi$), a circular paraboloid symmetric to the z-axis opening upward with vertex at the origin

105. $r = 7 \sin \theta \Rightarrow$ rectangular, $r = 7 \sin \theta \Rightarrow r = 7\left(\frac{y}{r}\right) \Rightarrow r^2 = 7y \Rightarrow x^2 + y^2 - 7y = 0 \Rightarrow x^2 + y^2 - 7y + \frac{49}{4}$ $= \frac{49}{4} \Rightarrow x^2 + \left(y - \frac{7}{2}\right)^2 = \frac{49}{4}$; spherical, $r = 7 \sin \theta \Rightarrow \rho \sin \phi = 7 \sin \theta$, a circular cylinder parallel to the z-axis generated by the circle

107. $\rho = 4 \Rightarrow$ rectangular, $\sqrt{x^2 + y^2 + z^2} = 4 \Rightarrow x^2 + y^2 + z^2 = 16$; cylindrical, $r^2 + z^2 = 16$, a sphere of radius 4 centered at the origin

109. $\phi = \frac{3\pi}{4} \Rightarrow$ cylindrical, $\tan^{-1}\left(\frac{r}{z}\right) = \frac{3\pi}{4} \Rightarrow \frac{r}{z} = -1 \Rightarrow z = -r$, $r \geq 0$; rectangular, $z = -\sqrt{x^2 + y^2}$, the lower nappe of a cone making an angle of $\frac{3\pi}{4}$ with the positive z-axis and having vertex at the origin

CHAPTER 10 ADDITIONAL EXERCISES–THEORY, EXAMPLES, APPLICATIONS

1. Information from ship A indicates the submarine is now on the line L_1: $x = 4 + 2t$, $y = 3t$, $z = -\frac{1}{3}t$; information from ship B indicates the submarine is now on the line L_2: $x = 18s$, $y = 5 - 6s$, $z = -s$. The current position of the sub is $\left(6, 3, -\frac{1}{3}\right)$ and occurs when the lines intersect at $t = 1$ and $s = \frac{1}{3}$. The straight line path of the submarine contains both points $P\left(2, -1, -\frac{1}{3}\right)$ and $Q\left(6, 3, -\frac{1}{3}\right)$; the line representing this path is L: $x = 2 + 4t$, $y = -1 + 4t$, $z = -\frac{1}{3}$. The submarine traveled the distance between P and Q in 4 minutes $\Rightarrow$ a speed of $\frac{\left|\overrightarrow{PQ}\right|}{4} = \frac{\sqrt{32}}{4} = \sqrt{2}$ thousand ft/min. In 20 minutes the submarine will move $20\sqrt{2}$ thousand ft from

Q along the line L $\Rightarrow 20\sqrt{2} = \sqrt{(2+4t-6)^2 + (-1+4t-3)^2 + 0^2} \Rightarrow 800 = 16(t-1)^2 + 16(t-1)^2 = 32(t-1)^2$

$\Rightarrow (t-1)^2 = \frac{800}{32} = 25 \Rightarrow t = 6 \Rightarrow$ the submarine will be located at $\left(26, 23, -\frac{1}{3}\right)$ in 20 minutes.

3. Work $= \mathbf{F} \cdot \overrightarrow{PQ} = |\mathbf{F}| \left|\overrightarrow{PQ}\right| \cos\theta = |160||250| \cos\frac{\pi}{6} = (40{,}000)\left(\frac{\sqrt{3}}{2}\right) \approx 34{,}641$ J

5. $|\mathbf{A}+\mathbf{B}|^2 = (\mathbf{A}+\mathbf{B}) \cdot (\mathbf{A}+\mathbf{B}) = \mathbf{A}\cdot\mathbf{A} + 2\mathbf{A}\cdot\mathbf{B} + \mathbf{B}\cdot\mathbf{B} \le |\mathbf{A}|^2 + 2|\mathbf{A}||\mathbf{B}| + |\mathbf{B}|^2 = (|\mathbf{A}| + |\mathbf{B}|)^2 \Rightarrow |\mathbf{A}+\mathbf{B}|$
$\le |\mathbf{A}| + |\mathbf{B}|$

7. Let α denote the angle between **C** and **A**, and β the angle between **C** and **B**. Let $a = |\mathbf{A}|$ and $b = |\mathbf{B}|$. Then

$$\cos\alpha = \frac{\mathbf{C}\cdot\mathbf{A}}{|\mathbf{C}||\mathbf{A}|} = \frac{(a\mathbf{B}+b\mathbf{A})\cdot\mathbf{A}}{|\mathbf{C}||\mathbf{A}|} = \frac{(a\mathbf{B}\cdot\mathbf{A}+b\mathbf{A}\cdot\mathbf{A})}{|\mathbf{C}||\mathbf{A}|} = \frac{(a\mathbf{B}\cdot\mathbf{A}+b\mathbf{A}\cdot\mathbf{A})}{|\mathbf{C}||\mathbf{A}|} = \frac{(a\mathbf{B}\cdot\mathbf{A}+ba^2)}{|\mathbf{C}|a} = \frac{\mathbf{B}\cdot\mathbf{A}+ba}{|\mathbf{C}|},$$

and likewise, $\cos\beta = \frac{\mathbf{A}\cdot\mathbf{B}+ba}{|\mathbf{C}|}$. Since the angle between A and B is always $\le \frac{\pi}{2}$ and $\cos\alpha = \cos\beta$, we have that $\alpha = \beta \Rightarrow$ **C** bisects the angle between **A** and **B**.

9. If $\mathbf{A} = a\mathbf{i} + b\mathbf{j} + c\mathbf{k}$, then $\mathbf{A}\cdot\mathbf{A} = a^2 + b^2 + c^2 \ge 0$ and $\mathbf{A}\cdot\mathbf{A} = 0$ iff $a = b = c = 0$.

11. If $\mathbf{A} = a\mathbf{i} + b\mathbf{j}$ and $\mathbf{B} = c\mathbf{i} + d\mathbf{j}$, then $\mathbf{A}\cdot\mathbf{B} = |\mathbf{A}||\mathbf{B}|\cos\theta \Rightarrow ac + bd = \sqrt{a^2+b^2}\sqrt{c^2+d^2}\cos\theta$
$\Rightarrow (ac+bd)^2 = (a^2+b^2)(c^2+d^2)\cos^2\theta \Rightarrow (ac+bd)^2 \le (a^2+b^2)(c^2+d^2)$, since $\cos^2\theta \le 1$.

13. (a) If $P(x,y,z)$ is a point in the plane determined by the three points $P_1(x_1,y_1,z_1)$, $P_2(x_2,y_2,z_2)$ and $P_3(x_3,y_3,z_3)$, then the vectors $\overrightarrow{PP_1}$, $\overrightarrow{PP_2}$ and $\overrightarrow{PP_3}$ all lie in the plane. Thus $\overrightarrow{PP_1}\cdot(\overrightarrow{PP_2}\times\overrightarrow{PP_3}) = 0$

$$\Rightarrow \begin{vmatrix} x_1-x & y_1-y & z_1-z \\ x_2-x & y_2-y & z_2-z \\ x_3-x & y_3-y & z_3-z \end{vmatrix} = 0$$ by the determinant formula for the triple scalar product in Section 10.4.

(b) Subtract row 1 from rows 2, 3, and 4 and evaluate the resulting determinant (which has the same value as the given determinant) by cofactor expansion about column 4. This expansion is exactly the determinant in part (a) so we have all points $P(x,y,z)$ in the plane determined by $P_1(x_1,y_1,z_1)$, $P_2(x_2,y_2,z_2)$, and $P_3(x_3,y_3,z_3)$.

15. If $Q(x,y)$ is a point on the line $ax + by = c$, then $\overrightarrow{P_1Q} = (x-x_1)\mathbf{i} + (y-y_1)\mathbf{j}$, and $\mathbf{n} = a\mathbf{i} + b\mathbf{j}$ is normal to the line. The distance is $\left|\text{proj}_{\mathbf{n}}\,\overrightarrow{P_1Q}\right| = \left|\frac{[(x-x_1)\mathbf{i}+(y-y_1)\mathbf{j}]\cdot(a\mathbf{i}+b\mathbf{j})}{\sqrt{a^2+b^2}}\right| = \frac{|a(x-x_1)+b(y-y_1)|}{\sqrt{a^2+b^2}}$

$= \frac{|ax_1+by_1-c|}{\sqrt{a^2+b^2}}$, since $c = ax + by$.

17. (a) If (x_1,y_1,z_1) is on the plane $Ax + By + Cz = D_1$, then the distance d between the planes is

$d = \frac{|Ax_1+By_1+Cz_1-D_2|}{\sqrt{A^2+B^2+C^2}} = \frac{|D_1-D_2|}{|A\mathbf{i}+B\mathbf{j}+C\mathbf{k}|}$, since $Ax_1+By_1+Cz_1 = D_1$, by Exercise 16(a).

(b) $d = \frac{|12-6|}{\sqrt{4+9+1}} = \frac{6}{\sqrt{14}}$

(c) $\frac{|2(3)+(-1)(2)+2(-1)+4|}{\sqrt{14}} = \frac{|2(3)+(-1)(2)+2(-1)+D|}{\sqrt{14}} \Rightarrow D = -8$ or $4 \Rightarrow$ the desired plane is $2x - y + 2x = 8$

(d) Choose the point $(2, 0, 1)$ on the plane. Then $\frac{|3-D|}{\sqrt{6}} = 5 \Rightarrow D = 3 \pm 5\sqrt{6} \Rightarrow$ the desired planes are $x - 2y + z = 3 + 5\sqrt{6}$ and $x - 2y + z = 3 - 5\sqrt{6}$.

19. (a) $\mathbf{A}\times\mathbf{B} = 4\mathbf{i}\times\mathbf{j} = 4\mathbf{k} \Rightarrow (\mathbf{A}\times\mathbf{B})\times\mathbf{C} = \mathbf{0}$; $(\mathbf{A}\cdot\mathbf{C})\mathbf{B} - (\mathbf{B}\cdot\mathbf{C})\mathbf{A} = 0\mathbf{B} - 0\mathbf{A} = \mathbf{0}$; $\mathbf{B}\times\mathbf{C} = 4\mathbf{i} \Rightarrow \mathbf{A}\times(\mathbf{B}\times\mathbf{C}) = \mathbf{0}$; $(\mathbf{A}\cdot\mathbf{C})\mathbf{B} - (\mathbf{A}\cdot\mathbf{B})\mathbf{C} = 0\mathbf{B} - 0\mathbf{C} = \mathbf{0}$

(b) $\mathbf{A}\times\mathbf{B} = \begin{vmatrix} \mathbf{i} & \mathbf{j} & \mathbf{k} \\ 1 & -1 & 1 \\ 2 & 1 & -2 \end{vmatrix} = \mathbf{i} + 4\mathbf{j} + 3\mathbf{k} \Rightarrow (\mathbf{A}\times\mathbf{B})\times\mathbf{C} = \begin{vmatrix} \mathbf{i} & \mathbf{j} & \mathbf{k} \\ 1 & 4 & 3 \\ -1 & 2 & -1 \end{vmatrix} = -10\mathbf{i} - 2\mathbf{j} + 6\mathbf{k}$;

$(\mathbf{A}\cdot\mathbf{C})\mathbf{B} - (\mathbf{B}\cdot\mathbf{C})\mathbf{A} = -4(2\mathbf{i}+\mathbf{j}-2\mathbf{k}) - 2(\mathbf{i}-\mathbf{j}+\mathbf{k}) = -10\mathbf{i} - 2\mathbf{j} + 6\mathbf{k}$;

$\mathbf{B}\times\mathbf{C} = \begin{vmatrix} \mathbf{i} & \mathbf{j} & \mathbf{k} \\ 2 & 1 & -2 \\ -1 & 2 & -1 \end{vmatrix} = 3\mathbf{i} + 4\mathbf{j} + 5\mathbf{k} \Rightarrow \mathbf{A}\times(\mathbf{B}\times\mathbf{C}) = \begin{vmatrix} \mathbf{i} & \mathbf{j} & \mathbf{k} \\ 1 & -1 & 1 \\ 3 & 4 & 5 \end{vmatrix} = -9\mathbf{i} - 2\mathbf{j} + 7\mathbf{k}$;

$(\mathbf{A}\cdot\mathbf{C})\mathbf{B} - (\mathbf{A}\cdot\mathbf{B})\mathbf{C} = -4(2\mathbf{i}+\mathbf{j}-2\mathbf{k}) - (-1)(-\mathbf{i}+2\mathbf{j}-\mathbf{k}) = -9\mathbf{i} - 2\mathbf{j} + 7\mathbf{k}$

(c) $\mathbf{A}\times\mathbf{B} = \begin{vmatrix} \mathbf{i} & \mathbf{j} & \mathbf{k} \\ 2 & 1 & 0 \\ 2 & -1 & 1 \end{vmatrix} = \mathbf{i} - 2\mathbf{j} - 4\mathbf{k} \Rightarrow (\mathbf{A}\times\mathbf{B})\times\mathbf{C} = \begin{vmatrix} \mathbf{i} & \mathbf{j} & \mathbf{k} \\ 1 & -2 & -4 \\ 1 & 0 & 2 \end{vmatrix} = -4\mathbf{i} - 6\mathbf{j} + 2\mathbf{k}$;

$(\mathbf{A}\cdot\mathbf{C})\mathbf{B} - (\mathbf{B}\cdot\mathbf{C})\mathbf{A} = 2(2\mathbf{i}-\mathbf{j}+\mathbf{k}) - 4(2\mathbf{i}+\mathbf{j}) = -4\mathbf{i} - 6\mathbf{j} + 2\mathbf{k}$;

$\mathbf{B}\times\mathbf{C} = \begin{vmatrix} \mathbf{i} & \mathbf{j} & \mathbf{k} \\ 2 & -1 & 1 \\ 1 & 0 & 2 \end{vmatrix} = -2\mathbf{i} - 3\mathbf{j} + \mathbf{k} \Rightarrow \mathbf{A}\times(\mathbf{B}\times\mathbf{C}) = \begin{vmatrix} \mathbf{i} & \mathbf{j} & \mathbf{k} \\ 2 & 1 & 0 \\ -2 & -3 & 1 \end{vmatrix} = \mathbf{i} - 2\mathbf{j} - 4\mathbf{k}$;

$(\mathbf{A}\cdot\mathbf{C})\mathbf{B} - (\mathbf{A}\cdot\mathbf{B})\mathbf{C} = 2(2\mathbf{i}-\mathbf{j}+\mathbf{k}) - 3(\mathbf{i}+2\mathbf{k}) = \mathbf{i} - 2\mathbf{j} - 4\mathbf{k}$

(d) $\mathbf{A}\times\mathbf{B} = \begin{vmatrix} \mathbf{i} & \mathbf{j} & \mathbf{k} \\ 1 & 1 & -2 \\ -1 & 0 & -1 \end{vmatrix} = -\mathbf{i} + 3\mathbf{j} + \mathbf{k} \Rightarrow (\mathbf{A}\times\mathbf{B})\times\mathbf{C} = \begin{vmatrix} \mathbf{i} & \mathbf{j} & \mathbf{k} \\ -1 & 3 & 1 \\ 2 & 4 & -2 \end{vmatrix} = -10\mathbf{i} - 10\mathbf{k}$;

$(\mathbf{A}\cdot\mathbf{C})\mathbf{B} - (\mathbf{B}\cdot\mathbf{C})\mathbf{A} = 10(-\mathbf{i}-\mathbf{k}) - 0(\mathbf{i}+\mathbf{j}-2\mathbf{k}) = -10\mathbf{i} - 10\mathbf{k}$;

$\mathbf{B}\times\mathbf{C} = \begin{vmatrix} \mathbf{i} & \mathbf{j} & \mathbf{k} \\ -1 & 0 & -1 \\ 2 & 4 & -2 \end{vmatrix} = 4\mathbf{i} - 4\mathbf{j} - 4\mathbf{k} \Rightarrow \mathbf{A}\times(\mathbf{B}\times\mathbf{C}) = \begin{vmatrix} \mathbf{i} & \mathbf{j} & \mathbf{k} \\ 1 & 1 & -2 \\ 4 & -4 & -4 \end{vmatrix} = -12\mathbf{i} - 4\mathbf{j} - 8\mathbf{k}$;

$(\mathbf{A}\cdot\mathbf{C})\mathbf{B} - (\mathbf{A}\cdot\mathbf{B})\mathbf{C} = 10(-\mathbf{i}-\mathbf{k}) - 1(2\mathbf{i}+4\mathbf{j}-2\mathbf{k}) = -12\mathbf{i} - 4\mathbf{j} - 8\mathbf{k}$

21. The formula is always true; $\mathbf{A} \times [\mathbf{A} \times (\mathbf{A} \times \mathbf{B})] \cdot \mathbf{C} = \mathbf{A} \times [(\mathbf{A} \cdot \mathbf{B})\mathbf{A} - (\mathbf{A} \cdot \mathbf{A})\mathbf{B}] \cdot \mathbf{C}$
$= [(\mathbf{A} \cdot \mathbf{B})\mathbf{A} \times \mathbf{A} - (\mathbf{A} \cdot \mathbf{A})\mathbf{A} \times \mathbf{B}] \cdot \mathbf{C} = -|\mathbf{A}|^2 \mathbf{A} \times \mathbf{B} \cdot \mathbf{C} = -|\mathbf{A}|^2 \mathbf{A} \cdot \mathbf{B} \times \mathbf{C}$

23. $\text{proj}_{\mathbf{z}}\, \mathbf{w} = -\text{proj}_{\mathbf{z}}\, \mathbf{v}$ and $\mathbf{w} - \text{proj}_{\mathbf{z}}\, \mathbf{w} = \mathbf{v} - \text{proj}_{\mathbf{z}}\, \mathbf{v}$ lies along the line L $\Rightarrow \mathbf{w} = (\mathbf{w} - \text{proj}_{\mathbf{z}}\, \mathbf{w}) + \text{proj}_{\mathbf{z}}\, \mathbf{w}$
$= (\mathbf{v} - \text{proj}_{\mathbf{z}}\, \mathbf{v}) + \text{proj}_{\mathbf{z}}\, \mathbf{w} = \mathbf{v} - 2\,\text{proj}_{\mathbf{z}}\, \mathbf{v} = \mathbf{v} - 2\left(\frac{\mathbf{v} \cdot \mathbf{z}}{|\mathbf{z}|^2}\right)\mathbf{z}$

25. (a) The vector from $(0, d)$ to $(kd, 0)$ is $\mathbf{r}_k = kd\mathbf{i} - d\mathbf{j} \Rightarrow |\mathbf{r}_k|^3 = \frac{1}{d^3(k^2+1)^{3/2}} \Rightarrow \frac{\mathbf{r}_k}{|\mathbf{r}_k|^3} = \frac{k\mathbf{i} - \mathbf{j}}{d^2(k^2+1)^{3/2}}$. The total force on the mass $(0, d)$ due to the masses Q_k for $k = -n, -n+1, \ldots, n-1, n$ is

$$\mathbf{F} = \frac{GMm}{d^2}(-\mathbf{j}) + \frac{GMm}{2d^2}\left(\frac{\mathbf{i}-\mathbf{j}}{\sqrt{2}}\right) + \frac{GMm}{5d^2}\left(\frac{2\mathbf{i}-\mathbf{j}}{\sqrt{5}}\right) + \ldots + \frac{GMm}{(n^2+1)d^2}\left(\frac{n\mathbf{i}-\mathbf{j}}{\sqrt{n^2+1}}\right) + \frac{GMm}{2d^2}\left(\frac{-\mathbf{i}-\mathbf{j}}{\sqrt{2}}\right)$$
$$+ \frac{GMm}{5d^2}\left(\frac{-2\mathbf{i}-\mathbf{j}}{\sqrt{5}}\right) + \ldots + \frac{GMm}{(n^2+1)d^2}\left(\frac{-n\mathbf{i}-\mathbf{j}}{\sqrt{n^2+1}}\right)$$

The $\mathbf{i}$ components cancel, giving

$$\mathbf{F} = \frac{GMm}{d^2}\left(-1 - \frac{2}{2\sqrt{2}} - \frac{2}{5\sqrt{5}} - \ldots - \frac{2}{(n^2+1)(n^2+1)^{1/2}}\right)\mathbf{j}$$ $\Rightarrow$ the magnitude of the force is

$$|\mathbf{F}| = \frac{GMm}{d^2}\left(1 + \sum_{i=1}^{n} \frac{2}{(i^2+1)^{3/2}}\right).$$

(b) Yes, it is finite: $\lim_{n\to\infty} |\mathbf{F}| = \frac{GMm}{d^2}\left(1 + \sum_{i=1}^{\infty} \frac{2}{(i^2+1)^{3/2}}\right)$ is finite since $\sum_{i=1}^{\infty} \frac{2}{(i^2+1)^{3/2}}$ converges.

CHAPTER 11 VECTOR-VALUED FUNCTIONS AND MOTION IN SPACE

11.1 VECTOR-VALUED FUNCTIONS AND SPACE CURVES

1. $x = t + 1$ and $y = t^2 - 1 \Rightarrow y = (x-1)^2 - 1 = x^2 - 2x$; $\mathbf{v} = \frac{d\mathbf{r}}{dt} = \mathbf{i} + 2t\mathbf{j} \Rightarrow \mathbf{a} = \frac{d\mathbf{v}}{dt} = 2\mathbf{j} \Rightarrow \mathbf{v} = \mathbf{i} + 2\mathbf{j}$ and $\mathbf{a} = 2\mathbf{j}$ at $t = 1$

3. $x = e^t$ and $y = \frac{2}{9}e^{2t} \Rightarrow y = \frac{2}{9}x^2$; $\mathbf{v} = \frac{d\mathbf{r}}{dt} = e^t\mathbf{i} + \frac{4}{9}e^{2t}\mathbf{j} \Rightarrow \mathbf{a} = e^t\mathbf{i} + \frac{8}{9}e^{2t}\mathbf{j} \Rightarrow \mathbf{v} = 3\mathbf{i} + 4\mathbf{j}$ and $\mathbf{a} = 3\mathbf{i} + 8\mathbf{j}$ at $t = \ln 3$

5. $\mathbf{v} = \frac{d\mathbf{r}}{dt} = (\cos t)\mathbf{i} - (\sin t)\mathbf{j}$ and $\mathbf{a} = \frac{d\mathbf{v}}{dt} = -(\sin t)\mathbf{i} - (\cos t)\mathbf{j}$
$\Rightarrow$ for $t = \frac{\pi}{4}$, $\mathbf{v}\left(\frac{\pi}{4}\right) = \frac{\sqrt{2}}{2}\mathbf{i} - \frac{\sqrt{2}}{2}\mathbf{j}$ and $\mathbf{a}\left(\frac{\pi}{4}\right) = -\frac{\sqrt{2}}{2}\mathbf{i} - \frac{\sqrt{2}}{2}\mathbf{j}$;
for $t = \frac{\pi}{2}$, $\mathbf{v}\left(\frac{\pi}{2}\right) = -\mathbf{j}$ and $\mathbf{a}\left(\frac{\pi}{2}\right) = -\mathbf{i}$

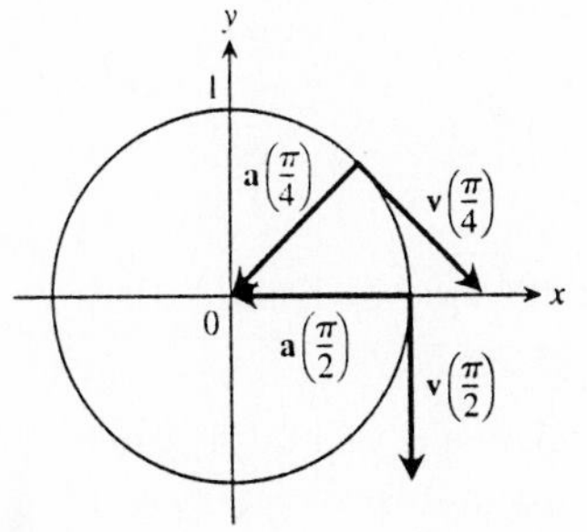

7. $\mathbf{v} = \frac{d\mathbf{r}}{dt} = (1 - \cos t)\mathbf{i} + (\sin t)\mathbf{j}$ and $\mathbf{a} = \frac{d\mathbf{v}}{dt} = (\sin t)\mathbf{i} + (\cos t)\mathbf{j}$
$\Rightarrow$ for $t = \pi$, $\mathbf{v}(\pi) = 2\mathbf{i}$ and $\mathbf{a}(\pi) = -\mathbf{j}$; for $t = \frac{3\pi}{2}$,
$\mathbf{v}\left(\frac{3\pi}{2}\right) = \mathbf{i} - \mathbf{j}$ and $\mathbf{a}\left(\frac{3\pi}{2}\right) = -\mathbf{i}$

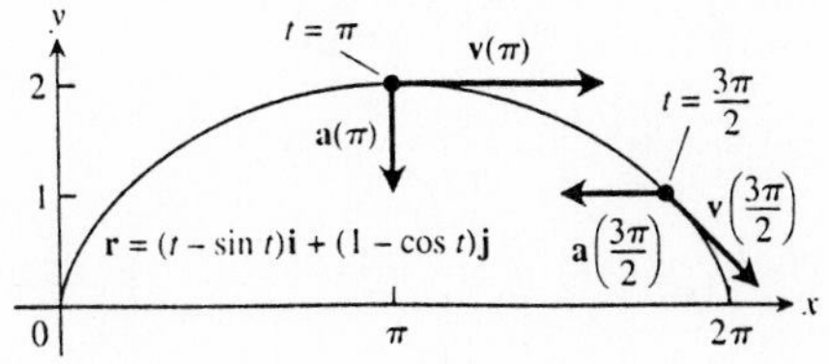

9. $\mathbf{r} = (t+1)\mathbf{i} + (t^2 - 1)\mathbf{j} + 2t\mathbf{k} \Rightarrow \mathbf{v} = \frac{d\mathbf{r}}{dt} = \mathbf{i} + 2t\mathbf{j} + 2\mathbf{k} \Rightarrow \mathbf{a} = \frac{d^2\mathbf{r}}{dt^2} = 2\mathbf{j}$; Speed: $|\mathbf{v}(1)| = \sqrt{1^2 + (2(1))^2 + 2^2} = 3$;
Direction: $\frac{\mathbf{v}(1)}{|\mathbf{v}(1)|} = \frac{\mathbf{i} + 2(1)\mathbf{j} + 2\mathbf{k}}{3} = \frac{1}{3}\mathbf{i} + \frac{2}{3}\mathbf{j} + \frac{2}{3}\mathbf{k} \Rightarrow \mathbf{v}(1) = 3\left(\frac{1}{3}\mathbf{i} + \frac{2}{3}\mathbf{j} + \frac{2}{3}\mathbf{k}\right)$

11. $\mathbf{r} = (2\cos t)\mathbf{i} + (3\sin t)\mathbf{j} + 4t\mathbf{k} \Rightarrow \mathbf{v} = \frac{d\mathbf{r}}{dt} = (-2\sin t)\mathbf{i} + (3\cos t)\mathbf{j} + 4\mathbf{k} \Rightarrow \mathbf{a} = \frac{d^2\mathbf{r}}{dt^2} = (-2\cos t)\mathbf{i} - (3\sin t)\mathbf{j}$;
Speed: $\left|\mathbf{v}\left(\frac{\pi}{2}\right)\right| = \sqrt{\left(-2\sin\frac{\pi}{2}\right)^2 + \left(3\cos\frac{\pi}{2}\right)^2 + 4^2} = 2\sqrt{5}$; Direction: $\frac{\mathbf{v}\left(\frac{\pi}{2}\right)}{\left|\mathbf{v}\left(\frac{\pi}{2}\right)\right|}$
$= \left(-\frac{2}{2\sqrt{5}}\sin\frac{\pi}{2}\right)\mathbf{i} + \left(\frac{3}{2\sqrt{5}}\cos\frac{\pi}{2}\right)\mathbf{j} + \frac{4}{2\sqrt{5}}\mathbf{k} = -\frac{1}{\sqrt{5}}\mathbf{i} + \frac{2}{\sqrt{5}}\mathbf{k} \Rightarrow \mathbf{v}\left(\frac{\pi}{2}\right) = 2\sqrt{5}\left(-\frac{1}{\sqrt{5}}\mathbf{i} + \frac{2}{\sqrt{5}}\mathbf{k}\right)$

13. $\mathbf{r} = (2\ln(t+1))\mathbf{i} + t^2\mathbf{j} + \frac{t^2}{2}\mathbf{k} \Rightarrow \mathbf{v} = \frac{d\mathbf{r}}{dt} = \left(\frac{2}{t+1}\right)\mathbf{i} + 2t\mathbf{j} + t\mathbf{k} \Rightarrow \mathbf{a} = \frac{d^2\mathbf{r}}{dt^2} = \left[\frac{-2}{(t+1)^2}\right]\mathbf{i} + 2\mathbf{j} + \mathbf{k}$;

Speed: $|\mathbf{v}(1)| = \sqrt{\left(\frac{2}{1+1}\right)^2 + (2(1))^2 + 1^2} = \sqrt{6}$; Direction: $\frac{\mathbf{v}(1)}{|\mathbf{v}(1)|} = \frac{\left(\frac{2}{1+1}\right)\mathbf{i} + 2(1)\mathbf{j} + (1)\mathbf{k}}{\sqrt{6}}$

$= \frac{1}{\sqrt{6}}\mathbf{i} + \frac{2}{\sqrt{6}}\mathbf{j} + \frac{1}{\sqrt{6}}\mathbf{k} \Rightarrow \mathbf{v}(1) = \sqrt{6}\left(\frac{1}{\sqrt{6}}\mathbf{i} + \frac{2}{\sqrt{6}}\mathbf{j} + \frac{1}{\sqrt{6}}\mathbf{k}\right)$

15. $\mathbf{v} = 3\mathbf{i} + \sqrt{3}\mathbf{j} + 2t\mathbf{k}$ and $\mathbf{a} = 2\mathbf{k} \Rightarrow \mathbf{v}(0) = 3\mathbf{i} + \sqrt{3}\mathbf{j}$ and $\mathbf{a}(0) = 2\mathbf{k} \Rightarrow |\mathbf{v}(0)| = \sqrt{3^2 + (\sqrt{3})^2 + 0^2} = \sqrt{12}$ and $|\mathbf{a}(0)| = \sqrt{2^2} = 2$; $\mathbf{v}(0)\cdot\mathbf{a}(0) = 0 \Rightarrow \cos\theta = 0 \Rightarrow \theta = \frac{\pi}{2}$

17. $\mathbf{v} = \left(\frac{2t}{t^2+1}\right)\mathbf{i} + \left(\frac{1}{t^2+1}\right)\mathbf{j} + t(t^2+1)^{-1/2}\mathbf{k}$ and $\mathbf{a} = \left[\frac{-2t^2+2}{(t^2+1)^2}\right]\mathbf{i} - \left[\frac{2t}{(t^2+1)^2}\right]\mathbf{j} + \left[\frac{1}{(t^2+1)^{3/2}}\right]\mathbf{k} \Rightarrow \mathbf{v}(0) = \mathbf{j}$ and $\mathbf{a}(0) = 2\mathbf{i} + \mathbf{k} \Rightarrow |\mathbf{v}(0)| = 1$ and $|\mathbf{a}(0)| = \sqrt{2^2+1^2} = \sqrt{5}$; $\mathbf{v}(0)\cdot\mathbf{a}(0) = 0 \Rightarrow \cos\theta = 0 \Rightarrow \theta = \frac{\pi}{2}$

19. $\mathbf{v} = (1-\cos t)\mathbf{i} + (\sin t)\mathbf{j}$ and $\mathbf{a} = (\sin t)\mathbf{i} + (\cos t)\mathbf{j} \Rightarrow \mathbf{v}\cdot\mathbf{a} = (\sin t)(1-\cos t) + (\sin t)(\cos t) = \sin t$. Thus, $\mathbf{v}\cdot\mathbf{a} = 0 \Rightarrow \sin t = 0 \Rightarrow t = 0, \pi$, or 2π

21. $\int_0^1 [t^3\mathbf{i} + 7\mathbf{j} + (t+1)\mathbf{k}]\,dt = \left[\frac{t^4}{4}\right]_0^1\mathbf{i} + [7t]_0^1\mathbf{j} + \left[\frac{t^2}{2} + t\right]_0^1\mathbf{k} = \frac{1}{4}\mathbf{i} + 7\mathbf{j} + \frac{3}{2}\mathbf{k}$

23. $\int_{-\pi/4}^{\pi/4} \left[(\sin t)\mathbf{i} + (1+\cos t)\mathbf{j} + (\sec^2 t)\mathbf{k}\right]dt = [-\cos t]_{-\pi/4}^{\pi/4}\mathbf{i} + [t + \sin t]_{-\pi/4}^{\pi/4}\mathbf{j} + [\tan t]_{-\pi/4}^{\pi/4}\mathbf{k}$

$= \left(\frac{\pi + 2\sqrt{2}}{2}\right)\mathbf{j} + 2\mathbf{k}$

25. $\int_1^4 \left(\frac{1}{t}\mathbf{i} + \frac{1}{5-t}\mathbf{j} + \frac{1}{2t}\mathbf{k}\right)dt = \;= [\ln t]_1^4\mathbf{i} + [-\ln(5-t)]_1^4\mathbf{j} + \left[\frac{1}{2}\ln t\right]_1^4\mathbf{k} = (\ln 4)\mathbf{i} + (\ln 4)\mathbf{j} + (\ln 2)\mathbf{k}$

27. $\mathbf{r} = \int(-t\mathbf{i} - t\mathbf{j} - t\mathbf{k})\,dt = -\frac{t^2}{2}\mathbf{i} - \frac{t^2}{2}\mathbf{j} - \frac{t^2}{2}\mathbf{k} + \mathbf{C}$; $\mathbf{r}(0) = 0\mathbf{i} - 0\mathbf{j} - 0\mathbf{k} + \mathbf{C} = \mathbf{i} + 2\mathbf{j} + 3\mathbf{k} \Rightarrow \mathbf{C} = \mathbf{i} + 2\mathbf{j} + 3\mathbf{k}$

$\Rightarrow \mathbf{r} = \left(-\frac{t^2}{2} + 1\right)\mathbf{i} + \left(-\frac{t^2}{2} + 2\right)\mathbf{j} + \left(-\frac{t^2}{2} + 3\right)\mathbf{k}$

29. $\mathbf{r} = \int\left[\left(\frac{3}{2}(t+1)^{1/2}\right)\mathbf{i} + e^{-t}\mathbf{j} + \left(\frac{1}{t+1}\right)\mathbf{k}\right]dt = (t+1)^{3/2}\mathbf{i} - e^{-t}\mathbf{j} + \ln(t+1)\mathbf{k} + \mathbf{C}$;

$\mathbf{r}(0) = (0+1)^{3/2}\mathbf{i} - e^{-0}\mathbf{j} + \ln(0+1)\mathbf{k} + \mathbf{C} = \mathbf{k} \Rightarrow \mathbf{C} = -\mathbf{i} + \mathbf{j} + \mathbf{k}$

$\Rightarrow \mathbf{r} = \left[(t+1)^{3/2} - 1\right]\mathbf{i} + (1 - e^{-t})\mathbf{j} + [1 + \ln(t+1)]\mathbf{k}$

31. $\frac{d\mathbf{r}}{dt} = \int (-32\mathbf{k})\,dt = -32t\mathbf{k} + \mathbf{C}_1$; $\frac{d\mathbf{r}}{dt}(0) = 8\mathbf{i} + 8\mathbf{j} \Rightarrow -32(0)\mathbf{k} + \mathbf{C}_1 = 8\mathbf{i} + 8\mathbf{j} \Rightarrow \mathbf{C}_1 = 8\mathbf{i} + 8\mathbf{j}$

$\Rightarrow \frac{d\mathbf{r}}{dt} = 8\mathbf{i} + 8\mathbf{j} - 32t\mathbf{k}$; $\mathbf{r} = \int (8\mathbf{i} + 8\mathbf{j} - 32t\mathbf{k})\,dt = 8t\mathbf{i} + 8t\mathbf{j} - 16t^2\mathbf{k} + \mathbf{C}_2$; $\mathbf{r}(0) = 100\mathbf{k}$

$\Rightarrow 8(0)\mathbf{i} + 8(0)\mathbf{j} - 16(0)^2\mathbf{k} + \mathbf{C}_2 = 100\mathbf{k} \Rightarrow \mathbf{C}_2 = 100\mathbf{k} \Rightarrow \mathbf{r} = 8t\mathbf{i} + 8t\mathbf{j} + (100 - 16t^2)\mathbf{k}$

33. $\mathbf{r}(t) = (\sin t)\mathbf{i} + (t^2 - \cos t)\mathbf{j} + e^t\mathbf{k} \Rightarrow \mathbf{v}(t) = (\cos t)\mathbf{i} + (2t + \sin t)\mathbf{j} + e^t\mathbf{k}$; $t_0 = 0 \Rightarrow \mathbf{v}(0) = \mathbf{i} + \mathbf{k}$ and $\mathbf{r}(0) = P_0 = (0, -1, 1) \Rightarrow x = 0 + t = t$, $y = -1$, and $z = 1 + t$ are parametric equations of the tangent line

35. $\mathbf{r}(t) = (a \sin t)\mathbf{i} + (a \cos t)\mathbf{j} + bt\mathbf{k} \Rightarrow \mathbf{v}(t) = (a \cos t)\mathbf{i} - (a \sin t)\mathbf{j} + b\mathbf{k}$; $t_0 = 2\pi \Rightarrow \mathbf{v}(0) = a\mathbf{i} + b\mathbf{k}$ and $\mathbf{r}(0) = P_0 = (0, a, 2b\pi) \Rightarrow x = 0 + at = at$, $y = a$, and $z = 2\pi b + bt$ are parametric equations of the tangent line

37. (a) $\mathbf{v}(t) = -(\sin t)\mathbf{i} + (\cos t)\mathbf{j} \Rightarrow \mathbf{a}(t) = -(\cos t)\mathbf{i} - (\sin t)\mathbf{j}$;

(i) $|\mathbf{v}(t)| = \sqrt{(-\sin t)^2 + (\cos t)^2} = 1 \Rightarrow$ constant speed;

(ii) $\mathbf{v} \cdot \mathbf{a} = (\sin t)(\cos t) - (\cos t)(\sin t) = 0 \Rightarrow$ yes, orthogonal;

(iii) counterclockwise movement;

(iv) yes, $\mathbf{r}(0) = \mathbf{i} + 0\mathbf{j}$

(b) $\mathbf{v}(t) = -(2 \sin 2t)\mathbf{i} + (2 \cos 2t)\mathbf{j} \Rightarrow \mathbf{a}(t) = -(4 \cos 2t)\mathbf{i} - (4 \sin 2t)\mathbf{j}$;

(i) $|\mathbf{v}(t)| = \sqrt{4 \sin^2 2t + 4 \cos^2 2t} = 2 \Rightarrow$ constant speed;

(ii) $\mathbf{v} \cdot \mathbf{a} = 8 \sin 2t \cos 2t - 8 \cos 2t \sin 2t = 0 \Rightarrow$ yes, orthogonal;

(iii) counterclockwise movement;

(iv) yes, $\mathbf{r}(0) = \mathbf{i} + 0\mathbf{j}$

(c) $\mathbf{v}(t) = -\sin\left(t - \frac{\pi}{2}\right)\mathbf{i} + \cos\left(t - \frac{\pi}{2}\right)\mathbf{j} \Rightarrow \mathbf{a}(t) = -\cos\left(t - \frac{\pi}{2}\right)\mathbf{i} - \sin\left(t - \frac{\pi}{2}\right)\mathbf{j}$;

(i) $|\mathbf{v}(t)| = \sqrt{\sin^2\left(t - \frac{\pi}{2}\right) + \cos^2\left(t - \frac{\pi}{2}\right)} = 1 \Rightarrow$ constant speed;

(ii) $\mathbf{v} \cdot \mathbf{a} = \sin\left(t - \frac{\pi}{2}\right)\cos\left(t - \frac{\pi}{2}\right) - \cos\left(t - \frac{\pi}{2}\right)\sin\left(t - \frac{\pi}{2}\right) = 0 \Rightarrow$ yes, orthogonal;

(iii) counterclockwise movement;

(iv) no, $\mathbf{r}(0) = 0\mathbf{i} - \mathbf{j}$ instead of $\mathbf{i} + 0\mathbf{j}$

(d) $\mathbf{v}(t) = -(\sin t)\mathbf{i} - (\cos t)\mathbf{j} \Rightarrow \mathbf{a}(t) = -(\cos t)\mathbf{i} + (\sin t)\mathbf{j}$;

(i) $|\mathbf{v}(t)| = \sqrt{(-\sin t)^2 + (-\cos t)^2} = 1 \Rightarrow$ constant speed;

(ii) $\mathbf{v} \cdot \mathbf{a} = (\sin t)(\cos t) - (\cos t)(\sin t) = 0 \Rightarrow$ yes, orthogonal;

(iii) clockwise movement;

(iv) yes, $\mathbf{r}(0) = \mathbf{i} - 0\mathbf{j}$

(e) $\mathbf{v}(t) = -(2t \sin t)\mathbf{i} + (2t \cos t)\mathbf{j} \Rightarrow \mathbf{a}(t) = -(2 \sin t + 2t \cos t)\mathbf{i} + (2 \cos t - 2t \sin t)\mathbf{j}$;

(i) $|\mathbf{v}(t)| = 2\sqrt{(\sin t + t \cos t)^2 + (\cos t - t \sin t)^2}$

$= 2\sqrt{\sin^2 t + \cos^2 t + 2t \sin t \cos t - 2t \sin t \cos t + t^2 \cos^2 t + t^2 \sin^2 t}$

$= 2\sqrt{1 + t^2} \Rightarrow$ variable speed;

(ii) $\mathbf{v}\cdot\mathbf{a} = 4\left(t\sin^2 t + t^2 \sin t\cos t\right) + 4\left(t\cos^2 t - t^2\cos t\sin t\right) = 4t \neq 0$ in general $\Rightarrow$ not orthogonal in general;

(iii) counterclockwise movement;

(iv) yes, $\mathbf{r}(0) = \mathbf{i} + 0\mathbf{j}$

39. $\frac{d\mathbf{v}}{dt} = \mathbf{a} = 3\mathbf{i} - \mathbf{j} + \mathbf{k} \Rightarrow \mathbf{v}(t) = 3t\mathbf{i} - t\mathbf{j} + t\mathbf{k} + \mathbf{C}_1$; the particle travels in the direction of the vector $(4-1)\mathbf{i} + (1-2)\mathbf{j} + (4-3)\mathbf{k} = 3\mathbf{i} - \mathbf{j} + \mathbf{k}$ (since it travels in a straight line), and at time $t = 0$ it has speed $2 \Rightarrow \mathbf{v}(0) = \frac{2}{\sqrt{9+1+1}}(3\mathbf{i} - \mathbf{j} + \mathbf{k}) = \mathbf{C}_1 \Rightarrow \frac{d\mathbf{r}}{dt} = \mathbf{v}(t) = \left(3t + \frac{6}{\sqrt{11}}\right)\mathbf{i} - \left(t + \frac{2}{\sqrt{11}}\right)\mathbf{j} + \left(t + \frac{2}{\sqrt{11}}\right)\mathbf{k}$

$\Rightarrow \mathbf{r}(t) = \left(\frac{3}{2}t^2 + \frac{6}{\sqrt{11}}t\right)\mathbf{i} - \left(\frac{1}{2}t^2 + \frac{2}{\sqrt{11}}t\right)\mathbf{j} + \left(\frac{1}{2}t^2 + \frac{2}{\sqrt{22}}t\right)\mathbf{k} + \mathbf{C}_2$; $\mathbf{r}(0) = \mathbf{i} + 2\mathbf{j} + 3\mathbf{k} = \mathbf{C}_2$

$\Rightarrow \mathbf{r}(t) = \left(\frac{3}{2}t^2 + \frac{6}{\sqrt{11}}t + 1\right)\mathbf{i} - \left(\frac{1}{2}t^2 + \frac{2}{\sqrt{11}}t - 2\right)\mathbf{j} + \left(\frac{1}{2}t^2 + \frac{2}{\sqrt{11}}t + 3\right)\mathbf{k}$

$= \left(\frac{1}{2}t^2 + \frac{2}{\sqrt{11}}t\right)(3\mathbf{i} - \mathbf{j} + \mathbf{k}) + (\mathbf{i} + 2\mathbf{j} + 3\mathbf{k})$

41. The velocity vector is tangent to the graph of $y^2 = 2x$ at the point $(2,2)$, has length 5, and a positive $\mathbf{i}$ component. Now, $y^2 = 2x \Rightarrow 2y\frac{dy}{dx} = 2 \Rightarrow \left.\frac{dy}{dx}\right|_{(2,2)} = \frac{2}{2\cdot 2} = \frac{1}{2} \Rightarrow$ the tangent vector lies in the direction of the vector $\mathbf{i} + \frac{1}{2}\mathbf{j} \Rightarrow$ the velocity vector is $\mathbf{v} = \frac{5}{\sqrt{1+\frac{1}{4}}}\left(\mathbf{i} + \frac{1}{2}\mathbf{j}\right) = \frac{5}{\left(\frac{\sqrt{5}}{2}\right)}\left(\mathbf{i} + \frac{1}{2}\mathbf{j}\right) = 2\sqrt{5}\mathbf{i} + \sqrt{5}\mathbf{j}$

43. $\mathbf{v} = (-3\sin t)\mathbf{j} + (2\cos t)\mathbf{k}$ and $\mathbf{a} = (-3\cos t)\mathbf{j} - (2\sin t)\mathbf{k}$; $|\mathbf{v}|^2 = 9\sin^2 t + 4\cos^2 t \Rightarrow \frac{d}{dt}\left(|\mathbf{v}|^2\right)$ $= 18\sin t\cos t - 8\cos t\sin t = 10\sin t\cos t$; $\frac{d}{dt}\left(|\mathbf{v}|^2\right) = 0 \Rightarrow 10\sin t\cos t = 0 \Rightarrow \sin t = 0$ or $\cos t = 0$ $\Rightarrow t = 0, \pi$ or $t = \frac{\pi}{2}, \frac{3\pi}{2}$. When $t = 0, \pi$, $|\mathbf{v}|^2 = 4 \Rightarrow |\mathbf{v}| = \sqrt{4} = 2$; when $t = \frac{\pi}{2}, \frac{3\pi}{2}$, $|\mathbf{v}| = \sqrt{9} = 3$. Therefore max $|\mathbf{v}|$ is 3 when $t = \frac{\pi}{2}, \frac{3\pi}{2}$, and min $|\mathbf{v}| = 2$ when $t = 0, \pi$. Next, $|\mathbf{a}|^2 = 9\cos^2 t + 4\sin^2 t$ $\Rightarrow \frac{d}{dt}\left(|\mathbf{a}|^2\right) = -18\cos t\sin t + 8\sin t\cos t = -10\sin t\cos t$; $\frac{d}{dt}\left(|\mathbf{a}|^2\right) = 0 \Rightarrow -10\sin t\cos t = 0 \Rightarrow \sin t = 0$ or $\cos t = 0 \Rightarrow t = 0, \pi$ or $t = \frac{\pi}{2}, \frac{3\pi}{2}$. When $t = 0, \pi$, $|\mathbf{a}|^2 = 9 \Rightarrow |\mathbf{a}| = 3$; when $t = \frac{\pi}{2}, \frac{3\pi}{2}$, $|\mathbf{a}|^2 = 4 \Rightarrow |\mathbf{a}| = 2$. Therefore, max $|\mathbf{a}| = 3$ when $t = 0, \pi$, and min $|\mathbf{a}| = 2$ when $t = \frac{\pi}{2}, \frac{3\pi}{2}$.

45. $\frac{d}{dt}(\mathbf{v}\cdot\mathbf{v}) = \mathbf{v}\cdot\frac{d\mathbf{v}}{dt} + \frac{d\mathbf{v}}{dt}\cdot\mathbf{v} = 2\mathbf{v}\cdot\frac{d\mathbf{v}}{dt} = 2\cdot 0 = 0 \Rightarrow \mathbf{v}\cdot\mathbf{v}$ is a constant $\Rightarrow |\mathbf{v}| = \sqrt{\mathbf{v}\cdot\mathbf{v}}$ is constant

47. $\frac{d}{dt}\left[\mathbf{r}\cdot\left(\frac{d\mathbf{r}}{dt}\times\frac{d^2\mathbf{r}}{dt^2}\right)\right] = \frac{d\mathbf{r}}{dt}\cdot\left(\frac{d\mathbf{r}}{dt}\times\frac{d^2\mathbf{r}}{dt^2}\right) + \mathbf{r}\cdot\left(\frac{d^2\mathbf{r}}{dt^2}\times\frac{d^2\mathbf{r}}{dt^2}\right) + \mathbf{r}\cdot\left(\frac{d\mathbf{r}}{dt}\times\frac{d^3\mathbf{r}}{dt^3}\right) = \mathbf{r}\cdot\left(\frac{d\mathbf{r}}{dt}\times\frac{d^3\mathbf{r}}{dt^3}\right)$, since $\mathbf{A}\cdot(\mathbf{A}\times\mathbf{B}) = 0$ and $\mathbf{A}\cdot(\mathbf{B}\times\mathbf{B}) = 0$ for any vectors $\mathbf{A}$ and $\mathbf{B}$

49. (a) $\mathbf{u} = f(t)\mathbf{i} + g(t)\mathbf{j} + h(t)\mathbf{k} \Rightarrow c\mathbf{u} = cf(t)\mathbf{i} + cg(t)\mathbf{j} + ch(t)\mathbf{k} \Rightarrow \frac{d}{dt}(c\mathbf{u}) = c\frac{df}{dt}\mathbf{i} + c\frac{dg}{dt}\mathbf{j} + c\frac{dh}{dt}\mathbf{k}$ $= c\left(\frac{df}{dt}\mathbf{i} + \frac{dg}{dt}\mathbf{j} + \frac{dh}{dt}\mathbf{k}\right) = c\frac{d\mathbf{u}}{dt}$

(b) $f\mathbf{u} = f\text{f}(t)\mathbf{i} + f\text{g}(t)\mathbf{j} + f\text{h}(t)\mathbf{k} \Rightarrow \frac{d}{dt}(f\mathbf{u}) = \left[\frac{df}{dt}\text{f}(t) + f\frac{d\text{f}}{dt}\right]\mathbf{i} + \left[\frac{df}{dt}\text{g}(t) + f\frac{dg}{dt}\right]\mathbf{j} + \left[\frac{df}{dt}\text{h}(t) + f\frac{dh}{dt}\right]\mathbf{k}$

$= \frac{df}{dt}[\text{f}(t)\mathbf{i} + \text{g}(t)\mathbf{j} + \text{h}(t)\mathbf{k}] + f\left[\frac{d\text{f}}{dt}\mathbf{i} + \frac{dg}{dt}\mathbf{j} + \frac{dh}{dt}\mathbf{k}\right] = \frac{df}{dt}\mathbf{u} + f\frac{d\mathbf{u}}{dt}$

51. Suppose $\mathbf{r}$ is continuous at $t = t_0$. Then $\lim\limits_{t \to t_0} \mathbf{r}(t) = \mathbf{r}(t_0) \Leftrightarrow \lim\limits_{t \to t_0} [f(t)\mathbf{i} + g(t)\mathbf{j} + h(t)\mathbf{k}]$ $= f(t_0)\mathbf{i} + g(t_0)\mathbf{j} + h(t_0)\mathbf{k} \Leftrightarrow \lim\limits_{t \to t_0} f(t) = f(t_0)$, $\lim\limits_{t \to t_0} g(t) = g(t_0)$, and $\lim\limits_{t \to t_0} h(t) = h(t_0) \Leftrightarrow$ f, g, and h are continuous at $t = t_0$.

53. $\mathbf{r}'(t_0)$ exists $\Rightarrow f'(t_0)\mathbf{i} + g'(t_0)\mathbf{j} + h'(t_0)\mathbf{k}$ exists $\Rightarrow f'(t_0)$, $g'(t_0)$, $h'(t_0)$ all exist $\Rightarrow$ f, g, and h are continuous at $t = t_0 \Rightarrow \mathbf{r}(t)$ is continuous at $t = t_0$

55. (a) Let u and $\mathbf{r}$ be continuous on [a, b]. Then $\lim\limits_{t \to t_0} u(t)\mathbf{r}(t) = \lim\limits_{t \to t_0} [u(t)f(t)\mathbf{i} + u(t)g(t)\mathbf{j} + u(t)h(t)\mathbf{k}]$

$= u(t_0)f(t_0)\mathbf{i} + u(t_0)g(t_0)\mathbf{j} + u(t_0)h(t_0)\mathbf{k} = u(t_0)\mathbf{r}(t_0) \Rightarrow u\mathbf{r}$ is continuous for every t_0 in [a, b].

(b) Let u and $\mathbf{r}$ be differentiable. Then $\frac{d}{dt}(u\mathbf{r}) = \frac{d}{dt}[u(t)f(t)\mathbf{i} + u(t)g(t)\mathbf{j} + u(t)h(t)\mathbf{k}]$

$= \left(\frac{du}{dt}f(t) + u(t)\frac{df}{dt}\right)\mathbf{i} + \left(\frac{du}{dt}g(t) + u(t)\frac{dg}{dt}\right)\mathbf{j} + \left(\frac{du}{dt}h(t) + u(t)\frac{dh}{dt}\right)\mathbf{k}$

$= [f(t)\mathbf{i} + g(t)\mathbf{j} + h(t)\mathbf{k}]\frac{du}{dt} + u(t)\left(\frac{df}{dt}\mathbf{i} + \frac{dg}{dt}\mathbf{j} + \frac{dh}{dt}\mathbf{k}\right) = \mathbf{r}\frac{du}{dt} + u\frac{d\mathbf{r}}{dt}$

57. $\frac{d}{dt}\int_a^t \mathbf{r}(\tau)\,d\tau = \frac{d}{dt}\int_a^t [f(\tau)\mathbf{i} + g(\tau)\mathbf{j} + h(\tau)\mathbf{k}]\,d\tau = \frac{d}{dt}\int_a^t f(\tau)\,d\tau\,\mathbf{i} + \frac{d}{dt}\int_a^t g(\tau)\,d\tau\,\mathbf{j} + \frac{d}{dt}\int_a^t h(\tau)\,d\tau\,\mathbf{k}$

$= f(t)\mathbf{i} + g(t)\mathbf{j} + h(t)\mathbf{k} = \mathbf{r}(t)$. Since $\frac{d}{dt}\int_a^t \mathbf{r}(\tau)\,d\tau = \mathbf{r}(t)$, we have that $\int_a^t \mathbf{r}(\tau)\,d\tau$ is an antiderivative of $\mathbf{r}$. If $\mathbf{R}$ is any antiderivative of $\mathbf{r}$, then $\mathbf{R}(t) = \int_a^t \mathbf{r}(\tau)\,d\tau + \mathbf{C}$ by Exercise 56(b). Then $\mathbf{R}(a) = \int_a^a \mathbf{r}(\tau)\,d\tau + \mathbf{C}$ $= \mathbf{0} + \mathbf{C} \Rightarrow \mathbf{C} = \mathbf{R}(a) \Rightarrow \int_a^t \mathbf{r}(\tau)\,d\tau = \mathbf{R}(t) - \mathbf{C} = \mathbf{R}(t) - \mathbf{R}(a) \Rightarrow \int_a^b \mathbf{r}(\tau)\,d\tau = \mathbf{R}(b) - \mathbf{R}(a)$.

11.2 MODELING PROJECTILE MOTION

1. $x = (v_0 \cos\alpha)t \Rightarrow (21\text{ km})\left(\frac{1000\text{ m}}{1\text{ km}}\right) = (840\text{ m/s})(\cos 60°)t \Rightarrow t = \frac{21{,}000\text{ m}}{(840\text{ m/s})(\cos 60°)} = 50$ seconds

3. (a) $t = \frac{2v_0 \sin\alpha}{g} = \frac{2(500\text{ m/s})(\sin 45°)}{9.8\text{ m/s}^2} = 72.2$ seconds; $R = \frac{v_0^2}{g}\sin 2\alpha = \frac{(500\text{ m/s})^2}{9.8\text{ m/s}^2}(\sin 90°) = 25{,}510.2$ m

(b) $x = (v_0 \cos\alpha)t \Rightarrow 5000\text{ m} = (500\text{ m/s})(\cos 45°)t \Rightarrow t = \frac{5000\text{ m}}{(500\text{ m/s})(\cos 45°)} \approx 14.14$ s; thus,

$y = (v_0 \sin\alpha)t - \frac{1}{2}gt^2 \Rightarrow y \approx (500\text{ m/s})(\sin 45°)(14.14\text{ s}) - \frac{1}{2}(9.8\text{ m/s}^2)(14.14\text{ s})^2 \approx 4020$ m

(c) $y_{max} = \frac{(v_0 \sin \alpha)^2}{2g} = \frac{((500 \text{ m/s})(\sin 45°))^2}{2(9.8 \text{ m/s}^2)} = 6378$ m

5. $x = x_0 + (v_0 \cos \alpha)t = 0 + (44 \cos 45°)t = 22\sqrt{2}t$ and $y = y_0 + (v_0 \sin \alpha)t - \frac{1}{2}gt^2 = 6.5 + (44 \sin 45°)t - 16t^2$ $= 6.5 + 22\sqrt{2}t - 16t^2$; the shot lands when $y = 0 \Rightarrow t = \frac{22\sqrt{2} \pm \sqrt{968 + 416}}{32} \approx 2.135$ sec since $t > 0$; thus $x = 22\sqrt{2}t \approx (22\sqrt{2})(2.134839) \approx 66.42$ ft

7. $R = \frac{v_0^2}{g} \sin 2\alpha \Rightarrow 10 \text{ m} = \left(\frac{v_0^2}{9.8 \text{ m/s}^2}\right)(\sin 90°) \Rightarrow v_0^2 = 98 \text{ m}^2\text{s}^2 \Rightarrow v_0 \approx 9.9$ m/s;

$6\text{m} \approx \frac{(9.9 \text{ m/s})^2}{9.8 \text{ m/s}^2}(\sin 2\alpha) \Rightarrow \sin 2\alpha \approx 0.59999 \Rightarrow 2\alpha \approx 36.87°$ or $143.12° \Rightarrow \alpha \approx 18.4°$ or $71.6°$

9. $R = \frac{v_0^2}{g} \sin 2\alpha \Rightarrow 3(248.8) \text{ ft} = \left(\frac{v_0^2}{32 \text{ ft/sec}^2}\right)(\sin 18°) \Rightarrow v_0^2 \approx 77{,}292.84 \text{ ft}^2/\text{sec}^2 \Rightarrow v_0 \approx 278.01 \text{ ft/sec} \approx 190$ mph

11. $x = (v_0 \cos \alpha)t \Rightarrow 135 \text{ ft} = (90 \text{ ft/sec})(\cos 30°)t \Rightarrow t \approx 1.732$ sec; $y = (v_0 \sin \alpha)t - \frac{1}{2}gt^2$ $\Rightarrow y \approx (90 \text{ ft/sec})(\sin 30°)(1.732 \text{ sec}) - \frac{1}{2}(32 \text{ ft/sec}^2)(1.732 \text{ sec})^2 \Rightarrow y \approx 29.94 \text{ ft} \Rightarrow$ the golf ball will clip the leaves at the top

13. $x = x_0 + (v_0 \cos \alpha)t = 0 + (v_0 \cos 40°)t \approx 0.766\, v_0 t$ and $y = y_0 + (v_0 \sin \alpha)t - \frac{1}{2}gt^2 = 6.5 + (v_0 \sin 40°)t - 16t^2$ $\approx 6.5 + 0.643\, v_0 t - 16t^2$; now the shot went 73.833 ft $\Rightarrow 73.833 = 0.766\, v_0 t \Rightarrow t \approx \frac{96.383}{v_0}$ sec; the shot lands when $y = 0 \Rightarrow 0 = 6.5 + (0.643)(96.383) - 16\left(\frac{96.383}{v_0}\right)^2 \Rightarrow 0 \approx 68.474 - \frac{148{,}634}{v_0^2} \Rightarrow v_0 \approx \sqrt{\frac{148{,}634}{68.474}}$ ≈ 46.6 ft/sec, the shot's initial speed

15. $R = \frac{v_0^2}{g} \sin 2\alpha = \frac{v_0^2}{g}(2 \sin \alpha \cos \alpha) = \frac{v_0^2}{g}[2 \cos(90° - \alpha) \sin(90° - \alpha)] = \frac{v_0^2}{g}[\sin 2(90° - \alpha)]$

17. $R = \frac{(2v_0)^2}{g} \sin 2\alpha = \frac{4v_0^2}{g} \sin 2\alpha = 4\left(\frac{v_0^2}{g} \sin \alpha\right)$ or 4 times the original range. Now, let the initial range be $R = \frac{v_0^2}{g} \sin 2\alpha$. Then we want the factor p so that pv_0 will double the range $\Rightarrow \frac{(pv_0)^2}{g} \sin 2\alpha = 2\left(\frac{v_0^2}{g} \sin 2\alpha\right)$ $\Rightarrow p^2 = 2 \Rightarrow p = \sqrt{2}$ or about 141%. The same percentage will approximately double the height.

19. $\frac{d\mathbf{r}}{dt} = \int (-g\mathbf{j})\, dt = -gt\mathbf{j} + \mathbf{C}_1$ and $\frac{d\mathbf{r}}{dt}(0) = (v_0 \cos \alpha)\mathbf{i} + (v_0 \sin \alpha)\mathbf{j} \Rightarrow -g(0)\mathbf{j} + \mathbf{C}_1 = (v_0 \cos \alpha)\mathbf{i} + (v_0 \sin \alpha)\mathbf{j}$ $\Rightarrow \mathbf{C}_1 = (v_0 \cos \alpha)\mathbf{i} + (v_0 \sin \alpha)\mathbf{j} \Rightarrow \frac{d\mathbf{r}}{dt} = (v_0 \cos \alpha)\mathbf{i} + (v_0 \sin \alpha - gt)\mathbf{j}$; $\mathbf{r} = \int [(v_0 \cos \alpha)\mathbf{i} + (v_0 \sin \alpha - gt)\mathbf{j}]\, dt$ $= (v_0 t \cos \alpha)\mathbf{i} + \left(v_0 t \sin \alpha - \frac{1}{2}gt^2\right)\mathbf{j} + \mathbf{C}_2$ and $\mathbf{r}(0) = x_0\mathbf{i} + y_0\mathbf{j} \Rightarrow [v_0(0) \cos \alpha]\mathbf{i} + \left[v_0(0) \sin \alpha - \frac{1}{2}g(0)^2\right]\mathbf{j} + \mathbf{C}_2$

$= x_0\mathbf{i} + y_0\mathbf{j} \Rightarrow \mathbf{C}_2 = x_0\mathbf{i} + y_0\mathbf{j} \Rightarrow \mathbf{r} = (x_0 + v_0 t\cos\alpha)\mathbf{i} + \left(y_0 + v_0 t\sin\alpha - \frac{1}{2}gt^2\right)\mathbf{j} \Rightarrow x = x_0 + v_0 t\cos\alpha$ and

$y = y_0 + v_0 t\sin\alpha - \frac{1}{2}gt^2$

21. The horizontal distance from Rebollo to the center of the cauldron is 90 ft $\Rightarrow$ the horizontal distance to the nearest rim is $x = 90 - \frac{1}{2}(12) = 84 \Rightarrow 84 = x_0 + (v_0\cos\alpha)t \approx 0 + \left(\frac{90g}{v_0\sin\alpha}\right)t \Rightarrow 84 = \frac{(90)(32)}{\sqrt{(68)(64)}}t$

$\Rightarrow t = 1.92$ sec. The vertical distance at this time is $y = y_0 + (v_0\sin\alpha)t - \frac{1}{2}gt^2$

$\approx 6 + \sqrt{(68)(64)}\,(1.92) - 16(1.92)^2 \approx 73.7$ ft $\Rightarrow$ the arrow clears the rim by 3.7 ft

23. When marble A is located R units downrange, we have $x = (v_0\cos\alpha)t \Rightarrow R = (v_0\cos\alpha)t \Rightarrow t = \frac{R}{v_0\cos\alpha}$. At that time the height of marble A is $y = y_0 + (v_0\sin\alpha)t - \frac{1}{2}gt^2 = (v_0\sin\alpha)\left(\frac{R}{v_0\cos\alpha}\right) - \frac{1}{2}g\left(\frac{R}{v_0\cos\alpha}\right)^2$

$\Rightarrow y = R\tan\alpha - \frac{1}{2}g\left(\frac{R^2}{v_0^2\cos^2\alpha}\right)$. The height of marble B at the same time $t = \frac{R}{v_0\cos\alpha}$ seconds is

$h = R\tan\alpha - \frac{1}{2}gt^2 = R\tan\alpha - \frac{1}{2}g\left(\frac{R^2}{v_0^2\cos^2\alpha}\right)$. Since the heights are the same, the marbles collide regardless of the initial velocity v_0.

25. $\mathbf{a}(t) = -g\mathbf{k} \Rightarrow \mathbf{v}(t) = -gt\mathbf{k} + \mathbf{C}$; $\mathbf{v}(0) = \mathbf{v}_0 \Rightarrow \mathbf{C} = \mathbf{v}_0 \Rightarrow \mathbf{v}(t) = -gt\mathbf{k} + \mathbf{v}_0 \Rightarrow \mathbf{r}(t) = -\frac{1}{2}gt^2\mathbf{k} + \mathbf{v}_0 t + \mathbf{C}_1$;

$\mathbf{r}(0) = \mathbf{0} \Rightarrow \mathbf{C}_1 = \mathbf{0} \Rightarrow \mathbf{r}(t) = -\frac{1}{2}gt^2\mathbf{k} + \mathbf{v}_0 t$

27. From Eq. (7) in the text, the maximum height is $y = \frac{(v_0\sin\alpha)^2}{2g}$ and this occurs for $x = \frac{v_0^2}{2g}\sin 2\alpha$ $= \frac{v_0^2\sin\alpha\cos\alpha}{g}$. These equations describe parametrically the points on a curve in the xy-plane associated with the maximum heights on the parabolic trajectories in terms of the parameter (launch angle) α.

Eliminating the parameter α, we have $x^2 = \frac{v_0^4\sin^2\alpha\cos^2\alpha}{g^2} = \frac{\left(v_0^4\sin^2\alpha\right)\left(1-\sin^2\alpha\right)}{g^2} = \frac{v_0^4\sin^2\alpha}{g^2} - \frac{v_0^4\sin^4\alpha}{g^2}$

$= \frac{v_0^2}{g}(2y) - (2y)^2 \Rightarrow x^2 + 4y^2 - \left(\frac{2v_0^2}{g}\right)y = 0 \Rightarrow x^2 + 4\left[y^2 - \left(\frac{v_0^2}{2g}\right)y + \frac{v_0^4}{16g^2}\right] = \frac{v_0^4}{16g^2} \Rightarrow x^2 + 4\left(y - \frac{v_0^2}{4g}\right)^2 = \frac{v_0^4}{16g^2}$, where $x \ge 0$.

11.3 ARC LENGTH AND THE UNIT TANGENT VECTOR T

1. $\mathbf{r} = (2\cos t)\mathbf{i} + (2\sin t)\mathbf{j} + \sqrt{5}t\mathbf{k} \Rightarrow \mathbf{v} = (-2\sin t)\mathbf{i} + (2\cos t)\mathbf{j} + \sqrt{5}\mathbf{k}$

$\Rightarrow |\mathbf{v}| = \sqrt{(-2\sin t)^2 + (2\cos t)^2 + \left(\sqrt{5}\right)^2} = \sqrt{4\sin^2 t + 4\cos^2 t + 5} = 3$; $\mathbf{T} = \frac{\mathbf{v}}{|\mathbf{v}|}$

$= \left(-\frac{2}{3}\sin t\right)\mathbf{i} + \left(\frac{2}{3}\cos t\right)\mathbf{j} + \frac{\sqrt{5}}{3}\mathbf{k}$ and Length $= \int_0^{\pi} |\mathbf{v}|\,dt = \int_0^{\pi} 3\,dt = [3t]_0^{\pi} = 3\pi$

3. $\mathbf{r} = t\mathbf{i} + \frac{2}{3}t^{3/2}\mathbf{k} \Rightarrow \mathbf{v} = \mathbf{i} + t^{1/2}\mathbf{k} \Rightarrow |\mathbf{v}| = \sqrt{1^2 + \left(t^{1/2}\right)^2} = \sqrt{1+t}$; $\mathbf{T} = \frac{\mathbf{v}}{|\mathbf{v}|} = \frac{1}{\sqrt{1+t}}\mathbf{i} + \frac{\sqrt{t}}{\sqrt{1+t}}\mathbf{k}$

and Length $= \int_0^8 \sqrt{1+t}\,dt = \left[\frac{2}{3}(1+t)^{3/2}\right]_0^8 = \frac{52}{3}$

5. $\mathbf{r} = \left(\cos^3 t\right)\mathbf{j} + \left(\sin^3 t\right)\mathbf{k} \Rightarrow \mathbf{v} = \left(-3\cos^2 t \sin t\right)\mathbf{j} + \left(3\sin^2 t \cos t\right)\mathbf{k} \Rightarrow |\mathbf{v}|$

$= \sqrt{\left(-3\cos^2 t \sin t\right)^2 + \left(3\sin^2 t \cos t\right)^2} = \sqrt{\left(9\cos^2 t \sin^2 t\right)\left(\cos^2 t + \sin^2 t\right)} = 3\,|\cos t \sin t|$;

$\mathbf{T} = \frac{\mathbf{v}}{|\mathbf{v}|} = \frac{-3\cos^2 t \sin t}{3\,|\cos t \sin t|}\mathbf{j} + \frac{3\sin^2 t \cos t}{3\,|\cos t \sin t|}\mathbf{k} = (-\cos t)\mathbf{j} + (\sin t)\mathbf{k}$, if $0 \le t \le \frac{\pi}{2}$, and

Length $= \int_0^{\pi/2} 3\,|\cos t \sin t|\,dt = \int_0^{\pi/2} 3\cos t \sin t\,dt = \int_0^{\pi/2} \frac{3}{2}\sin 2t\,dt = \left[-\frac{3}{4}\cos 2t\right]_0^{\pi/2} = \frac{3}{2}$

7. $\mathbf{r} = (t\cos t)\mathbf{i} + (t\sin t)\mathbf{j} + \frac{2\sqrt{2}}{3}t^{3/2}\mathbf{k} \Rightarrow \mathbf{v} = (\cos t - t\sin t)\mathbf{i} + (\sin t + t\cos t)\mathbf{j} + \left(\sqrt{2}\,t^{1/2}\right)\mathbf{k}$

$\Rightarrow |\mathbf{v}| = \sqrt{(\cos t - t\sin t)^2 + (\sin t + t\cos t)^2 + \left(\sqrt{2t}\right)^2} = \sqrt{1 + t^2 + 2t} = \sqrt{(t+1)^2} = |t+1| = t+1$, if $t \ge 0$;

$\mathbf{T} = \frac{\mathbf{v}}{|\mathbf{v}|} = \left(\frac{\cos t - t\sin t}{t+1}\right)\mathbf{i} + \left(\frac{\sin t + t\cos t}{t+1}\right)\mathbf{j} + \left(\frac{\sqrt{2}\,t^{1/2}}{t+1}\right)\mathbf{k}$ and Length $= \int_0^{\pi} (t+1)\,dt = \left[\frac{t^2}{2} + t\right]_0^{\pi} = \frac{\pi^2}{2} + \pi$

9. Let $P(t_0)$ denote the point. Then $\mathbf{v} = (5\cos t)\mathbf{i} - (5\sin t)\mathbf{j} + 12\mathbf{k}$ and $26\pi = \int_0^{t_0} \sqrt{25\cos^2 t + 25\sin^2 t + 144}\,dt$

$= \int_0^{t_0} 13\,dt = 13t_0 \Rightarrow t_0 = 2\pi$, and the point is $P(2\pi) = (5\sin 2\pi, 5\cos 2\pi, 24\pi) = (5, 0, 24\pi)$

11. $\mathbf{r} = (4\cos t)\mathbf{i} + (4\sin t)\mathbf{j} + 3t\mathbf{k} \Rightarrow \mathbf{v} = (-4\sin t)\mathbf{i} + (4\cos t)\mathbf{j} + 3\mathbf{k} \Rightarrow |\mathbf{v}| = \sqrt{(-4\sin t)^2 + (4\cos t)^2 + 3^2}$

$= \sqrt{25} = 5 \Rightarrow s(t) = \int_0^t 5\,d\tau = 5t \Rightarrow$ Length $= s\left(\frac{\pi}{2}\right) = \frac{5\pi}{2}$

13. $\mathbf{r} = \left(e^t \cos t\right)\mathbf{i} + \left(e^t \sin t\right)\mathbf{j} + e^t\mathbf{k} \Rightarrow \mathbf{v} = \left(e^t \cos t - e^t \sin t\right)\mathbf{i} + \left(e^t \sin t + e^t \cos t\right)\mathbf{j} + e^t\mathbf{k}$

$\Rightarrow |\mathbf{v}| = \sqrt{\left(e^t \cos t - e^t \sin t\right)^2 + \left(e^t \sin t + e^t \cos t\right)^2 + \left(e^t\right)^2} = = \sqrt{3e^{2t}} = \sqrt{3}\,e^t \Rightarrow s(t) = \int_0^t \sqrt{3}\,e^{\tau}\,d\tau$

$= \sqrt{3}\,e^t - \sqrt{3} \Rightarrow$ Length $= s(0) - s(-\ln 4) = 0 - \left(\sqrt{3}\,e^{-\ln 4} - \sqrt{3}\right) = \frac{3\sqrt{3}}{4}$

15. $\mathbf{r} = \left(\sqrt{2}t\right)\mathbf{i} + \left(\sqrt{2}t\right)\mathbf{j} + \left(1 - t^2\right)\mathbf{k} \Rightarrow \mathbf{v} = \sqrt{2}\mathbf{i} + \sqrt{2}\mathbf{j} - 2t\mathbf{k} \Rightarrow |\mathbf{v}| = \sqrt{\left(\sqrt{2}\right)^2 + \left(\sqrt{2}\right)^2 + (-2t)^2} = \sqrt{4 + 4t^2}$

$= 2\sqrt{1+t^2} \Rightarrow$ Length $= \int_0^1 2\sqrt{1+t^2}\,dt = \left[2\left(\frac{t}{2}\sqrt{1+t^2} + \frac{1}{2}\ln\left(t + \sqrt{1+t^2}\right)\right)\right]_0^1 = \sqrt{2} + \ln\left(1 + \sqrt{2}\right)$

17. (a) $\mathbf{r} = (\cos t)\mathbf{i} + (\sin t)\mathbf{j} + (1 - \cos t)\mathbf{k}$, $0 \le t \le 2\pi \Rightarrow x = \cos t$, $y = \sin t$, $z = 1 - \cos t \Rightarrow x^2 + y^2 = \cos^2 t + \sin^2 t = 1$, a right circular cylinder with the z-axis as the axis and radius $= 1$. Therefore $P(\cos t, \sin t, 1 - \cos t)$ lies on the cylinder $x^2 + y^2 = 1$; $t = 0 \Rightarrow P(1,0,0)$ is on the curve; $t = \frac{\pi}{2} \Rightarrow Q(0,1,1)$ is on the curve; $t = \pi \Rightarrow R(-1,0,2)$ is on the curve. Then $\overrightarrow{PQ} = -\mathbf{i} + \mathbf{j} + \mathbf{k}$ and $\overrightarrow{PR} = -2\mathbf{i} + 2\mathbf{k}$

$$\Rightarrow \overrightarrow{PQ} \times \overrightarrow{PR} = \begin{vmatrix} \mathbf{i} & \mathbf{j} & \mathbf{k} \\ -1 & 1 & 1 \\ -2 & 0 & 2 \end{vmatrix} = 2\mathbf{i} + 2\mathbf{k}$$ is a vector normal to the plane of P, Q, and R. Then the plane containing P, Q, and R has an equation $2x + 2z = 2(1) + 2(0)$ or $x + z = 1$. Any point on the curve will satisfy this equation since $x + z = \cos t + (1 - \cos t) = 1$. Therefore, any point on the curve lies on the intersection of the cylinder $x^2 + y^2 = 1$ and the plane $x + z = 1 \Rightarrow$ the curve is an ellipse.

(b) $\mathbf{v} = (-\sin t)\mathbf{i} + (\cos t)\mathbf{j} + (\sin t)\mathbf{k} \Rightarrow |\mathbf{v}| = \sqrt{\sin^2 t + \cos^2 t + \sin^2 t} = \sqrt{1 + \sin^2 t} \Rightarrow \mathbf{T} = \frac{\mathbf{v}}{|\mathbf{v}|}$

$$= \frac{(-\sin t)\mathbf{i} + (\cos t)\mathbf{j} + (\sin t)\mathbf{k}}{\sqrt{1 + \sin^2 t}} \Rightarrow \mathbf{T}(0) = \mathbf{j},\ \mathbf{T}\left(\frac{\pi}{2}\right) = \frac{-\mathbf{i} + \mathbf{k}}{\sqrt{2}},\ \mathbf{T}(\pi) = -\mathbf{j},\ \mathbf{T}\left(\frac{3\pi}{2}\right) = \frac{\mathbf{i} - \mathbf{k}}{\sqrt{2}}$$

(c) $\mathbf{a} = (-\cos t)\mathbf{i} - (\sin t)\mathbf{j} + (\cos t)\mathbf{k}$; $\mathbf{n} = \mathbf{i} + \mathbf{k}$ is normal to the plane $x + z = 1 \Rightarrow \mathbf{n} \cdot \mathbf{a} = -\cos t + \cos t = 0 \Rightarrow \mathbf{a}$ is orthogonal to $\mathbf{n} \Rightarrow \mathbf{a}$ is parallel to the plane; $\mathbf{a}(0) = -\mathbf{i} + \mathbf{k}$, $\mathbf{a}\left(\frac{\pi}{2}\right) = -\mathbf{j}$, $\mathbf{a}\left(\frac{3\pi}{2}\right) = \mathbf{j}$

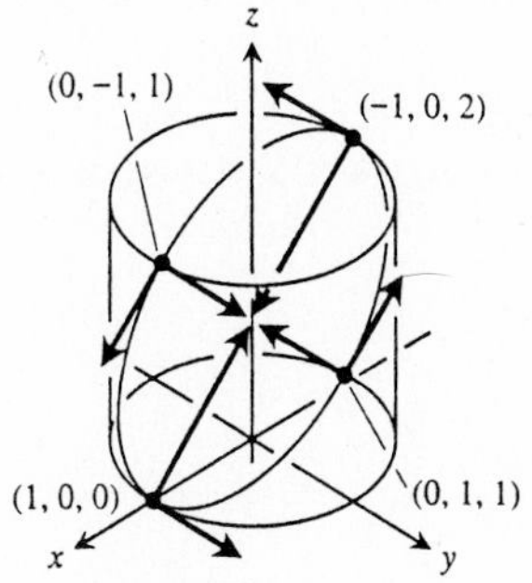

(d) $|\mathbf{v}| = \sqrt{1 + \sin^2 t}$ (See part (b) $\Rightarrow L = \int_0^{2\pi} \sqrt{1 + \sin^2 t}\, dt$

(e) $L \approx 7.64$ (by *Mathematica*)

11.4 CURVATURE, TORSION, AND THE TNB FRAME

1. $\mathbf{r} = t\mathbf{i} + \ln(\cos t)\mathbf{j} \Rightarrow \mathbf{v} = \mathbf{i} + \left(\frac{-\sin t}{\cos t}\right)\mathbf{j} = \mathbf{i} - (\tan t)\mathbf{j} \Rightarrow |\mathbf{v}| = \sqrt{1^2 + (-\tan t)^2} = \sqrt{\sec^2 t} = |\sec t| = \sec t$, since $-\frac{\pi}{2} < t < \frac{\pi}{2} \Rightarrow \mathbf{T} = \frac{\mathbf{v}}{|\mathbf{v}|} = \left(\frac{1}{\sec t}\right)\mathbf{i} - \left(\frac{\tan t}{\sec t}\right)\mathbf{j} = (\cos t)\mathbf{i} - (\sin t)\mathbf{j}$; $\frac{d\mathbf{T}}{dt} = (-\sin t)\mathbf{i} - (\cos t)\mathbf{j}$

$$\Rightarrow \left|\frac{d\mathbf{T}}{dt}\right| = \sqrt{(-\sin t)^2 + (-\cos t)^2} = 1 \Rightarrow \mathbf{N} = \frac{\left(\frac{d\mathbf{T}}{dt}\right)}{\left|\frac{d\mathbf{T}}{dt}\right|} = (-\sin t)\mathbf{i} - (\cos t)\mathbf{j};\ \mathbf{a} = \left(-\sec^2 t\right)\mathbf{j}$$

$$\Rightarrow \mathbf{v}\times\mathbf{a} = \begin{vmatrix} \mathbf{i} & \mathbf{j} & \mathbf{k} \\ 1 & -\tan t & 0 \\ 0 & -\sec^2 t & 0 \end{vmatrix} = \left(-\sec^2 t\right)\mathbf{k} \Rightarrow |\mathbf{v}\times\mathbf{a}| = \sqrt{\left(-\sec^2 t\right)^2} = \sec^2 t \Rightarrow \kappa = \frac{|\mathbf{v}\times\mathbf{a}|}{|\mathbf{v}|^3} = \frac{\sec^2 t}{\sec^3 t} = \cos t$$

3. $\mathbf{r} = (2t+3)\mathbf{i} + \left(5-t^2\right)\mathbf{j} \Rightarrow \mathbf{v} = 2\mathbf{i} - 2t\mathbf{j} \Rightarrow |\mathbf{v}| = \sqrt{2^2+(-2t)^2} = 2\sqrt{1+t^2} \Rightarrow \mathbf{T} = \frac{\mathbf{v}}{|\mathbf{v}|} = \frac{2}{2\sqrt{1+t^2}}\mathbf{i} + \frac{-2t}{2\sqrt{1+t^2}}\mathbf{j}$

$$= \frac{1}{\sqrt{1+t^2}}\mathbf{i} - \frac{t}{\sqrt{1+t^2}}\mathbf{j};\ \frac{d\mathbf{T}}{dt} = \frac{-t}{\left(\sqrt{1+t^2}\right)^3} - \frac{1}{\left(\sqrt{1+t^2}\right)^3}\mathbf{j} \Rightarrow \left|\frac{d\mathbf{T}}{dt}\right| = \sqrt{\left(\frac{-t}{\left(\sqrt{1+t^2}\right)^3}\right)^2 + \left(-\frac{1}{\left(\sqrt{1+t^2}\right)^3}\right)^2}$$

$$= \sqrt{\frac{1}{\left(1+t^2\right)^2}} = \frac{1}{1+t^2} \Rightarrow \mathbf{N} = \frac{\left(\frac{d\mathbf{T}}{dt}\right)}{\left|\frac{d\mathbf{T}}{dt}\right|} = \frac{-t}{\sqrt{1+t^2}}\mathbf{i} - \frac{1}{\sqrt{1+t^2}}\mathbf{j};\ \mathbf{a} = -2\mathbf{j} \Rightarrow \mathbf{v}\times\mathbf{a} = \begin{vmatrix} \mathbf{i} & \mathbf{j} & \mathbf{k} \\ 2 & -2t & 0 \\ 0 & -2 & 0 \end{vmatrix} = -4\mathbf{k}$$

$$\Rightarrow |\mathbf{v}\times\mathbf{a}| = \sqrt{(-4)^2} = 4 \Rightarrow \kappa = \frac{|\mathbf{v}\times\mathbf{a}|}{|\mathbf{v}|^3} = \frac{4}{\left(2\sqrt{1+t^2}\right)^3} = \frac{1}{2\left(\sqrt{1+t^2}\right)^3}$$

5. $\mathbf{r} = (2t+3)\mathbf{i} + \left(t^2-1\right)\mathbf{j} \Rightarrow \mathbf{v} = 2\mathbf{i} + 2t\mathbf{j} \Rightarrow |\mathbf{v}| = \sqrt{2^2+(2t)^2} = 2\sqrt{1+t^2} \Rightarrow a_T = \frac{d}{dt}|\mathbf{v}| = 2\left(\frac{1}{2}\right)\left(1+t^2\right)^{-1/2}(2t)$

$$= \frac{2t}{\sqrt{1+t^2}};\ \mathbf{a} = 2\mathbf{j} \Rightarrow |\mathbf{a}| = 2 \Rightarrow a_N = \sqrt{|\mathbf{a}|^2 - a_T^2} = \sqrt{2^2 - \left(\frac{2t}{\sqrt{1+t^2}}\right)^2} = \frac{2}{\sqrt{1+t^2}};\ \mathbf{a} = \frac{2t}{\sqrt{1+t^2}}\mathbf{T} + \frac{2}{\sqrt{1+t^2}}\mathbf{N}$$

7. (a) $\mathbf{r} = x\mathbf{i} + f(x)\mathbf{j} \Rightarrow \mathbf{v} = \mathbf{i} + f'(x)\mathbf{j} \Rightarrow \mathbf{a} = f''(x)\mathbf{j} \Rightarrow \mathbf{v}\times\mathbf{a} = \begin{vmatrix} \mathbf{i} & \mathbf{j} & \mathbf{k} \\ 1 & f'(x) & 0 \\ 0 & f''(x) & 0 \end{vmatrix} = f''(x)\mathbf{k}$

$$\Rightarrow |\mathbf{v}\times\mathbf{a}| = \sqrt{\left(f''(x)\right)^2} = |f''(x)| \text{ and } |\mathbf{v}| = \sqrt{1^2+[f'(x)]^2} = \sqrt{1+[f'(x)]^2} \Rightarrow \kappa = \frac{|\mathbf{v}\times\mathbf{a}|}{|\mathbf{v}|^3}$$

$$= \frac{|f''(x)|}{\left[1+\left(f'(x)\right)^2\right]^{3/2}}$$

(b) $y = \ln(\cos x) \Rightarrow \frac{dy}{dx} = \left(\frac{1}{\cos x}\right)(-\sin x) = -\tan x \Rightarrow \frac{d^2y}{dx^2} = -\sec^2 x \Rightarrow \kappa = \frac{|-\sec^2 x|}{\left[1+(-\tan x)^2\right]^{3/2}} = \frac{\sec^2 x}{|\sec^3 x|}$

$= \frac{1}{\sec x} = \cos x$, since $-\frac{\pi}{2} < x < \frac{\pi}{2}$

(c) $x = x_0$ gives a point of inflection $\Rightarrow f''(x_0) = 0$ (since f is twice differentiable) $\Rightarrow \kappa = 0$

9. (a) $\mathbf{r}(t) = f(t)\mathbf{i} + g(t)\mathbf{j} \Rightarrow \mathbf{v} = f'(t)\mathbf{i} + g'(t)\mathbf{j}$ is tangent to the curve at the point $(f(t), g(t))$;

$\mathbf{n}\cdot\mathbf{v} = \left[-g'(t)\mathbf{i} + f'(t)\mathbf{j}\right]\cdot\left[f'(t)\mathbf{i} + g'(t)\mathbf{j}\right] = -g'(t)f'(t) + f'(t)g'(t) = 0$; $-\mathbf{n}\cdot\mathbf{v} = -(\mathbf{n}\cdot\mathbf{v}) = 0$; thus,

$\mathbf{n}$ and $-\mathbf{n}$ are both normal to the curve at the point

(b) $\mathbf{r}(t) = t\mathbf{i} + e^{2t}\mathbf{j} \Rightarrow \mathbf{v} = \mathbf{i} + 2e^{2t}\mathbf{j} \Rightarrow \mathbf{n} = -2e^{2t}\mathbf{i} + \mathbf{j}$ points toward the concave side of the curve; $\mathbf{N} = \frac{\mathbf{n}}{|\mathbf{n}|}$ and $|\mathbf{n}| = \sqrt{4e^{4t}+1} \Rightarrow \mathbf{N} = \frac{-2e^{2t}}{\sqrt{1+4e^{4t}}}\mathbf{i} + \frac{1}{\sqrt{1+4e^{4t}}}\mathbf{j}$

(c) $\mathbf{r}(t) = \sqrt{4-t^2}\,\mathbf{i} + t\mathbf{j} \Rightarrow \mathbf{v} = \frac{-t}{\sqrt{4-t^2}}\mathbf{i} + \mathbf{j} \Rightarrow \mathbf{n} = -\mathbf{i} - \frac{t}{\sqrt{4-t^2}}\mathbf{j}$ points toward the concave side of the curve; $\mathbf{N} = \frac{\mathbf{n}}{|\mathbf{n}|}$ and $|\mathbf{n}| = \sqrt{1 + \frac{t^2}{4-t^2}} = \frac{2}{\sqrt{4-t^2}} \Rightarrow \mathbf{N} = -\frac{1}{2}\left(\sqrt{4-t^2}\,\mathbf{i} + t\mathbf{j}\right)$

11. $\mathbf{r} = (3\sin t)\mathbf{i} + (3\cos t)\mathbf{j} + 4t\mathbf{k} \Rightarrow \mathbf{v} = (3\cos t)\mathbf{i} + (-3\sin t)\mathbf{j} + 4\mathbf{k} \Rightarrow |\mathbf{v}| = \sqrt{(3\cos t)^2 + (-3\sin t)^2 + 4^2}$ $= \sqrt{25} = 5 \Rightarrow \mathbf{T} = \frac{\mathbf{v}}{|\mathbf{v}|} = \left(\frac{3}{5}\cos t\right)\mathbf{i} - \left(\frac{3}{5}\sin t\right)\mathbf{j} + \frac{4}{5}\mathbf{k} \Rightarrow \frac{d\mathbf{T}}{dt} = \left(-\frac{3}{5}\sin t\right)\mathbf{i} - \left(\frac{3}{5}\cos t\right)\mathbf{j}$ $\Rightarrow \left|\frac{d\mathbf{T}}{dt}\right| = \sqrt{\left(-\frac{3}{5}\sin t\right)^2 + \left(-\frac{3}{5}\cos t\right)^2} = \frac{3}{5} \Rightarrow \mathbf{N} = \frac{\left(\frac{d\mathbf{T}}{dt}\right)}{\left|\frac{d\mathbf{T}}{dt}\right|} = (-\sin t)\mathbf{i} - (\cos t)\mathbf{j}$; $\mathbf{a} = (-3\sin t)\mathbf{i} + (-3\cos t)\mathbf{j}$

$$\Rightarrow \mathbf{v}\times\mathbf{a} = \begin{vmatrix} \mathbf{i} & \mathbf{j} & \mathbf{k} \\ 3\cos t & -3\sin t & 4 \\ -3\sin t & -3\cos t & 0 \end{vmatrix} = (12\cos t)\mathbf{i} - (12\sin t)\mathbf{j} - 9\mathbf{k} \Rightarrow |\mathbf{v}\times\mathbf{a}|$$

$= \sqrt{(12\cos t)^2 + (-12\sin t)^2 + (-9)^2} = \sqrt{225} = 15 \Rightarrow \kappa = \frac{|\mathbf{v}\times\mathbf{a}|}{|\mathbf{v}|^3} = \frac{15}{5^3} = \frac{3}{25}$; $\mathbf{B} = \mathbf{T}\times\mathbf{N}$

$$= \begin{vmatrix} \mathbf{i} & \mathbf{j} & \mathbf{k} \\ \frac{3}{5}\cos t & -\frac{3}{5}\sin t & \frac{4}{5} \\ -\sin t & -\cos t & 0 \end{vmatrix} = \left(\frac{4}{5}\cos t\right)\mathbf{i} - \left(\frac{4}{5}\sin t\right)\mathbf{j} - \frac{3}{5}\mathbf{k};\ \frac{d\mathbf{a}}{dt} = (-3\cos t)\mathbf{i} + (3\sin t)\mathbf{j}$$

$$\Rightarrow \tau = \frac{\begin{vmatrix} 3\cos t & -3\sin t & 4 \\ -3\sin t & -3\sin t & 0 \\ -3\cos t & 3\sin t & 0 \end{vmatrix}}{|\mathbf{v}\times\mathbf{a}|^2} = \frac{-36\sin^2 t - 36\cos^2 t}{15^2} = -\frac{4}{25}$$

13. $\mathbf{r} = \left(e^t\cos t\right)\mathbf{i} + \left(e^t\sin t\right)\mathbf{j} + 2\mathbf{k} \Rightarrow \mathbf{v} = \left(e^t\cos t - e^t\sin t\right)\mathbf{i} + \left(e^t\sin t + e^t\cos t\right)\mathbf{j} \Rightarrow$ $|\mathbf{v}| = \sqrt{\left(e^t\cos t - e^t\sin t\right)^2 + \left(e^t\sin t + e^t\cos t\right)^2} = \sqrt{2e^{2t}} = e^t\sqrt{2}$;

$\mathbf{T} = \frac{\mathbf{v}}{|\mathbf{v}|} = \left(\frac{\cos t - \sin t}{\sqrt{2}}\right)\mathbf{i} + \left(\frac{\sin t + \cos t}{\sqrt{2}}\right)\mathbf{j} \Rightarrow \frac{d\mathbf{T}}{dt} = \left(\frac{-\sin t - \cos t}{\sqrt{2}}\right)\mathbf{i} + \left(\frac{\cos t - \sin t}{\sqrt{2}}\right)\mathbf{j}$

$\Rightarrow \left|\frac{d\mathbf{T}}{dt}\right| = \sqrt{\left(\frac{-\sin t - \cos t}{\sqrt{2}}\right)^2 + \left(\frac{\cos t - \sin t}{\sqrt{2}}\right)^2} = 1 \Rightarrow \mathbf{N} = \frac{\left(\frac{d\mathbf{T}}{dt}\right)}{\left|\frac{d\mathbf{T}}{dt}\right|} = \left(\frac{-\cos t - \sin t}{\sqrt{2}}\right)\mathbf{i} + \left(\frac{-\sin t + \cos t}{\sqrt{2}}\right)\mathbf{j}$;

$$\mathbf{a} = \left(-2e^t \sin t\right)\mathbf{i} + \left(2e^t \cos t\right)\mathbf{j} \Rightarrow \mathbf{v}\times\mathbf{a} = \begin{vmatrix} \mathbf{i} & \mathbf{j} & \mathbf{k} \\ e^t \cos t - e^t \sin t & e^t \sin t + e^t \cos t & 0 \\ -2e^t \sin t & 2e^t \cos t & 0 \end{vmatrix} = 2e^{2t}\mathbf{k}$$

$$\Rightarrow |\mathbf{v}\times\mathbf{a}| = \sqrt{\left(2e^{2t}\right)^2} = 2e^{2t} \Rightarrow \kappa = \frac{|\mathbf{v}\times\mathbf{a}|}{|\mathbf{v}|^3} = \frac{2e^{2t}}{\left(e^t\sqrt{2}\right)^3} = \frac{1}{e^t\sqrt{2}};$$

$$\mathbf{B} = \mathbf{T}\times\mathbf{N} = \begin{vmatrix} \mathbf{i} & \mathbf{j} & \mathbf{k} \\ \frac{\cos t - \sin t}{\sqrt{2}} & \frac{\sin t + \cos t}{\sqrt{2}} & 0 \\ \frac{-\cos t - \sin t}{\sqrt{2}} & \frac{-\sin t + \cos t}{\sqrt{2}} & 0 \end{vmatrix}$$

$$= \left[\tfrac{1}{2}(\cos t - \sin t)(-\sin t + \cos t) - \tfrac{1}{2}(-\cos t - \sin t)(\sin t + \cos t)\right]\mathbf{k}$$

$$= \left[\tfrac{1}{2}\left(\cos^2 t - 2\cos t \sin t + \sin^2 t\right) + \tfrac{1}{2}\left(\cos^2 t + 2\sin t\cos t + \sin^2 t\right)\right]\mathbf{k} = \mathbf{k};$$

$$\frac{d\mathbf{a}}{dt} = \left(-2e^t \sin t - 2e^t \cos t\right)\mathbf{i} + \left(2e^t \cos t - 2e^t \sin t\right)\mathbf{j}$$

$$\Rightarrow \tau = \frac{\begin{vmatrix} e^t \cos t - e^t \sin t & e^t \sin t + e^t \cos t & 0 \\ -2e^t \sin t & 2e^t \cos t & 0 \\ -2e^t \sin t - 2e^t \cos t & 2e^t \cos t - 2e^t \sin t & 0 \end{vmatrix}}{|\mathbf{v}\times\mathbf{a}|^2} = 0$$

15. $\mathbf{r} = \left(\frac{t^3}{3}\right)\mathbf{i} + \left(\frac{t^2}{2}\right)\mathbf{j},\ t > 0 \Rightarrow \mathbf{v} = t^2\mathbf{i} + t\mathbf{j} \Rightarrow |\mathbf{v}| = \sqrt{t^4 + t^2} = t\sqrt{t^2+1}$, since $t > 0 \Rightarrow \mathbf{T} = \frac{\mathbf{v}}{|\mathbf{v}|}$

$$= \frac{t}{\sqrt{t^2+t}}\mathbf{i} + \frac{1}{\sqrt{t^2+1}}\mathbf{j} \Rightarrow \frac{d\mathbf{T}}{dt} = \frac{1}{\left(t^2+1\right)^{3/2}}\mathbf{i} - \frac{t}{\left(t^2+1\right)^{3/2}}\mathbf{j} \Rightarrow \left|\frac{d\mathbf{T}}{dt}\right| = \sqrt{\left(\frac{1}{\left(t^2+1\right)^{3/2}}\right)^2 + \left(\frac{-t}{\left(t^2+1\right)^{3/2}}\right)^2}$$

$$= \sqrt{\frac{1+t^2}{\left(t^2+1\right)^3}} = \frac{1}{t^2+1} \Rightarrow \mathbf{N} = \frac{\left(\frac{d\mathbf{T}}{dt}\right)}{\left|\frac{d\mathbf{T}}{dt}\right|} = \frac{1}{\sqrt{t^2+1}}\mathbf{i} - \frac{t}{\sqrt{t^2+1}}\mathbf{j};\ \mathbf{a} = 2t\mathbf{i} + \mathbf{j} \Rightarrow \mathbf{v}\times\mathbf{a} = \begin{vmatrix} \mathbf{i} & \mathbf{j} & \mathbf{k} \\ t^2 & t & 0 \\ 2t & 1 & 0 \end{vmatrix} = -t^2\mathbf{k}$$

$$\Rightarrow |\mathbf{v}\times\mathbf{a}| = \sqrt{\left(-t^2\right)^2} = t^2 \Rightarrow \kappa = \frac{|\mathbf{v}\times\mathbf{a}|}{|\mathbf{v}|^3} = \frac{t^2}{\left(t\sqrt{t^2+1}\right)^3} = \frac{1}{t\left(t^2+1\right)^{3/2}};$$

$$\mathbf{B}=\mathbf{T}\times\mathbf{N}=\begin{vmatrix}\mathbf{i} & \mathbf{j} & \mathbf{k}\\ \frac{t}{\sqrt{t^2+1}} & \frac{1}{\sqrt{t^2+1}} & 0\\ \frac{1}{\sqrt{t^2+1}} & \frac{-t}{\sqrt{t^2+1}} & 0\end{vmatrix}=-\mathbf{k};\ \frac{d\mathbf{a}}{dt}=2\mathbf{i}\Rightarrow\tau=\frac{\begin{vmatrix}t^2 & t & 0\\ 2t & 1 & 0\\ 2 & 0 & 0\end{vmatrix}}{|\mathbf{v}\times\mathbf{a}|^2}=0$$

17. $\mathbf{r}=t\mathbf{i}+\left(a\cosh\frac{t}{a}\right)\mathbf{j},\ a>0\Rightarrow\mathbf{v}=\mathbf{i}+\left(\sinh\frac{t}{a}\right)\mathbf{j}\Rightarrow|\mathbf{v}|=\sqrt{1+\sinh^2\left(\frac{t}{a}\right)}=\sqrt{\cosh^2\left(\frac{t}{a}\right)}=\cosh\frac{t}{a}$

$\Rightarrow\mathbf{T}=\frac{\mathbf{v}}{|\mathbf{v}|}=\left(\operatorname{sech}\frac{t}{a}\right)\mathbf{i}+\left(\tanh\frac{t}{a}\right)\mathbf{j}\Rightarrow\frac{d\mathbf{T}}{dt}=\left(-\frac{1}{a}\operatorname{sech}\frac{t}{a}\tanh\frac{t}{a}\right)\mathbf{i}+\left(\frac{1}{a}\operatorname{sech}^2\frac{t}{a}\right)\mathbf{j}$

$\Rightarrow\left|\frac{d\mathbf{T}}{dt}\right|=\sqrt{\frac{1}{a^2}\operatorname{sech}^2\left(\frac{t}{a}\right)\tanh^2\left(\frac{t}{a}\right)+\frac{1}{a^2}\operatorname{sech}^4\left(\frac{t}{a}\right)}=\frac{1}{a}\operatorname{sech}\left(\frac{t}{a}\right)\Rightarrow\mathbf{N}=\frac{\left(\frac{d\mathbf{T}}{dt}\right)}{\left|\frac{d\mathbf{T}}{dt}\right|}=\left(-\tanh\frac{t}{a}\right)\mathbf{i}+\left(\operatorname{sech}\frac{t}{a}\right)\mathbf{j};$

$$\mathbf{a}=\left(\frac{1}{a}\cosh\frac{t}{a}\right)\mathbf{j}\Rightarrow\mathbf{v}\times\mathbf{a}=\begin{vmatrix}\mathbf{i} & \mathbf{j} & \mathbf{k}\\ 1 & \sinh\left(\frac{t}{a}\right) & 0\\ 0 & \frac{1}{a}\cosh\left(\frac{t}{a}\right) & 0\end{vmatrix}=\left(\frac{1}{a}\cosh\frac{t}{a}\right)\mathbf{k}\Rightarrow|\mathbf{v}\times\mathbf{a}|=\frac{1}{a}\cosh\left(\frac{t}{a}\right)\Rightarrow\kappa=\frac{|\mathbf{v}\times\mathbf{a}|}{|\mathbf{v}|^3}$$

$$=\frac{\frac{1}{a}\cosh\left(\frac{t}{a}\right)}{\cosh^3\left(\frac{t}{a}\right)}=\frac{1}{a}\operatorname{sech}^2\left(\frac{t}{a}\right);\ \mathbf{B}=\mathbf{T}\times\mathbf{N}=\begin{vmatrix}\mathbf{i} & \mathbf{j} & \mathbf{k}\\ \operatorname{sech}\left(\frac{t}{a}\right) & \tanh\left(\frac{t}{a}\right) & 0\\ -\tanh\left(\frac{t}{a}\right) & \operatorname{sech}\left(\frac{t}{a}\right) & 0\end{vmatrix}=\mathbf{k};\ \frac{d\mathbf{a}}{dt}=\frac{1}{a^2}\sinh\left(\frac{t}{a}\right)\mathbf{j}$$

$$\tau=\frac{\begin{vmatrix}1 & \sinh\left(\frac{t}{a}\right) & 0\\ 0 & \frac{1}{a}\cosh\left(\frac{t}{a}\right) & 0\\ 0 & \frac{1}{a^2}\sinh\left(\frac{t}{a}\right) & 0\end{vmatrix}}{|\mathbf{v}\times\mathbf{a}|^2}=0$$

19. $\mathbf{r}=(a\cos t)\mathbf{i}+(a\sin t)\mathbf{j}+bt\mathbf{k}\Rightarrow\mathbf{v}=(-a\sin t)\mathbf{i}+(a\cos t)\mathbf{j}+b\mathbf{k}\Rightarrow|\mathbf{v}|=\sqrt{(-a\sin t)^2+(a\cos t)^2+b^2}$

$=\sqrt{a^2+b^2}\Rightarrow a_T=\frac{d}{dt}|\mathbf{v}|=0;\ \mathbf{a}=(-a\cos t)\mathbf{i}+(-a\sin t)\mathbf{j}\Rightarrow|\mathbf{a}|=\sqrt{(-a\cos t)^2+(-a\sin t)^2}=\sqrt{a^2}=|a|$

$\Rightarrow a_N=\sqrt{|\mathbf{a}|^2-a_T^2}=\sqrt{|\mathbf{a}|^2-0^2}=|\mathbf{a}|=|a|\Rightarrow\mathbf{a}=(0)\mathbf{T}+|a|\mathbf{N}=|a|\mathbf{N}$

21. $\mathbf{r}=(t+1)\mathbf{i}+2t\mathbf{j}+t^2\mathbf{k}\Rightarrow\mathbf{v}=\mathbf{i}+2\mathbf{j}+2t\mathbf{k}\Rightarrow|\mathbf{v}|=\sqrt{1^2+2^2+(2t)^2}=\sqrt{5+4t^2}\Rightarrow a_T=\frac{1}{2}\left(5+4t^2\right)^{-1/2}(8t)$

$=4t\left(5+4t^2\right)^{-1/2}\Rightarrow a_T(1)=\frac{4}{\sqrt{9}}=\frac{4}{3};\ \mathbf{a}=2\mathbf{k}\Rightarrow\mathbf{a}(1)=2\mathbf{k}\Rightarrow|\mathbf{a}(1)|=2\Rightarrow a_N=\sqrt{|\mathbf{a}|^2-a_T^2}=\sqrt{2^2-\left(\frac{4}{3}\right)^2}$

$= \sqrt{\frac{20}{9}} = \frac{2\sqrt{5}}{3} \Rightarrow \mathbf{a}(1) = \frac{4}{3}\mathbf{T} + \frac{2\sqrt{5}}{3}\mathbf{N}$

23. $\mathbf{r} = t^2\mathbf{i} + \left(t + \frac{1}{3}t^3\right)\mathbf{j} + \left(t - \frac{1}{3}t^3\right)\mathbf{k} \Rightarrow \mathbf{v} = 2t\mathbf{i} + (1 + t^2)\mathbf{j} + (1 - t^2)\mathbf{k} \Rightarrow |\mathbf{v}| = \sqrt{(2t)^2 + (1 + t^2)^2 + (1 - t^2)^2}$

$= \sqrt{2(t^4 + 2t^2 + 1)} = \sqrt{2}(1 + t^2) \Rightarrow a_T = 2t\sqrt{2} \Rightarrow a_T(0) = 0$; $\mathbf{a} = 2\mathbf{i} + 2t\mathbf{j} - 2t\mathbf{k} \Rightarrow \mathbf{a}(0) = 2\mathbf{i} \Rightarrow |\mathbf{a}(0)| = 2$

$\Rightarrow a_N = \sqrt{|\mathbf{a}|^2 - a_T^2} = \sqrt{2^2 - 0^2} = 2 \Rightarrow \mathbf{a}(0) = (0)\mathbf{T} + 2\mathbf{N} = 2\mathbf{N}$

25. $\mathbf{r} = (\cos t)\mathbf{i} + (\sin t)\mathbf{j} - \mathbf{k} \Rightarrow \mathbf{v} = (-\sin t)\mathbf{i} + (\cos t)\mathbf{j} \Rightarrow |\mathbf{v}| = \sqrt{(-\sin t)^2 + (\cos t)^2} = 1 \Rightarrow \mathbf{T} = \frac{\mathbf{v}}{|\mathbf{v}|}$

$= (-\sin t)\mathbf{i} + (\cos t)\mathbf{j} \Rightarrow \mathbf{T}\left(\frac{\pi}{4}\right) = -\frac{\sqrt{2}}{2}\mathbf{i} + \frac{\sqrt{2}}{2}\mathbf{j}$; $\frac{d\mathbf{T}}{dt} = (-\cos t)\mathbf{i} - (\sin t)\mathbf{j} \Rightarrow \left|\frac{d\mathbf{T}}{dt}\right| = \sqrt{(-\cos t)^2 + (-\sin t)^2}$

$= 1 \Rightarrow \mathbf{N} = \frac{\left(\frac{d\mathbf{T}}{dt}\right)}{\left|\frac{d\mathbf{T}}{dt}\right|} = (-\cos t)\mathbf{i} - (\sin t)\mathbf{j} \Rightarrow \mathbf{N}\left(\frac{\pi}{4}\right) = -\frac{\sqrt{2}}{2}\mathbf{i} - \frac{\sqrt{2}}{2}\mathbf{j}$; $\mathbf{B} = \mathbf{T} \times \mathbf{N} = \begin{vmatrix} \mathbf{i} & \mathbf{j} & \mathbf{k} \\ -\sin t & \cos t & 0 \\ -\cos t & -\sin t & 0 \end{vmatrix} = \mathbf{k}$

$\Rightarrow \mathbf{B}\left(\frac{\pi}{4}\right) = \mathbf{k}$, the normal to the osculating plane; $\mathbf{r}\left(\frac{\pi}{4}\right) = \frac{\sqrt{2}}{2}\mathbf{i} + \frac{\sqrt{2}}{2}\mathbf{j} - \mathbf{k} \Rightarrow P = \left(\frac{\sqrt{2}}{2}, \frac{\sqrt{2}}{2}, -1\right)$ lies on the osculating plane $\Rightarrow 0\left(x - \frac{\sqrt{2}}{2}\right) + 0\left(y - \frac{\sqrt{2}}{2}\right) + (z - (-1)) = 0 \Rightarrow z = -1$ is the osculating plane; $\mathbf{T}$ is normal to the normal plane $\Rightarrow \left(-\frac{\sqrt{2}}{2}\right)\left(x - \frac{\sqrt{2}}{2}\right) + \left(\frac{\sqrt{2}}{2}\right)\left(y - \frac{\sqrt{2}}{2}\right) + 0(z - (-1)) = 0 \Rightarrow -\frac{\sqrt{2}}{2}x + \frac{\sqrt{2}}{2}y = 0$

$\Rightarrow -x + y = 0$ is the normal plane; $\mathbf{N}$ is normal to the rectifying plane

$\Rightarrow \left(-\frac{\sqrt{2}}{2}\right)\left(x - \frac{\sqrt{2}}{2}\right) + \left(-\frac{\sqrt{2}}{2}\right)\left(y - \frac{\sqrt{2}}{2}\right) + 0(z - (-1)) = 0 \Rightarrow -\frac{\sqrt{2}}{2}x - \frac{\sqrt{2}}{2}y = -1 \Rightarrow x + y = \sqrt{2}$ is the rectifying plane

27. Yes. If the car is moving along a curved path, then $\kappa \neq 0$ and $a_N = \kappa|\mathbf{v}|^2 \neq 0 \Rightarrow \mathbf{a} = a_T\mathbf{T} + a_N\mathbf{N} \neq \mathbf{0}$.

29. $\mathbf{a} \perp \mathbf{v} \Rightarrow \mathbf{a} \perp \mathbf{T} \Rightarrow a_T = 0 \Rightarrow \frac{d}{dt}|\mathbf{v}| = 0 \Rightarrow |\mathbf{v}|$ is constant

31. $\mathbf{a} = a_T\mathbf{T} + a_N\mathbf{N}$, where $a_T = \frac{d}{dt}|\mathbf{v}| = \frac{d}{dt}(\text{constant}) = 0$ and $a_N = \kappa|\mathbf{v}|^2 \Rightarrow \mathbf{F} = m\mathbf{a} = m\kappa|\mathbf{v}|^2\mathbf{N} \Rightarrow |\mathbf{F}| = m\kappa|\mathbf{v}|^2$

$= \left(m|\mathbf{v}|^2\right)\kappa$, a constant multiple of the curvature κ of the trajectory

33. $y = ax^2 \Rightarrow y' = 2ax \Rightarrow y'' = 2a$; from Exercise 7(a), $\kappa(x) = \frac{|2a|}{(1 + 4a^2x^2)^{3/2}} = |2a|(1 + 4a^2x^2)^{-3/2}$

$\Rightarrow \kappa'(x) = -\frac{3}{2}|2a|(1 + 4a^2x^2)^{-5/2}(8a^2x)$; thus, $\kappa'(x) = 0 \Rightarrow x = 0$. Now, $\kappa'(x) > 0$ for $x < 0$ and $\kappa'(x) < 0$ for $x > 0$ so that $\kappa(x)$ has an absolute maximum at $x = 0$ which is the vertex of the parabola. Since $x = 0$ is the only critical point for $\kappa(x)$, the curvature has no minimum value.

35. $\kappa = \frac{a}{a^2+b^2} \Rightarrow \frac{d\kappa}{da} = \frac{-a^2+b^2}{(a^2+b^2)^2}$; $\frac{d\kappa}{da} = 0 \Rightarrow a^2+b^2 = 0 \Rightarrow a = \pm b \Rightarrow a = b$ since a, b > 0. Now, $\frac{d\kappa}{da} > 0$ if $a < b$ and $\frac{d\kappa}{da} < 0$ if $a > b \Rightarrow \kappa$ is at a maximum for a = b and $\kappa(b) = \frac{b}{b^2+b^2} = \frac{1}{2b}$ is the maximum value

37. $\mathbf{r} = (x_0 + At)\mathbf{i} + (y_0 + Bt)\mathbf{j} + (z_0 + Ct)\mathbf{k} \Rightarrow \mathbf{v} = A\mathbf{i} + B\mathbf{j} + C\mathbf{k} \Rightarrow \mathbf{a} = \mathbf{0} \Rightarrow \mathbf{v} \times \mathbf{a} = \mathbf{0} \Rightarrow \kappa = 0$. Since the curve is a plane curve, $\tau = 0$.

39. (a) From Exercise 36, $\kappa = \frac{1}{t}$ and $|\mathbf{v}| = t \Rightarrow K = \int_a^b \left(\frac{1}{t}\right)(t)\,dt = b - a$

(b) $y = x^2 \Rightarrow x = t$ and $y = t^2$, $-\infty < t < \infty \Rightarrow \mathbf{r}(t) = t\mathbf{i} + t^2\mathbf{j} \Rightarrow \mathbf{v} = \mathbf{i} + 2t\mathbf{j} \Rightarrow |\mathbf{v}| = \sqrt{1+4t^2}$; also $\mathbf{a} = 2\mathbf{j}$

$$\Rightarrow \mathbf{v} \times \mathbf{a} = \begin{vmatrix} \mathbf{i} & \mathbf{j} & \mathbf{k} \\ 1 & 2t & 0 \\ 0 & 2 & 0 \end{vmatrix} = 2\mathbf{k} \Rightarrow |\mathbf{v} \times \mathbf{a}| = 2 \Rightarrow \kappa = \frac{|\mathbf{v} \times \mathbf{a}|}{|\mathbf{v}|^3} = \frac{2}{\left(\sqrt{1+4t^2}\right)^3}.$$ Then

$$K = \int_{-\infty}^{\infty} \frac{2}{\left(\sqrt{1+4t^2}\right)^3}\left(\sqrt{1+4t^2}\right) dt = \int_{-\infty}^{\infty} \frac{2}{1+4t^2}\,dt = \lim_{a\to-\infty} \int_a^0 \frac{2}{1+4t^2}\,dt + \lim_{b\to\infty} \int_0^b \frac{2}{1+4t^2}\,dt$$

$$= \lim_{a\to-\infty} \left[\tan^{-1} 2t\right]_a^0 + \lim_{b\to\infty} \left[\tan^{-1} 2t\right]_0^b = \lim_{a\to-\infty} \left(-\tan^{-1} 2a\right) + \lim_{b\to\infty} \left(\tan^{-1} 2b\right) = \frac{\pi}{2} + \frac{\pi}{2} = \pi$$

41. If a plane curve is sufficiently differentiable the torsion is zero as the following argument shows:
$\mathbf{r} = f(t)\mathbf{i} + g(t)\mathbf{j} \Rightarrow \mathbf{v} = f'(t)\mathbf{i} + g'(t)\mathbf{j} \Rightarrow \mathbf{a} = f''(t)\mathbf{i} + g''(t)\mathbf{j} \Rightarrow \frac{d\mathbf{a}}{dt} = f'''(t)\mathbf{i} + g'''(t)\mathbf{j}$

$$\Rightarrow \tau = \frac{\begin{vmatrix} f'(t) & g'(t) & 0 \\ f''(t) & g''(t) & 0 \\ f'''(t) & g'''(t) & 0 \end{vmatrix}}{|\mathbf{v} \times \mathbf{a}|^2} = 0$$

43. $\mathbf{r}(t) = f(t)\mathbf{i} + g(t)\mathbf{j} + h(t)\mathbf{k} \Rightarrow \mathbf{v} = f'(t)\mathbf{i} + g'(t)\mathbf{j} + h'(t)\mathbf{k}$; $\mathbf{v} \cdot \mathbf{k} = 0 \Rightarrow h'(t) = 0 \Rightarrow h(t) = C$
$\Rightarrow \mathbf{r}(t) = f(t)\mathbf{i} + g(t)\mathbf{j} + C\mathbf{k}$ and $\mathbf{r}(a) = f(a)\mathbf{i} + g(a)\mathbf{j} + C\mathbf{k} = \mathbf{0} \Rightarrow f(a) = 0$, $g(a) = 0$ and $C = 0 \Rightarrow h(t) = 0$.

45. $y = x^2 \Rightarrow f'(x) = 2x$ and $f''(x) = 2$

$$\Rightarrow \kappa = \frac{|2|}{\left(1+(2x)^2\right)^{3/2}} = \frac{2}{\left(1+4x^2\right)^{3/2}}$$

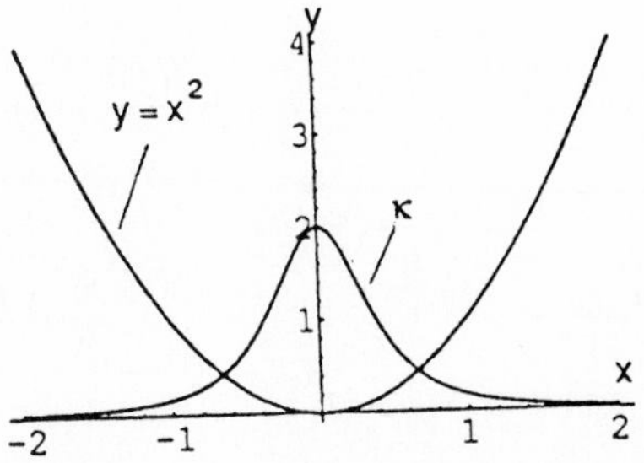

47. $y = \sin x \Rightarrow f'(x) = \cos x$ and $f''(x) = -\sin x$

$$\Rightarrow \kappa = \frac{|-\sin x|}{\left(1+\cos^2 x\right)^{3/2}} = \frac{|\sin x|}{\left(1+\cos^2 x\right)^{3/2}}$$

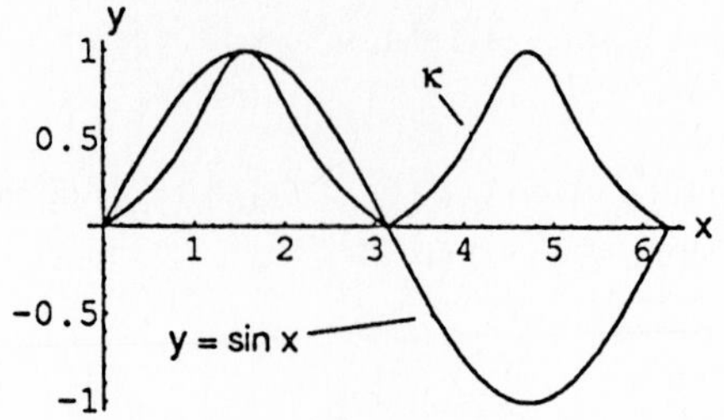

11.5 PLANETARY MOTION AND SATELLITES

1. $\frac{T^2}{a^3} = \frac{4\pi^2}{GM} \Rightarrow T^2 = \frac{4\pi^2}{GM}a^3 \Rightarrow T^2 = \frac{4\pi^2}{(6.6720 \times 10^{-11}\ \text{Nm}^2\text{kg}^{-2})(5.975 \times 10^{24}\ \text{kg})}(6{,}808{,}000\ \text{m})^3$

$\approx 3.125 \times 10^7\ \text{sec}^2 \Rightarrow T \approx \sqrt{3125 \times 10^4\ \text{sec}^2} \approx 55.90 \times 10^2\ \text{sec} \approx 93.2\ \text{min}$

3. $92.25\ \text{min} = 5535\ \text{sec}$ and $\frac{T^2}{a^3} = \frac{4\pi^2}{GM} \Rightarrow a^3 = \frac{GM}{4\pi^2}T^2$

$\Rightarrow a^3 = \frac{(6.6720 \times 10^{-11}\ \text{Nm}^2\text{kg}^{-2})(5.975 \times 10^{24}\ \text{kg})}{4\pi^2}(5535\ \text{sec})^2 = 3.094 \times 10^{20}\ \text{m}^3 \Rightarrow a \approx \sqrt[3]{3.094 \times 10^{20}\ \text{m}^3}$

$= 6.763 \times 10^6\ \text{m} \approx 6763\ \text{km}$; the mean distance from center of the Earth $= \frac{12{,}757\ \text{km} + 183\ \text{km} + 589\ \text{km}}{2}$

$= 6765\ \text{km}$

5. $a = 22{,}030\ \text{km} = 2.203 \times 10^7\ \text{m}$ and $T^2 = \frac{4\pi^2}{GM}a^3$

$\Rightarrow T^2 = \frac{4\pi^2}{(6.6720 \times 10^{-11}\ \text{Nm}^2\text{kg}^{-2})(6.418 \times 10^{23}\ \text{kg})}(2.203 \times 10^7\ \text{sec})^3 \approx 9.857 \times 10^9\ \text{sec}^2$

$\Rightarrow T \approx \sqrt{9.857 \times 10^8\ \text{sec}^2} \approx 9.928 \times 10^4\ \text{sec} \approx 1655\ \text{min}$

7. $T = 1477.4\ \text{min} = 88{,}644\ \text{sec} \Rightarrow a^3 = \frac{GMT^2}{4\pi^2}$

$= \frac{(6.6720 \times 10^{-11}\ \text{Nm}^2\text{kg}^{-2})(6.418 \times 10^{23}\ \text{kg})(88{,}644\ \text{sec})^2}{4\pi^2} = 8.523 \times 10^{21}\ \text{m}^3 \Rightarrow a \approx \sqrt[3]{8.523 \times 10^{21}\ \text{m}^3}$

$\approx 2.043 \times 10^7\ \text{m} = 20{,}430\ \text{km}$

9. $r = \frac{GM}{v^2} \Rightarrow v^2 = \frac{GM}{r} \Rightarrow |v| = \sqrt{\frac{GM}{r}} = \sqrt{\frac{(6.6720 \times 10^{-11}\ \text{Nm}^2\text{kg}^{-2})(5.975 \times 10^{24}\ \text{kg})}{r}} \approx 1.9966 \times 10^7 r^{-1/2}\ \text{m/sec}$

11. $e = \frac{r_0 v_0^2}{GM} - 1 \Rightarrow v_0^2 = \frac{GM(e+1)}{r_0} \Rightarrow v_0 = \sqrt{\frac{GM(e+1)}{r_0}}$;

Circle: $e = 0 \Rightarrow v_0 = \sqrt{\frac{GM}{r_0}}$

Ellipse: $0 < e < 1 \Rightarrow \sqrt{\frac{GM}{r_0}} < v_0 < \sqrt{\frac{2GM}{r_0}}$

Parabola: $e = 1 \Rightarrow v_0 = \sqrt{\frac{2GM}{r_0}}$

Hyperbola: $e > 1 \Rightarrow v_0 > \sqrt{\frac{2GM}{r_0}}$

13. $\Delta A = \frac{1}{2}\left|\mathbf{r}(t+\Delta t) \times \mathbf{r}(t)\right| \Rightarrow \frac{\Delta A}{\Delta t} = \frac{1}{2}\left|\frac{\mathbf{r}(t+\Delta t)}{\Delta t} \times \mathbf{r}(t)\right| = \frac{1}{2}\left|\frac{\mathbf{r}(t+\Delta t) - \mathbf{r}(t) + \mathbf{r}(t)}{\Delta t} \times \mathbf{r}(t)\right|$

$= \frac{1}{2}\left|\frac{\mathbf{r}(t+\Delta t) - \mathbf{r}(t)}{\Delta t} \times \mathbf{r}(t) + \frac{1}{\Delta t}\mathbf{r}(t) \times \mathbf{r}(t)\right| = \frac{1}{2}\left|\frac{\mathbf{r}(t+\Delta t) - \mathbf{r}(t)}{\Delta t} \times \mathbf{r}(t)\right| \Rightarrow \frac{dA}{dt} = \lim_{\Delta t \to 0} \frac{1}{2}\left|\frac{\mathbf{r}(t+\Delta t) - \mathbf{r}(t)}{\Delta t} \times \mathbf{r}(t)\right|$

$= \frac{1}{2}\left|\frac{d\mathbf{r}}{dt} \times \mathbf{r}(t)\right| = \frac{1}{2}\left|\mathbf{r}(t) \times \frac{d\mathbf{r}}{dt}\right| = \frac{1}{2}\left|\mathbf{r} \times \dot{\mathbf{r}}\right|$

15. (a) Let $\mathbf{r}_{AB}(t)$ denote the vector from planet A to planet B at time t. Then $\mathbf{r}_{AB}(t) = \mathbf{r}_B(t) - \mathbf{r}_A(t)$
$= [3\cos(\pi t) - 2\cos(2\pi t)]\mathbf{i} + [3\sin(\pi t) - 2\sin(2\pi t)]\mathbf{j}$

$= \left[3\cos(\pi t) - 2\left(\cos^2(\pi t) - \sin^2(\pi t)\right)\right]\mathbf{i} + [3\sin(\pi t) - 4\sin(\pi t)\cos(\pi t)]\mathbf{j}$

$= \left[3\cos(\pi t) - 4\cos^2(\pi t) + 2\right]\mathbf{i} + [(3 - 4\cos(\pi t))\sin(\pi t)]\mathbf{j} \Rightarrow$ parametric equations for the path are $x(t) = 2 + [3 - 4\cos(\pi t)]\cos(\pi t)$ and $y(t) = [3 - 4\cos(\pi t)]\sin(\pi t)$

(b) Setting $\theta = \pi t$ and $r = 3 - 4\cos\theta$, we see that $x - 2 = r\cos\theta$ and $y = r\sin\theta \Rightarrow$ the graph of the path of planet B is the limaçon $r = 3 - 4\cos\theta$ shown at the right. The planet A is located at $x = -2$.

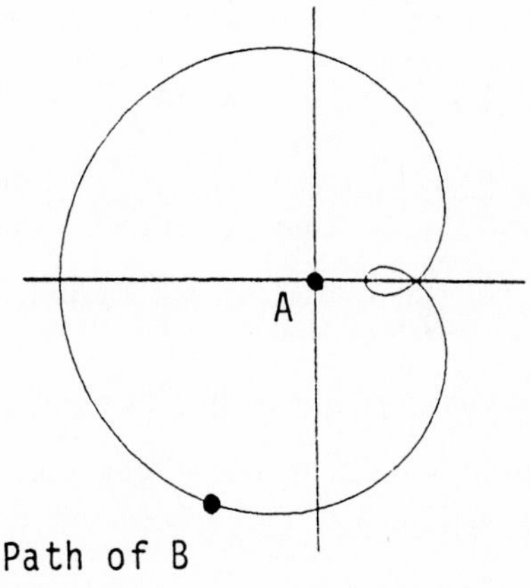

CHAPTER 11 PRACTICE EXERCISES

1. $\mathbf{r}(t) = (4\cos t)\mathbf{i} + \left(\sqrt{2}\sin t\right)\mathbf{j} \Rightarrow x = \cos t$ and $y = \sqrt{2}\sin t \Rightarrow \frac{x^2}{16} + \frac{y^2}{2} = 1$;

$\mathbf{v} = (-4\sin t)\mathbf{i} + \left(\sqrt{2}\cos t\right)\mathbf{j}$ and

$\mathbf{a} = (-4\cos t)\mathbf{i} - \left(\sqrt{2}\sin t\right)\mathbf{j}$; $\mathbf{r}(0) = \mathbf{i}$, $\mathbf{v}(0) = \sqrt{2}\mathbf{j}$,

$\mathbf{a}(0) = -4\mathbf{i}$; $\mathbf{r}\left(\frac{\pi}{4}\right) = 2\sqrt{2}\mathbf{i} + \mathbf{j}$, $\mathbf{v}\left(\frac{\pi}{4}\right) = -2\sqrt{2}\mathbf{i} + \mathbf{j}$,

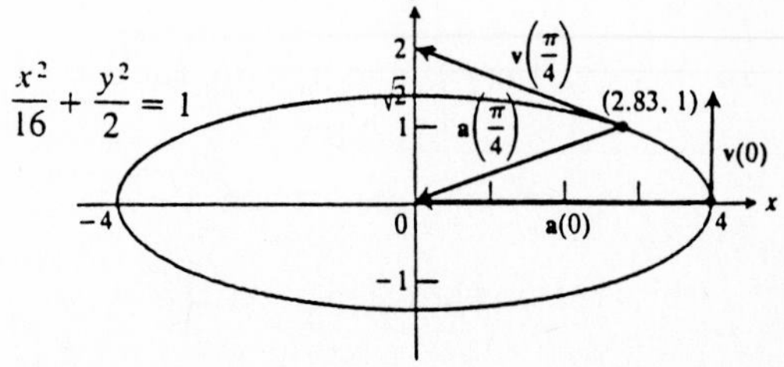

$\mathbf{a}\left(\frac{\pi}{4}\right) = -2\sqrt{2}\mathbf{i} - \mathbf{j}$; $|\mathbf{v}| = \sqrt{16\sin^2 t + 2\cos^2 t}$

$\Rightarrow a_T = \frac{d}{dt}|\mathbf{v}| = \frac{14\sin t\cos t}{\sqrt{16\sin^2 t + 2\cos^2 t}}$; at $t = 0$: $a_T = 0$, $a_N = \sqrt{|\mathbf{a}|^2 - 0} = 4$, $\kappa = \frac{a_N}{|\mathbf{v}|^2} = \frac{4}{2} = 2$;

at $t = \frac{\pi}{4}$: $a_T = \frac{7}{\sqrt{8+1}} = \frac{7}{3}$, $a_N = \sqrt{9 - \frac{49}{9}} = \frac{4\sqrt{2}}{3}$, $\kappa = \frac{a_N}{|\mathbf{v}|^2} = \frac{4\sqrt{2}}{27}$

3. $\mathbf{r} = \frac{1}{\sqrt{1+t^2}}\mathbf{i} + \frac{t}{\sqrt{1+t^2}}\mathbf{j} \Rightarrow \mathbf{v} = -t(1+t^2)^{-3/2}\mathbf{i} + (1+t^2)^{-3/2}\mathbf{j}$

$\Rightarrow |\mathbf{v}| = \sqrt{\left[-t(1+t^2)^{-3/2}\right]^2 + \left[(1+t^2)^{-3/2}\right]^2} = \frac{1}{1+t^2}$. We want to maximize $|\mathbf{v}|$: $\frac{d|\mathbf{v}|}{dt} = \frac{-2t}{(1+t^2)^2}$ and

$\frac{d|\mathbf{v}|}{dt} = 0 \Rightarrow \frac{-2t}{(1+t^2)^2} = 0 \Rightarrow t = 0$. For $t < 0$, $\frac{-2t}{(1+t^2)^2} > 0$; for $t > 0$, $\frac{-2t}{(1+t^2)^2} < 0 \Rightarrow |\mathbf{v}|_{max}$ occurs when

$t = 0 \Rightarrow |\mathbf{v}|_{max} = 1$

5. $\mathbf{v} = 3\mathbf{i} + 4\mathbf{j}$ and $\mathbf{a} = 5\mathbf{i} + 15\mathbf{j} \Rightarrow \mathbf{v}\times\mathbf{a} = \begin{vmatrix} \mathbf{i} & \mathbf{j} & \mathbf{k} \\ 3 & 4 & 0 \\ 5 & 15 & 0 \end{vmatrix} = 25\mathbf{k} \Rightarrow |\mathbf{v}\times\mathbf{a}| = 25$; $|\mathbf{v}| = \sqrt{3^2 + 4^2} = 5$

$\Rightarrow \kappa = \frac{|\mathbf{v}\times\mathbf{a}|}{|\mathbf{v}|^3} = \frac{25}{5^3} = \frac{1}{5}$

7. $\mathbf{r} = x\mathbf{i} + y\mathbf{j} \Rightarrow \mathbf{v} = \frac{dx}{dt}\mathbf{i} + \frac{dy}{dt}\mathbf{j}$ and $\mathbf{v}\cdot\mathbf{i} = y \Rightarrow \frac{dx}{dt} = y$. Since the particle moves around the unit circle $x^2 + y^2 = 1$, $2x\frac{dx}{dt} + 2y\frac{dy}{dt} = 0 \Rightarrow \frac{dy}{dt} = -\frac{x}{y}\frac{dx}{dt} \Rightarrow \frac{dy}{dt} = -\frac{x}{y}(y) = -x$. Since $\frac{dx}{dt} = y$ and $\frac{dy}{dt} = -x$, we have $\mathbf{v} = y\mathbf{i} - x\mathbf{j} \Rightarrow$ at $(1,0)$, $\mathbf{v} = -\mathbf{j}$ and the motion is clockwise.

9. $\frac{d\mathbf{r}}{dt}$ orthogonal to $\mathbf{r} \Rightarrow 0 = \frac{d\mathbf{r}}{dt}\cdot\mathbf{r} = \frac{1}{2}\frac{d\mathbf{r}}{dt}\cdot\mathbf{r} + \frac{1}{2}\mathbf{r}\cdot\frac{d\mathbf{r}}{dt} = \frac{1}{2}\frac{d}{dt}(\mathbf{r}\cdot\mathbf{r}) \Rightarrow \mathbf{r}\cdot\mathbf{r} = K$, a constant. If $\mathbf{r} = x\mathbf{i} + y\mathbf{j}$, where x and y are differentiable functions of t, then $\mathbf{r}\cdot\mathbf{r} = x^2 + y^2 \Rightarrow x^2 + y^2 = K$, which is the equation of a circle centered at the origin.

11. $y = y_0 + (v_0\sin\alpha)t - \frac{1}{2}gt^2 \Rightarrow y = 6.5 + (44\text{ ft/sec})(\sin 45°)(3\text{ sec}) - \frac{1}{2}(32\text{ ft/sec}^2)(3\text{ sec})^2 = 6.5 + 66\sqrt{2} - 144$ ≈ -41.36 ft $\Rightarrow$ the shot put is on the ground. Now, $y = 0 \Rightarrow 6.5 + 22\sqrt{2}t - 16t^2 = 0 \Rightarrow t \approx 2.13$ sec (the positive root) $\Rightarrow x \approx (44\text{ ft/sec})(\cos 45°)(2.13\text{ sec}) \approx 66.42$ ft or about 66 ft, 5 in. from the stopboard

13. $x = (v_0\cos\alpha)t$ and $y = (v_0\sin\alpha)t - \frac{1}{2}gt^2 \Rightarrow \tan\phi = \frac{y}{x} = \frac{(v_0\sin\alpha)t - \frac{1}{2}gt^2}{(v_0\cos\alpha)t} = \frac{(v_0\sin\alpha) - \frac{1}{2}gt}{v_0\cos\alpha}$

$\Rightarrow v_0\cos\alpha\tan\phi = v_0\sin\alpha - \frac{1}{2}gt \Rightarrow t = \frac{2v_0\sin\alpha - 2v_0\cos\alpha\tan\phi}{g}$, which is the time when the golf ball hits the upward slope. At this time

$x = (v_0 \cos \alpha)\left(\frac{2v_0 \sin \alpha - 2v_0 \cos \alpha \tan \phi}{g}\right)$

$= \left(\frac{2}{g}\right)\left(v_0^2 \sin \alpha \cos \alpha - v_0^2 \cos^2 \alpha \tan \phi\right)$. Now

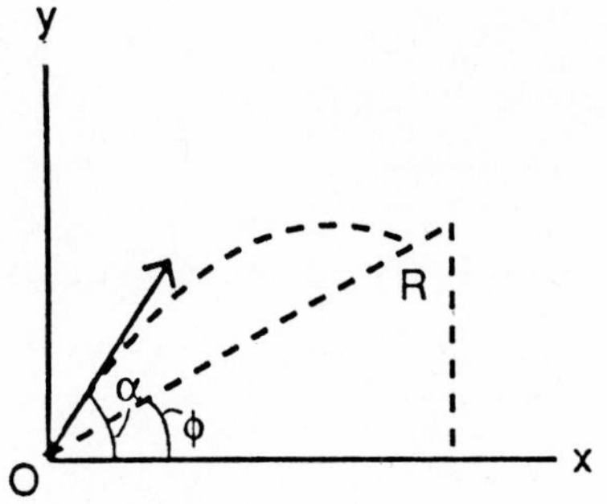

$OR = \frac{x}{\cos \phi} \Rightarrow OR = \left(\frac{2}{g}\right)\left(\frac{v_0^2 \sin \alpha \cos \alpha - v_0^2 \cos^2 \alpha \tan \phi}{\cos \phi}\right)$

$= \left(\frac{2v_0^2 \cos \alpha}{g}\right)\left(\frac{\sin \alpha}{\cos \phi} - \frac{\cos \alpha \tan \phi}{\cos \phi}\right)$

$= \left(\frac{2v_0^2 \cos \alpha}{g}\right)\left(\frac{\sin \alpha \cos \phi - \cos \alpha \sin \phi}{\cos^2 \phi}\right)$

$= \left(\frac{2v_0^2 \cos \alpha}{g \cos^2 \phi}\right)[\sin (\alpha - \phi)]$. The distance OR is maximized when x is maximized:

$\frac{dx}{d\alpha} = \left(\frac{2v_0^2}{g}\right)(\cos 2\alpha + \sin 2\alpha \tan \phi) = 0 \Rightarrow (\cos 2\alpha + \sin 2\alpha \tan \phi) = 0 \Rightarrow \cot 2\alpha + \tan \phi = 0$

$\Rightarrow \cot 2\alpha = \tan (-\phi) \Rightarrow 2\alpha = \frac{\pi}{2} + \phi \Rightarrow \alpha = \frac{\phi}{2} + \frac{\pi}{4}$

15. (a) $R = \frac{v_0^2}{g} \sin 2\alpha \Rightarrow 109.5 \text{ ft} = \left(\frac{v_0^2}{32 \text{ ft/sec}^2}\right)(\sin 90°) \Rightarrow v_0^2 = 3504 \text{ ft}^2/\text{sec}^2 \Rightarrow v_0 = \sqrt{3504 \text{ ft}^2/\text{sec}^2}$

≈ 59.19 ft/sec

(b) $x = (v_0 \cos \alpha)t$ and $y = 4 + (v_0 \sin \alpha)t - \frac{1}{2}gt^2$; when the cork hits the ground, $x = 177.75$ ft and $y = 0$

$\Rightarrow 177.75 = \left(v_0 \frac{1}{\sqrt{2}}\right)t$ and $0 = 4 + \left(v_0 \frac{1}{\sqrt{2}}\right)t - 16t^2 \Rightarrow 16t^2 = 4 + 177.75 \Rightarrow t = \frac{\sqrt{181.75}}{4}$

$\Rightarrow v_0 = \frac{(177.75)\sqrt{2}}{t} = \frac{4(177.75)\sqrt{2}}{\sqrt{181.75}} \approx 74.58$ ft/sec

17. $x^2 = \left(v_0^2 \cos^2 \alpha\right)t^2$ and $\left(y + \frac{1}{2}gt^2\right)^2 = \left(v_0^2 \sin^2 \alpha\right)t^2 \Rightarrow x^2 + \left(y + \frac{1}{2}gt^2\right) = v_0^2 t^2$

19. $\mathbf{r}(t) = \left[\int_0^t \cos\left(\frac{1}{2}\pi\theta^2\right) d\theta\right]\mathbf{i} + \left[\int_0^t \sin\left(\frac{1}{2}\pi\theta^2\right) d\theta\right]\mathbf{j} \Rightarrow \mathbf{v}(t) = \cos\left(\frac{\pi t^2}{2}\right)\mathbf{i} + \sin\left(\frac{\pi t^2}{2}\right)\mathbf{j} \Rightarrow |\mathbf{v}| = 1;$

$\mathbf{a}(t) = -\pi t \sin\left(\frac{\pi t^2}{2}\right)\mathbf{i} + \pi t \cos\left(\frac{\pi t^2}{2}\right)\mathbf{j} \Rightarrow \mathbf{v} \times \mathbf{a} = \begin{vmatrix} \mathbf{i} & \mathbf{j} & \mathbf{k} \\ \cos\left(\frac{\pi t^2}{2}\right) & \sin\left(\frac{\pi t^2}{2}\right) & 0 \\ -\pi t \sin\left(\frac{\pi t^2}{2}\right) & \pi t \cos\left(\frac{\pi t^2}{2}\right) & 0 \end{vmatrix}$

$= \pi t \mathbf{k} \Rightarrow \kappa = \frac{|\mathbf{v} \times \mathbf{a}|}{|\mathbf{v}|^3} = \pi t; \ |\mathbf{v}(t)| = \frac{ds}{dt} = 1 \Rightarrow s = t + C; \ \mathbf{r}(0) = \mathbf{0} \Rightarrow s(0) = 0 \Rightarrow C = 0 \Rightarrow \kappa = \pi s$

21. $\mathbf{r} = (2\cos t)\mathbf{i} + (2\sin t)\mathbf{j} + t^2\mathbf{k} \Rightarrow \mathbf{v} = (-2\sin t)\mathbf{i} + (2\cos t)\mathbf{j} + 2t\mathbf{k} \Rightarrow |\mathbf{v}| = \sqrt{(-2\sin t)^2 + (2\cos t)^2 + (2t)^2}$

$$= 2\sqrt{1+t^2} \Rightarrow \text{Length} = \int_0^{\pi/4} 2\sqrt{1+t^2}\,dt = \left[t\sqrt{1+t^2} + \ln\left|t + \sqrt{1+t^2}\right|\right]_0^{\pi/4} = \frac{\pi}{4}\sqrt{1+\frac{\pi^2}{16}} + \ln\left(\frac{\pi}{4} + \sqrt{1+\frac{\pi^2}{16}}\right)$$

23. $\mathbf{r} = \frac{4}{9}(1+t)^{3/2}\mathbf{i} + \frac{4}{9}(1-t)^{3/2}\mathbf{j} + \frac{1}{3}t\mathbf{k} \Rightarrow \mathbf{v} = \frac{2}{3}(1+t)^{1/2}\mathbf{i} - \frac{2}{3}(1-t)^{1/2}\mathbf{j} + \frac{1}{3}\mathbf{k}$

$$\Rightarrow |\mathbf{v}| = \sqrt{\left[\frac{2}{3}(1+t)^{1/2}\right]^2 + \left[-\frac{2}{3}(1-t)^{1/2}\right]^2 + \left(\frac{1}{3}\right)^2} = 1 \Rightarrow \mathbf{T} = \frac{2}{3}(1+t)^{1/2}\mathbf{i} - \frac{2}{3}(1-t)^{1/2}\mathbf{j} + \frac{1}{3}\mathbf{k}$$

$$\Rightarrow \mathbf{T}(0) = \frac{2}{3}\mathbf{i} - \frac{2}{3}\mathbf{j} + \frac{1}{3}\mathbf{k};\ \frac{d\mathbf{T}}{dt} = \frac{1}{3}(1+t)^{-1/2}\mathbf{i} + \frac{1}{3}(1-t)^{-1/2}\mathbf{j} \Rightarrow \frac{d\mathbf{T}}{dt}(0) = \frac{1}{3}\mathbf{i} + \frac{1}{3}\mathbf{j} \Rightarrow \left|\frac{d\mathbf{T}}{dt}(0)\right| = \frac{\sqrt{2}}{3}$$

$$\Rightarrow \mathbf{N}(0) = \frac{1}{\sqrt{2}}\mathbf{i} + \frac{1}{\sqrt{2}}\mathbf{j};\ \mathbf{B}(0) = \mathbf{T}(0) \times \mathbf{N}(0) = \begin{vmatrix} \mathbf{i} & \mathbf{j} & \mathbf{k} \\ \frac{2}{3} & -\frac{2}{3} & \frac{1}{3} \\ \frac{1}{\sqrt{2}} & \frac{1}{\sqrt{2}} & 0 \end{vmatrix} = -\frac{1}{3\sqrt{2}}\mathbf{i} + \frac{1}{3\sqrt{2}}\mathbf{j} + \frac{4}{3\sqrt{2}}\mathbf{k};$$

$$\mathbf{a} = \frac{1}{3}(1+t)^{-1/2}\mathbf{i} + \frac{1}{3}(1-t)^{-1/2}\mathbf{j} \Rightarrow \mathbf{a}(0) = \frac{1}{3}\mathbf{i} + \frac{1}{3}\mathbf{j} \text{ and } \mathbf{v}(0) = \frac{2}{3}\mathbf{i} - \frac{2}{3}\mathbf{j} + \frac{1}{3}\mathbf{k} \Rightarrow \mathbf{v}(0) \times \mathbf{a}(0)$$

$$= \begin{vmatrix} \mathbf{i} & \mathbf{j} & \mathbf{k} \\ \frac{2}{3} & -\frac{2}{3} & \frac{1}{3} \\ \frac{1}{3} & \frac{1}{3} & 0 \end{vmatrix} = -\frac{1}{9}\mathbf{i} + \frac{1}{9}\mathbf{j} + \frac{4}{9}\mathbf{k} \Rightarrow |\mathbf{v} \times \mathbf{a}| = \frac{\sqrt{2}}{3} \Rightarrow \kappa(0) = \frac{|\mathbf{v} \times \mathbf{a}|}{|\mathbf{v}|^3} = \frac{\left(\frac{\sqrt{2}}{3}\right)}{1^3} = \frac{\sqrt{2}}{3};$$

$$\dot{\mathbf{a}} = -\frac{1}{6}(1+t)^{-3/2}\mathbf{i} + \frac{1}{6}(1-t)^{-3/2}\mathbf{j} \Rightarrow \dot{\mathbf{a}}(0) = -\frac{1}{6}\mathbf{i} + \frac{1}{6}\mathbf{j} \Rightarrow \tau(0) = \frac{\begin{vmatrix} \frac{2}{3} & -\frac{2}{3} & \frac{1}{3} \\ \frac{1}{3} & \frac{1}{3} & 0 \\ -\frac{1}{6} & \frac{1}{6} & 0 \end{vmatrix}}{|\mathbf{v} \times \mathbf{a}|^2} = \frac{\left(\frac{1}{3}\right)\left(\frac{2}{18}\right)}{\left(\frac{\sqrt{2}}{3}\right)^2} = \frac{1}{6};$$

$t = 0 \Rightarrow \left(\frac{4}{9}, \frac{4}{9}, 0\right)$ is the point on the curve

25. $\mathbf{r} = t\mathbf{i} + \frac{1}{2}e^{2t}\mathbf{j} \Rightarrow \mathbf{v} = \mathbf{i} + e^{2t}\mathbf{j} \Rightarrow |\mathbf{v}| = \sqrt{1+e^{4t}} \Rightarrow \mathbf{T} = \frac{1}{\sqrt{1+e^{4t}}}\mathbf{i} + \frac{e^{2t}}{\sqrt{1+e^{4t}}}\mathbf{j} \Rightarrow \mathbf{T}(\ln 2) = \frac{1}{\sqrt{17}}\mathbf{i} + \frac{4}{\sqrt{17}}\mathbf{j};$

$$\frac{d\mathbf{T}}{dt} = \frac{-2e^{4t}}{\left(1+e^{4t}\right)^{3/2}}\mathbf{i} + \frac{2e^{2t}}{\left(1+e^{4t}\right)^{3/2}}\mathbf{j} \Rightarrow \frac{d\mathbf{T}}{dt}(\ln 2) = \frac{-32}{17\sqrt{17}}\mathbf{i} + \frac{8}{17\sqrt{17}}\mathbf{j} \Rightarrow \mathbf{N}(\ln 2) = -\frac{4}{\sqrt{17}}\mathbf{i} + \frac{1}{\sqrt{17}}\mathbf{j};$$

$$\mathbf{B}(\ln 2) = \mathbf{T}(\ln 2) \times \mathbf{N}(\ln 2) = \begin{vmatrix} \mathbf{i} & \mathbf{j} & \mathbf{k} \\ \frac{1}{\sqrt{17}} & \frac{4}{\sqrt{17}} & 0 \\ -\frac{4}{\sqrt{17}} & \frac{1}{\sqrt{17}} & 0 \end{vmatrix} = \mathbf{k};\ \mathbf{a} = 2e^{2t}\mathbf{j} \Rightarrow \mathbf{a}(\ln 2) = 8\mathbf{j} \text{ and } \mathbf{v}(\ln 2) = \mathbf{i} + 4\mathbf{j}$$

$$\Rightarrow \mathbf{v}(\ln 2)\times\mathbf{a}(\ln 2)=\begin{vmatrix}\mathbf{i}&\mathbf{j}&\mathbf{k}\\1&4&0\\0&8&0\end{vmatrix}=8\mathbf{k}\Rightarrow|\mathbf{v}\times\mathbf{a}|=8 \text{ and } |\mathbf{v}(\ln 2)|=\sqrt{17}\Rightarrow\kappa(\ln 2)=\frac{8}{17\sqrt{17}};\ \dot{\mathbf{a}}=4e^{2t}\mathbf{j}$$

$$\Rightarrow \dot{\mathbf{a}}(\ln 2)=16\mathbf{j}\Rightarrow\tau(\ln 2)=\frac{\begin{vmatrix}1&4&0\\0&8&0\\0&16&0\end{vmatrix}}{|\mathbf{v}\times\mathbf{a}|^2}=0;\ t=\ln 2\Rightarrow(\ln 2,2,0)\text{ is on the curve}$$

27. $\mathbf{r}=(2+3t+3t^2)\mathbf{i}+(4t+4t^2)\mathbf{j}-(6\cos t)\mathbf{k}\Rightarrow\mathbf{v}=(3+6t)\mathbf{i}+(4+8t)\mathbf{j}+(6\sin t)\mathbf{k}$

$$\Rightarrow|\mathbf{v}|=\sqrt{(3+6t)^2+(4+8t)^2+(6\sin t)^2}=\sqrt{25+100t+100t^2+36\sin^2 t}$$

$$\Rightarrow\frac{d|\mathbf{v}|}{dt}=\frac{1}{2}\left(25+100t+100t^2+36\sin^2 t\right)^{-1/2}(100+200t+72\sin t\cos t)\Rightarrow a_T(0)=\frac{d|\mathbf{v}|}{dt}(0)=10;$$

$$\mathbf{a}=6\mathbf{i}+8\mathbf{j}+(t\cos t)\mathbf{k}\Rightarrow|\mathbf{a}|=\sqrt{6^2+8^2+(6\cos t)^2}=\sqrt{100+36\cos^2 t}\Rightarrow|\mathbf{a}(0)|=\sqrt{136}$$

$$\Rightarrow a_N=\sqrt{|\mathbf{a}|^2-a_T^2}=\sqrt{136-10^2}=\sqrt{36}=6\Rightarrow\mathbf{a}(0)=10\mathbf{T}+6\mathbf{N}$$

29. $\mathbf{r}=(\sin t)\mathbf{i}+\left(\sqrt{2}\cos t\right)\mathbf{j}+(\sin t)\mathbf{k}\Rightarrow\mathbf{v}=(\cos t)\mathbf{i}-\left(\sqrt{2}\sin t\right)\mathbf{j}+(\cos t)\mathbf{k}$

$$\Rightarrow|\mathbf{v}|=\sqrt{(\cos t)^2+\left(-\sqrt{2}\sin t\right)^2+(\cos t)^2}=\sqrt{2}\Rightarrow\mathbf{T}=\frac{\mathbf{v}}{|\mathbf{v}|}=\left(\frac{1}{\sqrt{2}}\cos t\right)\mathbf{i}-(\sin t)\mathbf{j}+\left(\frac{1}{\sqrt{2}}\cos t\right)\mathbf{k};$$

$$\frac{d\mathbf{T}}{dt}=\left(-\frac{1}{\sqrt{2}}\sin t\right)\mathbf{i}-(\cos t)\mathbf{j}-\left(\frac{1}{\sqrt{2}}\sin t\right)\mathbf{k}\Rightarrow\left|\frac{d\mathbf{T}}{dt}\right|=\sqrt{\left(-\frac{1}{\sqrt{2}}\sin t\right)^2+(-\cos t)^2+\left(-\frac{1}{\sqrt{2}}\sin t\right)^2}=1$$

$$\Rightarrow\mathbf{N}=\frac{\left(\frac{d\mathbf{T}}{dt}\right)}{\left|\frac{d\mathbf{T}}{dt}\right|}=\left(-\frac{1}{\sqrt{2}}\sin t\right)\mathbf{i}-(\cos t)\mathbf{j}-\left(\frac{1}{\sqrt{2}}\sin t\right)\mathbf{k};\ \mathbf{B}=\mathbf{T}\times\mathbf{N}=\begin{vmatrix}\mathbf{i}&\mathbf{j}&\mathbf{k}\\\frac{1}{\sqrt{2}}\cos t&-\sin t&\frac{1}{\sqrt{2}}\cos t\\-\frac{1}{\sqrt{2}}\sin t&-\cos t&-\frac{1}{\sqrt{2}}\sin t\end{vmatrix}$$

$$=\frac{1}{\sqrt{2}}\mathbf{i}-\frac{1}{\sqrt{2}}\mathbf{k};\ \mathbf{a}=(-\sin t)\mathbf{i}-\left(\sqrt{2}\cos t\right)\mathbf{j}-(\sin t)\mathbf{k}\Rightarrow\mathbf{v}\times\mathbf{a}=\begin{vmatrix}\mathbf{i}&\mathbf{j}&\mathbf{k}\\\cos t&-\sqrt{2}\sin t&\cos t\\-\sin t&-\sqrt{2}\cos t&-\sin t\end{vmatrix}$$

$$=\sqrt{2}\,\mathbf{i}-\sqrt{2}\,\mathbf{k}\Rightarrow|\mathbf{v}\times\mathbf{a}|=\sqrt{4}=2\Rightarrow\kappa=\frac{|\mathbf{v}\times\mathbf{a}|}{|\mathbf{v}|^3}=\frac{2}{\left(\sqrt{2}\right)^3}=\frac{1}{\sqrt{2}};\ \dot{\mathbf{a}}=(-\cos t)\mathbf{i}+\left(\sqrt{2}\sin t\right)\mathbf{j}-(\cos t)\mathbf{k}$$

$$\Rightarrow\tau=\frac{\begin{vmatrix}\cos t&-\sqrt{2}\sin t&\cos t\\-\sin t&-\sqrt{2}\cos t&-\sin t\\-\cos t&\sqrt{2}\sin t&-\cos t\end{vmatrix}}{|\mathbf{v}\times\mathbf{a}|^2}=\frac{(\cos t)\left(\sqrt{2}\right)-\left(\sqrt{2}\sin t\right)(0)+(\cos t)\left(-\sqrt{2}\right)}{4}=0$$

31. $\mathbf{r} = 2\mathbf{i} + \left(4 \sin \frac{t}{2}\right)\mathbf{j} + \left(3 - \frac{t}{\pi}\right)\mathbf{k} \Rightarrow 0 = \mathbf{r} \cdot (\mathbf{i} - \mathbf{j}) = 2(1) + \left(4 \sin \frac{t}{2}\right)(-1) \Rightarrow 0 = 2 - 4 \sin \frac{t}{2} \Rightarrow \sin \frac{t}{2} = \frac{1}{2} \Rightarrow \frac{t}{2} = \frac{\pi}{6}$
$\Rightarrow t = \frac{\pi}{3}$ (for the first time)

33. $\mathbf{r} = e^t\mathbf{i} + (\sin t)\mathbf{j} + \ln(1-t)\mathbf{k} \Rightarrow \mathbf{v} = e^t\mathbf{i} + (\cos t)\mathbf{j} - \left(\frac{1}{1-t}\right)\mathbf{k} \Rightarrow \mathbf{v}(0) = \mathbf{i} + \mathbf{j} - \mathbf{k}$; $\mathbf{r}(0) = \mathbf{i} \Rightarrow (1,0,0)$ is on the line $\Rightarrow x = 1 + t$, $y = t$, and $z = -t$ are parametric equations of the line

35. $\Delta SOT \approx \Delta TOD \Rightarrow \frac{DO}{OT} = \frac{OT}{SO} \Rightarrow \frac{y_0}{6380} = \frac{6380}{6380 + 437}$
$\Rightarrow y_0 = \frac{6380^2}{6817} \Rightarrow y_0 \approx 5971$ km;

$$VA = \int_{5971}^{6380} 2\pi x \sqrt{1 + \left(\frac{dx}{dy}\right)^2}\, dy$$
$$= 2\pi \int_{5971}^{6817} \sqrt{6380^2 - y^2}\left(\frac{6380}{\sqrt{6380^2 - y^2}}\right) dy$$
$$= 2\pi \int_{5971}^{6817} 6380\, dy = 2\pi\left[6380y\right]_{5971}^{6817}$$
$= 16{,}395{,}469$ km$^2 \approx 1.639 \times 10^7$ km^2;

percentage visible $\approx \frac{16{,}395{,}469 \text{ km}^2}{4\pi(6380 \text{ km})^2} \approx 3.21\%$

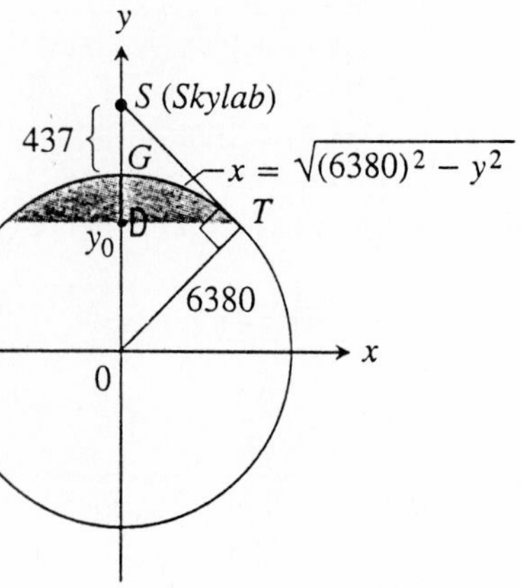

CHAPTER 11 ADDITIONAL EXERCISES–THEORY, EXAMPLES, APPLICATIONS

1. (a) The velocity of the boat at (x,y) relative to land is the sum of the velocity due to the rower and the velocity of the river, or $\mathbf{v} = \left[-\frac{1}{250}(y-50)^2 + 10\right]\mathbf{i} - 20\mathbf{j}$. Now, $\frac{dy}{dt} = -20 \Rightarrow y = -20t + c$; $y(0) = 100$
$\Rightarrow c = 100 \Rightarrow y = -20t + 100 \Rightarrow \mathbf{v} = \left[-\frac{1}{250}(-20t + 50)^2 + 10\right]\mathbf{i} - 20\mathbf{j} = \left(-\frac{8}{5}t^2 + 8t\right)\mathbf{i} - 20\mathbf{j}$
$\Rightarrow \mathbf{r}(t) = \left(-\frac{8}{15}t^3 + 4t^2\right)\mathbf{i} - 20t\mathbf{j} + \mathbf{C}_1$; $\mathbf{r}(0) = 0\mathbf{i} + 100\mathbf{j} \Rightarrow 100\mathbf{j} = \mathbf{C}_1 \Rightarrow \mathbf{r}(t)$
$= \left(-\frac{8}{15}t^3 + 4t^2\right)\mathbf{i} + (100 - 20t)\mathbf{j}$

(b) The boat reaches the shore when $y = 0 \Rightarrow 0 = -20t + 100$ from part (a) $\Rightarrow t = 5$
$\Rightarrow \mathbf{r}(5) = \left(-\frac{8}{15} \cdot 125 + 4 \cdot 25\right)\mathbf{i} + (100 - 20 \cdot 5)\mathbf{j} = \left(-\frac{200}{3} + 100\right)\mathbf{i} = \frac{100}{3}\mathbf{i}$; the distance downstream is therefore $\frac{100}{3}$ m

3. Let $\mathbf{a} = \mathbf{i} + \mathbf{j} + \mathbf{k}$ be the vector from O to A and $\mathbf{b} = \mathbf{i} + 3\mathbf{j} + 2\mathbf{k}$ be the vector from O to B. The vector $\mathbf{v}$ orthogonal to $\mathbf{a}$ and $\mathbf{b} \Rightarrow \mathbf{v}$ is parallel to $\mathbf{b} \times \mathbf{a}$ (since the rotation is clockwise). Now $\mathbf{b} \times \mathbf{a} = \mathbf{i} + \mathbf{j} - 2\mathbf{k}$; $\text{proj}_{\mathbf{a}}\, \mathbf{b} = \left(\frac{\mathbf{a} \cdot \mathbf{b}}{\mathbf{a} \cdot \mathbf{a}}\right)\mathbf{a} = 2\mathbf{i} + 2\mathbf{j} + 2\mathbf{k} \Rightarrow (2,2,2)$ is the center of the circular path $(1,3,2)$ takes $\Rightarrow$ radius $= \sqrt{1^2 + (-1)^2 + 0^2} = \sqrt{2} \Rightarrow$ arc length per second covered by the point is $\frac{3}{2}\sqrt{2}$ units/sec $= |\mathbf{v}|$ (velocity is

constant). A unit vector in the direction of $\mathbf{v}$ is $\dfrac{\mathbf{b} \times \mathbf{a}}{|\mathbf{b} \times \mathbf{a}|} = \dfrac{1}{\sqrt{6}}\mathbf{i} + \dfrac{1}{\sqrt{6}}\mathbf{j} - \dfrac{2}{\sqrt{6}}\mathbf{k} \Rightarrow \mathbf{v} = |\mathbf{v}|\left(\dfrac{\mathbf{b} \times \mathbf{a}}{|\mathbf{b} \times \mathbf{a}|}\right)$

$= \dfrac{3}{2}\sqrt{2}\left(\dfrac{1}{\sqrt{6}}\mathbf{i} + \dfrac{1}{\sqrt{6}}\mathbf{j} - \dfrac{2}{\sqrt{6}}\mathbf{k}\right) = \dfrac{\sqrt{3}}{2}\mathbf{i} + \dfrac{\sqrt{3}}{2}\mathbf{j} - \sqrt{3}\mathbf{k}$

5. (a) $\mathbf{r}(\theta) = (a\cos\theta)\mathbf{i} + (a\sin\theta)\mathbf{j} + b\theta\mathbf{k} \Rightarrow \dfrac{d\mathbf{r}}{dt} = [(-a\sin\theta)\mathbf{i} + (a\cos\theta)\mathbf{j} + b\mathbf{k}]\dfrac{d\theta}{dt}$; $|\mathbf{v}| = \sqrt{2gz} = \left|\dfrac{d\mathbf{r}}{dt}\right|$

$= \sqrt{a^2 + b^2}\,\dfrac{d\theta}{dt} \Rightarrow \dfrac{d\theta}{dt} = \sqrt{\dfrac{2gz}{a^2 + b^2}} = \sqrt{\dfrac{2gb\theta}{a^2 + b^2}} \Rightarrow \left.\dfrac{d\theta}{dt}\right|_{\theta = 2\pi} = \sqrt{\dfrac{4\pi gb}{a^2 + b^2}} = 2\sqrt{\dfrac{\pi gb}{a^2 + b^2}}$

(b) $\dfrac{d\theta}{dt} = \sqrt{\dfrac{2gb\theta}{a^2 + b^2}} \Rightarrow \dfrac{d\theta}{\sqrt{\theta}} = \sqrt{\dfrac{2gb}{a^2 + b^2}}\,dt \Rightarrow 2\theta^{1/2} = \sqrt{\dfrac{2gb}{a^2 + b^2}}\,t + C$; $t = 0 \Rightarrow \theta = 0 \Rightarrow C = 0$

$\Rightarrow 2\theta^{1/2} = \sqrt{\dfrac{2gb}{a^2 + b^2}}\,t \Rightarrow \theta = \dfrac{gbt^2}{2(a^2 + b^2)}$; $z = b\theta \Rightarrow z = \dfrac{gb^2 t^2}{2(a^2 + b^2)}$

(c) $\mathbf{v}(t) = \dfrac{d\mathbf{r}}{dt} = [(-a\sin\theta)\mathbf{i} + (a\cos\theta)\mathbf{j} + b\mathbf{k}]\dfrac{d\theta}{dt} = [(-a\sin\theta)\mathbf{i} + (a\cos\theta)\mathbf{j} + b\mathbf{k}]\left(\dfrac{gbt}{a^2 + b^2}\right)$, from part (b)

$\Rightarrow \mathbf{v}(t) = \left[\dfrac{(-a\sin\theta)\mathbf{i} + (a\cos\theta)\mathbf{j} + b\mathbf{k}}{\sqrt{a^2 + b^2}}\right]\left(\dfrac{gbt}{\sqrt{a^2 + b^2}}\right) = \dfrac{gbt}{\sqrt{a^2 + b^2}}\,\mathbf{T}$;

$\dfrac{d^2\mathbf{r}}{dt^2} = [(-a\cos\theta)\mathbf{i} - (a\sin\theta)\mathbf{j}]\left(\dfrac{d\theta}{dt}\right)^2 + [(-a\sin\theta)\mathbf{i} + (a\cos\theta)\mathbf{j} + b\mathbf{k}]\dfrac{d^2\theta}{dt^2}$

$= \left(\dfrac{gbt}{a^2 + b^2}\right)^2 [(-a\cos\theta)\mathbf{i} - (a\sin\theta)\mathbf{j}] + [(-a\sin\theta)\mathbf{i} + (a\cos\theta)\mathbf{j} + b\mathbf{k}]\left(\dfrac{gb}{a^2 + b^2}\right)$

$= \left[\dfrac{(-a\sin\theta)\mathbf{i} + (a\cos\theta)\mathbf{j} + b\mathbf{k}}{\sqrt{a^2 + b^2}}\right]\left(\dfrac{gb}{\sqrt{a^2 + b^2}}\right) + a\left(\dfrac{gbt}{a^2 + b^2}\right)^2 [(-\cos\theta)\mathbf{i} - (\sin\theta)\mathbf{j}]$

$= \dfrac{gb}{\sqrt{a^2 + b^2}}\,\mathbf{T} + a\left(\dfrac{gbt}{a^2 + b^2}\right)^2 \mathbf{N}$ (there is no component in the direction of $\mathbf{B}$).

7. $r = \dfrac{(1+e)r_0}{1 + e\cos\theta} \Rightarrow \dfrac{dr}{d\theta} = \dfrac{(1+e)r_0(e\sin\theta)}{(1 + e\cos\theta)^2}$; $\dfrac{dr}{d\theta} = 0 \Rightarrow \dfrac{(1+e)r_0(e\sin\theta)}{(1 + e\cos\theta)^2} = 0 \Rightarrow (1+e)r_0(e\sin\theta) = 0$

$\Rightarrow \sin\theta = 0 \Rightarrow \theta = 0$ or π. Note that $\dfrac{dr}{d\theta} > 0$ when $\sin\theta > 0$ and $\dfrac{dr}{d\theta} < 0$ when $\sin\theta < 0$. Since $\sin\theta < 0$ on

$-\pi < \theta < 0$ and $\sin\theta > 0$ on $0 < \theta < \pi$, r is a minimum when $\theta = 0$ and $r(0) = \dfrac{(1+e)r_0}{1 + e\cos 0} = r_0$

9. (a) $\mathbf{v} = \dfrac{dx}{dt}\mathbf{i} + \dfrac{dy}{dt}\mathbf{j}$ and $\mathbf{v} = \dfrac{dr}{dt}\mathbf{u}_r + r\dfrac{d\theta}{dt}\mathbf{u}_\theta = \left(\dfrac{dr}{dt}\right)[(\cos\theta)\mathbf{i} + (\sin\theta)\mathbf{j}] + \left(r\dfrac{d\theta}{dt}\right)[(-\sin\theta)\mathbf{i} + (\cos\theta)\mathbf{j}] \Rightarrow \mathbf{v} \cdot \mathbf{i}\,\dfrac{dx}{dt}$ and

$\mathbf{v} \cdot \mathbf{i} = \dfrac{dr}{dt}\cos\theta - r\dfrac{d\theta}{dt}\sin\theta \Rightarrow \dfrac{dx}{dt} = \dfrac{dr}{dt}\cos\theta - r\dfrac{d\theta}{dt}\sin\theta$; $\mathbf{v} \cdot \mathbf{j} = \dfrac{dy}{dt}$ and $\mathbf{v} \cdot \mathbf{j} = \dfrac{dr}{dt}\sin\theta + r\dfrac{d\theta}{dt}\cos\theta$

$\Rightarrow \dfrac{dy}{dt} = \dfrac{dr}{dt}\sin\theta + r\dfrac{d\theta}{dt}\cos\theta$

(b) $\mathbf{u}_r = (\cos\theta)\mathbf{i} + (\sin\theta)\mathbf{j} \Rightarrow \mathbf{v} \cdot \mathbf{u}_r = \dfrac{dx}{dt}\cos\theta + \dfrac{dy}{dt}\sin\theta$

$= \left(\dfrac{dr}{dt}\cos\theta - r\dfrac{d\theta}{dt}\sin\theta\right)(\cos\theta) + \left(\dfrac{dr}{dt}\sin\theta + r\dfrac{d\theta}{dt}\cos\theta\right)(\sin\theta)$ by part (a),

$\Rightarrow \mathbf{v} \cdot \mathbf{u}_r = \dfrac{dr}{dt}$; therefore, $\dfrac{dr}{dt} = \dfrac{dx}{dt}\cos\theta + \dfrac{dy}{dt}\sin\theta$;

$\mathbf{u}_\theta = -(\sin\theta)\mathbf{i} + (\cos\theta)\mathbf{j} \Rightarrow \mathbf{v}\cdot\mathbf{u}_\theta = -\frac{dx}{dt}\sin\theta + \frac{dy}{dt}\cos\theta$

$= \left(\frac{dr}{dt}\cos\theta - r\frac{d\theta}{dt}\sin\theta\right)(-\sin\theta) + \left(\frac{dr}{dt}\sin\theta + r\frac{d\theta}{dt}\cos\theta\right)(\cos\theta)$ by part (a) $\Rightarrow \mathbf{v}\cdot\mathbf{u}_\theta = r\frac{d\theta}{dt}$;

therefore, $r\frac{d\theta}{dt} = -\frac{dx}{dt}\sin\theta + \frac{dy}{dt}\cos\theta$

11. (a) Let $r = 2 - t$ and $\theta = 3t \Rightarrow \frac{dr}{dt} = -1$ and $\frac{d\theta}{dt} = 3 \Rightarrow \frac{d^2r}{dt^2} = \frac{d^2\theta}{dt^2} = 0$. The halfway point is $(1,3) \Rightarrow t = 1$;

$\mathbf{v} = \frac{dr}{dt}\mathbf{u}_r + r\frac{d\theta}{dt}\mathbf{u}_\theta \Rightarrow \mathbf{v}(1) = -\mathbf{u}_r + 3\mathbf{u}_\theta$; $\mathbf{a} = \left[\frac{d^2r}{dt^2} - r\left(\frac{d\theta}{dt}\right)^2\right]\mathbf{u}_r + \left[r\frac{d^2\theta}{dt^2} + 2\frac{dr}{dt}\frac{d\theta}{dt}\right]\mathbf{u}_\theta \Rightarrow \mathbf{a}(1) = -9\mathbf{u}_r - 6\mathbf{u}_\theta$

(b) It takes the beetle 2 min to crawl to the origin $\Rightarrow$ the rod has revolved 6 radians

$$\Rightarrow L = \int_0^6 \sqrt{[f(\theta)]^2 + [f'(\theta)]^2}\, d\theta = \int_0^6 \sqrt{\left(2 - \frac{\theta}{3}\right)^2 + \left(-\frac{1}{3}\right)^2}\, d\theta = \int_0^6 \sqrt{4 - \frac{4\theta}{3} + \frac{\theta^2}{9} + \frac{1}{9}}\, d\theta$$

$$= \int_0^6 \sqrt{\frac{37 - 12\theta + \theta^2}{9}}\, d\theta = \frac{1}{3}\int_0^6 \sqrt{(\theta-6)^2 + 1}\, d\theta = \frac{1}{3}\left[\frac{(\theta-6)}{2}\sqrt{(\theta-6)^2+1} + \frac{1}{2}\ln\left|\theta - 6 + \sqrt{(\theta-6)^2+1}\right|\right]_0^6$$

$$= \sqrt{37} - \frac{1}{6}\ln\left(\sqrt{37} - 6\right) \approx 6.5 \text{ in.}$$

13. (a) $\mathbf{u}_r \times \mathbf{u}_\theta = \begin{vmatrix} \mathbf{i} & \mathbf{j} & \mathbf{k} \\ \cos\theta & \sin\theta & 0 \\ -\sin\theta & \cos\theta & 0 \end{vmatrix} = \mathbf{k} \Rightarrow$ a right-handed frame of unit vectors

(b) $\frac{d\mathbf{u}_r}{d\theta} = (-\sin\theta)\mathbf{i} + (\cos\theta)\mathbf{j} = \mathbf{u}_\theta$ and $\frac{d\mathbf{u}_\theta}{d\theta} = (-\cos\theta)\mathbf{i} - (\sin\theta)\mathbf{j} = -\mathbf{u}_r$

(c) From Eq. (7), $\mathbf{v} = \dot{r}\mathbf{u}_r + r\dot{\theta}\mathbf{u}_\theta + \dot{z}\mathbf{k} \Rightarrow \mathbf{a} = \dot{\mathbf{v}} = \left(\ddot{r}\mathbf{u}_r + \dot{r}\,\dot{\mathbf{u}}_r\right) + \left(\dot{r}\,\dot{\theta}\mathbf{u}_\theta + r\ddot{\theta}\mathbf{u}_\theta + r\dot{\theta}\,\dot{\mathbf{u}}_\theta\right) + \ddot{z}\,\mathbf{k}$

$= \left(\ddot{r} - r\dot{\theta}^2\right)\mathbf{u}_r + \left(r\ddot{\theta} + 2\dot{r}\,\dot{\theta}\right)\mathbf{u}_\theta + \ddot{z}\,\mathbf{k}$

15. (a) $\mathbf{u}_\rho = (\sin\phi\cos\theta)\mathbf{i} + (\sin\phi\sin\theta)\mathbf{j} + (\cos\phi)\mathbf{k}$, $\mathbf{u}_\phi = (\cos\phi\cos\theta)\mathbf{i} + (\cos\phi\sin\theta)\mathbf{j} - (\sin\phi)\mathbf{k}$, and

$\mathbf{u}_\theta = \mathbf{u}_\rho \times \mathbf{u}_\phi = (-\sin\theta)\mathbf{i} + (\cos\theta)\mathbf{j}$ (we choose $\rho = 1$ for <u>unit</u> vectors)

(b) $\mathbf{u}_\rho \cdot \mathbf{u}_\phi = (\sin\phi\cos\theta)(\cos\phi\cos\theta) + (\sin\phi\sin\theta)(\cos\phi\sin\theta) + (\cos\phi)(-\sin\phi)$

$= (\sin\phi\cos\phi)\left(\cos^2\theta\right) + (\sin\phi\cos\phi)\left(\sin^2\theta\right) - \cos\phi\sin\phi = (\sin\phi\cos\phi) - (\cos\phi\sin\phi) = 0$

(c) $\mathbf{u}_\rho \times \mathbf{u}_\phi = \begin{vmatrix} \mathbf{i} & \mathbf{j} & \mathbf{k} \\ \sin\phi\cos\theta & \sin\phi\sin\theta & \cos\phi \\ \cos\phi\cos\theta & \cos\phi\sin\theta & -\sin\phi \end{vmatrix} = (-\sin\theta)\mathbf{i} + (\cos\theta)\mathbf{j} = \mathbf{u}_\theta$

(d) $\mathbf{u}_\rho \times \mathbf{u}_\phi = \mathbf{u}_\theta \Rightarrow$ right-handed frame

CHAPTER 12 MULTIVARIABLE FUNCTIONS AND PARTIAL DERIVATIVES

12.1 FUNCTIONS OF SEVERAL VARIABLES

1. (a) Domain: all points in the xy-plane
   (b) Range: all real numbers
   (c) level curves are straight lines $y - x = c$ parallel to the line $y = x$
   (d) no boundary points
   (e) both open and closed
   (f) unbounded

3. (a) Domain: all points in the xy-plane
   (b) Range: $z \geq 0$
   (c) level curves: for $f(x,y) = 0$, the origin; for $f(x,y) = c > 0$, ellipses with center $(0,0)$ and major and minor axes along the x- and y-axes, respectively
   (d) no boundary points
   (e) both open and closed
   (f) unbounded

5. (a) Domain: all points in the xy-plane
   (b) Range: all real numbers
   (c) level curves are hyperbolas with the x- and y-axes as asymptotes when $f(x,y) \neq 0$, and the x- and y-axes when $f(x,y) = 0$
   (d) no boundary points
   (e) both open and closed
   (f) unbounded

7. (a) Domain: all (x,y) satisfying $x^2 + y^2 < 16$
   (b) Range: $z \geq \frac{1}{4}$
   (c) level curves are circles centered at the origin with radii $r < 4$
   (d) boundary is the circle $x^2 + y^2 = 16$
   (e) open
   (f) bounded

9. (a) Domain: $(x,y) \neq (0,0)$
   (b) Range: all real numbers
   (c) level curves are circles with center $(0,0)$ and radii $r > 0$
   (d) boundary is the single point $(0,0)$
   (e) open
   (f) unbounded

11. (a) Domain: all (x,y) satisfying $-1 \leq y - x \leq 1$
    (b) Range: $-\frac{\pi}{2} \leq z \leq \frac{\pi}{2}$
    (c) level curves are straight lines of the form $y - x = c$ where $-1 \leq c \leq 1$
    (d) boundary is the two straight lines $y = 1 + x$ and $y = -1 + x$

(e) closed
(f) unbounded

13. f

15. a

17. d

19. (a)

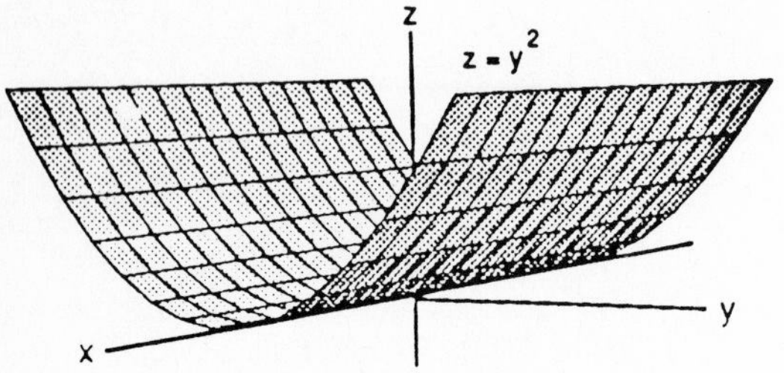

(b)

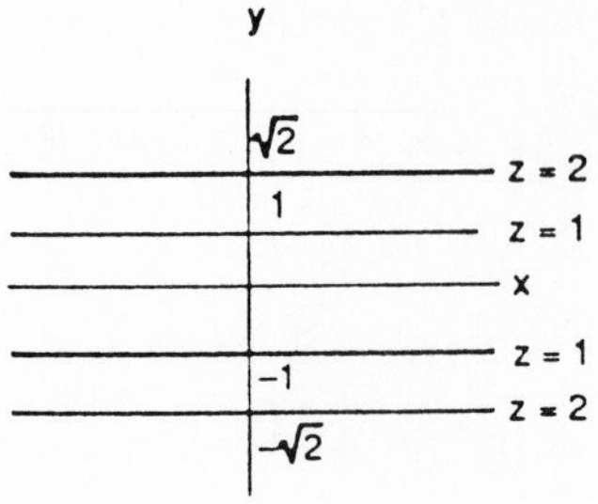

21. (a)

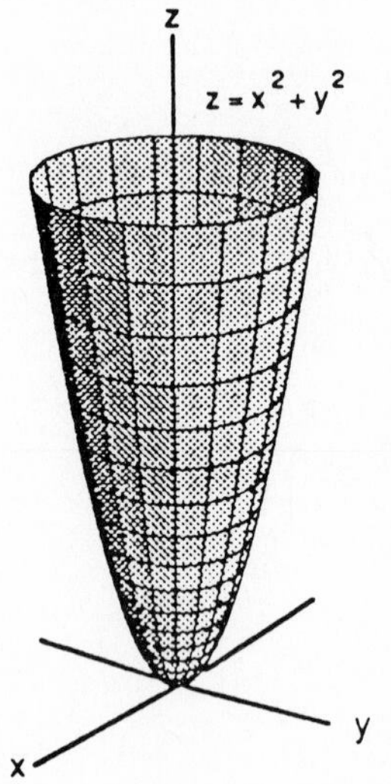

(b)

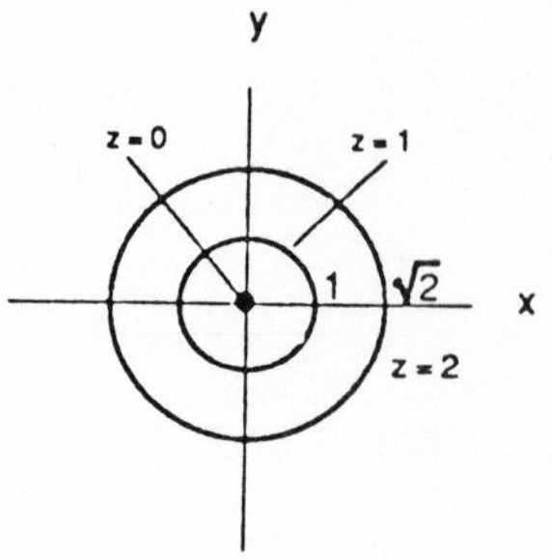

23. (a)

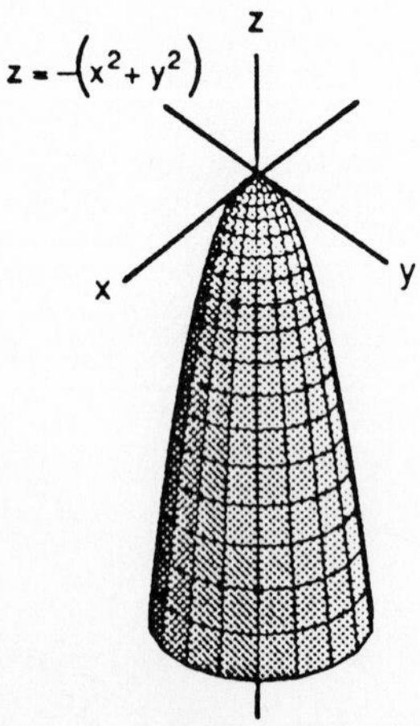

(b)

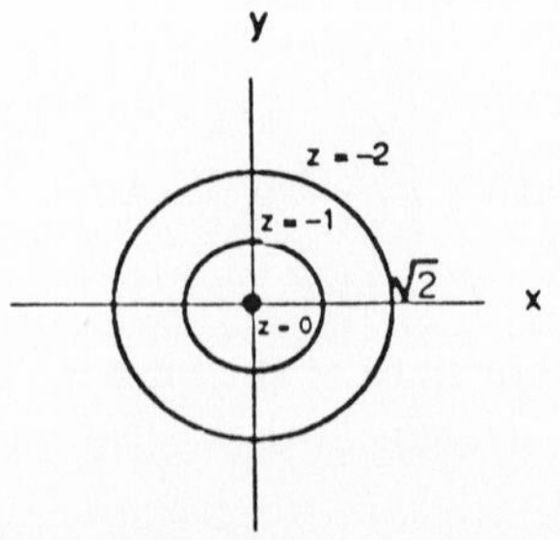

25. (a)

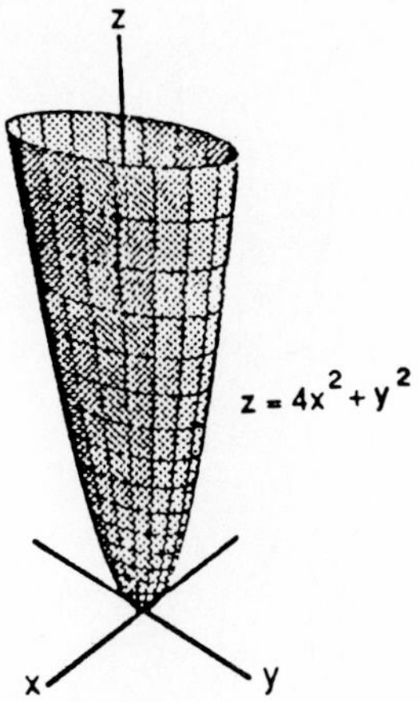

(b)

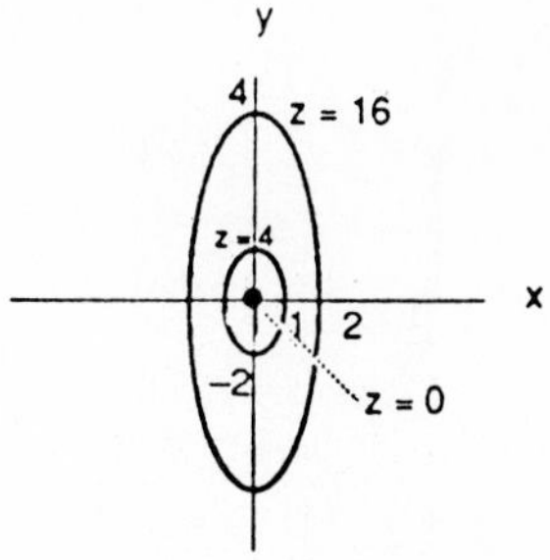

27. (a)

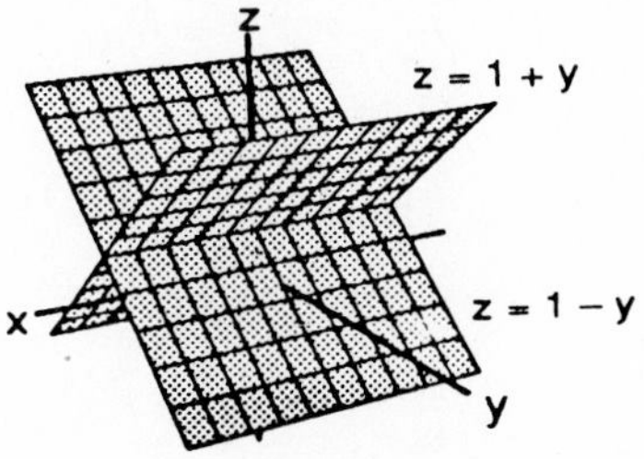

(b)

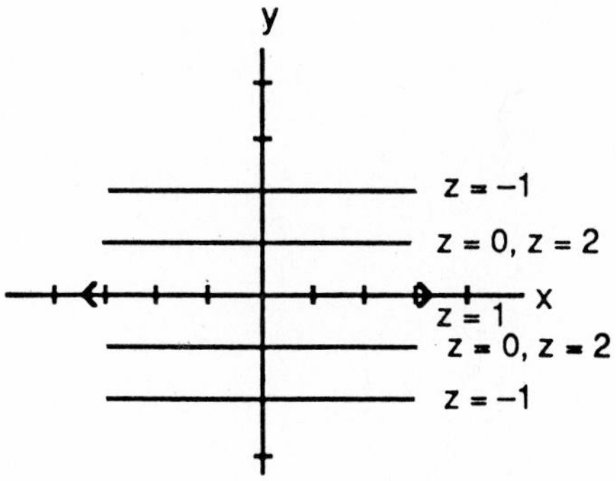

29.

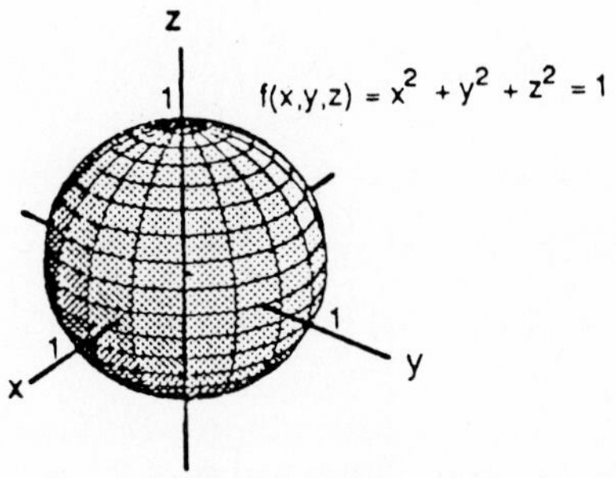

31.

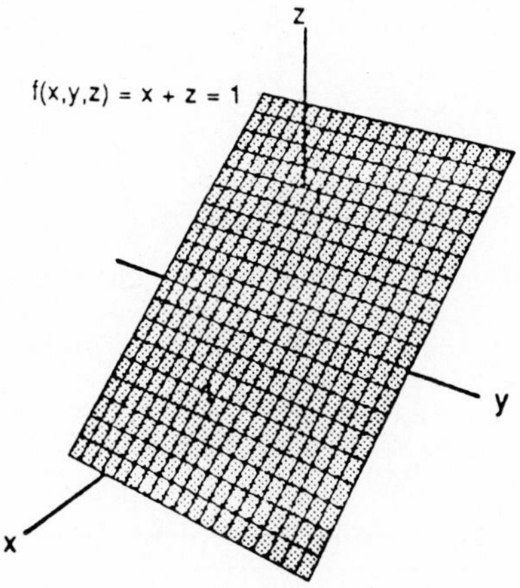

33.

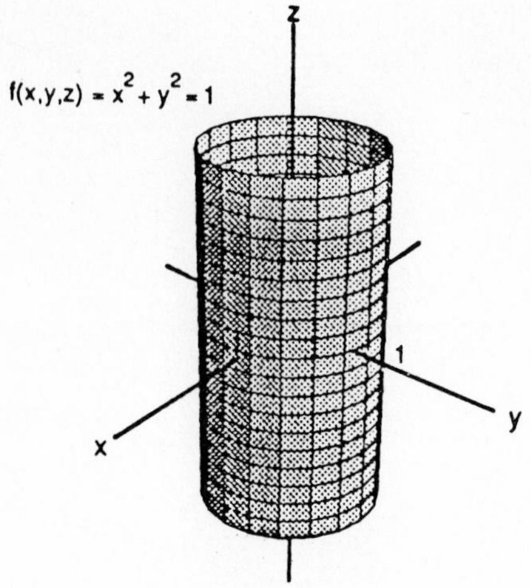

35.

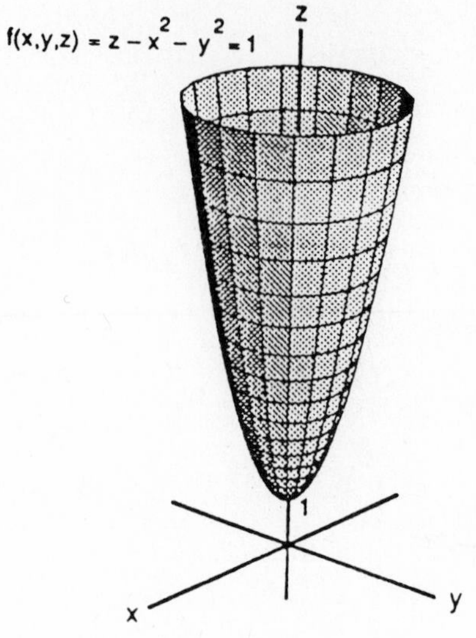

37. $f(x,y) = 16 - x^2 - y^2$ and $(2\sqrt{2}, \sqrt{2}) \Rightarrow z = 16 - (2\sqrt{2})^2 - (\sqrt{2})^2 = 6 \Rightarrow 6 = 16 - x^2 - y^2 \Rightarrow x^2 + y^2 = 10$

39. $f(x,y) = \int_x^y \frac{1}{1+t^2}\,dt$ at $(-\sqrt{2}, \sqrt{2}) \Rightarrow z = \tan^{-1} y - \tan^{-1} x$; at $(-\sqrt{2}, \sqrt{2}) \Rightarrow z = \tan^{-1}\sqrt{2} - \tan^{-1}(-\sqrt{2})$
$= 2\tan^{-1}\sqrt{2} \Rightarrow \tan^{-1} y - \tan^{-1} x = 2\tan^{-1}\sqrt{2}$

41. $f(x,y,z) = \sqrt{x-y} - \ln z$ at $(3,-1,1) \Rightarrow w = \sqrt{x-y} - \ln z$; at $(3,-1,1) \Rightarrow w = \sqrt{3-(-1)} - \ln 1 = 2$
$\Rightarrow \sqrt{x-y} - \ln z = 2$

43. $g(x,y,z) = \sum_{n=0}^{\infty} \frac{(x+y)^n}{n!\,z^n}$ at $(\ln 2, \ln 4, 3) \Rightarrow w = \sum_{n=0}^{\infty} \frac{(x+y)^n}{n!\,z^n} = e^{(x+y)/z}$; at $(\ln 2, \ln 4, 3) \Rightarrow w = e^{(\ln 2 + \ln 4)/3}$
$= e^{(\ln 8)/3} = e^{\ln 2} = 2 \Rightarrow 2 = e^{(x+y)/z} \Rightarrow \frac{x+y}{z} = \ln 2$

45. $f(x,y,z) = xyz$ and $x = 20 - t$, $y = t$, $z = 20 \Rightarrow w = (20-t)(t)(20)$ along the line $\Rightarrow w = 400t - 20t^2$
$\Rightarrow \frac{dw}{dt} = 400 - 40t$; $\frac{dw}{dt} = 0 \Rightarrow 400 - 40t = 0 \Rightarrow t = 10$ and $\frac{d^2w}{dt^2} = -40$ for all $t \Rightarrow$ yes, maximum at $t = 10$
$\Rightarrow x = 20 - 10 = 10$, $y = 10$, $z = 20 \Rightarrow$ maximum of f along the line is $f(10,10,20) = (10)(10)(20) = 2000$

47. $w = 4\left(\frac{Th}{d}\right)^{1/2} = 4\left[\frac{(290\text{ K})(16.8\text{ km})}{5\text{ K/km}}\right]^{1/2} \approx 124.86$ km $\Rightarrow$ must be $\frac{1}{2}(124.86) \approx 63$ km south of Nantucket

12.2 LIMITS AND CONTINUITY

1. $\lim\limits_{(x,y)\to(0,0)} \frac{3x^2 - y^2 + 5}{x^2 + y^2 + 2} = \frac{3(0)^2 - 0^2 + 5}{0^2 + 0^2 + 2} = \frac{5}{2}$

3. $\lim_{(x,y)\to(3,4)} \sqrt{x^2+y^2-1} = \sqrt{3^2+4^2-1} = \sqrt{24} = 2\sqrt{6}$

5. $\lim_{(x,y)\to\left(0,\frac{\pi}{4}\right)} \sec x \tan y = (\sec 0)\left(\tan \frac{\pi}{4}\right) = (1)(1) = 1$

7. $\lim_{(x,y)\to(0,\ln 2)} e^{x-y} = e^{0-\ln 2} = e^{\ln\left(\frac{1}{2}\right)} = \frac{1}{2}$

9. $\lim_{(x,y)\to(0,0)} \frac{e^y \sin x}{x} = \lim_{(x,y)\to(0,0)} (e^y)\left(\frac{\sin x}{x}\right) = e^0 \cdot \lim_{x\to 0} \left(\frac{\sin x}{x}\right) = 1 \cdot 1 = 1$

11. $\lim_{(x,y)\to(1,0)} \frac{x \sin y}{x^2+1} = \frac{1 \cdot \sin 0}{1^2+1} = \frac{0}{2} = 0$

13. $\lim_{\substack{(x,y)\to(1,1)\\ x\neq y}} \frac{x^2-2xy+y^2}{x-y} = \lim_{(x,y)\to(1,1)} \frac{(x-y)^2}{x-y} = \lim_{(x,y)\to(1,2)} (x-y) = (1-1) = 0$

15. $\lim_{\substack{(x,y)\to(1,1)\\ x\neq 1}} \frac{xy-y-2x+2}{x-1} = \lim_{\substack{(x,y)\to(1,1)\\ x\neq 1}} \frac{(x-1)(y-2)}{x-1} = \lim_{(x,y)\to(1,1)} (y-2) = (1-2) = -1$

17. $\lim_{\substack{(x,y)\to(0,0)\\ x\neq y}} \frac{x-y+2\sqrt{x}-2\sqrt{y}}{\sqrt{x}-\sqrt{y}} = \lim_{\substack{(x,y)\to(0,0)\\ x\neq y}} \frac{\left(\sqrt{x}-\sqrt{y}\right)\left(\sqrt{x}+\sqrt{y}+2\right)}{\sqrt{x}-\sqrt{y}} = \lim_{(x,y)\to(0,0)} \left(\sqrt{x}+\sqrt{y}+2\right)$

$= \left(\sqrt{0}+\sqrt{0}+2\right) = 2$

Note: (x,y) must approach $(0,0)$ through the first quadrant only with $x \neq y$.

19. $\lim_{\substack{(x,y)\to(2,0)\\ 2x-y\neq 4}} \frac{\sqrt{2x-y}-2}{2x-y-4} = \lim_{\substack{(x,y)\to(2,0)\\ 2x-y\neq 4}} \frac{\sqrt{2x-y}-2}{\left(\sqrt{2x-y}+2\right)\left(\sqrt{2x-y}-2\right)} = \lim_{(x,y)\to(2,0)} \frac{1}{\sqrt{2x-y}+2}$

$= \frac{1}{\sqrt{(2)(2)-0}+2} = \frac{1}{2+2} = \frac{1}{4}$

21. $\lim_{P\to(1,3,4)} \left(\frac{1}{x}+\frac{1}{y}+\frac{1}{z}\right) = \frac{1}{1}+\frac{1}{3}+\frac{1}{4} = \frac{12+4+3}{12} = \frac{19}{12}$

23. $\lim_{P\to(3,3,0)} \left(\sin^2 x+\cos^2 y+\sec^2 z\right) = \left(\sin^2 3+\cos^2 3\right)+\sec^2 0 = 1+1^2 = 2$

25. $\lim_{P\to(\pi,0,3)} ze^{-2y} \cos 2x = 3e^{-2(0)} \cos 2\pi = (3)(1)(1) = 3$

27. (a) All (x,y)
    (b) All (x,y) except $(0,0)$

29. (a) All (x,y) except where $x=0$ or $y=0$
    (b) All (x,y)

31. (a) All (x, y, z)
(b) All (x, y, z) except the interior of the cylinder $x^2 + y^2 = 1$

33. (a) All (x, y, z) with $z \neq 0$
(b) All (x, y, z) with $x^2 + z^2 \neq 1$

35. $\lim\limits_{\substack{(x,y)\to(0,0) \\ \text{along } y = x \\ x > 0}} -\frac{x}{\sqrt{x^2+y^2}} = \lim\limits_{x\to 0} -\frac{x}{\sqrt{x^2+x^2}} = \lim\limits_{x\to 0} -\frac{x}{\sqrt{2}\,|x|} = \lim\limits_{x\to 0} -\frac{x}{\sqrt{2}\,x} = \lim\limits_{x\to 0} -\frac{1}{\sqrt{2}} = -\frac{1}{\sqrt{2}}$;

$\lim\limits_{\substack{(x,y)\to(0,0) \\ \text{along } y = x \\ x < 0}} -\frac{x}{\sqrt{x^2+y^2}} = \lim\limits_{x\to 0} -\frac{x}{\sqrt{2}\,|x|} = \lim\limits_{x\to 0} -\frac{x}{\sqrt{2}(-x)} = \lim\limits_{x\to 0} \frac{1}{\sqrt{2}} = \frac{1}{\sqrt{2}}$

37. $\lim\limits_{\substack{(x,y)\to(0,0) \\ \text{along } y = kx^2}} \frac{x^4 - y^2}{x^4 + y^2} = \lim\limits_{x\to 0} \frac{x^4 - (kx^2)^2}{x^4 + (kx^2)^2} = \lim\limits_{x\to 0} \frac{x^4 - k^2x^4}{x^4 + k^2x^4} = \frac{1-k^2}{1+k^2} \Rightarrow$ different limits for different values of k

39. $\lim\limits_{\substack{(x,y)\to(0,0) \\ \text{along } y = kx \\ k \neq -1}} \frac{x - y}{x + y} = \lim\limits_{x\to 0} \frac{x - kx}{x + kx} = \frac{1-k}{1+k} \Rightarrow$ different limits for different values of k, $k \neq -1$

41. $\lim\limits_{\substack{(x,y)\to(0,0) \\ \text{along } y = kx^2 \\ k \neq 0}} \frac{x^2 + y}{y} = \lim\limits_{x\to 0} \frac{x^2 + kx^2}{kx^2} = \frac{1+k}{k} \Rightarrow$ different limits for different values of k, $k \neq 0$

43. No, the limit depends only on the values $f(x, y)$ has when $(x, y) \neq (x_0, y_0)$

45. $\lim\limits_{(x,y)\to(0,0)} \left(1 - \frac{x^2y^2}{3}\right) = 1$ and $\lim\limits_{(x,y)\to(0,0)} 1 = 1 \Rightarrow \lim\limits_{(x,y)\to(0,0)} \frac{\tan^{-1} xy}{xy} = 1$, by the Sandwich Theorem

47. The limit is 0 since $\left|\sin\left(\frac{1}{x}\right)\right| \le 1 \Rightarrow -1 \le \sin\left(\frac{1}{x}\right) \le 1 \Rightarrow -y \le y \sin\left(\frac{1}{x}\right) \le y$ for $y \ge 0$, and $-y \ge y \sin\left(\frac{1}{x}\right) \ge y$ for $y \le 0$. Thus as $(x, y) \to (0, 0)$, both $-y$ and y approach $0 \Rightarrow y \sin\left(\frac{1}{x}\right) \to 0$, by the Sandwich Theorem.

49. (a) $f(x, y)\big|_{y = mx} = \frac{2m}{1+m^2} = \frac{2 \tan \theta}{1 + \tan^2 \theta} = \sin 2\theta$. The value of $f(x, y) = \sin 2\theta$ varies with θ, which is the line's angle of inclination.

(b) Since $f(x, y)\big|_{y = mx} = \sin 2\theta$ and since $-1 \le \sin 2\theta \le 1$ for every θ, $\lim\limits_{(x,y)\to(0,0)} f(x, y)$ varies from -1 to 1 along $y = mx$.

51. $\lim\limits_{(x,y)\to(0,0)} \frac{x^3 - xy^2}{x^2 + y^2} = \lim\limits_{r\to 0} \frac{r^3 \cos^3 \theta - (r \cos \theta)(r^2 \sin^2 \theta)}{r^2 \cos^2 \theta + r^2 \sin^2 \theta} = \lim\limits_{r\to 0} \frac{r(\cos^3 \theta - \cos \theta \sin^2 \theta)}{1} = 0$

53. $\lim\limits_{(x,y)\to(0,0)} \frac{y^2}{x^2 + y^2} = \lim\limits_{r\to 0} \frac{r^2 \sin^2 \theta}{r^2} = \lim\limits_{r\to 0} (\sin^2 \theta) = \sin^2 \theta$; the limit does not exist since $\sin^2 \theta$ is between 0 and 1 depending on θ

55. $\lim\limits_{(x,y)\to(0,0)} \tan^{-1}\left[\frac{|x|+|y|}{x^2+y^2}\right] = \lim\limits_{r\to 0} \tan^{-1}\left[\frac{|r\cos\theta|+|r\sin\theta|}{r^2}\right] = \lim\limits_{r\to 0} \tan^{-1}\left[\frac{|r|\left(|\cos\theta|+|\sin\theta|\right)}{r^2}\right]$;

if $r \to 0^+$, then $\lim\limits_{r\to 0^+} \tan^{-1}\left[\frac{|r|\left(|\cos\theta|+|\sin\theta|\right)}{r^2}\right] = \lim\limits_{r\to 0^+} \tan^{-1}\left[\frac{|\cos\theta|+|\sin\theta|}{r}\right] = \frac{\pi}{2}$; if $r \to 0^-$, then

$\lim\limits_{r\to 0^-} \tan^{-1}\left[\frac{|r|\left(|\cos\theta|+|\sin\theta|\right)}{r^2}\right] = \lim\limits_{r\to 0^-} \tan^{-1}\left(\frac{|\cos\theta|+|\sin\theta|}{-r}\right) = \frac{\pi}{2} \Rightarrow$ the limit is $\frac{\pi}{2}$

57. $\lim\limits_{(x,y)\to(0,0)} \ln\left(\frac{3x^2 - x^2y^2 + 3y^2}{x^2+y^2}\right) = \lim\limits_{r\to 0} \ln\left(\frac{3r^2\cos^2\theta - r^4\cos^2\theta\sin^2\theta + 3r^2\sin^2\theta}{r^2}\right)$

$= \lim\limits_{r\to 0} \ln\left(3 - r^2\cos^2\theta\sin^2\theta\right) = \ln 3 \Rightarrow$ define $f(0,0) = \ln 3$

59. In Eq. (1), if the point (x,y) lies within a disk centered at (x_0,y_0) and radius less than δ, then $|f(x,y) - L| < \epsilon$; in Eq. (2), if the point (x,y) lies within a square centered at (x_0,y_0) with the side length less than 2δ, then $|f(x,y) - L| < \epsilon$. Since every circle of radius δ is circumscribed by a square of side length 2δ, $\sqrt{(x-x_0)^2 + (y-y_0)^2} < \delta \Rightarrow |x - x_0| < \delta$ and $|y - y_0| < \delta$; likewise, every square of side length $\frac{2\delta}{\sqrt{2}}$ is circumscribed by a circle of radius δ so that $|x - x_0| < \frac{\delta}{\sqrt{2}}$ and $|y - y_0| < \frac{\delta}{\sqrt{2}}$

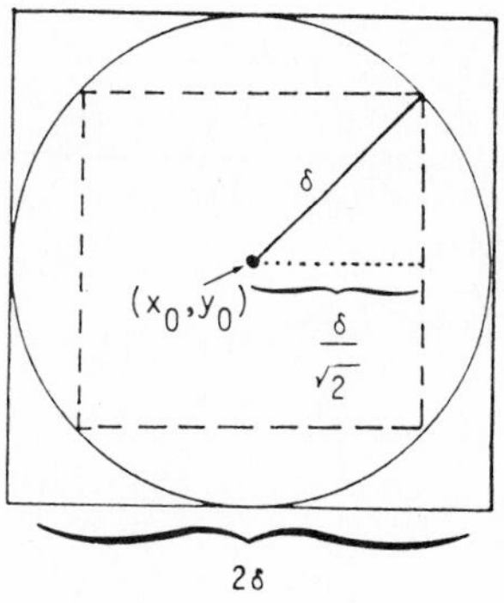

$\Rightarrow \sqrt{(x-x_0)^2 + (y-y_0)^2} < \delta$. Thus the requirements are equivalent: small circles give small inscribed squares, and small squares give small inscribed circles.

61. Let $\delta = 0.1$. Then $\sqrt{x^2+y^2} < \delta \Rightarrow \sqrt{x^2+y^2} < 0.1 \Rightarrow x^2+y^2 < 0.01 \Rightarrow |x^2+y^2-0| < 0.01 \Rightarrow |f(x,y) - f(0,0)| < 0.01 = \epsilon$.

63. Let $\delta = 0.005$. Then $|x| < \delta$ and $|y| < \delta \Rightarrow |f(x,y) - f(0,0)| = \left|\frac{x+y}{x^2+1} - 0\right| = \left|\frac{x+y}{x^2+1}\right| \le |x+y| < |x| + |y| < 0.005 + 0.005 = 0.01 = \epsilon$.

65. Let $\delta = \sqrt{0.015}$. Then $\sqrt{x^2+y^2+z^2} < \delta \Rightarrow |f(x,y,z) - f(0,0,0)| = |x^2+y^2+z^2-0| = |x^2+y^2+z^2|$ $= \left(\sqrt{x^2+t^2+x^2}\right)^2 < \left(\sqrt{0.015}\right)^2 = 0.015 = \epsilon$.

67. Let $\delta = 0.005$. Then $|x| < \delta$, $|y| < \delta$, and $|z| < \delta \Rightarrow |f(x,y,z) - f(0,0,0)| = \left|\frac{x+y+z}{x^2+y^2+z^2+1} - 0\right|$ $= \left|\frac{x+y+z}{x^2+y^2+z^2+1}\right| \le |x+y+z| \le |x| + |y| + |z| < 0.005 + 0.005 + 0.005 = 0.015 = \epsilon$.

69. $\lim_{(x,y,z)\to(x_0,y_0,z_0)} f(x,y,z) = \lim_{(x,y,z)\to(x_0,y_0,z_0)} (x+y+z) = x_0+y_0+z_0 = f(x_0,y_0,z_0) \Rightarrow$ f is continuous at every (x_0,y_0,z_0)

12.3 PARTIAL DERIVATIVES

1. $\frac{\partial f}{\partial x} = 4x,\ \frac{\partial f}{\partial y} = -3$

3. $\frac{\partial f}{\partial x} = 2x(y+2),\ \frac{\partial f}{\partial y} = x^2-1$

5. $\frac{\partial f}{\partial x} = 2y(xy-1),\ \frac{\partial f}{\partial y} = 2x(xy-1)$

7. $\frac{\partial f}{\partial x} = \frac{x}{\sqrt{x^2+y^2}},\ \frac{\partial f}{\partial y} = \frac{y}{\sqrt{x^2+y^2}}$

9. $\frac{\partial f}{\partial x} = -\frac{1}{(x+y)^2}\cdot\frac{\partial}{\partial x}(x+y) = -\frac{1}{(x+y)^2},\ \frac{\partial f}{\partial y} = -\frac{1}{(x+y)^2}\cdot\frac{\partial}{\partial y}(x+y) = -\frac{1}{(x+y)^2}$

11. $\frac{\partial f}{\partial x} = = \frac{(xy-1)(1)-(x+y)(y)}{(xy-1)^2} = \frac{-y^2-1}{(xy-1)^2},\ \frac{\partial f}{\partial y} = \frac{(xy-1)(1)-(x+y)(x)}{(xy-1)^2} = \frac{-x^2-1}{(xy-1)^2}$

13. $\frac{\partial f}{\partial x} = e^{(x+y+1)}\cdot\frac{\partial}{\partial x}(x+y+1) = e^{(x+y+1)},\ \frac{\partial f}{\partial y} = e^{(x+y+1)}\cdot\frac{\partial}{\partial y}(x+y+1) = e^{(x+y+1)}$

15. $\frac{\partial f}{\partial x} = \frac{1}{x+y}\cdot\frac{\partial}{\partial x}(x+y) = \frac{1}{x+y},\ \frac{\partial f}{\partial y} = \frac{1}{x+y}\cdot\frac{\partial}{\partial y}(x+y) = \frac{1}{x+y}$

17. $\frac{\partial f}{\partial x} = 2\sin(x-3y)\cdot\frac{\partial}{\partial x}\sin(x-3y) = 2\sin(x-3y)\cos(x-3y)\cdot\frac{\partial}{\partial x}(x-3y) = 2\sin(x-3y)\cos(x-3y)$,
$\frac{\partial f}{\partial y} = 2\sin(x-3y)\cdot\frac{\partial}{\partial y}\sin(x-3y) = 2\sin(x-3y)\cos(x-3y)\cdot\frac{\partial}{\partial y}(x-3y) = -6\sin(x-3y)\cos(x-3y)$

19. $\frac{\partial f}{\partial x} = yx^{y-1},\ \frac{\partial f}{\partial y} = x^y \ln x$

21. $\frac{\partial f}{\partial x} = -g(x),\ \frac{\partial f}{\partial y} = g(y)$

23. $f_x = 1+y^2,\ f_y = 2xy,\ f_z = -4z$

25. $f_x = 1,\ f_y = -\frac{y}{\sqrt{y^2+z^2}},\ f_z = -\frac{z}{\sqrt{y^2+z^2}}$

27. $f_x = \frac{yz}{\sqrt{1-x^2y^2z^2}},\ f_y = \frac{xz}{\sqrt{1-x^2y^2z^2}},\ f_z = \frac{xy}{\sqrt{1-x^2y^2z^2}}$

29. $f_x = \frac{1}{x+2y+3z},\ f_y = \frac{2}{x+2y+3z},\ f_z = \frac{3}{x+2y+3z}$

31. $f_x = -2xe^{-(x^2+y^2+z^2)},\ f_y = -2ye^{-(x^2+y^2+z^2)},\ f_z = -2ze^{-(x^2+y^2+z^2)}$

33. $f_x = \operatorname{sech}^2(x+2y+3z),\ f_y = 2\operatorname{sech}^2(x+2y+3z),\ f_z = 3\operatorname{sech}^2(x+2y+3z)$

35. $\frac{\partial f}{\partial t} = -2\pi\sin(2\pi t-\alpha),\ \frac{\partial f}{\partial \alpha} = \sin(2\pi t-\alpha)$

37. $\frac{\partial h}{\partial \rho} = \sin \phi \cos \theta,\ \frac{\partial h}{\partial \phi} = \rho \cos \phi \cos \theta,\ \frac{\partial h}{\partial \theta} = -\rho \sin \phi \sin \theta$

39. $W_P = V,\ W_V = P + \frac{\delta v^2}{2g},\ W_\delta = \frac{Vv^2}{2g},\ W_v = \frac{2V\delta v}{2g} = \frac{V\delta v}{g},\ W_g = -\frac{V\delta v^2}{2g^2}$

41. $\frac{\partial f}{\partial x} = 1 + y,\ \frac{\partial f}{\partial y} = 1 + x,\ \frac{\partial^2 f}{\partial x^2} = 0,\ \frac{\partial^2 f}{\partial y^2} = 0,\ \frac{\partial^2 f}{\partial y \partial x} = \frac{\partial^2 f}{\partial x \partial y} = 1$

43. $\frac{\partial g}{\partial x} = 2xy + y \cos x,\ \frac{\partial g}{\partial y} = x^2 - \sin y + \sin x,\ \frac{\partial^2 g}{\partial x^2} = 2y - y \sin x,\ \frac{\partial^2 g}{\partial y^2} = -\cos y,\ \frac{\partial^2 g}{\partial y \partial x} = \frac{\partial^2 g}{\partial x \partial y} = 2x + \cos x$

45. $\frac{\partial r}{\partial x} = \frac{1}{x+y},\ \frac{\partial r}{\partial y} = \frac{1}{x+y},\ \frac{\partial^2 r}{\partial x^2} = \frac{-1}{(x+y)^2},\ \frac{\partial^2 r}{\partial y^2} = \frac{-1}{(x+y)^2},\ \frac{\partial^2 r}{\partial y \partial x} = \frac{\partial^2 r}{\partial x \partial y} = \frac{-1}{(x+y)^2}$

47. $\frac{\partial w}{\partial x} = \frac{2}{2x+3y},\ \frac{\partial w}{\partial y} = \frac{3}{2x+3y},\ \frac{\partial^2 w}{\partial y \partial x} = \frac{-6}{(2x+3y)^2}$, and $\frac{\partial^2 w}{\partial x \partial y} = \frac{-6}{(2x+3y)^2}$

49. $\frac{\partial w}{\partial x} = y^2 + 2xy^3 + 3x^2y^4,\ \frac{\partial w}{\partial y} = 2xy + 3x^2y^2 + 4x^3y^3,\ \frac{\partial^2 w}{\partial y \partial x} = 2y + 6xy^2 + 12x^2y^3$, and
$\frac{\partial^2 w}{\partial x \partial y} = 2y + 6xy^2 + 12x^2y^3$

51. (a) x first (b) y first (c) x first (d) x first (e) y first (f) y first

53. $f_x(1,2) = \lim_{h\to 0} \frac{f(1+h,2) - f(1,2)}{h} = \lim_{h\to 0} \frac{[1-(1+h)+2-6(1+h)^2]-(2-6)}{h} = \lim_{h\to 0} \frac{-h-6(1+2h+h^2)+6}{h}$

$= \lim_{h\to 0} \frac{-13h-6h^2}{h} = \lim_{h\to 0} (-13-6h) = -13,$

$f_y(1,2) = \lim_{h\to 0} \frac{f(1,2+h) - f(1,2)}{h} = \lim_{h\to 0} \frac{[1-1+(2+h)-3(2+h)]-(2-6)}{h} = \lim_{h\to 0} \frac{(2-6-2h)-(2-6)}{h}$

$= \lim_{h\to 0} (-2) = -2$

55. $f_z(x_0,y_0,z_0) = \lim_{h\to 0} \frac{f(x_0,y_0,z_0+h) - f(x_0,y_0,z_0)}{h};$

$f_z(1,2,3) = \lim_{h\to 0} \frac{f(1,2,3+h) - f(1,2,3)}{h} = \lim_{h\to 0} \frac{2(3+h)^2 - 2(9)}{h} = \lim_{h\to 0} \frac{12h+2h^2}{h} = \lim_{h\to 0} (12+2h) = 12$

57. $y + \left(3z^2 \frac{\partial z}{\partial x}\right)x + z^3 - 2y\frac{\partial z}{\partial x} = 0 \Rightarrow (3xz^2 - 2y)\frac{\partial z}{\partial x} = -y - z^3 \Rightarrow$ at (1,1,1) we have $(3-2)\frac{\partial z}{\partial x} = -1 - 1$ or
$\frac{\partial z}{\partial x} = -2$

59. $a^2 = b^2 + c^2 - 2bc \cos A \Rightarrow 2a = (2bc \sin A)\frac{\partial A}{\partial a} \Rightarrow \frac{\partial A}{\partial a} = \frac{a}{bc \sin A}$; also $0 = 2b - 2c \cos A + (2bc \sin A)\frac{\partial A}{\partial b}$
$\Rightarrow 2c \cos A - 2b = (2bc \sin A)\frac{\partial A}{\partial b} \Rightarrow \frac{\partial A}{\partial b} = \frac{c \cos A - b}{bc \sin A}$

f − f₁
$\frac{dy}{dt} = \frac{\Delta y}{\Delta x}$
$L = z_1 + \frac{dz}{dt} = \frac{\Delta z}{\Delta x}$
$\frac{dz}{dy} = \frac{\Delta z}{\Delta y}$
$\frac{dy}{dt}\Delta x + f_1 = L$

61. Differentiating each equation implicitly gives $1 = v_x \ln u + \left(\frac{v}{u}\right)u_x$ and $0 = u_x \ln v + \left(\frac{u}{v}\right)v_x$ or

$$\left.\begin{array}{r}(\ln u)\, v_x \quad + \left(\frac{v}{u}\right)u_x = 1 \\ \left(\frac{u}{v}\right)v_x + (\ln v)\, u_x = 0\end{array}\right\} \Rightarrow v_x = \frac{\begin{vmatrix} 1 & \frac{v}{u} \\ 0 & \ln v \end{vmatrix}}{\begin{vmatrix} \ln u & \frac{v}{u} \\ \frac{u}{v} & \ln v \end{vmatrix}} = \frac{\ln v}{(\ln u)(\ln v) - 1}$$

63. $\frac{\partial f}{\partial x} = 2x,\ \frac{\partial f}{\partial y} = 2y,\ \frac{\partial f}{\partial z} = -4z \Rightarrow \frac{\partial^2 f}{\partial x^2} = 2,\ \frac{\partial^2 f}{\partial y^2} = 2,\ \frac{\partial^2 f}{\partial z^2} = -4 \Rightarrow \frac{\partial^2 f}{\partial x^2} + \frac{\partial^2 f}{\partial y^2} + \frac{\partial^2 f}{\partial z^2} = 2 + 2 + (-4) = 0$

65. $\frac{\partial f}{\partial x} = -2e^{-2y} \sin 2x,\ \frac{\partial f}{\partial y} = -e^{-2y} \cos 2x,\ \frac{\partial^2 f}{\partial x^2} = -4e^{-2y} \cos 2x,\ \frac{\partial^2 f}{\partial y^2} = 4e^{-2y} \cos 2x \Rightarrow \frac{\partial^2 f}{\partial x^2} + \frac{\partial^2 f}{\partial y^2}$
$= -4e^{-2y} \cos 2x + 4e^{-2y} \cos 2x = 0$

67. $\frac{\partial f}{\partial x} = -\frac{1}{2}\left(x^2 + y^2 + z^2\right)^{-3/2}(2x) = -x\left(x^2 + y^2 + z^2\right)^{-3/2},\ \frac{\partial f}{\partial y} = -\frac{1}{2}\left(x^2 + y^2 + z^2\right)^{-3/2}(2y)$
$= -y\left(x^2 + y^2 + z^2\right)^{-3/2},\ \frac{\partial f}{\partial z} = -\frac{1}{2}\left(x^2 + y^2 + z^2\right)^{-3/2}(2z) = -z\left(x^2 + y^2 + z^2\right)^{-3/2};$

$\frac{\partial^2 f}{\partial x^2} = -\left(x^2 + y^2 + z^2\right)^{-3/2} + 3x^2\left(x^2 + y^2 + z^2\right)^{-5/2},\ \frac{\partial^2 f}{\partial y^2} = -\left(x^2 + y^2 + z^2\right)^{-3/2} + 3y^2\left(x^2 + y^2 + z^2\right)^{-5/2},$
$\frac{\partial^2 f}{\partial z^2} = -\left(x^2 + y^2 + z^2\right)^{-3/2} + 3z^2\left(x^2 + y^2 + z^2\right)^{-5/2} \Rightarrow \frac{\partial^2 f}{\partial x^2} + \frac{\partial^2 f}{\partial y^2} + \frac{\partial^2 f}{\partial z^2}$
$= \left[-\left(x^2 + y^2 + z^2\right)^{-3/2} + 3x^2\left(x^2 + y^2 + z^2\right)^{-5/2}\right] + \left[-\left(x^2 + y^2 + z^2\right)^{-3/2} + 3y^2\left(x^2 + y^2 + z^2\right)^{-5/2}\right]$
$+ \left[-\left(x^2 + y^2 + z^2\right)^{-3/2} + 3z^2\left(x^2 + y^2 + z^2\right)^{-5/2}\right] = -3\left(x^2 + y^2 + z^2\right)^{-3/2} + \left(3x^2 + 3y^2 + 3z^2\right)\left(x^2 + y^2 + z^2\right)^{-5/2}$
$= 0$

69. $\frac{\partial w}{\partial x} = \cos(x + ct),\ \frac{\partial w}{\partial t} = c \cos(x + ct);\ \frac{\partial^2 w}{\partial x^2} = -\sin(x + ct),\ \frac{\partial^2 w}{\partial t^2} = -c^2 \sin(x + ct) \Rightarrow \frac{\partial^2 w}{\partial t^2} = c^2[-\sin(x + ct)]$
$= c^2 \frac{\partial^2 w}{\partial x^2}$

71. $\frac{\partial w}{\partial x} = \cos(x + ct) - 2 \sin(2x + 2ct),\ \frac{\partial w}{\partial t} = c \cos(x + ct) - 2c \sin(2x + 2ct);$
$\frac{\partial^2 w}{\partial x^2} = -\sin(x + ct) - 4 \cos(2x + 2ct),\ \frac{\partial^2 w}{\partial t^2} = -c^2 \sin(x + ct) - 4c^2 \cos(2x + 2ct)$
$\Rightarrow \frac{\partial^2 w}{\partial t^2} = c^2[-\sin(x + ct) - 4 \cos(2x + 2ct)] = c^2 \frac{\partial^2 w}{\partial x^2}$

73. $\frac{\partial w}{\partial x} = 2 \sec^2(2x - 2ct),\ \frac{\partial w}{\partial t} = -2c \sec^2(2x - 2ct);\ \frac{\partial^2 w}{\partial x^2} = 8 \sec^2(2x - 2ct) \tan(2x - 2ct),$

$\frac{\partial^2 w}{\partial t^2} = 8c^2 \sec^2(2x - 2ct) \tan(2x - 2ct) \Rightarrow \frac{\partial^2 w}{\partial t^2} = c^2[8 \sec^2(2x - 2ct) \tan(2x - 2ct)] = c^2 \frac{\partial^2 w}{\partial x^2}$

75. $\frac{\partial w}{\partial t} = \frac{\partial f}{\partial u}\frac{\partial u}{\partial t} = \frac{\partial f}{\partial u}(ac) \Rightarrow \frac{\partial^2 w}{\partial t^2} = (ac)\left(\frac{\partial^2 f}{\partial u^2}\right)(ac) = a^2c^2\,\frac{\partial^2 f}{\partial u^2};\ \frac{\partial w}{\partial x} = \frac{\partial f}{\partial u}\frac{\partial u}{\partial x} = \frac{\partial f}{\partial u}\cdot a \Rightarrow \frac{\partial^2 w}{\partial x^2} = \left(a\,\frac{\partial^2 f}{\partial u^2}\right)\cdot a$

$= a^2\,\frac{\partial^2 f}{\partial u^2} \Rightarrow \frac{\partial^2 w}{\partial t^2} = a^2c^2\,\frac{\partial^2 f}{\partial u^2} = c^2\left(a^2\,\frac{\partial^2 f}{\partial u^2}\right) = c^2\,\frac{\partial^2 w}{\partial x^2}$

12.4 DIFFERENTIABILITY, LINEARIZATION, AND DIFFERENTIALS

1. (a) $f(0,0) = 1,\ f_x(x,y) = 2x \Rightarrow f_x(0,0) = 0,\ f_y(x,y) = 2y \Rightarrow f_y(0,0) = 0 \Rightarrow L(x,y) = 1 + 0(x-0) + 0(y-0) = 1$

   (b) $f(1,1) = 3,\ f_x(1,1) = 2,\ f_y(1,1) = 2 \Rightarrow L(x,y) = 3 + 2(x-1) + 2(y-1) = 2x + 2y - 1$

3. (a) $f(0,0) = 5,\ f_x(x,y) = 3$ for all (x,y), $f_y(x,y) = -4$ for all $(x,y) \Rightarrow L(x,y) = 5 + 3(x-0) - 4(y-0)$

   $= 3x - 4y + 5$

   (b) $f(1,1) = 4,\ f_x(1,1) = 3,\ f_y(1,1) = -4 \Rightarrow L(x,y) = 4 + 3(x-1) - 4(y-1) = 3x - 4y + 5$

5. (a) $f(0,0) = 1,\ f_x(x,y) = e^x \cos y \Rightarrow f_x(0,0) = 1,\ f_y(x,y) = -e^x \sin y \Rightarrow f_y(0,0) = 0$

   $\Rightarrow L(x,y) = 1 + 1(x-0) + 0(y-0) = x + 1$

   (b) $f\left(0,\frac{\pi}{2}\right) = 0,\ f_x\left(0,\frac{\pi}{2}\right) = 0,\ f_y\left(0,\frac{\pi}{2}\right) = -1 \Rightarrow L(x,y) = 0 + 0(x-0) - 1\left(y - \frac{\pi}{2}\right) = -y + \frac{\pi}{2}$

7. $f(2,1) = 3,\ f_x(x,y) = 2x - 3y \Rightarrow f_x(2,1) = 1,\ f_y(x,y) = -3x \Rightarrow f_y(2,1) = -6 \Rightarrow L(x,y) = 3 + 1(x-2) - 6(y-1)$

   $= 7 + x - 6y;\ f_{xx}(x,y) = 2,\ f_{yy}(x,y) = 0,\ f_{xy}(x,y) = -3 \Rightarrow M = 3$; thus $|E(x,y)| \le \left(\frac{1}{2}\right)(3)\left(|x-2| + |y-1|\right)^2$

   $\le \left(\frac{3}{2}\right)(0.1 + 0.1)^2 = 0.06$

9. $f(0,0) = 1,\ f_x(x,y) = \cos y \Rightarrow f_x(0,0) = 1,\ f_y(x,y) = 1 - x\sin y \Rightarrow f_y(0,0) = 1$

   $\Rightarrow L(x,y) = 1 + 1(x-0) + 1(y-0) = x + y + 1;\ f_{xx}(x,y) = 0,\ f_{yy}(x,y) = -x\cos y,\ f_{xy}(x,y) = -\sin y \Rightarrow M = 1$;

   thus $|E(x,y)| \le \left(\frac{1}{2}\right)(1)\left(|x| + |y|\right)^2 \le \left(\frac{1}{2}\right)(0.2 + 0.2)^2 = 0.08$

11. $f(0,0) = 1,\ f_x(x,y) = e^x \cos y \Rightarrow f_x(0,0) = 1,\ f_y(x,y) = -e^x \sin y \Rightarrow f_y(0,0) = 0$

    $\Rightarrow L(x,y) = 1 + 1(x-0) + 0(y-0) = 1 + x;\ f_{xx}(x,y) = e^x \cos y,\ f_{yy}(x,y) = -e^x \cos y,\ f_{xy}(x,y) = -e^x \sin y$;

    $|x| \le 0.1 \Rightarrow -0.1 \le x \le 0.1$ and $|y| \le 0.1 \Rightarrow -0.1 \le y \le 0.1$; thus the max of $|f_{xx}(x,y)|$ on R is $e^{0.1}\cos(0.1)$

    ≤ 1.11, the max of $|f_{yy}(x,y)|$ on R is $e^{0.1}\cos(0.1) \le 1.11$, and the max of $|f_{xy}(x,y)|$ on R is $e^{0.1}\sin(0.1)$

    $\le 0.002 \Rightarrow M = 1.11$; thus $|E(x,y)| \le \left(\frac{1}{2}\right)(1.11)\left(|x| + |y|\right)^2 \le (0.555)(0.1 + 0.1)^2 = 0.0222$

13. $A = xy \Rightarrow dA = x\,dy + y\,dx$; if $x > y$ then a 1-unit change in y gives a greater change in dA than a 1-unit change in x. Thus, pay more attention to y which is the smaller of the two dimensions.

15. $T_x(x,y) = e^y + e^{-y}$ and $T_y(x,y) = x\left(e^y - e^{-y}\right) \Rightarrow dT = T_x(x,y)\,dx + T_y(x,y)\,dy$

    $= \left(e^y + e^{-y}\right)dx + x\left(e^y - e^{-y}\right)dy \Rightarrow dT|_{(2,\ln 2)} = 2.5\,dx + 3.0\,dy$. If $|dx| \le 0.1$ and $|dy| \le 0.02$, then the maximum possible error in the computed value of T is $(2.5)(0.1) + (3.0)(0.02) = 0.31$ in magnitude.

17. $V_r = 2\pi rh$ and $V_h = \pi r^2 \Rightarrow dV = V_r\, dr + V_h\, dh \Rightarrow dV = 2\pi rh\, dr + \pi r^2\, dh \Rightarrow dV|_{(5,12)} = 120\pi\, dr + 25\pi\, dh$; $|dr| \le 0.1$ cm and $|dh| \le 0.1$ cm $\Rightarrow dV \le (120\pi)(0.1) + (25\pi)(0.1) = 14.5\pi$ cm^3; $V(5,12) = 300\pi$ cm^3 $\Rightarrow$ maximum percentage error is $\pm\frac{14.5\pi}{300\pi} \times 100 = \pm 4.83\%$

19. $df = f_x(x,y)\, dx + f_y(x,y)\, dy = 3x^2y^4\, dx + 4x^3y^3\, dy \Rightarrow df|_{(1,1)} = 3\, dx + 4\, dy$; for a square, $dx = dy$ $\Rightarrow df = 7\, dx$ so that $|df| \le 0.1 \Rightarrow 7|dx| \le 0.1 \Rightarrow |dx| \le \frac{0.1}{7} \approx 0.014 \Rightarrow$ for the square, $|x-1| \le 0.014$ and $|y-1| \le 0.014$

21. From Exercise 20, $dR = \left(\frac{R}{R_1}\right)^2 dR_1 + \left(\frac{R}{R_2}\right)^2 dR_2$ so that R_1 changing from 20 to 20.1 ohms $\Rightarrow dR_1 = 0.1$ ohm and R_2 changing from 25 to 24.9 ohms $\Rightarrow dR_2 = -0.1$ ohms; $\frac{1}{R} = \frac{1}{R_1} + \frac{1}{R_2} \Rightarrow R = \frac{100}{9}$ ohms $\Rightarrow dR|_{(20,25)} = \frac{\left(\frac{100}{9}\right)^2}{(20)^2}(0.1) + \frac{\left(\frac{100}{9}\right)^2}{(25)^2}(-0.1) \approx 0.011$ ohms $\Rightarrow$ percentage change is $\left.\frac{dR}{R}\right|_{(20,25)} \times 100$ $= \frac{0.011}{\left(\frac{100}{9}\right)} \times 100 \approx 0.1\%$

23. (a) $f(1,1,1) = 3$, $f_x(1,1,1) = y + z|_{(1,1,1)} = 2$, $f_y(1,1,1) = x + z|_{(1,1,1)} = 2$, $f_z(1,1,1) = y + x|_{(1,1,1)} = 2$ $\Rightarrow L(x,y,z) = 3 + 2(x-1) + 2(y-1) + 2(z-1) = 2x + 2y + 2z - 3$

(b) $f(1,0,0) = 0$, $f_x(1,0,0) = 0$, $f_y(1,0,0) = 1$, $f_z(1,0,0) = 1 \Rightarrow L(x,y,z) = 0 + 0(x-1) + (y-0) + (z-0)$ $= y + z$

(c) $f(0,0,0) = 0$, $f_x(0,0,0) = 0$, $f_y(0,0,0) = 0$, $f_z(0,0,0) = 0 \Rightarrow L(x,y,z) = 0$

25. (a) $f(1,0,0) = 1$, $f_x(1,0,0) = \left.\frac{x}{\sqrt{x^2+y^2+z^2}}\right|_{(1,0,0)} = 1$, $f_y(1,0,0) = \left.\frac{y}{\sqrt{x^2+y^2+z^2}}\right|_{(1,0,0)} = 0$, $f_z(1,0,0) = \left.\frac{z}{\sqrt{x^2+y^2+z^2}}\right|_{(1,0,0)} = 0 \Rightarrow L(x,y,z) = 1 + 1(x-1) + 0(y-0) + 0(z-0) = x$

(b) $f(1,1,0) = \sqrt{2}$, $f_x(1,1,0) = \frac{1}{\sqrt{2}}$, $f_y(1,1,0) = \frac{1}{\sqrt{2}}$, $f_z(1,1,0) = 0$ $\Rightarrow L(x,y,z) = \sqrt{2} + \frac{1}{\sqrt{2}}(x-1) + \frac{1}{\sqrt{2}}(y-1) + 0(z-0) = \frac{1}{\sqrt{2}}x + \frac{1}{\sqrt{2}}y$

(c) $f(1,2,2) = 3$, $f_x(1,2,2) = \frac{1}{3}$, $f_y(1,2,2) = \frac{2}{3}$, $f_z(1,2,2) = \frac{2}{3} \Rightarrow L(x,y,z) = 3 + \frac{1}{3}(x-1) + \frac{2}{3}(y-2) + \frac{2}{3}(z-2)$ $= \frac{1}{3}x + \frac{2}{3}y + \frac{2}{3}z$

27. (a) $f(0,0,0) = 2$, $f_x(0,0,0) = e^x|_{(0,0,0)} = 1$, $f_y(0,0,0) = -\sin(y+z)|_{(0,0,0)} = 0$, $f_z(0,0,0) = -\sin(y+z)|_{(0,0,0)} = 0 \Rightarrow L(x,y,z) = 2 + 1(x-0) + 0(y-0) + 0(z-0) = 2 + x$

(b) $f\left(0,\frac{\pi}{2},0\right) = 1$, $f_x\left(0,\frac{\pi}{2},0\right) = 1$, $f_y\left(0,\frac{\pi}{2},0\right) = -1$, $f_z\left(0,\frac{\pi}{2},0\right) = -1 \Rightarrow L(x,y,z)$ $= 1 + 1(x-0) - 1\left(y - \frac{\pi}{2}\right) - 1(z-0) = x - y - z + \frac{\pi}{2} + 1$

(c) $f\left(0,\frac{\pi}{4},\frac{\pi}{4}\right)=1$, $f_x\left(0,\frac{\pi}{4},\frac{\pi}{4}\right)=1$, $f_y\left(0,\frac{\pi}{4},\frac{\pi}{4}\right)=-1$, $f_z\left(0,\frac{\pi}{4},\frac{\pi}{4}\right)=-1 \Rightarrow L(x,y,z)$
$= 1+1(x-0)-1\left(y-\frac{\pi}{4}\right)-1\left(z-\frac{\pi}{4}\right)=x-y-z+\frac{\pi}{2}+1$

29. $f(x,y,z)=xz-3yz+2$ at $P_0(1,1,2) \Rightarrow f(1,1,2)=-2$; $f_x=z$, $f_y=-3z$, $f_z=x-3y \Rightarrow L(x,y,z)$
$=-2+2(x-1)-6(y-1)-2(z-2)=2x-6y-2z+6$; $f_{xx}=0$, $f_{yy}=0$, $f_{zz}=0$, $f_{xy}=0$, $f_{yz}=-3$
$\Rightarrow M=3$; thus, $|E(x,y,z)| \le \left(\frac{1}{2}\right)(3)(0.01+0.01+0.02)^2=0.0024$

31. $f(x,y,z)=xy+2yz-3xz$ at $P_0(1,1,0) \Rightarrow f(1,1,0)=1$; $f_x=y-3z$, $f_y=x+2z$, $f_z=2y-3x$
$\Rightarrow L(x,y,z)=1+(x-1)+(y-1)-(z-0)=x+y-z-1$; $f_{xx}=0$, $f_{yy}=0$, $f_{zz}=0$, $f_{xy}=1$, $f_{xz}=-3$,
$f_{yz}=2 \Rightarrow M=3$; thus $|E(x,y,z)| \le \left(\frac{1}{2}\right)(3)(0.01+0.01+0.01)^2=0.00135$

33. (a) $dS=S_p\,dp+S_x\,dx+S_w\,dw+S_h\,dh=C\left(\frac{x^4}{wh^3}\,dp+\frac{4px^3}{wh^3}\,dx-\frac{px^4}{w^2h^3}\,dw-\frac{3px^4}{wh^4}\,dh\right)$
$=C\left(\frac{px^4}{wh^3}\right)\left(\frac{1}{p}\,dp+\frac{4}{x}\,dx-\frac{1}{w}\,dw-\frac{3}{h}\,dh\right)=S_0\left(\frac{1}{p_0}\,dp+\frac{4}{x_0}\,dx-\frac{1}{w_0}\,dw-\frac{3}{h_0}\,dh\right)$
$=S_0\left(\frac{1}{100}\,dp+dx-5\,dw-30\,dh\right)$, where $p_0=100$ N/m, $x_0=4$ m, $w_0=0.2$ m, $h_0=0.1$ m
(b) More sensitive to a change in height

35. $f(a,b,c,d)=\begin{vmatrix} a & b \\ c & d \end{vmatrix}=ad-bc \Rightarrow f_a=d$, $f_b=-c$, $f_c=-b$, $f_d=a \Rightarrow df=d\,da-c\,db-b\,dc+a\,dd$; since $|a|$ is much greater than $|b|$, $|c|$, and $|d|$, the function f is most sensitive to a change in d.

37. $V=lwh \Rightarrow V_l=wh$, $V_w=lh$, $V_h=lw \Rightarrow dV=wh\,dl+lh\,dw+lw\,dh \Rightarrow dV|_{(5,3,2)}=6\,dl+10\,dw+15\,dh$;
$dl=1$ in. $=\frac{1}{12}$ ft, $dw=1$ in. $=\frac{1}{12}$ ft, $dh=\frac{1}{2}$ in. $=\frac{1}{24}$ ft $\Rightarrow dV=6\left(\frac{1}{12}\right)+10\left(\frac{1}{12}\right)+15\left(\frac{1}{24}\right)=\frac{47}{24}$ ft^3

39. $u_x=e^y$, $u_y=xe^y+\sin z$, $u_z=y\cos z \Rightarrow du=e^y\,dx+(xe^y+\sin z)\,dy+(y\cos z)\,dz$
$\Rightarrow du|_{\left(2,\ln 3,\frac{\pi}{2}\right)}=3\,dx+7\,dy+0\,dz=3\,dx+7\,dy \Rightarrow$ magnitude of the maximum possible error
$\le 3(0.2)+7(0.6)=4.8$

41. If the first partial derivatives are continuous throughout an open region R, then by Eq. (3) in this section of the text, $f(x,y)=f(x_0,y_0)+f_x(x_0,y_0)\,\Delta x+f_y(x_0,y_0)\,\Delta y+\epsilon_1\Delta x+\epsilon_2\Delta y$, where ϵ_1, $\epsilon_2 \to 0$ as Δx, $\Delta y \to 0$. Then as $(x,y)\to(x_0,y_0)$, $\Delta x \to 0$ and $\Delta y \to 0 \Rightarrow \lim_{(x,y)\to(x_0,y_0)} f(x,y)=f(x_0,y_0) \Rightarrow$ f is continuous at every point (x_0,y_0) in R.

12.5 THE CHAIN RULE

1. (a) $\frac{\partial w}{\partial x} = 2x$, $\frac{\partial w}{\partial y} = 2y$, $\frac{dx}{dt} = -\sin t$, $\frac{dy}{dt} = \cos t \Rightarrow \frac{dw}{dt} = -2x \sin t + 2y \cos t = -2 \cos t \sin t + 2 \sin t \cos t$
$= 0$; $w = x^2 + y^2 = \cos^2 t + \sin^2 t = 1 \Rightarrow \frac{dw}{dt} = 0$

(b) $\frac{dw}{dt}(\pi) = 0$

3. (a) $\frac{\partial w}{\partial x} = \frac{1}{z}$, $\frac{\partial w}{\partial y} = \frac{1}{z}$, $\frac{\partial w}{\partial z} = \frac{-(x+y)}{z^2}$, $\frac{dx}{dt} = -2 \cos t \sin t$, $\frac{dy}{dt} = 2 \sin t \cos t$, $\frac{dz}{dt} = -\frac{1}{t^2}$
$\Rightarrow \frac{dw}{dt} = -\frac{2}{z} \cos t \sin t + \frac{2}{z} \sin t \cos t + \frac{x+y}{z^2 t^2} = \frac{\cos^2 t + \sin^2 t}{\left(\frac{1}{t^2}\right)(t^2)} = 1$; $w = \frac{x}{z} + \frac{y}{z} = \frac{\cos^2 t}{\left(\frac{1}{t}\right)} + \frac{\sin^2 t}{\left(\frac{1}{t}\right)} = t \Rightarrow \frac{dw}{dt} = 1$

(b) $\frac{dw}{dt}(3) = 1$

5. (a) $\frac{\partial w}{\partial x} = 2ye^x$, $\frac{\partial w}{\partial y} = 2e^x$, $\frac{\partial w}{\partial z} = -\frac{1}{z}$, $\frac{dx}{dt} = \frac{2t}{t^2+1}$, $\frac{dy}{dt} = \frac{1}{t^2+1}$, $\frac{dz}{dt} = e^t \Rightarrow \frac{dw}{dt} = \frac{4yte^x}{t^2+1} + \frac{2e^x}{t^2+1} - \frac{e^t}{z}$
$= \frac{(4t)(\tan^{-1} t)(t^2+1)}{t^2+1} + \frac{2(t^2+1)}{t^2+1} - \frac{e^t}{e^t} = 4t \tan^{-1} t + 1$; $w = 2ye^x - \ln z = (2 \tan^{-1} t)(t^2+1) - t$
$\Rightarrow \frac{dw}{dt} = \left(\frac{2}{t^2+1}\right)(t^2+1) + (2 \tan^{-1} t)(2t) - 1 = 4t \tan^{-1} t + 1$

(b) $\frac{dw}{dt}(1) = (4)(1)\left(\frac{\pi}{4}\right) + 1 = \pi + 1$

7. (a) $\frac{\partial z}{\partial r} = \frac{\partial z}{\partial x}\frac{\partial x}{\partial r} + \frac{\partial z}{\partial y}\frac{\partial y}{\partial r} = (4e^x \ln y)\left(\frac{\cos \theta}{r \cos \theta}\right) + \left(\frac{4e^x}{y}\right)(\sin \theta) = \frac{4e^x \ln y}{r} + \frac{4e^x \sin \theta}{y}$
$= \frac{4(r \cos \theta) \ln (r \sin \theta)}{r} + \frac{4(r \cos \theta)(\sin \theta)}{r \sin \theta} = (4 \cos \theta) \ln (r \sin \theta) + 4 \cos \theta$;

$\frac{\partial z}{\partial \theta} = \frac{\partial z}{\partial x}\frac{\partial x}{\partial \theta} + \frac{\partial z}{\partial y}\frac{\partial y}{\partial \theta} = (4e^x \ln y)\left(\frac{-r \sin \theta}{r \cos \theta}\right) + \left(\frac{4e^x}{y}\right)(r \cos \theta) = -(4e^x \ln y)(\tan \theta) + \frac{4e^x r \cos \theta}{y}$
$= [-4(r \cos \theta) \ln (r \sin \theta)](\tan \theta) + \frac{4(r \cos \theta)(r \cos \theta)}{r \sin \theta} = (-4r \sin \theta) \ln (r \sin \theta) + \frac{4r \cos^2 \theta}{\sin \theta}$;
$z = 4e^x \ln y = 4(r \cos \theta) \ln (r \sin \theta) \Rightarrow \frac{\partial z}{\partial r} = (4 \cos \theta) \ln (r \sin \theta) + 4(r \cos \theta)\left(\frac{\sin \theta}{r \sin \theta}\right)$
$= (4 \cos \theta) \ln (r \sin \theta) + 4 \cos \theta$; also $\frac{\partial z}{\partial \theta} = (-4r \sin \theta) \ln (r \sin \theta) + 4(r \cos \theta)\left(\frac{r \cos \theta}{r \sin \theta}\right)$
$= (-4r \sin \theta) \ln (r \sin \theta) + \frac{4r \cos^2 \theta}{\sin \theta}$

(b) At $\left(2, \frac{\pi}{4}\right)$: $\frac{\partial z}{\partial r} = 4 \cos \frac{\pi}{4} \ln\left(2 \sin \frac{\pi}{4}\right) + 4 \cos \frac{\pi}{4} = 2\sqrt{2} \ln \sqrt{2} + 2\sqrt{2} = \sqrt{2}(\ln 2 + 2)$;
$\frac{\partial z}{\partial \theta} = (-4)(2) \sin \frac{\pi}{4} \ln\left(2 \sin \frac{\pi}{4}\right) + \frac{(4)(2)\left(\cos^2 \frac{\pi}{4}\right)}{\left(\sin \frac{\pi}{4}\right)} = -4\sqrt{2} \ln \sqrt{2} + 4\sqrt{2} = -2\sqrt{2} \ln 2 + 4\sqrt{2}$

9. (a) $\frac{\partial w}{\partial u} = \frac{\partial w}{\partial x}\frac{\partial x}{\partial u} + \frac{\partial w}{\partial y}\frac{\partial y}{\partial u} + \frac{\partial w}{\partial z}\frac{\partial z}{\partial u} = (y+z)(1) + (x+z)(1) + (y+x)(v) = x + y + 2z + v(y+x)$
$= (u+v) + (u-v) + 2uv + v(2u) = 2u + 4uv$; $\frac{\partial w}{\partial v} = \frac{\partial w}{\partial x}\frac{\partial x}{\partial v} + \frac{\partial w}{\partial y}\frac{\partial y}{\partial v} + \frac{\partial w}{\partial z}\frac{\partial z}{\partial v}$
$= (y+z)(1) + (x+z)(-1) + (y+x)(u) = y - x + (y+x)u = -2v + (2u)u = -2v + 2u^2$;

$w = xy + yz + xz = (u^2 - v^2) + (u^2v - uv^2) + (u^2v + uv^2) = u^2 - v^2 + 2u^2v \Rightarrow \frac{\partial w}{\partial u} = 2u + 4uv$ and $\frac{\partial w}{\partial v} = -2v + 2u^2$

(b) At $\left(\frac{1}{2}, 1\right)$: $\frac{\partial w}{\partial u} = 2\left(\frac{1}{2}\right) + 4\left(\frac{1}{2}\right)(1) = 3$ and $\frac{\partial w}{\partial v} = -2(1) + 2\left(\frac{1}{2}\right)^2 = -\frac{3}{2}$

11. (a) $\frac{\partial u}{\partial x} = \frac{\partial u}{\partial p}\frac{\partial p}{\partial x} + \frac{\partial u}{\partial q}\frac{\partial q}{\partial x} + \frac{\partial u}{\partial r}\frac{\partial r}{\partial x} = \frac{1}{q-r} + \frac{r-p}{(q-r)^2} + \frac{p-q}{(q-r)^2} = \frac{q-r+r-p+p-q}{(q-r)^2} = 0$;

$\frac{\partial u}{\partial y} = \frac{\partial u}{\partial p}\frac{\partial p}{\partial y} + \frac{\partial u}{\partial q}\frac{\partial q}{\partial y} + \frac{\partial u}{\partial r}\frac{\partial r}{\partial y} = \frac{1}{q-r} - \frac{r-p}{(q-r)^2} + \frac{p-q}{(q-r)^2} = \frac{q-r-r+p+p-q}{(q-r)^2} = \frac{2p-2r}{(q-r)^2}$
$= \frac{(2x+2y+2z)-(2x+2y-2z)}{(2z-2y)^2} = \frac{z}{(z-y)^2}$; $\frac{\partial u}{\partial z} = \frac{\partial u}{\partial p}\frac{\partial p}{\partial z} + \frac{\partial u}{\partial q}\frac{\partial q}{\partial z} + \frac{\partial u}{\partial r}\frac{\partial r}{\partial z}$
$= \frac{1}{q-r} + \frac{r-p}{(q-r)^2} - \frac{p-q}{(q-r)^2} = \frac{q-r+r-p-p+q}{(q-r)^2} = \frac{2q-2p}{(q-r)^2} = \frac{-4y}{(2z-2y)^2} = -\frac{y}{(z-y)^2}$;

$u = \frac{p-q}{q-r} = \frac{2y}{2z-2y} = \frac{y}{z-y} \Rightarrow \frac{\partial u}{\partial x} = 0$, $\frac{\partial u}{\partial y} = \frac{(z-y)-y(-1)}{(z-y)^2} = \frac{z}{(z-y)^2}$, and $\frac{\partial u}{\partial z} = \frac{(z-y)(0)-y(1)}{(z-y)^2}$
$= -\frac{y}{(z-y)^2}$

(b) At $(\sqrt{3}, 2, 1)$: $\frac{\partial u}{\partial x} = 0$, $\frac{\partial u}{\partial y} = \frac{1}{(1-2)^2} = 1$, and $\frac{\partial u}{\partial z} = \frac{-2}{(1-2)^2} = -2$

13. $\frac{dz}{dt} = \frac{\partial z}{\partial x}\frac{dx}{dt} + \frac{\partial z}{\partial y}\frac{dy}{dt}$

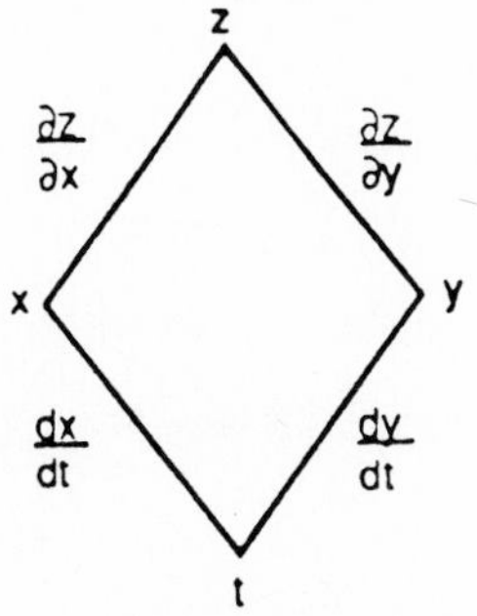

15. $\frac{\partial w}{\partial u} = \frac{\partial w}{\partial x}\frac{\partial x}{\partial u} + \frac{\partial w}{\partial y}\frac{\partial y}{\partial u} + \frac{\partial w}{\partial z}\frac{\partial z}{\partial u}$

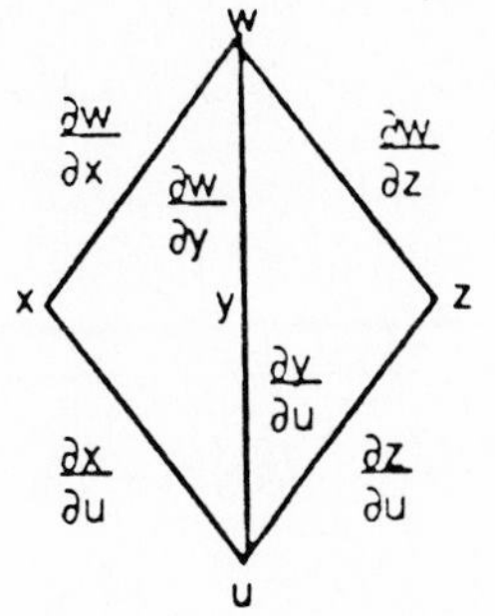

$\frac{\partial w}{\partial v} = \frac{\partial w}{\partial x}\frac{\partial x}{\partial v} + \frac{\partial w}{\partial y}\frac{\partial y}{\partial v} + \frac{\partial w}{\partial z}\frac{\partial z}{\partial v}$

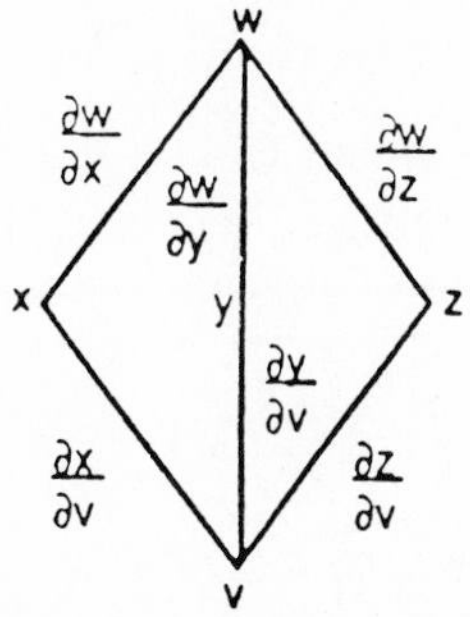

17. $\frac{\partial w}{\partial u} = \frac{\partial w}{\partial x}\frac{\partial x}{\partial u} + \frac{\partial w}{\partial y}\frac{\partial y}{\partial u}$

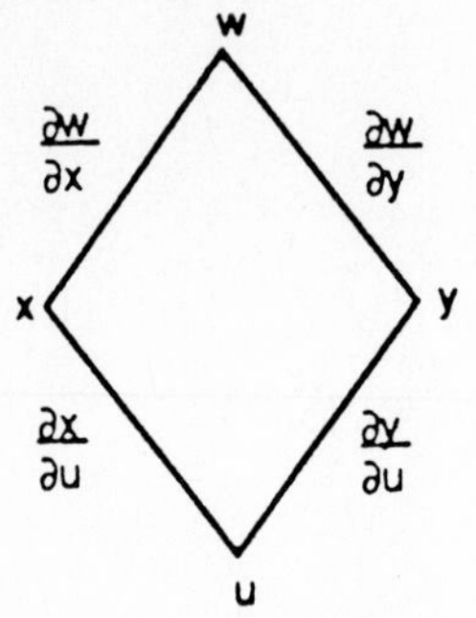

$\frac{\partial w}{\partial v} = \frac{\partial w}{\partial x}\frac{\partial x}{\partial v} + \frac{\partial w}{\partial y}\frac{\partial y}{\partial v}$

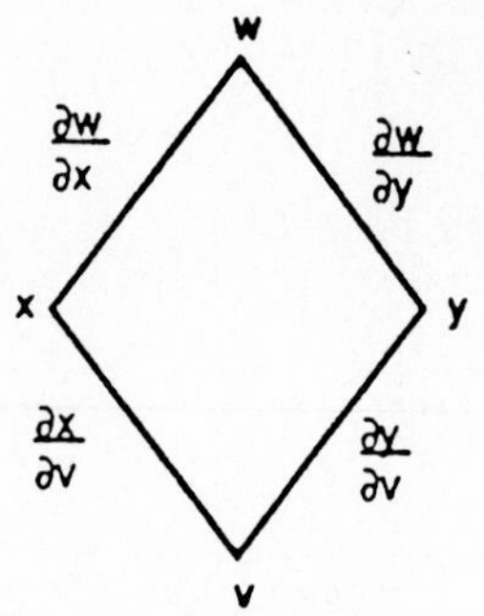

19. $\frac{\partial z}{\partial t} = \frac{\partial z}{\partial x}\frac{\partial x}{\partial t} + \frac{\partial z}{\partial y}\frac{\partial y}{\partial t}$

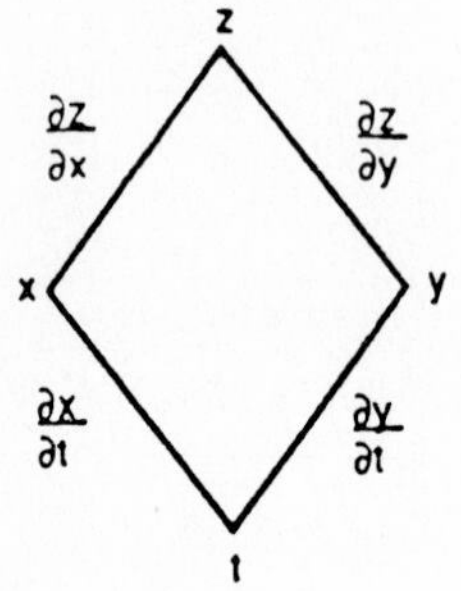

$\frac{\partial z}{\partial s} = \frac{\partial z}{\partial x}\frac{\partial x}{\partial s} + \frac{\partial z}{\partial y}\frac{\partial y}{\partial s}$

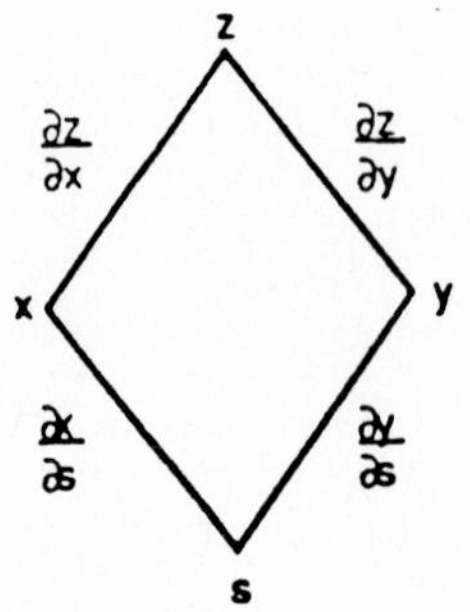

21. $\frac{\partial w}{\partial s} = \frac{dw}{du}\frac{\partial u}{\partial s}$

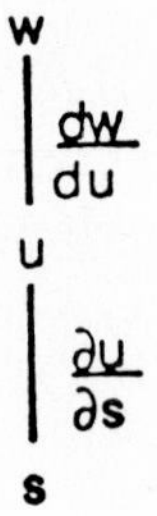

$\frac{\partial w}{\partial t} = \frac{dw}{du}\frac{\partial u}{\partial t}$

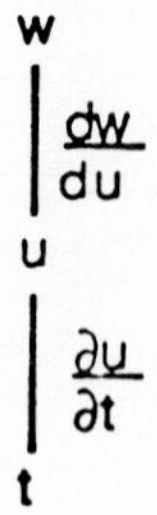

23. $\frac{\partial w}{\partial r} = \frac{\partial w}{\partial x}\frac{dx}{dr} + \frac{\partial w}{\partial y}\frac{dy}{dr} = \frac{\partial w}{\partial x}\frac{dx}{dr}$ since $\frac{dy}{dr} = 0$ $\qquad$ $\frac{\partial w}{\partial s} = \frac{\partial w}{\partial x}\frac{dx}{ds} + \frac{\partial w}{\partial y}\frac{dy}{ds} = \frac{\partial w}{\partial y}\frac{dy}{ds}$ since $\frac{dx}{ds} = 0$

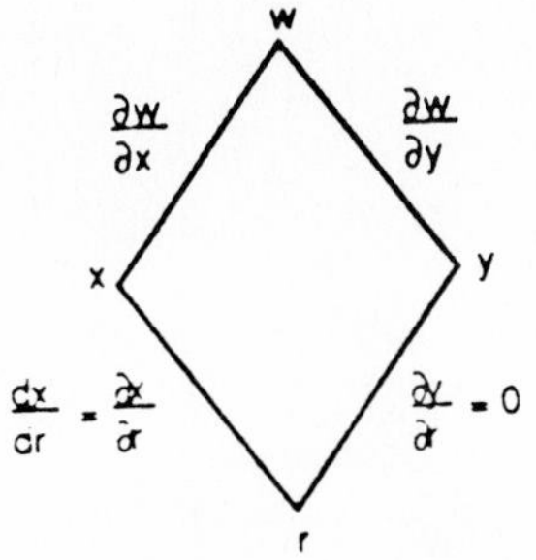

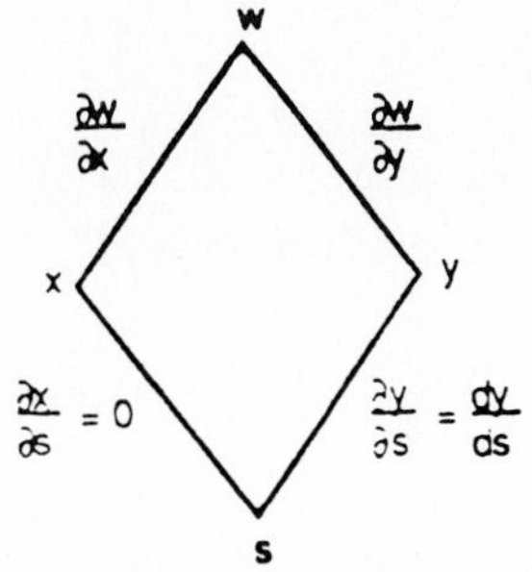

25. Let $F(x,y) = x^3 - 2y^2 + xy = 0 \Rightarrow F_x(x,y) = 3x^2 + y$ and $F_y(x,y) = -4y + x \Rightarrow \frac{dy}{dx} = -\frac{F_x}{F_y} = -\frac{3x^2+y}{(-4y+x)}$

$\Rightarrow \frac{dy}{dx}(1,1) = \frac{4}{3}$

27. Let $F(x,y) = x^2 + xy + y^2 - 7 = 0 \Rightarrow F_x(x,y) = 2x + y$ and $F_y(x,y) = x + 2y \Rightarrow \frac{dy}{dx} = -\frac{F_x}{F_y} = -\frac{2x+y}{x+2y}$

$\Rightarrow \frac{dy}{dx}(1,2) = -\frac{4}{5}$

29. Let $F(x,y,z) = z^3 - xy + yz + y^3 - 2 = 0 \Rightarrow F_x(x,y,z) = -y,\ F_y(x,y,z) = -x + z + 3y^2,\ F_z(x,y,z) = 3z^2 + y$

$\Rightarrow \frac{\partial z}{\partial x} = -\frac{F_x}{F_y} = -\frac{-y}{3x^2+y} = \frac{y}{3z^2+y} \Rightarrow \frac{\partial z}{\partial x}(1,1,1) = \frac{1}{4};\ \frac{\partial z}{\partial y} = -\frac{F_y}{F_z} = -\frac{-x+z+3y^2}{3z^2+y} = \frac{x-z-3y^2}{3z^2+y}$

$\Rightarrow \frac{\partial z}{\partial y}(1,1,1) = -\frac{3}{4}$

31. Let $F(x,y,z) = \sin(x+y) + \sin(y+z) + \sin(z+z) = 0 \Rightarrow F_x(x,y,z) = \cos(x+y) + \cos(x+z)$,

$F_y(x,y,z) = \cos(x+y) + \cos(y+z),\ F_z(x,y,z) = \cos(y+z) + \cos(x+z) \Rightarrow \frac{\partial z}{\partial x} = -\frac{F_x}{F_z}$

$= -\frac{\cos(x+y)+\cos(x+z)}{\cos(y+z)+\cos(x+z)} \Rightarrow \frac{\partial z}{\partial x}(\pi,\pi,\pi) = -1;\ \frac{\partial z}{\partial y} = -\frac{F_y}{F_z} = -\frac{\cos(x+y)+\cos(y+z)}{\cos(y+z)+\cos(x+z)} \Rightarrow \frac{\partial z}{\partial y}(\pi,\pi,\pi) = -1$

33. $\frac{\partial w}{\partial r} = \frac{\partial w}{\partial x}\frac{\partial x}{\partial r} + \frac{\partial w}{\partial y}\frac{\partial y}{\partial r} + \frac{\partial w}{\partial z}\frac{\partial z}{\partial r} = 2(x+y+z)(1) + 2(x+y+z)[-\sin(r+s)] + 2(x+y+z)[\cos(r+s)]$

$= 2(x+y+z)[1 - \sin(r+s) + \cos(r+s)] = 2[r - s + \cos(r+s) + \sin(r+s)][1 - \sin(r+s) + \cos(r+s)]$

$\Rightarrow \left.\frac{\partial w}{\partial r}\right|_{r=1,s=-1} = 2(3)(2) = 12$

35. $\frac{\partial w}{\partial v} = \frac{\partial w}{\partial x}\frac{\partial x}{\partial v} + \frac{\partial w}{\partial y}\frac{\partial y}{\partial v} = \left(2x - \frac{y}{x^2}\right)(-2) + \left(\frac{1}{x}\right)(1) = \left[2(u-2v+1) - \frac{2u+v-2}{(u-2v+1)^2}\right](-2) + \frac{1}{u-2v+1}$

$\Rightarrow \left.\frac{\partial w}{\partial v}\right|_{u=0,v=0} = -7$

37. $\frac{\partial z}{\partial u} = \frac{dz}{dx}\frac{\partial x}{\partial u} = \left(\frac{5}{1+x^2}\right)e^u = \left[\frac{5}{1+(e^u+\ln v)^2}\right]e^u \Rightarrow \left.\frac{\partial z}{\partial u}\right|_{u=\ln 2,v=1} = \left[\frac{5}{1+(2)^2}\right](2) = 2;$

$\frac{\partial z}{\partial v} = \frac{dz}{dx}\frac{\partial x}{\partial v} = \left(\frac{5}{1+x^2}\right)\left(\frac{1}{v}\right) = \left[\frac{5}{1+(e^u+\ln v)^2}\right]\left(\frac{1}{v}\right) \Rightarrow \left.\frac{\partial z}{\partial v}\right|_{u=\ln 2,v=1} = \left[\frac{5}{1+(2)^2}\right](1) = 1$

39. $V = IR \Rightarrow \frac{\partial V}{\partial I} = R$ and $\frac{\partial V}{\partial R} = I$; $\frac{dV}{dt} = \frac{\partial V}{\partial I}\frac{dI}{dt} + \frac{\partial V}{\partial R}\frac{dR}{dt} = R\frac{dI}{dt} + I\frac{dR}{dt} \Rightarrow -0.01$ volts/sec $= (600 \text{ ohms})\frac{dI}{dt} + (0.04 \text{ amps})(0.5 \text{ ohms/sec}) \Rightarrow \frac{dI}{dt} = -0.00005$ amps/sec

41. $\frac{\partial f}{\partial x} = \frac{\partial f}{\partial u}\frac{\partial u}{\partial x} + \frac{\partial f}{\partial v}\frac{\partial v}{\partial x} + \frac{\partial f}{\partial w}\frac{\partial w}{\partial x} = \frac{\partial f}{\partial u}(1) + \frac{\partial f}{\partial v}(0) + \frac{\partial f}{\partial w}(-1) = \frac{\partial f}{\partial u} - \frac{\partial f}{\partial w}$,
$\frac{\partial f}{\partial y} = \frac{\partial f}{\partial u}\frac{\partial u}{\partial y} + \frac{\partial f}{\partial v}\frac{\partial v}{\partial y} + \frac{\partial f}{\partial w}\frac{\partial w}{\partial y} = \frac{\partial f}{\partial u}(-1) + \frac{\partial f}{\partial v}(1) + \frac{\partial f}{\partial w}(0) = -\frac{\partial f}{\partial u} + \frac{\partial f}{\partial v}$, and
$\frac{\partial f}{\partial z} = \frac{\partial f}{\partial u}\frac{\partial u}{\partial z} + \frac{\partial f}{\partial v}\frac{\partial v}{\partial z} + \frac{\partial f}{\partial w}\frac{\partial w}{\partial z} = \frac{\partial f}{\partial u}(0) + \frac{\partial f}{\partial v}(-1) + \frac{\partial f}{\partial w}(1) = -\frac{\partial f}{\partial v} + \frac{\partial f}{\partial w} \Rightarrow \frac{\partial f}{\partial x} + \frac{\partial f}{\partial y} + \frac{\partial f}{\partial z} = 0$

43. $w_x = \frac{\partial w}{\partial x} = \frac{\partial w}{\partial u}\frac{\partial u}{\partial x} + \frac{\partial w}{\partial v}\frac{\partial v}{\partial x} = x\frac{\partial w}{\partial u} + y\frac{\partial w}{\partial v} \Rightarrow w_{xx} = \frac{\partial w}{\partial u} + x\frac{\partial}{\partial x}\left(\frac{\partial w}{\partial u}\right) + y\frac{\partial}{\partial x}\left(\frac{\partial w}{\partial v}\right)$
$= \frac{\partial w}{\partial u} + x\left(\frac{\partial^2 w}{\partial u^2}\frac{\partial u}{\partial x} + \frac{\partial^2 w}{\partial v \partial u}\frac{\partial v}{\partial x}\right) + y\left(\frac{\partial^2 w}{\partial u \partial v}\frac{\partial u}{\partial x} + \frac{\partial^2 w}{\partial v^2}\frac{\partial v}{\partial x}\right) = \frac{\partial w}{\partial u} + x\left(x\frac{\partial^2 w}{\partial u^2} + y\frac{\partial^2 w}{\partial v \partial u}\right) + y\left(x\frac{\partial^2 w}{\partial u \partial v} + y\frac{\partial^2 w}{\partial v^2}\right)$
$= \frac{\partial w}{\partial u} + x^2\frac{\partial^2 w}{\partial u^2} + 2xy\frac{\partial^2 w}{\partial v \partial u} + y^2\frac{\partial^2 w}{\partial v^2}$; $w_y = \frac{\partial w}{\partial y} = \frac{\partial w}{\partial u}\frac{\partial u}{\partial y} + \frac{\partial w}{\partial v}\frac{\partial v}{\partial y} = -y\frac{\partial w}{\partial u} + x\frac{\partial w}{\partial v}$
$\Rightarrow w_{yy} = -\frac{\partial w}{\partial u} - y\left(\frac{\partial^2 w}{\partial u^2}\frac{\partial u}{\partial y} + \frac{\partial^2 w}{\partial v \partial u}\frac{\partial v}{\partial y}\right) + x\left(\frac{\partial^2 w}{\partial u \partial v}\frac{\partial u}{\partial y} + \frac{\partial^2 w}{\partial v^2}\frac{\partial v}{\partial y}\right)$
$= -\frac{\partial w}{\partial u} - y\left(-y\frac{\partial^2 w}{\partial u^2} + x\frac{\partial^2 w}{\partial v \partial u}\right) + x\left(-y\frac{\partial^2 w}{\partial u \partial v} + x\frac{\partial^2 w}{\partial v^2}\right) = -\frac{\partial w}{\partial u} + y^2\frac{\partial^2 w}{\partial u^2} - 2xy\frac{\partial^2 w}{\partial v \partial u} + x^2\frac{\partial^2 w}{\partial v^2}$; thus
$w_{xx} + w_{yy} = (x^2 + y^2)\frac{\partial^2 w}{\partial u^2} + (x^2 + y^2)\frac{\partial^2 w}{\partial v^2} = (x^2 + y^2)(w_{uu} + w_{vv}) = 0$, since $w_{uu} + w_{vv} = 0$

45. $f_x(x,y,z) = \cos t$, $f_y(x,y,z) = \sin t$, and $f_z(x,y,z) = t^2 + t - 2 \Rightarrow \frac{df}{dt} = \frac{\partial f}{\partial x}\frac{dx}{dt} + \frac{\partial f}{\partial y}\frac{dy}{dt} + \frac{\partial f}{\partial z}\frac{dz}{dt}$
$= (\cos t)(-\sin t) + (\sin t)(\cos t) + (t^2 + t - 2)(1) = t^2 + t - 2$; $\frac{df}{dt} = 0 \Rightarrow t^2 + t - 2 = 0 \Rightarrow t = -2$ or $t = 1$; $t = -2 \Rightarrow x = \cos(-2)$, $y = \sin(-2)$, $z = -2$ for the point $(\cos(-2), \sin(-2), -2)$; $t = 1 \Rightarrow x = \cos 1$, $y = \sin 1$, $z = 1$ for the point $(\cos 1, \sin 1, 1)$

47. (a) $\frac{\partial T}{\partial x} = 8x - 4y$ and $\frac{\partial T}{\partial y} = 8y - 4x \Rightarrow \frac{dT}{dt} = \frac{\partial T}{\partial x}\frac{dx}{dt} + \frac{\partial T}{\partial y}\frac{dy}{dt} = (8x - 4y)(-\sin t) + (8y - 4x)(\cos t)$
$= (8\cos t - 4\sin t)(-\sin t) + (8\sin t - 4\cos t)(\cos t) = 4\sin^2 t - 4\cos^2 t \Rightarrow \frac{d^2T}{dt^2} = 16\sin t\cos t$;
$\frac{dT}{dt} = 0 \Rightarrow 4\sin^2 t - 4\cos^2 t = 0 \Rightarrow \sin^2 t = \cos^2 t \Rightarrow \sin t = \cos t$ or $\sin t = -\cos t \Rightarrow t = \frac{\pi}{4}, \frac{5\pi}{4}, \frac{3\pi}{4}, \frac{7\pi}{4}$ on the interval $0 \le t \le 2\pi$;

$\left.\frac{d^2T}{dt^2}\right|_{t=\frac{\pi}{4}} = 16\sin\frac{\pi}{4}\cos\frac{\pi}{4} > 0 \Rightarrow$ T has a minimum at $(x,y) = \left(\frac{\sqrt{2}}{2}, \frac{\sqrt{2}}{2}\right)$;

$\left.\frac{d^2T}{dt^2}\right|_{t=\frac{3\pi}{4}} = 16\sin\frac{3\pi}{4}\cos\frac{3\pi}{4} < 0 \Rightarrow$ T has a maximum at $(x,y) = \left(-\frac{\sqrt{2}}{2}, \frac{\sqrt{2}}{2}\right)$;

$\left.\frac{d^2T}{dt^2}\right|_{t=\frac{5\pi}{4}} = 16\sin\frac{5\pi}{4}\cos\frac{5\pi}{4} > 0 \Rightarrow$ T has a minimum at $(x,y) = \left(-\frac{\sqrt{2}}{2}, -\frac{\sqrt{2}}{2}\right)$;

$\left.\frac{d^2T}{dt^2}\right|_{t=\frac{7\pi}{4}} = 16\sin\frac{7\pi}{4}\cos\frac{7\pi}{4} < 0 \Rightarrow$ T has a maximum at $(x,y) = \left(\frac{\sqrt{2}}{2}, -\frac{\sqrt{2}}{2}\right)$

(b) $T = 4x^2 - 4xy + 4y^2 \Rightarrow \frac{\partial T}{\partial x} = 8x - 4y$, and $\frac{\partial T}{\partial y} = 8y - 4x$ so the extreme values occur at the four points found in part (a): $T\left(-\frac{\sqrt{2}}{2}, \frac{\sqrt{2}}{2}\right) = T\left(\frac{\sqrt{2}}{2}, -\frac{\sqrt{2}}{2}\right) = 4\left(\frac{1}{2}\right) - 4\left(-\frac{1}{2}\right) + 4\left(\frac{1}{2}\right) = 6$, the maximum and $T\left(\frac{\sqrt{2}}{2}, \frac{\sqrt{2}}{2}\right) = T\left(-\frac{\sqrt{2}}{2}, -\frac{\sqrt{2}}{2}\right) = 4\left(\frac{1}{2}\right) - 4\left(\frac{1}{2}\right) + 4\left(\frac{1}{2}\right) = 2$, the minimum

49. $G(u,x) = \int_a^u g(t,x)\,dt$ where $u = f(x) \Rightarrow \frac{dG}{dx} = \frac{\partial G}{\partial u}\frac{du}{dx} + \frac{\partial G}{\partial x}\frac{dx}{dx} = g(u,x)f'(x) + \int_a^u g_x(t,x)\,dt$; thus

$$F(x) = \int_0^{x^2} \sqrt{t^4 + x^3}\,dt \Rightarrow F'(x) = \sqrt{\left(x^2\right)^4 + x^3}\,(2x) + \int_0^{x^2} \frac{\partial}{\partial x}\sqrt{t^4 + x^3}\,dt = 2x\sqrt{x^8 + x^3} + \int_0^{x^2} \frac{3x^2}{2\sqrt{t^4 + x^3}}\,dt$$

12.6 PARTIAL DERIVATIVES WITH CONSTRAINED VARIABLES

1. $w = x^2 + y^2 + z^2$ and $z = x^2 + y^2$:

(a) $\begin{pmatrix} y \\ z \end{pmatrix} \to \begin{pmatrix} x = x(y,z) \\ y = y \\ z = z \end{pmatrix} \to w \Rightarrow \left(\frac{\partial w}{\partial y}\right)_z = \frac{\partial w}{\partial x}\frac{\partial x}{\partial y} + \frac{\partial w}{\partial y}\frac{\partial y}{\partial y} + \frac{\partial w}{\partial z}\frac{\partial z}{\partial y}$; $\frac{\partial z}{\partial y} = 0$ and $\frac{\partial z}{\partial y} = 2x\frac{\partial x}{\partial y} + 2y\frac{\partial y}{\partial y}$

$= 2x\frac{\partial x}{\partial y} + 2y \Rightarrow 0 = 2x\frac{\partial x}{\partial y} + 2y \Rightarrow \frac{\partial x}{\partial y} = -\frac{y}{x} \Rightarrow \left(\frac{\partial w}{\partial y}\right)_z = (2x)\left(-\frac{y}{x}\right) + (2y)(1) + (2z)(0) = -2y + 2y = 0$

(b) $\begin{pmatrix} x \\ z \end{pmatrix} \to \begin{pmatrix} x = x \\ y = y(x,z) \\ z = z \end{pmatrix} \to w \Rightarrow \left(\frac{\partial w}{\partial z}\right)_x = \frac{\partial w}{\partial x}\frac{\partial x}{\partial z} + \frac{\partial w}{\partial y}\frac{\partial y}{\partial z} + \frac{\partial w}{\partial z}\frac{\partial z}{\partial z}$; $\frac{\partial x}{\partial z} = 0$ and $\frac{\partial z}{\partial z} = 2x\frac{\partial x}{\partial z} + 2y\frac{\partial y}{\partial z}$

$\Rightarrow 1 = 2y\frac{\partial y}{\partial z} \Rightarrow \frac{\partial y}{\partial z} = \frac{1}{2y} \Rightarrow \left(\frac{\partial w}{\partial z}\right)_x = (2x)(0) + (2y)\left(\frac{1}{2y}\right) + (2z)(1) = 1 + 2z$

(c) $\begin{pmatrix} y \\ z \end{pmatrix} \to \begin{pmatrix} x = x(y,z) \\ y = y \\ z = z \end{pmatrix} \to w \Rightarrow \left(\frac{\partial w}{\partial z}\right)_y = \frac{\partial w}{\partial x}\frac{\partial x}{\partial z} + \frac{\partial w}{\partial y}\frac{\partial y}{\partial z} + \frac{\partial w}{\partial z}\frac{\partial z}{\partial z}$; $\frac{\partial y}{\partial z} = 0$ and $\frac{\partial z}{\partial z} = 2x\frac{\partial x}{\partial z} + 2y\frac{\partial y}{\partial z}$

$\Rightarrow 1 = 2x\frac{\partial x}{\partial z} \Rightarrow \frac{\partial x}{\partial z} = \frac{1}{2x} \Rightarrow \left(\frac{\partial w}{\partial z}\right)_y = (2x)\left(\frac{1}{2x}\right) + (2y)(0) + (2z)(1) = 1 + 2z$

3. $U = f(P,V,T)$ and $PV = nRT$

(a) $\begin{pmatrix} P \\ V \end{pmatrix} \to \begin{pmatrix} P = P \\ V = V \\ T = \frac{PV}{nR} \end{pmatrix} \to U \Rightarrow \left(\frac{\partial U}{\partial P}\right)_V = \frac{\partial U}{\partial P}\frac{\partial P}{\partial P} + \frac{\partial U}{\partial V}\frac{\partial V}{\partial P} + \frac{\partial U}{\partial T}\frac{\partial T}{\partial P} = \frac{\partial U}{\partial P} + \left(\frac{\partial U}{\partial V}\right)(0) + \left(\frac{\partial U}{\partial T}\right)\left(\frac{V}{nR}\right)$

$= \frac{\partial U}{\partial P} + \left(\frac{\partial U}{\partial T}\right)\left(\frac{V}{nR}\right)$

(b) $\begin{pmatrix} V \\ T \end{pmatrix} \to \begin{pmatrix} P = \frac{nRT}{V} \\ V = V \\ T = T \end{pmatrix} \to U \Rightarrow \left(\frac{\partial U}{\partial T}\right)_V = \frac{\partial U}{\partial P}\frac{\partial P}{\partial T} + \frac{\partial U}{\partial V}\frac{\partial V}{\partial T} + \frac{\partial U}{\partial T}\frac{\partial T}{\partial T} = \left(\frac{\partial U}{\partial P}\right)\left(\frac{nR}{V}\right) + \left(\frac{\partial U}{\partial V}\right)(0) + \frac{\partial U}{\partial T}$

$= \left(\frac{\partial U}{\partial P}\right)\left(\frac{nR}{V}\right) + \frac{\partial U}{\partial T}$

5. $w = x^2y^2 + yz - z^3$ and $x^2 + y^2 + z^2 = 6$

(a) $\begin{pmatrix} x \\ y \end{pmatrix} \to \begin{pmatrix} x = x \\ y = y \\ z = z(x,y) \end{pmatrix} \to w \Rightarrow \left(\frac{\partial w}{\partial y}\right)_x = \frac{\partial w}{\partial x}\frac{\partial x}{\partial y} + \frac{\partial w}{\partial y}\frac{\partial y}{\partial y} + \frac{\partial w}{\partial z}\frac{\partial z}{\partial y}$

$= (2xy^2)(0) + (2x^2y + z)(1) + (y - 3z^2)\frac{\partial z}{\partial y} = 2x^2y + z + (y - 3z^2)\frac{\partial z}{\partial y}$. Now $(2x)\frac{\partial x}{\partial y} + 2y + (2z)\frac{\partial z}{\partial y} = 0$ and

$\frac{\partial x}{\partial y} = 0 \Rightarrow 2y + (2z)\frac{\partial z}{\partial y} = 0 \Rightarrow \frac{\partial z}{\partial y} = -\frac{y}{z}$. At $(w,x,y,z) = (4,2,1,-1)$, $\frac{\partial z}{\partial y} = -\frac{1}{-1} = 1 \Rightarrow \left(\frac{\partial w}{\partial y}\right)_x \Big|_{(4,2,1,-1)}$

$= \left[(2)(2)^2(1) + (-1)\right] + \left[1 - 3(-1)^2\right](1) = 5$

(b) $\begin{pmatrix} y \\ z \end{pmatrix} \to \begin{pmatrix} x = x(y,z) \\ y = y \\ z = z \end{pmatrix} \to w \Rightarrow \left(\frac{\partial w}{\partial y}\right)_z = \frac{\partial w}{\partial x}\frac{\partial x}{\partial y} + \frac{\partial w}{\partial y}\frac{\partial y}{\partial y} + \frac{\partial w}{\partial z}\frac{\partial z}{\partial y}$

$= (2xy^2)\frac{\partial x}{\partial y} + (2x^2y + z)(1) + (y - 3z^2)(0) = (2x^2y)\frac{\partial x}{\partial y} + 2x^2y + z$. Now $(2x)\frac{\partial x}{\partial y} + 2y + (2z)\frac{\partial z}{\partial y} = 0$ and

$\frac{\partial z}{\partial y} = 0 \Rightarrow (2x)\frac{\partial x}{\partial y} + 2y = 0 \Rightarrow \frac{\partial x}{\partial y} = -\frac{y}{x}$. At $(w,x,y,z) = (4,2,1,-1)$, $\frac{\partial x}{\partial y} = -\frac{1}{2} \Rightarrow \left(\frac{\partial w}{\partial y}\right)_z \Big|_{(4,2,1,-1)}$

$= (2)(2)(1)^2\left(-\frac{1}{2}\right) + (2)(2)^2(1) + (-1) = 5$

7. (a) $\begin{pmatrix} r \\ \theta \end{pmatrix} \to \begin{pmatrix} x = r\cos\theta \\ y = r\sin\theta \end{pmatrix} \Rightarrow \left(\frac{\partial x}{\partial r}\right)_\theta = \cos\theta$

(b) $x^2 + y^2 = r^2 \Rightarrow 2x + 2y\frac{\partial y}{\partial x} = 2r\frac{\partial r}{\partial x}$ and $\frac{\partial y}{\partial x} = 0 \Rightarrow 2x = 2r\frac{\partial r}{\partial x} \Rightarrow \frac{\partial r}{\partial x} = \frac{x}{r} \Rightarrow \left(\frac{\partial r}{\partial x}\right)_y = \frac{x}{\sqrt{x^2 + y^2}}$

9. If x is a differentiable function of y and z, then $f(x,y,z) = 0 \Rightarrow \frac{\partial f}{\partial x}\frac{\partial x}{\partial x} + \frac{\partial f}{\partial y}\frac{\partial y}{\partial x} + \frac{\partial f}{\partial z}\frac{\partial z}{\partial x} = 0 \Rightarrow \frac{\partial f}{\partial x} + \frac{\partial f}{\partial y}\frac{\partial y}{\partial x} = 0$

$\Rightarrow \left(\frac{\partial x}{\partial y}\right)_z = -\frac{\partial f/\partial y}{\partial f/\partial z}$. Similarly, if y is a differentiable function of x and z, $\left(\frac{\partial y}{\partial z}\right)_x = -\frac{\partial f/\partial z}{\partial f/\partial x}$ and if z is a differentiable function of x and y, $\left(\frac{\partial z}{\partial x}\right)_y = -\frac{\partial f/\partial x}{\partial f/\partial y}$. Then $\left(\frac{\partial x}{\partial y}\right)_z\left(\frac{\partial y}{\partial z}\right)_x\left(\frac{\partial z}{\partial x}\right)_y$

$= \left(-\frac{\partial f/\partial y}{\partial f/\partial z}\right)\left(-\frac{\partial f/\partial z}{\partial f/\partial x}\right)\left(-\frac{\partial f/\partial x}{\partial f/\partial y}\right) = -1.$

11. If x and y are independent, then $g(x,y,z)=0 \Rightarrow \frac{\partial g}{\partial x}\frac{\partial x}{\partial y}+\frac{\partial g}{\partial y}\frac{\partial y}{\partial y}+\frac{\partial g}{\partial z}\frac{\partial z}{\partial y}=0$ and $\frac{\partial x}{\partial y}=0 \Rightarrow \frac{\partial g}{\partial y}+\frac{\partial g}{\partial z}\frac{\partial z}{\partial y}=0$

$\Rightarrow \left(\frac{\partial z}{\partial y}\right)_x=-\frac{\partial g/\partial y}{\partial g/\partial z}$, as claimed.

12.7 DIRECTIONAL DERIVATIVES, GRADIENT VECTORS, AND TANGENT PLANES

1. $\frac{\partial f}{\partial x}=-1,\ \frac{\partial f}{\partial y}=1 \Rightarrow \nabla f=-\mathbf{i}+\mathbf{j}$; $f(2,1)=-1$
$\Rightarrow -1=y-x$ is the level curve

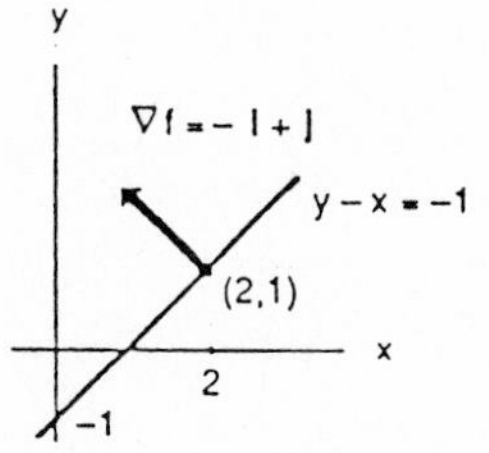

3. $\frac{\partial g}{\partial x}=-2x \Rightarrow \frac{\partial g}{\partial x}(-1,0)=2$; $\frac{\partial g}{\partial y}=1$

$\Rightarrow \nabla g=2\mathbf{i}+\mathbf{j}$; $g(-1,0)=-1$

$\Rightarrow -1=y-x^2$ is the level curve

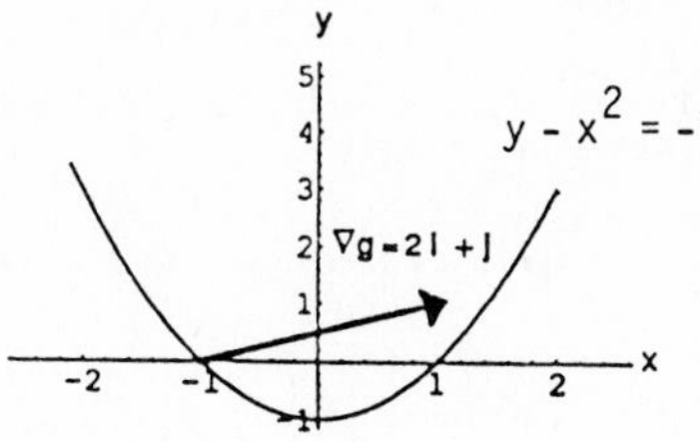

5. $\frac{\partial f}{\partial x}=2x+\frac{z}{x} \Rightarrow \frac{\partial f}{\partial x}(1,1,1)=3$; $\frac{\partial f}{\partial y}=2y \Rightarrow \frac{\partial f}{\partial y}(1,1,1)=2$; $\frac{\partial f}{\partial z}=-4z+\ln x \Rightarrow \frac{\partial f}{\partial z}(1,1,1)=-4$;
thus $\nabla f=3\mathbf{i}+2\mathbf{j}-4\mathbf{k}$

7. $\frac{\partial f}{\partial x}=-\frac{x}{(x^2+y^2+z^2)^{3/2}}+\frac{1}{x} \Rightarrow \frac{\partial f}{\partial x}(-1,2,-2)=-\frac{26}{27}$; $\frac{\partial f}{\partial y}=-\frac{y}{(x^2+y^2+z^2)^{3/2}}+\frac{1}{y} \Rightarrow \frac{\partial f}{\partial y}(-1,2,-2)=\frac{23}{54}$;
$\frac{\partial f}{\partial z}=-\frac{z}{(x^2+y^2+z^2)^{3/2}}+\frac{1}{z} \Rightarrow \frac{\partial f}{\partial z}(-1,2,-2)=-\frac{23}{54}$; thus $\nabla f=-\frac{26}{27}\mathbf{i}+\frac{23}{54}\mathbf{j}-\frac{23}{54}\mathbf{k}$

9. $\mathbf{u}=\frac{\mathbf{A}}{|\mathbf{A}|}=\frac{4\mathbf{i}+3\mathbf{j}}{\sqrt{4^2+3^2}}=\frac{4}{5}\mathbf{i}+\frac{3}{5}\mathbf{j}$; $f_x(x,y)=2y \Rightarrow f_x(5,5)=10$; $f_y(x,y)=2x-6y \Rightarrow f_y(5,5)=-20$

$\Rightarrow \nabla f=10\mathbf{i}-20\mathbf{j} \Rightarrow (D_{\mathbf{u}}f)_{P_0}=\nabla f\cdot\mathbf{u}=10\left(\frac{4}{5}\right)-20\left(\frac{3}{5}\right)=-4$

11. $\mathbf{u}=\frac{\mathbf{A}}{|\mathbf{A}|}=\frac{12\mathbf{i}+5\mathbf{j}}{\sqrt{12^2+5^2}}=\frac{12}{13}\mathbf{i}+\frac{5}{13}\mathbf{j}$; $g_x(x,y)=1+\frac{y^2}{x^2}+\frac{2y\sqrt{3}}{2xy\sqrt{4x^2y^2-1}} \Rightarrow g_x(1,1)=3$; $g_y(x,y)$

$=-\frac{2y}{x}+\frac{2x\sqrt{3}}{2xy\sqrt{4x^2y^2-1}} \Rightarrow g_y(1,1)=-1 \Rightarrow \nabla g=3\mathbf{i}-\mathbf{j} \Rightarrow (D_{\mathbf{u}}g)_{P_0}=\nabla g\cdot\mathbf{u}=\frac{36}{13}-\frac{5}{13}=\frac{31}{13}$

13. $\mathbf{u} = \frac{\mathbf{A}}{|\mathbf{A}|} = \frac{3\mathbf{i} + 6\mathbf{j} - 2\mathbf{k}}{\sqrt{3^2 + 6^2 + (-2)^2}} = \frac{3}{7}\mathbf{i} + \frac{6}{7}\mathbf{j} - \frac{2}{7}\mathbf{k}$; $f_x(x,y,z) = y + z \Rightarrow f_x(1,-1,2) = 1$; $f_y(x,y,z) = x + z$ $\Rightarrow f_y(1,-1,2) = 3$; $f_z(x,y,z) = y + x \Rightarrow f_z(1,-1,2) = 0 \Rightarrow \nabla f = \mathbf{i} + 3\mathbf{j} \Rightarrow (D_{\mathbf{u}}f)_{P_0} = \nabla f \cdot \mathbf{u} = \frac{3}{7} + \frac{18}{7} = 3$

15. $\mathbf{u} = \frac{\mathbf{A}}{|\mathbf{A}|} = \frac{2\mathbf{i} + \mathbf{j} - 2\mathbf{k}}{\sqrt{2^2 + 1^2 + (-2)^2}} = \frac{2}{3}\mathbf{i} + \frac{1}{3}\mathbf{j} - \frac{2}{3}\mathbf{k}$; $g_x(x,y,z) = 3e^x \cos yz \Rightarrow g_x(0,0,0) = 3$; $g_y(x,y,z) = -3ze^x \sin yz$ $\Rightarrow g_y(0,0,0) = 0$; $g_z(x,y,z) = -3ye^x \sin yz \Rightarrow g_z(0,0,0) = 0 \Rightarrow \nabla g = 3\mathbf{i} \Rightarrow (D_{\mathbf{u}}g)_{P_0} = \nabla g \cdot \mathbf{u} = 2$

17. $\nabla f = (2x + y)\mathbf{i} + (x + 2y)\mathbf{j} \Rightarrow \nabla f(-1,1) = -\mathbf{i} + \mathbf{j} \Rightarrow \mathbf{u} = \frac{\nabla f}{|\nabla f|} = \frac{-\mathbf{i} + \mathbf{j}}{\sqrt{(-1)^2 + 1^2}} = -\frac{1}{\sqrt{2}}\mathbf{i} + \frac{1}{\sqrt{2}}\mathbf{j}$; f increases most rapidly in the direction $\mathbf{u} = -\frac{1}{\sqrt{2}}\mathbf{i} + \frac{1}{\sqrt{2}}\mathbf{j}$ and decreases most rapidly in the direction $-\mathbf{u} = \frac{1}{\sqrt{2}}\mathbf{i} - \frac{1}{\sqrt{2}}\mathbf{j}$; $(D_{\mathbf{u}}f)_{P_0} = \nabla f \cdot \mathbf{u} = |\nabla f| = \sqrt{2}$ and $(D_{-\mathbf{u}}f)_{P_0} = -\sqrt{2}$

19. $\nabla f = \frac{1}{y}\mathbf{i} - \left(\frac{x}{y^2} + z\right)\mathbf{j} - y\mathbf{k} \Rightarrow \nabla f(4,1,1) = \mathbf{i} - 5\mathbf{j} - \mathbf{k} \Rightarrow \mathbf{u} = \frac{\nabla f}{|\nabla f|} = \frac{\mathbf{i} - 5\mathbf{j} - \mathbf{k}}{\sqrt{1^2 + (-5)^2 + (-1)^2}}$ $= \frac{1}{3\sqrt{3}}\mathbf{i} - \frac{5}{3\sqrt{3}}\mathbf{j} - \frac{1}{3\sqrt{3}}\mathbf{k}$; f increases most rapidly in the direction of $\mathbf{u} = \frac{1}{3\sqrt{3}}\mathbf{i} - \frac{5}{3\sqrt{3}}\mathbf{j} - \frac{1}{3\sqrt{3}}\mathbf{k}$ and decreases most rapidly in the direction $-\mathbf{u} = -\frac{1}{3\sqrt{3}}\mathbf{i} + \frac{5}{3\sqrt{3}}\mathbf{j} + \frac{1}{3\sqrt{3}}\mathbf{k}$; $(D_{\mathbf{u}}f)_{P_0} = \nabla f \cdot \mathbf{u} = |\nabla f| = 3\sqrt{3}$ and $(D_{-\mathbf{u}}f)_{P_0} = -3\sqrt{3}$

21. $\nabla f = \left(\frac{1}{x} + \frac{1}{x}\right)\mathbf{i} + \left(\frac{1}{y} + \frac{1}{y}\right)\mathbf{j} + \left(\frac{1}{z} + \frac{1}{z}\right)\mathbf{k} \Rightarrow \nabla f(1,1,1) = 2\mathbf{i} + 2\mathbf{j} + 2\mathbf{k} \Rightarrow \mathbf{u} = \frac{\nabla f}{|\nabla f|} = \frac{1}{\sqrt{3}}\mathbf{i} + \frac{1}{\sqrt{3}}\mathbf{j} + \frac{1}{\sqrt{3}}\mathbf{k}$; f increases most rapidly in the direction $\mathbf{u} = \frac{1}{\sqrt{3}}\mathbf{i} + \frac{1}{\sqrt{3}}\mathbf{j} + \frac{1}{\sqrt{3}}\mathbf{k}$; $(D_{\mathbf{u}}f)_{P_0} = \nabla f \cdot \mathbf{u} = |\nabla f| = 2\sqrt{3}$ and $(D_{-\mathbf{u}}f)_{P_0} = -2\sqrt{3}$

23. $\nabla f = \left(\frac{x}{x^2 + y^2 + z^2}\right)\mathbf{i} + \left(\frac{y}{x^2 + y^2 + z^2}\right)\mathbf{j} + \left(\frac{z}{x^2 + y^2 + z^2}\right)\mathbf{k} \Rightarrow \nabla f(3,4,12) = \frac{3}{169}\mathbf{i} + \frac{4}{169}\mathbf{j} + \frac{12}{169}\mathbf{k}$; $\mathbf{u} = \frac{\mathbf{A}}{|\mathbf{A}|} = \frac{3\mathbf{i} + 6\mathbf{j} - 2\mathbf{k}}{\sqrt{3^2 + 6^2 + (-2)^2}} = \frac{3}{7}\mathbf{i} + \frac{6}{7}\mathbf{j} - \frac{2}{7}\mathbf{k} \Rightarrow \nabla f \cdot \mathbf{u} = \frac{9}{1183}$ and $df = (\nabla f \cdot \mathbf{u})\, ds = \left(\frac{9}{1183}\right)(0.1) \approx 0.000760$

25. $\nabla g = (1 + \cos z)\mathbf{i} + (1 - \sin z)\mathbf{j} + (-x \sin z - y \cos z)\mathbf{k} \Rightarrow \nabla g(2,-1,0) = 2\mathbf{i} + \mathbf{j} + \mathbf{k}$; $\mathbf{A} = \overrightarrow{P_0P_1} = -2\mathbf{i} + 2\mathbf{j} + 2\mathbf{k}$ $\Rightarrow \mathbf{u} = \frac{\mathbf{A}}{|\mathbf{A}|} = \frac{-2\mathbf{i} + 2\mathbf{j} + 2\mathbf{k}}{\sqrt{(-2)^2 + 2^2 + 2^2}} = -\frac{1}{\sqrt{3}}\mathbf{i} + \frac{1}{\sqrt{3}}\mathbf{j} + \frac{1}{\sqrt{3}}\mathbf{k} \Rightarrow \nabla g \cdot \mathbf{u} = 0$ and $dg = (\nabla g \cdot \mathbf{u})\, ds = (0)(0.2) = 0$

27. $\nabla f = 2x\mathbf{i} + 2y\mathbf{j} + 2z\mathbf{k} \Rightarrow \nabla f(1,1,1) = 2\mathbf{i} + 2\mathbf{j} + 2\mathbf{k} \Rightarrow$ Tangent plane: $2(x-1) + 2(y-1) + 2(z-1) = 0$ $\Rightarrow x + y + z = 3$; Normal line: $x = 1 + 2t$, $y = 1 + 2t$, $z = 1 + 2t$

29. $\nabla f = -2x\mathbf{i} + 2\mathbf{k} \Rightarrow \nabla f(2,0,2) = -4\mathbf{i} + 2\mathbf{k} \Rightarrow$ Tangent plane: $-4(x-2) + 2(z-2) = 0 \Rightarrow -4x + 2z + 4 = 0$ $\Rightarrow -2x + z + 2 = 0$; Normal line: $x = 2 - 4t$, $y = 0$, $z = 2 + 2t$

31. $\nabla f = (-\pi \sin \pi x - 2xy + ze^{xz})\mathbf{i} + (-x^2 + z)\mathbf{j} + (xe^{xz} + y)\mathbf{k} \Rightarrow \nabla f(0,1,2) = 2\mathbf{i} + 2\mathbf{j} + \mathbf{k} \Rightarrow$ Tangent plane: $2(x-0) + 2(y-1) + 1(z-2) = 0 \Rightarrow 2x + 2y + z - 4 = 0$; Normal line: $x = 2t,\ y = 1 + 2t,\ z = 2 + t$

33. $\nabla f = \mathbf{i} + \mathbf{j} + \mathbf{k}$ for all points $\Rightarrow \nabla f(0,1,0) = \mathbf{i} + \mathbf{j} + \mathbf{k} \Rightarrow$ Tangent plane: $1(x-0) + 1(y-1) + 1(z-0) = 0$ $\Rightarrow x + y + z - 1 = 0$; Normal line: $x = t,\ y = 1 + t,\ z = t$

35. $z = f(x,y) = \ln(x^2 + y^2) \Rightarrow f_x(x,y) = \frac{2x}{x^2+y^2}$ and $f_y(x,y) = \frac{2y}{x^2+y^2} \Rightarrow f_x(1,0) = 2$ and $f_y(1,0) = 0 \Rightarrow$ from Eq. (10) the tangent plane at $(1,0,0)$ is $2(x-1) - z = 0$ or $2x - z - 2 = 0$

37. $z = f(x,y) = \sqrt{y-x} \Rightarrow f_x(x,y) = -\frac{1}{2}(y-x)^{-1/2}$ and $f_y(x,y) = \frac{1}{2}(y-x)^{-1/2} \Rightarrow f_x(1,2) = -\frac{1}{2}$ and $f_y(1,2) = \frac{1}{2}$ $\Rightarrow$ from Eq. (10) the tangent plane at $(1,2,1)$ is $-\frac{1}{2}(x-1) + \frac{1}{2}(y-2) - (z-1) = 0 \Rightarrow x - y + 2z - 1 = 0$

39. $\nabla f = 2x\mathbf{i} + 2y\mathbf{j} \Rightarrow \nabla f(\sqrt{2}, \sqrt{2}) = 2\sqrt{2}\,\mathbf{i} + 2\sqrt{2}\,\mathbf{j}$
$\Rightarrow$ Tangent line: $2\sqrt{2}(x - \sqrt{2}) + 2\sqrt{2}(y - \sqrt{2}) = 0$
$\Rightarrow \sqrt{2}x + \sqrt{2}y = 4$

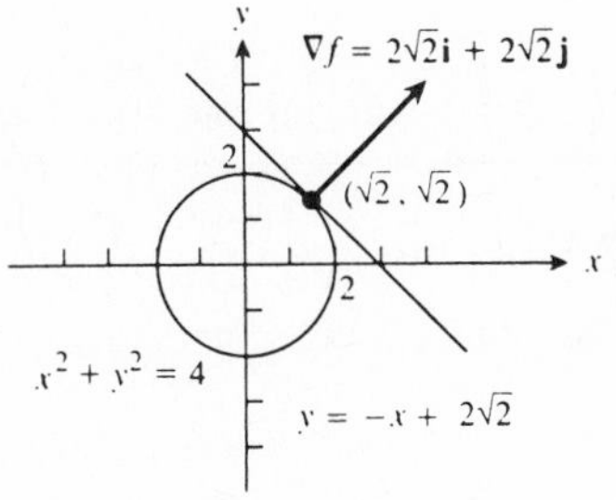

41. $\nabla f = y\mathbf{i} + x\mathbf{j} \Rightarrow \nabla f(2,-2) = -2\mathbf{i} + 2\mathbf{j}$
$\Rightarrow$ Tangent line: $-2(x-2) + 2(y+2) = 0$
$\Rightarrow y = x - 4$

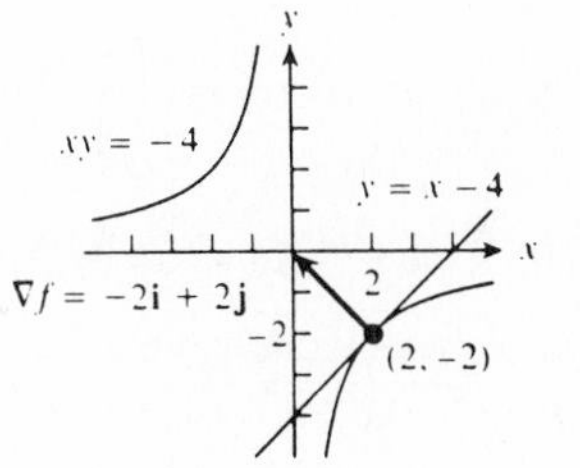

43. $\nabla f = \mathbf{i} + 2y\mathbf{j} + 2\mathbf{k} \Rightarrow \nabla f(1,1,1) = \mathbf{i} + 2\mathbf{j} + 2\mathbf{k}$ and $\nabla g = \mathbf{i}$ for all points; $\mathbf{v} = \nabla f \times \nabla g$

$$\Rightarrow \mathbf{v} = \begin{vmatrix} \mathbf{i} & \mathbf{j} & \mathbf{k} \\ 1 & 2 & 2 \\ 1 & 0 & 0 \end{vmatrix} = 2\mathbf{j} - 2\mathbf{k} \Rightarrow \text{Tangent line: } x = 1,\ y = 1 + 2t,\ z = 1 - 2t$$

45. $\nabla f = 2x\mathbf{i} + 2\mathbf{j} + 2\mathbf{k} \Rightarrow \nabla f\left(1, 1, \frac{1}{2}\right) = 2\mathbf{i} + 2\mathbf{j} + 2\mathbf{k}$ and $\nabla g = \mathbf{j}$ for all points; $\mathbf{v} = \nabla f \times \nabla g$

$$\Rightarrow \mathbf{v} = \begin{vmatrix} \mathbf{i} & \mathbf{j} & \mathbf{k} \\ 2 & 2 & 2 \\ 0 & 1 & 0 \end{vmatrix} = -2\mathbf{i} + 2\mathbf{k} \Rightarrow \text{Tangent line: } x = 1 - 2t,\ y = 1,\ z = \frac{1}{2} + 2t$$

47. $\nabla f = (3x^2 + 6xy^2 + 4y)\mathbf{i} + (6x^2y + 3y^2 + 4x)\mathbf{j} - 2z\mathbf{k} \Rightarrow \nabla f(1,1,3) = 13\mathbf{i} + 13\mathbf{j} - 6\mathbf{k}$; $\nabla g = 2x\mathbf{i} + 2y\mathbf{j} + 2z\mathbf{k}$

$\Rightarrow \nabla g(1,1,3) = 2\mathbf{i} + 2\mathbf{j} + 6\mathbf{k}$; $\mathbf{v} = \nabla f \times \nabla g \Rightarrow \mathbf{v} = \begin{vmatrix} \mathbf{i} & \mathbf{j} & \mathbf{k} \\ 13 & 13 & -6 \\ 2 & 2 & 6 \end{vmatrix} = 90\mathbf{i} - 90\mathbf{j} \Rightarrow$ Tangent line:

$x = 1 + 90t,\ y = 1 - 90t,\ z = 3$

49. $\nabla f = y\mathbf{i} + (x + 2y)\mathbf{j} \Rightarrow \nabla f(3,2) = 2\mathbf{i} + 7\mathbf{j}$; a vector orthogonal to ∇f is $\mathbf{A} = 7\mathbf{i} - 2\mathbf{j} \Rightarrow \mathbf{u} = \frac{\mathbf{A}}{|\mathbf{A}|} = \frac{7\mathbf{i} - 2\mathbf{j}}{\sqrt{7^2 + (-2)^2}}$
$= \frac{7}{\sqrt{53}}\mathbf{i} - \frac{2}{\sqrt{53}}\mathbf{j}$ and $-\mathbf{u} = -\frac{7}{\sqrt{53}}\mathbf{i} + \frac{2}{\sqrt{53}}\mathbf{j}$ are the directions where the derivative is zero

51. $\nabla f = (2x - 3y)\mathbf{i} + (-3x + 8y)\mathbf{j} \Rightarrow \nabla f(1,2) = -4\mathbf{i} + 13\mathbf{j} \Rightarrow |\nabla f(1,2)| = \sqrt{(-4)^2 + (13)^2} = \sqrt{185}$; no, the maximum rate of change is $\sqrt{185} < 14$

53. $\nabla f = f_x(1,2)\mathbf{i} + f_y(1,2)\mathbf{j}$ and $\mathbf{u}_1 = \frac{\mathbf{i} + \mathbf{j}}{\sqrt{1^2 + 1^2}} = \frac{1}{\sqrt{2}}\mathbf{i} + \frac{1}{\sqrt{2}}\mathbf{j} \Rightarrow (D_{\mathbf{u}_1}f)(1,2) = f_x(1,2)\left(\frac{1}{\sqrt{2}}\right) + f_y(1,2)\left(\frac{1}{\sqrt{2}}\right)$
$= 2\sqrt{2} \Rightarrow f_x(1,2) + f_y(1,2) = 4$; $\mathbf{u}_2 = -\mathbf{j} \Rightarrow (D_{\mathbf{u}_2}f)(1,2) = f_x(1,2)(0) + f_y(1,2)(-1) = -3 \Rightarrow -f_y(1,2) = -3$
$\Rightarrow f_y(1,2) = 3$; then $f_x(1,2) + 3 = 4 \Rightarrow f_x(1,2) = 1$; thus $\nabla f(1,2) = \mathbf{i} + 3\mathbf{j}$ and $\mathbf{u} = \frac{\mathbf{A}}{|\mathbf{A}|} = \frac{-\mathbf{i} - 2\mathbf{j}}{\sqrt{(-1)^2 + (-2)^2}}$
$= -\frac{1}{\sqrt{5}}\mathbf{i} - \frac{2}{\sqrt{5}}\mathbf{j} \Rightarrow (D_{\mathbf{u}}f)_{P_0} = \nabla f \cdot \mathbf{u} = -\frac{1}{\sqrt{5}} - \frac{6}{\sqrt{5}} = -\frac{7}{\sqrt{5}}$

55. (a) The unit tangent vector at $\left(\frac{1}{2}, \frac{\sqrt{3}}{2}\right)$ in the direction of motion is $\mathbf{u} = \frac{\sqrt{3}}{2}\mathbf{i} - \frac{1}{2}\mathbf{j}$;

$\nabla T = (\sin 2y)\mathbf{i} + (2x \cos 2y)\mathbf{j} \Rightarrow \nabla T\left(\frac{1}{2}, \frac{\sqrt{3}}{2}\right) = \left(\sin \sqrt{3}\right)\mathbf{i} + \left(\cos \sqrt{3}\right)\mathbf{j} \Rightarrow D_{\mathbf{u}}T\left(\frac{1}{2}, \frac{\sqrt{3}}{2}\right) = \nabla T \cdot \mathbf{u}$
$= \frac{\sqrt{3}}{2} \sin \sqrt{3} - \frac{1}{2} \cos \sqrt{3} \approx 0.935°$ C/ft

(b) $\mathbf{r}(t) = (\sin 2t)\mathbf{i} + (\cos 2t)\mathbf{j} \Rightarrow \mathbf{v}(t) = (2 \cos 2t)\mathbf{i} - (2 \sin 2t)\mathbf{j}$ and $|\mathbf{v}| = 2$; $\frac{dT}{dt} = \frac{\partial T}{\partial x}\frac{dx}{dt} + \frac{\partial T}{\partial y}\frac{dy}{dt}$
$= \nabla T \cdot \mathbf{v} = \left(\nabla T \cdot \frac{\mathbf{v}}{|\mathbf{v}|}\right)|\mathbf{v}| = (D_{\mathbf{u}}T)\,|\mathbf{v}|$, where $\mathbf{u} = \frac{\mathbf{v}}{|\mathbf{v}|}$; at $\left(\frac{1}{2}, \frac{\sqrt{3}}{2}\right)$ we have $\mathbf{u} = \frac{\sqrt{3}}{2}\mathbf{i} - \frac{1}{2}\mathbf{j}$ from part (a)
$\Rightarrow \frac{dT}{dt} = \left(\frac{\sqrt{3}}{2} \sin \sqrt{3} - \frac{1}{2} \cos \sqrt{3}\right) \cdot 2 = \sqrt{3} \sin \sqrt{3} - \cos \sqrt{3} \approx 1.87°$ C/sec

57. $\nabla f = 2x\mathbf{i} + 2y\mathbf{j} + 2z\mathbf{k} = (2 \cos t)\mathbf{i} + (2 \sin t)\mathbf{j} + 2t\mathbf{k}$ and $\mathbf{v} = (-\sin t)\mathbf{i} + (\cos t)\mathbf{j} + \mathbf{k} \Rightarrow \mathbf{u} = \frac{\mathbf{v}}{|\mathbf{v}|}$
$= \frac{(-\sin t)\mathbf{i} + (\cos t)\mathbf{j} + \mathbf{k}}{\sqrt{(\sin t)^2 + (\cos t)^2 + 1^2}} = \left(\frac{-\sin t}{\sqrt{2}}\right)\mathbf{i} + \left(\frac{\cos t}{\sqrt{2}}\right)\mathbf{j} + \frac{1}{\sqrt{2}}\mathbf{k} \Rightarrow (D_{\mathbf{u}}f)_{P_0} = \nabla f \cdot \mathbf{u}$
$= (2 \cos t)\left(\frac{-\sin t}{\sqrt{2}}\right) + (2 \sin t)\left(\frac{\cos t}{\sqrt{2}}\right) + (2t)\left(\frac{1}{\sqrt{2}}\right) = \frac{2t}{\sqrt{2}} \Rightarrow (D_{\mathbf{u}}f)\left(\frac{-\pi}{4}\right) = \frac{-\pi}{2\sqrt{2}}$, $(D_{\mathbf{u}}f)(0) = 0$ and
$(D_{\mathbf{u}}f)\left(\frac{\pi}{4}\right) = \frac{\pi}{2\sqrt{2}}$

59. If (x, y) is a point on the line, then $\mathbf{T}(x,y) = (x - x_0)\mathbf{i} + (y - y_0)\mathbf{j}$ is a vector parallel to the line $\Rightarrow \mathbf{T}\cdot\mathbf{N} = 0$ $\Rightarrow A(x - x_0) + B(y - y_0) = 0$, as claimed.

61. $x = g(t)$ and $y = h(t) \Rightarrow \mathbf{r} = g(t)\mathbf{i} + h(t)\mathbf{j} \Rightarrow \mathbf{v} = g'(t)\mathbf{i} + h'(t)\mathbf{j} \Rightarrow \mathbf{T} = \frac{\mathbf{v}}{|\mathbf{v}|} = \frac{g'(t)\mathbf{i} + h'(t)\mathbf{j}}{\sqrt{[g'(t)]^2 + [h'(t)]^2}}$;

$z = f(x,y) \Rightarrow \frac{df}{dt} = \frac{\partial f}{\partial x}\frac{dx}{dt} + \frac{\partial f}{\partial y}\frac{dy}{dt} = \frac{\partial f}{\partial x}g'(t) + \frac{\partial f}{\partial y}h'(t) = \nabla f\cdot\mathbf{T}$. If $f(g(t),h(t)) = c$, then $\frac{df}{dt} = 0$

$\Rightarrow \frac{\partial f}{\partial x}g'(t) + \frac{\partial f}{\partial y}h'(t) = 0 \Rightarrow \nabla f\cdot\mathbf{T} = 0 \Rightarrow \nabla f$ is normal to $\mathbf{T}$

63. The directional derivative is the scalar component. With ∇f evaluated at P_0, the scalar component of ∇f in the direction of $\mathbf{u}$ is $\nabla f\cdot\mathbf{u} = (D_{\mathbf{u}}f)_{P_0}$.

65. (a) $\nabla(kf) = \frac{\partial(kf)}{\partial x}\mathbf{i} + \frac{\partial(kf)}{\partial y}\mathbf{j} + \frac{\partial(kf)}{\partial z}\mathbf{k} = k\left(\frac{\partial f}{\partial x}\right)\mathbf{i} + k\left(\frac{\partial f}{\partial y}\right)\mathbf{j} + k\left(\frac{\partial f}{\partial z}\right)\mathbf{k} = k\left(\frac{\partial f}{\partial x}\mathbf{i} + \frac{\partial f}{\partial y}\mathbf{j} + \frac{\partial f}{\partial z}\mathbf{k}\right) = k\nabla f$

(b) $\nabla(f+g) = \frac{\partial(f+g)}{\partial x}\mathbf{i} + \frac{\partial(f+g)}{\partial y}\mathbf{j} + \frac{\partial(f+g)}{\partial z}\mathbf{k} = \left(\frac{\partial f}{\partial x} + \frac{\partial g}{\partial x}\right)\mathbf{i} + \left(\frac{\partial f}{\partial y} + \frac{\partial g}{\partial y}\right)\mathbf{j} + \left(\frac{\partial f}{\partial z} + \frac{\partial g}{\partial z}\right)\mathbf{k}$

$= \frac{\partial f}{\partial x}\mathbf{i} + \frac{\partial g}{\partial x}\mathbf{i} + \frac{\partial f}{\partial y}\mathbf{j} + \frac{\partial g}{\partial y}\mathbf{j} + \frac{\partial f}{\partial z}\mathbf{k} + \frac{\partial g}{\partial z}\mathbf{k} = \left(\frac{\partial f}{\partial x}\mathbf{i} + \frac{\partial f}{\partial y}\mathbf{j} + \frac{\partial f}{\partial z}\mathbf{k}\right) + \left(\frac{\partial g}{\partial x}\mathbf{i} + \frac{\partial g}{\partial y}\mathbf{j} + \frac{\partial g}{\partial z}\mathbf{k}\right) = \nabla f + \nabla g$

(c) $\nabla(f-g) = \nabla f - \nabla g$ (Substitute $-g$ for g in part (b) above)

(d) $\nabla(fg) = \frac{\partial(fg)}{\partial x}\mathbf{i} + \frac{\partial(fg)}{\partial y}\mathbf{j} + \frac{\partial(fg)}{\partial z}\mathbf{k} = \left(\frac{\partial f}{\partial x}g + \frac{\partial g}{\partial x}f\right)\mathbf{i} + \left(\frac{\partial f}{\partial y}g + \frac{\partial g}{\partial y}f\right)\mathbf{j} + \left(\frac{\partial f}{\partial z}g + \frac{\partial g}{\partial z}f\right)\mathbf{k}$

$= \left(\frac{\partial f}{\partial x}g\right)\mathbf{i} + \left(\frac{\partial g}{\partial x}f\right)\mathbf{i} + \left(\frac{\partial f}{\partial y}g\right)\mathbf{j} + \left(\frac{\partial g}{\partial y}f\right)\mathbf{j} + \left(\frac{\partial f}{\partial z}g\right)\mathbf{k} + \left(\frac{\partial g}{\partial z}f\right)\mathbf{k}$

$= f\left(\frac{\partial g}{\partial x}\mathbf{i} + \frac{\partial g}{\partial y}\mathbf{j} + \frac{\partial g}{\partial z}\mathbf{k}\right) + g\left(\frac{\partial f}{\partial x}\mathbf{i} + \frac{\partial f}{\partial y}\mathbf{j} + \frac{\partial f}{\partial z}\mathbf{k}\right) = f\nabla g + g\nabla f$

(e) $\nabla\left(\frac{f}{g}\right) = \frac{\partial\left(\frac{f}{g}\right)}{\partial x}\mathbf{i} + \frac{\partial\left(\frac{f}{g}\right)}{\partial y}\mathbf{j} + \frac{\partial\left(\frac{f}{g}\right)}{\partial z}\mathbf{k} = \left(\frac{g\frac{\partial f}{\partial x} - f\frac{\partial g}{\partial x}}{g^2}\right)\mathbf{i} + \left(\frac{g\frac{\partial f}{\partial y} - f\frac{\partial g}{\partial y}}{g^2}\right)\mathbf{j} + \left(\frac{g\frac{\partial f}{\partial z} - f\frac{\partial g}{\partial z}}{g^2}\right)\mathbf{k}$

$= \left(\frac{g\frac{\partial f}{\partial x}\mathbf{i} + g\frac{\partial f}{\partial y}\mathbf{j} + g\frac{\partial f}{\partial z}\mathbf{k}}{g^2}\right) - \left(\frac{f\frac{\partial g}{\partial x}\mathbf{i} + f\frac{\partial g}{\partial y}\mathbf{j} + f\frac{\partial g}{\partial z}\mathbf{k}}{g^2}\right) = \frac{g\left(\frac{\partial f}{\partial x}\mathbf{i} + \frac{\partial f}{\partial y}\mathbf{j} + \frac{\partial f}{\partial z}\mathbf{k}\right)}{g^2} - \frac{f\left(\frac{\partial g}{\partial x}\mathbf{i} + \frac{\partial g}{\partial y}\mathbf{j} + \frac{\partial g}{\partial z}\mathbf{k}\right)}{g^2}$

$= \frac{g\nabla f}{g^2} - \frac{f\nabla g}{g^2} = \frac{g\nabla f - f\nabla g}{g^2}$

12.8 EXTREME VALUES AND SADDLE POINTS

1. $f_x(x,y) = 2x + y + 3 = 0$ and $f_y(x,y) = x + 2y - 3 = 0 \Rightarrow x = -3$ and $y = 3 \Rightarrow$ critical point is $(-3,3)$; $f_{xx}(-3,3) = 2$, $f_{yy}(-3,3) = 2$, $f_{xy}(-3,3) = 1 \Rightarrow f_{xx}f_{yy} - f_{xy}^2 = 3 > 0$ and $f_{xx} > 0 \Rightarrow$ local minimum of $f(-3,3) = -5$

3. $f_x(x,y) = 2y - 10x + 4 = 0$ and $f_y(x,y) = 2x - 4y + 4 = 0 \Rightarrow x = \frac{2}{3}$ and $y = \frac{4}{3} \Rightarrow$ critical point is $\left(\frac{2}{3}, \frac{4}{3}\right)$; $f_{xx}\left(\frac{2}{3}, \frac{4}{3}\right) = -10$, $f_{yy}\left(\frac{2}{3}, \frac{4}{3}\right) = -4$, $f_{xy}\left(\frac{2}{3}, \frac{4}{3}\right) = 2 \Rightarrow f_{xx}f_{yy} - f_{xy}^2 = 36 > 0$ and $f_{xx} < 0 \Rightarrow$ local maximum of $f\left(\frac{2}{3}, \frac{4}{3}\right) = 0$

5. $f_x(x,y) = 2x + y + 3 = 0$ and $f_y(x,y) = x + 2 = 0 \Rightarrow x = -2$ and $y = 1 \Rightarrow$ critical point is $(-2, 1)$; $f_{xx}(-2,1) = 2$, $f_{yy}(-2,1) = 0$, $f_{xy}(-2,1) = 1 \Rightarrow f_{xx}f_{yy} - f_{xy}^2 = -1 < 0 \Rightarrow$ saddle point

7. $f_x(x,y) = 5y - 14x + 3 = 0$ and $f_y(x,y) = 5x - 6 = 0 \Rightarrow x = \frac{6}{5}$ and $y = \frac{69}{25} \Rightarrow$ critical point is $\left(\frac{6}{5}, \frac{69}{25}\right)$; $f_{xx}\left(\frac{6}{5}, \frac{69}{25}\right) = -14$, $f_{yy}\left(\frac{6}{5}, \frac{69}{25}\right) = 0$, $f_{xy}\left(\frac{6}{5}, \frac{69}{25}\right) = 5 \Rightarrow f_{xx}f_{yy} - f_{xy}^2 = -25 < 0 \Rightarrow$ saddle point

9. $f_x(x,y) = 2x - 4y = 0$ and $f_y(x,y) = -4x + 2y + 6 = 0 \Rightarrow x = 2$ and $y = 1 \Rightarrow$ critical point is $(2, 1)$; $f_{xx}(2,1) = 2$, $f_{yy}(2,1) = 2$, $f_{xy}(2,1) = -4 \Rightarrow f_{xx}f_{yy} - f_{xy}^2 = -12 < 0 \Rightarrow$ saddle point

11. $f_x(x,y) = 4x + 3y - 5 = 0$ and $f_y(x,y) = 3x + 8y + 2 = 0 \Rightarrow x = 2$ and $y = -1 \Rightarrow$ critical point is $(2, -1)$; $f_{xx}(2,-1) = 4$, $f_{yy}(2,-1) = 8$, $f_{xy}(2,-1) = 3 \Rightarrow f_{xx}f_{yy} - f_{xy}^2 = 23 > 0$ and $f_{xx} > 0 \Rightarrow$ local minimum of $f(2,-1) = -6$

13. $f_x(x,y) = 2x - 2 = 0$ and $f_y(x,y) = -2y + 4 = 0 \Rightarrow x = 1$ and $y = 2 \Rightarrow$ critical point is $(1, 2)$; $f_{xx}(1,2) = 2$, $f_{yy}(1,2) = -2$, $f_{xy}(1,2) = 0 \Rightarrow f_{xx}f_{yy} - f_{xy}^2 = -4 < 0 \Rightarrow$ saddle point

15. $f_x(x,y) = 2x + 2y = 0$ and $f_y(x,y) = 2x = 0 \Rightarrow x = 0$ and $y = 0 \Rightarrow$ critical point is $(0, 0)$; $f_{xx}(0,0) = 2$, $f_{yy}(0,0) = 0$, $f_{xy}(0,0) = 2 \Rightarrow f_{xx}f_{yy} - f_{xy}^2 = -4 < 0 \Rightarrow$ saddle point

17. $f_x(x,y) = 3x^2 - 2y = 0$ and $f_y(x,y) = -3y^2 - 2x = 0 \Rightarrow x = 0$ and $y = 0$, or $x = -\frac{2}{3}$ and $y = \frac{2}{3} \Rightarrow$ critical points are $(0,0)$ and $\left(-\frac{2}{3}, \frac{2}{3}\right)$; for $(0,0)$: $f_{xx}(0,0) = 6x\big|_{(0,0)} = 0$, $f_{yy}(0,0) = -6y\big|_{(0,0)} = 0$, $f_{xy}(0,0) = -2$ $\Rightarrow f_{xx}f_{yy} - f_{xy}^2 = -4 < 0 \Rightarrow$ saddle point; for $\left(-\frac{2}{3}, \frac{2}{3}\right)$: $f_{xx}\left(-\frac{2}{3}, \frac{2}{3}\right) = -4$, $f_{yy}\left(-\frac{2}{3}, \frac{2}{3}\right) = -4$, $f_{xy}\left(-\frac{2}{3}, \frac{2}{3}\right) = -2$ $\Rightarrow f_{xx}f_{yy} - f_{xy}^2 = 12 > 0$ and $f_{xx} < 0 \Rightarrow$ local maximum of $f\left(-\frac{2}{3}, \frac{2}{3}\right) = \frac{170}{27}$

19. $f_x(x,y) = 12x - 6x^2 + 6y = 0$ and $f_y(x,y) = 6y + 6x = 0 \Rightarrow x = 0$ and $y = 0$, or $x = 1$ and $y = -1 \Rightarrow$ critical points are $(0,0)$ and $(1,-1)$; for $(0,0)$: $f_{xx}(0,0) = 12 - 12x\big|_{(0,0)} = 12$, $f_{yy}(0,0) = 6$, $f_{xy}(0,0) = 6 \Rightarrow f_{xx}f_{yy} - f_{xy}^2$ $= 36 > 0$ and $f_{xx} > 0 \Rightarrow$ local minimum of $f(0,0) = 0$; for $(1,-1)$: $f_{xx}(1,-1) = 0$, $f_{yy}(1,-1) = 6$, $f_{xy}(1,-1) = 6 \Rightarrow f_{xx}f_{yy} - f_{xy}^2 = -36 < 0 \Rightarrow$ saddle point

21. $f_x(x,y) = 27x^2 - 4y = 0$ and $f_y(x,y) = y^2 - 4x = 0 \Rightarrow x = 0$ and $y = 0$, or $x = \frac{4}{9}$ and $y = \frac{4}{3} \Rightarrow$ critical points are $(0,0)$ and $\left(\frac{4}{9}, \frac{4}{3}\right)$; for $(0,0)$: $f_{xx}(0,0) = 54x\big|_{(0,0)} = 0$, $f_{yy}(0,0) = 2y\big|_{(0,0)} = 0$, $f_{xy}(0,0) = -4 \Rightarrow f_{xx}f_{yy} - f_{xy}^2$ $= -16 < 0 \Rightarrow$ saddle point; for $\left(\frac{4}{9}, \frac{4}{3}\right)$: $f_{xx}\left(\frac{4}{9}, \frac{4}{3}\right) = 24$, $f_{yy}\left(\frac{4}{9}, \frac{4}{3}\right) = \frac{8}{3}$, $f_{xy}\left(\frac{4}{9}, \frac{4}{3}\right) = -4 \Rightarrow f_{xx}f_{yy} - f_{xy}^2 = 48 > 0$

and $f_{xx} > 0 \Rightarrow$ local minimum of $f\left(\frac{4}{9}, \frac{4}{3}\right) = -\frac{64}{81}$

23. $f_x(x,y) = 3x^2 + 6x = 0 \Rightarrow x = 0$ or $x = -2$; $f_y(x,y) = 3y^2 - 6y = 0 \Rightarrow y = 0$ or $y = 2 \Rightarrow$ the critical points are $(0,0)$, $(0,2)$, $(-2,0)$, and $(-2,2)$; for $(0,0)$: $f_{xx}(0,0) = 6x + 6\big|_{(0,0)} = 6$, $f_{yy}(0,0) = 6y - 6\big|_{(0,0)} = -6$, $f_{xy}(0,0) = 0 \Rightarrow f_{xx}f_{yy} - f_{xy}^2 = -36 < 0 \Rightarrow$ saddle point; for $(0,2)$: $f_{xx}(0,2) = 6$, $f_{yy}(0,2) = 6$, $f_{xy}(0,2) = 0$ $\Rightarrow f_{xx}f_{yy} - f_{xy}^2 = 36 > 0$ and $f_{xx} > 0 \Rightarrow$ local minimum of $f(0,2) = -12$; for $(-2,0)$: $f_{xx}(-2,0) = -6$, $f_{yy}(-2,0) = -6$, $f_{xy}(-2,0) = 0 \Rightarrow f_{xx}f_{yy} - f_{xy}^2 = 36 > 0$ and $f_{xx} < 0 \Rightarrow$ local maximum of $f(-2,0) = -4$; for $(-2,2)$: $f_{xx}(-2,2) = -6$, $f_{yy}(-2,2) = 6$, $f_{xy}(-2,2) = 0 \Rightarrow f_{xx}f_{yy} - f_{xy}^2 = -36 < 0 \Rightarrow$ saddle point

25. $f_x(x,y) = 4y - 4x^3 = 0$ and $f_y(x,y) = 4x - 4y^3 = 0 \Rightarrow x = y \Rightarrow x(1 - x^2) = 0 \Rightarrow x = 0, 1, -1 \Rightarrow$ the critical points are $(0,0)$, $(1,1)$, and $(-1,-1)$; for $(0,0)$: $f_{xx}(0,0) = -12x^2\big|_{(0,0)} = 0$, $f_{yy}(0,0) = -12y^2\big|_{(0,0)} = 0$, $f_{xy}(0,0) = 4 \Rightarrow f_{xx}f_{yy} - f_{xy}^2 = -16 < 0 \Rightarrow$ saddle point; for $(1,1)$: $f_{xx}(1,1) = -12$, $f_{yy}(1,1) = -12$, $f_{xy}(1,1) = 4$ $\Rightarrow f_{xx}f_{yy} - f_{xy}^2 = 128 > 0$ and $f_{xx} < 0 \Rightarrow$ local maximum of $f(1,1) = 2$; for $(-1,-1)$: $f_{xx}(-1,-1) = -12$, $f_{yy}(-1,-1) = -12$, $f_{xy}(-1,-1) = 4 \Rightarrow f_{xx}f_{yy} - f_{xy}^2 = 128 > 0$ and $f_{xx} < 0 \Rightarrow$ local maximum of $f(-1,-1) = 2$

27. $f_x(x,y) = \frac{-2x}{(x^2+y^2-1)^2} = 0$ and $f_y(x,y) = \frac{-2y}{(x^2+y^2-1)^2} = 0 \Rightarrow x = 0$ and $y = 0 \Rightarrow$ the critical point is $(0,0)$; $f_{xx} = \frac{4x^2 - 2y^2 + 2}{(x^2+y^2-1)^3}$, $f_{yy} = \frac{-2x^2 + 4y^2 + 2}{(x^2+y^2-1)^3}$, $f_{xy} = \frac{6xy}{(x^2+y^2-1)^3}$; $f_{xx}(0,0) = -2$, $f_{yy}(0,0) = -2$, $f_{xy}(0,0) = 0$ $\Rightarrow f_{xx}f_{yy} - f_{xy}^2 = 4 > 0$ and $f_{xx} < 0 \Rightarrow$ local maximum of $f(0,0) = -1$

29. $f_x(x,y) = y \cos x = 0$ and $f_y(x,y) = \sin x = 0 \Rightarrow x = n\pi$, n an integer, and $y = 0 \Rightarrow$ the critical points are $(n\pi, 0)$, n an integer (Note: $\cos x$ and $\sin x$ cannot both be 0 for the same x, so $\sin x$ must be 0 and $y = 0$); $f_{xx} = -y \sin x$, $f_{yy} = 0$, $f_{xy} = \cos x$; $f_{xx}(n\pi,0) = 0$, $f_{yy}(n\pi,0) = 0$, $f_{xy}(n\pi,0) = 1$ if n is even and $f_{xy}(n\pi,0) = -1$ if n is odd $\Rightarrow f_{xx}f_{yy} - f_{xy}^2 = -1 < 0 \Rightarrow$ saddle point; $f(n\pi,0) = 0$ for every n

31. (i) On OA, $f(x,y) = f(0,y) = y^2 - 4y + 1$ on $0 \le y \le 2$;
$f'(0,y) = 2y - 4 = 0 \Rightarrow y = 2$;
$f(0,0) = 1$ and $f(0,2) = -3$

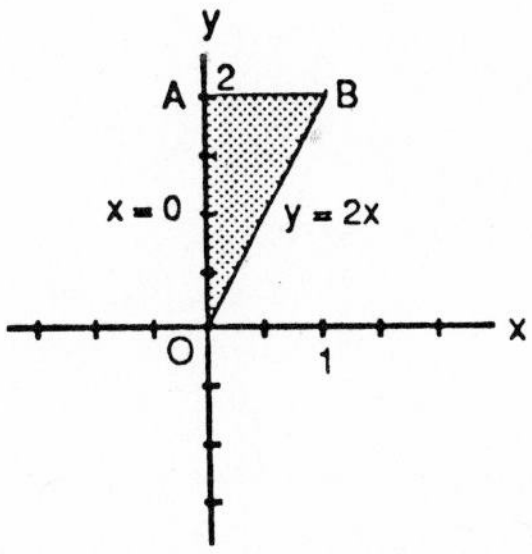

(ii) On AB, $f(x,y) = f(x,2) = 2x^2 - 4x - 3$ on $0 \le x \le 1$;
$f'(x,2) = 4x - 4 = 0 \Rightarrow x = 1$;
$f(0,2) = -3$ and $f(1,2) = -5$

(iii) On OB, $f(x,y) = f(x,2x) = 6x^2 - 12x + 1$ on $0 \le x \le 1$;
endpoint values have been found above; $f'(x,2x) = 12x - 12 = 0 \Rightarrow x = 1$ and $y = 2$, but $(1,2)$ is not an interior point of OB

(iv) For interior points of the triangular region,
$f_x(x,y) = 4x - 4 = 0$ and $f_y(x,y) = 2y - 4 = 0$
$\Rightarrow x = 1$ and $y = 2$, but $(1,2)$ is not an interior point of the region. Therefore, the absolute maximum is 1 at $(0,0)$ and the absolute minimum is -5 at $(1,2)$.

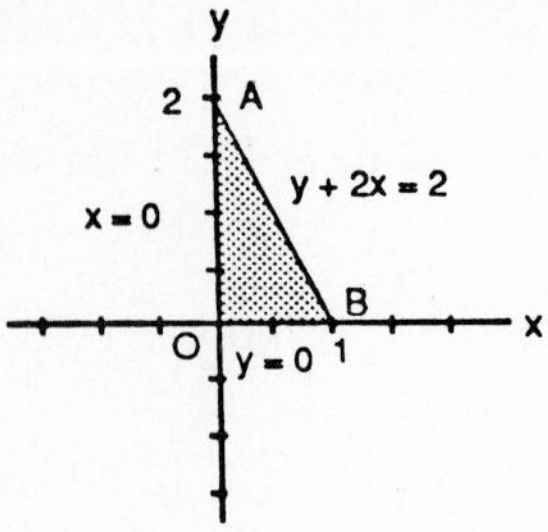

33. (i) On OA, $f(x,y) = f(0,y) = y^2$ on $0 \le y \le 2$; $f'(0,y) = 2y = 0$ $\Rightarrow y = 0$ and $x = 0$; $f(0,0) = 0$ and $f(0,2) = 4$

(ii) On OB, $f(x,y) = f(x,0) = x^2$ on $0 \le x \le 1$; $f'(x,0) = 2x = 0$ $\Rightarrow x = 0$ and $y = 0$; $f(0,0) = 0$ and $f(1,0) = 1$

(iii) On AB, $f(x,y) = f(x,-2x+2) = 5x^2 - 8x + 4$ on $0 \le x \le 1$; $f'(x,-2x+2) = 10x - 8 = 0 \Rightarrow x = \frac{4}{5}$ and $y = \frac{2}{5}$; $f\left(\frac{4}{5},\frac{2}{5}\right) = \frac{4}{5}$ and $f(0,2) = 4$

(iv) For interior points of the triangular region, $f_x(x,y) = 2x = 0$ and $f_y(x,y) = 2y = 0$ $\Rightarrow x = 0$ and $y = 0$, but $(0,0)$ is not an interior point of the region. Therefore the absolute maximum is 4 at $(0,2)$ and the absolute minimum is 0 at $(0,0)$.

35. (i) On OC, $T(x,y) = T(x,0) = x^2 - 6x + 2$ on $0 \le x \le 5$; $T'(x,0) = 2x - 6 = 0 \Rightarrow x = 3$ and $y = 0$; $T(3,0) = -7$, $T(0,0) = 2$, and $T(5,0) = -3$

(ii) On CB, $T(x,y) = T(5,y) = y^2 + 5y - 3$ on $-3 \le y \le 0$; $T'(5,y) = 2y + 5 = 0 \Rightarrow y = -\frac{5}{2}$ and $x = 6$; $T\left(5,-\frac{5}{2}\right) = -\frac{37}{4}$ and $T(5,-3) = -9$

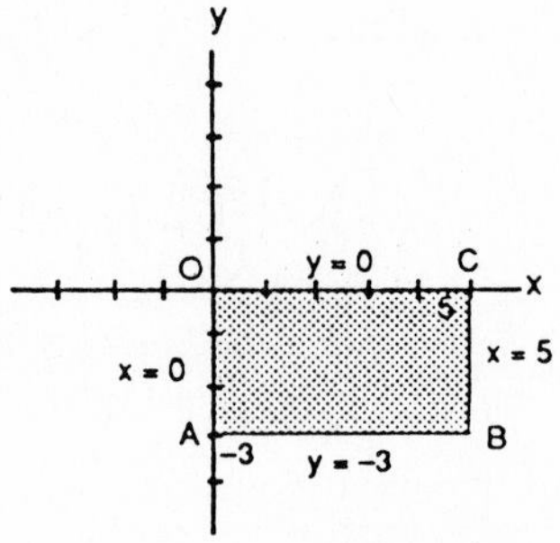

(iii) On AB, $T(x,y) = T(x,-3) = x^2 - 9x + 11$ on $0 \le x \le 5$; $T'(x,-3) = 2x - 9 = 0 \Rightarrow x = \frac{9}{2}$ and $y = -3$; $T\left(\frac{9}{2},-3\right) = -\frac{37}{4}$ and $T(0,-3) = 11$

(iv) On AO, $T(x,y) = T(0,y) = y^2 + 2$ on $-3 \le y \le 0$; $T'(0,y) = 2y = 0 \Rightarrow y = 0$ and $x = 0$, but $(0,0)$ is not an interior point of AO

(v) For interior points of the rectangular region, $T_x(x,y) = 2x + y - 6 = 0$ and $T_y(x,y) = x + 2y = 0 \Rightarrow x = 4$ and $y = -2$, an interior critical point with $T(4,-2) = -10$. Therefore the absolute maximum is 11 at $(0,-3)$ and the absolute minimum is -10 at $(4,-2)$.

37. (i) On AB, $f(x,y) = f(1,y) = 3\cos y$ on $-\frac{\pi}{4} \le y \le \frac{\pi}{4}$; $f'(1,y) = -3\sin y = 0 \Rightarrow y = 0$ and $x = 1$; $f(1,0) = 3$, $f\left(1,-\frac{\pi}{4}\right) = \frac{3\sqrt{2}}{2}$, and $f\left(1,\frac{\pi}{4}\right) = \frac{3\sqrt{2}}{2}$

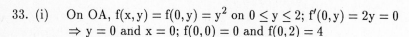

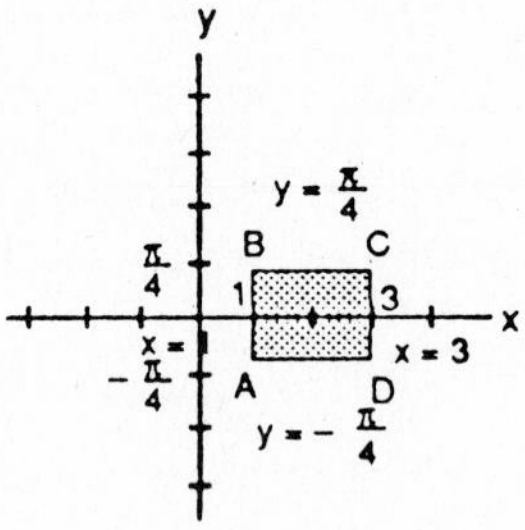

(ii) On CD, $f(x,y) = f(3,y) = 3\cos y$ on $-\frac{\pi}{4} \le y \le \frac{\pi}{4}$; $f'(3,y) = -3\sin y = 0 \Rightarrow y = 0$ and $x = 3$; $f(3,0) = 3$, $f\left(3,-\frac{\pi}{4}\right) = \frac{3\sqrt{2}}{2}$ and $f\left(3,\frac{\pi}{4}\right) = \frac{3\sqrt{2}}{2}$

(iii) On BC, $f(x,y) = f\left(x,\frac{\pi}{4}\right) = \frac{\sqrt{2}}{2}(4x - x^2)$ on $1 \le x \le 3$; $f'\left(x,\frac{\pi}{4}\right) = \sqrt{2}(2-x) = 0 \Rightarrow x = 2$ and $y = \frac{\pi}{4}$; $f\left(2,\frac{\pi}{4}\right) = 2\sqrt{2}$, $f\left(1,\frac{\pi}{4}\right) = \frac{3\sqrt{2}}{2}$, and $f\left(3,\frac{\pi}{4}\right) = \frac{3\sqrt{2}}{2}$

(iv) On AD, $f(x,y) = f\left(x,-\frac{\pi}{4}\right) = \frac{\sqrt{2}}{2}(4x - x^2)$ on $1 \le x \le 3$; $f'\left(x,-\frac{\pi}{4}\right) = \sqrt{2}(2-x) = 0 \Rightarrow x = 2$ and $y = -\frac{\pi}{4}$; $f\left(2,-\frac{\pi}{4}\right) = 2\sqrt{2}$, $f\left(1,-\frac{\pi}{4}\right) = \frac{3\sqrt{2}}{2}$, and $f\left(3,-\frac{\pi}{4}\right) = \frac{3\sqrt{2}}{2}$

(v) For interior points of the region, $f_x(x,y) = (4-2x)\cos y = 0$ and $f_y(x,y) = -(4x - x^2)\sin y = 0 \Rightarrow x = 2$ and $y = 0$, which is an interior critical point with $f(2,0) = 4$. Therefore the absolute maximum is 4 at $(2,0)$ and the absolute minimum is $\frac{3\sqrt{2}}{2}$ at $\left(3, -\frac{\pi}{4}\right)$, $\left(3, \frac{\pi}{4}\right)$, $\left(1, -\frac{\pi}{4}\right)$, and $\left(1, \frac{\pi}{4}\right)$.

39. Let $F(a,b) = \int_a^b (6 - x - x^2)\,dx$ where $a \le b$. The boundary of the domain of F is the line $a = b$ in the ab-plane, and $F(a,a) = 0$, so F is identically 0 on the boundary of its domain. For interior critical points we have: $\frac{\partial F}{\partial a} = -(6 - a - a^2) = 0 \Rightarrow a = -3,\ 2$ and $\frac{\partial F}{\partial b} = (6 - b - b^2) = 0 \Rightarrow b = -3,\ 2$. Since $a \le b$, there is only one interior critical point $(-3,2)$ and $F(-3,2) = \int_{-3}^{2} (6 - x - x^2)\,dx$ gives the area under the parabola $y = 6 - x - x^2$ that is above the x-axis. Therefore, $a = -3$ and $b = 2$.

41. $T_x(x,y) = 2x - 1 = 0$ and $T_y(x,y) = 4y = 0 \Rightarrow x = \frac{1}{2}$ and $y = 0$ with $T\left(\frac{1}{2}, 0\right) = -\frac{1}{4}$; on the boundary $x^2 + y^2 = 1$: $T(x,y) = -x^2 - x + 2$ for $-1 \le x \le 1 \Rightarrow T'(x,y) = -2x - 1 = 0 \Rightarrow x = -\frac{1}{2}$ and $y = \pm\frac{\sqrt{3}}{2}$; $T\left(-\frac{1}{2}, \frac{\sqrt{3}}{2}\right) = \frac{9}{4}$, $T\left(-\frac{1}{2}, -\frac{\sqrt{3}}{2}\right) = \frac{9}{4}$, $T(-1,0) = 2$, and $T(1,0) = 0 \Rightarrow$ the hottest is $2\frac{1}{4}^\circ$ at $\left(-\frac{1}{2}, \frac{\sqrt{3}}{2}\right)$ and $\left(-\frac{1}{2}, -\frac{\sqrt{3}}{2}\right)$; the coldest is $-\frac{1}{4}^\circ$ at $\left(\frac{1}{2}, 0\right)$.

43. (a) $f_x(x,y) = 2x - 4y = 0$ and $f_y(x,y) = 2y - 4x = 0 \Rightarrow x = 0$ and $y = 0$; $f_{xx}(0,0) = 2$, $f_{yy}(0,0) = 2$, $f_{xy}(0,0) = -4 \Rightarrow f_{xx}f_{yy} - f_{xy}^2 = -12 < 0 \Rightarrow$ saddle point at $(0,0)$

(b) $f_x(x,y) = 2x - 2 = 0$ and $f_y(x,y) = 2y - 4 = 0 \Rightarrow x = 1$ and $y = 2$; $f_{xx}(1,2) = 2$, $f_{yy}(1,2) = 2$, $f_{xy}(1,2) = 0 \Rightarrow f_{xx}f_{yy} - f_{xy}^2 = 4 > 0$ and $f_{xx} > 0 \Rightarrow$ local minimum at $(1,2)$

(c) $f_x(x,y) = 9x^2 - 9 = 0$ and $f_y(x,y) = 2y + 4 = 0 \Rightarrow x = \pm 1$ and $y = -2$; $f_{xx}(1,-2) = 18x\big|_{(1,-2)} = 18$, $f_{yy}(1,-2) = 2$, $f_{xy}(1,-2) = 0 \Rightarrow f_{xx}f_{yy} - f_{xy}^2 = 36 > 0$ and $f_{xx} > 0 \Rightarrow$ local minimum at $(1,-2)$; $f_{xx}(-1,-2) = -18$, $f_{yy}(-1,-2) = 2$, $f_{xy}(-1,-2) = 0 \Rightarrow f_{xx}f_{yy} - f_{xy}^2 = -36 < 0 \Rightarrow$ saddle point at $(-1,-2)$

45. If $k = 0$, then $f(x,y) = x^2 + y^2 \Rightarrow f_x(x,y) = 2x = 0$ and $f_y(x,y) = 2y = 0 \Rightarrow x = 0$ and $y = 0 \Rightarrow (0,0)$ is the only critical point. If $k \ne 0$, $f_x(x,y) = 2x + ky = 0 \Rightarrow y = -\frac{2}{k}x$; $f_y(x,y) = kx + 2y = 0 \Rightarrow kx + 2\left(-\frac{2}{k}x\right) = 0$ $\Rightarrow kx - \frac{4x}{k} = 0 \Rightarrow \left(k - \frac{4}{k}\right)x = 0 \Rightarrow x = 0$ or $k = \pm 2 \Rightarrow y = \left(-\frac{2}{k}\right)(0) = 0$ or $y = -x$; in any case $(0,0)$ is a critical point.

47. (a) No; for example $f(x,y) = xy$ has a saddle point at $(a,b) = (0,0)$ where $f_x = f_y = 0$.

(b) If $f_{xx}(a,b)$ and $f_{yy}(a,b)$ differ in sign, then $f_{xx}(a,b)\,f_{yy}(a,b) < 0$ so $f_{xx}f_{yy} - f_{xy}^2 < 0$. The surface must therefore have a saddle point at (a,b) by the second derivative test.

49. We want the point on $z = 10 - x^2 - y^2$ where the tangent plane is parallel to the plane $x + 2y + 3z = 0$. To find a normal vector to $z = 10 - x^2 - y^2$ let $w = z + x^2 + y^2 - 10$. Then $\nabla w = 2x\mathbf{i} + 2y\mathbf{j} + \mathbf{k}$ is normal to

$z = 10 - x^2 - y^2$ at (x, y). The vector ∇w is parallel to $\mathbf{i} + 2\mathbf{j} + 3\mathbf{k}$ which is normal to the plane $x + 2y + 3z = 0$ if $6x\mathbf{i} + 6y\mathbf{j} + 3\mathbf{k} = \mathbf{i} + 2\mathbf{j} + 3\mathbf{k}$ or $x = \frac{1}{6}$ and $y = \frac{1}{3}$. Thus the point is $\left(\frac{1}{6}, \frac{1}{3}, 10 - \frac{1}{36} - \frac{1}{9}\right)$ or $\left(\frac{1}{6}, \frac{1}{3}, \frac{355}{36}\right)$.

51. No, because the domain $x \ge 0$ and $y \ge 0$ is unbounded since x and y can be as large as we please. Absolute extrema are guaranteed for continuous functions defined over closed and bounded domains in the plane. Since the domain is unbounded, the continuous function $f(x, y) = x + y$ need not have an absolute maximum (although, in this case, it does have an absolute minimum value of $f(0, 0) = 0$).

53. (a) $\frac{df}{dt} = \frac{\partial f}{\partial x}\frac{dx}{dt} + \frac{\partial f}{\partial y}\frac{dy}{dt} = \frac{dx}{dt} + \frac{dy}{dt} = -2 \sin t + 2 \cos t = 0 \Rightarrow \cos t = \sin t \Rightarrow x = y$

(i) On the semicircle $x^2 + y^2 = 4$, $y \ge 0$, we have $t = \frac{\pi}{4}$ and $x = y = \sqrt{2} \Rightarrow f(\sqrt{2}, \sqrt{2}) = 2\sqrt{2}$. At the endpoints, $f(-2, 0) = -2$ and $f(2, 0) = 2$. Therefore the absolute minimum is $f(-2, 0) = -2$ when $t = \pi$; the absolute maximum is $f(\sqrt{2}, \sqrt{2}) = 2\sqrt{2}$ when $t = \frac{\pi}{4}$.

(ii) On the quartercircle $x^2 + y^2 = 4$, $x \ge 0$ and $y \ge 0$, the endpoints give $f(0, 2) = 2$ and $f(2, 0) = 2$. Therefore the absolute minimum is $f(2, 0) = 2$ and $f(0, 2) = 2$ when $t = 0, \frac{\pi}{2}$ respectively; the absolute maximum is $f(\sqrt{2}, \sqrt{2}) = 2\sqrt{2}$ when $t = \frac{\pi}{4}$.

(b) $\frac{dg}{dt} = \frac{\partial g}{\partial x}\frac{dx}{dt} + \frac{\partial g}{\partial y}\frac{dy}{dt} = y\frac{dx}{dt} + x\frac{dy}{dt} = -4 \sin^2 t + 4 \cos^2 t = 0 \Rightarrow \cos t = \pm \sin t \Rightarrow x = \pm y$.

(i) On the semicircle $x^2 + y^2 = 4$, $y \ge 0$, we obtain $x = y = \sqrt{2}$ at $t = \frac{\pi}{4}$ and $x = -\sqrt{2}$, $y = \sqrt{2}$ at $t = \frac{3\pi}{4}$. Then $g(\sqrt{2}, \sqrt{2}) = 2$ and $g(-\sqrt{2}, \sqrt{2}) = -2$. At the endpoints, $g(-2, 0) = g(2, 0) = 0$. Therefore the absolute minimum is $g(-\sqrt{2}, \sqrt{2}) = -2$ when $t = \frac{3\pi}{4}$; the absolute maximum is $g(\sqrt{2}, \sqrt{2}) = 2$ when $t = \frac{\pi}{4}$.

(ii) On the quartercircle $x^2 + y^2 = 4$, $x \ge 0$ and $y \ge 0$, the endpoints give $g(0, 2) = 0$ and $g(2, 0) = 0$. Therefore the absolute minimum is $g(2, 0) = 0$ and $g(0, 2) = 0$ when $t = 0, \frac{\pi}{2}$ respectively; the absolute maximum is $g(\sqrt{2}, \sqrt{2}) = 2$ when $t = \frac{\pi}{4}$.

(c) $\frac{dh}{dt} = \frac{\partial h}{\partial x}\frac{dx}{dt} + \frac{\partial h}{\partial y}\frac{dy}{dt} = 4x\frac{dx}{dt} + 2y\frac{dy}{dt} = (8 \cos t)(-2 \sin t) + (4 \sin t)(2 \cos t) = -8 \cos t \sin t = 0$ $\Rightarrow t = 0, \frac{\pi}{2}, \pi$ yielding the points $(2, 0)$, $(0, 2)$, and $(-2, 0)$, respectively.

(i) On the semicircle $x^2 + y^2 = 4$, $y \ge 0$ we have $h(2, 0) = 8$, $h(0, 2) = 4$, and $h(-2, 0) = 8$. Therefore, the absolute minimum is $h(0, 2) = 4$ when $t = \frac{\pi}{2}$; the absolute maximum is $h(2, 0) = 8$ and $h(-2, 0) = 8$ when $t = 0, \pi$ respectively.

(ii) On the quartercircle $x^2 + y^2 = 4$, $x \ge 0$ and $y \ge 0$ the absolute minimum is $h(0, 2) = 4$ when $t = \frac{\pi}{2}$; the absolute maximum is $h(2, 0) = 8$ when $t = 0$.

55. $\frac{df}{dt} = \frac{\partial f}{\partial x}\frac{dx}{dt} + \frac{\partial f}{\partial y}\frac{dy}{dt} = y\frac{dx}{dt} + x\frac{dy}{dt}$

(i) $x = 2t$ and $y = t + 1 \Rightarrow \frac{df}{dt} = (t + 1)(2) + (2t)(1) = 4t + 2 = 0 \Rightarrow t = -\frac{1}{2} \Rightarrow x = -1$ and $y = \frac{1}{2}$ with $f\left(-1, \frac{1}{2}\right) = -\frac{1}{2}$. The absolute minimum is $f\left(-1, \frac{1}{2}\right) = -\frac{1}{2}$ when $t = -\frac{1}{2}$; no absolute maximum.

(ii) For the endpoints: $t=-1 \Rightarrow x=-2$ and $y=0$ with $f(-2,0)=0$; $t=0 \Rightarrow x=0$ and $y=1$ with $f(0,1)=0$. The absolute minimum is $f\left(-1,\frac{1}{2}\right)=-\frac{1}{2}$ when $t=-\frac{1}{2}$; the absolute maximum is $f(0,1)=0$ and $f(-2,0)=0$ when $t=-1$, 0 respectively.

(iii) There are no interior critical points. for the endpoints: $t=0 \Rightarrow x=0$ and $y=1$ with $f(0,1)=0$; $t=1 \Rightarrow x=2$ and $y=2$ with $f(2,2)=4$. The absolute minimum is $f(0,1)=0$ when $t=0$; the absolute maximum is $f(2,2)=4$ when $t=1$.

57. $m=\frac{(2)(-1)-3(-14)}{(2)^2-3(10)}=-\frac{20}{13}$ and

$b=\frac{1}{3}\left[-1-\left(-\frac{20}{13}\right)(2)\right]=\frac{9}{13}$

$\Rightarrow y=-\frac{20}{13}x+\frac{9}{13};\ y\Big|_{x=4}=-\frac{71}{13}$

k	x_k	y_k	x_k^2	$x_k y_k$
1	−1	2	1	−2
2	0	1	0	0
3	3	−4	9	−12
Σ	2	−1	10	−14

59. $m=\frac{(3)(5)-3(8)}{(3)^2-3(5)}=\frac{3}{2}$ and

$b=\frac{1}{3}\left[5-\frac{3}{2}(3)\right]=\frac{1}{6}$

$\Rightarrow y=\frac{3}{2}x+\frac{1}{6};\ y\Big|_{x=4}=\frac{37}{6}$

k	x_k	y_k	x_k^2	$x_k y_k$
1	0	0	0	0
2	1	2	1	2
3	2	3	4	6
Σ	3	5	5	8

61. $m=\frac{(162)(41.32)-6(1192.8)}{(162)^2-6(5004)}\approx 0.122$ and

$b=\frac{1}{6}[41.32-(0.122)(162)]\approx 3.58$

$\Rightarrow y=0.122x+3.58$

k	x_k	y_k	x_k^2	$x_k y_k$
1	12	5.27	144	63.24
2	18	5.68	324	102.24
3	24	6.25	576	150
4	30	7.21	900	216.3
5	36	8.20	1296	295.2
6	42	8.71	1764	365.82
Σ	162	41.32	5004	1192.8

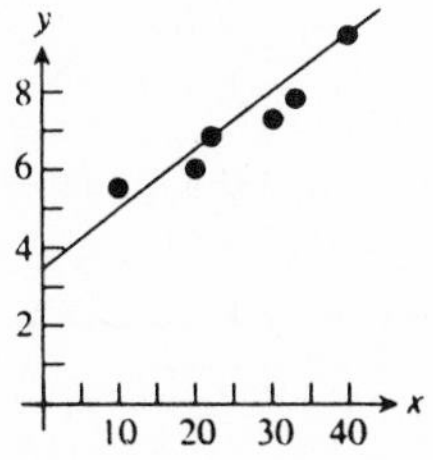

63. (b) $m = \frac{(3201)(17{,}785) - 10(5{,}710{,}292)}{(3201)^2 - 10(1{,}430{,}389)}$
≈ 0.0427 and $b = \frac{1}{10}[17{,}785 - (0.0427)(3201)]$
$\approx 1764.8 \Rightarrow y = 0.0427K + 1764.8$

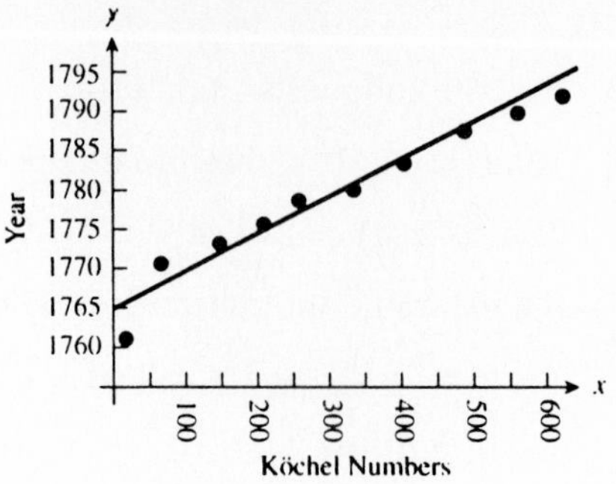

(c) $K = 364 \Rightarrow y = (0.0427)(364)$
$\Rightarrow y = (0.0427)(364) + 1764.8$
≈ 1780

k	K_k	y_k	K_k^2	$K_k y_k$
1	1	1761	1	1761
2	75	1771	5625	132,825
3	155	1772	24,025	274,660
4	219	1775	47,961	388,725
5	271	1777	73,441	481,567
6	351	1780	123,201	624,780
7	425	1783	180,625	757,775
8	503	1786	253,009	898,358
9	575	1789	330,625	1,028,675
10	626	1791	391,876	1,121,166
Σ	3201	17,785	1,430,389	5,710,292

12.9 LAGRANGE MULTIPLIERS

1. $\nabla f = y\mathbf{i} + x\mathbf{j}$ and $\nabla g = 2x\mathbf{i} + 4y\mathbf{j}$ so that $\nabla f = \lambda \nabla g \Rightarrow y\mathbf{i} + x\mathbf{j} = \lambda(2x\mathbf{i} + 4y\mathbf{j}) \Rightarrow y = 2x\lambda$ and $x = 4y\lambda$
$\Rightarrow x = 8x\lambda^2 \Rightarrow \lambda = \pm\frac{\sqrt{2}}{4}$ or $x = 0$.
CASE 1: If $x = 0$, then $y = 0$. But $(0,0)$ is not on the ellipse so $x \neq 0$.
CASE 2: $x \neq 0 \Rightarrow \lambda = \pm\frac{\sqrt{2}}{4} \Rightarrow x = \pm\sqrt{2}y \Rightarrow \left(\pm\sqrt{2}y\right)^2 + 2y^2 = 1 \Rightarrow y = \pm\frac{1}{2}$.
Therefore f takes on its extreme values at $\left(\pm\sqrt{2}, \frac{1}{2}\right)$ and $\left(\pm\sqrt{2}, -\frac{1}{2}\right)$. The extreme values of f on the ellipse are $\pm\frac{\sqrt{2}}{2}$.

3. $\nabla f = -2x\mathbf{i} - 2y\mathbf{j}$ and $\nabla g = \mathbf{i} + 3\mathbf{j}$ so that $\nabla f = \lambda \nabla g \Rightarrow -2x\mathbf{i} - 2y\mathbf{j} = \lambda(\mathbf{i} + 3\mathbf{j}) \Rightarrow x = -\frac{\lambda}{2}$ and $y = -\frac{3\lambda}{2}$
$\Rightarrow \left(-\frac{\lambda}{2}\right) + 3\left(-\frac{3\lambda}{2}\right) = 10 \Rightarrow \lambda = -2 \Rightarrow x = 1$ and $y = 3 \Rightarrow$ f takes on its extreme value at $(1,3)$ on the line.
The extreme value is $f(1,3) = 49 - 1 - 9 = 39$.

5. We optimize $f(x,y) = x^2 + y^2$, the square of the distance to the origin, subject to the constraint $g(x,y) = xy^2 - 54 = 0$. Thus $\nabla f = 2x\mathbf{i} + 2y\mathbf{j}$ and $\nabla g = y^2\mathbf{i} + 2xy\mathbf{j}$ so that $\nabla f = \lambda \nabla g \Rightarrow 2x\mathbf{i} + 2y\mathbf{j}$

$= \lambda(y^2\mathbf{i} + 2xy\mathbf{j}) \Rightarrow 2x = \lambda y^2$ and $2y = 2\lambda xy$.

CASE 1: If $y = 0$, then $x = 0$. But $(0,0)$ does not satisfy the constraint $xy^2 = 54$ so $y \neq 0$.

CASE 2: If $y \neq 0$, then $2 = 2\lambda x \Rightarrow x = \frac{1}{\lambda} \Rightarrow 2\left(\frac{1}{\lambda}\right) = \lambda y^2 \Rightarrow y^2 = \frac{2}{\lambda^2}$. Then $xy^2 = 54 \Rightarrow \left(\frac{1}{\lambda}\right)\left(\frac{2}{\lambda^2}\right) = 54$

$\Rightarrow \lambda^3 = \frac{1}{27} \Rightarrow \lambda = \pm\frac{1}{3} \Rightarrow x = \pm 3$. Since $xy^2 = 54$ we cannot have $x = -3$, so $x = 3$ and $y^2 = 18$

$\Rightarrow x = 3$ and $y = \pm 3\sqrt{2}$.

Therefore $(3, \pm 3\sqrt{2})$ are the points on the curve $xy^2 = 54$ nearest the origin (since $xy^2 = 54$ has points increasingly far away as y gets close to 0, no points are farthest away).

7. (a) $\nabla f = \mathbf{i} + \mathbf{j}$ and $\nabla g = y\mathbf{i} + x\mathbf{j}$ so that $\nabla f = \lambda \nabla g \Rightarrow \mathbf{i} + \mathbf{j} = \lambda(y\mathbf{i} + x\mathbf{j}) \Rightarrow 1 = \lambda y$ and $1 = \lambda x \Rightarrow y = \frac{1}{\lambda}$ and $x = \frac{1}{\lambda} \Rightarrow \frac{1}{\lambda^2} = 16 \Rightarrow \lambda = \pm\frac{1}{4}$. Use $\lambda = \frac{1}{4}$ since $x > 0$ and $y > 0$. Then $x = 4$ and $y = 4 \Rightarrow$ the minimum value is 8 at the point $(4,4)$. Now, $xy = 16$, $x > 0$, $y > 0$ is a branch of a hyperbola in the first quadrant with the x-and y-axes as asymptotes. The equations $x + y = c$ give a family of parallel lines with $m = -1$. As these lines move away from the origin, the number c increases. Thus the minimum value of c occurs where $x + y = c$ is tangent to the hyperbola's branch.

(b) $\nabla f = y\mathbf{i} + x\mathbf{j}$ and $\nabla g = \mathbf{i} + \mathbf{j}$ so that $\nabla f = \lambda \nabla g \Rightarrow y\mathbf{i} + x\mathbf{j} = \lambda(\mathbf{i} + \mathbf{j}) \Rightarrow y = \lambda = x \Rightarrow y + y = 16 \Rightarrow y = 8$ $\Rightarrow x = 8 \Rightarrow f(8,8) = 64$ is the maximum value. The equations $xy = c$ ($x > 0$ and $y > 0$ or $x < 0$ and $y < 0$ to get a maximum value) give a family of hyperbolas in the first and third quadrants with the x- and y-axes as asymptotes. The maximum value of c occurs where the hyperbola $xy = c$ is tangent to the line $x + y = 16$.

9. $V = \pi r^2 h \Rightarrow 16\pi = \pi r^2 h \Rightarrow 16 = r^2 h \Rightarrow g(r,h) = r^2 h - 16$; $S = 2\pi rh + 2\pi r^2 \Rightarrow \nabla S = (2\pi h + 4\pi r)\mathbf{i} + 2\pi r\mathbf{j}$ and $\nabla g = 2rh\mathbf{i} + r^2\mathbf{j}$ so that $\nabla S = \lambda \nabla g \Rightarrow (2\pi rh + 4\pi r)\mathbf{i} + 2\pi r\mathbf{j} = \lambda(2rh\mathbf{i} + r^2\mathbf{j}) \Rightarrow 2\pi rh + 4\pi r = 2rh\lambda$ and $2\pi r = \lambda r^2 \Rightarrow r = 0$ or $\lambda = \frac{2\pi}{r}$. But $r = 0$ gives no physical can, so $r \neq 0 \Rightarrow \lambda = \frac{2\pi}{r} \Rightarrow 2\pi h + 4\pi r$ $= 2rh\left(\frac{2\pi}{r}\right) \Rightarrow 2r = h \Rightarrow 16 = r^2(2r) \Rightarrow r = 2 \Rightarrow h = 4$; thus $r = 2$ cm and $h = 4$ cm give the only extreme surface area of 24π cm^2. Since $r = 4$ cm and $h = 1$ cm $\Rightarrow V = 16\pi$ cm^3 and $S = 40\pi$ cm^2, which is a larger surface area, then 24π cm^2 must be the minimum surface area.

11. $A = (2x)(2y) = 4xy$ subject to $g(x,y) = \frac{x^2}{16} + \frac{y^2}{9} - 1 = 0$; $\nabla A = 4y\mathbf{i} + 4x\mathbf{j}$ and $\nabla g = \frac{x}{8}\mathbf{i} + \frac{2y}{9}\mathbf{j}$ so that ∇A $= \lambda \nabla g \Rightarrow 4y\mathbf{i} + 4x\mathbf{j} = \lambda\left(\frac{x}{8}\mathbf{i} + \frac{2y}{9}\mathbf{j}\right) \Rightarrow 4y = \left(\frac{x}{8}\right)\lambda$ and $4x = \left(\frac{2y}{9}\right)\lambda \Rightarrow \lambda = \frac{32y}{x}$ and $4x = \left(\frac{2y}{9}\right)\left(\frac{32y}{x}\right)$

$\Rightarrow y = \pm\frac{3}{4}x \Rightarrow \frac{x^2}{16} + \frac{\left(\pm\frac{3}{4}x\right)^2}{9} = 1 \Rightarrow x^2 = 8 \Rightarrow x = \pm 2\sqrt{2}$. We use $x = 2\sqrt{2}$ since x represents distance.

Then $y = \frac{3}{4}(2\sqrt{2}) = \frac{3\sqrt{2}}{2}$, so the length is $2x = 4\sqrt{2}$ and the width is $2y = 3\sqrt{2}$.

13. $\nabla f = 2x\mathbf{i} + 2y\mathbf{j}$ and $\nabla g = (2x - 2)\mathbf{i} + (2y - 4)\mathbf{j}$ so that $\nabla f = \lambda \nabla g = 2x\mathbf{i} + 2y\mathbf{j} = \lambda[(2x - 2)\mathbf{i} + (2y - 4)\mathbf{j}]$ $\Rightarrow 2x = \lambda(2x - 2)$ and $2y = \lambda(2y - 4) \Rightarrow x = \frac{\lambda}{\lambda - 1}$ and $y = \frac{2\lambda}{\lambda - 1}$, $\lambda \neq 1 \Rightarrow y = 2x \Rightarrow x^2 - 2x + (2x)^2 - 4(2x)$ $= 0 \Rightarrow x = 0$ and $y = 0$, or $x = 2$ and $y = 4$. Therefore $f(0,0) = 0$ is the minimum value and $f(2,4) = 20$ is the maximum value. (Note that $\lambda = 1$ gives $2x = 2x - 2$ or $0 = -2$, which is impossible.)

15. $\nabla T = (8x - 4y)\mathbf{i} + (-4x + 2y)\mathbf{j}$ and $g(x,y) = x^2 + y^2 - 25 = 0 \Rightarrow \nabla g = 2x\mathbf{i} + 2y\mathbf{j}$ so that $\nabla T = \lambda \nabla g$ $\Rightarrow (8x - 4y)\mathbf{i} + (-4x + 2y)\mathbf{j} = \lambda(2x\mathbf{i} + 2y\mathbf{j}) \Rightarrow 8x - 4y = 2\lambda x$ and $-4x + 2y = 2\lambda y \Rightarrow y = \frac{-2x}{\lambda - 1}$, $\lambda \neq 1$ $\Rightarrow 8x - 4\left(\frac{-2x}{\lambda - 1}\right) = 2\lambda x \Rightarrow x = 0$, or $\lambda = 0$, or $\lambda = 5$.

CASE 1: $x = 0 \Rightarrow y = 0$; but $(0,0)$ is not on $x^2 + y^2 = 25$ so $x \neq 0$.

CASE 2: $\lambda = 0 \Rightarrow y = 2x \Rightarrow x^2 + (2x)^2 = 25 \Rightarrow x = \pm\sqrt{5}$ and $y = 2x$.

CASE 3: $\lambda = 5 \Rightarrow y = \frac{-2x}{4} = -\frac{x}{2} \Rightarrow x^2 + \left(-\frac{x}{2}\right)^2 = 25 \Rightarrow x = \pm 2\sqrt{5} \Rightarrow x = 2\sqrt{5}$ and $y = -\sqrt{5}$, or $x = -2\sqrt{5}$ and $y = \sqrt{5}$.

Therefore $T(\sqrt{5}, 2\sqrt{5}) = 0° = T(-\sqrt{5}, -2\sqrt{5})$ is the minimum value and $T(2\sqrt{5}, -\sqrt{5}) = 125°$ $= T(-2\sqrt{5}, \sqrt{5})$ is the maximum value. (Note: $\lambda = 1 \Rightarrow x = 0$ from the equation $-4x + 2y = 2\lambda y$; but we found $x \neq 0$ in CASE 1.)

17. Let $f(x,y,z) = (x-1)^2 + (y-1)^2 + (z-1)^2$ be the square of the distance from $(1,1,1)$. Then $\nabla f = 2(x-1)\mathbf{i} + 2(y-1)\mathbf{j} + 2(z-1)\mathbf{k}$ and $\nabla g = \mathbf{i} + 2\mathbf{j} + 3\mathbf{k}$ so that $\nabla f = \lambda \nabla g$ $\Rightarrow 2(x-1)\mathbf{i} + 2(y-1)\mathbf{j} + 2(z-1)\mathbf{k} = \lambda(\mathbf{i} + 2\mathbf{j} + 3\mathbf{k}) \Rightarrow 2(x-1) = \lambda,\ 2(y-1) = 2\lambda,\ 2(z-1) = 3\lambda$ $\Rightarrow 2(y-1) = 2[2(x-1)]$ and $2(z-1) = 3[2(x-1)] \Rightarrow x = \frac{y+1}{2} \Rightarrow z + 2 = 3\left(\frac{y+1}{2}\right)$ or $z = \frac{3y-1}{2}$; thus $\frac{y+1}{2} + 2y + 3\left(\frac{3y-1}{2}\right) - 13 = 0 \Rightarrow y = 2 \Rightarrow x = \frac{3}{2}$ and $z = \frac{5}{2}$. Therefore the point $\left(\frac{3}{2}, 2, \frac{5}{2}\right)$ is closest (since no point on the plane is farthest from the point $(1,1,1)$).

19. Let $f(x,y,z) = x^2 + y^2 + z^2$ be the square of the distance from the origin. Then $\nabla f = 2x\mathbf{i} + 2y\mathbf{j} + 2z\mathbf{k}$ and $\nabla g = 2x\mathbf{i} - 2y\mathbf{j} - 2z\mathbf{k}$ so that $\nabla f = \lambda \nabla g \Rightarrow 2x\mathbf{i} + 2y\mathbf{j} + 2z\mathbf{k} = \lambda(2x\mathbf{i} - 2y\mathbf{j} - 2z\mathbf{k}) \Rightarrow 2x = 2x\lambda,\ 2y = -2y\lambda$, and $2z = -2z\lambda \Rightarrow x = 0$ or $\lambda = 1$.

CASE 1: $\lambda = 1 \Rightarrow 2y = -2y \Rightarrow y = 0$; $2z = -2z \Rightarrow z = 0 \Rightarrow x^2 - 1 = 0 \Rightarrow x = \pm 1$ and $y = z = 0$.

CASE 2: $x = 0 \Rightarrow -y^2 - z^2 = 1$, which has no solution.

Therefore the points $(\pm 1, 0, 0)$ are closest to the origin $\Rightarrow$ the minimum distance from the surface to the origin is 1 (since there is no maximum distance from the surface to the origin).

21. Let $f(x,y,z) = x^2 + y^2 + z^2$ be the square of the distance to the origin. Then $\nabla f = 2x\mathbf{i} + 2y\mathbf{j} + 2z\mathbf{k}$ and $\nabla g = -y\mathbf{i} - x\mathbf{j} + 2z\mathbf{k}$ so that $\nabla f = \lambda \nabla g \Rightarrow 2x\mathbf{i} + 2y\mathbf{j} + 2z\mathbf{k} = \lambda(-y\mathbf{i} - x\mathbf{j} + 2z\mathbf{k}) \Rightarrow 2x = -y\lambda,\ 2y = -x\lambda$, and $2z = 2z\lambda \Rightarrow \lambda = 1$ or $z = 0$.

CASE 1: $\lambda = 1 \Rightarrow 2x = -y$ and $2y = -x \Rightarrow y = 0$ and $x = 0 \Rightarrow z^2 - 4 = 0 \Rightarrow z = \pm 2$ and $x = y = 0$.

CASE 2: $z = 0 \Rightarrow -xy - 4 = 0 \Rightarrow y = -\frac{4}{x}$. Then $2x = \frac{4}{x}\lambda \Rightarrow \lambda = \frac{x^2}{2}$, and $-\frac{8}{x} = -x\lambda \Rightarrow -\frac{8}{x} = -x\left(\frac{x^2}{2}\right)$ $\Rightarrow x^4 = 16 \Rightarrow x = \pm 2$. Thus, $x = 2$ and $y = -2$, or $x = -2$ and $y = 2$.

Therefore we get four points: $(2,-2,0)$, $(-2,2,0)$, $(0,0,2)$ and $(0,0,-2)$. But the points $(0,0,2)$ and $(0,0,-2)$ are closest to the origin since they are 2 units away and the others are $2\sqrt{2}$ units away.

23. $\nabla f = \mathbf{i} - 2\mathbf{j} + 5\mathbf{k}$ and $\nabla g = 2x\mathbf{i} + 2y\mathbf{j} + 2z\mathbf{k}$ so that $\nabla f = \lambda \nabla g \Rightarrow \mathbf{i} - 2\mathbf{j} + 5\mathbf{k} = \lambda(2x\mathbf{i} + 2y\mathbf{j} + 2z\mathbf{k}) \Rightarrow 1 = 2x\lambda$, $-2 = 2y\lambda$, and $5 = 2z\lambda \Rightarrow x = \frac{1}{2\lambda},\ y = -\frac{1}{\lambda} = -2x$, and $z = \frac{5}{2\lambda} = 5x \Rightarrow x^2 + (-2x)^2 + (5x)^2 = 30 \Rightarrow x = \pm 1$. Thus, $x = 1,\ y = -2,\ z = 5$ or $x = -1,\ y = 2,\ z = -5$. Therefore $f(1,-2,5) = 30$ is the maximum value and $f(-1,2,-5) = -30$ is the minimum value.

25. $f(x,y,z) = x^2+y^2+z^2$ and $g(x,y,z) = x+y+z-9 = 0 \Rightarrow \nabla f = 2x\mathbf{i}+2y\mathbf{j}+2z\mathbf{k}$ and $\nabla g = \mathbf{i}+\mathbf{j}+\mathbf{k}$ so that $\nabla f = \lambda \nabla g \Rightarrow 2x\mathbf{i}+2y\mathbf{j}+2z\mathbf{k} = \lambda(\mathbf{i}+\mathbf{j}+\mathbf{k}) \Rightarrow 2x = \lambda,\ 2y = \lambda,$ and $2z = \lambda \Rightarrow x = y = z \Rightarrow x+x+x-9 = 0$ $\Rightarrow x = 3,\ y = 3,$ and $z = 3$.

27. $V = 6xyz$ and $g(x,y,z) = x^2+y^2+z^2-1 = 0 \Rightarrow \nabla V = 6yz\mathbf{i}+6xz\mathbf{j}+6xy\mathbf{k}$ and $\nabla g = 2x\mathbf{i}+2y\mathbf{j}+2z\mathbf{k}$ so that $\nabla f = \lambda \nabla g \Rightarrow 3yz = \lambda x,\ 3xz = \lambda y,$ and $3xy = \lambda z \Rightarrow 3xyz = \lambda x^2$ and $3xyz = \lambda y^2 \Rightarrow y = \pm x \Rightarrow z = \pm x$ $\Rightarrow x^2+x^2+x^2 = 1 \Rightarrow x = \frac{1}{\sqrt{3}}$ since $x > 0 \Rightarrow$ the dimensions of the box are $\frac{2}{\sqrt{3}}$ by $\frac{2}{\sqrt{3}}$ by $\frac{2}{\sqrt{3}}$ for maximum volume. (Note that there is no minimum volume since the box could be made arbitrarily thin.)

29. $\nabla T = 16x\mathbf{i}+4z\mathbf{j}+(4y-16)\mathbf{k}$ and $\nabla g = 8x\mathbf{i}+2y\mathbf{j}+8z\mathbf{k}$ so that $\nabla T = \lambda \nabla g \Rightarrow 16x\mathbf{i}+4z\mathbf{j}+(4y-16)\mathbf{k}$ $= \lambda(8x\mathbf{i}+2y\mathbf{j}+8z\mathbf{k}) \Rightarrow 16x = 8x\lambda,\ 4z = 2y\lambda,$ and $4y-16 = 8z\lambda \Rightarrow \lambda = 2$ or $x = 0$.

CASE 1: $\lambda = 2 \Rightarrow 4z = 2y(2) \Rightarrow z = y$. Then $4z-16 = 16z \Rightarrow z = -\frac{4}{3} \Rightarrow y = -\frac{4}{3}$. Then $4x^2+\left(-\frac{4}{3}\right)^2+4\left(-\frac{4}{3}\right)^2 = 16 \Rightarrow x = \pm\frac{4}{3}$.

CASE 2: $x = 0 \Rightarrow \lambda = \frac{2z}{y} \Rightarrow 4y-16 = 8z\left(\frac{2z}{y}\right) \Rightarrow y^2-4y = 4z^2 \Rightarrow 4(0)^2+y^2+\left(y^2-4y\right)-16 = 0$ $\Rightarrow y^2-2y-8 = 0 \Rightarrow (y-4)(y+2) = 0 \Rightarrow y = 4$ or $y = -2$. Now $y = 4 \Rightarrow 4z^2 = 4^2-4(4)$ $\Rightarrow z = 0$ and $y = -2 \Rightarrow 4z^2 = (-2)^2-4(-2) \Rightarrow z = \pm\sqrt{3}$.

The temperatures are $T\left(\pm\frac{4}{3},-\frac{4}{3},-\frac{4}{3}\right) = 642\frac{2}{3}^{\circ}$, $T(0,4,0) = 600^{\circ}$, $T\left(0,-2,\sqrt{3}\right) = \left(600-24\sqrt{3}\right)^{\circ}$, and $T\left(0,-2,-\sqrt{3}\right) = \left(600+24\sqrt{3}\right)^{\circ} \approx 641.6^{\circ}$. Therefore $\left(\pm\frac{4}{3},-\frac{4}{3},-\frac{4}{3}\right)$ are the hottest points on the space probe.

31. $\nabla U = (y+2)\mathbf{i}+x\mathbf{j}$ and $\nabla g = 2\mathbf{i}+\mathbf{j}$ so that $\nabla U = \lambda \nabla g \Rightarrow (y+2)\mathbf{i}+x\mathbf{j} = \lambda(2\mathbf{i}+\mathbf{j}) \Rightarrow y+2 = 2\lambda$ and $x = \lambda \Rightarrow y+2 = 2x \Rightarrow y = 2x-2 \Rightarrow 2x+(2x-2) = 30 \Rightarrow x = 8$ and $y = 14$. Therefore $U(8,14) = \$128$ is the maximum value of U under the constraint.

33. Let $g_1(x,y,z) = 2x-y = 0$ and $g_2(x,y,z) = y+z = 0 \Rightarrow \nabla g_1 = 2\mathbf{i}-\mathbf{j},\ \nabla g_2 = \mathbf{j}+\mathbf{k}$, and $\nabla f = 2x\mathbf{i}+2\mathbf{j}-2z\mathbf{k}$ so that $\nabla f = \lambda \nabla g_1 + \mu \nabla g_2 \Rightarrow 2x\mathbf{i}+2\mathbf{j}-2z\mathbf{k} = \lambda(2\mathbf{i}-\mathbf{j})+\mu(\mathbf{j}+\mathbf{k}) \Rightarrow 2x\mathbf{i}+2\mathbf{j}-2z\mathbf{k} = 2\lambda\mathbf{i}+(\mu-\lambda)\mathbf{j}+\mu\mathbf{k}$ $\Rightarrow 2x = 2\lambda,\ 2 = \mu-\lambda,$ and $-2z = \mu \Rightarrow x = \lambda$. Then $2 = -2z-x \Rightarrow x = -2z-2$ so that $2x-y = 0$ $\Rightarrow 2(-2z-2)-y = 0 \Rightarrow -4z-4-y = 0$. This equation coupled with $y+z = 0$ implies $z = -\frac{4}{3}$ and $y = \frac{4}{3}$. Then $x = \frac{2}{3}$ so that $\left(\frac{2}{3},\frac{4}{3},-\frac{4}{3}\right)$ is the point that gives the maximum value $f\left(\frac{2}{3},\frac{4}{3},-\frac{4}{3}\right) = \left(\frac{2}{3}\right)^2+2\left(\frac{4}{3}\right)-\left(-\frac{4}{3}\right)^2$ $= \frac{4}{3}$.

35. Let $f(x,y,z) = x^2+y^2+z^2$ be the square of the distance from the origin. We want to minimize $f(x,y,z)$ subject to the constraints $g_1(x,y,z) = y+2z-12 = 0$ and $g_2(x,y,z) = x+y-6 = 0$. Thus $\nabla f = 2x\mathbf{i}+2y\mathbf{j}+2z\mathbf{k}$, $\nabla g_1 = \mathbf{j}+2\mathbf{k}$, and $\nabla g_2 = \mathbf{i}+\mathbf{j}$ so that $\nabla f = \lambda \nabla g_1 + \mu \nabla g_2 \Rightarrow 2x = \mu,\ 2y = \lambda+\mu,$ and $2z = 2\lambda$. Then $0 = y+2z-12 = \left(\frac{\lambda}{2}+\frac{\mu}{2}\right)+2\lambda-12 \Rightarrow \frac{5}{2}\lambda+\frac{1}{2}\mu = 12 \Rightarrow 5\lambda+\mu = 24;\ 0 = x+y-6 = \frac{\mu}{2}+\left(\frac{\lambda}{2}+\frac{\mu}{2}\right)-6$ $\Rightarrow \frac{1}{2}\lambda+\mu = 6 \Rightarrow \lambda+2\mu = 12$. Solving these two equations for λ and μ gives $\lambda = 4$ and $\mu = 4 \Rightarrow x = \frac{\mu}{2} = 2$, $y = \frac{\lambda+\mu}{2} = 4$, and $z = \lambda = 4$. The point $(2,4,4)$ on the line of intersection is closest to the origin. (There is no maximum distance from the origin since points on the line can be arbitrarily far away.)

37. Let $g_1(x,y,z) = z - 1 = 0$ and $g_2(x,y,z) = x^2 + y^2 + z^2 - 10 = 0 \Rightarrow \nabla g_1 = \mathbf{k}$, $\nabla g_2 = 2x\mathbf{i} + 2y\mathbf{j} + 2z\mathbf{k}$, and $\nabla f = 2xyz\mathbf{i} + x^2z\mathbf{j} + x^2y\mathbf{k}$ so that $\nabla f = \lambda \nabla g_1 + \mu \nabla g_2 \Rightarrow 2xyz\mathbf{i} + x^2z\mathbf{j} + x^2y\mathbf{k} = \lambda(\mathbf{k}) + \mu(2x\mathbf{i} + 2y\mathbf{j} + 2z\mathbf{k})$ $\Rightarrow 2xyz = 2x\mu$, $x^2z = 2y\mu$, and $x^2y = 2z\mu + \lambda \Rightarrow xyz = x\mu \Rightarrow x = 0$ or $yz = \mu \Rightarrow \mu = y$ since $z = 1$.

CASE 1: $x = 0$ and $z = 1 \Rightarrow y^2 - 9 = 0$ (from g_2) $\Rightarrow y = \pm 3$ yielding the points $(0, \pm 3, 1)$.

CASE 2: $\mu = y \Rightarrow x^2z = 2y^2 \Rightarrow x^2 = 2y^2$ (since $z = 1$) $\Rightarrow 2y^2 + y^2 + 1 - 10 = 0$ (from g_2) $\Rightarrow 3y^2 - 9 = 0$ $\Rightarrow y = \pm\sqrt{3} \Rightarrow x^2 = 2(\pm\sqrt{3})^2 \Rightarrow x = \pm\sqrt{6}$ yielding the points $(\pm\sqrt{6}, \pm\sqrt{3}, 1)$.

Now $f(0, \pm 3, 1) = 1$ and $f(\pm\sqrt{6}, \pm\sqrt{3}, 1) = 6(\pm\sqrt{3}) + 1 = 1 \pm 6\sqrt{3}$. Therefore the maximum of f is $1 + 6\sqrt{3}$ at $(\pm\sqrt{6}, \sqrt{3}, 1)$, and the minimum of f is $1 - 6\sqrt{3}$ at $(\pm\sqrt{6}, -\sqrt{3}, 1)$.

39. Let $g_1(x,y,z) = y - x = 0$ and $g_2(x,y,z) = x^2 + y^2 + z^2 - 4 = 0$. Then $\nabla f = y\mathbf{i} + x\mathbf{j} + 2z\mathbf{k}$, $\nabla g_1 = -\mathbf{i} + \mathbf{j}$, and $\nabla g_2 = 2x\mathbf{i} + 2y\mathbf{j} + 2z\mathbf{k}$ so that $\nabla f = \lambda \nabla g_1 + \mu \nabla g_2 \Rightarrow y\mathbf{i} + x\mathbf{j} + 2z\mathbf{k} = \lambda(-\mathbf{i} + \mathbf{j}) + \mu(2x\mathbf{i} + 2y\mathbf{j} + 2z\mathbf{k})$ $\Rightarrow y = -\lambda + 2x\mu$, $x = \lambda + 2y\mu$, and $2z = 2z\mu \Rightarrow z = 0$ or $\mu = 1$.

CASE 1: $z = 0 \Rightarrow x^2 + y^2 - 4 = 0 \Rightarrow 2x^2 - 4 = 0$ (since $x = y$) $\Rightarrow x = \pm\sqrt{2}$ and $y = \pm\sqrt{2}$ yielding the points $(\pm\sqrt{2}, \pm\sqrt{2}, 0)$.

CASE 2: $\mu = 1 \Rightarrow y = -\lambda + 2x$ and $x = \lambda + 2y \Rightarrow x + y = 2(x + y) \Rightarrow 2x = 2(2x)$ since $x = y \Rightarrow x = 0 \Rightarrow y = 0$ $\Rightarrow z^2 - 4 = 0 \Rightarrow z = \pm 2$ yielding the points $(0, 0, \pm 2)$.

Now, $f(0, 0, \pm 2) = 4$ and $f(\pm\sqrt{2}, \pm\sqrt{2}, 0) = 2$. Therefore the maximum value of f is 4 at $(0, 0, \pm 2)$ and the minimum value of f is 2 at $(\pm\sqrt{2}, \pm\sqrt{2}, 0)$.

41. $\nabla f = \mathbf{i} + \mathbf{j}$ and $\nabla g = y\mathbf{i} + x\mathbf{j}$ so that $\nabla f = \lambda \nabla g \Rightarrow \mathbf{i} + \mathbf{j} = \lambda(y\mathbf{i} + x\mathbf{j}) \Rightarrow 1 = y\lambda$ and $1 = x\lambda \Rightarrow y = x$ $\Rightarrow y^2 = 16 \Rightarrow y = \pm 4 \Rightarrow (4, 4)$ and $(-4, -4)$ are candidates for the location of extreme values. But as $x \to \infty$, $y \to \infty$ and $f(x,y) \to \infty$; as $x \to -\infty$, $y \to 0$ and $f(x,y) \to -\infty$. Therefore no maximum or minimum value exists subject to the constraint.

43. (a) Maximize $f(a,b,c) = a^2b^2c^2$ subject to $a^2 + b^2 + c^2 = r^2$. Thus $\nabla f = 2ab^2c^2\mathbf{i} + 2a^2bc^2\mathbf{j} + 2a^2b^2c\mathbf{k}$ and $\nabla g = 2a\mathbf{i} + 2b\mathbf{j} + 2c\mathbf{k}$ so that $\nabla f = \lambda \nabla g \Rightarrow 2ab^2c^2 = 2a\lambda$, $2a^2bc^2 = 2b\lambda$, and $2a^2b^2c = 2c\lambda$ $\Rightarrow 2a^2b^2c^2 = 2a^2\lambda = 2b^2\lambda = 2c^2\lambda \Rightarrow \lambda = 0$ or $a^2 = b^2 = c^2$.

CASE 1: $\lambda = 0 \Rightarrow a^2b^2c^2 = 0$.

CASE 2: $a^2 = b^2 = c^2 \Rightarrow f(a,b,c) = a^2a^2a^2$ and $3a^2 = r^2 \Rightarrow f(a,b,c) = \left(\frac{r^2}{3}\right)^3$ is the maximum value.

(b) The point $(\sqrt{a}, \sqrt{b}, \sqrt{c})$ is on the sphere if $a + b + c = r^2$. Moreover, by part (a), $abc = f(\sqrt{a}, \sqrt{b}, \sqrt{c})$ $\le \left(\frac{r^2}{3}\right)^3 \Rightarrow (abc)^{1/3} \le \frac{r^2}{3} = \frac{a+b+c}{3}$, as claimed.

12.10 TAYLOR'S FORMULA

1. $f(x,y) = xe^y \Rightarrow f_x = e^y$, $f_y = xe^y$, $f_{xx} = 0$, $f_{xy} = e^y$, $f_{yy} = xe^y$

$\Rightarrow f(x,y) \approx f(0,0) + xf_x(0,0) + yf_y(0,0) + \frac{1}{2}\left[x^2f_{xx}(0,0) + 2xyf_{xy}(0,0) + y^2f_{yy}(0,0)\right]$

$= 0 + x \cdot 1 + y \cdot 0 + \frac{1}{2}(x^2 \cdot 0 + 2xy \cdot 1 + y^2 \cdot 0) = x + xy$ quadratic approximation;

$f_{xxx} = 0,\ f_{xxy} = 0,\ f_{xyy} = e^y,\ f_{yyy} = xe^y$

$\Rightarrow f(x,y) \approx \text{quadratic} + \frac{1}{6}\left[x^3 f_{xxx}(0,0) + 3x^2 y f_{xxy}(0,0) + 3xy^2 f_{xyy}(0,0) + y^3 f_{yyy}(0,0)\right]$

$= x + xy + \frac{1}{6}\left(x^3 \cdot 0 + 3x^2 y \cdot 0 + 3xy^2 \cdot 1 + y^3 \cdot 0\right) = x + xy + \frac{1}{2}xy^2$, cubic approximation

3. $f(x,y) = y \sin x \Rightarrow f_x = y \cos x,\ f_y = \sin x,\ f_{xx} = -y \sin x,\ f_{xy} = \cos x,\ f_{yy} = 0$

$\Rightarrow f(x,y) \approx f(0,0) + x f_x(0,0) + y f_y(0,0) + \frac{1}{2}\left[x^2 f_{xx}(0,0) + 2xy f_{xy}(0,0) + y^2 f_{yy}(0,0)\right]$

$= 0 + x \cdot 0 + y \cdot 0 + \frac{1}{2}\left(x^2 \cdot 0 + 2xy \cdot 1 + y^2 \cdot 0\right) = xy$, quadratic approximation;

$f_{xxx} = -y \cos x,\ f_{xxy} = -\sin x,\ f_{xyy} = 0,\ f_{yyy} = 0$

$\Rightarrow f(x,y) \approx \text{quadratic} + \frac{1}{6}\left[x^3 f_{xxx}(0,0) + 3x^2 y f_{xxy}(0,0) + 3xy^2 f_{xyy}(0,0) + y^3 f_{yyy}(0,0)\right]$

$= xy + \frac{1}{6}\left(x^3 \cdot 0 + 3x^2 y \cdot 0 + 3xy^2 \cdot 0 + y^3 \cdot 0\right) = xy$, cubic approximation

5. $f(x,y) = e^x \ln(1+y) \Rightarrow f_x = e^x \ln(1+y),\ f_y = \frac{e^x}{1+y},\ f_{xx} = e^x \ln(1+y),\ f_{xy} = \frac{e^x}{1+y},\ f_{yy} = -\frac{e^x}{(1+y)^2}$

$\Rightarrow f(x,y) \approx f(0,0) + x f_x(0,0) + y f_y(0,0) + \frac{1}{2}\left[x^2 f_{xx}(0,0) + 2xy f_{xy}(0,0) + y^2 f_{yy}(0,0)\right]$

$= 0 + x \cdot 0 + y \cdot 1 + \frac{1}{2}\left[x^2 \cdot 0 + 2xy \cdot 1 + y^2 \cdot (-1)\right] = y + \frac{1}{2}\left(2xy - y^2\right)$, quadratic approximation;

$f_{xxx} = e^x \ln(1+y),\ f_{xxy} = \frac{e^x}{1+y},\ f_{xyy} = -\frac{e^x}{(1+y)^2},\ f_{yyy} = \frac{2e^x}{(1+y)^3}$

$\Rightarrow f(x,y) \approx \text{quadratic} + \frac{1}{6}\left[x^3 f_{xxx}(0,0) + 3x^2 y f_{xxy}(0,0) + 3xy^2 f_{xyy}(0,0) + y^3 f_{yyy}(0,0)\right]$

$= y + \frac{1}{2}\left(2xy - y^2\right) + \frac{1}{6}\left[x^3 \cdot 0 + 3x^2 y \cdot 1 + 3xy^2 \cdot (-1) + y^3 \cdot 2\right]$

$= y + \frac{1}{2}\left(2xy - y^2\right) + \frac{1}{6}\left(3x^2 y - 3xy^2 + 2y^3\right)$, cubic approximation

7. $f(x,y) = \sin\left(x^2 + y^2\right) \Rightarrow f_x = 2x \cos\left(x^2 + y^2\right),\ f_y = 2y \cos\left(x^2 + y^2\right),\ f_{xx} = 2 \cos\left(x^2 + y^2\right) - 4x^2 \sin\left(x^2 + y^2\right)$,
$f_{xy} = -4xy \sin\left(x^2 + y^2\right),\ f_{yy} = 2 \cos\left(x^2 + y^2\right) - 4y^2 \sin\left(x^2 + y^2\right)$

$\Rightarrow f(x,y) \approx f(0,0) + x f_x(0,0) + y f_y(0,0) + \frac{1}{2}\left[x^2 f_{xx}(0,0) + 2xy f_{xy}(0,0) + y^2 f_{yy}(0,0)\right]$

$= 0 + x \cdot 0 + y \cdot 0 + \frac{1}{2}\left(x^2 \cdot 2 + 2xy \cdot 0 + y^2 \cdot 2\right) = x^2 + y^2$, quadratic approximation;

$f_{xxx} = -12x \sin\left(x^2 + y^2\right) - 8x^3 \cos\left(x^2 + y^2\right),\ f_{xxy} = -4y \sin\left(x^2 + y^2\right) - 8x^2 y \cos\left(x^2 + y^2\right)$,
$f_{xyy} = -4x \sin\left(x^2 + y^2\right) - 8xy^2 \cos\left(x^2 + y^2\right),\ f_{yyy} = -12y \sin\left(x^2 + y^2\right) - 8y^3 \cos\left(x^2 + y^2\right)$

$\Rightarrow f(x,y) \approx \text{quadratic} + \frac{1}{6}\left[x^3 f_{xxx}(0,0) + 3x^2 y f_{xxy}(0,0) + 3xy^2 f_{xyy}(0,0) + y^3 f_{yyy}(0,0)\right]$

$= x^2 + y^2 + \frac{1}{6}\left(x^3 \cdot 0 + 3x^2 y \cdot 0 + 3xy^2 \cdot 0 + y^3 \cdot 0\right) = x^2 + y^2$, cubic approximation

9. $f(x,y) = \frac{1}{1-x-y} \Rightarrow f_x = \frac{1}{(1-x-y)^2} = f_y,\ f_{xx} = \frac{2}{(1-x-y)^3} = f_{xy} = f_{yy}$

$\Rightarrow f(x,y) \approx f(0,0) + x f_x(0,0) + y f_y(0,0) + \frac{1}{2}\left[x^2 f_{xx}(0,0) + 2xy f_{xy}(0,0) + y^2 f_{yy}(0,0)\right]$

$= 1 + x \cdot 1 + y \cdot 1 + \frac{1}{2}(x^2 \cdot 2 + 2xy \cdot 2 + y^2 \cdot 2) = 1 + (x + y) + (x^2 + 2xy + y^2)$

$= 1 + (x + y) + (x + y)^2$, quadratic approximation; $f_{xxx} = \frac{6}{(1 - x - y)^4} = f_{xxy} = f_{xyy} = f_{yyy}$

$\Rightarrow f(x, y) \approx \text{quadratic} + \frac{1}{6}\left[x^3 f_{xxx}(0,0) + 3x^2 y f_{xxy}(0,0) + 3xy^2 f_{xyy}(0,0) + y^3 f_{yyy}(0,0)\right]$

$= 1 + (x + y) + (x + y)^2 + \frac{1}{6}(x^3 \cdot 6 + 3x^2 y \cdot 6 + 3xy^2 \cdot 6 + y^3 \cdot 6)$

$= 1 + (x + y) + (x + y)^2 + (x^3 + 3x^2 y + 3xy^2 + y^3) = 1 + (x + y) + (x + y)^2 + (x + y)^3$, cubic approximation

11. $f(x, y) = \cos x \cos y \Rightarrow f_x = -\sin x \cos y$, $f_y = -\cos x \sin y$, $f_{xx} = -\cos x \cos y$, $f_{xy} = \sin x \sin y$,

$f_{yy} = -\cos x \cos y \Rightarrow f(x, y) \approx f(0,0) + x f_x(0,0) + y f_y(0,0) + \frac{1}{2}\left[x^2 f_{xx}(0,0) + 2xy f_{xy}(0,0) + y^2 f_{yy}(0,0)\right]$

$= 1 + x \cdot 0 + y \cdot 0 + \frac{1}{2}[x^2 \cdot (-1) + 2xy \cdot 0 + y^2 \cdot (-1)] = 1 - \frac{x^2}{2} - \frac{y^2}{2}$, quadratic approximation. Since all partial derivatives of f are products of sines and cosines, the absolute value of these derivatives is less than or equal to 1 $\Rightarrow E(x, y) \le \frac{1}{6}[(0.1)^3 + 3(0.1)^3 + 3(0.1)^3 + 0.1)^3] \le 0.00134$.

CHAPTER 12 PRACTICE EXERCISES

1. Domain: All points in the xy-plane

Range: $z \ge 0$

Level curves are ellipses with major axis along the y-axis and minor axis along the x-axis.

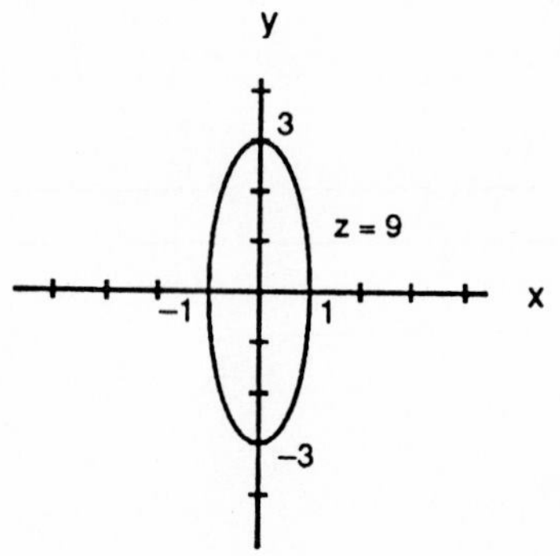

3. Domain: All (x, y) such that $x \ne 0$ and $y \ne 0$

Range: $z \ne 0$

Level curves are hyperbolas with the x- and y-axes as asymptotes.

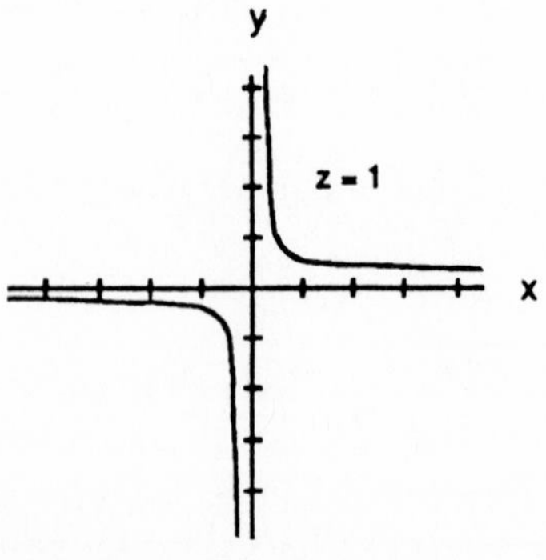

5. Domain: All (x, y, z) such that $(x, y, z) \neq (0,0,0)$

Range: All real numbers

Level surfaces are paraboloids of revolution with the z-axis as axis.

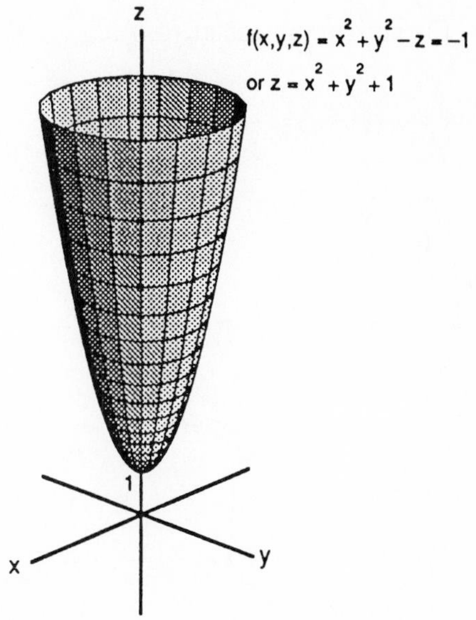

7. Domain: All (x, y, z) such that $(x, y, z) \neq (0,0,0)$

Range: Positive real numbers

Level surfaces are spheres with center $(0,0,0)$ and radius $r > 0$.

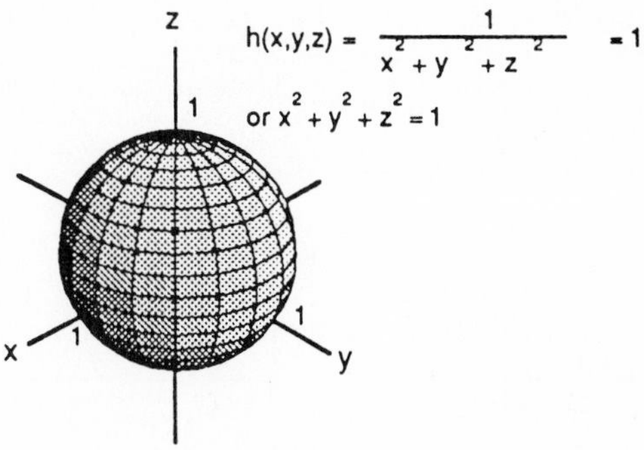

9. $\lim\limits_{(x,y)\to(\pi, \ln 2)} e^y \cos x = e^{\ln 2} \cos \pi = (2)(-1) = -2$

11. $\lim\limits_{\substack{(x,y)\to(1,1) \\ x \neq \pm y}} \frac{x - y}{x^2 - y^2} = \lim\limits_{\substack{(x,y)\to(1,1) \\ x \neq \pm y}} \frac{x - y}{(x-y)(x+y)} = \lim\limits_{(x,y)\to(1,1)} \frac{1}{x+y} = \frac{1}{1+1} = \frac{1}{2}$

13. $\lim\limits_{P\to(1,-1,e)} \ln|x+y+z| = \ln|1+(-1)+e| = \ln e = 1$

15. Let $y = kx^2$, $k \neq 1$. Then $\lim\limits_{\substack{(x,y)\to(0,0) \\ y \neq x^2}} \frac{y}{x^2 - y} = \lim\limits_{(x, kx^2)\to(0,0)} \frac{kx^2}{x^2 - kx^2} = \frac{k}{1-k^2}$ which gives different limits for different values of $k \Rightarrow$ the limit does not exist.

17. (a) Let $y = kx$. Then $\lim\limits_{(x,y)\to(0,0)} \frac{x^2 - y^2}{x^2 + y^2} = \frac{x^2 - k^2x^2}{x^2 + k^2x^2} = \frac{1-k^2}{1+k^2}$ which gives different limits for different values of $k \Rightarrow$ the limit does not exist so $f(0,0)$ cannot be defined in a way that makes f continuous at the origin.

(b) Along the x-axis, y = 0 and $\lim\limits_{(x,y)\to(0,0)} \frac{\sin(x-y)}{|x+y|} = \lim\limits_{x\to 0} \frac{\sin x}{|x|} = \begin{cases} 1, & x>0 \\ -1, & x<0 \end{cases}$, so the limit fails to exist $\Rightarrow$ f is not continuous at (0, 0).

19. $\frac{\partial g}{\partial r} = \cos\theta + \sin\theta$, $\frac{\partial g}{\partial \theta} = -r\sin\theta + r\cos\theta$

21. $\frac{\partial f}{\partial R_1} = -\frac{1}{R_1^2}$, $\frac{\partial f}{\partial R_2} = -\frac{1}{R_2^2}$, $\frac{\partial f}{\partial R_3} = -\frac{1}{R_3^2}$

23. $\frac{\partial P}{\partial n} = \frac{RT}{V}$, $\frac{\partial P}{\partial R} = \frac{nT}{V}$, $\frac{\partial P}{\partial T} = \frac{nR}{V}$, $\frac{\partial P}{\partial V} = -\frac{nRT}{V^2}$

25. $\frac{\partial g}{\partial x} = \frac{1}{y}$, $\frac{\partial g}{\partial y} = 1 - \frac{x}{y^2} \Rightarrow \frac{\partial^2 g}{\partial x^2} = 0$, $\frac{\partial^2 g}{\partial y^2} = \frac{2x}{y^3}$, $\frac{\partial^2 g}{\partial y \partial x} = \frac{\partial^2 g}{\partial x \partial y} = -\frac{1}{y^2}$

27. $\frac{\partial f}{\partial x} = 1 + y - 15x^2 + \frac{2x}{x^2+1}$, $\frac{\partial f}{\partial y} = x \Rightarrow \frac{\partial^2 f}{\partial x^2} = -30x + \frac{2-2x^2}{(x^2+1)^2}$, $\frac{\partial^2 f}{\partial y^2} = 0$, $\frac{\partial^2 f}{\partial y \partial x} = \frac{\partial^2 f}{\partial x \partial y} = 1$

29. $f\left(\frac{\pi}{4},\frac{\pi}{4}\right) = \frac{1}{2}$, $f_x\left(\frac{\pi}{4},\frac{\pi}{4}\right) = \cos x \cos y\,\big|_{(\pi/4,\pi/4)} = \frac{1}{2}$, $f_y\left(\frac{\pi}{4},\frac{\pi}{4}\right) = -\sin x \sin y\,\big|_{(\pi/4,\pi/4)} = -\frac{1}{2}$
$\Rightarrow L(x,y) = \frac{1}{2} + \frac{1}{2}\left(x - \frac{\pi}{4}\right) - \frac{1}{2}\left(y - \frac{\pi}{4}\right) = \frac{1}{2} + \frac{1}{2}x - \frac{1}{2}y$; $f_{xx}(x,y) = -\sin x \cos y$, $f_{yy}(x,y) = -\sin x \cos y$, and $f_{xy}(x,y) = -\cos x \sin y$. Thus an upper bound for E depends on the bound M used for $|f_{xx}|$, $|f_{xy}|$, and $|f_{yy}|$.
With $M = \frac{\sqrt{2}}{2}$ we have $|E(x,y)| \le \frac{1}{2}\left(\frac{\sqrt{2}}{2}\right)\left(\left|x - \frac{\pi}{4}\right| + \left|y - \frac{\pi}{4}\right|\right)^2 \le \frac{\sqrt{2}}{4}(0.2)^2 \le 0.0142$;
with M = 1, $|E(x,y)| \le \frac{1}{2}(1)\left(\left|x - \frac{\pi}{4}\right| + \left|y - \frac{\pi}{4}\right|\right)^2 = \frac{1}{2}(0.2)^2 = 0.02$.

31. $f(1,0,0) = 0$, $f_x(1,0,0) = y - 3z\,\big|_{(1,0,0)} = 0$, $f_y(1,0,0) = x + 2z\,\big|_{(1,0,0)} = 1$, $f_z(1,0,0) = 2y - 3x\,\big|_{(1,0,0)} = -3$
$\Rightarrow L(x,y,z) = 0(x-1) + (y-0) - 3(z-0) = y - 3z$; $f(1,1,0) = 1$, $f_x(1,1,0) = 1$, $f_y(1,1,0) = 1$, $f_z(1,1,0) = -1$
$\Rightarrow L(x,y,z) = 1 + (x-1) + (y-1) - 1(z-0) = x + y - z - 1$

33. $V = \pi r^2 h \Rightarrow dV = 2\pi rh\,dr + \pi r^2\,dh \Rightarrow dV\,\big|_{(1.5,\,5280)} = 2\pi(1.5)(5280)\,dr + \pi(1.5)^2\,dh = 15{,}840\pi\,dr + 2.25\pi\,dh$.
You should be more careful with the diameter since it has a greater effect on dV.

35. $dI = \frac{1}{R}dV - \frac{V}{R^2}dR \Rightarrow dI\,\Big|_{(24,100)} = \frac{1}{100}dV - \frac{24}{100^2}dR \Rightarrow dI\,\Big|_{dV=-1,\,dR=-20} = -0.01 + (480)(.0001) = 0.038$,
or increases by 0.038 amps; % change in $V = (100)\left(-\frac{1}{24}\right) \approx -4.17\%$; % change in $R = \left(-\frac{20}{100}\right)(100) = -20\%$;
$I = \frac{24}{100} = 0.24 \Rightarrow$ estimated % change in $I = \frac{dI}{I} \times 100 = \frac{0.038}{0.24} \times 100 \approx 15.83\%$

37. (a) $y = uv \Rightarrow dy = v\,du + u\,dv$; percentage change in $u \le 2\% \Rightarrow |du| \le 0.02$, and percentage change in $v \le 3\%$
$\Rightarrow |dv| \le 0.03$; $\frac{dy}{y} = \frac{v\,du + u\,dv}{uv} = \frac{du}{u} + \frac{dv}{v} \Rightarrow \left|\frac{dy}{y} \times 100\right| = \left|\frac{du}{u} \times 100 + \frac{dv}{v} \times 100\right| \le \left|\frac{du}{u} \times 100\right| + \left|\frac{dv}{v} \times 100\right|$
$\le 2\% + 3\% = 5\%$

(b) $z = u + v \Rightarrow \frac{dz}{z} = \frac{du + dv}{u + v} = \frac{du}{u + v} + \frac{dv}{u + v} \le \frac{du}{u} + \frac{dv}{v}$ (since $u > 0$, $v > 0$)

$\Rightarrow \left|\frac{dz}{z} \times 100\right| \le \left|\frac{du}{u} \times 100 + \frac{dv}{v} \times 100\right| = \left|\frac{dy}{y} \times 100\right|$

39. $\frac{\partial w}{\partial x} = y \cos(xy + \pi)$, $\frac{\partial w}{\partial y} = x \cos(xy + \pi)$, $\frac{dx}{dt} = e^t$, $\frac{dy}{dt} = \frac{1}{t+1}$

$\Rightarrow \frac{dw}{dt} = [y \cos(xy + \pi)]e^t + [x \cos(xy + \pi)]\left(\frac{1}{t+1}\right)$; $t = 0 \Rightarrow x = 1$ and $y = 0$

$\Rightarrow \left.\frac{dw}{dt}\right|_{t=0} = 0 \cdot 1 + [1 \cdot (-1)]\left(\frac{1}{0+1}\right) = -1$

41. $\frac{\partial w}{\partial x} = 2 \cos(2x - y)$, $\frac{\partial w}{\partial y} = -\cos(2x - y)$, $\frac{\partial x}{\partial r} = 1$, $\frac{\partial x}{\partial s} = \cos s$, $\frac{\partial y}{\partial r} = s$, $\frac{\partial y}{\partial s} = r$

$\Rightarrow \frac{\partial w}{\partial r} = [2 \cos(2x - y)](1) + [-\cos(2x - y)](s)$; $r = \pi$ and $s = 0 \Rightarrow x = \pi$ and $y = 0$

$\Rightarrow \left.\frac{\partial w}{\partial r}\right|_{(\pi,0)} = (2 \cos 2\pi) - (\cos 2\pi)(0) = 2$; $\frac{\partial w}{\partial s} = [2 \cos(2x - y)](\cos s) + [-\cos(2x - y)](r)$

$\Rightarrow \left.\frac{\partial w}{\partial s}\right|_{(\pi,0)} = (2 \cos 2\pi)(\cos 0) - (\cos 2\pi)(\pi) = 2 - \pi$

43. $\frac{\partial f}{\partial x} = y + z$, $\frac{\partial f}{\partial y} = x + z$, $\frac{\partial f}{\partial z} = y + x$, $\frac{dx}{dt} = -\sin t$, $\frac{dy}{dt} = \cos t$, $\frac{dz}{dt} = -2 \sin 2t$

$\Rightarrow \frac{df}{dt} = -(y + z)(\sin t) + (x + z)(\cos t) - 2(y + x)(\sin 2t)$; $t = 1 \Rightarrow x = \cos 1$, $y = \sin 1$, and $z = \cos 2$

$\Rightarrow \left.\frac{df}{dt}\right|_{t=1} = -(\sin 1 + \cos 2)(\sin 1) + (\cos 1 + \cos 2)(\cos 1) - 2(\sin 1 + \cos 1)(\sin 2)$

45. $F(x,y) = 1 - x - y^2 - \sin xy \Rightarrow F_x = -1 - y \cos xy$ and $F_y = -2y - x \cos xy \Rightarrow \frac{dy}{dx} = -\frac{F_x}{F_y} = \frac{-1 - y \cos xy}{-2y - x \cos xy}$

$= \frac{1 + y \cos xy}{-2y - x \cos xy} \Rightarrow$ at $(x,y) = (0,1)$ we have $\left.\frac{dy}{dx}\right|_{(0,1)} = \frac{1+1}{-2} = -1$

47. (a) y, z are independent with $w = x^2 e^{yz}$ and $z = x^2 - y^2 \Rightarrow \frac{\partial w}{\partial y} = \frac{\partial w}{\partial x}\frac{\partial x}{\partial y} + \frac{\partial w}{\partial y}\frac{\partial y}{\partial y} + \frac{\partial w}{\partial z}\frac{\partial z}{\partial y}$

$= \left(2xe^{yz}\right)\frac{\partial x}{\partial y} + \left(zx^2e^{yz}\right)(1) + \left(yx^2e^{yz}\right)(0)$; $z = x^2 - y^2 \Rightarrow 0 = 2x\frac{\partial x}{\partial y} - 2y \Rightarrow \frac{\partial x}{\partial y} = \frac{y}{x}$; therefore,

$\left(\frac{\partial w}{\partial y}\right)_z = \left(2xe^{yz}\right)\left(\frac{y}{x}\right) + zx^2e^{yz} = \left(2y + zx^2\right)e^{yz}$

(b) z, x are independent with $w = x^2 e^{yz}$ and $z = x^2 - y^2 \Rightarrow \frac{\partial w}{\partial z} = \frac{\partial w}{\partial x}\frac{\partial x}{\partial z} + \frac{\partial w}{\partial y}\frac{\partial y}{\partial z} + \frac{\partial w}{\partial z}\frac{\partial z}{\partial z}$

$= \left(2xe^{yz}\right)(0) + \left(zx^2e^{yz}\right)\frac{\partial y}{\partial z} + \left(yx^2e^{yz}\right)(1)$; $z = x^2 - y^2 \Rightarrow 1 = 0 - 2y\frac{\partial y}{\partial z} \Rightarrow \frac{\partial y}{\partial z} = -\frac{1}{2y}$; therefore,

$\left(\frac{\partial w}{\partial z}\right)_x = \left(zx^2e^{yz}\right)\left(-\frac{1}{2y}\right) + yx^2e^{yz} = x^2e^{yz}\left(y - \frac{z}{2y}\right)$

(c) z, y are independent with $w = x^2 e^{yz}$ and $z = x^2 - y^2 \Rightarrow \frac{\partial w}{\partial z} = \frac{\partial w}{\partial x}\frac{\partial x}{\partial z} + \frac{\partial w}{\partial y}\frac{\partial y}{\partial z} + \frac{\partial w}{\partial z}\frac{\partial z}{\partial z}$

$= \left(2xe^{yz}\right)\frac{\partial x}{\partial z} + \left(zx^2e^{yz}\right)(0) + \left(yx^2e^{yz}\right)(1)$; $z = x^2 - y^2 \Rightarrow 1 = 2x\frac{\partial x}{\partial z} - 0 \Rightarrow \frac{\partial x}{\partial z} = \frac{1}{2x}$; therefore,

$\left(\frac{\partial w}{\partial z}\right)_y = \left(2xe^{yz}\right)\left(\frac{1}{2x}\right) + yx^2e^{yz} = \left(1 + x^2y\right)e^{yz}$

49. $\nabla f = (-\sin x \cos y)\mathbf{i} - (\cos x \sin y)\mathbf{j} \Rightarrow \nabla f|_{\left(\frac{\pi}{4},\frac{\pi}{4}\right)} = -\frac{1}{2}\mathbf{i} - \frac{1}{2}\mathbf{j} \Rightarrow |\nabla f| = \sqrt{\left(-\frac{1}{2}\right)^2 + \left(-\frac{1}{2}\right)^2} = \frac{1}{\sqrt{2}} = \frac{\sqrt{2}}{2}$;

$\mathbf{u} = \frac{\nabla f}{|\nabla f|} = -\frac{\sqrt{2}}{2}\mathbf{i} - \frac{\sqrt{2}}{2}\mathbf{j} \Rightarrow$ f increases most rapidly in the direction $\mathbf{u} = -\frac{\sqrt{2}}{2}\mathbf{i} - \frac{\sqrt{2}}{2}\mathbf{j}$ and decreases most rapidly in the direction $-\mathbf{u} = \frac{\sqrt{2}}{2}\mathbf{i} + \frac{\sqrt{2}}{2}\mathbf{j}$; $(D_{\mathbf{u}}f)_{P_0} = |\nabla f| = \frac{\sqrt{2}}{2}$ and $(D_{-\mathbf{u}}f)_{P_0} = -\frac{\sqrt{2}}{2}$;

$\mathbf{u}_1 = \frac{\mathbf{A}}{|\mathbf{A}|} = \frac{3\mathbf{i} + 4\mathbf{j}}{\sqrt{3^2 + 4^2}} = \frac{3}{5}\mathbf{i} + \frac{4}{5}\mathbf{j} \Rightarrow (D_{\mathbf{u}_1}f)_{P_0} = \nabla f \cdot \mathbf{u}_1 = \left(-\frac{1}{2}\right)\left(\frac{3}{5}\right) + \left(-\frac{1}{2}\right)\left(\frac{4}{5}\right) = -\frac{7}{10}$

51. $\nabla f = \left(\frac{2}{2x+3y+6z}\right)\mathbf{i} + \left(\frac{3}{2x+3y+6z}\right)\mathbf{j} + \left(\frac{6}{2x+3y+6z}\right)\mathbf{k} \Rightarrow \nabla f|_{(-1,-1,1)} = 2\mathbf{i} + 3\mathbf{j} + 6\mathbf{k}$;

$\mathbf{u} = \frac{\nabla f}{|\nabla f|} = \frac{2\mathbf{i} + 3\mathbf{j} + 6\mathbf{k}}{\sqrt{2^2 + 3^2 + 6^2}} = \frac{2}{7}\mathbf{i} + \frac{3}{7}\mathbf{j} + \frac{6}{7}\mathbf{k} \Rightarrow$ f increases most rapidly in the direction $\mathbf{u} = \frac{2}{7}\mathbf{i} + \frac{3}{7}\mathbf{j} + \frac{6}{7}\mathbf{k}$ and decreases most rapidly in the direction $-\mathbf{u} = -\frac{2}{7}\mathbf{i} - \frac{3}{7}\mathbf{j} - \frac{6}{7}\mathbf{k}$; $(D_{\mathbf{u}}f)_{P_0} = |\nabla f| = 7$, $(D_{-\mathbf{u}}f)_{P_0} = -7$;

$\mathbf{u}_1 = \frac{\mathbf{A}}{|\mathbf{A}|} = \frac{2}{7}\mathbf{i} + \frac{3}{7}\mathbf{j} + \frac{6}{7}\mathbf{k} \Rightarrow (D_{\mathbf{u}_1}f)_{P_0} = (D_{\mathbf{u}}f)_{P_0} = 7$

53. $\mathbf{r} = (\cos 3t)\mathbf{i} + (\sin 3t)\mathbf{j} + 3t\mathbf{k} \Rightarrow \mathbf{v}(t) = (-3\sin 3t)\mathbf{i} + (3\cos 3t)\mathbf{j} + 3\mathbf{k} \Rightarrow \mathbf{v}\left(\frac{\pi}{3}\right) = -3\mathbf{j} + 3\mathbf{k}$

$\Rightarrow \mathbf{u} = -\frac{1}{\sqrt{2}}\mathbf{j} + \frac{1}{\sqrt{2}}\mathbf{k}$; $f(x,y,z) = xyz \Rightarrow \nabla f = yz\mathbf{i} + xz\mathbf{j} + xy\mathbf{k}$; $t = \frac{\pi}{3}$ yields the point on the helix $(-1, 0, \pi)$

$\Rightarrow \nabla f|_{(1,0,\pi)} = -\pi\mathbf{j} \Rightarrow \nabla f \cdot \mathbf{u} = (-\pi\mathbf{j}) \cdot \left(-\frac{1}{\sqrt{2}}\mathbf{j} + \frac{1}{\sqrt{2}}\mathbf{k}\right) = \frac{\pi}{\sqrt{2}}$

55. (a) Let $\nabla f = a\mathbf{i} + b\mathbf{j}$ at $(1,2)$. The direction toward $(2,2)$ is determined by $\mathbf{v}_1 = (2-1)\mathbf{i} + (2-2)\mathbf{j} = \mathbf{i} = \mathbf{u}$ so that $\nabla f \cdot \mathbf{u} = 2 \Rightarrow a = 2$. The direction toward $(1,1)$ is determined by $\mathbf{v}_2 + (1-1)\mathbf{i} + (1-2)\mathbf{j} = -\mathbf{j} = \mathbf{u}$ so that $\nabla f \cdot \mathbf{u} = -2 \Rightarrow -b = -2 \Rightarrow b = 2$. Therefore $\nabla f = 2\mathbf{i} + 2\mathbf{j}$.

(b) The direction toward $(4,6)$ is determined by $\mathbf{v}_3 = (4-1)\mathbf{i} + (6-2)\mathbf{j} = 3\mathbf{i} + 4\mathbf{j} \Rightarrow \mathbf{u} = \frac{3}{5}\mathbf{i} + \frac{4}{5}\mathbf{j}$

$\Rightarrow \nabla f \cdot \mathbf{u} = \frac{14}{5}$.

57. $\nabla f = 2x\mathbf{i} + \mathbf{j} + 2z\mathbf{k} \Rightarrow$

$\nabla f|_{(0,-1,-1)} = \mathbf{j} - 2\mathbf{k}$,

$\nabla f|_{(0,0,0)} = \mathbf{j}$,

$\nabla f|_{(0,-1,1)} = \mathbf{j} + 2\mathbf{k}$

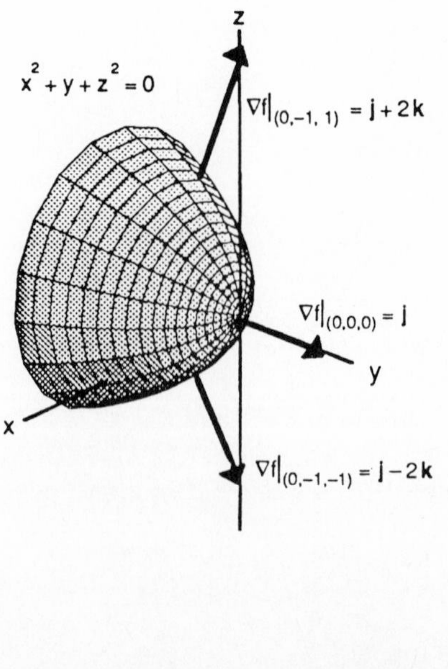

59. $\nabla f = 2x\mathbf{i} - \mathbf{j} - 5\mathbf{k} \Rightarrow \nabla f|_{(2,-1,1)} = 4\mathbf{i} - \mathbf{j} - 5\mathbf{k} \Rightarrow$ Tangent Plane: $4(x-2) - (y+1) - 5(z-1) = 0$
$\Rightarrow 4x - y - 5z = 4$; Normal Line: $x = 2 + 4t$, $y = -1 - t$, $z = 1 - 5t$

61. $\frac{\partial z}{\partial x} = \frac{2x}{x^2+y^2} \Rightarrow \left.\frac{\partial z}{\partial x}\right|_{(0,1,0)} = 0$ and $\frac{\partial z}{\partial y} = \frac{2y}{x^2+y^2} \Rightarrow \left.\frac{\partial z}{\partial y}\right|_{(0,1,0)} = 2$; thus the tangent plane is
$2(y-1) - (z-0) = 0$ or $2y - z - 2 = 0$

63. $\nabla f = (-\cos x)\mathbf{i} + \mathbf{j} \Rightarrow \nabla f|_{(\pi,1)} = \mathbf{i} + \mathbf{j} \Rightarrow$ the tangent line is $(x - \pi) + (y - 1) = 0 \Rightarrow x + y = \pi + 1$; the normal line is $y - 1 = 1(x - \pi) \Rightarrow y = x - \pi + 1$

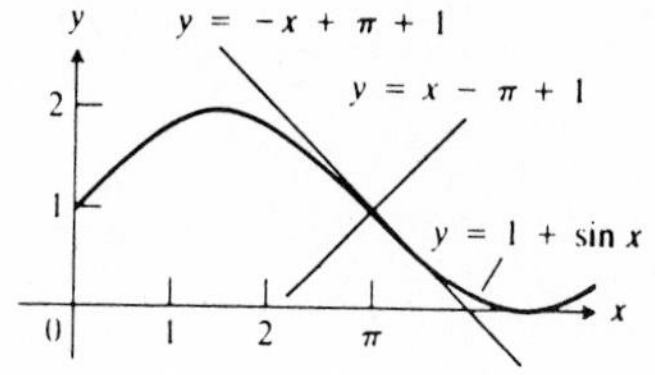

65. Let $f(x,y,z) = x^2 + 2y + 2z - 4$ and $g(x,y,z) = y - 1$. Then $\nabla f = 2x\mathbf{i} + 2\mathbf{j} + 2\mathbf{k}|_{\left(1,1,\frac{1}{2}\right)} = 2\mathbf{i} + 2\mathbf{j} + 2\mathbf{k}$

and $\nabla g = \mathbf{j} \Rightarrow \nabla f \times \nabla g = \begin{vmatrix} \mathbf{i} & \mathbf{j} & \mathbf{k} \\ 2 & 2 & 2 \\ 0 & 1 & 0 \end{vmatrix} = -2\mathbf{i} + 2\mathbf{k} \Rightarrow$ the line is $x = 1 - 2t$, $y = 1$, $z = \frac{1}{2} + 2t$

67. $f_x(x,y) = 2x - y + 2 = 0$ and $f_y(x,y) = -x + 2y + 2 = 0 \Rightarrow x = -2$ and $y = -2 \Rightarrow (-2,-2)$ is the critical point; $f_{xx}(-2,-2) = 2$, $f_{yy}(-2,-2) = 2$, $f_{xy}(-2,-2) = -1 \Rightarrow f_{xx}f_{yy} - f_{xy}^2 = 3 > 0$ and $f_{xx} > 0 \Rightarrow$ local minimum value of $f(-2,-2) = -8$

69. $f_x(x,y) = 6x^2 + 3y = 0$ and $f_y(x,y) = 3x + 6y^2 = 0 \Rightarrow y = -2x^2$ and $3x + 6(4x^4) = 0 \Rightarrow x(1 + 8x^3) = 0$ $\Rightarrow x = 0$ and $y = 0$, or $x = -\frac{1}{2}$ and $y = -\frac{1}{2} \Rightarrow$ the critical points are $(0,0)$ and $\left(-\frac{1}{2}, -\frac{1}{2}\right)$. For $(0,0)$: $f_{xx}(0,0) = 12x|_{(0,0)} = 0$, $f_{yy}(0,0) = 12y|_{(0,0)} = 0$, $f_{xy}(0,0) = 3 \Rightarrow f_{xx}f_{yy} - f_{xy}^2 = -9 < 0 \Rightarrow$ saddle point with $f(0,0) = 0$. For $\left(-\frac{1}{2}, -\frac{1}{2}\right)$: $f_{xx} = -6$, $f_{yy} = -6$, $f_{xy} = 3 \Rightarrow f_{xx}f_{yy} - f_{xy}^2 = 27 > 0$ and $f_{xx} < 0 \Rightarrow$ local maximum value of $f\left(-\frac{1}{2}, -\frac{1}{2}\right) = \frac{1}{4}$

71. $f_x(x,y) = 3x^2 + 6x = 0$ and $f_y(x,y) = 3y^2 - 6y = 0 \Rightarrow x(x+2) = 0$ and $y(y-2) = 0 \Rightarrow x = 0$ or $x = -2$ and $y = 0$ or $y = 2 \Rightarrow$ the critical points are $(0,0)$, $(0,2)$, $(-2,0)$, and $(-2,2)$. For $(0,0)$: $f_{xx}(0,0) = 6x + 6|_{(0,0)}$ $= 6$, $f_{yy}(0,0) = 6y - 6|_{(0,0)} = -6$, $f_{xy}(0,0) = 0 \Rightarrow f_{xx}f_{yy} - f_{xy}^2 = -36 < 0 \Rightarrow$ saddle point with $f(0,0) = 0$. For $(0,2)$: $f_{xx}(0,2) = 6$, $f_{yy}(0,2) = 6$, $f_{xy}(0,2) = 0 \Rightarrow f_{xx}f_{yy} - f_{xy}^2 = 36 > 0$ and $f_{xx} > 0 \Rightarrow$ local minimum value of $f(0,2) = -4$. For $(-2,0)$: $f_{xx}(-2,0) = -6$, $f_{yy}(-2,0) = -6$, $f_{xy}(-2,0) = 0 \Rightarrow f_{xx}f_{yy} - f_{xy}^2 = 36 > 0$ and $f_{xx} < 0$ $\Rightarrow$ local maximum value of $f(-2,0) = 4$. For $(-2,2)$: $f_{xx}(-2,2) = -6$, $f_{yy}(-2,2) = 6$, $f_{xy}(-2,2) = 0$ $\Rightarrow f_{xx}f_{yy} - f_{xy}^2 = -36 < 0 \Rightarrow$ saddle point with $f(-2,2) = 0$.

73. (i) On OA, $f(x,y) = f(0,y) = y^2 + 3y$ for $0 \le y \le 4$ $\Rightarrow f'(0,y) = 2y + 3 = 0 \Rightarrow y = -\frac{3}{2}$. But $\left(0, -\frac{3}{2}\right)$ is not in the region.

Endpoints: $f(0,0) = 0$ and $f(0,4) = 28$.

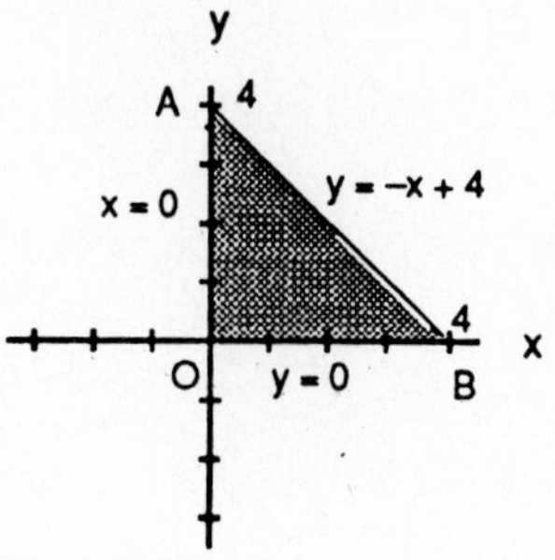

(ii) On AB, $f(x,y) = f(x, -x+4) = x^2 - 10x + 28$ for $0 \le x \le 4 \Rightarrow f'(x, -x+4) = 2x - 10 = 0$ $\Rightarrow x = 5,\ y = -1$. But $(5,-1)$ is not in the region.

Endpoints: $f(4,0) = 4$ and $f(0,4) = 28$.

(iii) On OB, $f(x,y) = f(x,0) = x^2 - 3x$ for $0 \le x \le 4 \Rightarrow f'(x,0) = 2x - 3 \Rightarrow x = \frac{3}{2}$ and $y = 0 \Rightarrow \left(\frac{3}{2}, 0\right)$ is a critical point with $f\left(\frac{3}{2}, 0\right) = -\frac{9}{4}$. Endpoints: $f(0,0) = 0$ and $f(4,0) = 4$.

(iv) For the interior of the triangular region, $f_x(x,y) = 2x + y - 3 = 0$ and $f_y(x,y) = x + 2y + 3 = 0 \Rightarrow x = 3$ and $y = -3$. But $(3,-3)$ is not in the region. Therefore the absolute maximum is 28 at $(0,4)$ and the absolute minimum is $-\frac{9}{4}$ at $\left(\frac{3}{2}, 0\right)$.

75. (i) On AB, $f(x,y) = f(-2,y) = y^2 - y - 4$ for $-2 \le y \le 2$ $\Rightarrow f'(-2,y) = 2y - 1 \Rightarrow y = \frac{1}{2}$ and $x = -2 \Rightarrow \left(-2, \frac{1}{2}\right)$ is an interior critical point in AB with $f\left(-2, \frac{1}{2}\right)$ $= -\frac{17}{4}$.

Endpoints: $f(-2,-2) = 2$ and $f(2,2) = -2$.

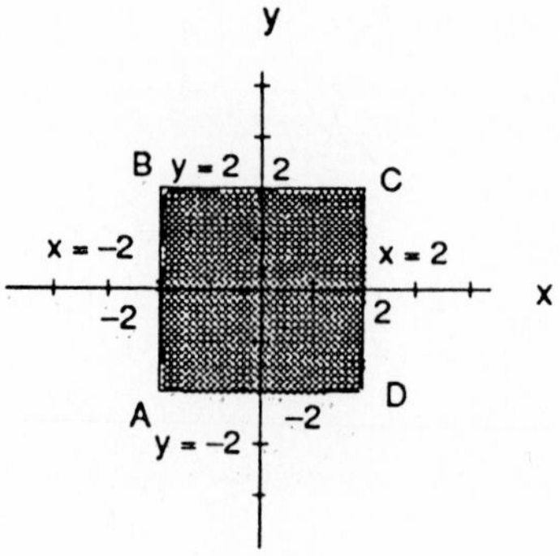

(ii) On BC, $f(x,y) = f(x,2) = -2$ for $-2 \le x \le 2$ $\Rightarrow f'(x,2) = 0 \Rightarrow$ no critical points in the interior of BC.

Endpoints: $f(-2,2) = -2$ and $f(2,2) = -2$.

(iii) On CD, $f(x,y) = f(2,y) = y^2 - 5y + 4$ for $-2 \le y \le 2$ $\Rightarrow f'(2,y) = 2y - 5 = 0 \Rightarrow y = \frac{5}{2}$ and $x = 2$. But $\left(2, \frac{5}{2}\right)$ is not in the region.

Endpoints: $f(2,-2) = 18$ and $f(2,2) = -2$.

(iv) On AD, $f(x,y) = f(x,-2) = 4x + 10$ for $-2 \le x \le 2 \Rightarrow f'(x,-2) = 4 \Rightarrow$ no critical points in the interior of AD. Endpoints: $f(-2,-2) = 2$ and $f(2,-2) = 18$.

(v) For the interior of the square, $f_x(x,y) = -y + 2 = 0$ and $f_y(x,y) = 2y - x - 3 = 0 \Rightarrow y = 2$ and $x = 1$ $\Rightarrow (1,2)$ is an interior critical point of the square with $f(1,2) = -2$. Therefore the absolute maximum is 18 at $(2,-2)$ and the absolute minimum is $-\frac{17}{4}$ at $\left(-2, \frac{1}{2}\right)$.

77. (i) On AB, $f(x,y) = f(x,x+2) = -2x+4$ for $-2 \le x \le 2$

$\Rightarrow f'(x,x+2) = -2 = 0 \Rightarrow$ no critical points in the interior of AB.

Endpoints: $f(-2,0) = 8$ and $f(2,4) = 0$.

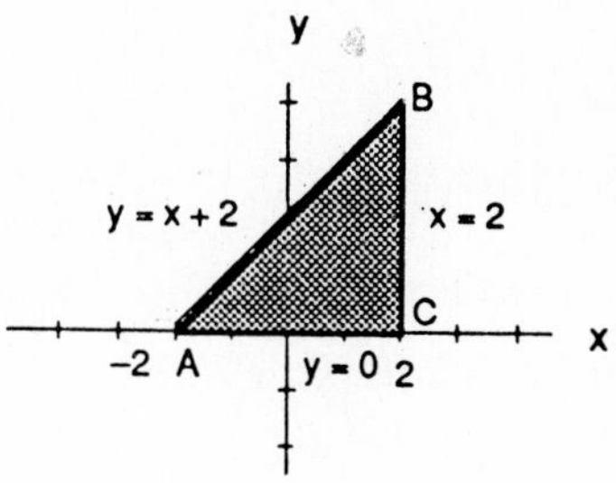

(ii) On BC, $f(x,y) = f(2,y) = -y^2 + 4y$ for $0 \le y \le 4$

$\Rightarrow f'(2,y) = -2y + 4 = 0 \Rightarrow y = 2$ and $x = 2 \Rightarrow (2,2)$ is an interior critical point of BC with $f(2,2) = 4$.

Endpoints: $f(2,0) = 0$ and $f(2,4) = 0$.

(iii) On AC, $f(x,y) = f(x,0) = x^2 - 2x$ for $-2 \le x \le 2$

$\Rightarrow f'(x,0) = 2x - 2 \Rightarrow x = 1$ and $y = 0 \Rightarrow (1,0)$ is an interior critical point of AC with $f(1,0) = -1$.

Endpoints: $f(-2,0) = 8$ and $f(2,0) = 0$.

(iv) For the interior of the triangular region, $f_x(x,y) = 2x - 2 = 0$ and $f_y(x,y) = -2y + 4 = 0 \Rightarrow x = 1$ and $y = 2 \Rightarrow (1,2)$ is an interior critical point of the region with $f(1,2) = 3$. Therefore the absolute maximum is 8 at $(-2,0)$ and the absolute minimum is -1 at $(1,0)$.

79. (i) On AB, $f(x,y) = f(-1,y) = y^3 - 3y^2 + 2$ for $-1 \le y \le 1$

$\Rightarrow f'(-1,y) = 3y^2 - 6y = 0 \Rightarrow y = 0$ and $x = -1$, or $y = 2$ and $x = -1 \Rightarrow (-1,0)$ is an interior critical point of AB with $f(-1,0) = 2$; $(-1,2)$ is outside the boundary.

Endpoints: $f(-1,-1) = -2$ and $f(-1,1) = 0$.

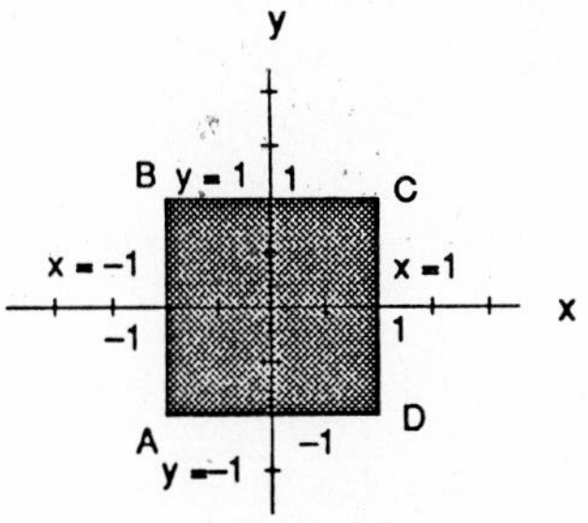

(ii) On BC, $f(x,y) = f(x,1) = x^3 + 3x^2 - 2$ for $-1 \le x \le 1$

$\Rightarrow f'(x,1) = 3x^2 + 6x = 0 \Rightarrow x = 0$ and $y = 1$, or $x = 2$ and $y = 1 \Rightarrow (0,1)$ is an interior critical point of BC with $f(0,1) = -2$; $(2,1)$ is outside the boundary. Endpoints: $f(-1,1) = 0$ and $f(1,1) = 2$.

(iii) On CD, $f(x,y) = f(1,y) = y^3 - 3y^2 + 4$ for $-1 \le y \le 1 \Rightarrow f'(1,y) = 3y^2 - 6y = 0 \Rightarrow y = 0$ and $x = 1$, or $y = 2$ and $x = 1 \Rightarrow (1,0)$ is an interior critical point of CD with $f(1,0) = 4$; $(1,2)$ is outside the boundary.

Endpoints: $f(1,1) = 2$ and $f(1,-1) = 0$.

(iv) On AD, $f(x,y) = f(x,-1) = x^3 + 3x^2 - 4$ for $-1 \le x \le 1 \Rightarrow f'(x,-1) = 3x^2 + 6x = 0 \Rightarrow x = 0$ and $y = -1$, or $x = -2$ and $y = -1 \Rightarrow (0,-1)$ is an interior point of AD with $f(0,-1) = -4$; $(-2,-1)$ is outside the boundary. Endpoints: $f(-1,-1) = -2$ and $f(1,-1) = 0$.

(v) For the interior of the square, $f_x(x,y) = 3x^2 + 6x = 0$ and $f_y(x,y) = 3y^2 - 6y = 0 \Rightarrow x = 0$ or $x = -2$, and $y = 0$ or $y = 2 \Rightarrow (0,0)$ is an interior critical point of the square region with $f(0,0) = 0$; the points $(0,2)$, $(-2,0)$, and $(-2,2)$ are outside the region. Therefore the absolute maximum is 4 at $(1,0)$ and the absolute minimum is -4 at $(0,-1)$.

81. $\nabla f = 3x^2\mathbf{i} + 2y\mathbf{j}$ and $\nabla g = 2x\mathbf{i} + 2y\mathbf{j}$ so that $\nabla f = \lambda \nabla g \Rightarrow 3x^2\mathbf{i} + 2y\mathbf{j} = \lambda(2x\mathbf{i} + 2y\mathbf{j}) \Rightarrow 3x^2 = 2x\lambda$ and $2y = 2y\lambda \Rightarrow \lambda = 1$ or $y = 0$.

CASE 1: $\lambda = 1 \Rightarrow 3x^2 = 2x \Rightarrow x = 0$ or $x = \frac{2}{3}$; $x = 0 \Rightarrow y = \pm 1$ yielding the points $(0,1)$ and $(0,-1)$; $x = \frac{2}{3}$

$\Rightarrow y = \pm\frac{\sqrt{5}}{3}$ yielding the points $\left(\frac{2}{3}, \frac{\sqrt{5}}{3}\right)$ and $\left(\frac{2}{3}, -\frac{\sqrt{5}}{3}\right)$.

CASE 2: $y = 0 \Rightarrow x^2 - 1 = 0 \Rightarrow x = \pm 1$ yielding the points $(1,0)$ and $(-1,0)$.

Evaluations give $f(0, \pm 1) = 1$, $f\left(\frac{2}{3}, \pm\frac{\sqrt{5}}{3}\right) = \frac{23}{27}$, $f(1,0) = 1$, and $f(-1,0) = -1$. Therefore the absolute maximum is 1 at $(0, \pm 1)$ and $(1,0)$, and the absolute minimum is -1 at $(-1,0)$.

83. (i) $f(x,y) = x^2 + 3y^2 + 2y$ on $x^2 + y^2 = 1 \Rightarrow \nabla f = 2x\mathbf{i} + (6y+2)\mathbf{j}$ and $\nabla g = 2x\mathbf{i} + 2y\mathbf{j}$ so that $\nabla f = \lambda \nabla g$ $\Rightarrow 2x\mathbf{i} + (6y+2)\mathbf{j} = \lambda(2x\mathbf{i} + 2y\mathbf{j}) \Rightarrow 2x = 2x\lambda$ and $6y + 2 = 2y\lambda \Rightarrow \lambda = 1$ or $x = 0$.

CASE 1: $\lambda = 1 \Rightarrow 6y + 2 = 2y \Rightarrow y = -\frac{1}{2}$ and $x = \pm\frac{\sqrt{3}}{2}$ yielding the points $\left(\pm\frac{\sqrt{3}}{2}, -\frac{1}{2}\right)$.

CASE 2: $x = 0 \Rightarrow y^2 = 1 \Rightarrow y = \pm 1$ yielding the points $(0, \pm 1)$.

Evaluations give $f\left(\pm\frac{\sqrt{3}}{2}, -\frac{1}{2}\right) = \frac{1}{2}$, $f(0,1) = 5$, and $f(0,-1) = 1$. Therefore $\frac{1}{2}$ and 5 are the extreme values on the boundary of the disk.

(ii) For the interior of the disk, $f_x(x,y) = 2x = 0$ and $f_y(x,y) = 6y + 2 = 0 \Rightarrow x = 0$ and $y = -\frac{1}{3}$ $\Rightarrow \left(0, -\frac{1}{3}\right)$ is an interior critical point with $f\left(0, -\frac{1}{3}\right) = -\frac{1}{3}$. Therefore the absolute maximum of f on the disk is 5 at $(0,1)$ and the absolute minimum of f on the disk is $-\frac{1}{3}$ at $\left(0, -\frac{1}{3}\right)$.

85. $\nabla f = \mathbf{i} - \mathbf{j} + \mathbf{k}$ and $\nabla g = 2x\mathbf{i} + 2y\mathbf{j} + 2z\mathbf{k}$ so that $\nabla f = \lambda \nabla g \Rightarrow \mathbf{i} - \mathbf{j} + \mathbf{k} = \lambda(2x\mathbf{i} + 2y\mathbf{j} + 2z\mathbf{k}) \Rightarrow 1 = 2x\lambda$, $-1 = 2y\lambda$, $1 = 2z\lambda \Rightarrow x = -y = z = \frac{1}{\lambda}$. Thus $x^2 + y^2 + z^2 = 1 \Rightarrow 3x^2 = 1 \Rightarrow x = \pm\frac{1}{\sqrt{3}}$ yielding the points $\left(\frac{1}{\sqrt{3}}, -\frac{1}{\sqrt{3}}, \frac{1}{\sqrt{3}}\right)$ and $\left(-\frac{1}{\sqrt{3}}, \frac{1}{\sqrt{3}}, -\frac{1}{\sqrt{3}}\right)$. Evaluations give the absolute maximum value of $f\left(\frac{1}{\sqrt{3}}, -\frac{1}{\sqrt{3}}, \frac{1}{\sqrt{3}}\right) = \frac{3}{\sqrt{3}} = \sqrt{3}$ and the absolute minimum value of $f\left(-\frac{1}{\sqrt{3}}, \frac{1}{\sqrt{3}}, -\frac{1}{\sqrt{3}}\right) = -\sqrt{3}$.

87. The cost is $f(x,y,z) = 2axy + 2bxz + 2cyz$ subject to the constraint $xyz = V$. Then $\nabla f = \lambda \nabla g$ $\Rightarrow 2ay + 2bz = \lambda yz$, $2ax + 2cz = \lambda xz$, and $2bx + 2cy = \lambda xy \Rightarrow 2axy + 2bxz = \lambda xyz$, $2axy + 2cyz = \lambda xyz$, and $2bxz + 2cyz = \lambda xyz \Rightarrow 2axy + 2bxz = 2axy + 2cyz \Rightarrow y = \left(\frac{b}{c}\right)x$. Also $2axy + 2bxz = 2bxz + 2cyz \Rightarrow z = \left(\frac{a}{c}\right)x$.

Then $x\left(\frac{b}{c}x\right)\left(\frac{a}{c}x\right) = V \Rightarrow x^3 = \frac{c^2V}{ab} \Rightarrow \text{width} = x = \left(\frac{c^2V}{ab}\right)^{1/3}$, $\text{Depth} = y = \left(\frac{b}{c}\right)\left(\frac{c^2V}{ab}\right)^{1/3} = \left(\frac{b^2V}{ac}\right)^{1/3}$, and

$\text{Height} = z = \left(\frac{a}{c}\right)\left(\frac{c^2V}{ab}\right)^{1/3} = \left(\frac{a^2V}{bc}\right)^{1/3}$

89. $\nabla f = (y+z)\mathbf{i} + x\mathbf{j} + x\mathbf{k}$, $\nabla g = 2x\mathbf{i} + 2y\mathbf{j}$, and $\nabla h = z\mathbf{i} + x\mathbf{k}$ so that $\nabla f = \lambda \nabla g + \mu \nabla h$ $\Rightarrow (y+z)\mathbf{i} + x\mathbf{j} + x\mathbf{k} = \lambda(2x\mathbf{i} + 2y\mathbf{j}) + \mu(z\mathbf{i} + x\mathbf{k}) \Rightarrow y + z = 2\lambda x + \mu z$, $x = 2\lambda y$, $x = \mu x \Rightarrow x = 0$ or $\mu = 1$.

CASE 1: $x = 0$ which is impossible since $xz = 1$.

CASE 2: $\mu = 1 \Rightarrow y + z = 2\lambda x + z \Rightarrow y = 2\lambda x$ and $x = 2\lambda y \Rightarrow y = (2\lambda)(2\lambda y) \Rightarrow y = 0$ or $4\lambda^2 = 1$. If $y = 0$, then $x^2 = 1 \Rightarrow x = \pm 1$ so with $xz = 1$ we obtain the points $(1,0,1)$ and $(-1,0,-1)$. If $4\lambda^2 = 1$, then $\lambda = \pm\frac{1}{2}$. For $\lambda = -\frac{1}{2}$, $y = -x$ so $x^2 + y^2 = 1 \Rightarrow x^2 = \frac{1}{2}$ $\Rightarrow x = \pm\frac{1}{\sqrt{2}}$ with $xz = 1 \Rightarrow z = \pm\sqrt{2}$, and we obtain the points $\left(\frac{1}{\sqrt{2}}, -\frac{1}{\sqrt{2}}, \sqrt{2}\right)$ and $\left(-\frac{1}{\sqrt{2}}, \frac{1}{\sqrt{2}}, -\sqrt{2}\right)$. For $\lambda = \frac{1}{2}$, $y = x \Rightarrow x^2 = \frac{1}{2} \Rightarrow x = \pm\frac{1}{\sqrt{2}}$ with $xz = 1 \Rightarrow z = \pm\sqrt{2}$, and we obtain the points $\left(\frac{1}{\sqrt{2}}, \frac{1}{\sqrt{2}}, \sqrt{2}\right)$ and $\left(-\frac{1}{\sqrt{2}}, -\frac{1}{\sqrt{2}}, -\sqrt{2}\right)$.

Evaluations give $f(1,0,1) = 1$, $f(-1,0,-1) = 1$, $f\left(\frac{1}{\sqrt{2}}, -\frac{1}{\sqrt{2}}, \sqrt{2}\right) = \frac{1}{2}$, $f\left(-\frac{1}{\sqrt{2}}, \frac{1}{\sqrt{2}}, -\sqrt{2}\right) = \frac{1}{2}$, $f\left(\frac{1}{\sqrt{2}}, \frac{1}{\sqrt{2}}, \sqrt{2}\right) = \frac{3}{2}$, and $f\left(-\frac{1}{\sqrt{2}}, -\frac{1}{\sqrt{2}}, -\sqrt{2}\right) = \frac{3}{2}$. Therefore the absolute maximum is $\frac{3}{2}$ at $\left(\frac{1}{\sqrt{2}}, \frac{1}{\sqrt{2}}, \sqrt{2}\right)$ and $\left(-\frac{1}{\sqrt{2}}, -\frac{1}{\sqrt{2}}, -\sqrt{2}\right)$, and the absolute minimum is $\frac{1}{2}$ at $\left(-\frac{1}{\sqrt{2}}, \frac{1}{\sqrt{2}}, -\sqrt{2}\right)$ and $\left(\frac{1}{\sqrt{2}}, -\frac{1}{\sqrt{2}}, \sqrt{2}\right)$.

91. Note that $x = r\cos\theta$ and $y = r\sin\theta \Rightarrow r = \sqrt{x^2 + y^2}$ and $\theta = \tan^{-1}\left(\frac{y}{x}\right)$. Thus,

$$\frac{\partial w}{\partial x} = \frac{\partial w}{\partial r}\frac{\partial r}{\partial x} + \frac{\partial w}{\partial \theta}\frac{\partial \theta}{\partial x} = \left(\frac{\partial w}{\partial r}\right)\left(\frac{x}{\sqrt{x^2+y^2}}\right) + \left(\frac{\partial w}{\partial \theta}\right)\left(\frac{-y}{x^2+y^2}\right) = (\cos\theta)\frac{\partial w}{\partial r} - \left(\frac{\sin\theta}{r}\right)\frac{\partial w}{\partial \theta};$$

$$\frac{\partial w}{\partial y} = \frac{\partial w}{\partial r}\frac{\partial r}{\partial y} + \frac{\partial w}{\partial \theta}\frac{\partial \theta}{\partial y} = \left(\frac{\partial w}{\partial r}\right)\left(\frac{y}{\sqrt{x^2+y^2}}\right) + \left(\frac{\partial w}{\partial \theta}\right)\left(\frac{x}{x^2+y^2}\right) = (\sin\theta)\frac{\partial w}{\partial r} + \left(\frac{\cos\theta}{r}\right)\frac{\partial w}{\partial \theta}$$

93. $\frac{\partial u}{\partial y} = b$ and $\frac{\partial u}{\partial x} = a \Rightarrow \frac{\partial w}{\partial x} = \frac{dw}{du}\frac{\partial u}{\partial x} = a\frac{dw}{du}$ and $\frac{\partial w}{\partial y} = \frac{dw}{du}\frac{\partial u}{\partial y} = b\frac{dw}{du} \Rightarrow \frac{1}{a}\frac{\partial w}{\partial x} = \frac{dw}{du}$ and $\frac{1}{b}\frac{\partial w}{\partial y} = \frac{dw}{du}$ $\Rightarrow \frac{1}{a}\frac{\partial w}{\partial x} = \frac{1}{b}\frac{\partial w}{\partial y} \Rightarrow b\frac{\partial w}{\partial x} = a\frac{\partial w}{\partial y}$

95. $e^u\cos v - x = 0 \Rightarrow (e^u\cos v)\frac{\partial u}{\partial x} - (e^u\sin v)\frac{\partial v}{\partial x} = 1$; $e^u\sin v - y = 0 \Rightarrow (e^u\sin v)\frac{\partial u}{\partial x} + (e^u\cos v)\frac{\partial v}{\partial x} = 0$. Solving this system yields $\frac{\partial u}{\partial x} = e^{-u}\cos v$ and $\frac{\partial v}{\partial x} = -e^{-u}\sin v$. Similarly, $e^u\cos v - x = 0$ $\Rightarrow (e^u\cos v)\frac{\partial u}{\partial y} - (e^u\sin v)\frac{\partial v}{\partial y} = 0$ and $e^u\sin v - y = 0 \Rightarrow (e^u\sin v)\frac{\partial u}{\partial y} + (e^u\cos v)\frac{\partial v}{\partial y} = 1$. Solving this second system yields $\frac{\partial u}{\partial y} = e^{-u}\sin v$ and $\frac{\partial v}{\partial y} = e^{-u}\cos v$. Therefore $\left(\frac{\partial u}{\partial x}\mathbf{i} + \frac{\partial u}{\partial y}\mathbf{j}\right)\cdot\left(\frac{\partial v}{\partial x}\mathbf{i} + \frac{\partial v}{\partial y}\mathbf{j}\right)$ $= [(e^{-u}\cos v)\mathbf{i} + (e^{-u}\sin v)\mathbf{j}]\cdot[(-e^{-u}\sin v)\mathbf{i} + (e^{-u}\cos v)\mathbf{j}] = 0 \Rightarrow$ the vectors are orthogonal $\Rightarrow$ the angle between the vectors is the constant $\frac{\pi}{2}$.

97. $(y+z)^2 + (z-x)^2 = 16 \Rightarrow \nabla f = -2(z-x)\mathbf{i} + 2(y+z)\mathbf{j} + 2(y+2z-x)\mathbf{k}$; if the normal line is parallel to the yz-plane, then x is constant $\Rightarrow \frac{\partial f}{\partial x} = 0 \Rightarrow -2(z-x) = 0 \Rightarrow z = x \Rightarrow (y+z)^2 + (z-z)^2 = 16 \Rightarrow y + z = \pm 4$. Let $x = t \Rightarrow z = t \Rightarrow y = -t \pm 4$. Therefore the points are $(t, -t \pm 4, t)$, t a real number.

99. $\nabla f = \lambda(x\mathbf{i} + y\mathbf{j} + z\mathbf{k}) \Rightarrow \frac{\partial f}{\partial x} = \lambda x \Rightarrow f(x,y,z) = \frac{1}{2}\lambda x^2 + g(y,z)$ for some function $g \Rightarrow \lambda y = \frac{\partial f}{\partial y} = \frac{\partial g}{\partial y}$ $\Rightarrow g(y,z) = \frac{1}{2}\lambda y^2 + h(z)$ for some function $h \Rightarrow yz = \frac{\partial f}{\partial z} = \frac{\partial g}{\partial z} = h'(z) \Rightarrow h(z) = \frac{1}{2}\lambda z^2 + C$ for some arbitrary constant $C \Rightarrow g(y,z) = \frac{1}{2}\lambda y^2 + \left(\frac{1}{2}\lambda z^2 + C\right) \Rightarrow f(x,y,z) = \frac{1}{2}\lambda x^2 + \frac{1}{2}\lambda y^2 + \frac{1}{2}\lambda z^2 + C \Rightarrow f(0,0,a) = \frac{1}{2}\lambda a^2 + C$ and $f(0,0,-a) = \frac{1}{2}\lambda(-a)^2 + C \Rightarrow f(0,0,a) = f(0,0,-a)$ for any constant a, as claimed.

101. Let $f(x,y,z) = xy + z - 2 \Rightarrow \nabla f = y\mathbf{i} + z\mathbf{j} + \mathbf{k}$. At $(1,1,1)$, we have $\nabla f = \mathbf{i} + \mathbf{j} + \mathbf{k} \Rightarrow$ the normal line is $x = 1 + t$, $y = 1 + t$, $z = 1 + t$, so at $t = -1 \Rightarrow x = 0$, $y = 0$, $z = 0$ and the normal line passes through the origin.

CHAPTER 12 ADDITIONAL EXERCISES–THEORY, EXAMPLES, APPLICATIONS

1. By definition, $f_{xy}(0,0) = \lim_{h\to 0} \frac{f_x(0,h) - f_x(0,0)}{h}$ so we need to calculate the first partial derivatives in the numerator. For $(x,y) \neq (0,0)$ we calculate $f_x(x,y)$ by applying the differentiation rules to the formula for $f(x,y)$: $f_x(x,y) = \frac{x^2y - y^3}{x^2 + y^2} + (xy)\frac{(x^2+y^2)(2x) - (x^2-y^2)(2x)}{(x^2+y^2)^2} = \frac{x^2y - y^3}{x^2+y^2} + \frac{4x^2y^3}{(x^2+y^2)^2} \Rightarrow f_x(0,h) = -\frac{h^3}{h^2} = -h$. For $(x,y) = (0,0)$ we apply the definition: $f_x(0,0) = \lim_{h\to 0} \frac{f(h,0) - f(0,0)}{h} = \lim_{h\to 0} \frac{0-0}{h} = 0$. Then by definition $f_{xy}(0,0) = \lim_{h\to 0} \frac{-h-0}{h} = -1$. Similarly, $f_{yx}(0,0) = \lim_{h\to 0} \frac{f_y(h,0) - f_y(0,0)}{h}$, so for $(x,y) \neq (0,0)$ we have $f_y(x,y) = \frac{x^3 - xy^2}{x^2+y^2} - \frac{4x^3y^2}{(x^2+y^2)^2} \Rightarrow f_y(h,0) = \frac{h^3}{h^2} = h$; for $(x,y) = (0,0)$ we obtain $f_y(0,0) = \lim_{h\to 0} \frac{f(0,h) - f(0,0)}{h}$ $= \lim_{h\to 0} \frac{0-0}{h} = 0$. Then by definition $f_{yx}(0,0) = \lim_{h\to 0} \frac{h-0}{h} = 1$. Note that $f_{xy}(0,0) \neq f_{yx}(0,0)$ in this case.

3. Substitution of $u + u(x)$ and $v = v(x)$ in $g(u,v)$ gives $g(u(x),v(x))$ which is a function of the independent variable x. Then, $g(u,v) = \int_u^v f(t)\,dt \Rightarrow \frac{dg}{dx} = \frac{\partial g}{\partial u}\frac{du}{dx} + \frac{\partial g}{\partial v}\frac{dv}{dx} = \left(\frac{\partial}{\partial u}\int_u^v f(t)\,dt\right)\frac{du}{dx} + \left(\frac{\partial}{\partial v}\int_u^v f(t)\,dt\right)\frac{dv}{dx}$

$$= \left(-\frac{\partial}{\partial u}\int_v^u f(t)\,dt\right)\frac{du}{dx} + \left(\frac{\partial}{\partial v}\int_u^v f(t)\,dt\right)\frac{dv}{dx} = -f(u(x))\frac{du}{dx} + f(v(x))\frac{dv}{dx} = f(v(x))\frac{dv}{dx} - f(u(x))\frac{du}{dx}$$

5. (a) Let $u = tx$, $v = ty$, and $w = f(u,v) = f(u(t,x), v(t,y)) = f(tx,ty) = t^n f(x,y)$, where t, x, and y are independent variables. Then $nt^{n-1}f(x,y) = \frac{\partial w}{\partial t} = \frac{\partial w}{\partial u}\frac{\partial u}{\partial t} + \frac{\partial w}{\partial v}\frac{\partial v}{\partial t} = x\frac{\partial w}{\partial u} + y\frac{\partial w}{\partial v}$. Now, $\frac{\partial w}{\partial x} = \frac{\partial w}{\partial u}\frac{\partial u}{\partial x} + \frac{\partial w}{\partial v}\frac{\partial v}{\partial x} = \left(\frac{\partial w}{\partial u}\right)(t) + \left(\frac{\partial w}{\partial v}\right)(0) = t\frac{\partial w}{\partial u} \Rightarrow \frac{\partial w}{\partial u} = \left(\frac{1}{t}\right)\left(\frac{\partial w}{\partial x}\right)$. Likewise, $\frac{\partial w}{\partial y} = \frac{\partial w}{\partial u}\frac{\partial u}{\partial y} + \frac{\partial w}{\partial v}\frac{\partial v}{\partial y} = \left(\frac{\partial w}{\partial u}\right)(0) + \left(\frac{\partial w}{\partial v}\right)(t) \Rightarrow \frac{\partial w}{\partial v} = \left(\frac{1}{t}\right)\left(\frac{\partial w}{\partial y}\right)$. Therefore, $nt^{n-1}f(x,y) = x\frac{\partial w}{\partial u} + y\frac{\partial w}{\partial v} = \left(\frac{x}{t}\right)\left(\frac{\partial w}{\partial x}\right) + \left(\frac{y}{t}\right)\left(\frac{\partial w}{\partial y}\right) = u\frac{\partial w}{\partial x} + v\frac{\partial w}{\partial y}$. When $t = 1$, $u = x$, $v = y$, and $w = f(x,y) \Rightarrow \frac{\partial w}{\partial x} = \frac{\partial f}{\partial x}$ and $\frac{\partial w}{\partial y} = \frac{\partial f}{\partial x} \Rightarrow nf(x,y) = x\frac{\partial f}{\partial x} + y\frac{\partial f}{\partial y}$, as claimed.

(b) From part (a), $nt^{n-1}f(x,y) = x\frac{\partial w}{\partial u} + y\frac{\partial w}{\partial v}$. Differentiating with respect to t again we obtain

$$n(n-1)t^{n-2}f(x,y) = x\frac{\partial^2 w}{\partial u^2}\frac{\partial u}{\partial t} + x\frac{\partial^2 w}{\partial v \partial w}\frac{\partial v}{\partial t} + y\frac{\partial^2 w}{\partial u \partial v}\frac{\partial u}{\partial t} + y\frac{\partial^2 w}{\partial v^2}\frac{\partial v}{\partial t} = x^2\frac{\partial^2 w}{\partial u^2} + 2xy\frac{\partial^2 w}{\partial u \partial v} + y^2\frac{\partial^2 w}{\partial v^2}.$$

Also from part (a), $\frac{\partial^2 w}{\partial x^2} = \frac{\partial}{\partial x}\left(\frac{\partial w}{\partial x}\right) = \frac{\partial}{\partial x}\left(t\frac{\partial w}{\partial u}\right) = t\frac{\partial^2 w}{\partial u^2}\frac{\partial u}{\partial x} + t\frac{\partial^2 w}{\partial v \partial u}\frac{\partial v}{\partial x} = t^2\frac{\partial^2 w}{\partial u^2}$, $\frac{\partial^2 w}{\partial y^2} = \frac{\partial}{\partial y}\left(\frac{\partial w}{\partial y}\right)$

$= \frac{\partial}{\partial y}\left(t\frac{\partial w}{\partial v}\right) = t\frac{\partial^2 w}{\partial u \partial v}\frac{\partial u}{\partial y} + t\frac{\partial^2 w}{\partial v^2}\frac{\partial v}{\partial y} = t^2\frac{\partial^2 w}{\partial v^2}$, and $\frac{\partial^2 w}{\partial y \partial x} = \frac{\partial}{\partial y}\left(\frac{\partial w}{\partial x}\right) = \frac{\partial}{\partial y}\left(t\frac{\partial w}{\partial u}\right) = t\frac{\partial^2 w}{\partial u^2}\frac{\partial u}{\partial y} + t\frac{\partial^2 w}{\partial v \partial u}\frac{\partial v}{\partial y}$

$= t^2\frac{\partial^2 w}{\partial v \partial u} \Rightarrow \left(\frac{1}{t^2}\right)\frac{\partial^2 w}{\partial x^2} = \frac{\partial^2 w}{\partial u^2}$, $\left(\frac{1}{t^2}\right)\frac{\partial^2 w}{\partial y^2} = \frac{\partial^2 w}{\partial v^2}$, and $\left(\frac{1}{t^2}\right)\frac{\partial^2 w}{\partial y \partial x} = \frac{\partial^2 w}{\partial v \partial u}$

$\Rightarrow n(n-1)t^{n-2}f(x,y) = \left(\frac{x^2}{t^2}\right)\left(\frac{\partial^2 w}{\partial x^2}\right) + \left(\frac{2xy}{t^2}\right)\left(\frac{\partial^2 w}{\partial y \partial x}\right) + \left(\frac{y^2}{t^2}\right)\left(\frac{\partial^2 w}{\partial y^2}\right)$ for $t \neq 0$. When $t = 1$, $w = f(x,y)$ and

we have $n(n-1)f(x,y) = x^2\left(\frac{\partial^2 f}{\partial x^2}\right) + 2xy\left(\frac{\partial^2 f}{\partial x \partial y}\right) + y^2\left(\frac{\partial^2 f}{\partial y^2}\right)$ as claimed.

7. (a) $\mathbf{r} = x\mathbf{i} + y\mathbf{j} + z\mathbf{k} \Rightarrow r = |\mathbf{r}| = \sqrt{x^2+y^2+z^2}$ and $\nabla r = \frac{x}{\sqrt{x^2+y^2+z^2}}\mathbf{i} + \frac{y}{\sqrt{x^2+y^2+z^2}}\mathbf{j} + \frac{z}{\sqrt{x^2+y^2+z^2}}\mathbf{k}$

$= \frac{\mathbf{r}}{r}$

(b) $r^n = \left(\sqrt{x^2+y^2+z^2}\right)^n$

$\Rightarrow \nabla(r^n) = nx\left(x^2+y^2+z^2\right)^{(n/2)-1}\mathbf{i} + ny\left(x^2+y^2+z^2\right)^{(n/2)-1}\mathbf{j} + nz\left(x^2+y^2+z^2\right)^{(n/2)-1}\mathbf{k}$

$= nr^{n-2}\mathbf{r}$

(c) Let $n = 2$ in part (b). Then $\frac{1}{2}\nabla\left(r^2\right) = \mathbf{r} \Rightarrow \nabla\left(\frac{1}{2}r^2\right) = \mathbf{r} \Rightarrow \frac{r^2}{2} = \frac{1}{2}\left(x^2+y^2+z^2\right)$ is the function.

(d) $d\mathbf{r} = dx\mathbf{i} + dy\mathbf{j} + dz\mathbf{k} \Rightarrow \mathbf{r} \cdot d\mathbf{r} = x\,dx + y\,dy + z\,dz$, and $dr = r_x\,dx + r_y\,dy + r_z\,dz = \frac{x}{r}\,dx + \frac{y}{r}\,dy + \frac{z}{r}\,dz$

$\Rightarrow r\,dr = x\,dx + y\,dy + z\,dz = \mathbf{r} \cdot d\mathbf{r}$

(e) $\mathbf{A} = a\mathbf{i} + b\mathbf{j} + c\mathbf{k} \Rightarrow \mathbf{A} \cdot \mathbf{r} = ax + by + cz \Rightarrow \nabla(\mathbf{A} \cdot \mathbf{r}) = a\mathbf{i} + b\mathbf{j} + c\mathbf{k} = \mathbf{A}$

9. $f(x,y,z) = xz^2 - yz + \cos xy - 1 \Rightarrow \nabla f = \left(z^2 - y \sin xy\right)\mathbf{i} + (-z - x \sin xy)\mathbf{j} + (2xz - y)\mathbf{k} \Rightarrow \nabla f(0,0,1) = \mathbf{i} - \mathbf{j}$ $\Rightarrow$ the tangent plane is $x - y = 0$; $\mathbf{r} = (\ln t)\mathbf{i} + (t \ln t)\mathbf{j} + t\mathbf{k} \Rightarrow \mathbf{r}' = \left(\frac{1}{t}\right)\mathbf{i} + (\ln t + 1)\mathbf{j} + \mathbf{k}$; $x = y = 0$, $z = 1$ $\Rightarrow t = 1 \Rightarrow \mathbf{r}'(1) = \mathbf{i} + \mathbf{j} + \mathbf{k}$. Since $(\mathbf{i}+\mathbf{j}+\mathbf{k}) \cdot (\mathbf{i}-\mathbf{j}) = \mathbf{r}'(1) \cdot \nabla f = 0$, $\mathbf{r}$ is parallel to the plane, and $\mathbf{r}(1) = 0\mathbf{i} + 0\mathbf{j} + \mathbf{k} \Rightarrow \mathbf{r}$ is contained in the plane.

11. $\frac{\partial w}{\partial r} = \frac{\partial w}{\partial x}\frac{\partial x}{\partial r} + \frac{\partial w}{\partial y}\frac{\partial y}{\partial r} = \frac{\partial w}{\partial x}(\cos\theta) + \frac{\partial w}{\partial y}(\sin\theta)$; $\frac{\partial w}{\partial \theta} = \frac{\partial w}{\partial x}\frac{\partial x}{\partial \theta} + \frac{\partial w}{\partial y}\frac{\partial y}{\partial \theta} = \frac{\partial w}{\partial x}(-r\sin\theta) + \frac{\partial w}{\partial y}(r\cos\theta)$;

$\frac{\partial w}{\partial r}\mathbf{u}_r = \left[\frac{\partial w}{\partial x}(\cos\theta) + \frac{\partial w}{\partial y}(\sin\theta)\right]\left[(\cos\theta)\mathbf{i} + (\sin\theta)\mathbf{j}\right]$ and

$\frac{1}{r}\frac{\partial w}{\partial \theta}\mathbf{u}_\theta = \left[\frac{\partial w}{\partial x}(-\sin\theta) + \frac{\partial w}{\partial y}(\cos\theta)\right]\left[(-\sin\theta)\mathbf{i} + (\cos\theta)\mathbf{j}\right] \Rightarrow \frac{\partial w}{\partial r}\mathbf{u}_r + \frac{1}{r}\frac{\partial w}{\partial \theta}\mathbf{u}_\theta + \frac{\partial w}{\partial z}\mathbf{k}$

$= \frac{\partial w}{\partial x}\left(\cos^2\theta + \sin^2\theta\right)\mathbf{i} + \frac{\partial w}{\partial y}\left(\cos^2\theta + \sin^2\theta\right)\mathbf{j} + \frac{\partial w}{\partial z}\mathbf{k} = \frac{\partial w}{\partial x}\mathbf{i} + \frac{\partial w}{\partial y}\mathbf{j} + \frac{\partial w}{\partial z}\mathbf{k} = \nabla w$

13. $\frac{\partial z}{\partial x} = 3x^2 - 9y = 0$ and $\frac{\partial z}{\partial y} = 3y^2 - 9x = 0 \Rightarrow y = \frac{1}{3}x^2$ and $3\left(\frac{1}{3}x^2\right)^2 - 9x = 0 \Rightarrow \frac{1}{3}x^4 - 9x = 0$ $\Rightarrow x(x^3 - 27) = 0 \Rightarrow x = 0$ or $x = 3$. Now $x = 0 \Rightarrow y = 0$ or $(0,0)$ and $x = 3 \Rightarrow y = 3$ or $(3,3)$. Next $\frac{\partial^2 z}{\partial x^2} = 6x$, $\frac{\partial^2 z}{\partial y^2} = 6y$, and $\frac{\partial^2 z}{\partial x \partial y} = -9$. For $(0;0)$, $\frac{\partial^2 z}{\partial x^2}\frac{\partial^2 z}{\partial y^2} - \left(\frac{\partial^2 z}{\partial x \partial y}\right)^2 = -81 \Rightarrow$ no extremum (a saddle point), and for $(3,3)$, $\frac{\partial^2 z}{\partial x^2}\frac{\partial^2 z}{\partial y^2} - \left(\frac{\partial^2 z}{\partial x \partial y}\right)^2 = 243 > 0$ and $\frac{\partial^2 z}{\partial x^2} = 18 > 0 \Rightarrow$ a local minimum.

15. Let $f(x,y,z) = \frac{x^2}{a^2} + \frac{y^2}{b^2} + \frac{z^2}{c^2} - 1 \Rightarrow \nabla f = \frac{2x}{a^2}\mathbf{i} + \frac{2y}{b^2}\mathbf{j} + \frac{2z}{c^2}\mathbf{k} \Rightarrow$ an equation of the plane tangent at the point $P_0(x_0, y_0, y_0)$ is $\left(\frac{2x_0}{a^2}\right)x + \left(\frac{2y_0}{b^2}\right)y + \left(\frac{2z_0}{c^2}\right)z = \frac{2x_0^2}{a^2} + \frac{2y_0^2}{b^2} + \frac{2z_0^2}{c^2} = 2$ or $\left(\frac{x_0}{a^2}\right)x + \left(\frac{y_0}{b^2}\right)y + \left(\frac{z_0}{c^2}\right)z = 1$.

The intercepts of the plane are $\left(\frac{a^2}{x_0}, 0, 0\right)$, $\left(0, \frac{b^2}{y_0}, 0\right)$ and $\left(0, 0, \frac{c^2}{z_0}\right)$. The volume of the tetrahedron formed by the plane and the coordinate planes is $V = \left(\frac{1}{3}\right)\left(\frac{1}{2}\right)\left(\frac{a^2}{x_0}\right)\left(\frac{b^2}{y_0}\right)\left(\frac{c^2}{z_0}\right) \Rightarrow$ we need to maximize $V(x,y,z) = \frac{(abc)^2}{2}(xyz)^{-1}$ subject to the constraint $f(x,y,z) = \frac{x^2}{a^2} + \frac{y^2}{b^2} + \frac{z^2}{c^2} = 1$. Thus,

$\left[-\frac{(abc)^2}{6}\right]\left(\frac{1}{x^2yz}\right) = \frac{2x}{a^2}\lambda$, $\left[-\frac{(abc)^2}{6}\right]\left(\frac{1}{xy^2z}\right) = \frac{2y}{b^2}\lambda$, and $\left[-\frac{(abc)^2}{6}\right]\left(\frac{1}{xyz^2}\right) = \frac{2z}{c^2}\lambda$. Multiply the first equation by a^2yz, the second by b^2xz, and the third by c^2xy. Then equate the first and second $\Rightarrow a^2y^2 = b^2x^2$ $\Rightarrow y = \frac{b}{a}x$, $x > 0$; equate the first and third $\Rightarrow a^2z^2 = c^2x^2 \Rightarrow z = \frac{c}{a}x$, $x > 0$; substitute into $f(x,y,z) = 0$ $\Rightarrow x = \sqrt{\frac{a}{3}} \Rightarrow y = \sqrt{\frac{b}{3}} \Rightarrow z = \sqrt{\frac{c}{3}} \Rightarrow V = \frac{\sqrt{3}}{2}abc$.

17. Let (x_0, y_0) be any point in R. We must show $\lim_{(x,y)\to(x_0,y_0)} f(x,y) = f(x_0,y_0)$ or, equivalently that $\lim_{(h,k)\to(0,0)} |f(x_0+h, y_0+k) - f(x_0,y_0)| = 0$. Consider $f(x_0+h, y_0+k) - f(x_0,y_0)$ $= [f(x_0+h, y_0+k) - f(x_0, y_0+k)] + [f(x_0, y_0+k) - f(x_0,y_0)]$. Let $F(x) = f(x, y_0+k)$ and apply the Mean Value Theorem: there exists ξ with $x_0 < \xi < x_0 + h$ such that $F'(\xi)h = F(x_0+h) - F(x_0) \Rightarrow hf_x(\xi, y_0+k)$ $= f(x_0+h, y_0+k) - f(x_0, y_0+k)$. Similarly, $k\,f_y(x_0, \eta) = f(x_0, y_0+k) - f(x_0, y_0)$ for some η with $y_0 < \eta < y_0 + k$. Then $|f(x_0+h, y_0+k) - f(x_0,y_0)| \le |hf_x(\xi, y_0+k)| + |kf_y(x_0,\eta)|$. If M, N are positive real numbers such that $|f_x| \le M$ and $|f_y| \le N$ for all (x,y) in the xy-plane, then $|f(x_0+h, y_0+k) - f(x_0,y_0)|$ $\le M|h| + N|k|$. As $(h,k) \to 0$, $|f(x_0+h, y_0+k) - f(x_0,y_0)| \to 0 \Rightarrow \lim_{(h,k)\to(0,0)} |f(x_0+h, y_0+k) - f(x_0,y_0)|$ $= 0 \Rightarrow f$ is continuous at (x_0, y_0).

19. $\frac{\partial f}{\partial x} = 0 \Rightarrow f(x,y) = h(y)$ is a function of y only. Also, $\frac{\partial g}{\partial y} = \frac{\partial f}{\partial x} = 0 \Rightarrow g(x,y) = k(x)$ is a function of x only. Moreover, $\frac{\partial f}{\partial y} = \frac{\partial g}{\partial x} \Rightarrow h'(y) = k'(x)$ for all x and y. This can happen only if $h'(y) = k'(x) = c$ is a constant. Integration gives $h(y) = cy + c_1$ and $k(x) = cx + c_2$, where c_1 and c_2 are constants. Therefore $f(x,y) = cy + c_1$ and $g(x,y) = cx + c_2$. Then $f(1,2) = g(1,2) = 5 \Rightarrow 5 = 2c + c_1 = c + c_2$, and $f(0,0) = 4 \Rightarrow c_1 = 4 \Rightarrow c = \frac{1}{2}$ $\Rightarrow c_2 = \frac{9}{2}$. Thus, $f(x,y) = \frac{1}{2}y + 4$ and $g(x,y) = \frac{1}{2}x + \frac{9}{2}$.

21. Since the particle is heat-seeking, at each point (x,y) it moves in the direction of maximal temperature increase, that is in the direction of $\nabla T(x,y) = \left(e^{-2y} \sin x\right)\mathbf{i} + \left(2e^{-2y} \cos x\right)\mathbf{j}$. Since $\nabla T(x,y)$ is parallel to the particle's velocity vector, it is tangent to the path $y = f(x)$ of the particle $\Rightarrow f'(x) = \frac{2e^{-2y} \cos x}{e^{-2y} \sin x} = 2 \cot x$. Integration gives $f(x) = 2 \ln|\sin x| + C$ and $f\left(\frac{\pi}{4}\right) = 0 \Rightarrow 0 = 2 \ln\left|\sin \frac{\pi}{4}\right| + C \Rightarrow C = -2 \ln \frac{\sqrt{2}}{2} = \ln\left(\frac{2}{\sqrt{2}}\right)^2$ $= \ln 2$. Therefore, the path of the particle is the graph of $y = 2 \ln|\sin x| + \ln 2$.

23. (a) $\mathbf{k}$ is a vector normal to $z = 10 - x^2 - y^2$ at the point $(0,0,10)$. So directions tangential to S at $(0,0,10)$ will be unit vectors $\mathbf{u} = a\mathbf{i} + b\mathbf{j}$. Also, $\nabla T(x,y,z) = (2xy + 4)\mathbf{i} + (x^2 + 2yz + 14)\mathbf{j} + (y^2 + 1)\mathbf{k}$ $\Rightarrow \nabla T(0,0,10) = 4\mathbf{i} + 14\mathbf{j} + \mathbf{k}$. We seek the unit vector $\mathbf{u} = a\mathbf{i} + b\mathbf{j}$ such that $D_{\mathbf{u}}T(0,0,10)$ $= (4\mathbf{i} + 14\mathbf{j} + \mathbf{k}) \cdot (a\mathbf{i} + b\mathbf{j}) = (4\mathbf{i} + 14\mathbf{j}) \cdot (a\mathbf{i} + b\mathbf{j})$ is a maximum. The maximum will occur when $a\mathbf{i} + b\mathbf{j}$ has the same direction as $4\mathbf{i} + 14\mathbf{j}$, or $\mathbf{u} = \frac{1}{\sqrt{53}}(2\mathbf{i} + 7\mathbf{j})$.

(b) A vector normal to S at $(1,1,8)$ is $\mathbf{n} = 2\mathbf{i} + 2\mathbf{j} + \mathbf{k}$. Now, $\nabla T(1,1,8) = 6\mathbf{i} + 31\mathbf{j} + 2\mathbf{k}$ and we seek the unit vector $\mathbf{u}$ such that $D_{\mathbf{u}}T(1,1,8) = \nabla T \cdot \mathbf{u}$ has its largest value. Now write $\nabla T = \mathbf{v} + \mathbf{w}$, where $\mathbf{v}$ is parallel to ∇T and $\mathbf{w}$ is orthogonal to ∇T. Then $D_{\mathbf{u}}T = \nabla T \cdot \mathbf{u} = (\mathbf{v} + \mathbf{w}) \cdot \mathbf{u} = \mathbf{v} \cdot \mathbf{u} + \mathbf{w} \cdot \mathbf{u} = \mathbf{w} \cdot \mathbf{u}$. Thus $D_{\mathbf{u}}T(1,1,8)$ is a maximum when $\mathbf{u}$ has the same direction as $\mathbf{w}$. Now, $\mathbf{w} = \nabla T - \left(\frac{\nabla T \cdot \mathbf{n}}{|\mathbf{n}|^2}\right)\mathbf{n}$

$$= (6\mathbf{i} + 31\mathbf{j} + 2\mathbf{k}) - \left(\frac{12 + 62 + 2}{4 + 4 + 1}\right)(2\mathbf{i} + 2\mathbf{j} + \mathbf{k}) = \left(6 - \frac{152}{9}\right)\mathbf{i} + \left(31 - \frac{152}{9}\right)\mathbf{j} + \left(2 - \frac{76}{9}\right)\mathbf{k}$$

$$= -\frac{98}{9}\mathbf{i} + \frac{127}{9}\mathbf{j} - \frac{58}{9}\mathbf{k} \Rightarrow \mathbf{u} = \frac{\mathbf{w}}{|\mathbf{w}|} = -\frac{1}{\sqrt{29{,}097}}(98\mathbf{i} - 127\mathbf{j} + 58\mathbf{k}).$$

25. $w = e^{rt} \sin \pi x \Rightarrow w_t = re^{rt} \sin \pi x$ and $w_x = \pi e^{rt} \cos \pi x \Rightarrow w_{xx} = -\pi^2 e^{rt} \sin \pi x$; $w_{xx} = \frac{1}{c^2} w_t$, where c^2 is the positive constant determined by the material of the rod $\Rightarrow -\pi^2 e^{rt} \sin \pi x = \frac{1}{c^2}\left(re^{rt} \sin \pi x\right)$ $\Rightarrow \left(r + c^2\pi^2\right)e^{rt} \sin \pi x = 0 \Rightarrow r = -c^2\pi^2 \Rightarrow w = e^{-c^2\pi^2 t} \sin \pi x$

NOTES:

CHAPTER 13 MULTIPLE INTEGRALS

13.1 DOUBLE INTEGRALS

1. $\int_0^3 \int_0^2 (4 - y^2)\, dy\, dx = \int_0^3 \left[4y - \frac{y^3}{3}\right]_0^2 dx = \frac{16}{3} \int_0^3 dx = 16$

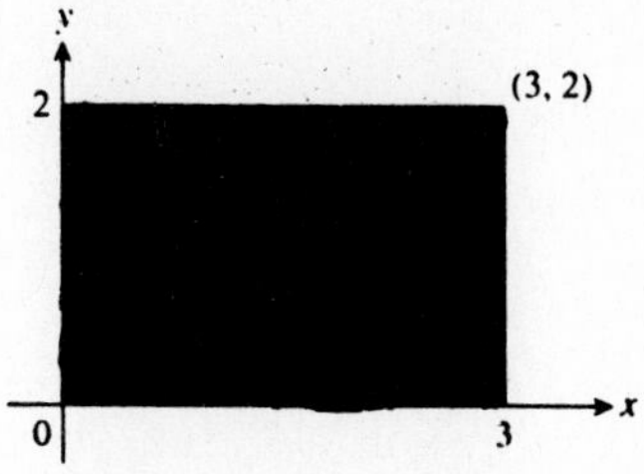

3. $\int_{-1}^0 \int_{-1}^1 (x + y + 1)\, dx\, dy = \int_{-1}^0 \left[\frac{x^2}{2} + yx + x\right]_{-1}^1 dy$

$= \int_{-1}^0 (2y + 2)\, dy = \left[y^2 + 2y\right]_{-1}^0 = 1$

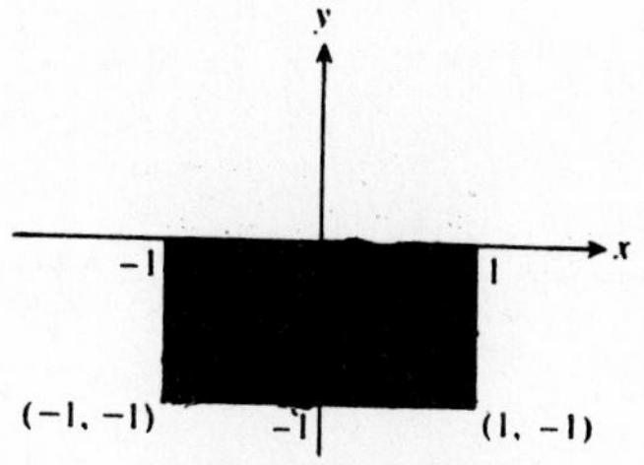

5. $\int_0^\pi \int_0^x (x \sin y)\, dy\, dx = \int_0^\pi [-x \cos y]_0^x\, dx$

$= \int_0^\pi (x - x \cos x)\, dx = \left[\frac{x^2}{2} - (\cos x + x \sin x)\right]_0^\pi = \frac{\pi^2}{2} + 2$

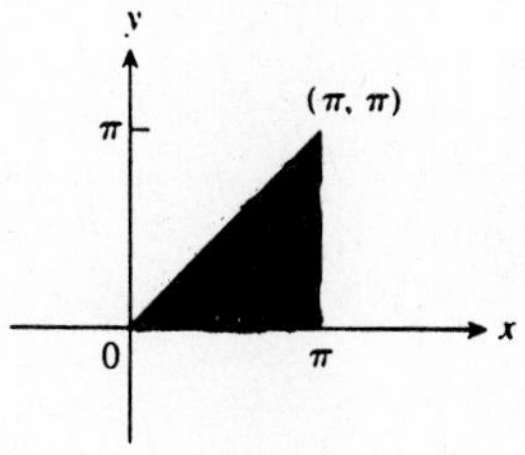

7. $\int_1^{\ln 8} \int_0^{\ln y} e^{x+y}\, dx\, dy = \int_1^{\ln 8} \left[e^{x+y}\right]_0^{\ln y} dy = \int_1^{\ln 8} (ye^y - e^y)\, dy$

$= \left[(y-1)e^y - e^y\right]_1^{\ln 8} = 8(\ln 8 - 1) - 8 + e = 8 \ln 8 - 16 + e$

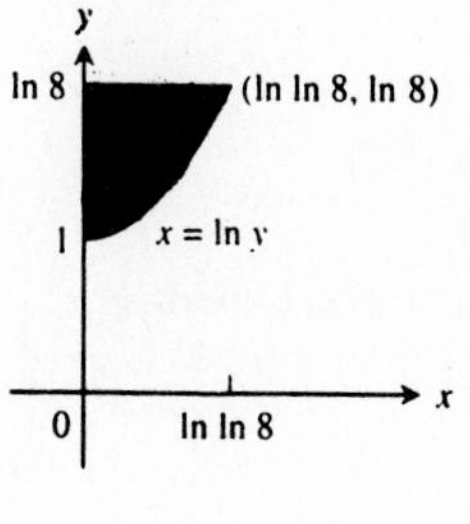

9. $\int_0^1 \int_0^{y^2} 3y^3 e^{xy}\, dx\, dy = \int_0^1 \left[3y^2 e^{xy}\right]_0^{y^2} dy$

$= \int_0^1 \left(3y^2 e^{y^3} - 3y^2\right) dy = \left[e^{y^3} - y^3\right]_0^1 = e - 2$

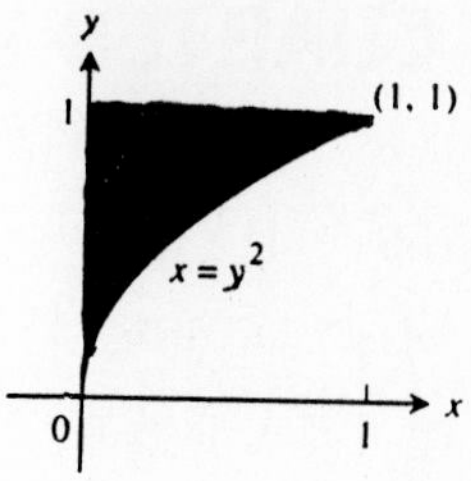

11. $\int_1^2 \int_x^{2x} \frac{x}{y}\, dy\, dx = \int_1^2 \left[x \ln y\right]_x^{2x} dx = (\ln 2) \int_1^2 x\, dx = \frac{3}{2} \ln 2$

13. $\int_0^1 \int_0^{1-x} (x^2 + y^2)\, dy\, dx = \int_0^1 \left[x^2 y + \frac{y^3}{3}\right]_0^{1-x} dx = \int_0^1 \left[x^2(1-x) + \frac{(1-x)^3}{3}\right] dx = \int_0^1 \left[x^2 - x^3 + \frac{(1-x)^3}{3}\right] dx$

$= \left[\frac{x^3}{3} - \frac{x^4}{4} - \frac{(1-x)^4}{12}\right]_0^1 = \left(\frac{1}{3} - \frac{1}{4} - 0\right) - \left(0 - 0 - \frac{1}{12}\right) = \frac{1}{6}$

15. $\int_0^1 \int_0^{1-u} (v - \sqrt{u})\, dv\, du = \int_0^1 \left[\frac{v^2}{2} - v\sqrt{u}\right]_0^{1-u} du = \int_0^1 \left[\frac{1 - 2u + u^2}{2} - \sqrt{u}(1-u)\right] du$

$= \int_0^1 \left(\frac{1}{2} - u + \frac{u^2}{2} - u^{1/2} + u^{3/2}\right) du = \left[\frac{u}{2} - \frac{u^2}{2} + \frac{u^3}{6} - \frac{2}{3}u^{3/2} + \frac{2}{5}u^{5/2}\right]_0^1 = \frac{1}{2} - \frac{1}{2} + \frac{1}{6} - \frac{2}{3} + \frac{2}{5} = -\frac{1}{2} + \frac{2}{5} = -\frac{1}{10}$

17. $\int_{-2}^0 \int_v^{-v} 2\, dp\, dv = \int_{-2}^0 [p]_v^{-v}\, dv = 2 \int_{-2}^0 -2v\, dv$

$= -2\left[v^2\right]_{-2}^0 = 8$

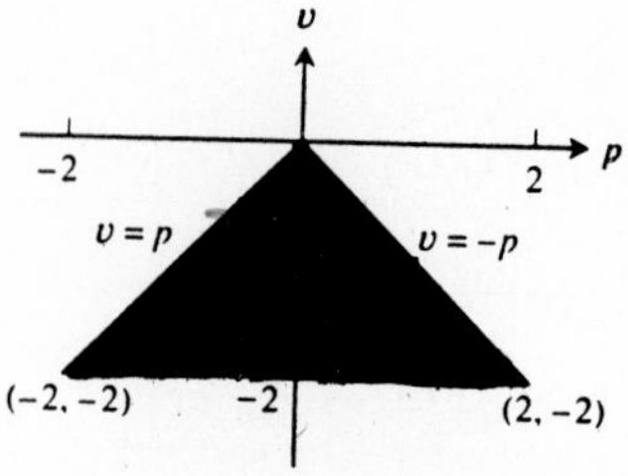

19. $\int_{-\pi/3}^{\pi/3} \int_0^{\sec t} 3 \cos t\, du\, dt = \int_{-\pi/3}^{\pi/3} \left[(3 \cos t)u\right]_0^{\sec t}$

$= \int_{-\pi/3}^{\pi/3} 3\, dt = 2\pi$

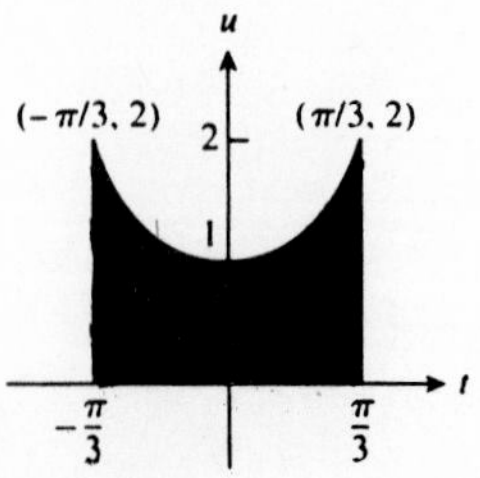

21. $\int_2^4 \int_0^{(4-y)/2} dx\, dy$

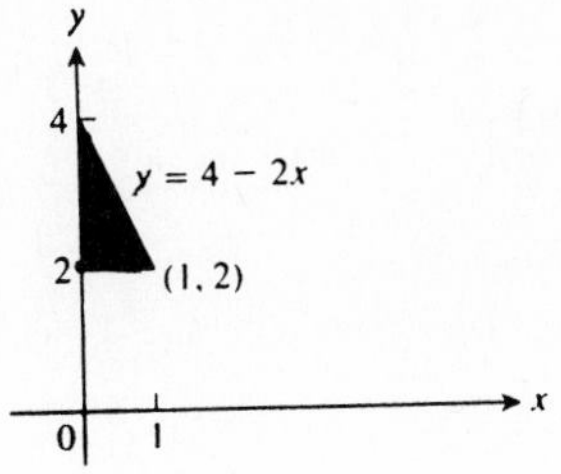

23. $\int_0^1 \int_{x^2}^{x} dy\, dx$

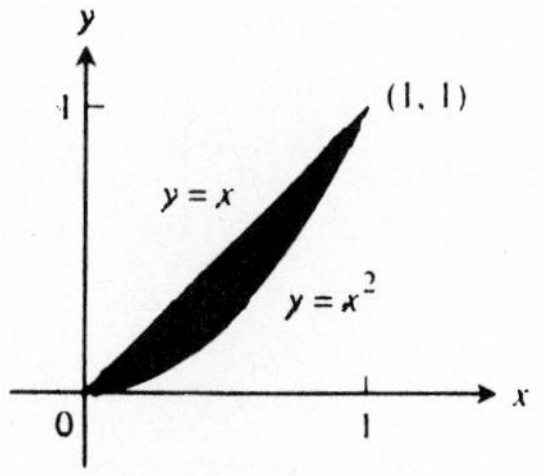

25. $\int_1^e \int_{\ln y}^{1} dx\, dy$

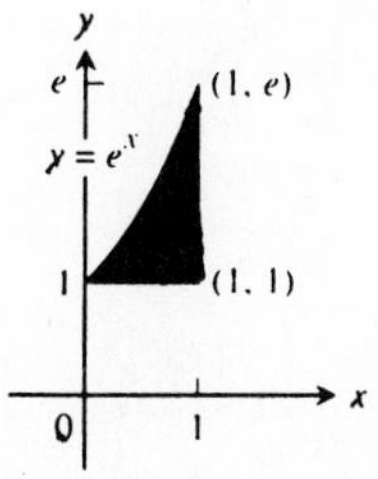

27. $\int_0^9 \int_0^{\frac{1}{2}\sqrt{9-y}} 16x\, dx\, dy$

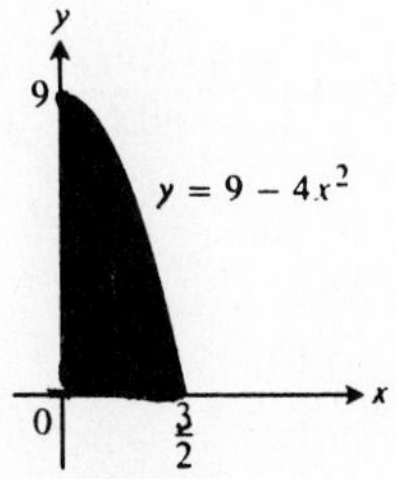

29. $\int_{-1}^1 \int_0^{\sqrt{1-x^2}} 3y\, dy\, dx$

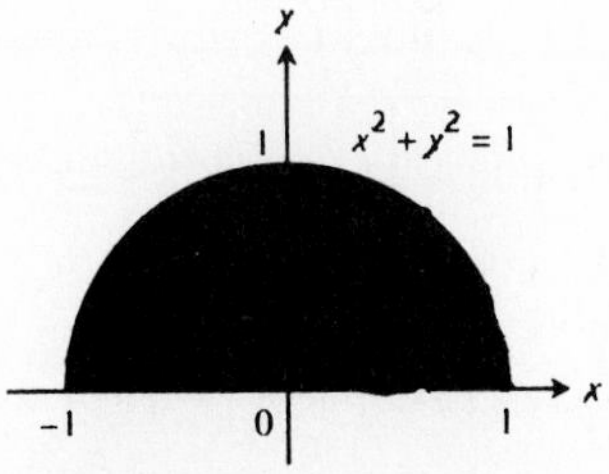

31. $\displaystyle\int_0^{\pi}\int_x^{\pi} \frac{\sin y}{y}\,dy\,dx = \int_0^{\pi}\int_0^{y} \frac{\sin y}{y}\,dx\,dy = \int_0^{\pi} \sin y\,dy = 2$

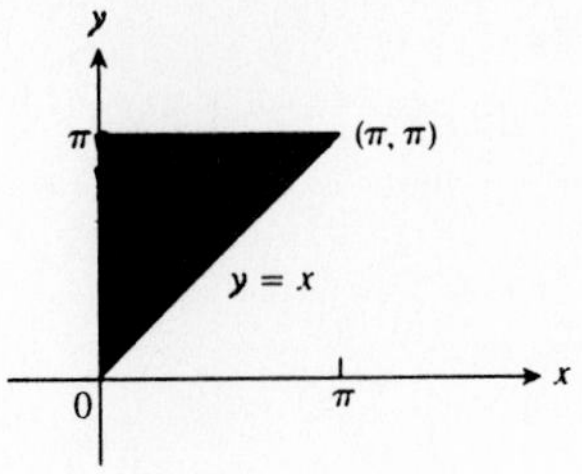

33. $\displaystyle\int_0^{1}\int_y^{1} x^2e^{xy}\,dx\,dy = \int_0^{1}\int_0^{x} x^2e^{xy}\,dy\,dx = \int_0^{1} \left[xe^{xy}\right]_0^x\,dx$

$\displaystyle= \int_0^{1} \left(xe^{x^2} - x\right)dx = \left[\frac{1}{2}e^{x^2} - \frac{x^2}{2}\right]_0^1 = \frac{e-2}{2}$

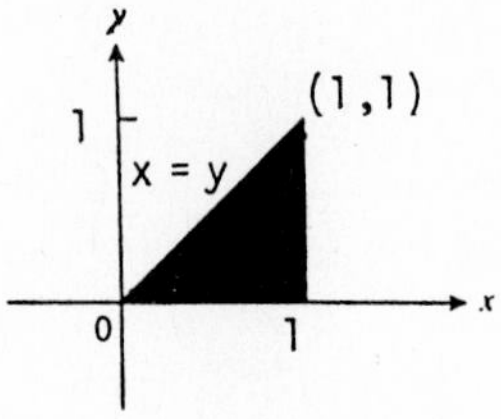

35. $\displaystyle\int_0^{2\sqrt{\ln 3}}\int_{y/2}^{\sqrt{\ln 3}} e^{x^2}\,dx\,dy = \int_0^{\sqrt{\ln 3}}\int_0^{2x} e^{x^2}\,dy\,dx$

$\displaystyle= \int_0^{\sqrt{\ln 3}} 2xe^{x^2}\,dx = \left[e^{x^2}\right]_0^{\sqrt{\ln 3}} = e^{\ln 3} - 1 = 2$

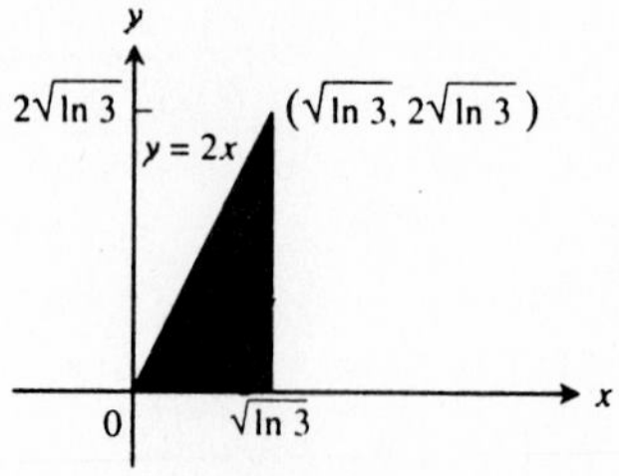

37. $\displaystyle\int_0^{1/16}\int_{y^{1/4}}^{1/2} \cos\left(16\pi x^5\right)dx\,dy = \int_0^{1/2}\int_0^{x^4} \cos\left(16\pi x^5\right)dy\,dx$

$\displaystyle= \int_0^{1/2} x^4\cos\left(16\pi x^5\right)dx = \left[\frac{\sin\left(16\pi x^5\right)}{80\pi}\right]_0^{1/2} = \frac{1}{80\pi}$

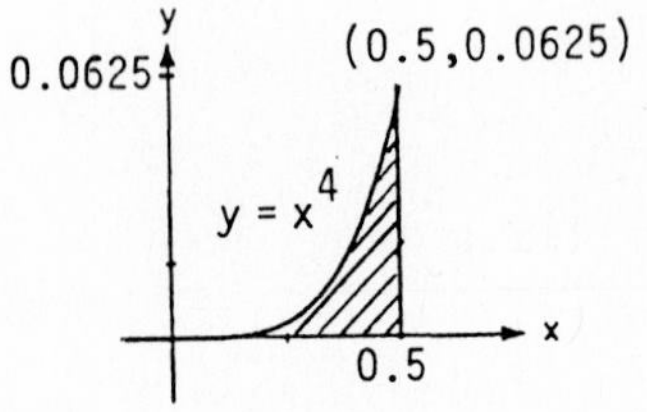

39. $\displaystyle\iint_R \left(y - 2x^2\right)dA = \int_{-1}^{0}\int_{-x-1}^{x+1} \left(y - 2x^2\right)dy\,dx + \int_0^{1}\int_{x-1}^{1-x} \left(y - 2x^2\right)dy\,dx$

$\displaystyle= \int_{-1}^{0} \left[\frac{1}{2}y^2 - 2x^2y\right]_{-x-1}^{x+1} dx + \int_0^{1} \left[\frac{1}{2}y^2 - 2x^2y\right]_{x-1}^{1-x} dx$

$\displaystyle= \int_{-1}^{0} \left[\frac{1}{2}(x+1)^2 - 2x^2(x+1) - \frac{1}{2}(-x-1)^2 + 2x^2(-x-1)\right]dx$

$\displaystyle+ \int_0^{1} \left[\frac{1}{2}(1-x)^2 - 2x^2(1-x) - \frac{1}{2}(x-1)^2 + 2x^2(x-1)\right]dx$

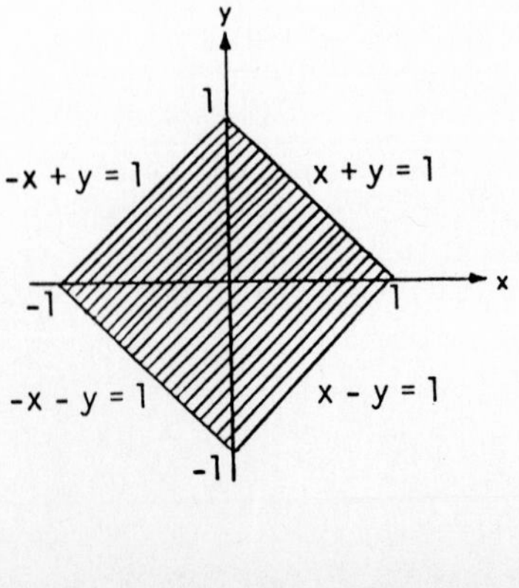

$$= -4\int_{-1}^{0}(x^3+x^2)\,dx + 4\int_{0}^{1}(x^3-x^2)\,dx = -4\left[\frac{x^4}{4}+\frac{x^3}{3}\right]_{-1}^{0} + 4\left[\frac{x^4}{4}-\frac{x^3}{3}\right]_{0}^{1}$$

$$= 4\left[\frac{(-1)^4}{4}+\frac{(-1)^3}{3}\right] + 4\left(\frac{1}{4}-\frac{1}{3}\right) = 8\left(\frac{3}{12}-\frac{4}{12}\right) = -\frac{8}{12} = -\frac{2}{3}$$

41. $$V = \int_{0}^{1}\int_{x}^{2-x}(x^2+y^2)\,dy\,dx = \int_{0}^{1}\left[x^2y+\frac{y^3}{3}\right]_{x}^{2-x}dx = \int_{0}^{1}\left[2x^2-\frac{7x^3}{3}+\frac{(2-x)^3}{3}\right]dx = \left[\frac{2x^3}{3}-\frac{7x^4}{12}-\frac{(2-x)^4}{12}\right]_{0}^{1}$$

$$= \left(\frac{2}{3}-\frac{7}{12}-\frac{1}{12}\right)-\left(0-0-\frac{16}{12}\right) = \frac{4}{3}$$

43. $$V = \int_{-4}^{1}\int_{3x}^{4-x^2}(x+4)\,dy\,dx = \int_{-4}^{1}\left[xy+4y\right]_{3x}^{4-x^2}dx = \int_{-4}^{1}\left[x(4-x^2)+4(4-x^2)-3x^2-12x\right]dx$$

$$= \int_{-4}^{1}(-x^3-7x^2-8x+16)\,dx = \left[-\frac{1}{4}x^4-\frac{7}{3}x^3-4x^2+16x\right]_{-4}^{1} = \left(-\frac{1}{4}-\frac{7}{3}+12\right)-\left(\frac{64}{3}-64\right)$$

$$= \frac{157}{3}-\frac{1}{4} = \frac{625}{12}$$

45. $$V = \int_{0}^{2}\int_{0}^{3}(4-y^2)\,dx\,dy = \int_{0}^{2}\left[4x-y^2x\right]_{0}^{3}dy = \int_{0}^{2}(12-3y^2)\,dy = \left[12y-y^3\right]_{0}^{2} = 24-8 = 16$$

47. $$V = \int_{0}^{2}\int_{0}^{2-x}(12-3y^2)\,dy\,dx = \int_{0}^{2}\left[12y-y^3\right]_{0}^{2-x}dx = \int_{0}^{2}\left[24-12x-(2-x)^3\right]dx$$

$$= \left[24x-6x^2+\frac{(2-x)^4}{4}\right]_{0}^{2} = 20$$

49. $$V = \int_{1}^{2}\int_{-1/x}^{1/x}(x+1)\,dy\,dx = \int_{1}^{2}\left[xy+y\right]_{-1/x}^{1/x}dx = \int_{1}^{2}\left[1+\frac{1}{x}-\left(-1-\frac{1}{x}\right)\right] = 2\int_{1}^{2}\left(1+\frac{1}{x}\right)dx$$

$$= 2\left[x+\ln x\right]_{1}^{2} = 2(1+\ln 2)$$

51. $$\int_{1}^{\infty}\int_{e^{-x}}^{1}\frac{1}{x^3y}\,dy\,dx = \int_{1}^{\infty}\left[\frac{\ln y}{x^3}\right]_{e^{-x}}^{1}dx = \int_{1}^{\infty}-\left(\frac{-x}{x^3}\right)dx = -\lim_{b\to\infty}\left[\frac{1}{x}\right]_{1}^{b} = -\lim_{b\to\infty}\left(\frac{1}{b}-1\right) = 1$$

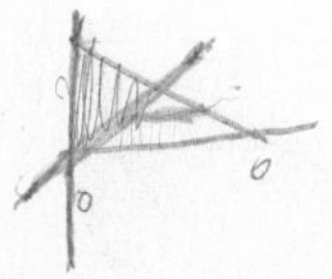

53. $\int_{-\infty}^{\infty}\int_{-\infty}^{\infty} \frac{1}{(x^2+1)(y^2+1)}\,dx\,dy = 2\int_0^{\infty}\left(\frac{2}{y^2+1}\right)\left(\lim_{b\to\infty}\tan^{-1}b-\tan^{-1}0\right)dy = 2\pi\lim_{b\to\infty}\int_0^b \frac{1}{y^2+1}\,dy$

$= 2\pi\left(\lim_{b\to\infty}\tan^{-1}b-\tan^{-1}0\right) = (2\pi)\left(\frac{\pi}{2}\right) = \pi^2$

55. $\iint_R f(x,y)\,dA \approx \frac{1}{4}f\left(-\frac{1}{2},0\right)+\frac{1}{8}f(0,0)+\frac{1}{8}f\left(\frac{1}{4},0\right)+\frac{1}{4}f\left(\frac{1}{2},0\right)+\frac{1}{4}f\left(-\frac{1}{2},\frac{1}{2}\right)+\frac{1}{8}f\left(0,\frac{1}{2}\right)+\frac{1}{8}f\left(\frac{1}{4},\frac{1}{2}\right)$

$= \frac{1}{4}\left(-\frac{1}{2}+\frac{1}{2}+0\right)+\frac{1}{8}\left(0+\frac{1}{4}+\frac{1}{2}+\frac{3}{4}\right) = \frac{3}{16}$

57. The ray $\theta = \frac{\pi}{6}$ meets the circle $x^2+y^2=4$ at the point $(\sqrt{3},1) \Rightarrow$ the ray is represented by the line $y = \frac{x}{\sqrt{3}}$.

Thus, $\iint_R f(x,y)\,dA = \int_0^{\sqrt{3}}\int_{x/\sqrt{3}}^{\sqrt{4-x^2}} \sqrt{4-x^2}\,dy\,dx = \int_0^{\sqrt{3}}\left[(4-x^2)-\frac{x}{3}\sqrt{4-x^2}\right]dx = \left[4x-\frac{x^3}{3}+\frac{(4-x^2)^{3/2}}{3\sqrt{3}}\right]_0^{\sqrt{3}}$

$= \frac{20\sqrt{3}}{9}$

59. $V = \int_0^1\int_x^{2-x}(x^2+y^2)\,dy\,dx = \int_0^1\left[x^2y+\frac{y^3}{3}\right]_x^{2-x}dx$

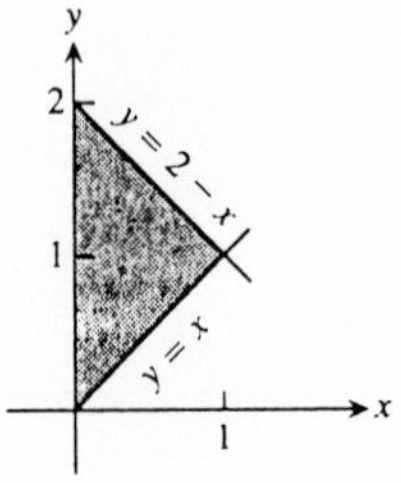

$= \int_0^1\left[2x^2-\frac{7x^3}{3}+\frac{(2-x)^3}{3}\right]dx = \left[\frac{2x^3}{3}-\frac{7x^4}{12}-\frac{(2-x)^4}{12}\right]_0^1$

$= \left(\frac{2}{3}-\frac{7}{12}-\frac{1}{12}\right)-\left(0-0-\frac{16}{12}\right) = \left(\frac{2}{3}+\frac{8}{12}\right) = \frac{4}{3}$

61. To maximize the integral, we want the domain to include all points where the integrand is positive and to exclude all points where the integrand is negative. These criteria are met by the points (x,y) such that $4-x^2-2y^2 \ge 0$ or $x^2+2y^2 \le 4$, which is the ellipse $x^2+2y^2=4$ together with its interior.

63. No, it is not all right. By Fubini's theorem, the two orders of integration must give the same result.

65. $\int_{-b}^{b}\int_{-b}^{b} e^{-x^2-y^2}\,dx\,dy = \int_{-b}^{b}\int_{-b}^{b} e^{-y^2}e^{-x^2}\,dx\,dy = \int_{-b}^{b} e^{-y^2}\left(\int_{-b}^{b} e^{-x^2}\,dx\right)dy = \left(\int_{-b}^{b} e^{-x^2}\,dx\right)\left(\int_{-b}^{b} e^{-y^2}\,dy\right)$

$= \left(\int_{-b}^{b} e^{-x^2}\,dx\right)^2 = \left(2\int_0^b e^{-x^2}\,dx\right)^2 = 4\left(\int_0^b e^{-x^2}\,dx\right)^2$; taking limits as $b \to \infty$ gives the stated result.

67. $\int_1^3\int_1^x \frac{1}{xy}\,dy\,dx \approx 0.603$

69. $\int_0^1\int_0^1 \tan^{-1}xy\,dy\,dx \approx 0.233$

13.2 AREAS, MOMENTS, AND CENTERS OF MASS

1. $\int_0^2 \int_0^{2-x} dy\, dx = \int_0^2 (2-x)\, dx = \left[2x - \frac{x^2}{2}\right]_0^2 = 2,$

or $\int_0^2 \int_0^{2-y} dx\, dy = \int_0^2 (2-y)\, dy = 2$

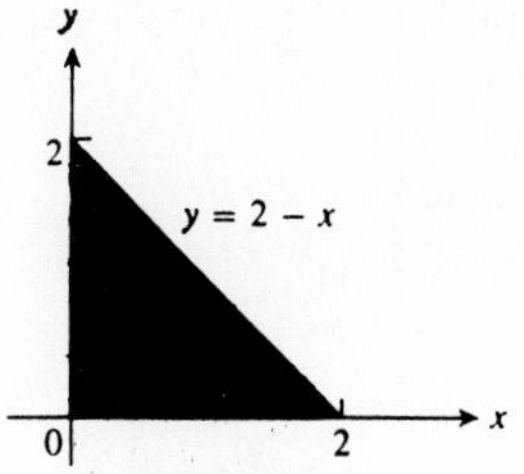

3. $\int_{-2}^{1} \int_{y-2}^{-y^2} dx\, dy = \int_{-2}^{1} \left(-y^2 - y + 2\right) dy = \left[-\frac{y^3}{3} - \frac{y^2}{2} + 2y\right]_{-2}^{1}$

$= \left(-\frac{1}{3} - \frac{1}{2} + 2\right) - \left(\frac{8}{3} - 2 - 4\right) = \frac{9}{2}$

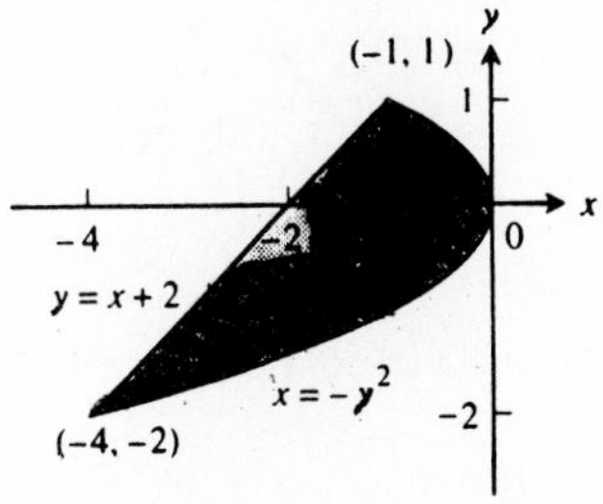

5. $\int_0^{\ln 2} \int_0^{e^x} dy\, dx = \int_0^{\ln 2} e^x\, dx = \left[e^x\right]_0^{\ln 2} = 2 - 1 = 1$

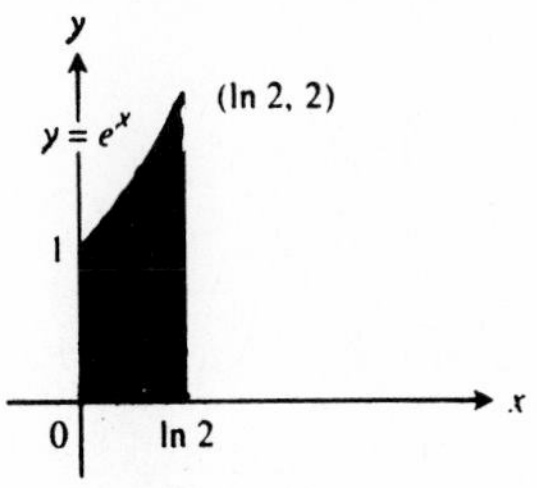

7. $\int_0^1 \int_{y^2}^{2y-y^2} dx\, dy = \int_0^1 \left(2y - 2y^2\right) dy = \left[y^2 - \frac{2}{3}y^3\right]_0^1 = \frac{1}{3}$

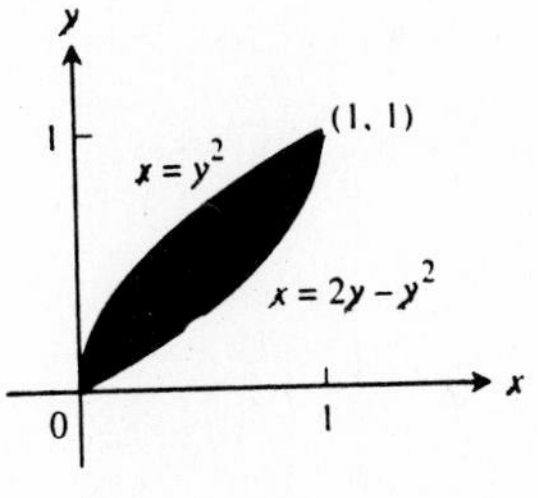

9. $\int_0^6 \int_{y^2/3}^{2y} dx\, dy = \int_0^6 \left(2y - \frac{y^2}{3}\right) dy = \left[y^2 - \frac{y^3}{9}\right]_0^6$

$= 36 - \frac{216}{9} = 12$

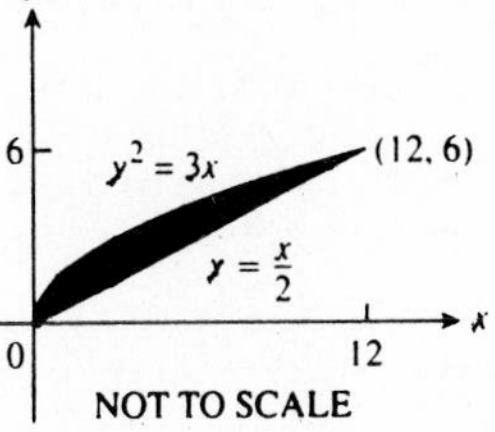

11. $\int_0^{\pi/4}\int_{\sin x}^{\cos x} dy\,dx = \int_0^{\pi/4} (\cos x - \sin x)\,dx = [\sin x + \cos x]_0^{\pi/4}$

$= \left(\frac{\sqrt{2}}{2} + \frac{\sqrt{2}}{2}\right) - (0+1) = \sqrt{2} - 1$

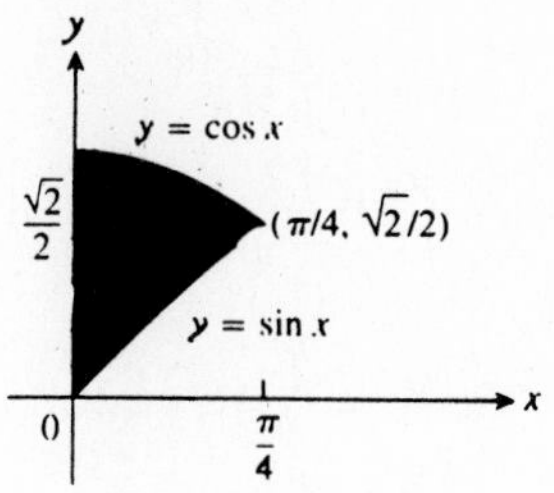

13. $\int_{-1}^{0}\int_{-2x}^{1-x} dy\,dx + \int_0^2\int_{-x/2}^{1-x} dy\,dx$

$= \int_{-1}^{0} (1+x)\,dx + \int_0^2 \left(1 - \frac{x}{2}\right) dx$

$= \left[x + \frac{x^2}{2}\right]_{-1}^{0} + \left[x - \frac{x^2}{4}\right]_0^2 = -\left(-1 + \frac{1}{2}\right) + (2-1) = \frac{3}{2}$

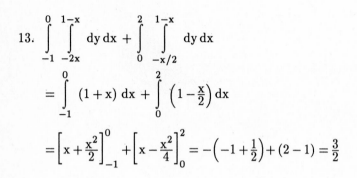

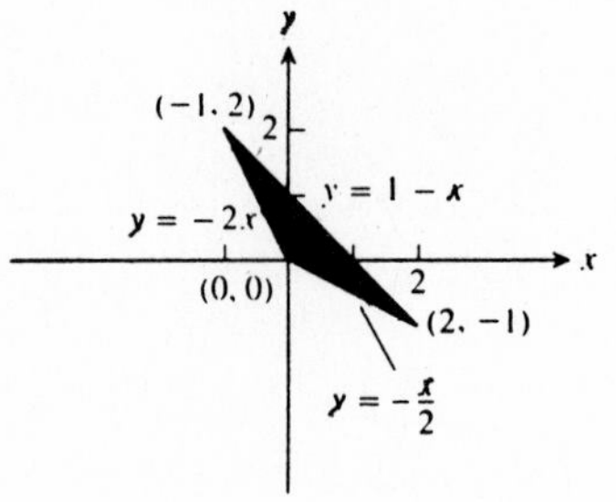

15. (a) average $= \frac{1}{\pi^2}\int_0^{\pi}\int_0^{\pi} \sin(x+y)\,dy\,dx = \frac{1}{\pi^2}\int_0^{\pi} [-\cos(x+y)]_0^{\pi}\,dx = \frac{1}{\pi^2}\int_0^{\pi} [-\cos(x+\pi) + \cos x]\,dx$

$= \frac{1}{\pi^2}[-\sin(x+\pi) + \sin x]_0^{\pi} = \frac{1}{\pi^2}[(-\sin 0 + \sin \pi) - (-\sin \pi + \sin 0)] = 0$

(b) average $= \frac{1}{\left(\frac{\pi^2}{2}\right)}\int_0^{\pi}\int_0^{\pi/2} \sin(x+y)\,dy\,dx = \frac{2}{\pi^2}\int_0^{\pi} [-\cos(x+y)]_0^{\pi/2}\,dx = \frac{2}{\pi^2}\int_0^{\pi} \left[-\cos\left(x + \frac{\pi}{2}\right) + \cos x\right] dx$

$= \frac{2}{\pi^2}\left[-\sin\left(x + \frac{\pi}{2}\right) + \sin x\right]_0^{\pi} = \frac{2}{\pi^2}\left[\left(-\sin\frac{3\pi}{2} + \sin \pi\right) - \left(-\sin\frac{\pi}{2} + \sin 0\right)\right] = \frac{4}{\pi^2}$

17. average height $= \frac{1}{4}\int_0^2\int_0^2 (x^2+y^2)\,dy\,dx = \frac{1}{4}\int_0^2 \left[x^2y + \frac{y^3}{3}\right]_0^2 dx = \frac{1}{4}\int_0^2 \left(2x^2 + \frac{8}{3}\right) dx = \frac{1}{2}\left[\frac{x^3}{3} + \frac{4x}{3}\right]_0^2 = \frac{8}{3}$

19. $M = \int_0^1\int_x^{2-x^2} 3\,dy\,dx = 3\int_0^1 (2 - x^2 - x)\,dx = \frac{7}{2}$; $M_y = \int_0^1\int_x^{2-x^2} 3x\,dy\,dx = 3\int_0^1 [xy]_x^{2-x^2}\,dx$

$= 3\int_0^1 (2x - x^3 - x^2)\,dx = \frac{5}{4}$; $M_x = \int_0^1\int_x^{2-x^2} 3y\,dy\,dx = \frac{3}{2}\int_0^1 [y^2]_x^{2-x^2}\,dx = \frac{3}{2}\int_0^1 (4 - 5x^2 + x^4)\,dx = \frac{19}{5}$

$\Rightarrow \bar{x} = \frac{5}{14}$ and $\bar{y} = \frac{38}{35}$

21. $M = \int_0^2 \int_{y^2/2}^{4-y} dx\,dy = \int_0^2 \left(4 - y - \frac{y^2}{2}\right) dy = \frac{14}{3};\ M_y = \int_0^2 \int_{y^2/2}^{4-y} x\,dx\,dy = \frac{1}{2}\int_0^2 \left[x^2\right]_{y^2/2}^{4-y} dy$

$= \frac{1}{2}\int_0^2 \left(16 - 8y + y^2 - \frac{y^4}{4}\right) dy = \frac{128}{15};\ M_x = \int_0^2 \int_{y^2/2}^{4-y} y\,dx\,dy = \int_0^2 \left(4y - y^2 - \frac{y^3}{2}\right) dy = \frac{10}{3}$

$\Rightarrow \overline{x} = \frac{64}{35}$ and $\overline{y} = \frac{5}{7}$

23. $M = 2\int_0^1 \int_0^{\sqrt{1-x^2}} dy\,dx = 2\int_0^1 \sqrt{1-x^2}\,dx = 2\left(\frac{\pi}{4}\right) = \frac{\pi}{2};\ M_x = 2\int_0^1 \int_0^{\sqrt{1-x^2}} y\,dy\,dx = \int_0^1 \left[y^2\right]_0^{\sqrt{1-x^2}} dx$

$= \int_0^1 (1 - x^2)\,dx = \left[x - \frac{x^3}{3}\right]_0^1 = \frac{2}{3} \Rightarrow \overline{y} = \frac{4}{3\pi}$ and $\overline{x} = 0$, by symmetry

25. $M = \int_0^a \int_0^{\sqrt{a^2-x^2}} dy\,dx = \frac{\pi a^2}{4};\ M_y = \int_0^a \int_0^{\sqrt{a^2-x^2}} x\,dy\,dx = \int_0^a \left[xy\right]_0^{\sqrt{a^2-x^2}} dx = \int_0^a x\sqrt{a^2 - x^2}\,dx = \frac{a^3}{3}$

$\Rightarrow \overline{x} = \overline{y} = \frac{4a}{3\pi}$, by symmetry

27. $M = \int_0^\pi \int_0^{\sin x} dy\,dx = \int_0^\pi \sin x\,dx = 2;\ M_x = \int_0^\pi \int_0^{\sin x} y\,dy\,dx = \frac{1}{2}\int_0^\pi \left[y^2\right]_0^{\sin x} dx = \frac{1}{2}\int_0^\pi \sin^2 x\,dx$

$= \frac{1}{4}\int_0^\pi (1 - \cos 2x)\,dx = \frac{\pi}{4} \Rightarrow \overline{x} = \frac{\pi}{2}$ and $\overline{y} = \frac{\pi}{8}$

29. $M = \int_{-\infty}^0 \int_0^{e^x} dy\,dx = \int_{-\infty}^0 e^x\,dx = \lim_{b\to-\infty} \int_b^0 e^x\,dx = 1 - \lim_{b\to-\infty} e^b = 1;\ M_y = \int_{-\infty}^0 \int_0^{e^x} x\,dy\,dx = \int_{-\infty}^0 xe^x\,dx$

$= \lim_{b\to-\infty} \int_b^0 xe^x\,dx = \lim_{b\to-\infty} \left[xe^x - e^x\right]_b^0 = -1 - \lim_{b\to-\infty} \left(be^b - e^b\right) = -1;\ M_x = \int_{-\infty}^0 \int_0^{e^x} y\,dy\,dx$

$= \frac{1}{2}\int_{-\infty}^0 e^{2x}\,dx = \frac{1}{2}\lim_{b\to-\infty} \int_b^0 e^{2x}\,dx = \frac{1}{4} \Rightarrow \overline{x} = -1$ and $\overline{y} = \frac{1}{4}$

31. $M = \int_0^2 \int_{-y}^{y-y^2} (x + y)\,dx\,dy = \int_0^2 \left[\frac{x^2}{2} + xy\right]_{-y}^{y-y^2} dy = \int_0^2 \left(\frac{y^4}{2} - 2y^3 + 2y^2\right) dy = \left[\frac{y^5}{10} - \frac{y^4}{2} + \frac{2y^3}{3}\right]_0^2 = \frac{8}{15};$

$I_x = \int_0^2 \int_{-y}^{y-y^2} y^2(x + y)\,dx\,dy = \int_0^2 \left[\frac{x^2y^2}{2} + xy^3\right]_{-y}^{y-y^2} dy = \int_0^2 \left(\frac{y^6}{2} - 2y^5 + 2y^4\right) dy = \frac{64}{105};$

$R_x = \sqrt{\frac{I_x}{M}} = \sqrt{\frac{8}{7}} = 2\sqrt{\frac{2}{7}}$

33. $M = \int_0^1 \int_x^{2-x} (6x + 3y + 3)\, dy\, dx = \int_0^1 \left[6xy + \frac{3}{2}y^2 + 3y\right]_x^{2-x} dx = \int_0^1 \left(12 - 12x^2\right) dx = 8;$

$M_y = \int_0^1 \int_x^{2-x} x(6x + 3y + 3)\, dy\, dx = \int_0^1 \left(12x - 12x^3\right) dx = 3;\ M_x = \int_0^1 \int_x^{2-x} y(6x + 3y + 3)\, dy\, dx$

$= \int_0^1 \left(14 - 6x - 6x^2 - 2x^3\right) dx = \frac{17}{2} \Rightarrow \bar{x} = \frac{3}{8}$ and $\bar{y} = \frac{17}{16}$

35. $M = \int_0^1 \int_0^6 (x + y + 1)\, dx\, dy = \int_0^1 (6y + 24)\, dy = 27;\ M_x = \int_0^1 \int_0^6 y(x + y + 1)\, dx\, dy = \int_0^1 y(6y + 24)\, dy = 14;$

$M_y = \int_0^1 \int_0^6 x(x + y + 1)\, dx\, dy = \int_0^1 (18y + 90)\, dy = 99 \Rightarrow \bar{x} = \frac{11}{3}$ and $\bar{y} = \frac{14}{27};\ I_y = \int_0^1 \int_0^6 x^2(x + y + 1)\, dx\, dy$

$= 216 \int_0^1 \left(\frac{y}{3} + \frac{11}{6}\right) dy = 432;\ R_y = \sqrt{\frac{I_y}{M}} = 4$

37. $M = \int_{-1}^1 \int_0^{x^2} (7y + 1)\, dy\, dx = \int_{-1}^1 \left(\frac{7x^4}{2} + x^2\right) dx = \frac{31}{15};\ M_x = \int_{-1}^1 \int_0^{x^2} y(7y + 1)\, dy\, dx = \int_{-1}^1 \left(\frac{7x^6}{3} + \frac{x^4}{2}\right) dx = \frac{13}{15};$

$M_y = \int_{-1}^1 \int_0^{x^2} x(7y + 1)\, dy\, dx = \int_{-1}^1 \left(\frac{7x^5}{2} + x^3\right) dx = 0 \Rightarrow \bar{x} = 0$ and $\bar{y} = \frac{13}{31};\ I_y = \int_{-1}^1 \int_0^{x^2} x^2(7y + 1)\, dy\, dx$

$= \int_{-1}^1 \left(\frac{7x^6}{2} + x^4\right) dx = \frac{7}{5};\ R_y = \sqrt{\frac{I_y}{M}} = \sqrt{\frac{21}{31}}$

39. $M = \int_0^1 \int_{-y}^{y} (y + 1)\, dx\, dy = \int_0^1 \left(2y^2 + 2y\right) dy = \frac{5}{3};\ M_x = \int_0^1 \int_{-y}^{y} y(y + 1)\, dx\, dy = 2\int_0^1 \left(y^3 + y^2\right) dy = \frac{7}{6};$

$M_y = \int_0^1 \int_{-y}^{y} x(y + 1)\, dx\, dy = \int_0^1 0\, dy = 0 \Rightarrow \bar{x} = 0$ and $\bar{y} = \frac{7}{10};\ I_x = \int_0^1 \int_{-y}^{y} y^2(y + 1)\, dx\, dy = \int_0^1 \left(2y^4 + 2y^3\right) dy$

$= \frac{9}{10} \Rightarrow R_x = \sqrt{\frac{I_x}{M}} = \frac{3\sqrt{6}}{10};\ I_y = \int_0^1 \int_{-y}^{y} x^2(y + 1)\, dx\, dy = \frac{1}{3}\int_0^1 \left(2y^4 + 2y^3\right) dy = \frac{3}{10} \Rightarrow R_y = \sqrt{\frac{I_y}{M}} = \frac{3\sqrt{2}}{10};$

$I_o = I_x + I_y = \frac{6}{5} \Rightarrow R_0 = \sqrt{\frac{I_o}{M}} = \frac{3\sqrt{2}}{5}$

41. $\int_{-5}^{5}\int_{-2}^{0} \frac{10{,}000e^y}{1+\frac{|x|}{2}}\,dy\,dx = 10{,}000(1-e^{-2})\int_{-5}^{5}\frac{dx}{1+\frac{|x|}{2}} = 10{,}000(1-e^{-2})\left[\int_{-5}^{0}\frac{dx}{1-\frac{x}{2}}+\int_{0}^{5}\frac{dx}{1+\frac{x}{2}}\right]$

$= 10{,}000(1-e^{-2})\left[-2\ln\left(1-\frac{x}{2}\right)\right]_{-5}^{0} + 10{,}000(1-e^{-2})\left[2\ln\left(1+\frac{x}{2}\right)\right]_{0}^{5}$

$= 10{,}000(1-e^{-2})\left[2\ln\left(1+\frac{5}{2}\right)\right] + 10{,}000(1-e^{-2})\left[2\ln\left(1+\frac{5}{2}\right)\right] = 40{,}000(1-e^{-2})\ln\left(\frac{7}{2}\right) \approx 43{,}329$

43. $M = \int_{-1}^{1}\int_{0}^{a(1-x^2)} dy\,dx = 2a\int_{0}^{1}(1-x^2)\,dx = 2a\left[x-\frac{x^3}{3}\right]_0^1 = \frac{4a}{3}$; $M_x = \int_{-1}^{1}\int_{0}^{a(1-x^2)} y\,dy\,dx$

$= \frac{2a^2}{2}\int_0^1 (1-2x^2+x^4)\,dx = a^2\left[x-\frac{2x^3}{3}+\frac{x^5}{5}\right]_0^1 = \frac{8a^2}{15} \Rightarrow \bar{y} = \frac{M_x}{M} = \frac{\left(\frac{8a^2}{15}\right)}{\left(\frac{4a}{3}\right)} = \frac{2a}{5}$. The angle θ between the x-axis and the line segment from the fulcrum to the center of mass on the y-axis plus 45° must be no more than 90° if the center of mass is to lie on the left side of the line $x = 1 \Rightarrow \theta + \frac{\pi}{4} \le \frac{\pi}{2} \Rightarrow \tan^{-1}\left(\frac{2a}{5}\right) \le \frac{\pi}{4} \Rightarrow a \le \frac{5}{2}$. Thus, if $0 < a \le \frac{5}{2}$, then the appliance will have to be tipped more than 45° to fall over.

45. $M = \int_0^1\int_{-1/\sqrt{1-x^2}}^{1/\sqrt{1-x^2}} dy\,dx = \int_0^1 \frac{2}{\sqrt{1-x^2}}\,dx = \left[2\sin^{-1}x\right]_0^1 = 2\left(\frac{\pi}{2}-0\right) = \pi$; $M_y = \int_0^1\int_{-1/\sqrt{1-x^2}}^{1/\sqrt{1-x^2}} x\,dy\,dx$

$= \int_0^1 \frac{2x}{\sqrt{1-x^2}}\,dx = \left[-2(1-x^2)^{1/2}\right]_0^1 = 2 \Rightarrow \bar{x} = \frac{2}{\pi}$ and $\bar{y} = 0$ by symmetry

47. (a) $\frac{1}{2} = M = \int_0^1\int_{y^2}^{2y-y^2} \delta\,dx\,dy = 2\delta\int_0^1 (y-y^2)\,dy = 2\delta\left[\frac{y^2}{2}-\frac{y^3}{3}\right]_0^1 = 2\delta\left(\frac{1}{6}\right) = \frac{\delta}{3} \Rightarrow \delta = \frac{3}{2}$

(b) average value $= \frac{\int_0^1\int_{y^2}^{2y-y^2}(y+1)\,dx\,dy}{\int_0^1\int_{y^2}^{2y-y^2}dx\,dy} = \frac{\left(\frac{1}{2}\right)}{\left(\frac{1}{3}\right)} = \frac{3}{2} = \delta$, so the values are the same

49. (a) $\bar{x} = \frac{M_y}{M} = 0 \Rightarrow M_y = \iint_R x\delta(x,y)\,dy\,dx = 0$

(b) $I_L = \iint_R (x-h)^2\,\delta(x,y)\,dA = \iint_R x^2\,\delta(x,y)\,dA - \iint_R 2hx\,\delta(x,y)\,dA + \iint_R h^2\,\delta(x,y)\,dA$

$= I_y - 0 + h^2\iint_R \delta(x,y)\,dA = I_{c.m.} + mh^2$

51. $M_{x_{P_1 \cup P_2}} = \iint_{R_1} y\, dA_1 + \iint_{R_2} y\, dA_2 = M_{x_1} + M_{x_2} \Rightarrow \bar{x} = \frac{M_{x_1} + M_{x_2}}{m_1 + m_2}$; likewise, $\bar{y} = \frac{M_{y_1} + M_{y_2}}{m_1 + m_2}$;

thus $\mathbf{c} = \bar{x}\mathbf{i} + \bar{y}\mathbf{j} = \frac{1}{m_1 + m_2}\left[\left(M_{x_1} + M_{x_2}\right)\mathbf{i} + \left(M_{y_1} + M_{y_2}\right)\mathbf{j}\right] = \frac{1}{m_1 + m_2}\left[(m_1\bar{x}_1 + m_2\bar{x}_2)\mathbf{i} + (m_1\bar{y}_1 + m_2\bar{y}_2)\mathbf{j}\right]$

$= \frac{1}{m_1 + m_2}\left[m_1(\bar{x}_1\mathbf{i} + \bar{y}_1\mathbf{j}) + m_2(\bar{x}_2\mathbf{i} + \bar{y}_2\mathbf{j})\right] = \frac{m_1\mathbf{c}_1 + m_2\mathbf{c}_2}{m_1 + m_2}$

53. (a) $\mathbf{c} = \frac{8(\mathbf{i} + 3\mathbf{j}) + 2(3\mathbf{i} + 3.5\mathbf{j})}{8 + 2} = \frac{14\mathbf{j} + 31\mathbf{k}}{10} \Rightarrow \bar{x} = \frac{7}{5}$ and $\bar{y} = \frac{31}{10}$

(b) $\mathbf{c} = \frac{8(\mathbf{i} + 3\mathbf{j}) + 6(5\mathbf{i} + 2\mathbf{j})}{14} = \frac{38\mathbf{i} + 36\mathbf{j}}{14} \Rightarrow \bar{x} = \frac{19}{7}$ and $\bar{y} = \frac{18}{7}$

(c) $\mathbf{c} = \frac{2(3\mathbf{i} + 3.5\mathbf{j}) + 6(5\mathbf{i} + 2\mathbf{j})}{8} = \frac{36\mathbf{i} + 19\mathbf{j}}{8} \Rightarrow \bar{x} = \frac{9}{2}$ and $\bar{y} = \frac{19}{8}$

(d) $\mathbf{c} = \frac{8(\mathbf{i} + 3\mathbf{j}) + 2(3\mathbf{i} + 3.5\mathbf{j}) + 6(5\mathbf{i} + 2\mathbf{j})}{16} = \frac{44\mathbf{i} + 43\mathbf{j}}{16} \Rightarrow \bar{x} = \frac{11}{4}$ and $\bar{y} = \frac{43}{16}$

55. Place the midpoint of the triangle's base at the origin and above the semicircle. Then the center of mass of the triangle is $\left(0, \frac{h}{3}\right)$, and the center of mass of the disk is $\left(0, -\frac{4a}{3\pi}\right)$ from Exercise 25. From Pappus's formula, $\mathbf{c} = \frac{(ah)\left(\frac{h}{3}\mathbf{j}\right) + \left(\frac{\pi a^2}{2}\right)\left(-\frac{4a}{3\pi}\mathbf{j}\right)}{\left(ah + \frac{\pi a^2}{2}\right)} = \frac{\left(\frac{ah^2 - 2a^3}{3}\right)\mathbf{j}}{\left(ah + \frac{\pi a^2}{2}\right)}$, so the centroid is on the boundary if $ah^2 - 2a^3 = 0 \Rightarrow h^2 = 2a^2 \Rightarrow h = a\sqrt{2}$. In order for the center of mass to be inside T we must have $ah^2 - 2a^3 > 0$ or $h > a\sqrt{2}$.

13.3 DOUBLE INTEGRALS IN POLAR FORM

1. $\int_{-1}^{1} \int_{0}^{\sqrt{1-x^2}} dy\, dx = \int_0^{\pi} \int_0^1 r\, dr\, d\theta = \frac{1}{2} \int_0^{\pi} d\theta = \frac{\pi}{2}$

3. $\int_0^1 \int_0^{\sqrt{1-y^2}} (x^2 + y^2)\, dx\, dy = \int_0^{\pi/2} \int_0^1 r^3\, dr\, d\theta = \frac{1}{4} \int_0^{\pi/2} d\theta = \frac{\pi}{8}$

5. $\int_{-a}^{a} \int_{-\sqrt{a^2-x^2}}^{\sqrt{a^2-x^2}} dy\, dx = \int_0^{2\pi} \int_0^a r\, dr\, d\theta = \frac{a^2}{2} \int_0^{2\pi} d\theta = \pi a^2$

7. $\int_0^6 \int_0^y x \, dx \, dy = \int_{\pi/4}^{\pi/2} \int_0^{6 \csc \theta} r^2 \cos \theta \, dr \, d\theta = 72 \int_{\pi/4}^{\pi/2} \cot \theta \csc^2 \theta \, d\theta = -36 \left[\cot^2 \theta\right]_{\pi/4}^{\pi/2} = 36$

9. $\int_{-1}^{0} \int_{-\sqrt{1-x^2}}^{0} \frac{2}{1+\sqrt{x^2+y^2}} \, dy \, dx = \int_{\pi}^{3\pi/2} \int_0^1 \frac{2r}{1+r} \, dr \, d\theta = 2 \int_{\pi}^{3\pi/2} \int_0^1 \left(1 - \frac{1}{1+r}\right) dr \, d\theta = 2 \int_{\pi}^{3\pi/2} (1 - \ln 2) \, d\theta$

$= (1 - \ln 2)\pi$

11. $\int_0^{\ln 2} \int_0^{\sqrt{(\ln 2)^2 - y^2}} e^{\sqrt{x^2+y^2}} \, dx \, dy = \int_0^{\pi/2} \int_0^{\ln 2} r e^r \, dr \, d\theta = \int_0^{\pi/2} (2 \ln 2 - 1) \, d\theta = \frac{\pi}{2}(2 \ln 2 - 1)$

13. $\int_0^2 \int_0^{\sqrt{1-(x-1)^2}} \frac{x+y}{x^2+y^2} \, dy \, dx = \int_0^{\pi/2} \int_0^{2 \cos \theta} \frac{r(\cos \theta + \sin \theta)}{r^2} r \, dr \, d\theta = \int_0^{\pi/2} \left(2 \cos^2 \theta + 2 \sin \theta \cos \theta\right) d\theta$

$= \left[\theta + \frac{\sin 2\theta}{2} + \sin^2 \theta\right]_0^{\pi/2} = \frac{\pi + 2}{2} = \frac{\pi}{2} + 1$

15. $\int_{-1}^{1} \int_{-\sqrt{1-y^2}}^{\sqrt{1-y^2}} \ln\left(x^2 + y^2 + 1\right) dx \, dy = 4 \int_0^{\pi/2} \int_0^1 \ln\left(r^2 + 1\right) r \, dr \, d\theta = 2 \int_0^{\pi/2} (\ln 4 - 1) \, d\theta = \pi(\ln 4 - 1)$

$\left[\frac{r^3}{3} \cos \theta\right]_0^{6 \csc \theta}$

17. $\int_0^{\pi/2} \int_0^{2\sqrt{2 - \sin 2\theta}} r \, dr \, d\theta = 2 \int_0^{\pi/2} (2 - \sin 2\theta) \, d\theta = 2(\pi - 1)$

19. $A = 2 \int_0^{\pi/6} \int_0^{12 \cos 3\theta} r \, dr \, d\theta = 144 \int_0^{\pi/6} \cos^2 3\theta \, d\theta = 12\pi$

21. $A = \int_0^{\pi/2} \int_0^{1+\sin \theta} r \, dr \, d\theta = \frac{1}{2} \int_0^{\pi/2} \left(\frac{3}{2} + 2 \sin \theta - \frac{\cos 2\theta}{2}\right) d\theta = \frac{3\pi}{8} + 1$

23. $M_x = \int_0^{\pi} \int_0^{1-\cos \theta} 3r^2 \sin \theta \, dr \, d\theta = 2 \int_0^{\pi} (1 - \cos \theta)^3 \sin \theta \, d\theta = 4$

25. $M = 2 \int_{\pi/6}^{\pi/2} \int_3^{6 \sin \theta} dr \, d\theta = 2 \int_{\pi/6}^{\pi/2} (6 \sin \theta - 3) \, d\theta = 6[-2 \cos \theta - \theta]_{\pi/6}^{\pi/2} = 6\sqrt{3} - 2\pi$

27. $M = 2\int_0^{\pi}\int_0^{1+\cos\theta} r\,dr\,d\theta = \int_0^{\pi}(1+\cos\theta)^2\,d\theta = \frac{3\pi}{2}$; $M_y = 2\int_0^{2\pi}\int_0^{1+\cos\theta} r^2\cos\theta\,dr\,d\theta$

$= \int_0^{2\pi}\left(\frac{4\cos\theta}{3}+\frac{15}{24}+\cos 2\theta-\sin^2\theta\cos\theta+\frac{\cos 4\theta}{4}\right)d\theta = \frac{5\pi}{4} \Rightarrow \bar{x}=\frac{5}{6}$ and $\bar{y}=0$, by symmetry

29. average $= \frac{4}{\pi a^2}\int_0^{\pi/2}\int_0^{a} r\sqrt{a^2-r^2}\,dr\,d\theta = \frac{4}{3\pi a^2}\int_0^{\pi/2} a^3\,d\theta = \frac{2a}{3}$

31. average $= \frac{1}{\pi a^2}\int_{-a}^{a}\int_{-\sqrt{a^2-x^2}}^{\sqrt{a^2-x^2}} \sqrt{x^2+y^2}\,dy\,dx = \frac{1}{\pi a^2}\int_0^{2\pi}\int_0^{a} r^2\,dr\,d\theta = \frac{a}{3\pi}\int_0^{2\pi} d\theta = \frac{2a}{3}$

33. $\int_0^{2\pi}\int_1^{\sqrt{e}}\left(\frac{\ln r^2}{r}\right) r\,dr\,d\theta = \int_0^{2\pi}\int_1^{\sqrt{e}} 2\ln r\,dr\,d\theta = 2\int_0^{2\pi}[r\ln r - r]_1^{e^{1/2}}\,d\theta = 2\int_0^{2\pi}\sqrt{e}\left[\left(\frac{1}{2}-1\right)+1\right]d\theta = 2\pi\sqrt{e}$

35. $V = 2\int_0^{\pi/2}\int_1^{1+\cos\theta} r^2\cos\theta\,dr\,d\theta = \frac{2}{3}\int_0^{\pi/2}\left(3\cos^2\theta+3\cos^3\theta+\cos^4\theta\right)d\theta$

$= \frac{2}{3}\left[\frac{15\theta}{8}+\sin 2\theta+3\sin\theta-\sin^3\theta+\frac{\sin 4\theta}{32}\right]_0^{\pi/2} = \frac{4}{3}+\frac{5\pi}{8}$

37. (a) $I^2 = \int_0^{\infty}\int_0^{\infty} e^{-\left(x^2+y^2\right)}\,dx\,dy = \int_0^{\pi/2}\int_0^{\infty}\left(e^{-r^2}\right) r\,dr\,d\theta = \int_0^{\pi/2}\left[\lim_{b\to\infty}\int_0^{b} re^{-r^2}\,dr\right]d\theta$

$= -\frac{1}{2}\int_0^{\pi/2}\lim_{b\to\infty}\left(e^{-b^2}-1\right)d\theta = \frac{1}{2}\int_0^{\pi/2} d\theta = \frac{\pi}{4} \Rightarrow I = \frac{\sqrt{\pi}}{2}$

(b) $\lim_{x\to\infty}\int_0^{x}\frac{2e^{-t^2}}{\sqrt{\pi}}\,dt = \frac{2}{\sqrt{\pi}}\int_0^{\infty} e^{-t^2}\,dt = \left(\frac{2}{\sqrt{\pi}}\right)\left(\frac{\sqrt{\pi}}{2}\right) = 1$, from part (a)

39. Over the disk $x^2+y^2 \le \frac{3}{4}$: $\iint_R \frac{1}{1-x^2-y^2}\,dA = \int_0^{2\pi}\int_0^{\sqrt{3}/2}\frac{r}{1-r^2}\,dr\,d\theta = \int_0^{2\pi}\left[-\frac{1}{2}\ln\left(1-r^2\right)\right]_0^{\sqrt{3}/2} d\theta$

$= \int_0^{2\pi}\left(-\frac{1}{2}\ln\frac{1}{4}\right)d\theta = (\ln 2)\int_0^{2\pi} d\theta = \pi\ln 4$

Over the disk $x^2+y^2 \le 1$: $\iint_R \frac{1}{1-x^2-y^2}\,dA = \int_0^{2\pi}\int_0^{1}\frac{r}{1-r^2}\,dr\,d\theta = \int_0^{2\pi}\left[\lim_{a\to 1^-}\int_0^{a}\frac{r}{1-r^2}\,dr\right]d\theta$

$= \int_0^{2\pi} \lim_{a \to 1^-} \left[-\frac{1}{2}\ln(1-a^2)\right] d\theta = 2\pi \cdot \lim_{a \to 1^-} \left[-\frac{1}{2}\ln(1-a^2)\right] = 2\pi \cdot \infty$, so the integral does not exist over

$x^2 + y^2 \le 1$

41. $\text{average} = \frac{1}{\pi a^2} \int_0^{2\pi} \int_0^a \left[(r\cos\theta - h)^2 + r^2 \sin^2\theta\right] r\, dr\, d\theta = \frac{1}{\pi a^2} \int_0^{2\pi} \int_0^a \left(r^3 - 2r^2 h\cos\theta + rh^2\right) dr\, d\theta$

$$= \frac{1}{\pi a^2} \int_0^{2\pi} \left(\frac{a^4}{4} - \frac{2a^3 h\cos\theta}{3} + \frac{a^2h^2}{2}\right) d\theta = \frac{1}{\pi} \int_0^{2\pi} \left(\frac{a^2}{4} - \frac{2ah\cos\theta}{3} + \frac{h^2}{2}\right) d\theta = \frac{1}{\pi}\left[\frac{a^2\theta}{4} - \frac{2ah\sin\theta}{3} + \frac{h^2\theta}{2}\right]_0^{2\pi}$$

$$= \frac{1}{2}\left(a^2 + 2h^2\right)$$

13.4 TRIPLE INTEGRALS IN RECTANGULAR COORDINATES

1. $\int_0^1 \int_0^{1-z} \int_0^2 dx\, dy\, dz = 2 \int_0^1 \int_0^{1-z} dy\, dz = 2 \int_0^1 (1-z)\, dz = 2\left[z - \frac{z^2}{2}\right]_0^1 = 2\left(1 - \frac{1}{2}\right) = 1$

3. $\int_0^1 \int_0^{2-2x} \int_0^{3-3x-3y/2} dz\, dy\, dx = \int_0^1 \int_0^{2-2x} \left(3 - 3x - \frac{3}{2}y\right) dy\, dx$

$$= \int_0^1 \left[3(1-x)\cdot 2(1-x) - \frac{3}{4}\cdot 4(1-x)^2\right] dx$$

$$= 3 \int_0^1 (1-x)^2\, dx = \left[-(1-x)^3\right]_0^1 = 1,$$

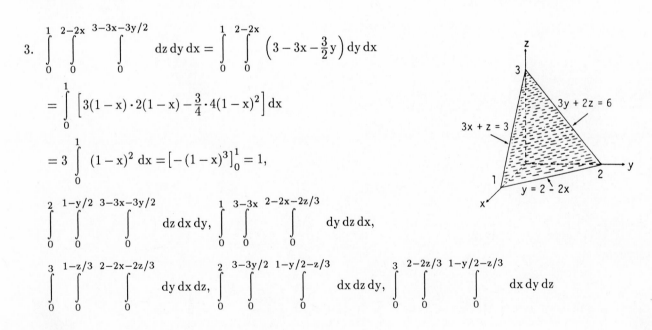

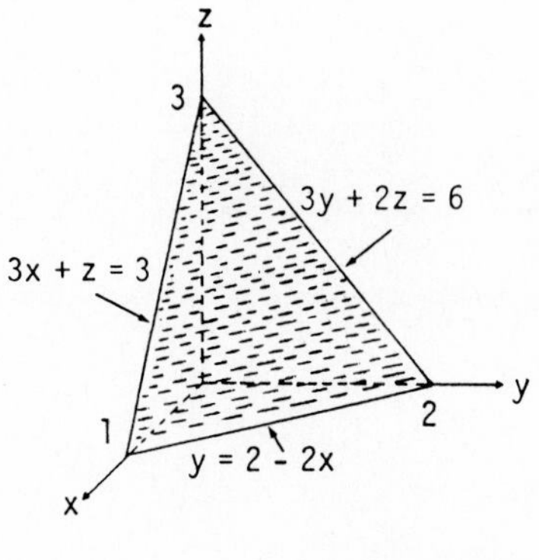

$\int_0^2 \int_0^{1-y/2} \int_0^{3-3x-3y/2} dz\, dx\, dy, \quad \int_0^1 \int_0^{3-3x} \int_0^{2-2x-2z/3} dy\, dz\, dx,$

$\int_0^3 \int_0^{1-z/3} \int_0^{2-2x-2z/3} dy\, dx\, dz, \quad \int_0^2 \int_0^{3-3y/2} \int_0^{1-y/2-z/3} dx\, dz\, dy, \quad \int_0^3 \int_0^{2-2z/3} \int_0^{1-y/2-z/3} dx\, dy\, dz$

5. $$\int_{-2}^{2}\int_{-\sqrt{4-x^2}}^{\sqrt{4-x^2}}\int_{x^2+y^2}^{8-x^2-y^2} dz\,dy\,dx = 4\int_{0}^{2}\int_{0}^{\sqrt{4-x^2}}\int_{x^2+y^2}^{8-x^2-y^2} dz\,dy\,dx$$

$$= 4\int_{0}^{2}\int_{0}^{\sqrt{4-x^2}} \left[8-2\left(x^2+y^2\right)\right] dy\,dx$$

$$= 8\int_{0}^{2}\int_{0}^{\sqrt{4-x^2}} \left(4-x^2-y^2\right) dy\,dx = 8\int_{0}^{\pi/2}\int_{0}^{2} \left(4-r^2\right) r\,dr\,d\theta$$

$$= 8\int_{0}^{\pi/2} \left[2r^2-\frac{r^4}{4}\right]_0^2 d\theta = 32\int_{0}^{\pi/2} d\theta = 32\left(\frac{\pi}{2}\right) = 16\pi,$$

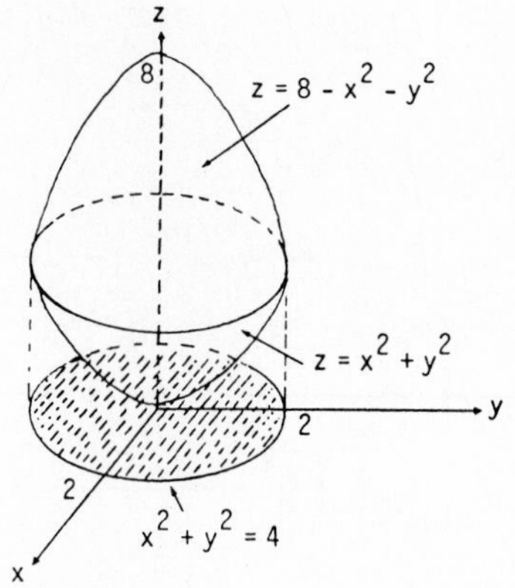

$$\int_{-2}^{2}\int_{-\sqrt{4-y^2}}^{\sqrt{4-y^2}}\int_{x^2+y^2}^{8-x^2-y^2} dz\,dx\,dy, \quad \int_{-2}^{2}\int_{y^2}^{4}\int_{-\sqrt{z-y^2}}^{\sqrt{z-y^2}} dx\,dz\,dy + \int_{-2}^{2}\int_{4}^{8-y^2}\int_{-\sqrt{8-z-y^2}}^{\sqrt{8-z-y^2}} dx\,dz\,dy,$$

$$\int_{0}^{4}\int_{-\sqrt{z}}^{\sqrt{z}}\int_{-\sqrt{z-y^2}}^{\sqrt{z-y^2}} dx\,dy\,dz + \int_{4}^{8}\int_{-\sqrt{8-z}}^{\sqrt{8-z}}\int_{-\sqrt{8-z-y^2}}^{\sqrt{8-z-y^2}} dx\,dy\,dz,$$

$$\int_{-2}^{2}\int_{x^2}^{4}\int_{-\sqrt{z-x^2}}^{\sqrt{z-x^2}} dy\,dz\,dx + \int_{-2}^{2}\int_{4}^{8-x^2}\int_{-\sqrt{8-z-x^2}}^{\sqrt{8-z-x^2}} dy\,dz\,dx,$$

$$\int_{0}^{4}\int_{-\sqrt{z}}^{\sqrt{z}}\int_{-\sqrt{z-x^2}}^{\sqrt{z-x^2}} dy\,dx\,dz + \int_{4}^{8}\int_{-\sqrt{8-z}}^{\sqrt{8-z}}\int_{-\sqrt{8-z-x^2}}^{\sqrt{8-z-x^2}} dy\,dx\,dz$$

7. $$\int_{0}^{1}\int_{0}^{1}\int_{0}^{1} \left(x^2+y^2+z^2\right) dz\,dy\,dx = \int_{0}^{1}\int_{0}^{1} \left(x^2+y^2+\frac{1}{3}\right) dy\,dx = \int_{0}^{1} \left(x^2+\frac{2}{3}\right) dx = 1$$

9. $$\int_{1}^{e}\int_{1}^{e}\int_{1}^{e} \frac{1}{xyz}\,dx\,dy\,dz = \int_{1}^{e}\int_{1}^{e} \left[\frac{\ln x}{yz}\right]_1^e dy\,dz = \int_{1}^{e}\int_{1}^{e} \frac{1}{yz}\,dy\,dz = \int_{1}^{e} \left[\frac{\ln y}{z}\right]_1^e dz = \int_{1}^{e} \frac{1}{z}\,dz = 1$$

11. $$\int_{0}^{1}\int_{0}^{\pi}\int_{0}^{\pi} y\sin z\,dx\,dy\,dz = \int_{0}^{1}\int_{0}^{\pi} \pi y\sin z\,dy\,dz = \frac{\pi^3}{2}\int_{0}^{1} \sin z\,dz = \frac{\pi^3}{2}(1-\cos 1)$$

13. $\int_0^3 \int_0^{\sqrt{9-x^2}} \int_0^{\sqrt{9-x^2}} dz\,dy\,dx = \int_0^3 \int_0^{\sqrt{9-x^2}} \sqrt{9-x^2}\,dy\,dx = \int_0^3 (9-x^2)\,dx = \left[9x - \frac{x^3}{3}\right]_0^3 = 18$

15. $\int_0^1 \int_0^{2-x} \int_0^{2-x-y} dz\,dy\,dx = \int_0^1 \int_0^{2-x} (2-x-y)\,dy\,dx = \int_0^1 \left[(2-x)^2 - \frac{1}{2}(2-x)^2\right] dx = \frac{1}{2}\int_0^1 (2-x)^2\,dx$

$= \left[-\frac{1}{6}(2-x)^3\right]_0^1 = -\frac{1}{6} + \frac{8}{6} = \frac{7}{6}$

17. $\int_0^\pi \int_0^\pi \int_0^\pi \cos(u+v+w)\,du\,dv\,dw = \int_0^\pi \int_0^\pi [\sin(w+v+\pi) - \sin(w+v)]\,dv\,dw$

$= \int_0^\pi [(-\cos(w+2\pi) + \cos(w+\pi)) + (\cos(w+\pi) - \cos w)]\,dw$

$= [-\sin(w+2\pi) + \sin(w+\pi) - \sin w + \sin(w+\pi)]_0^\pi = 0$

19. $\int_0^{\pi/4} \int_0^{\ln \sec v} \int_{-\infty}^{2t} e^x\,dx\,dt\,dv = \int_0^{\pi/4} \int_0^{\ln \sec v} \lim_{b\to-\infty} \left(e^{2t} - e^b\right) dt\,dv = \int_0^{\pi/4} \int_0^{\ln \sec v} e^{2t}\,dt\,dv = \int_0^{\pi/4} \frac{1}{2}e^{2\ln \sec v}\,dv$

$= \int_0^{\pi/4} \frac{\sec^2 v}{2}\,dv = \left[\frac{\tan v}{2}\right]_0^{\pi/4} = \frac{1}{2}$

21. (a) $\int_{-1}^1 \int_0^{1-x^2} \int_{x^2}^{1-z} dy\,dz\,dx$ (b) $\int_0^1 \int_{-\sqrt{1-z}}^{\sqrt{1-z}} \int_{x^2}^{1-z} dy\,dx\,dz$ (c) $\int_0^1 \int_0^{1-z} \int_{-\sqrt{y}}^{\sqrt{y}} dx\,dy\,dz$

(d) $\int_0^1 \int_0^{1-y} \int_{-\sqrt{y}}^{\sqrt{y}} dx\,dz\,dy$ (e) $\int_0^1 \int_{-\sqrt{y}}^{\sqrt{y}} \int_0^{1-y} dz\,dx\,dy$

23. $V = \int_0^1 \int_{-1}^1 \int_0^{y^2} dz\,dy\,dx = \int_0^1 \int_{-1}^1 y^2\,dy\,dx = \frac{2}{3}\int_0^1 dx = \frac{2}{3}$

25. $V = \int_0^4 \int_0^{\sqrt{4-x}} \int_0^{2-y} dz\,dy\,dx = \int_0^4 \int_0^{\sqrt{4-x}} (2-y)\,dy\,dx = \int_0^4 \left[2\sqrt{4-x} - \left(\frac{4-x}{2}\right)\right] dx$

$= \left[-\frac{4}{3}(4-x)^{3/2} + \frac{1}{4}(4-x)^2\right]_0^4 = \frac{4}{3}(4)^{3/2} - \frac{1}{4}(16) = \frac{32}{3} - 4 = \frac{20}{3}$

27. $V = \int_0^1 \int_0^{2-2x} \int_0^{3-3x-3y/2} dz\,dy\,dx = \int_0^1 \int_0^{2-2x} \left(3 - 3x - \frac{3}{2}y\right) dy\,dx = \int_0^1 \left[6(1-x)^2 - \frac{3}{4}\cdot 4(1-x)^2\right] dx$

$= \int_0^1 3(1-x)^2\,dx = \left[-(1-x)^3\right]_0^1 = 1$

29. $V = 8 \int_0^1 \int_0^{\sqrt{1-x^2}} \int_0^{\sqrt{1-x^2}} dz\,dy\,dx = 8 \int_0^1 \int_0^{\sqrt{1-x^2}} \sqrt{1-x^2}\,dy\,dx = 8 \int_0^1 (1-x^2)\,dx = \frac{16}{3}$

31. $V = \int_0^4 \int_0^{(\sqrt{16-y^2})/2} \int_0^{4-y} dx\,dz\,dy = \int_0^4 \int_0^{(\sqrt{16-y^2})/2} (4-y)\,dz\,dy = \int_0^4 \frac{\sqrt{16-y^2}}{2}(4-y)\,dy$

$= \int_0^4 2\sqrt{16-y^2}\,dy - \frac{1}{2}\int_0^4 y\sqrt{16-y^2}\,dy = \left[y\sqrt{16-y^2} + 16\sin^{-1}\frac{y}{4}\right]_0^4 + \left[\frac{1}{6}(16-y^2)^{3/2}\right]_0^4$

$= 16\left(\frac{\pi}{2}\right) - \frac{1}{6}(16)^{3/2} = 8\pi - \frac{32}{3}$

33. $\int_0^2 \int_0^{2-x} \int_{(2-x-y)/2}^{4-2x-2y} dz\,dy\,dx = \int_0^2 \int_0^{2-x} \left(3 - \frac{3x}{2} - \frac{3y}{2}\right) dy\,dx$

$= \int_0^2 \left[3\left(1-\frac{x}{2}\right)(2-x) - \frac{3}{4}(2-x)^2\right] dx$

$= \int_0^2 \left[6 - 6x + \frac{3x^2}{2} - \frac{3(2-x)^2}{4}\right] dx$

$= \left[6x - 3x^2 + \frac{x^3}{2} + \frac{(2-x)^3}{4}\right]_0^2 = (12 - 12 + 4 + 0) - \frac{2^3}{4} = 2$

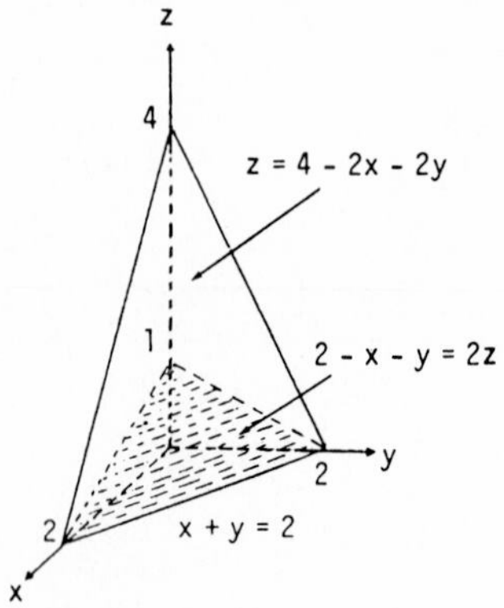

35. $V = 2\int_{-2}^{2} \int_0^{\sqrt{4-x^2}/2} \int_0^{x+2} dz\,dy\,dx = 2\int_{-2}^{2} \int_0^{\sqrt{4-x^2}/2} (x+2)\,dy\,dx = \int_{-2}^{2} (x+2)\sqrt{4-x^2}\,dx$

$= \int_{-2}^{2} 2\sqrt{4-x^2}\,dx + \int_{-2}^{2} x\sqrt{4-x^2}\,dx = \left[x\sqrt{4-x^2} + 4\sin^{-1}\frac{x}{2}\right]_{-2}^{2} + \left[-\frac{1}{3}(4-x^2)^{3/2}\right]_{-2}^{2}$

$= 4\left(\frac{\pi}{2}\right) - 4\left(-\frac{\pi}{2}\right) = 4\pi$

37. $\text{average} = \frac{1}{8}\int_0^2 \int_0^2 \int_0^2 (x^2+9)\,dz\,dy\,dx = \frac{1}{8}\int_0^2 \int_0^2 (2x^2+18)\,dy\,dx = \frac{1}{8}\int_0^2 (4x^2+36)\,dx = \frac{31}{3}$

39. average $= \int_0^1 \int_0^1 \int_0^1 (x^2+y^2+z^2)\,dz\,dy\,dx = \int_0^1 \int_0^1 \left(x^2+y^2+\frac{1}{3}\right) dy\,dx = \int_0^1 \left(x^2+\frac{2}{3}\right) dx = 1$

41. $\int_0^4 \int_0^1 \int_{2y}^2 \frac{4\cos(x^2)}{2\sqrt{z}}\,dx\,dy\,dz = \int_0^4 \int_0^2 \int_0^{x/2} \frac{4\cos(x^2)}{2\sqrt{z}}\,dy\,dx\,dz = \int_0^4 \int_0^2 \frac{x\cos(x^2)}{2\sqrt{z}}\,dx\,dz = \int_0^4 \left(\frac{\sin 4}{2}\right) z^{-1/2}\,dz$

$= \left[(\sin 4)z^{1/2}\right]_0^4 = 2 \sin 4$

43. $\int_0^1 \int_{\sqrt[3]{z}}^1 \int_0^{\ln 3} \frac{\pi e^{2x}\sin(\pi y^2)}{y^2}\,dx\,dy\,dz = \int_0^1 \int_{\sqrt[3]{z}}^1 \frac{4\pi \sin(\pi y^2)}{y^2}\,dy\,dz = \int_0^1 \int_0^{y^3} \frac{4\pi \sin(\pi y^2)}{y^2}\,dz\,dy$

$= \int_0^1 4\pi y \sin(\pi y^2)\,dy = \left[-2\cos(\pi y^2)\right]_0^1 = -2(-1)+2(1) = 4$

45. $\int_0^1 \int_0^{4-a-x^2} \int_a^{4-x^2-y} dz\,dy\,dx = \frac{4}{15} \Rightarrow \int_0^1 \int_0^{4-a-x^2} (4-x^2-y-a)\,dy\,dx = \frac{4}{15}$

$\Rightarrow \int_0^1 \left[(4-a-x^2)^2 - \frac{1}{2}(4-a-x^2)^2\right] dx = \frac{4}{15} \Rightarrow \frac{1}{2}\int_0^1 (4-a-x^2)^2\,dx = \frac{4}{15} \Rightarrow \int_0^1 \left[(4-a)^2 - 2x^2(4-a) + x^4\right] dx$

$= \frac{8}{15} \Rightarrow \left[(4-a)^2 x - \frac{2}{3}x^3(4-a) + \frac{x^5}{5}\right]_0^1 = \frac{8}{15} \Rightarrow (4-a)^2 - \frac{2}{3}(4-a) + \frac{1}{5} = \frac{8}{15} \Rightarrow 15(4-a)^2 - 10(4-a) - 5 = 0$

$\Rightarrow 3(4-a)^2 - 2(4-a) - 1 = 0 \Rightarrow [3(4-a)+1][(4-a)-1] = 0 \Rightarrow 4-a = -\frac{1}{3}$ or $4-a = 1 \Rightarrow a = \frac{13}{3}$ or $a = 3$

47. To minimize the integral, we want the domain to include all points where the integrand is negative and to exclude all points where it is positive. These criteria are met by the points (x,y,z) such that $4x^2+4y^2+z^2-4 \le 0$ or $4x^2+4y^2+z^2 \le 4$, which is a solid ellipsoid centered at the origin.

13.5 MASSES AND MOMENTS IN THREE DIMENSIONS

1. $I_x = \int_{-c/2}^{c/2} \int_{-b/2}^{b/2} \int_{-a/2}^{a/2} (y^2+z^2)\,dx\,dy\,dz = a \int_{-c/2}^{c/2} \int_{-b/2}^{b/2} (y^2+z^2)\,dy\,dz = a \int_{-c/2}^{c/2} \left[\frac{y^3}{3} + yz^2\right]_{-b/2}^{b/2} dz$

$= a \int_{-c/2}^{c/2} \left(\frac{b^3}{12} + bz^2\right) dz = ab\left[\frac{b^2}{12}z + \frac{z^3}{3}\right]_{-c/2}^{c/2} = ab\left(\frac{b^2 c}{12} + \frac{c^3}{12}\right) = \frac{abc}{12}(b^2+c^2) = \frac{M}{12}(b^2+c^2);$

$R_x = \sqrt{\frac{b^2+c^2}{12}}$; likewise $R_y = \sqrt{\frac{a^2+c^2}{12}}$ and $R_z = \sqrt{\frac{a^2+b^2}{12}}$, by symmetry

3. $I_x = \int_0^a \int_0^b \int_0^c (y^2+z^2)\,dz\,dy\,dx = \int_0^a \int_0^b \left(cy^2 + \frac{c^3}{3}\right) dy\,dx = \int_0^a \left(\frac{cb^3}{3} + \frac{c^3b}{3}\right) dx = \frac{abc(b^2+c^2)}{3}$

$= \frac{M}{3}(b^2+c^2)$ where $M = abc$; $I_y = \frac{M}{3}(a^2+c^2)$ and $I_z = \frac{M}{3}(a^2+b^2)$, by symmetry

5. $M = 4\int_0^1 \int_0^1 \int_{4y^2}^4 dz\,dy\,dx = 4\int_0^1 \int_0^1 (4-4y^2)\,dy\,dx = 16\int_0^1 \frac{2}{3}\,dx = \frac{32}{3}$; $M_{xy} = 4\int_0^1 \int_0^1 \int_{4y^2}^4 z\,dz\,dy\,dx$

$= 2\int_0^1 \int_0^1 (16-16y^4)\,dy\,dx = \frac{128}{5}\int_0^1 dx = \frac{128}{5} \Rightarrow \bar{z} = \frac{12}{5}$, and $\bar{x} = \bar{y} = 0$, by symmetry;

$I_x = 4\int_0^1 \int_0^1 \int_{4y^2}^4 (y^2+z^2)\,dz\,dy\,dx = 4\int_0^1 \int_0^1 \left[\left(4y^2 + \frac{64}{3}\right) - \left(4y^4 + \frac{64y^6}{3}\right)\right] dy\,dx = 4\int_0^1 \frac{1976}{105}\,dx = \frac{7904}{105}$;

$I_y = 4\int_0^1 \int_0^1 \int_{4y^2}^4 (x^2+z^2)\,dz\,dy\,dx = 4\int_0^1 \int_0^1 \left[\left(4x^2 + \frac{64}{3}\right) - \left(4x^2y^2 + \frac{64y^6}{3}\right)\right] dy\,dx = 4\int_0^1 \left(\frac{8}{3}x^2 + \frac{128}{7}\right) dx$

$= \frac{4832}{63}$; $I_z = 4\int_0^1 \int_0^1 \int_{4y^2}^4 (x^2+y^2)\,dz\,dy\,dx = 16\int_0^1 \int_0^1 (x^2 - x^2y^2 + y^2 - y^4)\,dy\,dx$

$= 16\int_0^1 \left(\frac{2x^2}{3} + \frac{2}{15}\right) dx = \frac{256}{45}$

7. (a) $M = 4\int_0^2 \int_0^{\sqrt{4-x^2}} \int_{x^2+y^2}^4 dz\,dy\,dx = 4\int_0^{\pi/2} \int_0^2 \int_{r^2}^4 r\,dz\,dr\,d\theta = 4\int_0^{\pi/2} \int_0^2 (4r - r^3)\,dr\,d\theta = 4\int_0^{\pi/2} 4\,d\theta = 8\pi$;

$M_{xy} = \int_0^{2\pi} \int_0^2 \int_{r^2}^4 zr\,dz\,dr\,d\theta = \int_0^{2\pi} \int_0^2 \frac{r}{2}(16 - r^4)\,dr\,d\theta = \frac{32}{3}\int_0^{2\pi} d\theta = \frac{64\pi}{3} \Rightarrow \bar{z} = \frac{8}{3}$, and $\bar{x} = \bar{y} = 0$,
by symmetry

(b) $M = 8\pi \Rightarrow 4\pi = \int_0^{2\pi} \int_0^{\sqrt{c}} \int_{r^2}^{c} r\,dz\,dr\,d\theta = \int_0^{2\pi} \int_0^{\sqrt{c}} (cr - r^3)\,dr\,d\theta = \int_0^{2\pi} \frac{c^2}{4}\,d\theta = \frac{c^2\pi}{2} \Rightarrow c^2 = 8 \Rightarrow c = 2\sqrt{2}$,
since $c > 0$

9. The plane $y + 2z = 2$ is the top of the wedge $\Rightarrow I_L = \int_{-2}^2 \int_{-2}^4 \int_{-1}^{(2-y)/2} \left[(y-6)^2 + z^2\right] dz\,dy\,dx$

$= \int_{-2}^2 \int_{-2}^4 \left[\frac{(y-6)^2(4-y)}{2} + \frac{(2-y)^3}{24} + \frac{1}{3}\right] dy\,dx$; let $t = 2 - y \Rightarrow I_L = 4\int_{-2}^4 \left(\frac{13t^3}{24} + 5t^2 + 16t + \frac{49}{3}\right) dt = 1386$;

$M = \frac{1}{2}(3)(6)(4) = 36 \Rightarrow R_L = \sqrt{\frac{I_L}{M}} = \sqrt{\frac{77}{2}}$

11. $M = 8$; $I_L = \int_0^4 \int_0^2 \int_0^1 \left[z^2 + (y-2)^2\right] dz\,dy\,dx = \int_0^4 \int_0^2 \left(y^2 - 4y + \frac{13}{3}\right) dy\,dx = \frac{10}{3} \int_0^4 dx = \frac{40}{3}$

$\Rightarrow R_L = \sqrt{\frac{I_L}{M}} = \sqrt{\frac{5}{3}}$

13. (a) $M = \int_0^2 \int_0^{2-x} \int_0^{2-x-y} 2x\,dz\,dy\,dx = \int_0^2 \int_0^{2-x} \left(4x - 2x^2 - 2xy\right) dy\,dx = \int_0^2 \left(x^3 - 4x^2 + 4x\right) dx = \frac{4}{3}$

(b) $M_{xy} = \int_0^2 \int_0^{2-x} \int_0^{2-x-y} 2xz\,dz\,dy\,dx = \int_0^2 \int_0^{2-x} x(2-x-y)^2\,dy\,dx = \int_0^2 \frac{x(2-x)^3}{3}\,dx = \frac{8}{15}$; $M_{xz} = \frac{8}{15}$ by symmetry; $M_{yz} = \int_0^2 \int_0^{2-x} \int_0^{2-x-y} 2x^2\,dz\,dy\,dx = \int_0^2 \int_0^{2-x} 2x^2(2-x-y)\,dy\,dx = \int_0^2 \left(2x - x^2\right)^2 dx = \frac{16}{15}$

$\Rightarrow \overline{x} = \frac{4}{5}$, and $\overline{y} = \overline{z} = \frac{2}{5}$

15. (a) $M = \int_0^1 \int_0^1 \int_0^1 (x+y+z+1)\,dz\,dy\,dx = \int_0^1 \int_0^1 \left(x + y + \frac{3}{2}\right) dy\,dx = \int_0^1 (x+2)\,dx = \frac{5}{2}$

(b) $M_{xy} = \int_0^1 \int_0^1 \int_0^1 z(x+y+z+1)\,dz\,dy\,dx = \frac{1}{2} \int_0^1 \int_0^1 \left(x + y + \frac{5}{3}\right) dy\,dx = \frac{1}{2} \int_0^1 \left(x + \frac{13}{6}\right) dx = \frac{4}{3}$

$\Rightarrow M_{xy} = M_{yz} = M_{xz} = \frac{4}{3}$, by symmetry $\Rightarrow \overline{x} = \overline{y} = \overline{z} = \frac{8}{15}$

(c) $I_z = \int_0^1 \int_0^1 \int_0^1 \left(x^2 + y^2\right)(x+y+z+1)\,dz\,dy\,dx = \int_0^1 \int_0^1 \left(x^2+y^2\right)\left(x + y + \frac{3}{2}\right) dy\,dx$

$= \int_0^1 \left(x^3 + 2x^2 + \frac{1}{3}x + \frac{3}{4}\right) dx = \frac{11}{6} \Rightarrow I_x = I_y = I_z = \frac{11}{6}$, by symmetry

(d) $R_x = R_y = R_z = \sqrt{\frac{I_z}{M}} = \sqrt{\frac{11}{15}}$

17. $M = \int_0^1 \int_{z-1}^{1-z} \int_0^{\sqrt{z}} (2y+5)\,dy\,dx\,dz = \int_0^1 \int_{z-1}^{1-z} \left(z + 5\sqrt{z}\right) dx\,dz = \int_0^1 2\left(z + 5\sqrt{z}\right)(1-z)\,dz$

$= 2 \int_0^1 \left(5z^{1/2} + z - 5z^{3/2} - z^2\right) dz = 2\left[\frac{10}{3} z^{3/2} + \frac{1}{2} z^2 - 2z^{5/2} - \frac{1}{3} z^3\right]_0^1 = 2\left(\frac{9}{3} - \frac{3}{2}\right) = 3$

19. (a) Let ΔV_i be the volume of the ith piece, and let (x_i, y_i, z_i) be a point in the ith piece. Then the work done by gravity in moving the ith piece to the xy-plane is approximately $W_i = m_i g z_i = (x_i + y_i + z_i + 1)g\,\Delta V_i\, z_i$ $\Rightarrow$ the total work done is the triple integral $W = \int_0^1 \int_0^1 \int_0^1 (x + y + z + 1)gz\, dz\, dy\, dx$

$$= g \int_0^1 \int_0^1 \left[\tfrac{1}{2}xz^2 + \tfrac{1}{2}yz^2 + \tfrac{1}{3}z^3 + \tfrac{1}{2}z^2\right]_0^1 dy\, dx = g \int_0^1 \int_0^1 \left(\tfrac{1}{2}x + \tfrac{1}{2}y + \tfrac{5}{6}\right) dy\, dx = g \int_0^1 \left[\tfrac{1}{2}xy + \tfrac{1}{4}y^2 + \tfrac{5}{6}y\right]_0^1 dx$$

$$= g \int_0^1 \left(\tfrac{1}{2}x + \tfrac{13}{12}\right) dx = g\left[\frac{x^2}{4} + \frac{13}{12}x\right]_0^1 = g\left(\frac{16}{12}\right) = \tfrac{4}{3}g$$

(b) From Exercise 15 the center of mass is $\left(\frac{8}{15}, \frac{8}{15}, \frac{8}{15}\right)$ and the mass of the liquid is $\frac{5}{2}$ $\Rightarrow$ the work done by gravity in moving the center of mass to the xy-plane is $W = mgd = \left(\frac{5}{2}\right)(g)\left(\frac{8}{15}\right) = \frac{4}{3}g$, which is the same as the work done in part (a).

21. (a) $\bar{x} = \frac{M_{yz}}{M} = 0 \Rightarrow \iiint_R x\delta(x, y, z)\, dx\, dy\, dz = 0 \Rightarrow M_{yz} = 0$

(b) $I_L = \iiint_R |\mathbf{v} - h\mathbf{i}|^2\, dm = \iiint_R |(x - h)\mathbf{i} + y\mathbf{j}|^2\, dm = \iiint_R (x^2 - 2xh + h^2 + y^2)\, dm$

$$= \iiint_R (x^2 + y^2)\, dm - 2h \iiint_R x\, dm + h^2 \iiint_R dm = I_x - 0 + h^2 m = I_{c.m.} + h^2 m$$

23. (a) $(\bar{x}, \bar{y}, \bar{z}) = \left(\frac{a}{2}, \frac{b}{2}, \frac{c}{2}\right) \Rightarrow I_z = I_{c.m.} + abc\left(\sqrt{\frac{a^2}{4} + \frac{b^2}{4}}\right)^2 \Rightarrow I_{c.m.} = I_z - \frac{abc(a^2 + b^2)}{4}$

$$= \frac{abc(a^2 + b^2)}{3} - \frac{abc(a^2 + b^2)}{4} = \frac{abc(a^2 + b^2)}{12};\ R_{c.m.} = \sqrt{\frac{I_{c.m.}}{M}} = \sqrt{\frac{a^2 + b^2}{12}}$$

(b) $I_L = I_{c.m.} + abc\left(\sqrt{\frac{a^2}{4} + \left(\frac{b}{2} - 2b\right)^2}\right)^2 = \frac{abc(a^2 + b^2)}{12} + \frac{abc(a^2 + 9b^2)}{4} = \frac{abc(4a^2 + 28b^2)}{12}$

$$= \frac{abc(a^2 + 7b^2)}{3};\ R_L = \sqrt{\frac{I_L}{M}} = \sqrt{\frac{a^2 + 7b^2}{3}}$$

25. $M_{yz_{B_1 \cup B_2}} = \iiint_{B_1} x\, dV_1 + \iiint_{B_2} x\, dV_2 = M_{(yz)_1} + M_{(yz)_2} \Rightarrow \bar{x} = \frac{M_{(yz)_1} + M_{(yz)_2}}{m_1 + m_2}$; similarly,

$\bar{y} = \frac{M_{(xz)_1} + M_{(xz)_2}}{m_1 + m_2}$ and $\bar{z} = \frac{M_{(xy)_1} + M_{(xy)_2}}{m_1 + m_2}$ $\Rightarrow \mathbf{c} = \bar{x}\mathbf{i} + \bar{y}\mathbf{j} + \bar{z}\mathbf{k}$

$$= \frac{1}{m_1 + m_2}\left[\left(M_{(yz)_1} + M_{(yz)_2}\right)\mathbf{i} + \left(M_{(xz)_1} + M_{(xz)_2}\right)\mathbf{j} + \left(M_{(xy)_1} + M_{(xy)_2}\right)\mathbf{k}\right]$$

$= \frac{1}{m_1 + m_2}[(m_1\bar{x}_1 + m_2\bar{x}_2)\mathbf{i} + (m_1\bar{y}_1 + m_2\bar{y}_2)\mathbf{j} + (m_1\bar{z}_1 + m_2\bar{z}_2)\mathbf{k}]$

$= \frac{1}{m_1 + m_2}[m_1(\bar{x}_1\mathbf{i} + \bar{y}_1\mathbf{j} + \bar{z}_1\mathbf{k}) + m_2(\bar{x}_2\mathbf{i} + \bar{y}_2\mathbf{j} + \bar{z}_2\mathbf{k})] = \frac{m_1\mathbf{c}_1 + m_2\mathbf{c}_2}{m_1 + m_2}$

27. (a) $\mathbf{c} = \frac{\left(\frac{\pi a^2 h}{3}\right)\left(\frac{h}{4}\mathbf{k}\right) + \left(\frac{2\pi a^3}{3}\right)\left(-\frac{3a}{8}\mathbf{k}\right)}{m_1 + m_2} = \frac{\left(\frac{a^2\pi}{3}\right)\left(\frac{h^2 - 3a^2}{4}\mathbf{k}\right)}{m_1 + m_2}$, where $m_1 = \frac{\pi a^2 h}{3}$ and $m_2 = \frac{2\pi a^3}{3}$; if

$\frac{h^2 - 3a^2}{4} = 0$, or $h = a\sqrt{3}$, then the centroid is on the common base

(b) See the solution to Exercise 55, Section 13.2, to see that $h = a\sqrt{2}$.

13.6 TRIPLE INTEGRALS IN CYLINDRICAL AND SPHERICAL COORDINATES

1. $\int_0^{2\pi}\int_0^1\int_r^{\sqrt{2-r^2}} dz\, r\, dr\, d\theta = \int_0^{2\pi}\int_0^1 \left[r(2-r^2)^{1/2} - r^2\right] dr\, d\theta = \int_0^{2\pi}\left[-\frac{1}{3}(2-r^2)^{3/2} - \frac{r^3}{3}\right]_0^1 d\theta$

$= \int_0^{2\pi}\left(\frac{2^{3/2}}{3} - \frac{2}{3}\right) d\theta = \frac{4\pi(\sqrt{2}-1)}{3}$

3. $\int_0^{2\pi}\int_0^{\theta/2\pi}\int_0^{3+24r^2} dz\, r\, dr\, d\theta = \int_0^{2\pi}\int_0^{\theta/2\pi} (3r + 24r^3)\, dr\, d\theta = \int_0^{2\pi}\left[\frac{3}{2}r^2 + 6r^4\right]_0^{\theta/2\pi} d\theta = \frac{3}{2}\int_0^{2\pi}\left(\frac{\theta^2}{4\pi^2} + \frac{4\theta^4}{16\pi^4}\right) d\theta$

$= \frac{3}{2}\left[\frac{\theta^3}{12\pi^2} + \frac{\theta^5}{5\pi^4}\right]_0^{2\pi} = \frac{17\pi}{5}$

5. $\int_0^{2\pi}\int_0^1\int_r^{(2-r^2)^{-1/2}} 3\, dz\, r\, dr\, d\theta = 3\int_0^{2\pi}\int_0^1 \left[r(2-r^2)^{-1/2} - r^2\right] dr\, d\theta = 3\int_0^{2\pi}\left[-(2-r^2)^{1/2} - \frac{r^3}{3}\right]_0^1 d\theta$

$= 3\int_0^{2\pi}\left(\sqrt{2} - \frac{4}{3}\right) d\theta = \pi(6\sqrt{2} - 8)$

7. $\int_0^{2\pi}\int_0^3\int_0^{z/3} r^3\, dr\, dz\, d\theta = \int_0^{2\pi}\int_0^3 \frac{z^4}{324}\, dz\, d\theta = \int_0^{2\pi} \frac{3}{20}\, d\theta = \frac{3\pi}{10}$

9. $\int_0^1\int_0^{\sqrt{z}}\int_0^{2\pi} (r^2\cos^2\theta + z^2)\, r\, d\theta\, dr\, dz = \int_0^1\int_0^{\sqrt{z}} \left[\frac{r^2\theta}{2} + \frac{r^2\sin 2\theta}{4} + z^2\theta\right]_0^{2\pi} r\, dr\, dz = \int_0^1\int_0^{\sqrt{z}} (\pi r^3 + 2\pi r z^2)\, dr\, dz$

$= \int_0^1 \left[\frac{\pi r^4}{4} + \pi r^2 z^2\right]_0^{\sqrt{z}} dz = \int_0^1 \left(\frac{\pi z^2}{4} + \pi z^3\right) dz = \left[\frac{\pi z^3}{12} + \frac{\pi z^4}{4}\right]_0^1 = \frac{\pi}{3}$

11. (a) $\int_0^{2\pi}\int_0^1\int_0^{\sqrt{4-r^2}} dz\ r\ dr\, d\theta$

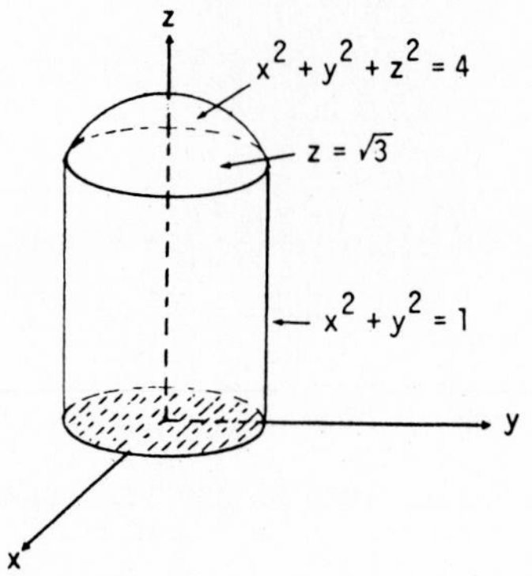

(b) $\int_0^{2\pi}\int_0^{\sqrt{3}}\int_0^1 r\ dr\, dz\, d\theta + \int_0^{2\pi}\int_{\sqrt{3}}^2\int_0^{\sqrt{4-z^2}} r\ dr\, dz\, d\theta$

(c) $\int_0^1\int_0^{\sqrt{4-r^2}}\int_0^{2\pi} r\ d\theta\, dz\, dr$

13. $\int_{-\pi/2}^{\pi/2}\int_0^{\cos\theta}\int_0^{3r^2} f(r,\theta,z)\ dz\ r\ dr\, d\theta$

15. $\int_0^{\pi}\int_0^{2\sin\theta}\int_0^{4-r\sin\theta} f(r,\theta,z)\ dz\ r\ dr\, d\theta$

17. $\int_{-\pi/2}^{\pi/2}\int_1^{1+\cos\theta}\int_0^4 f(r,\theta,z)\ dz\ r\ dr\, d\theta$

19. $\int_0^{\pi/4}\int_0^{\sec\theta}\int_0^{2-r\sin\theta} f(r,\theta,z)\ dz\ r\ dr\, d\theta$

21. $\int_0^{\pi}\int_0^{\pi}\int_0^{2\sin\phi} \rho^2\sin\phi\ d\rho\, d\phi\, d\theta = \frac{8}{3}\int_0^{\pi}\int_0^{\pi}\sin^4\phi\ d\phi\ d\theta = \frac{8}{3}\int_0^{\pi}\left(\left[-\frac{\sin^3\phi\cos\phi}{4}\right]_0^{\pi} + \frac{3}{4}\int_0^{\pi}\sin^2\phi\ d\phi\right) d\theta$

$= 2\int_0^{\pi}\int_0^{\pi}\sin^2\phi\ d\phi\, d\theta = \int_0^{\pi}\left[\theta - \frac{\sin 2\theta}{2}\right]_0^{\pi} d\theta = \int_0^{\pi}\pi\ d\theta = \pi^2$

23. $\int_0^{2\pi}\int_0^{\pi}\int_0^{(1-\cos\phi)/2} \rho^2\sin\phi\ d\rho\, d\phi\, d\theta = \frac{1}{24}\int_0^{2\pi}\int_0^{\pi}(1-\cos\phi)^3\sin\phi\ d\phi\ d\theta = \frac{1}{96}\int_0^{2\pi}\left[(1-\cos\phi)^4\right]_0^{\pi} d\theta$

$= \frac{1}{96}\int_0^{2\pi}\left(2^4 - 0\right) d\theta = \frac{16}{96}\int_0^{2\pi} d\theta = \frac{1}{6}(2\pi) = \frac{\pi}{3}$

25. $\int_0^{2\pi}\int_0^{\pi/3}\int_{\sec\phi}^{2} 3\rho^2\sin\phi\ d\rho\, d\phi\ d\theta = \int_0^{2\pi}\int_0^{\pi/3}\left(8-\sec^3\phi\right)\sin\phi\ d\phi\, d\theta = \int_0^{2\pi}\left[-8\cos\phi - \frac{1}{2}\sec^2\phi\right]_0^{\pi/3} d\theta$

$= \int_0^{2\pi}\left[(-4-2) - \left(-8 - \frac{1}{2}\right)\right] d\theta = \frac{5}{2}\int_0^{2\pi} d\theta = 5\pi$

27. $\int_0^2 \int_{-\pi}^0 \int_{\pi/4}^{\pi/2} \rho^3 \sin 2\phi \, d\phi \, d\theta \, d\rho = \int_0^2 \int_{-\pi}^0 \rho^3 \left[-\frac{\cos 2\phi}{2}\right]_{\pi/4}^{\pi/2} d\theta \, d\rho = \int_0^2 \int_{-\pi}^0 \frac{\rho^3}{2} \, d\theta \, d\rho = \int_0^2 \frac{\rho^3 \pi}{2} \, d\rho$

$= \left[\frac{\pi \rho^4}{8}\right]_0^2 = 2\pi$

29. $\int_0^1 \int_0^\pi \int_0^{\pi/4} 12\rho \sin^3 \phi \, d\phi \, d\theta \, d\rho = \int_0^1 \int_0^\pi \left(12\rho\left[\frac{-\sin^2 \phi \cos \phi}{3}\right]_0^{\pi/4} + 8\rho \int_0^{\pi/4} \sin \phi \, d\phi\right) d\theta \, d\rho$

$= \int_0^1 \int_0^\pi \left(-\frac{2\rho}{\sqrt{2}} - 8\rho\left[\cos \phi\right]_0^{\pi/4}\right) d\theta \, d\rho = \int_0^1 \int_0^\pi \left(8\rho - \frac{10\rho}{\sqrt{2}}\right) d\theta \, d\rho = \pi \int_0^1 \left(8\rho - \frac{10\rho}{\sqrt{2}}\right) d\rho = \pi\left[4\rho^2 - \frac{5\rho^2}{\sqrt{2}}\right]_0^1$

$= \frac{(4\sqrt{2}-5)\,\pi}{\sqrt{2}}$

31. (a) $x^2 + y^2 = 1 \Rightarrow \rho^2 \sin^2 \phi = 1$, and $\rho \sin \phi = 1 \Rightarrow \rho = \csc \phi$; thus

$$\int_0^{2\pi} \int_0^{\pi/6} \int_0^2 \rho^2 \sin \phi \, d\rho \, d\phi \, d\theta + \int_0^{2\pi} \int_{\pi/6}^{\pi/2} \int_0^{\csc \phi} \rho^2 \sin \phi \, d\rho \, d\phi \, d\theta$$

(b) $\int_0^{2\pi} \int_1^2 \int_{\pi/6}^{\sin^{-1}(1/\rho)} \rho^2 \sin \phi \, d\phi \, d\rho \, d\theta + \int_0^{2\pi} \int_0^2 \int_0^{\pi/6} \rho^2 \sin \phi \, d\phi \, d\rho \, d\theta$

33. $V = \int_0^{2\pi} \int_0^{\pi/2} \int_{\cos \phi}^2 \rho^2 \sin \phi \, d\rho \, d\phi \, d\theta = \frac{1}{3} \int_0^{2\pi} \int_0^{\pi/2} (8 - \cos^3 \phi) \sin \phi \, d\phi \, d\theta$

$= \frac{1}{3} \int_0^{2\pi} \left[-8 \cos \phi + \frac{\cos^4 \phi}{4}\right]_0^{\pi/2} d\theta = \frac{1}{3} \int_0^{2\pi} \left(8 - \frac{1}{4}\right) d\theta = \left(\frac{31}{12}\right)(2\pi) = \frac{31\pi}{6}$

35. $V = \int_0^{2\pi} \int_0^\pi \int_0^{1-\cos \phi} \rho^2 \sin \phi \, d\rho \, d\phi \, d\theta = \frac{1}{3} \int_0^{2\pi} \int_0^\pi (1 - \cos \phi)^3 \sin \phi \, d\phi \, d\theta = \frac{1}{3} \int_0^{2\pi} \left[\frac{(1-\cos \phi)^4}{4}\right]_0^\pi d\theta$

$= \frac{1}{12}(2)^4 \int_0^{2\pi} d\theta = \frac{4}{3}(2\pi) = \frac{8\pi}{3}$

37. $V = \int_0^{2\pi} \int_{\pi/4}^{\pi/2} \int_0^{2 \cos \phi} \rho^2 \sin \phi \, d\rho \, d\phi \, d\theta = \frac{8}{3} \int_0^{2\pi} \int_{\pi/4}^{\pi/2} \cos^3 \phi \sin \phi \, d\phi \, d\theta = \frac{8}{3} \int_0^{2\pi} \left[-\frac{\cos^4 \phi}{4}\right]_{\pi/4}^{\pi/2} d\theta$

$= \left(\frac{8}{3}\right)\left(\frac{1}{16}\right) \int_0^{2\pi} d\theta = \frac{1}{6}(2\pi) = \frac{\pi}{3}$

39. (a) $8\int_0^{\pi/2}\int_0^{\pi/2}\int_0^{2} \rho^2 \sin\phi \, d\rho \, d\phi \, d\theta$ (b) $8\int_0^{\pi/2}\int_0^{2}\int_0^{\sqrt{4-r^2}} dz \, r \, dr \, d\theta$

(c) $8\int_0^{2}\int_0^{\sqrt{4-x^2}}\int_0^{\sqrt{4-x^2-y^2}} dz \, dy \, dx$

41. (a) $V=\int_0^{2\pi}\int_0^{\pi/3}\int_{\sec\phi}^{2} \rho^2 \sin\phi \, d\rho \, d\phi \, d\theta$ (b) $V=\int_0^{2\pi}\int_0^{\sqrt{3}}\int_1^{\sqrt{4-r^2}} dz \, r \, dr \, d\theta$

(c) $V=\int_{-\sqrt{3}}^{\sqrt{3}}\int_{-\sqrt{3-x^2}}^{\sqrt{3-x^2}}\int_1^{\sqrt{4-x^2-y^2}} dz \, dy \, dx$

(d) $V=\int_0^{2\pi}\int_0^{\sqrt{3}} \left[r\left(4-r^2\right)^{1/2}-r\right] dr \, d\theta = \int_0^{2\pi}\left[-\frac{\left(4-r^2\right)^{3/2}}{3}-\frac{r^2}{2}\right]_0^{\sqrt{3}} d\theta = \int_0^{2\pi}\left(-\frac{1}{3}-\frac{3}{2}+\frac{4^{3/2}}{3}\right) d\theta$

$=\frac{5}{6}\int_0^{2\pi} d\theta = \frac{5\pi}{3}$

43. $V=4\int_0^{\pi/2}\int_0^{1}\int_{r^4-1}^{4-4r^2} dz \, r \, dr \, d\theta = 4\int_0^{\pi/2}\int_0^{1}\left(5r-4r^3-r^5\right) dr \, d\theta = 4\int_0^{\pi/2}\left(\frac{5}{2}-1-\frac{1}{6}\right) d\theta$

$=4\int_0^{\pi/2}\frac{8}{6} \, d\theta = \frac{8\pi}{3}$

45. $V=\int_{3\pi/2}^{2\pi}\int_0^{3\cos\theta}\int_0^{-r\sin\theta} dz \, r \, dr \, d\theta = \int_{3\pi/2}^{2\pi}\int_0^{3\cos\theta} -r^2\sin\theta \, dr \, d\theta = \int_{3\pi/2}^{2\pi}\left(-9\cos^3\theta\right)(\sin\theta) \, d\theta$

$=\left[\frac{9}{4}\cos^4\theta\right]_{3\pi/2}^{2\pi} = \frac{9}{4}-0=\frac{9}{4}$

47. $V=\int_0^{\pi/2}\int_0^{\sin\theta}\int_0^{\sqrt{1-r^2}} dz \, r \, dr \, d\theta = \int_0^{\pi/2}\int_0^{\sin\theta} r\sqrt{1-r^2} \, dr \, d\theta = \int_0^{\pi/2}\left[-\frac{1}{3}\left(1-r^2\right)^{3/2}\right]_0^{\sin\theta} d\theta$

$=-\frac{1}{3}\int_0^{\pi/2}\left[\left(1-\sin^2\theta\right)^{3/2}-1\right] d\theta = -\frac{1}{3}\int_0^{\pi/2}\left(\cos^3\theta-1\right) d\theta = -\frac{1}{3}\left(\left[\frac{\cos^2\theta\sin\theta}{3}\right]_0^{\pi/2}+\frac{2}{3}\int_0^{\pi/2}\cos\theta \, d\theta\right)+\left[\frac{\theta}{3}\right]_0^{\pi/2}$

$=-\frac{2}{9}[\sin\theta]_0^{\pi/2}+\frac{\pi}{6}=\frac{-4+3\pi}{18}$

49. $V = \int_0^{2\pi}\int_{\pi/3}^{2\pi/3}\int_0^{a} \rho^2 \sin \phi \, d\rho \, d\phi \, d\theta = \int_0^{2\pi}\int_{\pi/3}^{2\pi/3} \frac{a^3}{3} \sin \phi \, d\phi \, d\theta = \frac{a^3}{3}\int_0^{2\pi} \left[-\cos \phi\right]_{\pi/3}^{2\pi/3} d\theta = \frac{a^3}{3}\int_0^{2\pi} \left(\frac{1}{2}+\frac{1}{2}\right) d\theta = \frac{2\pi a^3}{3}$

51. $V = \int_0^{2\pi}\int_0^{\pi/3}\int_{\sec \phi}^{2} \rho^2 \sin \phi \, d\rho \, d\phi \, d\theta$

$= \frac{1}{3}\int_0^{2\pi}\int_0^{\pi/3} \left(8 \sin \phi - \tan \phi \sec^2 \phi\right) d\phi \, d\theta$

$= \frac{1}{3}\int_0^{2\pi} \left[-8 \cos \phi - \frac{1}{2} \tan^2 \phi\right]_0^{\pi/3} d\theta$

$= \frac{1}{3}\int_0^{2\pi} \left[-4 - \frac{1}{2}(3) + 8\right] d\theta = \frac{1}{3}\int_0^{2\pi} \frac{5}{2} \, d\theta = \frac{5}{6}(2\pi) = \frac{5\pi}{3}$

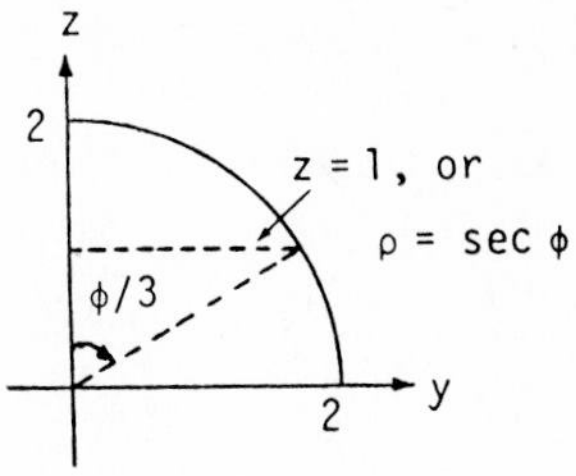

53. $V = 4\int_0^{\pi/2}\int_0^{1}\int_0^{r^2} dz \, r \, dr \, d\theta = 4\int_0^{\pi/2}\int_0^{1} r^3 \, dr \, d\theta = \int_0^{\pi/2} d\theta = \frac{\pi}{2}$

55. $V = 8\int_0^{\pi/2}\int_1^{\sqrt{2}}\int_0^{r} dz \, r \, dr \, d\theta = 8\int_0^{\pi/2}\int_1^{\sqrt{2}} r^2 \, dr \, d\theta = 8\left(\frac{2\sqrt{2}-1}{3}\right)\int_0^{\pi/2} d\theta = \frac{4\pi\left(2\sqrt{2}-1\right)}{3}$

57. $V = \int_0^{2\pi}\int_0^{2}\int_0^{4-r\sin\theta} dz \, r \, dr \, d\theta = \int_0^{2\pi}\int_0^{2} \left(4r - r^2 \sin \theta\right) dr \, d\theta = 8\int_0^{2\pi} \left(1 - \frac{\sin \theta}{3}\right) d\theta = 16\pi$

59. The paraboloids intersect when $4x^2 + 4y^2 = 5 - x^2 - y^2 \Rightarrow x^2 + y^2 = 1$ and $z = 4$

$\Rightarrow V = 4\int_0^{\pi/2}\int_0^{1}\int_{4r^2}^{5-r^2} dz \, r \, dr \, d\theta = 4\int_0^{\pi/2}\int_0^{1} \left(5r - 5r^3\right) dr \, d\theta = 20\int_0^{\pi/2} \left[\frac{r^2}{2} - \frac{r^4}{4}\right]_0^1 d\theta = 5\int_0^{\pi/2} d\theta = \frac{5\pi}{2}$

61. $V = 8\int_0^{\pi/2}\int_0^{1}\int_0^{\sqrt{4-r^2}} dz \, r \, dr \, d\theta = 8\int_0^{\pi/2}\int_0^{1} r\left(4-r^2\right)^{1/2} dr \, d\theta = 8\int_0^{\pi/2} \left[-\frac{1}{3}\left(4-r^2\right)^{3/2}\right]_0^1 d\theta$

$= -\frac{8}{3}\int_0^{\pi/2} \left(3^{3/2} - 8\right) d\theta = \frac{4\pi\left(8 - 3\sqrt{3}\right)}{3}$

63. $\text{average} = \frac{1}{2\pi}\int_0^{2\pi}\int_0^{1}\int_{-1}^{1} r^2 \, dz \, dr \, d\theta = \frac{1}{2\pi}\int_0^{2\pi}\int_0^{1} 2r^2 \, dr \, d\theta = \frac{1}{3\pi}\int_0^{2\pi} d\theta = \frac{2}{3}$

65. average $= \frac{1}{\left(\frac{4\pi}{3}\right)} \int_0^{2\pi} \int_0^{\pi} \int_0^{1} \rho^3 \sin\phi \, d\rho \, d\phi \, d\theta = \frac{3}{16\pi} \int_0^{2\pi} \int_0^{\pi} \sin\phi \, d\phi \, d\theta = \frac{3}{8\pi} \int_0^{2\pi} d\theta = \frac{3}{4}$

67. $M = 4 \int_0^{\pi/2} \int_0^{1} \int_0^{r} dz \, r \, dr \, d\theta = 4 \int_0^{\pi/2} \int_0^{1} r^2 \, dr \, d\theta = \frac{4}{3} \int_0^{\pi/2} d\theta = \frac{2\pi}{3}$; $M_{xy} = \int_0^{2\pi} \int_0^{1} \int_0^{r} z \, dz \, r \, dr \, d\theta$

$= \frac{1}{2} \int_0^{2\pi} \int_0^{1} r^3 \, dr \, d\theta = \frac{1}{8} \int_0^{2\pi} d\theta = \frac{\pi}{4} \Rightarrow \bar{z} = \frac{M_{xy}}{M} = \left(\frac{\pi}{4}\right)\left(\frac{3}{2\pi}\right) = \frac{3}{8}$, and $\bar{x} = \bar{y} = 0$, by symmetry

69. $M = \frac{8\pi}{3}$; $M_{xy} = \int_0^{2\pi} \int_{\pi/3}^{\pi/2} \int_0^{2} z\rho^2 \sin\phi \, d\rho \, d\phi \, d\theta = \int_0^{2\pi} \int_{\pi/3}^{\pi/2} \int_0^{2} \rho^3 \cos\phi \sin\phi \, d\rho \, d\phi \, d\theta = 4 \int_0^{2\pi} \int_{\pi/3}^{\pi/2} \cos\phi \sin\phi \, d\phi \, d\theta$

$= 4 \int_0^{2\pi} \left[\frac{\sin^2\phi}{2}\right]_{\pi/3}^{\pi/2} d\theta = 4 \int_0^{2\pi} \left(\frac{1}{2} - \frac{3}{8}\right) d\theta = \frac{1}{2} \int_0^{2\pi} d\theta = \pi \Rightarrow \bar{z} = \frac{M_{xy}}{M} = (\pi)\left(\frac{3}{8\pi}\right) = \frac{3}{8}$, and $\bar{x} = \bar{y} = 0$,
by symmetry

71. $M = \int_0^{2\pi} \int_0^{4} \int_0^{\sqrt{r}} dz \, r \, dr \, d\theta = \int_0^{2\pi} \int_0^{4} r^{3/2} \, dr \, d\theta = \frac{64}{5} \int_0^{2\pi} d\theta = \frac{128\pi}{5}$; $M_{xy} = \int_0^{2\pi} \int_0^{4} \int_0^{\sqrt{r}} z \, dz \, r \, dr \, d\theta$

$= \frac{1}{2} \int_0^{2\pi} \int_0^{4} r^2 \, dr \, d\theta = \frac{32}{3} \int_0^{2\pi} d\theta = \frac{64\pi}{3} \Rightarrow \bar{z} = \frac{M_{xy}}{M} = \frac{5}{6}$, and $\bar{x} = \bar{y} = 0$, by symmetry

73. $I_z = \int_0^{2\pi} \int_1^{2} \int_0^{4} (x^2 + y^2) \, dz \, r \, dr \, d\theta = 4 \int_0^{2\pi} \int_r^{2} r^3 \, dr \, d\theta = \int_0^{2\pi} 15 \, d\theta = 30\pi$; $M = \int_0^{2\pi} \int_1^{2} \int_0^{4} dz \, r \, dr \, d\theta$

$= \int_0^{2\pi} \int_1^{2} 4r \, dr \, d\theta = \int_0^{2\pi} 6 \, d\theta = 12\pi \Rightarrow R_z = \sqrt{\frac{I_z}{M}} = \sqrt{\frac{5}{2}}$

75. We orient the cone with its vertex at the origin and axis along the z-axis $\Rightarrow \phi = \frac{\pi}{4}$. We use the the x-axis which is through the vertex and parallel to the base of the cone $\Rightarrow I_x = \int_0^{2\pi} \int_0^{1} \int_r^{1} (r^2 \sin^2\theta + z^2) \, dz \, r \, dr \, d\theta$

$= \int_0^{2\pi} \int_0^{1} \left(r^3 \sin^2\theta - r^4 \sin^2\theta + \frac{r}{3} - \frac{r^4}{4}\right) dr \, d\theta = \int_0^{2\pi} \left(\frac{\sin^2\theta}{20} + \frac{1}{10}\right) d\theta = \left[\frac{\theta}{40} - \frac{\sin 2\theta}{80} + \frac{\theta}{10}\right]_0^{2\pi} = \frac{\pi}{20} + \frac{\pi}{5} = \frac{\pi}{4}$

7. $\int_{-3}^{3}\int_{0}^{(1/2)\sqrt{9-x^2}} y\,dy\,dx = \int_{-3}^{3}\left[\frac{y^2}{2}\right]_0^{(1/2)\sqrt{9-x^2}} dx$

$= \int_{-3}^{3} \frac{1}{8}(9-x^2)\,dx = \left[\frac{9x}{8} - \frac{x^3}{24}\right]_{-3}^{3}$

$= \left(\frac{27}{8} - \frac{27}{24}\right) - \left(-\frac{27}{8} + \frac{27}{24}\right) = \frac{27}{6} = \frac{9}{2}$

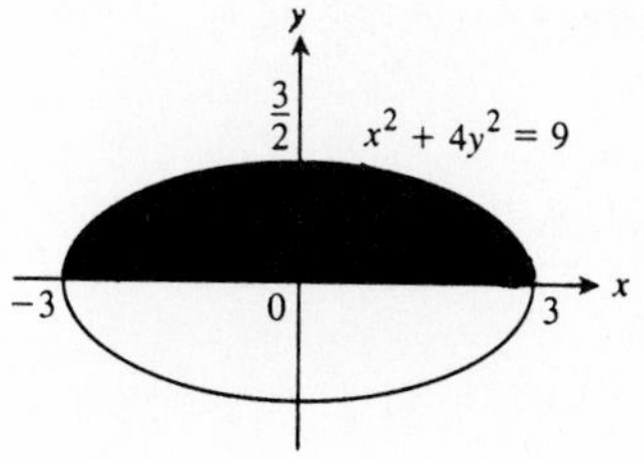

9. $\int_0^1\int_{2y}^2 4\cos(x^2)\,dx\,dy = \int_0^2\int_0^{\pi/2} 4\cos(x^2)\,dy\,dx = \int_0^2 2x\cos(x^2)\,dx = \left[\sin(x^2)\right]_0^2 = \sin 4$

11. $\int_0^8\int_{\sqrt[3]{x}}^2 \frac{1}{y^4+1}\,dy\,dx = \int_0^2\int_0^{y^3} \frac{1}{y^4+1}\,dx\,dy = \frac{1}{4}\int_0^2 \frac{4y^3}{y^4+1}\,dy = \frac{\ln 17}{4}$

13. $A = \int_{-2}^{0}\int_{2x+4}^{4-x^2} dy\,dx = \int_{-2}^{0}\left(-x^2-2x\right)dx = \frac{4}{3}$

15. $V = \int_0^1\int_x^{2-x}(x^2+y^2)\,dy\,dx = \int_0^1\left[x^2y + \frac{y^3}{3}\right]_x^{2-x} dx = \int_0^1\left[2x^2 + \frac{(2-x)^3}{3} - \frac{7x^3}{3}\right]dx = \left[\frac{2x^3}{3} - \frac{(2-x)^4}{12} - \frac{7x^4}{12}\right]_0^1$

$= \left(\frac{2}{3} - \frac{1}{12} - \frac{7}{12}\right) + \frac{2^4}{12} = \frac{4}{3}$

17. average value $= \int_0^1\int_0^1 xy\,dy\,dx = \int_0^1\left[\frac{xy^2}{2}\right]_0^1 dx = \int_0^1 \frac{x}{2}\,dx = \frac{1}{4}$

19. $M = \int_1^2\int_{2/x}^2 dy\,dx = \int_1^2\left(2 - \frac{2}{x}\right)dx = 2 - \ln 4$; $M_y = \int_1^2\int_{2/x}^2 x\,dy\,dx = \int_1^2 x\left(2-\frac{2}{x}\right)dx = 1$;

$M_x = \int_1^2\int_{2/x}^2 y\,dy\,dx = \int_1^2\left(2 - \frac{2}{x^2}\right)dx = 1 \Rightarrow \bar{x} = \bar{y} = \frac{1}{2-\ln 4}$

21. $I_o = \int_0^2\int_{2x}^4 (x^2+y^2)(3)\,dy\,dx = 3\int_0^2\left(4x^2 + \frac{64}{3} - \frac{14x^3}{3}\right)dx = 104$

23. $M = \delta\int_0^3\int_0^{2x/3} dy\,dx = \delta\int_0^3 \frac{2x}{3}\,dx = 3\delta$; $I_x = \delta\int_0^3\int_0^{2x/3} y^2\,dy\,dx = \frac{8\delta}{81}\int_0^3 x^3\,dx = \left(\frac{8\delta}{81}\right)\left(\frac{3^4}{4}\right) = 2\delta \Rightarrow R_x = \sqrt{\frac{2}{3}}$

25. $M = \int_{-1}^{1}\int_{-1}^{1} \left(x^2+y^2+\frac{1}{3}\right) dy\,dx = \int_{-1}^{1} \left(2x^2+\frac{4}{3}\right) dx = 4$; $M_x = \int_{-1}^{1}\int_{-1}^{1} y\left(x^2+y^2+\frac{1}{3}\right) dy\,dx = \int_{-1}^{1} 0\,dx = 0$;

$M_y = \int_{-1}^{1}\int_{-1}^{1} x\left(x^2+y^2+\frac{1}{3}\right) dy\,dx = \int_{-1}^{1} \left(2x^3+\frac{4}{3}x\right) dx = 0$

27. $\int_{-1}^{1}\int_{-\sqrt{1-x^2}}^{\sqrt{1-x^2}} \frac{2}{\left(1+x^2+y^2\right)^2}\,dy\,dx = \int_{0}^{2\pi}\int_{0}^{1} \frac{2r}{\left(1+r^2\right)^2}\,dr\,d\theta = \int_{0}^{2\pi} \left[-\frac{1}{1+r^2}\right]_0^1 d\theta = \frac{1}{2}\int_{0}^{2\pi} d\theta = \pi$

29. $M = \int_{-\pi/3}^{\pi/3}\int_{0}^{3} r\,dr\,d\theta = \frac{9}{2}\int_{-\pi/3}^{\pi/3} d\theta = 3\pi$; $M_y = \int_{-\pi/3}^{\pi/3}\int_{0}^{3} r^2 \cos\theta\,dr\,d\theta = 9\int_{-\pi/3}^{\pi/3} \cos\theta\,d\theta = 9\sqrt{3} \Rightarrow \bar{x} = \frac{3\sqrt{3}}{\pi}$,

and $\bar{y} = 0$ by symmetry

31. (a) $M = 2\int_{0}^{\pi/2}\int_{1}^{1+\cos\theta} r\,dr\,d\theta$ (b)

$= \int_{0}^{\pi/2} \left(2\cos\theta + \frac{1+\cos 2\theta}{2}\right) d\theta = \frac{8+\pi}{4}$;

$M_y = \int_{-\pi/2}^{\pi/2}\int_{1}^{1+\cos\theta} (r\cos\theta)\,r\,dr\,d\theta$

$= \int_{-\pi/2}^{\pi/2} \left(\cos^2\theta + \cos^3\theta + \frac{\cos^4\theta}{3}\right) d\theta$

$= \frac{32+15\pi}{24} \Rightarrow \bar{x} = \frac{15\pi+32}{6\pi+48}$, and

$\bar{y} = 0$ by symmetry

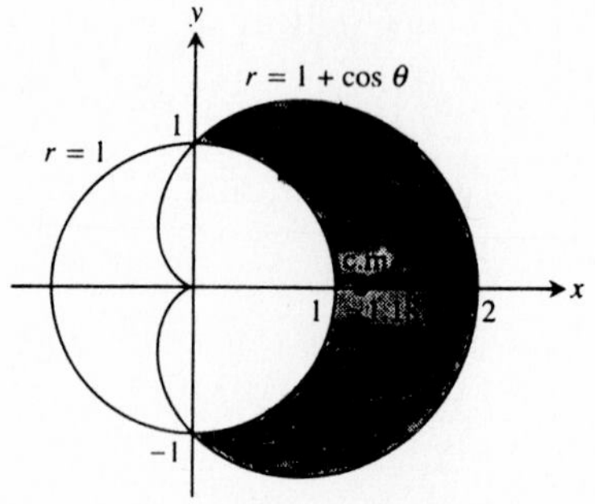

33. $\left(x^2+y^2\right)^2 - \left(x^2-y^2\right) = 0 \Rightarrow r^4 - r^2\cos 2\theta = 0 \Rightarrow r^2 = \cos 2\theta$ so the integral is $\int_{-\pi/4}^{\pi/4}\int_{0}^{\sqrt{\cos 2\theta}} \frac{r}{\left(1+r^2\right)^2}\,dr\,d\theta$

$= \int_{-\pi/4}^{\pi/4} \left[-\frac{1}{2\left(1+r^2\right)}\right]_0^{\sqrt{\cos 2\theta}} d\theta = \frac{1}{2}\int_{-\pi/4}^{\pi/4} \left(1-\frac{1}{1+\cos 2\theta}\right) d\theta = \frac{1}{2}\int_{-\pi/4}^{\pi/4} \left(1-\frac{1}{2\cos^2\theta}\right) d\theta$

$= \frac{1}{2}\int_{-\pi/4}^{\pi/4} \left(1-\frac{\sec^2\theta}{2}\right) d\theta = \frac{1}{2}\left[\theta - \frac{\tan\theta}{2}\right]_{-\pi/4}^{\pi/4} = \frac{\pi-2}{4}$

35. $\int_0^{\pi}\int_0^{\pi}\int_0^{\pi} \cos(x+y+z)\,dx\,dy\,dz = \int_0^{\pi}\int_0^{\pi} [\sin(z+y+\pi) - \sin(z+y)]\,dy\,dz$

$= \int_0^{\pi} [-\cos(z+2\pi) + \cos(z+\pi) + \cos z - \cos(z+\pi)]\,dz = 0$

37. $\int_0^1\int_0^{x^2}\int_0^{x+y} (2x-y-z)\,dz\,dy\,dx = \int_0^1\int_0^{x^2} \left(\frac{3x^2}{2} - \frac{3y^2}{2}\right) dy\,dx = \int_0^1 \left(\frac{3x^4}{2} - \frac{x^6}{2}\right) dx = \frac{8}{35}$

39. $V = 2\int_0^{\pi/2}\int_{-\cos y}^{0}\int_0^{-2x} dz\,dx\,dy = 2\int_0^{\pi/2}\int_{-\cos y}^{0} -2x\,dx\,dy = 2\int_0^{\pi/2} \cos^2 y\,dy = 2\left[\frac{y}{2} + \frac{\sin 2y}{4}\right]_0^{\pi/2} = \frac{\pi}{2}$

41. $\text{average} = \frac{1}{3}\int_0^1\int_0^3\int_0^1 30xz\sqrt{x^2+y}\,dz\,dy\,dx = \frac{1}{3}\int_0^1\int_0^3 15x\sqrt{x^2+y}\,dy\,dx = \frac{1}{3}\int_0^3\int_0^1 15x\sqrt{x^2+y}\,dx\,dy$

$= \frac{1}{3}\int_0^3 \left[5\left(x^2+y\right)^{3/2}\right]_0^1 dy = \frac{1}{3}\int_0^3 \left[5(1+y)^{3/2} - 5y^{3/2}\right] dy = \frac{1}{3}\left[2(1+y)^{5/2} - 2y^{5/2}\right]_0^3 = \frac{1}{3}\left[2(4)^{5/2} - 2(3)^{5/2} - 2\right]$

$= \frac{1}{3}\left[2\left(31 - 3^{5/2}\right)\right]$

43. (a) $\int_{-\sqrt{2}}^{\sqrt{2}}\int_{-\sqrt{2-y^2}}^{\sqrt{2-y^2}}\int_{\sqrt{x^2+y^2}}^{\sqrt{4-x^2-y^2}} 3\,dz\,dx\,dy$

(b) $\int_0^{2\pi}\int_0^{\pi/4}\int_0^{2} 3\rho^2 \sin\phi\,d\rho\,d\phi\,d\theta$

(c) $\int_0^{2\pi}\int_0^{\sqrt{2}}\int_r^{\sqrt{4-r^2}} 3\,dz\,r\,dr\,d\theta = 3\int_0^{2\pi}\int_0^{\sqrt{2}} \left[r\left(4-r^2\right)^{1/2} - r^2\right] dr\,d\theta = 3\int_0^{2\pi} \left[-\frac{1}{3}\left(4-r^2\right)^{3/2} - \frac{r^3}{3}\right]_0^{\sqrt{2}} d\theta$

$= \int_0^{2\pi} \left(-2^{3/2} - 2^{3/2} + 4^{3/2}\right) d\theta = \left(8 - 4\sqrt{2}\right)\int_0^{2\pi} d\theta = 2\pi\left(8 - 4\sqrt{2}\right)$

45. (a) $\int_0^{2\pi}\int_0^{\pi/4}\int_0^{\sec\phi} \rho^2 \sin\phi\,d\rho\,d\phi\,d\theta$

(b) $\int_0^{2\pi}\int_0^{\pi/4}\int_0^{\sec\phi} \rho^2 \sin\phi\,d\rho\,d\phi\,d\theta = \frac{1}{3}\int_0^{2\pi}\int_0^{\pi/4} (\sec\phi)(\sec\phi\tan\phi)\,d\phi\,d\theta = \frac{1}{3}\int_0^{2\pi} \left[\frac{1}{2}\tan^2\phi\right]_0^{\pi/4} d\theta = \frac{1}{6}\int_0^{2\pi} d\theta = \frac{\pi}{3}$

47. $\int_0^1 \int_{\sqrt{1-x^2}}^{\sqrt{3-x^2}} \int_1^{\sqrt{4-x^2-y^2}} z^2yx\, dz\, dy\, dx + \int_1^{\sqrt{3}} \int_0^{\sqrt{3-x^2}} \int_1^{\sqrt{4-x^2-y^2}} z^2yx\, dz\, dy\, dx$

49. (a) $\int_{-\sqrt{3}}^{\sqrt{3}} \int_{-\sqrt{3-x^2}}^{\sqrt{3-x^2}} \int_1^{\sqrt{4-x^2-y^2}} dz\, dy\, dx$ (b) $\int_0^{2\pi} \int_0^{\sqrt{3}} \int_1^{\sqrt{4-r^2}} dz\, r\, dr\, d\theta$

(c) $\int_0^{2\pi} \int_0^{\pi/3} \int_{\sec\phi}^{2} \rho^2 \sin\phi\, d\rho\, d\phi\, d\theta$

51. (a) $V = \int_0^{2\pi} \int_0^2 \int_2^{\sqrt{8-r^2}} dz\, r\, dr\, d\theta = \int_0^{2\pi} \int_0^2 \left(r\sqrt{8-r^2} - 2r\right) dr\, d\theta = \int_0^{2\pi} \left[-\frac{1}{3}\left(8-r^2\right)^{3/2} - r^2\right]_0^2 d\theta$

$= \int_0^{2\pi} \left[-\frac{1}{3}(4)^{3/2} - 4 + \frac{1}{3}(8)^{3/2}\right] d\theta = \int_0^{2\pi} \frac{4}{3}\left(-2-3+2\sqrt{8}\right) d\theta = \frac{4}{3}\left(4\sqrt{2}-5\right) \int_0^{2\pi} d\theta = \frac{8\pi\left(4\sqrt{2}-5\right)}{3}$

(b) $V = \int_0^{2\pi} \int_0^{\pi/4} \int_{2\sec\phi}^{\sqrt{8}} \rho^2 \sin\phi\, d\rho\, d\phi\, d\theta = \frac{8}{3} \int_0^{2\pi} \int_0^{\pi/4} \left(2\sqrt{2}\sin\phi - \sec^3\phi\sin\phi\right) d\phi\, d\theta$

$= \frac{8}{3} \int_0^{2\pi} \int_0^{\pi/4} \left(2\sqrt{2}\sin\phi - \tan\phi\sec^2\phi\right) d\phi\, d\theta = \frac{8}{3} \int_0^{2\pi} \left[-2\sqrt{2}\cos\phi - \frac{1}{2}\tan^2\phi\right]_0^{\pi/4} d\theta$

$= \frac{8}{3} \int_0^{2\pi} \left(-2 - \frac{1}{2} + 2\sqrt{2}\right) d\theta = \frac{8}{3} \int_0^{2\pi} \left(\frac{-5+4\sqrt{2}}{2}\right) d\theta = \frac{8\pi\left(4\sqrt{2}-5\right)}{3}$

53. With the centers of the spheres at the origin, $I_z = \int_0^{2\pi} \int_0^{\pi} \int_a^b \delta(\rho\sin\phi)^2 \left(\rho^2\sin\phi\right) d\rho\, d\phi\, d\theta$

$= \frac{\delta\left(b^5-a^5\right)}{5} \int_0^{2\pi} \int_0^{\pi} \sin^3\phi\, d\phi\, d\theta = \frac{\delta\left(b^5-a^5\right)}{5} \int_0^{2\pi} \int_0^{\pi} \left(\sin\phi - \cos^2\phi\sin\phi\right) d\phi\, d\theta$

$= \frac{\delta\left(b^5-a^5\right)}{5} \int_0^{2\pi} \left[-\cos\phi + \frac{\cos^3\phi}{3}\right]_0^{\pi} d\theta = \frac{4\delta\left(b^5-a^5\right)}{15} \int_0^{2\pi} d\theta = \frac{8\pi\delta\left(b^5-a^5\right)}{15}$

55. $x = u + y$ and $y = v \Rightarrow x = u + v$ and $y = v$

$\Rightarrow J(u, v) = \begin{vmatrix} 1 & 1 \\ 0 & 1 \end{vmatrix} = 1$; the boundary of the image G is obtained from the boundary of R as follows:

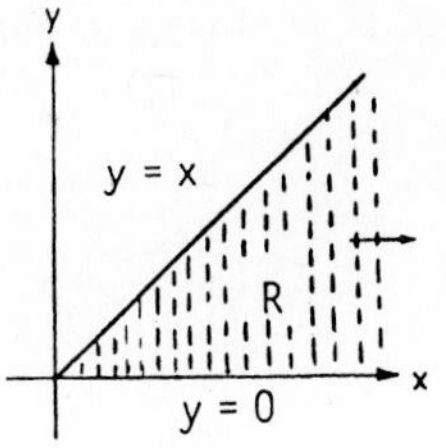

xy-equations for the boundary of R	Corresponding uv-equations for the boundary of G	Simplified uv-equations
$y = x$	$v = u + v$	$u = 0$
$y = 0$	$v = 0$	$v = 0$

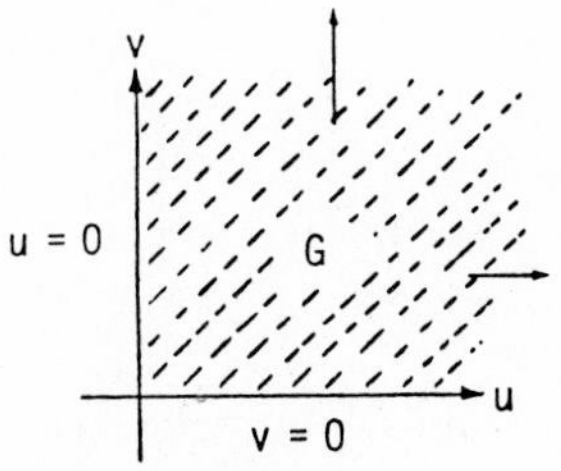

$$\Rightarrow \int_0^{\infty}\int_0^{x} e^{-sx} f(x - y, y)\, dy\, dx = \int_0^{\infty}\int_0^{\infty} e^{-s(u+v)} f(u, v)\, du\, dv$$

CHAPTER 13 ADDITIONAL EXERCISES–THEORY, EXAMPLES, APPLICATIONS

1. (a) $V = \int_{-3}^{2}\int_{x}^{6-x^2} x^2\, dy\, dx$ (b) $V = \int_{-3}^{2}\int_{x}^{6-x^2}\int_{0}^{x^2} dz\, dy\, dx$

(c) $V = \int_{-3}^{2}\int_{x}^{6-x^2} x^2\, dy\, dx = \int_{-3}^{2} \left(6x^2 - x^4 - x^3\right) dx = \left[2x^3 - \frac{x^5}{5} - \frac{x^4}{4}\right]_{-3}^{2} = \frac{125}{4}$

3. Using cylindrical coordinates, $V = \int_0^{2\pi}\int_0^1\int_0^{2-r(\cos\theta+\sin\theta)} dz\, r\, dr\, d\theta = \int_0^{2\pi}\int_0^1 \left(2r - r^2\cos\theta - r^2\sin\theta\right) dr\, d\theta$

$$= \int_0^{2\pi} \left(1 - \frac{1}{3}\cos\theta - \frac{1}{3}\sin\theta\right) d\theta = \left[\theta + \frac{1}{3}\sin\theta - \frac{1}{3}\cos\theta\right]_0^{2\pi} = 2\pi$$

5. The surfaces intersect when $3 - x^2 - y^2 = 2x^2 + 2y^2 \Rightarrow x^2 + y^2 = 1$. Thus the volume is

$$V = 4\int_0^1\int_0^{\sqrt{1-x^2}}\int_{2x^2+2y^2}^{3-x^2-y^2} dz\, dy\, dx = 4\int_0^{\pi/2}\int_0^1\int_{2r^2}^{3-r^2} dz\, r\, dr\, d\theta = 4\int_0^{\pi/2}\int_0^1 \left(3r - 3r^3\right) dr\, d\theta = 3\int_0^{\pi/2} d\theta = \frac{3\pi}{2}$$

7. (a) The radius of the hole is 1, and the radius of the sphere is 2.

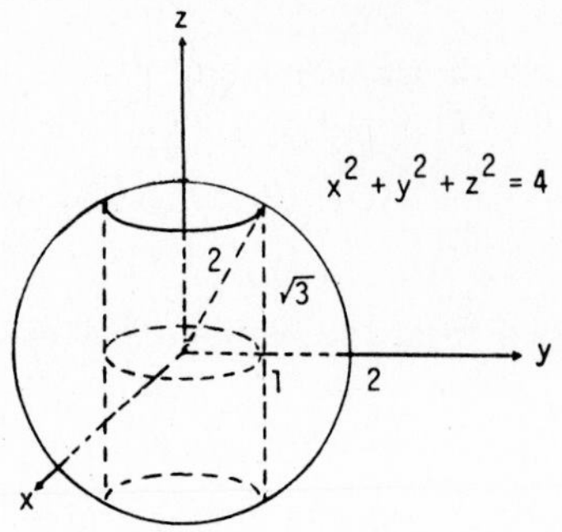

(b) $V = 2\int_0^{2\pi}\int_0^{\sqrt{3}}\int_1^{\sqrt{4-z^2}} r\,dr\,dz\,d\theta = \int_0^{2\pi}\int_0^{\sqrt{3}} (3-z^2)\,dz\,d\theta = 2\sqrt{3}\int_0^{2\pi} d\theta = 4\sqrt{3}\pi$

9. The surfaces intersect when $x^2+y^2 = \frac{x^2+y^2+1}{2} \Rightarrow x^2+y^2 = 1$. Thus the volume in cylindrical coordinates is $V = 4\int_0^{\pi/2}\int_0^1\int_{r^2}^{(r^2+1)/2} dz\; r\,dr\,d\theta = 4\int_0^{\pi/2}\int_0^1 \left(\frac{r}{2}-\frac{r^3}{2}\right) dr\,d\theta = 4\int_0^{\pi/2}\left[\frac{r^2}{4}-\frac{r^4}{8}\right]_0^1 d\theta$

$= \frac{1}{2}\int_0^{\pi/2} d\theta = \frac{\pi}{4}$

11. $\int_0^1\int_y^{\sqrt{y}} f(x,y)\,dx\,dy$

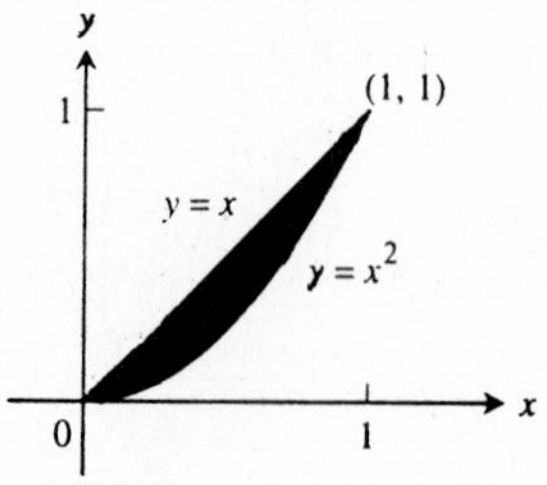

13. $\int_0^\infty \frac{e^{-ax}-e^{-bx}}{x}\,dx = \int_0^\infty\int_a^b e^{-xy}\,dy\,dx = \int_a^b\int_0^\infty e^{-xy}\,dx\,dy = \int_a^b\left(\lim_{t\to\infty}\int_0^t e^{-xy}\,dx\right)dy$

$= \int_a^b \lim_{t\to\infty}\left[-\frac{e^{-xy}}{y}\right]_0^t dy = \int_a^b \lim_{t\to\infty}\left(\frac{1}{y}-\frac{e^{-yt}}{y}\right) dy = \int_a^b \frac{1}{y}\,dy = [\ln y]_a^b = \ln\left(\frac{b}{a}\right)$

15. $\int_0^x\int_0^u e^{m(x-t)} f(t)\,dt\,du = \int_0^x\int_1^x e^{m(x-t)} f(t)\,du\,dt = \int_0^x (x-t)e^{m(x-t)} f(t)\,dt$; also

$\int_0^x\int_0^v\int_0^u e^{m(x-t)} f(t)\,dt\,du\,dv = \int_0^x\int_t^x\int_t^v e^{m(x-t)} f(t)\,du\,dv\,dt = \int_0^x\int_t^x (v-t)e^{m(x-t)} f(t)\,dv\,dt$

$$= \int_0^x \left[\frac{1}{2}(v-t)^2 e^{m(x-t)} f(t)\right]_t^x dt = \int_0^x \frac{(x-t)^2}{2} e^{m(x-t)} f(t)\, dt$$

17. $I_o(a) = \int_0^a \int_0^{x/a^2} (x^2+y^2)\, dy\, dx = \int_0^a \left[x^2 y + \frac{y^3}{3}\right]_0^{x/a^2} dx = \int_0^a \left(\frac{x^3}{a^2} + \frac{x^3}{3a^6}\right) dx = \left[\frac{x^4}{4a^2} + \frac{x^4}{12a^6}\right]_0^a$

$= \frac{a^2}{4} + \frac{1}{12}a^{-2}$; $I_o'(a) = \frac{1}{2}a - \frac{1}{6}a^{-3} = 0 \Rightarrow a^4 = \frac{1}{3} \Rightarrow a = \sqrt[4]{\frac{1}{3}} = \frac{1}{\sqrt[4]{3}}$. Since $I_o''(a) = \frac{1}{2} + \frac{1}{2}a^{-4} > 0$, the value of a does provide a minimum for the polar moment of inertia $I_o(a)$.

19. $M = 2\int_0^{\pi/2} \int_1^{1+\cos\theta} r\, dr\, d\theta = \int_0^{\pi/2} \left(2\cos\theta + \frac{1+\cos 2\theta}{2}\right) d\theta = \frac{8+\pi}{4}$;

$M_y = \int_{-\pi/2}^{\pi/2} \int_1^{1+\cos\theta} r^2 \cos\theta\, dr\, d\theta = \int_{-\pi/2}^{\pi/2} \left(\cos^2\theta + \cos^3\theta + \frac{\cos^4\theta}{3}\right) d\theta$

$= \left[\frac{\theta}{2} + \frac{\sin 2\theta}{4} + \frac{\cos^2\theta \sin\theta}{3} + \frac{\cos^3\theta \sin\theta}{12}\right]_{-\pi/2}^{\pi/2} + \int_{-\pi/2}^{\pi/2} \left(\frac{2}{3}\cos\theta + \frac{1}{4}\cos^2\theta\right) d\theta$

$= \frac{\pi}{2} + \left[\frac{2}{3}\sin\theta + \frac{1}{4}\left(\frac{\theta}{2} + \frac{\sin 2\theta}{4}\right)\right]_{-\pi/2}^{\pi/2} = \frac{\pi}{2} + \frac{4}{3} + \frac{\pi}{8} = \frac{5\pi}{8} + \frac{4}{3} = \frac{32+15\pi}{24} \Rightarrow \bar{x} = \frac{15\pi+32}{6\pi+48}$, and

$\bar{y} = 0$ by symmetry

21. $M = \int_{-\theta}^{\theta} \int_{b\sec\theta}^{a} r\, dr\, d\theta = \int_{-\theta}^{\theta} \left(\frac{a^2}{2} - \frac{b^2}{2}\sec^2\theta\right) d\theta$

$= a^2\theta - b^2\tan\theta = a^2\cos^{-1}\left(\frac{b}{a}\right) - b^2\left(\frac{\sqrt{a^2-b^2}}{b}\right)$

$= a^2\cos^{-1}\left(\frac{b}{a}\right) - b\sqrt{a^2-b^2}$; $I_o = \int_{-\theta}^{\theta} \int_{b\sec\theta}^{a} r^3\, dr\, d\theta$

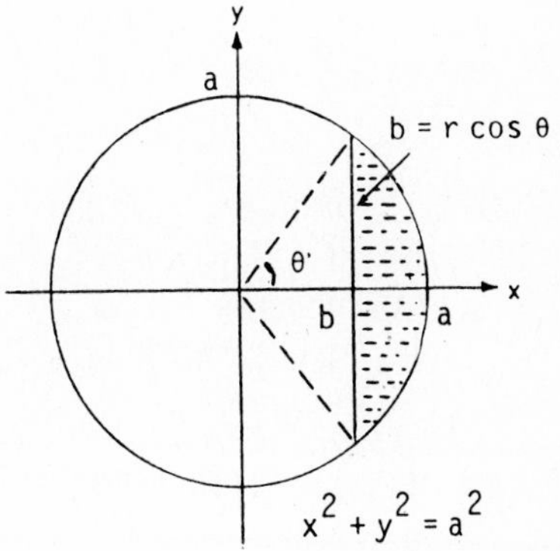

$= \frac{1}{4}\int_{-\theta}^{\theta} (a^4 + b^4\sec^4\theta)\, d\theta = \frac{1}{4}\int_{-\theta}^{\theta} \left[a^4 + b^4(1+\tan^2\theta)(\sec^2\theta)\right] d\theta$

$= \frac{1}{4}\left[a^4\theta - b^4\tan\theta - \frac{b^4\tan^3\theta}{3}\right]_{-\theta}^{\theta} = \frac{a^4\theta}{2} - \frac{b^4\tan\theta}{2} - \frac{b^4\tan^3\theta}{6}$

$= \frac{1}{2}a^4\cos^{-1}\left(\frac{b}{a}\right) - \frac{1}{2}b^3\sqrt{a^2-b^2} - \frac{1}{6}b^3(a^2-b^2)^{3/2}$

23. (a) $M = \int_0^{2\pi}\int_0^1\int_0^{r^2} z\,dz\,r\,dr\,d\theta = \int_0^{2\pi}\int_0^1 \frac{1}{2}r^5\,dr\,d\theta = \frac{1}{12}\int_0^{2\pi} d\theta = \frac{\pi}{6}$; $M_{xy} = \int_0^{2\pi}\int_0^1\int_0^{r^2} z^2\,dz\,r\,dr\,d\theta$

$= \frac{1}{3}\int_0^{2\pi}\int_0^1 r^7\,dr\,d\theta = \frac{1}{24}\int_0^{2\pi} d\theta = \frac{\pi}{12} \Rightarrow \bar{z} = \frac{1}{2}$, and $\bar{x} = \bar{y} = 0$ by symmetry; $I_z = \int_0^{2\pi}\int_0^1\int_0^{r^2} zr^3\,dz\,dr\,d\theta$

$= \frac{1}{2}\int_0^{2\pi}\int_0^1 r^7\,dr\,d\theta = \frac{1}{16}\int_0^{2\pi} d\theta = \frac{\pi}{8} \Rightarrow R_z = \sqrt{\frac{I_z}{M}} = \frac{\sqrt{3}}{2}$

(b) $M = \int_0^{2\pi}\int_0^1\int_0^{r^2} r^2\,dz\,dr\,d\theta = \int_0^{2\pi}\int_0^1 r^4\,dr\,d\theta = \frac{1}{5}\int_0^{2\pi} d\theta = \frac{2\pi}{5}$; $M_{xy} = \int_0^{2\pi}\int_0^1\int_0^{r^2} zr^2\,dz\,dr\,d\theta$

$= \frac{1}{2}\int_0^{2\pi}\int_0^1 r^6\,dr\,d\theta = \frac{1}{14}\int_0^{2\pi} d\theta = \frac{\pi}{7} \Rightarrow \bar{z} = \frac{5}{14}$, and $\bar{x} = \bar{y} = 0$ by symmetry; $I_z = \int_0^{2\pi}\int_0^1\int_0^{r^2} r^4\,dz\,dr\,d\theta$

$= \int_0^{2\pi}\int_0^1 r^6\,dr\,d\theta = \frac{1}{7}\int_0^{2\pi} d\theta = \frac{2\pi}{7} \Rightarrow R_z = \sqrt{\frac{I_z}{M}} = \sqrt{\frac{5}{7}}$

25. $M = \frac{2}{3}\pi a^3$; $M_{xy} = \int_0^{2\pi}\int_0^{\pi/2}\int_0^a (\rho^3 \sin\phi\cos\phi)\,d\rho\,d\phi\,d\theta = \frac{a^4}{4}\int_0^{2\pi}\int_0^{\pi/2} \sin\phi\cos\phi\,d\phi\,d\theta = \frac{a^4}{8}\int_0^{2\pi} d\theta = \frac{a^4\pi}{4}$

$\Rightarrow \bar{z} = \frac{3a}{8}$, and $\bar{x} = \bar{y} = 0$, by symmetry

27. $\int_0^a\int_0^b e^{\max(b^2x^2,\,a^2y^2)}\,dy\,dx = \int_0^a\int_0^{bx/a} e^{b^2x^2}\,dy\,dx + \int_0^b\int_0^{ay/b} e^{a^2y^2}\,dx\,dy$

$= \int_0^a \left(\frac{b}{a}x\right)e^{b^2x^2}\,dx + \int_0^b \left(\frac{a}{b}y\right)e^{a^2y^2}\,dy = \left[\frac{1}{2ab}e^{b^2x^2}\right]_0^a + \left[\frac{1}{2ba}e^{a^2y^2}\right]_0^b = \frac{1}{2ab}\left(e^{b^2a^2}-1\right) + \frac{1}{2ab}\left(e^{a^2b^2}-1\right)$

$= \frac{1}{ab}\left(e^{a^2b^2}-1\right)$

29. (a) (i) Fubini's Theorem
(ii) Treating $G(y)$ as a constant
(iii) Algebraic rearrangement
(iv) The definite integral is a constant number

(b) $\int_0^{\ln 2}\int_0^{\pi/2} e^x \cos y\,dy\,dx = \left(\int_0^{\ln 2} e^x\,dx\right)\left(\int_0^{\pi/2} \cos y\,dy\right) = \left(e^{\ln 2} - e^0\right)\left(\sin\frac{\pi}{2} - \sin 0\right) = (1)(1) = 1$

(c) $\int_1^2\int_{-1}^1 \frac{x}{y^2}\,dx\,dy = \left(\int_1^2 \frac{1}{y^2}\,dy\right)\left(\int_{-1}^1 x\,dx\right) = \left[-\frac{1}{y}\right]_1^2\left[\frac{x^2}{2}\right]_{-1}^1 = \left(-\frac{1}{2}+1\right)\left(\frac{1}{2}-\frac{1}{2}\right) = 0$

31. (a) $I^2 = \int_0^\infty \int_0^\infty e^{-(x^2+y^2)}\,dx\,dy = \int_0^{\pi/2} \int_0^\infty \left(e^{-r^2}\right) r\,dr\,d\theta = \int_0^{\pi/2} \left[\lim_{b\to\infty} \int_0^b re^{-r^2}\,dr\right] d\theta$

$= -\frac{1}{2} \int_0^{\pi/2} \lim_{b\to\infty} \left(e^{-b^2} - 1\right) d\theta = \frac{1}{2} \int_0^{\pi/2} d\theta = \frac{\pi}{4} \Rightarrow I = \frac{\sqrt{\pi}}{2}$

(b) $\Gamma\left(\frac{1}{2}\right) = \int_0^\infty t^{-1/2} e^{-t}\,dt = \int_0^\infty \left(y^2\right)^{-1/2} e^{-y^2}(2y)\,dy = 2\int_0^\infty e^{-y^2}\,dy = 2\left(\frac{\sqrt{\pi}}{2}\right) = \sqrt{\pi}$, where $y = \sqrt{t}$

33. For a height h in the bowl the volume of water is $V = \int_{-\sqrt{h}}^{\sqrt{h}} \int_{-\sqrt{h-x^2}}^{\sqrt{h-x^2}} \int_{x^2+y^2}^{h} dz\,dy\,dx$

$= \int_{-\sqrt{h}}^{\sqrt{h}} \int_{-\sqrt{h-x^2}}^{\sqrt{h-x^2}} \left(h - x^2 - y^2\right) dy\,dx = \int_0^{2\pi} \int_0^{\sqrt{h}} \left(h - r^2\right) r\,dr\,d\theta = \int_0^{2\pi} \left[\frac{hr^2}{2} - \frac{r^4}{4}\right]_0^{\sqrt{h}} d\theta = \int_0^{2\pi} \frac{h^2}{4}\,d\theta = \frac{h^2\pi}{2}.$

Since the top of the bowl has area 10π, then we calibrate the bowl by comparing it to a right circular cylinder whose cross sectional area is 10π from $z = 0$ to $z = 10$. If such a cylinder contains $\frac{h^2\pi}{2}$ cubic inches of water to a depth w then we have $10\pi w = \frac{h^2\pi}{2} \Rightarrow w = \frac{h^2}{20}$. So for 1 inch of rain, $w = 1$ and $h = \sqrt{20}$; for 3 inches of rain, $w = 3$ and $h = \sqrt{60}$.

35. Using cylindrical coordinates with $x = r\cos\theta$, $y = y$, and $z = r\sin\theta$, we have $V = \int_a^b \int_0^{2\pi} \int_0^{f(r)} r\,dy\,d\theta\,dr$

$= \int_a^b \int_0^{2\pi} [yr]_0^{f(r)}\, r\,d\theta\,dr = \int_a^b \int_0^{2\pi} r\,f(r)\,d\theta\,dr$

$= \int_a^b 2\pi r\,f(r)\,dr = 2\pi \int_a^b x\,f(x)\,dx.$

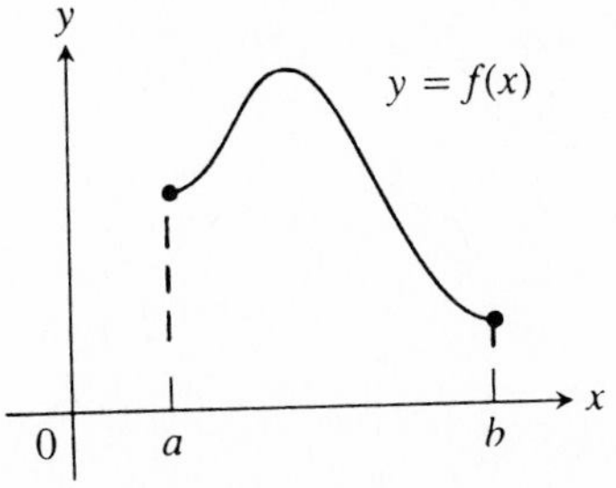

NOTES:

CHAPTER 14 INTEGRATION IN VECTOR FIELDS

14.1 LINE INTEGRALS

1. $\mathbf{r} = t\mathbf{i} + (1-t)\mathbf{j} \Rightarrow x = t$ and $y = 1 - t \Rightarrow y = 1 - x \Rightarrow$ (c)

3. $\mathbf{r} = (2\cos t)\mathbf{i} + (2\sin t)\mathbf{j} \Rightarrow x = 2\cos t$ and $y = 2\sin t \Rightarrow x^2 + y^2 = 4 \Rightarrow$ (g)

5. $\mathbf{r} = t\mathbf{i} + t\mathbf{j} + t\mathbf{k} \Rightarrow x = t,\ y = t,$ and $z = t \Rightarrow$ (d)

7. $\mathbf{r} = \left(t^2 - 1\right)\mathbf{j} + 2t\mathbf{k} \Rightarrow y = t^2 - 1$ and $z = 2t \Rightarrow y = \frac{z^2}{4} - 1 \Rightarrow$ (f)

9. $\mathbf{r}(t) = t\mathbf{i} + (1-t)\mathbf{j},\ 0 \le t \le 1 \Rightarrow \frac{d\mathbf{r}}{dt} = \mathbf{i} - \mathbf{j} \Rightarrow \left|\frac{d\mathbf{r}}{dt}\right| = \sqrt{2}\mathbf{j}$; $x = t$ and $y = 1 - t \Rightarrow x + y = t + (1-t) = 1$

$$\Rightarrow \int_C f(x,y,z)\, ds = \int_0^1 f(t, 1-t, 0)\, \frac{d\mathbf{r}}{dt}\, dt = \int_0^1 (1)\left(\sqrt{2}\right) dt = \left[\sqrt{2}\,t\right]_0^1 = \sqrt{2}$$

11. $\mathbf{r}(t) = 2t\mathbf{i} + t\mathbf{j} + (2-2t)\mathbf{k},\ 0 \le t \le 1 \Rightarrow \frac{d\mathbf{r}}{dt} = 2\mathbf{i} + \mathbf{j} - 2\mathbf{k} \Rightarrow \left|\frac{d\mathbf{r}}{dt}\right| = \sqrt{4+1+4} = 3$; $xy + y + z$

$$= (2t)t + t + (2-2t) \Rightarrow \int_C f(x,y,z)\, ds = \int_0^1 \left(2t^2 - t + 2\right) 3\, dt = 3\left[\frac{2}{3}t^3 - \frac{1}{2}t^2 + 2t\right]_0^1 = 3\left(\frac{2}{3} - \frac{1}{2} + 2\right) = \frac{13}{2}$$

13. $\mathbf{r}(t) = (\mathbf{i} + 2\mathbf{j} + 3\mathbf{k}) + t(-\mathbf{i} - 3\mathbf{j} - 2\mathbf{k}) = (1-t)\mathbf{i} + (2-3t)\mathbf{j} + (3-2t)\mathbf{k},\ 0 \le t \le 1 \Rightarrow \frac{d\mathbf{r}}{dt} = -\mathbf{i} - 3\mathbf{j} - 2\mathbf{k}$

$$\Rightarrow \left|\frac{d\mathbf{r}}{dt}\right| = \sqrt{1+9+4} = \sqrt{14};\ x + y + z = (1-t) + (2-3t) + (3-2t) = 6 - 6t \Rightarrow \int_C f(x,y,z)\, ds$$

$$= \int_0^1 (6-6t)\,\sqrt{14}\, dt = 6\sqrt{14}\left[1 - \frac{t^2}{2}\right]_0^1 = \left(6\sqrt{14}\right)\left(\frac{1}{2}\right) = 3\sqrt{14}$$

15. C_1: $\mathbf{r}(t) = t\mathbf{i} + t^2\mathbf{j},\ 0 \le t \le 1 \Rightarrow \frac{d\mathbf{r}}{dt} = \mathbf{i} + 2t\mathbf{j} \Rightarrow \left|\frac{d\mathbf{r}}{dt}\right| = \sqrt{1+4t^2}$; $x + \sqrt{y} - z^2 = t + \sqrt{t^2} - 0 = t + |t| = 2t$

$$\Rightarrow \int_{C_1} f(x,y,z)\, ds = \int_0^1 2t\sqrt{1+4t^2}\, dt = \left[\frac{1}{6}\left(1+4t^2\right)^{3/2}\right]_0^1 = \frac{1}{6}(5)^{3/2} - \frac{1}{6} = \frac{1}{6}\left(5\sqrt{5} - 1\right);$$

C_2: $\mathbf{r}(t) = \mathbf{i} + \mathbf{j} + t\mathbf{k},\ 0 \le t \le 1 \Rightarrow \frac{d\mathbf{r}}{dt} = \mathbf{k} \Rightarrow \left|\frac{d\mathbf{r}}{dt}\right| = 1$; $x + \sqrt{y} - z^2 = 1 + \sqrt{1} - t^2 = 2 - t^2$

$$\Rightarrow \int_{C_2} f(x,y,z)\, ds = \int_0^1 \left(2 - t^2\right)(1)\, dt = \left[2t - \frac{1}{3}t^3\right]_0^1 = 2 - \frac{1}{3} = \frac{5}{3};\ \text{therefore} \int_C f(x,y,z)\, ds$$

$= \int_{C_1} f(x,y,z)\, ds + \int_{C_2} f(x,y,z)\, ds = \frac{5}{6}\sqrt{5} + \frac{3}{2}$

17. $\mathbf{r}(t) = t\mathbf{i} + t\mathbf{j} + t\mathbf{k},\ 0 < a \le t \le b \Rightarrow \frac{d\mathbf{r}}{dt} = \mathbf{i} + \mathbf{j} + \mathbf{k} \Rightarrow \left|\frac{d\mathbf{r}}{dt}\right| = \sqrt{3};\ \frac{x+y+z}{x^2+y^2+z^2} = \frac{t+t+t}{t^2+t^2+t^2} = \frac{1}{t}$

$\Rightarrow \int_C f(x,y,z)\, ds = \int_a^b \left(\frac{1}{t}\right)\sqrt{3}\, dt = \left[\sqrt{3}\ln|t|\right]_a^b = \sqrt{3}\ln\left(\frac{b}{a}\right)$, since $0 < a \le b$

19. $\mathbf{r}(x) = x\mathbf{i} + y\mathbf{j} = x\mathbf{i} + \frac{x^2}{2}\mathbf{j},\ 0 \le x \le 2 \Rightarrow \frac{d\mathbf{r}}{dx} = \mathbf{i} + x\mathbf{j} \Rightarrow \left|\frac{d\mathbf{r}}{dx}\right| = \sqrt{1+x^2};\ f(x,y) = f\left(x, \frac{x^2}{2}\right) = \frac{x^3}{\left(\frac{x^2}{2}\right)} = 2x \Rightarrow \int_C f\, ds$

$= \int_0^2 (2x)\sqrt{1+x^2}\, dx = \left[\frac{2}{3}\left(1+x^2\right)^{3/2}\right]_0^2 = \frac{2}{3}\left(5^{3/2} - 1\right) = \frac{10\sqrt{5}-2}{3}$

21. $\mathbf{r}(t) = (2\cos t)\mathbf{i} + (2\sin t)\mathbf{j},\ 0 \le t \le \frac{\pi}{2} \Rightarrow \frac{d\mathbf{r}}{dt} = (-2\sin t)\mathbf{i} + (2\cos t)\mathbf{j} \Rightarrow \left|\frac{d\mathbf{r}}{dt}\right| = 2;\ f(x,y) = f(2\cos t, 2\sin t)$

$= 2\cos t + 2\sin t \Rightarrow \int_C f\, ds = \int_0^{\pi/2} (2\cos t + 2\sin t)(2)\, dt = [4\sin t - 4\cos t]_0^{\pi/2} = 4 - (-4) = 8$

23. $\mathbf{r}(t) = \left(t^2-1\right)\mathbf{j} + 2t\mathbf{k},\ 0 \le t \le 1 \Rightarrow \frac{d\mathbf{r}}{dt} = 2t\mathbf{j} + 2\mathbf{k} \Rightarrow \left|\frac{d\mathbf{r}}{dt}\right| = 2\sqrt{t^2+1};\ M = \int_C \delta(x,y,z)\, ds = \int_0^1 \delta(t)\left(2\sqrt{t^2+1}\right) dt$

$= \int_0^1 \left(\frac{3}{2}t\right)\left(2\sqrt{t^2+1}\right) dt = \left[\left(t^2+1\right)\right]_0^1 = 2^{3/2} - 1 = 2\sqrt{2} - 1$

25. $\mathbf{r}(t) = \sqrt{2}t\,\mathbf{i} + \sqrt{2}t\,\mathbf{j} + \left(4-t^2\right)\mathbf{k},\ 0 \le t \le 1 \Rightarrow \frac{d\mathbf{r}}{dt} = \sqrt{2}\mathbf{i} + \sqrt{2}\mathbf{j} - 2t\mathbf{k} \Rightarrow \left|\frac{d\mathbf{r}}{dt}\right| = \sqrt{2+2+4t^2} = 2\sqrt{1+t^2};$

(a) $M = \int_C \delta\, ds = \int_0^1 (3t)\left(2\sqrt{1+t^2}\right) dt = \left[2\left(1+t^2\right)^{3/2}\right]_0^1 = 2\left(2^{3/2}-1\right) = 4\sqrt{2} - 2$

(b) $M = \int_C \delta\, ds = \int_0^1 (1)\left(2\sqrt{1+t^2}\right) dt = \left[t\sqrt{1+t^2} + \ln\left(t + \sqrt{1+t^2}\right)\right]_0^1 = \left[\sqrt{2} + \ln\left(1+\sqrt{2}\right)\right] - (0 + \ln 1)$

$= \sqrt{2} + \ln\left(1+\sqrt{2}\right)$

27. Let $x = a\cos t$ and $y = a\sin t$, $0 \le t \le 2\pi$. Then $\frac{dx}{dt} = -a\sin t$, $\frac{dy}{dt} = a\cos t$, $\frac{dz}{dt} = 0$

$\Rightarrow \sqrt{\left(\frac{dx}{dt}\right)^2 + \left(\frac{dy}{dt}\right)^2 + \left(\frac{dz}{dt}\right)^2}\, dt = a\, dt;\ I_z = \int_C \left(x^2+y^2\right)\delta\, ds = \int_0^{2\pi} \left(a^2\sin^2 t + a^2\cos^2 t\right)a\delta\, dt$

$= \int_0^{2\pi} a^3\delta\, dt = 2\pi\delta a^3;\ M = \int_C \delta(x,y,z)\, ds = \int_0^{2\pi} \delta a\, dt = 2\pi\delta a \Rightarrow R_z = \sqrt{\frac{I_z}{M}} = \sqrt{\frac{2\pi a^3\delta}{2\pi a\delta}} = a.$

29. $\mathbf{r}(t) = (\cos t)\mathbf{i} + (\sin t)\mathbf{j} + t\mathbf{k},\ 0 \le t \le 2\pi \Rightarrow \frac{d\mathbf{r}}{dt} = (-\sin t)\mathbf{i} + (\cos t)\mathbf{j} + \mathbf{k} \Rightarrow \left|\frac{d\mathbf{r}}{dt}\right| = \sqrt{\sin^2 t + \cos^2 t + 1} = \sqrt{2};$

(a) $M = \int_C \delta\, ds = \int_0^{2\pi} \delta\sqrt{2}\, dt = 2\pi\delta\sqrt{2};\ I_z = \int_C (x^2+y^2)\delta\, ds = \int_0^{2\pi} (\cos^2 t + \sin^2 t)\delta\sqrt{2}\, dt = 2\pi\delta\sqrt{2}$

$\Rightarrow R_z = \sqrt{\frac{I_z}{M}} = 1$

(b) $M = \int_C \delta(x,y,z)\, ds = \int_0^{4\pi} \delta\sqrt{2}\, dt = 4\pi\delta\sqrt{2}$ and $I_z = \int_C (x^2+y^2)\delta\, ds = \int_0^{4\pi} \delta\sqrt{2}\, dt = 4\pi\delta\sqrt{2}$

$\Rightarrow R_z = \sqrt{\frac{I_z}{M}} = 1$

31. $\delta(x,y,z) = 2 - z$ and $\mathbf{r}(t) = (\cos t)\mathbf{j} + (\sin t)\mathbf{k},\ 0 \le t \le \pi \Rightarrow M = 2\pi - 2$ as found in Example 4 of the text;

also $\left|\frac{d\mathbf{r}}{dt}\right| = 1;\ I_x = \int_C (y^2+z^2)\delta\, ds = \int_0^{\pi} (\cos^2 t + \sin^2 t)(2 - \sin t)\, dt = \int_0^{\pi} (2 - \sin t)\, dt = 2\pi - 2 \Rightarrow R_x = \sqrt{\frac{I_x}{M}}$

$= 1$

14.2 VECTOR FIELDS, WORK, CIRCULATION, AND FLUX

1. $f(x,y,z) = (x^2+y^2+z^2)^{-1/2} \Rightarrow \frac{\partial f}{\partial x} = -\frac{1}{2}(x^2+y^2+z^2)^{-3/2}(2x) = -x(x^2+y^2+z^2)^{-3/2}$; similarly,

$\frac{\partial f}{\partial y} = -y(x^2+y^2+z^2)^{-3/2}$ and $\frac{\partial f}{\partial z} = -z(x^2+y^2+z^2)^{-3/2} \Rightarrow \nabla f = \frac{-x\mathbf{i} - y\mathbf{j} - z\mathbf{k}}{(x^2+y^2+z^2)^{3/2}}$

3. $g(x,y,z) = e^x - \ln(x^2+y^2) \Rightarrow \frac{\partial g}{\partial x} = -\frac{2x}{x^2+y^2},\ \frac{\partial g}{\partial y} = -\frac{2y}{x^2+y^2}$ and $\frac{\partial g}{\partial z} = e^z$

$\Rightarrow \nabla g = \left(\frac{-2x}{x^2+y^2}\right)\mathbf{i} - \left(\frac{2y}{x^2+y^2}\right)\mathbf{j} + e^z\mathbf{k}$

5. $|\mathbf{F}|$ inversely proportional to the square of the distance from (x,y) to the origin $\Rightarrow \sqrt{(M(x,y))^2 + (N(x,y))^2}$

$= \frac{k}{x^2+y^2},\ k > 0$; $\mathbf{F}$ points toward the origin $\Rightarrow \mathbf{F}$ is in the direction of $\mathbf{n} = \frac{-x}{\sqrt{x^2+y^2}}\mathbf{i} - \frac{y}{\sqrt{x^2+y^2}}\mathbf{j}$

$\Rightarrow \mathbf{F} = a\mathbf{n}$, for some constant $a > 0$. Then $M(x,y) = \frac{-ax}{\sqrt{x^2+y^2}}$ and $N(x,y) = \frac{-ay}{\sqrt{x^2+y^2}}$

$\Rightarrow \sqrt{(M(x,y))^2 + (N(x,y))^2} = a \Rightarrow a = \frac{k}{x^2+y^2} \Rightarrow \mathbf{F} = \frac{-kx}{(x^2+y^2)^{3/2}}\mathbf{i} - \frac{ky}{(x^2+y^2)^{3/2}}\mathbf{j}$, for any constant $k > 0$

7. Substitute the parametric representations for $\mathbf{r}(t) = x(t)\mathbf{i} + y(t)\mathbf{j} + z(t)\mathbf{k}$ representing each path into the vector field $\mathbf{F}$, and calculate the work $W = \int_C \mathbf{F} \cdot \frac{d\mathbf{r}}{dt}$.

(a) $\mathbf{F} = 3t\mathbf{i} + 2t\mathbf{j} + 4t\mathbf{k}$ and $\frac{d\mathbf{r}}{dt} = \mathbf{i} + \mathbf{j} + \mathbf{k} \Rightarrow \mathbf{F} \cdot \frac{d\mathbf{r}}{dt} = 9t \Rightarrow W = \int_0^1 9t\,dt = \frac{9}{2}$

(b) $\mathbf{F} = 3t^2\mathbf{i} + 2t\mathbf{j} + 4t^4\mathbf{k}$ and $\frac{d\mathbf{r}}{dt} = \mathbf{i} + 2t\mathbf{j} + 4t^3\mathbf{k} \Rightarrow \mathbf{F} \cdot \frac{d\mathbf{r}}{dt} = 7t^2 + 16t^7 \Rightarrow W = \int_0^1 (7t^2 + 16t^7)\,dt = \left[\frac{7}{3}t^3 + 2t^8\right]_0^1$
$= \frac{7}{3} + 2 = \frac{13}{3}$

(c) $\mathbf{r}_1 = t\mathbf{i} + t\mathbf{j}$ and $\mathbf{r}_2 = \mathbf{i} + \mathbf{j} + t\mathbf{k}$; $\mathbf{F}_1 = 3t\mathbf{i} + 2t\mathbf{j}$ and $\frac{d\mathbf{r}_1}{dt} = \mathbf{i} + \mathbf{j} \Rightarrow \mathbf{F}_1 \cdot \frac{d\mathbf{r}_1}{dt} = 5t \Rightarrow W_1 = \int_0^1 5t\,dt = \frac{5}{2}$;
$\mathbf{F}_2 = 3\mathbf{i} + 2\mathbf{j} + 4t\mathbf{k}$ and $\frac{d\mathbf{r}_2}{dt} = \mathbf{k} \Rightarrow \mathbf{F}_2 \cdot \frac{d\mathbf{r}_2}{dt} = 4t \Rightarrow W_2 = \int_0^1 4t\,dt = 2 \Rightarrow W = W_1 + W_2 = \frac{9}{2}$

9. Substitute the parametric representation for $\mathbf{r}(t) = x(t)\mathbf{i} + y(t)\mathbf{j} + z(t)\mathbf{k}$ representing each path into the vector field $\mathbf{F}$, and calculate the work $W = \int_C \mathbf{F} \cdot \frac{d\mathbf{r}}{dt}$.

(a) $\mathbf{F} = \sqrt{t}\mathbf{i} - 2t\mathbf{j} + \sqrt{t}\mathbf{k}$ and $\frac{d\mathbf{r}}{dt} = \mathbf{i} + \mathbf{j} + \mathbf{k} \Rightarrow \mathbf{F} \cdot \frac{d\mathbf{r}}{dt} = 2\sqrt{t} - 2t \Rightarrow W = \int_0^1 (2\sqrt{t} - 2t)dt = \left[\frac{4}{3}t^{3/2} - t^2\right]_0^1 = \frac{1}{3}$

(b) $\mathbf{F} = t^2\mathbf{i} - 2t\mathbf{j} + t\mathbf{k}$ and $\frac{d\mathbf{r}}{dt} = \mathbf{i} + 2t\mathbf{j} + 4t^3\mathbf{k} \Rightarrow \mathbf{F} \cdot \frac{d\mathbf{r}}{dt} = 4t^4 - 3t^2 \Rightarrow W = \int_0^1 (4t^4 - 3t^2)\,dt = \left[\frac{4}{5}t^5 - t^3\right]_0^1 = -\frac{1}{5}$

(c) $\mathbf{r}_1 = t\mathbf{i} + t\mathbf{j}$ and $\mathbf{r}_2 = \mathbf{i} + \mathbf{j} + t\mathbf{k}$; $\mathbf{F}_1 = -2t\mathbf{j} + \sqrt{t}\,\mathbf{k}$ and $\frac{d\mathbf{r}_1}{dt} = \mathbf{i} + \mathbf{j} \Rightarrow \mathbf{F}_1 \cdot \frac{d\mathbf{r}_1}{dt} = -2t \Rightarrow W_1 = \int_0^1 -2t\,dt$
$= -1$; $\mathbf{F}_2 = \sqrt{t}\mathbf{i} - 2\mathbf{j} + \mathbf{k}$ and $\frac{d\mathbf{r}_2}{dt} = \mathbf{k} \Rightarrow \mathbf{F}_2 \cdot \frac{d\mathbf{r}_2}{dt} = 1 \Rightarrow W_2 = \int_0^1 dt = 1 \Rightarrow W = W_1 + W_2 = 0$

11. Substitute the parametric representation for $\mathbf{r}(t) = x(t)\mathbf{i} + y(t)\mathbf{j} + z(t)\mathbf{k}$ representing each path into the vector field $\mathbf{F}$, and calculate the work $W = \int_C \mathbf{F} \cdot \frac{d\mathbf{r}}{dt}$.

(a) $\mathbf{F} = (3t^2 - 3t)\mathbf{i} + 3t\mathbf{j} + \mathbf{k}$ and $\frac{d\mathbf{r}}{dt} = \mathbf{i} + \mathbf{j} + \mathbf{k} \Rightarrow \mathbf{F} \cdot \frac{d\mathbf{r}}{dt} = 3t^2 + 1 \Rightarrow W = \int_0^1 (3t^2 + 1)\,dt = \left[t^3 + t\right]_0^1 = 2$

(b) $\mathbf{F} = (3t^2 - 3t)\mathbf{i} + 3t^4\mathbf{j} + \mathbf{k}$ and $\frac{d\mathbf{r}}{dt} = \mathbf{i} + 2t\mathbf{j} + 4t^3\mathbf{k} \Rightarrow \mathbf{F} \cdot \frac{d\mathbf{r}}{dt} = 6t^5 + 4t^3 + 3t^2 - 3t$
$\Rightarrow W = \int_0^1 (6t^5 + 4t^3 + 3t^2 - 3t)\,dt = \left[t^6 + t^4 + t^3 - \frac{3}{2}t^2\right]_0^1 = \frac{3}{2}$

(c) $\mathbf{r}_1 = t\mathbf{i} + t\mathbf{j}$ and $\mathbf{r}_2 = \mathbf{i} + \mathbf{j} + t\mathbf{k}$; $\mathbf{F}_1 = (3t^2 - 3t)\mathbf{i} + \mathbf{k}$ and $\frac{d\mathbf{r}_1}{dt} = \mathbf{i} + \mathbf{j} \Rightarrow \mathbf{F}_1 \cdot \frac{d\mathbf{r}_1}{dt} = 3t^2 - 3t$

$\Rightarrow W_1 = \int_0^1 (3t^2 - 3t)\, dt = \left[t^3 - \frac{3}{2}t^2\right]_0^1 = -\frac{1}{2}$; $\mathbf{F}_2 = 3t\mathbf{j} + \mathbf{k}$ and $\frac{d\mathbf{r}_2}{dt} = \mathbf{k} \Rightarrow \mathbf{F}_2 \cdot \frac{d\mathbf{r}_2}{dt} = 1 \Rightarrow W_2 = \int_0^1 dt = 1$

$\Rightarrow W = W_1 + W_2 = \frac{1}{2}$

13. $\mathbf{r} = t\mathbf{i} + t^2\mathbf{j} + t\mathbf{k}$, $0 \le t \le 1$, and $\mathbf{F} = xy\mathbf{i} + y\mathbf{j} - yz\mathbf{k} \Rightarrow \mathbf{F} = t^3\mathbf{i} + t^2\mathbf{j} - t^3\mathbf{k}$ and $\frac{d\mathbf{r}}{dt} = \mathbf{i} + 2t\mathbf{j} + \mathbf{k}$

$\Rightarrow \mathbf{F} \cdot \frac{d\mathbf{r}}{dt} = 2t^3 \Rightarrow \text{work} = \int_0^1 2t^3\, dt = \frac{1}{2}$

15. $\mathbf{r} = (\sin t)\mathbf{i} + (\cos t)\mathbf{j} + t\mathbf{k}$, $0 \le t \le 2\pi$, and $\mathbf{F} = z\mathbf{i} + x\mathbf{j} + y\mathbf{k} \Rightarrow \mathbf{F} = t\mathbf{i} + (\sin t)\mathbf{j} + (\cos t)\mathbf{k}$ and

$\frac{d\mathbf{r}}{dt} = (\cos t)\mathbf{i} - (\sin t)\mathbf{j} + \mathbf{k} \Rightarrow \mathbf{F} \cdot \frac{d\mathbf{r}}{dt} = t \cos t - \sin^2 t + \cos t \Rightarrow \text{work} = \int_0^{2\pi} (t \cos t - \sin^2 t + \cos t)\, dt$

$= \left[\cos t + t \sin t - \frac{t}{2} + \frac{\sin 2t}{4} + \sin t\right]_0^{2\pi} = -\pi$

17. $x = t$ and $y = x^2 = t^2 \Rightarrow \mathbf{r} = t\mathbf{i} + t^2\mathbf{j}$, $-1 \le t \le 2$, and $\mathbf{F} = xy\mathbf{i} + (x+y)\mathbf{j} \Rightarrow \mathbf{F} = t^3\mathbf{i} + (t + t^2)\mathbf{j}$ and

$\frac{d\mathbf{r}}{dt} = \mathbf{i} + 2t\mathbf{j} \Rightarrow \mathbf{F} \cdot \frac{d\mathbf{r}}{dt} = t^3 + (2t^2 + 2t^3) = 3t^3 + 2t^2 \Rightarrow \int_C xy\, dx + (x+y)\, dy = \int_C \mathbf{F} \cdot \frac{d\mathbf{r}}{dt}\, dt = \int_{-1}^{2} (3t^3 + 2t^2)\, dt$

$= \left[\frac{3}{4}t^4 + \frac{2}{3}t^3\right]_{-1}^2 = \left(12 + \frac{16}{3}\right) - \left(\frac{3}{4} - \frac{2}{3}\right) = \frac{45}{4} + \frac{18}{3} = \frac{207}{12}$

19. $\mathbf{r} = x\mathbf{i} + y\mathbf{j} = y^2\mathbf{i} + y\mathbf{j}$, $2 \ge y \ge -1$, and $\mathbf{F} = x^2\mathbf{i} - y\mathbf{j} = y^4\mathbf{i} - y\mathbf{j} \Rightarrow \frac{d\mathbf{r}}{dy} = 2y\mathbf{i} + \mathbf{j}$ and $\mathbf{F} \cdot \frac{d\mathbf{r}}{dy} = 2y^5 - y$

$\Rightarrow \int_C \mathbf{F} \cdot \mathbf{T}\, ds = \int_2^{-1} \mathbf{F} \cdot \frac{d\mathbf{r}}{dy}\, dy = \int_2^{-1} (2y^5 - y)\, dy = \left[\frac{1}{3}y^6 - \frac{1}{2}y^2\right]_2^{-1} = \left(\frac{1}{3} - \frac{1}{2}\right) - \left(\frac{64}{3} - \frac{4}{2}\right) = \frac{3}{2} - \frac{63}{3} = -\frac{39}{2}$

21. $\mathbf{r} = (\mathbf{i} + \mathbf{j}) + t(\mathbf{i} + 2\mathbf{j}) = (1+t)\mathbf{i} + (1+2t)\mathbf{j}$, $0 \le t \le 1$, and $\mathbf{F} = xy\mathbf{i} + (y - x)\mathbf{j} \Rightarrow \mathbf{F} = (1 + 3t + 2t^2)\mathbf{i} + t\mathbf{j}$ and

$\frac{d\mathbf{r}}{dt} = \mathbf{i} + 2\mathbf{j} \Rightarrow \mathbf{F} \cdot \frac{d\mathbf{r}}{dt} = 1 + 5t + 2t^2 \Rightarrow \text{work} = \int_C \mathbf{F} \cdot \frac{d\mathbf{r}}{dt}\, dt = \int_0^1 (1 + 5t + 2t^2)\, dt = \left[t + \frac{5}{2}t^2 + \frac{2}{3}t^3\right]_0^1 = \frac{25}{6}$

23. (a) $\mathbf{r} = (\cos t)\mathbf{i} + (\sin t)\mathbf{j}$, $0 \le t \le 2\pi$, $\mathbf{F}_1 = x\mathbf{i} + y\mathbf{j}$, and $\mathbf{F}_2 = -y\mathbf{i} + x\mathbf{j} \Rightarrow \frac{d\mathbf{r}}{dt} = (-\sin t)\mathbf{i} + (\cos t)\mathbf{j}$,

$\mathbf{F}_1 = (\cos t)\mathbf{i} + (\sin t)\mathbf{j}$, and $\mathbf{F}_2 = (-\sin t)\mathbf{i} + (\cos t)\mathbf{j} \Rightarrow \mathbf{F}_1 \cdot \frac{d\mathbf{r}}{dt} = 0$ and $\mathbf{F}_2 \cdot \frac{d\mathbf{r}}{dt} = \sin^2 t + \cos^2 t = 1$

$\Rightarrow \text{Circ}_1 = \int_0^{2\pi} 0\, dt = 0$ and $\text{Circ}_2 = \int_0^{2\pi} dt = 2\pi$; $\mathbf{n} = (\cos t)\mathbf{i} + (\sin t)\mathbf{j} \Rightarrow \mathbf{F}_1 \cdot \mathbf{n} = \cos^2 t + \sin^2 t = 1$ and

$\mathbf{F}_2 \cdot \mathbf{n} = 0 \Rightarrow \text{Flux}_1 = \int_0^{2\pi} dt = 2\pi$ and $\text{Flux}_2 = \int_0^{2\pi} 0\, dt = 0$

(b) $\mathbf{r} = (\cos t)\mathbf{i} + (4 \sin t)\mathbf{j},\ 0 \le t \le 2\pi \Rightarrow \frac{d\mathbf{r}}{dt} = (-\sin t)\mathbf{i} + (4 \cos t)\mathbf{j},\ \mathbf{F}_1 = (\cos t)\mathbf{i} + (4 \sin t)\mathbf{j}$, and

$\mathbf{F}_2 = (-4 \sin t)\mathbf{i} + (\cos t)\mathbf{j} \Rightarrow \mathbf{F}_1 \cdot \frac{d\mathbf{r}}{dt} = 15 \sin t \cos t$ and $\mathbf{F}_2 \cdot \frac{d\mathbf{r}}{dt} = 4 \Rightarrow \text{Circ}_1 = \int_0^{2\pi} 15 \sin t \cos t\, dt$

$= \left[\frac{15}{2} \sin^2 t\right]_0^{2\pi} = 0$ and $\text{Circ}_2 = \int_0^{2\pi} 4\, dt = 8\pi$; $\mathbf{n} = \left(\frac{4}{\sqrt{17}} \cos t\right)\mathbf{i} + \left(\frac{1}{\sqrt{17}} \sin t\right)\mathbf{j} \Rightarrow \mathbf{F}_1 \cdot \mathbf{n}$

$= \frac{4}{\sqrt{17}} \cos^2 t + \frac{4}{\sqrt{17}} \sin^2 t$ and $\mathbf{F}_2 \cdot \mathbf{n} = -\frac{15}{\sqrt{17}} \sin t \cos t \Rightarrow \text{Flux}_1 = \int_0^{2\pi} (\mathbf{F}_1 \cdot \mathbf{n})\,|\mathbf{v}|\, dt = \int_0^{2\pi} \left(\frac{4}{\sqrt{17}}\right)\sqrt{17}\, dt$

$= 8\pi$ and $\text{Flux}_2 = \int_0^{2\pi} (\mathbf{F}_2 \cdot \mathbf{n})\,|\mathbf{v}|\, dt = \int_0^{2\pi} \left(-\frac{15}{\sqrt{17}} \sin t \cos t\right) \sqrt{17}\, dt = \left[-\frac{15}{2} \sin^2 t\right]_0^{2\pi} = 0$

25. $\mathbf{F}_1 = (a \cos t)\mathbf{i} + (a \sin t)\mathbf{j},\ \frac{d\mathbf{r}_1}{dt} = (-a \sin t)\mathbf{i} + (a \cos t)\mathbf{j} \Rightarrow \mathbf{F}_1 \cdot \frac{d\mathbf{r}_1}{dt} = 0 \Rightarrow \text{Circ}_1 = 0$; $M_1 = a \cos t$,

$N_1 = a \sin t$, $dx = -a \sin t\, dt$, $dy = a \cos t\, dt \Rightarrow \text{Flux}_1 = \int_C M_1\, dy - N_1\, dx = \int_0^{\pi} \left(a^2 \cos^2 t + a^2 \sin^2 t\right) dt$

$= \int_0^{\pi} a^2\, dt = a^2\pi$;

$\mathbf{F}_2 = t\mathbf{i},\ \frac{d\mathbf{r}_2}{dt} = \mathbf{i} \Rightarrow \mathbf{F}_2 \cdot \frac{d\mathbf{r}_2}{dt} = t \Rightarrow \text{Circ}_2 = \int_{-a}^{a} t\, dt = 0$; $M_2 = t$, $N_2 = 0$, $dx = dt$, $dy = 0 \Rightarrow \text{Flux}_2$

$= \int_C M_2\, dy - N_2\, dx = \int_{-a}^{a} 0\, dt = 0$; therefore, $\text{Circ} = \text{Circ}_1 + \text{Circ}_2 = 0$ and $\text{Flux} = \text{Flux}_1 + \text{Flux}_2 = a^2\pi$

27. $\mathbf{F}_1 = (-a \sin t)\mathbf{i} + (a \cos t)\mathbf{j},\ \frac{d\mathbf{r}_1}{dt} = (-a \sin t)\mathbf{i} + (a \cos t)\mathbf{j} \Rightarrow \mathbf{F}_1 \cdot \frac{d\mathbf{r}_1}{dt} = a^2 \sin^2 t + a^2 \cos^2 t = a^2$

$\Rightarrow \text{Circ}_1 = \int_0^{\pi} a^2\, dt = a^2\pi$; $M_1 = -a \sin t$, $N_1 = a \cos t$, $dx = -a \sin t\, dt$, $dy = a \cos t\, dt$

$\Rightarrow \text{Flux}_1 = \int_C M_1\, dy - N_1\, dx = \int_0^{\pi} \left(-a^2 \sin t \cos t + a^2 \sin t \cos t\right) dt = 0$; $\mathbf{F}_2 = t\mathbf{j},\ \frac{d\mathbf{r}_2}{dt} = \mathbf{i} \Rightarrow \mathbf{F}_2 \cdot \frac{d\mathbf{r}_2}{dt} = 0$

$\Rightarrow \text{Circ}_2 = 0$; $M_2 = 0$, $N_2 = t$, $dx = dt$, $dy = 0 \Rightarrow \text{Flux}_2 = \int_C M_2\, dy - N_2\, dx = \int_{-a}^{a} -t\, dt = 0$; therefore,

$\text{Circ} = \text{Circ}_1 + \text{Circ}_2 = a^2\pi$ and $\text{Flux} = \text{Flux}_1 + \text{Flux}_2 = 0$

29. (a) $\mathbf{r} = (\cos t)\mathbf{i} + (\sin t)\mathbf{j},\ 0 \le t \le \pi$, and $\mathbf{F} = (x + y)\mathbf{i} - (x^2 + y^2)\mathbf{j} \Rightarrow \frac{d\mathbf{r}}{dt} = (-\sin t)\mathbf{i} + (\cos t)\mathbf{j}$ and

$\mathbf{F} = (\cos t + \sin t)\mathbf{i} - \left(\cos^2 t + \sin^2 t\right)\mathbf{j} \Rightarrow \mathbf{F} \cdot \frac{d\mathbf{r}}{dt} = -\sin t \cos t - \sin^2 t - \cos t \Rightarrow \int_C \mathbf{F} \cdot \mathbf{T}\, ds$

$$= \int_0^{\pi} \left(-\sin t \cos t - \sin^2 t - \cos t\right) dt = \left[-\tfrac{1}{2}\sin^2 t - \tfrac{t}{2} + \frac{\sin 2t}{4} - \sin t\right]_0^{\pi} = -\tfrac{\pi}{2}$$

(b) $\mathbf{r} = (1-2t)\mathbf{i}$, $0 \le t \le 1$, and $\mathbf{F} = (x+y)\mathbf{i} - (x^2+y^2)\mathbf{j} \Rightarrow \frac{d\mathbf{r}}{dt} = -\mathbf{i}$ and $\mathbf{F} = (1-2t)\mathbf{i} - (1-2t)^2\mathbf{j} \Rightarrow$

$$\mathbf{F}\cdot\frac{d\mathbf{r}}{dt} = 2t - 1 \Rightarrow \int_C \mathbf{F}\cdot\mathbf{T}\, ds = \int_0^1 (2t-1)\, dt = \left[t^2 - t\right]_0^1 = 0$$

(c) $\mathbf{r}_1 = (1-t)\mathbf{i} - t\mathbf{j}$, $0 \le t \le 1$, and $\mathbf{F} = (x+y)\mathbf{i} - (x^2+y^2)\mathbf{j} \Rightarrow \frac{d\mathbf{r}_1}{dt} = -\mathbf{i} - \mathbf{j}$ and $\mathbf{F} = (1-2t)\mathbf{i} - \left(1 - 2t + 2t^2\right)\mathbf{j}$

$$\Rightarrow \mathbf{F}\cdot\frac{d\mathbf{r}_1}{dt} = (2t-1) + \left(1 - 2t + 2t^2\right) = 2t^2 \Rightarrow \text{Flow}_1 = \int_{C_1} \mathbf{F}\cdot\frac{d\mathbf{r}_1}{dt} = \int_0^1 2t^2\, dt = \tfrac{2}{3};\ \mathbf{r}_2 = -t\mathbf{i} + (t-1)\mathbf{j},$$

$0 \le t \le 1$, and $\mathbf{F} = (x+y)\mathbf{i} - (x^2+y^2)\mathbf{j} \Rightarrow \frac{d\mathbf{r}_2}{dt} = -\mathbf{i} + \mathbf{j}$ and $\mathbf{F} = -\mathbf{i} - \left(t^2 + t^2 - 2t + 1\right)\mathbf{j}$

$$= -\mathbf{i} - \left(2t^2 - 2t + 1\right)\mathbf{j} \Rightarrow \mathbf{F}\cdot\frac{d\mathbf{r}_2}{dt} = 1 - \left(2t^2 - 2t + 1\right) = 2t - 2t^2 \Rightarrow \text{Flow}_2 = \int_{C_2} \mathbf{F}\cdot\frac{d\mathbf{r}_2}{dt} = \int_0^1 \left(2t - 2t^2\right) dt$$

$$= \left[t^2 - \tfrac{2}{3}t^3\right]_0^1 = \tfrac{1}{3} \Rightarrow \text{Flow} = \text{Flow}_1 + \text{Flow}_2 = \tfrac{2}{3} + \tfrac{1}{3} = 1$$

31. $\mathbf{F} = -\frac{y}{\sqrt{x^2+y^2}}\mathbf{i} + \frac{x}{\sqrt{x^2+y^2}}\mathbf{j}$ on $x^2 + y^2 = 4$;

at $(2,0)$, $\mathbf{F} = \mathbf{j}$; at $(0,2)$, $\mathbf{F} = -\mathbf{i}$; at $(-2,0)$,

$\mathbf{F} = -\mathbf{j}$; at $(0,-2)$, $\mathbf{F} = \mathbf{i}$; at $\left(1, \sqrt{3}\right)$, $\mathbf{F} = -\frac{\sqrt{3}}{2}\mathbf{i} + \frac{1}{2}\mathbf{j}$;

at $\left(1, -\sqrt{3}\right)$, $\mathbf{F} = \frac{\sqrt{3}}{2}\mathbf{i} + \frac{1}{2}\mathbf{j}$; at $\left(-1, \sqrt{3}\right)$,

$\mathbf{F} = -\frac{\sqrt{3}}{2}\mathbf{i} - \frac{1}{2}\mathbf{j}$; at $\left(-1, -\sqrt{3}\right)$, $\mathbf{F} = \frac{\sqrt{3}}{2}\mathbf{i} - \frac{1}{2}\mathbf{j}$

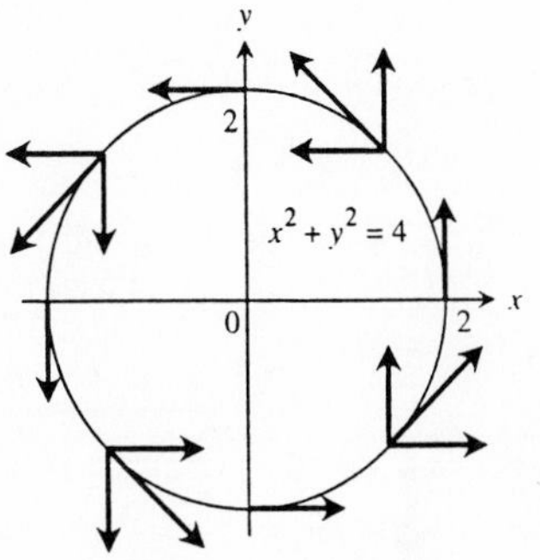

33. (a) $\mathbf{G} = P(x,y)\mathbf{i} + Q(x,y)\mathbf{j}$ is to have a magnitude $\sqrt{a^2+b^2}$ and to be tangent to $x^2 + y^2 = a^2 + b^2$ in a counterclockwise direction. Thus $x^2 + y^2 = a^2 + b^2 \Rightarrow 2x + 2yy' = 0 \Rightarrow y' = -\frac{x}{y}$ is the slope of the tangent line at any point on the circle $\Rightarrow y' = -\frac{a}{b}$ at (a,b). Let $\mathbf{v} = -b\mathbf{i} + a\mathbf{j} \Rightarrow |\mathbf{v}| = \sqrt{a^2+b^2}$, with $\mathbf{v}$ in a counterclockwise direction and tangent to the circle. Then let $P(x,y) = -y$ and $Q(x,y) = x$ $\Rightarrow \mathbf{G} = -y\mathbf{i} + x\mathbf{j} \Rightarrow$ for (a,b) on $x^2 + y^2 = a^2 + b^2$ we have $\mathbf{G} = -b\mathbf{i} + a\mathbf{j}$ and $|\mathbf{G}| = \sqrt{a^2+b^2}$.

(b) $\mathbf{G} = \left(\sqrt{x^2+y^2}\right)\mathbf{F} = \left(\sqrt{a^2+b^2}\right)\mathbf{F}$, since $x^2 + y^2 = a^2 + b^2$

35. The slope of the line through (x,y) and the origin is $\frac{y}{x} \Rightarrow \mathbf{v} = x\mathbf{i} + y\mathbf{j}$ is a vector parallel to that line and pointing away from the origin $\Rightarrow \mathbf{F} = -\frac{x\mathbf{i} + y\mathbf{j}}{\sqrt{x^2+y^2}}$ is the unit vector pointing toward the origin.

37. $\mathbf{F} = -4t^3\mathbf{i} + 8t^2\mathbf{j} + 2\mathbf{k}$ and $\frac{d\mathbf{r}}{dt} = \mathbf{i} + 2t\mathbf{j} \Rightarrow \mathbf{F} \cdot \frac{d\mathbf{r}}{dt} = 12t^3 \Rightarrow \text{Flow} = \int_0^2 12t^3\, dt = \left[3t^4\right]_0^2 = 48$

39. $\mathbf{F} = (\cos t - \sin t)\mathbf{i} + (\cos t)\mathbf{k}$ and $\frac{d\mathbf{r}}{dt} = (-\sin t)\mathbf{i} + (\cos t)\mathbf{k} \Rightarrow \mathbf{F} \cdot \frac{d\mathbf{r}}{dt} = -\sin t \cos t + 1$

$$\Rightarrow \text{Flow} = \int_0^\pi (-\sin t \cos t + 1)\, dt = \left[\frac{1}{2}\cos^2 t + t\right]_0^\pi = \left(\frac{1}{2} + \pi\right) - \left(\frac{1}{2} + 0\right) = \pi$$

41. C_1: $\mathbf{r} = (\cos t)\mathbf{i} + (\sin t)\mathbf{j} + t\mathbf{k}$, $0 \le t \le \frac{\pi}{2} \Rightarrow \mathbf{F} = (2\cos t)\mathbf{i} + 2t\mathbf{j} + (2\sin t)\mathbf{k}$ and $\frac{d\mathbf{r}}{dt} = (-\sin t)\mathbf{i} + (\cos t)\mathbf{j} + \mathbf{k}$

$$\Rightarrow \mathbf{F} \cdot \frac{d\mathbf{r}}{dt} = -2\cos t \sin t + 2t\cos t + 2\sin t = -\sin 2t + 2t\cos t + 2\sin t$$

$$\Rightarrow \text{Flow}_1 = \int_0^{\pi/2} (-\sin 2t + 2t\cos t + 2\sin t)\, dt = \left[\frac{1}{2}\cos 2t + 2t\sin t + 2\cos t - 2\cos t\right]_0^{\pi/2} = -1 + \pi;$$

C_2: $\mathbf{r} = \mathbf{j} + \frac{\pi}{2}(1 - t)\mathbf{k}$, $0 \le t \le 1 \Rightarrow \mathbf{F} = \pi(1 - t)\mathbf{j} + 2\mathbf{k}$ and $\frac{d\mathbf{r}}{dt} = -\frac{\pi}{2}\mathbf{k} \Rightarrow \mathbf{F} \cdot \frac{d\mathbf{r}}{dt} = -\pi$

$$\Rightarrow \text{Flow}_2 = \int_0^1 -\pi\, dt = [-\pi t]_0^1 = -\pi;$$

C_3: $\mathbf{r} = t\mathbf{i} + (1 - t)\mathbf{j}$, $0 \le t \le 1 \Rightarrow \mathbf{F} = 2t\mathbf{i} + 2(1 - t)\mathbf{j}$ and $\frac{d\mathbf{r}}{dt} = \mathbf{i} - \mathbf{j} \Rightarrow \mathbf{F} \cdot \frac{d\mathbf{r}}{dt} = 2t$

$$\Rightarrow \text{Flow}_3 = \int_0^1 2t\, dt = \left[t^2\right]_0^1 = 1 \Rightarrow \text{Circulation} = (-1 + \pi) - \pi + 1 = 0$$

43. Let $x = t$ be the parameter $\Rightarrow y = x^2 = t^2$ and $z = x = t \Rightarrow \mathbf{r} = t\mathbf{i} + t^2\mathbf{j} + t\mathbf{k}$, $0 \le t \le 1$ from $(0, 0, 0)$ to $(1, 1, 1)$

$\Rightarrow \frac{d\mathbf{r}}{dt} = \mathbf{i} + 2t\mathbf{j} + \mathbf{k}$ and $\mathbf{F} = xy\mathbf{i} + y\mathbf{j} - yz\mathbf{k} = t^3\mathbf{i} + t^2\mathbf{j} - t^3\mathbf{k} \Rightarrow \mathbf{F} \cdot \frac{d\mathbf{r}}{dt} = t^3 + 2t^3 - t^3 = 2t^3 \Rightarrow \text{Flow} = \int_0^1 2t^3\, dt$

$= \frac{1}{2}$

45. Yes. The work and area have the same numerical value because work $= \int_C \mathbf{F} \cdot d\mathbf{r} = \int_C y\mathbf{i} \cdot d\mathbf{r}$

$= \int_b^a [f(t)\mathbf{i}] \cdot \left[\mathbf{i} + \frac{df}{dt}\mathbf{j}\right] dt$ [On the path, y equals f(t)]

$= \int_a^b f(t)\, dt$ = Area under the curve [because $f(t) > 0$]

14.3 PATH INDEPENDENCE, POTENTIAL FUNCTIONS, AND CONSERVATIVE FIELDS

1. $\frac{\partial P}{\partial y} = x = \frac{\partial N}{\partial z}, \frac{\partial M}{\partial z} = y = \frac{\partial P}{\partial x}, \frac{\partial N}{\partial x} = z = \frac{\partial M}{\partial y} \Rightarrow$ Conservative

3. $\frac{\partial P}{\partial y} = -1 \neq 1 = \frac{\partial N}{\partial z} \Rightarrow$ Not Conservative

5. $\frac{\partial N}{\partial x} = 0 \neq 1 = \frac{\partial M}{\partial y} \Rightarrow$ Not Conservative

7. $\frac{\partial f}{\partial x} = 2x \Rightarrow f(x,y,z) = x^2 + g(y,z) \Rightarrow \frac{\partial f}{\partial y} = \frac{\partial g}{\partial y} = 3y \Rightarrow g(y,z) = \frac{3y^2}{2} + h(z) \Rightarrow f(x,y,z) = x^2 + \frac{3y^2}{2} + h(z)$
$\Rightarrow \frac{\partial f}{\partial z} = h'(z) = 4z \Rightarrow h(z) = 2z^2 + C \Rightarrow f(x,y,z) = x^2 + \frac{3y^2}{2} + 2z^2 + C$

9. $\frac{\partial f}{\partial x} = e^{y+2z} \Rightarrow f(x,y,z) = xe^{y+2z} + g(y,z) \Rightarrow \frac{\partial f}{\partial y} = xe^{y+2z} + \frac{\partial g}{\partial y} = xe^{y+2z} \Rightarrow \frac{\partial g}{\partial y} = 0 \Rightarrow f(x,y,z)$
$= xe^{y+2z} + h(z) \Rightarrow \frac{\partial f}{\partial z} = 2xe^{y+2z} + h'(z) = 2xe^{y+2z} \Rightarrow h'(z) = 0 \Rightarrow h(z) = C \Rightarrow f(x,y,z) = xe^{y+2z} + C$

11. $\frac{\partial f}{\partial z} = \frac{z}{y^2+z^2} \Rightarrow f(x,y,z) = \frac{1}{2}\ln(y^2+z^2) + g(x,y) \Rightarrow \frac{\partial f}{\partial x} = \frac{\partial g}{\partial x} = \ln x + \sec^2(x+y) \Rightarrow g(x,y)$
$= (x \ln x - x) + \tan(x+y) + h(y) \Rightarrow f(x,y,z) = \frac{1}{2}\ln(y^2+z^2) + (x \ln x - x) + \tan(x+y) + h(y)$
$\Rightarrow \frac{\partial f}{\partial y} = \frac{y}{y^2+z^2} + \sec^2(x+y) + h'(y) = \sec^2(x+y) + \frac{y}{y^2+z^2} \Rightarrow h'(y) = 0 \Rightarrow h(y) = C \Rightarrow f(x,y,z)$
$= \frac{1}{2}\ln(y^2+z^2) + (x \ln x - x) + \tan(x+y) + C$

13. Let $\mathbf{F}(x,y,z) = 2x\mathbf{i} + 2y\mathbf{j} + 2z\mathbf{k} \Rightarrow \frac{\partial P}{\partial y} = 0 = \frac{\partial N}{\partial z}, \frac{\partial M}{\partial z} = 0 = \frac{\partial P}{\partial x}, \frac{\partial N}{\partial x} = 0 = \frac{\partial M}{\partial y} \Rightarrow M\,dx + N\,dy + P\,dz$ is exact; $\frac{\partial f}{\partial x} = 2x \Rightarrow f(x,y,z) = x^2 + g(y,z) \Rightarrow \frac{\partial f}{\partial y} = \frac{\partial g}{\partial y} = 2y \Rightarrow g(y,z) = y^2 + h(z) \Rightarrow f(x,y,z) = x^2 + y^2 = h(z)$
$\Rightarrow \frac{\partial f}{\partial z} = h'(z) = 2z \Rightarrow h(z) = z^2 + C \Rightarrow f(x,y,z) = x^2 + y^2 + z^2 + C \Rightarrow \int_{(0,0,0)}^{(2,3,-6)} 2x\,dx + 2y\,dy + 2z\,dz$
$= f(2,3,-6) - f(0,0,0) = 2^2 + 3^2 + (-6)^2 = 49$

15. Let $\mathbf{F}(x,y,z) = 2xy\mathbf{i} + (x^2 - z^2)\mathbf{j} - 2yz\mathbf{k} \Rightarrow \frac{\partial P}{\partial y} = -2z = \frac{\partial N}{\partial z}, \frac{\partial M}{\partial z} = 0 = \frac{\partial P}{\partial x}, \frac{\partial N}{\partial x} = 2x = \frac{\partial M}{\partial y}$
$\Rightarrow M\,dx + N\,dy + P\,dz$ is exact; $\frac{\partial f}{\partial x} = 2xy \Rightarrow f(x,y,z) = x^2y + g(y,z) \Rightarrow \frac{\partial f}{\partial y} = x^2 + \frac{\partial g}{\partial y} = x^2 - z^2 \Rightarrow \frac{\partial g}{\partial y} = -z^2$
$\Rightarrow g(y,z) = -yz^2 + h(z) \Rightarrow f(x,y,z) = x^2y - yz^2 + h(z) \Rightarrow \frac{\partial f}{\partial z} = -2yz + h'(z) = -2yz \Rightarrow h'(z) = 0 \Rightarrow h(z) = C$
$\Rightarrow f(x,y,z) = x^2y - yz^2 + C \Rightarrow \int_{(0,0,0)}^{(1,2,3)} 2xy\,dx + (x^2 - z^2)\,dy - 2yz\,dz = f(1,2,3) - f(0,0,0) = 2 - 2(3)^2 = -16$

17. Let $\mathbf{F}(x,y,z) = (\sin y \cos x)\mathbf{i} + (\cos y \sin x)\mathbf{j} + \mathbf{k} \Rightarrow \frac{\partial P}{\partial y} = 0 = \frac{\partial N}{\partial z}, \frac{\partial M}{\partial z} = 0 = \frac{\partial P}{\partial x}, \frac{\partial N}{\partial x} = \cos y \cos x = \frac{\partial M}{\partial y}$
$\Rightarrow M\,dx + N\,dy + P\,dz$ is exact; $\frac{\partial f}{\partial x} = \sin y \cos x \Rightarrow f(x,y,z) = \sin y \sin x + g(y,z) \Rightarrow \frac{\partial f}{\partial y} = \cos y \sin x + \frac{\partial g}{\partial y}$

$= \cos y \sin x \Rightarrow \frac{\partial g}{\partial y} = 0 \Rightarrow g(y,z) = h(z) \Rightarrow f(x,y,z) = \sin y \sin x + h(z) \Rightarrow \frac{\partial f}{\partial z} = h'(z) = 1 \Rightarrow h(z) = z + C$

$\Rightarrow f(x,y,z) = \sin y \sin x + z + C \Rightarrow \int_{(1,0,0)}^{(0,1,1)} \sin y \cos x\, dx + \cos y \sin x\, dy + dz = f(0,1,1) - f(1,0,0)$

$= (0+1+C) - (0+0+C) = 1$

19. Let $\mathbf{F}(x,y,z) = 3x^2\mathbf{i} + \left(\frac{z^2}{y}\right)\mathbf{j} + (2z \ln y)\mathbf{k} \Rightarrow \frac{\partial P}{\partial y} = \frac{2z}{y} = \frac{\partial N}{\partial z}, \frac{\partial M}{\partial z} = 0 = \frac{\partial P}{\partial x}, \frac{\partial N}{\partial x} = 0 = \frac{\partial M}{\partial y}$

$\Rightarrow M\,dx + N\,dy + P\,dz$ is exact; $\frac{\partial f}{\partial x} = 3x^2 \Rightarrow f(x,y,z) = x^3 + g(y,z) \Rightarrow \frac{\partial f}{\partial y} = \frac{\partial g}{\partial y} = \frac{z^2}{y} \Rightarrow g(y,z) = z^2 \ln y + h(z)$

$\Rightarrow f(x,y,z) = x^3 + z^2 \ln y + h(z) \Rightarrow \frac{\partial f}{\partial z} = 2z \ln y + h'(z) = 2z \ln y \Rightarrow h'(z) = 0 \Rightarrow h(z) = C \Rightarrow f(x,y,z)$

$= x^3 + z^2 \ln y + C \Rightarrow \int_{(1,1,1)}^{(1,2,3)} 3x^2\, dx + \frac{z^2}{y}\, dy + 2z \ln y\, dz = f(1,2,3) - f(1,1,1)$

$= (1 + 9 \ln 2 + C) - (1 + 0 + C) = 9 \ln 2$

21. Let $\mathbf{F}(x,y,z) = \left(\frac{1}{y}\right)\mathbf{i} + \left(\frac{1}{z} - \frac{x}{y^2}\right)\mathbf{j} - \left(\frac{y}{z^2}\right)\mathbf{k} \Rightarrow \frac{\partial P}{\partial y} = -\frac{1}{z^2} = \frac{\partial N}{\partial z}, \frac{\partial M}{\partial z} = 0 = \frac{\partial P}{\partial x}, \frac{\partial N}{\partial x} = -\frac{1}{y^2} = \frac{\partial M}{\partial y}$

$\Rightarrow M\,dx + N\,dy + P\,dz$ is exact; $\frac{\partial f}{\partial x} = \frac{1}{y} \Rightarrow f(x,y,z) = \frac{x}{y} + g(y,z) \Rightarrow \frac{\partial f}{\partial y} = -\frac{x}{y^2} + \frac{\partial g}{\partial y} = \frac{1}{z} - \frac{x}{y^2}$

$\Rightarrow \frac{\partial g}{\partial y} = \frac{1}{z} \Rightarrow g(y,z) = \frac{y}{z} + h(z) \Rightarrow f(x,y,z) = \frac{x}{y} + \frac{y}{z} + h(z) \Rightarrow \frac{\partial f}{\partial z} = -\frac{y}{z^2} + h'(z) = -\frac{y}{z^2} \Rightarrow h'(z) = 0 \Rightarrow h(z) = C$

$\Rightarrow f(x,y,z) = \frac{x}{y} + \frac{y}{z} + C \Rightarrow \int_{(1,1,1)}^{(2,2,2)} \frac{1}{y}\, dx + \left(\frac{1}{z} - \frac{x}{y^2}\right) dy - \frac{y}{z^2}\, dz = f(2,2,2) - f(1,1,1) = \left(\frac{2}{2} + \frac{2}{2} + C\right) - \left(\frac{1}{1} + \frac{1}{1} + C\right)$

$= 0$

23. $\mathbf{r} = (\mathbf{i} + \mathbf{j} + \mathbf{k}) + t(\mathbf{i} + 2\mathbf{j} - 2\mathbf{k}) = (1+t)\mathbf{i} + (1+2t)\mathbf{j} + (1-2t)\mathbf{k} \Rightarrow dx = dt,\ dy = 2\,dt,\ dz = -2\,dt$

$\Rightarrow \int_{(1,1,1)}^{(2,3,-1)} y\, dx + x\, dy + 4\, dz = \int_0^1 (2t+1)\, dt + (t+1)(2\,dt) + 4(-2)\, dt = \int_0^1 (4t-5)\, dt = \left[2t^2 - 5t\right]_0^1 = -3$

25. $\frac{\partial P}{\partial y} = 0 = \frac{\partial N}{\partial z}, \frac{\partial M}{\partial z} = 2z = \frac{\partial P}{\partial x}, \frac{\partial N}{\partial x} = 0 = \frac{\partial M}{\partial y} \Rightarrow M\,dx + N\,dy + P\,dz$ is exact $\Rightarrow \mathbf{F}$ is conservative

$\Rightarrow$ path independence

27. $\frac{\partial P}{\partial y} = 0 = \frac{\partial N}{\partial z}, \frac{\partial M}{\partial z} = 2z = \frac{\partial P}{\partial x}, \frac{\partial N}{\partial x} = -\frac{2x}{y^2} = \frac{\partial M}{\partial y} \Rightarrow \mathbf{F}$ is conservative $\Rightarrow$ there exists an f so that $\mathbf{F} = \nabla f$;

$\frac{\partial f}{\partial x} = \frac{2x}{y} \Rightarrow f(x,y) = \frac{x^2}{y} + g(y) \Rightarrow \frac{\partial f}{\partial y} = -\frac{x^2}{y^2} + g'(y) = \frac{1-x^2}{y^2} \Rightarrow g'(y) = \frac{1}{y^2} \Rightarrow g(y) = -\frac{1}{y} + C$

$\Rightarrow f(x,y) = \frac{x^2}{y} - \frac{1}{y} + C \Rightarrow \mathbf{F} = \nabla\left(\frac{x^2-1}{y}\right)$

29. $\frac{\partial P}{\partial y} = 0 = \frac{\partial N}{\partial z}, \frac{\partial M}{\partial z} = 0 = \frac{\partial P}{\partial x}, \frac{\partial N}{\partial x} = 1 = \frac{\partial M}{\partial y} \Rightarrow \mathbf{F}$ is conservative $\Rightarrow$ there exists an f so that $\mathbf{F} = \nabla f$;

$\frac{\partial f}{\partial x} = x^2 + y \Rightarrow f(x,y,z) = \frac{1}{3}x^3 + xy + g(y,z) \Rightarrow \frac{\partial f}{\partial y} = x + \frac{\partial g}{\partial y} = y^2 + x \Rightarrow \frac{\partial g}{\partial y} = y^2 \Rightarrow g(y,z) = \frac{1}{3}y^3 + h(z)$

$\Rightarrow f(x,y,z) = \frac{1}{3}x^3 + xy + \frac{1}{3}y^3 + h(z) \Rightarrow \frac{\partial f}{\partial z} = h'(z) = ze^z \Rightarrow h(z) = ze^z - e^z + C \Rightarrow f(x,y,z)$

$= \frac{1}{3}x^3 + xy + \frac{1}{3}y^3 + ze^z - e^z + C \Rightarrow \mathbf{F} = \nabla\left(\frac{1}{3}x^3 + xy + \frac{1}{3}y^3 + ze^z - e^z\right)$

(a) work $= \int_A^B \mathbf{F} \cdot \frac{d\mathbf{r}}{dt}\, dt = \int_A^B \mathbf{F} \cdot d\mathbf{r} = \left[\frac{1}{3}x^3 + xy + \frac{1}{3}y^3 + ze^z - e^z\right]_{(1,0,0)}^{(1,0,1)} = \left(\frac{1}{3} + 0 + 0 + e - e\right) - \left(\frac{1}{3} + 0 + 0 - 1\right)$

$= 1$

(b) work $= \int_A^B \mathbf{F} \cdot d\mathbf{r} = \left[\frac{1}{3}x^3 + xy + \frac{1}{3}y^3 + ze^z - e^z\right]_{(1,0,0)}^{(1,0,1)} = 1$

(c) work $= \int_A^B \mathbf{F} \cdot d\mathbf{r} = \left[\frac{1}{3}x^3 + xy + \frac{1}{3}y^3 + ze^z - e^z\right]_{(1,0,0)}^{(1,0,1)} = 1$

Note: Since $\mathbf{F}$ is conservative, $\int_A^B \mathbf{F} \cdot d\mathbf{r}$ is independent of the path from $(1,0,0)$ to $(1,0,1)$.

31. (a) $\mathbf{F} = \nabla(x^3y^2) \Rightarrow \mathbf{F} = 3x^2y^2\mathbf{i} + 2x^3y\mathbf{j}$; let C_1 be the path from $(-1,1)$ to $(0,0) \Rightarrow x = t - 1$ and $y = -t + 1,\ 0 \le t \le 1 \Rightarrow \mathbf{F} = 3(t-1)^2(-t+1)^2\mathbf{i} + 2(t-1)^3(-t+1)\mathbf{j} = 3(t-1)^4\mathbf{i} - 2(t-1)^4\mathbf{j}$

and $\mathbf{r}_1 = (t-1)\mathbf{i} + (-t+1)\mathbf{j} \Rightarrow d\mathbf{r}_1 = dt\,\mathbf{i} - dt\,\mathbf{j} \Rightarrow \int_{C_1} \mathbf{F} \cdot d\mathbf{r}_1 = \int_0^1 \left[3(t-1)^4 + 2(t-1)^4\right] dt$

$= \int_0^1 5(t-1)^4\, dt = \left[(t-1)^5\right]_0^1 = 1$; let C_2 be the path from $(0,0)$ to $(1,1) \Rightarrow x = t$ and $y = t$,

$0 \le t \le 1 \Rightarrow \mathbf{F} = 3t^4\mathbf{i} + 2t^4\mathbf{j}$ and $\mathbf{r}_2 = t\mathbf{i} + t\mathbf{j} \Rightarrow d\mathbf{r}_2 = dt\,\mathbf{i} + dt\,\mathbf{j} \Rightarrow \int_{C_2} \mathbf{F} \cdot d\mathbf{r}_2 = \int_0^1 \left(3t^4 + 2t^4\right) dt$

$= \int_0^1 5t^4\, dt = 1 \Rightarrow \int_C \mathbf{F} \cdot d\mathbf{r} = \int_{C_1} \mathbf{F} \cdot d\mathbf{r}_1 + \int_{C_2} \mathbf{F} \cdot d\mathbf{r}_2 = 2$

(b) Since $f(x,y) = x^3y^2$ is a potential function for $\mathbf{F}$, $\int_{(-1,1)}^{(1,1)} \mathbf{F} \cdot d\mathbf{r} = f(1,1) - f(-1,1) = 2$

33. Let $-GmM = C \Rightarrow \mathbf{F} = C\left[\frac{x}{\left(x^2+y^2+z^2\right)^{3/2}}\mathbf{i} + \frac{y}{\left(x^2+y^2+z^2\right)^{3/2}}\mathbf{j} + \frac{z}{\left(x^2+y^2+z^2\right)^{3/2}}\mathbf{k}\right]$

$\Rightarrow \frac{\partial P}{\partial y} = \frac{-3yzC}{\left(x^2+y^2+z^2\right)^{5/2}} = \frac{\partial N}{\partial z}, \frac{\partial M}{\partial z} = \frac{-3xzC}{\left(x^2+y^2+z^2\right)^{5/2}} = \frac{\partial P}{\partial x}, \frac{\partial N}{\partial x} = \frac{-3xyC}{\left(x^2+y^2+z^2\right)^{5/2}} = \frac{\partial M}{\partial y} \Rightarrow \mathbf{F} = \nabla f$ for some f; $\frac{\partial f}{\partial x} = \frac{xC}{(x^2+y^2+z^2)^{3/2}} \Rightarrow f(x,y,z) = -\frac{C}{(x^2+y^2+z^2)^{1/2}} + g(y,z) \Rightarrow \frac{\partial f}{\partial y} = \frac{yC}{(x^2+y^2+z^2)^{3/2}} + \frac{\partial g}{\partial y}$ $= \frac{yC}{(x^2+y^2+z^2)^{3/2}} \Rightarrow \frac{\partial g}{\partial y} = 0 \Rightarrow g(y,z) = h(z) \Rightarrow \frac{\partial f}{\partial z} = \frac{zC}{(x^2+y^2+z^2)^{3/2}} + h'(z) = \frac{zC}{(x^2+y^2+z^2)^{3/2}}$ $\Rightarrow h(z) = C_1 \Rightarrow f(x,y,z) = -\frac{C}{(x^2+y^2+z^2)^{1/2}} + C_1$. Let $C_1 = 0 \Rightarrow f(x,y,z) = \frac{GmM}{(x^2+y^2+z^2)^{1/2}}$ is a potential function for **F**.

35. (a) If the differential form is exact, then $\frac{\partial P}{\partial y} = \frac{\partial N}{\partial z} \Rightarrow 2ay = cy$ for all y $\Rightarrow 2a = c$, $\frac{\partial M}{\partial z} = \frac{\partial P}{\partial x} \Rightarrow 2cx = 2cx$ for all x, and $\frac{\partial N}{\partial x} = \frac{\partial M}{\partial y} \Rightarrow by = 2ay$ for all y $\Rightarrow b = 2a$ and $c = 2a$

(b) $\mathbf{F} = \nabla f \Rightarrow$ the differential form with $a = 1$ in part (a) is exact $\Rightarrow b = 2$ and $c = 2$

37. The path will not matter; the work along any path will be the same because the field is conservative.

14.4 GREEN'S THEOREM IN THE PLANE

1. $M = -y = -a \sin t$, $N = x = a \cos t$, $dx = -a \sin t\, dt$, $dy = a \cos t\, dt \Rightarrow \frac{\partial M}{\partial x} = 0$, $\frac{\partial M}{\partial y} = -1$, $\frac{\partial N}{\partial x} = 1$, and $\frac{\partial N}{\partial y} = 0$;

Equation (11): $\oint_C M\, dy - N\, dx = \int_0^{2\pi} [(-a \sin t)(a \cos t) - (a \cos t)(-a \sin t)]\, dt = \int_0^{2\pi} 0\, dt = 0$;

$\iint_R \left(\frac{\partial M}{\partial x} + \frac{\partial N}{\partial y}\right) dx\, dy = \iint_R 0\, dx\, dy = 0$, Flux

Equation (12): $\oint_C M\, dx + N\, dy = \int_0^{2\pi} [(-a \sin t)(-a \sin t) - (a \cos t)(a \cos t)]\, dt = \int_0^{2\pi} a^2\, dt = 2\pi a^2$;

$\iint_R \left(\frac{\partial N}{\partial x} - \frac{\partial M}{\partial y}\right) dx\, dy = \int_{-a}^{a} \int_{-\sqrt{a^2-x^2}}^{\sqrt{a^2-x^2}} 2\, dy\, dx = \int_{-a}^{a} 4\sqrt{a^2-x^2}\, dx = 4\left[\frac{x}{2}\sqrt{a^2-x^2} + \frac{a^2}{2} \sin^{-1} \frac{x}{a}\right]_{-a}^{a}$

$= 2a^2\left(\frac{\pi}{2} + \frac{\pi}{2}\right) = 2a^2\pi$, Circulation

3. $M = 2x = 2a \cos t$, $N = -3y = -3a \sin t$, $dx = -a \sin t\, dt$, $dy = a \cos t\, dt \Rightarrow \frac{\partial M}{\partial x} = 2$, $\frac{\partial M}{\partial y} = 0$, $\frac{\partial N}{\partial x} = 0$, and $\frac{\partial N}{\partial y} = -3$;

Equation (11): $\oint_C M\, dy - N\, dx = \int_0^{2\pi} [(2a \cos t)(a \cos t) + (3a \sin t)(-a \sin t)]\, dt$

$$= \int_0^{2\pi} \left(2a^2 \cos^2 t - 3a^2 \sin^2 t\right) dt = 2a^2\left[\frac{t}{2} + \frac{\sin 2t}{4}\right]_0^{2\pi} - 3a^2\left[\frac{t}{2} - \frac{\sin 2t}{4}\right]_0^{2\pi} = 2\pi a^2 - 3\pi a^2 = -\pi a^2;$$

$$\iint_R \left(\frac{\partial M}{\partial x} + \frac{\partial N}{\partial y}\right) = \iint_R -1\, dx\, dy = \int_0^{2\pi}\int_0^{a} -r\, dr\, d\theta = \int_0^{2\pi} -\frac{a^2}{2}\, d\theta = -\pi a^2, \text{ Flux}$$

Equation (12): $\oint_C M\, dx + N\, dy = \int_0^{2\pi} [(2a \cos t)(-a \sin t) + (-3a \sin t)(a \cos t)]\, dt$

$$= \int_0^{2\pi} \left(-2a^2 \sin t \cos t - 3a^2 \sin t \cos t\right) dt = -5a^2\left[\frac{1}{2}\sin^2 t\right]_0^{2\pi} = 0; \quad \iint_R 0\, dx\, dy = 0, \text{ Circulation}$$

5. $M = x - y,\ N = y - x \Rightarrow \frac{\partial M}{\partial x} = 1,\ \frac{\partial M}{\partial y} = -1,\ \frac{\partial N}{\partial x} = -1,\ \frac{\partial N}{\partial y} = 1 \Rightarrow \text{Flux} = \iint_R 2\, dx\, dy = \int_0^1\int_0^1 2\, dx\, dy = 2;$

$$\text{Circ} = \iint_R [-1 - (-1)]\, dx\, dy = 0$$

7. $M = y^2 - x^2,\ N = x^2 + y^2 \Rightarrow \frac{\partial M}{\partial x} = -2x,\ \frac{\partial M}{\partial y} = 2y,\ \frac{\partial N}{\partial x} = 2x,\ \frac{\partial N}{\partial y} = 2y \Rightarrow \text{Flux} = \iint_R (-2x + 2y)\, dx\, dy$

$$= \int_0^3\int_0^x (-2x + 2y)\, dy\, dx = \int_0^3 \left(-2x^2 + x^2\right) dx = \left[-\frac{1}{3}x^3\right]_0^3 = -9; \text{ Circ} = \iint_R (2x - 2y)\, dx\, dy$$

$$= \int_0^3\int_0^x (2x - 2y)\, dy\, dx = \int_0^3 x^2\, dx = 9$$

9. $M = x + e^x \sin y,\ N = x + e^x \cos y \Rightarrow \frac{\partial M}{\partial x} = 1 + e^x \sin y,\ \frac{\partial M}{\partial y} = e^x \cos y,\ \frac{\partial N}{\partial x} = 1 + e^x \cos y,\ \frac{\partial N}{\partial y} = -e^x \sin y$

$$\Rightarrow \text{Flux} = \iint_R dx\, dy = \int_{-\pi/4}^{\pi/4}\int_0^{\sqrt{\cos 2\theta}} r\, dr\, d\theta = \int_{-\pi/4}^{\pi/4} \left(\frac{1}{2}\cos 2\theta\right) d\theta = \left[\frac{1}{4}\sin 2\theta\right]_{-\pi/4}^{\pi/4} = \frac{1}{2};$$

$$\text{Circ} = \iint_R \left(1 + e^x \cos y - e^x \cos y\right) dx\, dy = \iint_R dx\, dy = \int_{-\pi/4}^{\pi/4}\int_0^{\sqrt{\cos 2\theta}} r\, dr\, d\theta = \int_{-\pi/4}^{\pi/4} \left(\frac{1}{2}\cos 2\theta\right) d\theta = \frac{1}{2}$$

11. $M = xy,\ N = y^2 \Rightarrow \frac{\partial M}{\partial x} = y,\ \frac{\partial M}{\partial y} = x,\ \frac{\partial N}{\partial x} = 0,\ \frac{\partial N}{\partial y} = 2y \Rightarrow \text{Flux} = \iint_R (y + 2y)\, dy\, dx = \int_0^1\int_{x^2}^{x} 3y\, dy\, dx$

$$= \int_0^1 \left(\frac{3x^2}{2} - \frac{3x^4}{2}\right) dx = \frac{1}{5}; \text{ Circ} = \iint_R -x\, dy\, dx = \int_0^1\int_{x^2}^{x} -x\, dy\, dx = \int_0^1 \left(-x^2 + x^3\right) dx = -\frac{1}{12}$$

13. $M = 3xy - \frac{x}{1+y^2}$, $N = e^x + \tan^{-1} y \Rightarrow \frac{\partial M}{\partial x} = 3y - \frac{1}{1+y^2}$, $\frac{\partial N}{\partial y} = \frac{1}{1+y^2}$

$$\Rightarrow \text{Flux} = \iint_R \left(3y - \frac{1}{1+y^2} + \frac{1}{1+y^2}\right) dx\,dy = \iint_R 3y\,dx\,dy = \int_0^{2\pi} \int_0^{a(1+\cos\theta)} (3r\sin\theta)\,r\,dr\,d\theta$$

$$= \int_0^{2\pi} a^3(1+\cos\theta)^3(\sin\theta)\,d\theta = \left[-\frac{a^3}{4}(1+\cos\theta)^4\right]_0^{2\pi} = -4a^3 - \left(-4a^3\right) = 0$$

15. $M = 2xy^3$, $N = 4x^2y^2 \Rightarrow \frac{\partial M}{\partial y} = 6xy^2$, $\frac{\partial N}{\partial x} = 8xy^2 \Rightarrow \text{work} = \oint_C 2xy^3\,dx + 4x^2y^2\,dy = \iint_R \left(8xy^2 - 6xy^2\right) dx\,dy$

$$= \int_0^1 \int_0^{x^3} 2xy^2\,dy\,dx = \int_0^1 \tfrac{2}{3}x^{10}\,dx = \frac{2}{33}$$

17. $M = y^2$, $N = x^2 \Rightarrow \frac{\partial M}{\partial y} = 2y$, $\frac{\partial N}{\partial x} = 2x \Rightarrow \oint_C y^2\,dx + x^2\,dy = \iint_R (2x - 2y)\,dy\,dx$

$$= \int_0^1 \int_0^{1-x} (2x-2y)\,dy\,dx = \int_0^1 \left(-3x^2 + 4x - 1\right) dx = \left[-x^3 + 2x^2 - x\right]_0^1 = -1 + 2 - 1 = 0$$

19. $M = 6y + x$, $N = y + 2x \Rightarrow \frac{\partial M}{\partial y} = 6$, $\frac{\partial N}{\partial x} = 2 \Rightarrow \oint_C (6y + x)\,dx + (y + 2x)\,dy = \iint_R (2-6)\,dy\,dx$

$$= -4(\text{Area of the circle}) = -16\pi$$

21. $M = x = a\cos t$, $N = y = a\sin t \Rightarrow dx = -a\sin t\,dt$, $dy = a\cos t\,dt \Rightarrow \text{Area} = \frac{1}{2}\oint_C x\,dy - y\,dx$

$$= \frac{1}{2}\int_0^{2\pi} \left(a^2\cos^2 t + a^2\sin^2 t\right) dt = \frac{1}{2}\int_0^{2\pi} a^2\,dt = \pi a^2$$

23. $M = x = a\cos^3 t$, $N = y = \sin^3 t \Rightarrow dx = -3\cos^2 t\sin t\,dt$, $dy = 3\sin^2 t\cos t\,dt \Rightarrow \text{Area} = \frac{1}{2}\oint_C x\,dy - y\,dx$

$$= \frac{1}{2}\int_0^{2\pi} \left(3\sin^2 t\cos^2 t\right)\left(\cos^2 t + \sin^2 t\right) dt = \frac{1}{2}\int_0^{2\pi} \left(3\sin^2 t\cos^2 t\right) dt = \frac{3}{8}\int_0^{2\pi} \sin^2 2t\,dt = \frac{3}{16}\int_0^{4\pi} \sin^2 u\,du$$

$$= \frac{3}{16}\left[\frac{u}{2} - \frac{\sin 2u}{4}\right]_0^{4\pi} = \frac{3}{8}\pi$$

25. (a) $M = f(x)$, $N = g(y) \Rightarrow \frac{\partial M}{\partial y} = 0$, $\frac{\partial N}{\partial x} = 0 \Rightarrow \oint_C f(x)\,dx + g(y)\,dy = \iint_R \left(\frac{\partial N}{\partial x} - \frac{\partial M}{\partial y}\right) dx\,dy$

$$= \iint_R 0\,dx\,dy = 0$$

(b) $M = ky,\ N = hx \Rightarrow \frac{\partial M}{\partial y} = k,\ \frac{\partial N}{\partial x} = h \Rightarrow \oint_C ky\ dx + hx\ dy = \iint_R \left(\frac{\partial N}{\partial x} - \frac{\partial M}{\partial y}\right) dx\,dy$

$= \iint_R (h - k)\ dx\,dy = (h-k)(\text{Area of the region})$

27. The integral is 0 for any simple closed plane curve C. The reasoning: By the tangential form of Green's Theorem, with $M = 4x^3y$ and $N = x^4$, $\oint_C 4x^3y\ x + x^4\ dy = \iint_R \left[\frac{\partial}{\partial x}(x^4) - \frac{\partial}{\partial y}(4x^3y)\right] dx\,dy$

$= \iint_R \underbrace{\left(4x^3 - 4x^3\right)}_{0} dx\,dy = 0.$

29. Let $M = x$ and $N = 0 \Rightarrow \frac{\partial M}{\partial x} = 1$ and $\frac{\partial N}{\partial y} = 0 \Rightarrow \oint M\ dy - N\ dx = \iint_R \left(\frac{\partial M}{\partial x} + \frac{\partial N}{\partial y}\right) dx\,dy \Rightarrow \oint_C x\ dy$

$= \iint_R (1+0)\ dx\,dy \Rightarrow \text{Area of R} = \iint_R dx\,dy = \oint_C x\ dy$; similarly, $M = y$ and $N = 0 \Rightarrow \frac{\partial M}{\partial y} = 1$ and

$\frac{\partial N}{\partial x} = 0 \Rightarrow \oint_C M\ dx + N\ dy = \iint_R \left(\frac{\partial N}{\partial x} + \frac{\partial M}{\partial y}\right) dy\,dx \Rightarrow \oint_C y\ dx = \iint_R (0-1)\ dy\,dx \Rightarrow -\oint_C y\ dx$

$= \iint_R dx\,dy = \text{Area of R}$

31. Let $\delta(x,y) = 1 \Rightarrow \bar{x} = \frac{M_y}{M} = \frac{\iint_R x\,\delta(x,y)\ dA}{\iint_R \delta(x,y)\ dA} = \frac{\iint_R x\ dA}{\iint_R dA} = \frac{\iint_R x\ dA}{A} \Rightarrow A\bar{x} = \iint_R x\ dA = \iint_R (x+0)\ dx\,dy$

$= \oint_C \frac{x^2}{2}\ dy,\ A\bar{x} = \iint_R x\ dA = \iint_R (0+x)\ dx\,dy = -\oint_C xy\ dx$, and $A\bar{x} = \iint_R x\ dA = \iint_R \left(\frac{2}{3}x + \frac{1}{3}x\right) dx\,dy$

$= \oint_C \frac{1}{3}x^2\ dy - \frac{1}{3}xy\ dx \Rightarrow \frac{1}{2}\oint_C x^2\ dy = -\oint_C xy\ dx = \frac{1}{3}\oint_C x^2\ dy - xy\ dx = A\bar{x}$

33. $M = \frac{\partial f}{\partial y},\ N = -\frac{\partial f}{\partial x} \Rightarrow \frac{\partial M}{\partial y} = \frac{\partial^2 f}{\partial y^2},\ \frac{\partial N}{\partial x} = -\frac{\partial^2 f}{\partial x^2} \Rightarrow \oint_C \frac{\partial f}{\partial y}\ dx - \frac{\partial f}{\partial x}\ dy = \iint_R \left(-\frac{\partial^2 f}{\partial x^2} - \frac{\partial^2 f}{\partial y^2}\right) dx\,dy = 0$ for such curves C

35. (a) $\nabla f = \left(\frac{2x}{x^2+y^2}\right)\mathbf{i} + \left(\frac{2y}{x^2+y^2}\right)\mathbf{j} \Rightarrow M = \frac{2x}{x^2+y^2},\ N = \frac{2y}{x^2+y^2}$; since M, N are discontinuous at (0,0), we cannot apply Green's Theorem over C. Thus, let C_h be the circle $x = h\cos\theta,\ y = h\sin\theta,\ 0 < h \le a$ and

let C_1 be the circle $x = a\cos t$, $y = a\sin t$, $a > 0$. Then $\oint_C \mathbf{F}\cdot\mathbf{n}\,ds = \oint_{C_1} M\,dy - N\,dx + \oint_{C_h} M\,dy - N\,dx$

$= \oint_{C_1} \frac{2x}{x^2+y^2}\,dx - \frac{2y}{x^2+y^2}\,dx + \oint_{C_h} \frac{2x}{x^2+y^2}\,dy - \frac{2y}{x^2+y^2}\,dx$. In the first integral, let $x = a\cos t$, $y = a\sin t$

$\Rightarrow dx = -a\sin t\,dt$, $dy = a\cos t\,dt$, $M = 2a\cos t$, $N = 2a\sin t$, $0 \le \ \le 2\pi$. In the second integral, let

$x = h\cos\theta$, $y = h\sin\theta \Rightarrow dx = -h\sin\theta\,d\theta$, $dy = h\cos\theta\,d\theta$, $M = 2h\cos\theta$, $N = 2h\sin\theta$, $0 \le \theta \le 2\pi$.

Then $\oint_C \mathbf{F}\cdot\mathbf{n}\,ds = \oint_{C_1} \frac{2x}{x^2+y^2}\,dy - \frac{2y}{x^2+y^2}\,dx + \oint_{C_h} \frac{2x}{x^2+y^2}\,dy - \frac{2y}{x^2+y^2}\,dx$

$$= \oint_{C_1} \frac{(2a\cos t)(a\cos t)\,dt}{a^2} - \frac{(2a\sin t)(-a\sin t)\,dt}{a^2} + \oint_{C_h} \frac{(2h\cos\theta)(h\cos\theta)\,d\theta}{h^2} - \frac{(2h\sin\theta)(-h\sin\theta)\,d\theta}{h^2}$$

$$= \int_0^{2\pi} 2\,dt + \int_{2\pi}^{0} 2\,d\theta = 0 \text{ for every } h$$

(b) If K is any simple closed curve surrounding C_h (K contains (0,0)), then $\oint_C \mathbf{F}\cdot\mathbf{n}\,ds$

$= \oint_{C_1} M\,dy - N\,dx + \oint_{C_h} M\,dy - N\,dx$, and in polar coordinates, $\nabla f\cdot\mathbf{n} = M\,dy - N\,dx$

$= \left(\frac{2r\cos\theta}{r^2}\right)(r\cos\theta\,d\theta + \sin\theta\,dr) - \left(\frac{2r\sin\theta}{r^2}\right)(-r\sin\theta\,d\theta + \cos\theta\,dr) = \frac{2r^2}{r^2}\,d\theta = 2\,d\theta$. Now,

2θ increases by 4π as K is traversed once counterclockwise from $\theta = 0$ to $\theta = 2\pi \Rightarrow \oint_C \mathbf{F}\cdot\mathbf{n}\,ds = 0$

(since $\oint_{C_h} M\,dy - N\,dx = -4\pi$) when (0,0) is in the region, but $\oint_K \mathbf{F}\cdot\mathbf{n}\,ds = 4\pi$ when (0,0) is not in the region.

37. $\int_{g_1(y)}^{g_2(y)} \frac{\partial N}{\partial x}\,dx\,dy = N(g_2(y),y) - N(g_1(y),y) \Rightarrow \int_c^d \int_{g_1(y)}^{g_2(y)} \left(\frac{\partial N}{\partial x}\,dx\right) dy = \int_c^d [N(g_2(y),y) - N(g_1(y),y)]\,dy$

$= \int_c^d N(g_2(y),y)\,dy - \int_c^d N(g_1(y),y)\,dy = \int_c^d N(g_2(y),y)\,dy + \int_d^c N(g_1(y),y)\,dy = \int_{C_2} N\,dy + \int_{C_1} N\,dy$

$= \oint_C dy \Rightarrow \oint_C N\,dy = \iint_R \frac{\partial N}{\partial x}\,dx\,dy$

39. The curl of a conservative two-dimensional field is zero. The reasoning: A two-dimensional field $\mathbf{F} = M\mathbf{i} + N\mathbf{j}$ can be considered to be the restriction to the xy-plane of a three-dimensional field whose k component is zero, and whose $\mathbf{i}$ and $\mathbf{j}$ components are independent of z. For such a field to be conservative, we must have $\frac{\partial N}{\partial x} = \frac{\partial M}{\partial y}$ by the component test in Section 14.3 $\Rightarrow$ curl $\mathbf{F} = \frac{\partial N}{\partial x} - \frac{\partial M}{\partial y} = 0$.

14.5 SURFACE AREA AND SURFACE INTEGRALS

1. $\mathbf{p}=\mathbf{k}$, $\nabla f = 2x\mathbf{i}+2y\mathbf{j}-\mathbf{k} \Rightarrow |\nabla f| = \sqrt{(2x)^2+(2y)^2+(-1)^2} = \sqrt{4x^2+4y^2+1}$ and $|\nabla f\cdot\mathbf{p}| = 1$;

$z=2 \Rightarrow x^2+y^2=2$; thus $S = \iint_R \frac{|\nabla f|}{|\nabla f\cdot\mathbf{p}|}\,dA = \iint_R \sqrt{4x^2+4y^2+1}\,dx\,dy$

$= \iint_R \sqrt{4r^2\cos^2\theta+4r^2\sin^2\theta+1}\;r\,dr\,d\theta = \int_0^{2\pi}\int_0^{\sqrt{2}} \sqrt{4r^2+1}\;r\,dr\,d\theta = \int_0^{2\pi}\left[\frac{1}{12}(4r^2+1)^{3/2}\right]_0^{\sqrt{2}} d\theta$

$= \int_0^{2\pi} \frac{13}{6}\,d\theta = \frac{13}{3}\pi$

3. $\mathbf{p}=\mathbf{k}$, $\nabla f = \mathbf{i}+2\mathbf{j}+2\mathbf{k} \Rightarrow |\nabla f| = 3$ and $|\nabla f\cdot\mathbf{p}| = 2$; $x=y^2$ and $x=2-y^2$ intersect at $(1,1)$ and $(1,-1)$

$\Rightarrow S = \iint_R \frac{|\nabla f|}{|\nabla f\cdot\mathbf{p}|}\,dA = \iint_R \frac{3}{2}\,dx\,dy = \int_{-1}^{1}\int_{y^2}^{2-y^2} \frac{3}{2}\,dx\,dy = \int_{-1}^{1}(3-3y^2)\,dy = 4$

5. $\mathbf{p}=\mathbf{k}$, $\nabla f = 2x\mathbf{i}-2\mathbf{j}-2\mathbf{k} \Rightarrow |\nabla f| = \sqrt{(2x)^2+(-2)^2+(-2)^2} = \sqrt{4x^2+8} = 2\sqrt{x^2+2}$ and $|\nabla f\cdot\mathbf{p}| = 2$

$\Rightarrow S = \iint_R \frac{|\nabla f|}{|\nabla f\cdot\mathbf{p}|}\,dA = \iint_R \frac{2\sqrt{x^2+2}}{2}\,dx\,dy = \int_0^2\int_0^{3x} \sqrt{x^2+2}\,dy\,dx = \int_0^2 3x\sqrt{x^2+2}\,dx = \left[(x^2+2)^{3/2}\right]_0^2$

$= 6\sqrt{6}-2\sqrt{2}$

7. $\mathbf{p}=\mathbf{k}$, $\nabla f = c\mathbf{i}-\mathbf{k} \Rightarrow |\nabla f| = \sqrt{c^2+1}$ and $|\nabla f\cdot\mathbf{p}| = 1 \Rightarrow S = \iint_R \frac{|\nabla f|}{|\nabla f\cdot\mathbf{p}|}\,dA = \iint_R \sqrt{c^2+1}\,dx\,dy$

$= \int_0^{2\pi}\int_0^1 \sqrt{c^2+1}\;r\,dr\,d\theta = \int_0^{2\pi} \frac{\sqrt{c^2+1}}{2}\,d\theta = \pi\sqrt{c^2+1}$

9. $\mathbf{p}=\mathbf{i}$, $\nabla f = \mathbf{i}+2y\mathbf{j}+2z\mathbf{k} \Rightarrow |\nabla f| = \sqrt{1^2+(2y)^2+(2z)^2} = \sqrt{1+4y^2+4z^2}$ and $|\nabla f\cdot\mathbf{p}| = 1$; $1 \le y^2+z^2 \le 4$

$\Rightarrow S = \iint_R \frac{|\nabla f|}{|\nabla f\cdot\mathbf{p}|}\,dA = \iint_R \sqrt{1+4y^2+4z^2}\,dy\,dz = \int_0^{2\pi}\int_1^2 \sqrt{1+4r^2\cos^2\theta+4r^2\sin^2\theta}\;r\,dr\,d\theta$

$= \int_0^{2\pi}\int_1^2 \sqrt{1+4r^2}\;r\,dr\,d\theta = \int_0^{2\pi}\left[\frac{1}{12}(1+4r^2)^{3/2}\right]_1^2 d\theta = \int_0^{2\pi}\frac{1}{12}\left(17\sqrt{17}-5\sqrt{5}\right)d\theta = \frac{\pi}{6}\left(17\sqrt{17}-5\sqrt{5}\right)$

11. $\mathbf{p}=\mathbf{k}$, $\nabla f=\left(2x-\frac{2}{x}\right)\mathbf{i}+\sqrt{15}\,\mathbf{j}-\mathbf{k}\Rightarrow|\nabla f|=\sqrt{\left(2x-\frac{2}{x}\right)^2+\left(\sqrt{15}\right)^2+(-1)^2}=\sqrt{4x^2+8+\frac{4}{x^2}}=\sqrt{\left(2x+\frac{2}{x}\right)^2}$

$=2x+\frac{2}{x}$, on $1\le x\le 2$ and $|\nabla f\cdot\mathbf{p}|=1\Rightarrow S=\iint_R\frac{|\nabla f|}{|\nabla f\cdot\mathbf{p}|}\,dA=\iint_R\left(2x+2x^{-1}\right)dx\,dy$

$=\int_0^1\int_1^2\left(2x+2x^{-1}\right)dx\,dy=\int_0^1\left[x^2+2\ln x\right]_1^2 dy=\int_0^1(3+2\ln 2)\,dy=3+2\ln 2$

13. The bottom face S of the cube is in the xy-plane $\Rightarrow z=0\Rightarrow g(x,y,0)=x+y$ and $f(x,y,z)=z=0\Rightarrow\mathbf{p}=\mathbf{k}$ and $\nabla f=\mathbf{k}\Rightarrow|\nabla f|=1$ and $|\nabla f\cdot\mathbf{p}|=1\Rightarrow d\sigma=dx\,dy\Rightarrow\iint_S g\,d\sigma=\iint_R(x+y)\,dx\,dy$

$=\int_0^a\int_0^a(x+y)\,dx\,dy=\int_0^a\left(\frac{a^2}{2}+ay\right)dy=a^3$. Because of symmetry, we also get a^3 over the face of the cube in the xz-plane and a^3 over the face of the cube in the yz-plane. Next, on the top of the cube, $g(x,y,z)$ $=g(x,y,a)=x+y+a$ and $f(x,y,z)=z=a\Rightarrow\mathbf{p}=\mathbf{k}$ and $\nabla f=\mathbf{k}\Rightarrow|\nabla f|=1$ and $|\nabla f\cdot\mathbf{p}|=1\Rightarrow d\sigma=dx\,dy$

$\iint_S g\,d\sigma=\iint_R(x+y+a)\,dx\,dy=\int_0^a\int_0^a(x+y+a)\,dx\,dy=\int_0^a\int_0^a(x+y)\,dx\,dy+\int_0^a\int_0^a a\,dx\,dy=2a^3.$

Because of symmetry, the integral is also $2a^3$ over each of the other two faces. Therefore,

$\iint_{\text{cube}}(x+y+z)\,d\sigma=3\left(a^3+2a^3\right)=9a^3.$

15. On the faces in the coordinate planes, $g(x,y,z)=0\Rightarrow$ the integral over these faces is 0.

On the face $x=a$, we have $f(x,y,z)=x=a$ and $g(x,y,z)=g(a,y,z)=ayz\Rightarrow\mathbf{p}=\mathbf{i}$ and $\nabla f=\mathbf{i}\Rightarrow|\nabla f|=1$ and $|\nabla f\cdot\mathbf{p}|=1\Rightarrow d\sigma=dy\,dz\Rightarrow\iint_S g\,d\sigma=\iint_S ayz\,d\sigma=\int_0^c\int_0^b ayz\,dy\,dz=\frac{ab^2c^2}{4}$.

On the face $y=b$, we have $f(x,y,z)=y=b$ and $g(x,y,z)=g(x,b,z)=bxz\Rightarrow\mathbf{p}=\mathbf{j}$ and $\nabla f=\mathbf{j}\Rightarrow|\nabla f|=1$ and $|\nabla f\cdot\mathbf{p}|=1\Rightarrow d\sigma=dx\,dz\Rightarrow\iint_S g\,d\sigma=\iint_S bxz\,d\sigma=\int_0^c\int_0^a bxz\,dz\,dx=\frac{a^2bc^2}{4}$.

On the face $z=c$, we have $f(x,y,z)=z=c$ and $g(x,y,z)=g(x,y,c)=cxy\Rightarrow\mathbf{p}=\mathbf{k}$ and $\nabla f=\mathbf{k}\Rightarrow|\nabla f|=1$ and $|\nabla f\cdot\mathbf{p}|=1\Rightarrow d\sigma=dy\,dx\Rightarrow\iint_S g\,d\sigma=\iint_S cxy\,d\sigma=\int_0^b\int_0^a cxy\,dx\,dy=\frac{a^2b^2c}{4}$. Therefore,

$\iint_S g(x,y,z)\,d\sigma=\frac{abc(ab+ac+bc)}{4}.$

17. $f(x,y,z) = 2x + 2y + z = 2 \Rightarrow \nabla f = 2\mathbf{i} + 2\mathbf{j} + \mathbf{k}$ and $g(x,y,z) = x + y + (2 - 2x - 2y) = 2 - x - y \Rightarrow \mathbf{p} = \mathbf{k}$,

$|\nabla f| = 3$ and $|\nabla f \cdot \mathbf{p}| = 1 \Rightarrow d\sigma = 3\, dy\, dx$; $z = 0 \Rightarrow 2x + 2y = 2 \Rightarrow y = 1 - x \Rightarrow \iint_S g\, d\sigma = \iint_S (2 - x - y)\, d\sigma$

$$= 3 \int_0^1 \int_0^{1-x} (2 - x - y)\, dy\, dx = 3 \int_0^1 \left[(2-x)(1-x) - \tfrac{1}{2}(1-x)^2\right] dx = 3 \int_0^1 \left(\tfrac{3}{2} - 2x + \tfrac{x^2}{2}\right) dx = 2$$

19. $g(x,y,z) = z$, $\mathbf{p} = \mathbf{k} \Rightarrow \nabla g = \mathbf{k} \Rightarrow |\nabla g| = 1$ and $|\nabla g \cdot \mathbf{p}| = 1 \Rightarrow \text{Flux} = \iint_S \mathbf{F} \cdot \mathbf{n}\, d\sigma = \iint_R (\mathbf{F} \cdot \mathbf{k})\, dA$

$$= \int_0^2 \int_0^3 3\, dy\, dx = 18$$

21. $\nabla g = 2x\mathbf{i} + 2y\mathbf{j} + 2z\mathbf{k} \Rightarrow |\nabla g| = \sqrt{4x^2 + 4y^2 + 4z^2} = 2a$; $\mathbf{n} = \dfrac{2x\mathbf{i} + 2y\mathbf{j} + 2z\mathbf{k}}{2\sqrt{x^2 + y^2 + z^2}} = \dfrac{x\mathbf{i} + y\mathbf{j} + z\mathbf{k}}{a} \Rightarrow \mathbf{F} \cdot \mathbf{n} = \frac{z^2}{a}$;

$|\nabla g \cdot \mathbf{k}| = 2z \Rightarrow d\sigma = \frac{2a}{2z}\, dA \Rightarrow \text{Flux} = \iint_R \left(\frac{z^2}{a}\right)\left(\frac{a}{z}\right) dA = \iint_R z\, dA = \iint_R \sqrt{a^2 - (x^2 + y^2)}\, dx\, dy$

$$= \int_0^{\pi/2} \int_0^a \sqrt{a^2 - r^2}\, r\, dr\, d\theta = \frac{\pi a^3}{6}$$

23. From Exercise 21, $\mathbf{n} = \dfrac{x\mathbf{i} + y\mathbf{j} + z\mathbf{k}}{a}$ and $d\sigma = \frac{a}{z}\, dA \Rightarrow \mathbf{F} \cdot \mathbf{n} = \frac{xy}{a} - \frac{xy}{a} + \frac{z}{a} = \frac{z}{a} \Rightarrow \text{Flux} = \iint_R \left(\frac{z}{a}\right)\left(\frac{a}{z}\right) dA$

$$= \iint_R 1\, dA = \frac{\pi a^2}{4}$$

25. From Exercise 21, $\mathbf{n} = \dfrac{x\mathbf{i} + y\mathbf{j} + z\mathbf{k}}{a}$ and $d\sigma = \frac{a}{z}\, dA \Rightarrow \mathbf{F} \cdot \mathbf{n} = \frac{x^2}{a} + \frac{y^2}{a} + \frac{z^2}{a} = a \Rightarrow \text{Flux}$

$$= \iint_R a\left(\frac{a}{z}\right) dA = \iint_R \frac{a^2}{z}\, dA = \iint_R \frac{a^2}{\sqrt{a^2 - (x^2 + y^2)}}\, dA = \int_0^{\pi/2} \int_0^a \frac{a^2}{\sqrt{a^2 - r^2}}\, r\, dr\, d\theta$$

$$= \int_0^{\pi/2} a^2\left[-\sqrt{a^2 - r^2}\right]_0^a d\theta = \frac{\pi a^3}{2}$$

27. $g(x,y,z) = y^2 + z = 4 \Rightarrow \nabla g = 2y\mathbf{j} + \mathbf{k} \Rightarrow |\nabla g| = \sqrt{4y^2 + 1} \Rightarrow \mathbf{n} = \dfrac{2y\mathbf{j} + \mathbf{k}}{\sqrt{4y^2 + 1}}$

$\Rightarrow \mathbf{F} \cdot \mathbf{n} = \dfrac{2xy - 3z}{\sqrt{4y^2 + 1}}$; $\mathbf{p} = \mathbf{k} \Rightarrow |\nabla g \cdot \mathbf{p}| = 1 \Rightarrow d\sigma = \sqrt{4y^2 + 1}\, dA \Rightarrow \text{Flux}$

$$= \iint_R \left(\frac{2xy - 3z}{\sqrt{4y^2 + 1}}\right)\sqrt{4y^2 + 1}\, dA = \iint_R (2xy - 3z)\, dA;\ z = 0 \text{ and } z = 4 - y^2 \Rightarrow y^2 = 4$$

$$\Rightarrow \text{Flux} = \iint_R \left[2xy - 3\left(4 - y^2\right)\right] dA = \int_0^1 \int_{-2}^{2} \left(2xy - 12 + 3y^2\right) dy\, dx = \int_0^1 \left[xy^2 - 12y + y^3\right]_{-2}^{2} dx$$

$$= \int_0^1 -32\, dx = -32$$

29. $g(x,y,z) = y - e^x = 0 \Rightarrow \nabla g = -e^x\mathbf{i} + \mathbf{j} \Rightarrow |\nabla g| = \sqrt{e^{2x}+1} \Rightarrow \mathbf{n} = \frac{e^x\mathbf{i} - \mathbf{j}}{\sqrt{e^{2x}+1}} \Rightarrow \mathbf{F}\cdot\mathbf{n} = \frac{-2e^x - 2y}{\sqrt{e^{2x}+1}}$; $\mathbf{p} = \mathbf{i}$

$$\Rightarrow |\nabla g\cdot\mathbf{p}| = e^x \Rightarrow d\sigma = \frac{\sqrt{e^{2x}+1}}{e^x}\, dA \Rightarrow \text{Flux} = \iint_R \left(\frac{-2e^x - 2y}{\sqrt{e^{2x}+1}}\right)\left(\frac{\sqrt{e^{2x}+1}}{e^x}\right) dA = \iint_R \frac{-2e^x - 2e^x}{e^x}\, dA$$

$$= \iint_R -4\, dA = \int_0^1 \int_1^2 -4\, dy\, dz = -4$$

31. On the face $z = a$: $g(x,y,z) = z \Rightarrow \nabla g = \mathbf{k} \Rightarrow |\nabla g| = 1$; $\mathbf{n} = \mathbf{k} \Rightarrow \mathbf{F}\cdot\mathbf{n} = 2xz = 2ax$ since $z = a$;

$$d\sigma = dx\, dy \Rightarrow \text{Flux} = \iint_R 2ax\, dx\, dy = \int_0^a \int_0^a 2ax\, dx\, dy = a^4.$$

On the face $z = 0$: $g(x,y,z) = z \Rightarrow \nabla g = \mathbf{k} \Rightarrow |\nabla g| = 1$; $\mathbf{n} = -\mathbf{k} \Rightarrow \mathbf{F}\cdot\mathbf{n} = -2xz = 0$ since $z = 0$;

$$d\sigma = dx\, dy \Rightarrow \text{Flux} = \iint_R 0\, dx\, dy = 0.$$

On the face $x = a$: $g(x,y,z) = x \Rightarrow \nabla g = \mathbf{i} \Rightarrow |\nabla g| = 1$; $\mathbf{n} = \mathbf{i} \Rightarrow \mathbf{F}\cdot\mathbf{n} = 2xy = 2ay$ since $x = a$;

$$d\sigma = dy\, dz \Rightarrow \text{Flux} = \int_0^a \int_0^a 2ay\, dy\, dz = a^4.$$

On the face $x = 0$: $g(x,y,z) = x \Rightarrow \nabla g = \mathbf{i} \Rightarrow |\nabla g| = 1$; $\mathbf{n} = -\mathbf{i} \Rightarrow \mathbf{F}\cdot\mathbf{n} = -2xy = 0$ since $x = 0$ $\Rightarrow \text{Flux} = 0$.

On the face $y = a$: $g(x,y,z) = y \Rightarrow \nabla g = \mathbf{j} \Rightarrow |\nabla g| = 1$; $\mathbf{n} = \mathbf{j} \Rightarrow \mathbf{F}\cdot\mathbf{n} = 2yz = 2az$ since $y = a$;

$$d\sigma = dz\, dx \Rightarrow \text{Flux} = \int_0^a \int_0^a 2az\, dz\, dx = a^4.$$

On the face $y = 0$: $g(x,y,z) = y \Rightarrow \nabla g = \mathbf{j} \Rightarrow |\nabla g| = 1$; $\mathbf{n} = -\mathbf{j} \Rightarrow \mathbf{F}\cdot\mathbf{n} = -2yz = 0$ since $y = 0$ $\Rightarrow \text{Flux} = 0$. Therefore, Total Flux $= 3a^4$.

33. $\nabla f = 2x\mathbf{i} + 2y\mathbf{j} + 2z\mathbf{k} \Rightarrow |\nabla f| = \sqrt{4x^2 + 4y^2 + 4z^2} = 2a$; $\mathbf{p} = \mathbf{k} \Rightarrow |\nabla f\cdot\mathbf{p}| = 2z$ since $z \ge 0 \Rightarrow d\sigma = \frac{2a}{2z}\, dA$

$$= \frac{a}{z}\, dA;\ M = \iint_S \delta\, d\sigma = \frac{\delta}{8}(\text{surface area of sphere}) = \frac{\delta\pi a^2}{2};\ M_{xy} = \iint_S z\delta\, d\sigma = \delta \iint_R z\left(\frac{a}{z}\right) dA$$

$$= a\delta \iint_R dA = a\delta \int_0^{\pi/2} \int_0^a r\, dr\, d\theta = \frac{\delta\pi a^3}{4} \Rightarrow \bar{z} = \frac{M_{xy}}{M} = \left(\frac{\delta\pi a^3}{4}\right)\left(\frac{2}{\delta\pi a^2}\right) = \frac{a}{2}.$$ Because of symmetry, $\bar{x} = \bar{y}$

$= \frac{a}{2} \Rightarrow$ the centroid is $\left(\frac{a}{2}, \frac{a}{2}, \frac{a}{2}\right)$.

35. Because of symmetry, $\bar{x} = \bar{y} = 0$; $M = \iint_S \delta \, d\sigma = \delta \iint_S d\sigma = (\text{Area of S})\delta = 3\pi\sqrt{2}\,\delta$; $\nabla f = 2x\mathbf{i} + 2y\mathbf{j} - 2z\mathbf{k}$

$\Rightarrow |\nabla f| = \sqrt{4x^2 + 4y^2 + 4z^2} = 2\sqrt{x^2 + y^2 + z^2}$; $\mathbf{p} = \mathbf{k} \Rightarrow |\nabla f \cdot \mathbf{p}| = 2z \Rightarrow d\sigma = \frac{2\sqrt{x^2 + y^2 + z^2}}{2z}\, dA$

$= \frac{\sqrt{x^2 + y^2 + (x^2 + y^2)}}{z}\, dA = \frac{\sqrt{2}\sqrt{x^2 + y^2}}{z}\, dA \Rightarrow M_{xy} = \delta \iint_S z\left(\frac{\sqrt{2}\sqrt{x^2 + y^2}}{z}\right) dA$

$= \delta \iint_S \sqrt{2}\sqrt{x^2 + y^2}\, dA = \delta \int_0^{2\pi}\int_1^2 \sqrt{2}\, r^2 \, dr\, d\theta = \frac{14\pi\sqrt{2}}{3}\delta \Rightarrow \bar{z} = \frac{\left(\frac{14\pi\sqrt{2}}{3}\delta\right)}{3\pi\sqrt{2}\,\delta} = \frac{14}{9}$

$\Rightarrow (\bar{x}, \bar{y}, \bar{z}) = \left(0, 0, \frac{14}{9}\right)$. Next, $I_z = \iint_S (x^2 + y^2)\delta \, d\sigma = \iint_S (x^2 + y^2)\left(\frac{\sqrt{2}\sqrt{x^2 + y^2}}{z}\right) \delta \, dA$

$= \delta\sqrt{2} \iint_S (x^2 + y^2)\, dA = \delta\sqrt{2} \int_0^{2\pi}\int_1^2 r^3 \, dr\, d\theta = \frac{15\pi\sqrt{2}}{2}\delta \Rightarrow R_z = \sqrt{\frac{I_z}{M}} = \frac{\sqrt{10}}{2}$

37. (a) Let the diameter lie on the z-axis and let $f(x, y, z) = x^2 + y^2 + z^2 = a^2$, $z \ge 0$ be the upper hemisphere

$\Rightarrow \nabla f = 2x\mathbf{i} + 2y\mathbf{j} + 2z\mathbf{k} \Rightarrow |\nabla f| = \sqrt{4x^2 + 4y^2 + 4z^2} = 2a$, $a > 0$; $\mathbf{p} = \mathbf{k} \Rightarrow |\nabla f \cdot \mathbf{p}| = 2z$ since $z \ge 0$

$\Rightarrow d\sigma = \frac{a}{z}\, dA \Rightarrow I_z = \iint_S \delta(x^2 + y^2)\left(\frac{a}{z}\right) d\sigma = a\delta \iint_R \frac{x^2 + y^2}{\sqrt{a^2 - (x^2 + y^2)}}\, dA = a\delta \int_0^{2\pi}\int_0^a \frac{r^2}{\sqrt{a^2 - r^2}}\, r\, dr\, d\theta$

$= a\delta \int_0^{2\pi} \left[-r^2\sqrt{a^2 - r^2} - \frac{2}{3}\left(a^2 - r^2\right)^{3/2}\right]_0^a d\theta = a\delta \int_0^{2\pi} \frac{2}{3}a^3 \, d\theta = \frac{4\pi}{3}a^4\delta \Rightarrow$ the moment of inertia is $\frac{8\pi}{3}a^4\delta$ for the whole sphere

(b) $I_L = I_{c.m.} + mh^2$, where m is the mass of the body and h is the distance between the parallel lines; now, $I_{c.m.} = \frac{8\pi}{3}a^4\delta$ (from part a) and $\frac{m}{2} = \iint_S \delta \, d\sigma = \delta \iint_R \left(\frac{a}{z}\right) dA = a\delta \iint_R \frac{1}{\sqrt{a^2 - (x^2 + y^2)}}\, dy\, dx$

$= a\delta \int_0^{2\pi}\int_0^a \frac{1}{\sqrt{a^2 - r^2}}\, r\, dr\, d\theta = a\delta \int_0^{2\pi} \left[-\sqrt{a^2 - r^2}\right]_0^a d\theta = a\delta \int_0^{2\pi} a\, d\theta = 2\pi a^2\delta$ and $h = a$

$\Rightarrow I_L = \frac{8\pi}{3}a^4\delta + 4\pi a^2\delta a^2 = \frac{20\pi}{3}a^4\delta$

39. $f_x(x,y) = 2x,\ f_y(x,y) = 2y \Rightarrow \sqrt{f_x^2 + f_y^2 + 1} = \sqrt{4x^2 + 4y^2 + 1} \Rightarrow \text{Area} = \iint_R \sqrt{4x^2 + 4y^2 + 1}\, dx\, dy$

$= \int_0^{2\pi} \int_0^{\sqrt{3}} \sqrt{4r^2 + 1}\ r\, dr\, d\theta = \frac{\pi}{6}\left(13\sqrt{13} - 1\right)$

41. $f_x(x,y) = \frac{x}{\sqrt{x^2 + y^2}},\ f_y(x,y) = \frac{y}{\sqrt{x^2 + y^2}} \Rightarrow \sqrt{f_x^2 + f_y^2 + 1} = \sqrt{\frac{x^2}{x^2 + y^2} + \frac{y^2}{x^2 + y^2} + 1} = \sqrt{2}$

$\Rightarrow \text{Area} = \iint_{R_{xy}} \sqrt{2}\, dx\, dy = \sqrt{2}(\text{Area between the ellipse and the circle}) = \sqrt{2}(6\pi - \pi) = 5\pi\sqrt{2}$

43. $y = \frac{2}{3}z^{3/2} \Rightarrow f_x(x,z) = 0,\ f_z(x,z) = z^{1/2} \Rightarrow \sqrt{f_x^2 + f_z^2 + 1} = \sqrt{z + 1};\ y = \frac{16}{3} \Rightarrow \frac{16}{3} = \frac{2}{3}z^{3/2} \Rightarrow z = 4$

$\Rightarrow \text{Area} = \int_0^4 \int_0^1 \sqrt{z + 1}\, dx\, dz = \int_0^4 \sqrt{z + 1}\, dz = \frac{2}{3}\left(5\sqrt{5} - 1\right)$

14.6 PARAMETRIZED SURFACES

1. In cylindrical coordinates, let $x = r\cos\theta$, $y = r\sin\theta$, $z = \left(\sqrt{x^2 + y^2}\right)^2 = r^2$. Then $\mathbf{r}(r,\theta) = (r\cos\theta)\mathbf{i} + (r\sin\theta)\mathbf{j} + r^2\mathbf{k}$, $0 \le r \le 2$, $0 \le \theta \le 2\pi$.

3. In cylindrical coordinates, let $x = r\cos\theta$, $y = r\sin\theta$, $z = \frac{\sqrt{x^2 + y^2}}{2} \Rightarrow z = \frac{r}{2}$. Then $\mathbf{r}(r,\theta) = (r\cos\theta)\mathbf{i} + (r\sin\theta)\mathbf{j} + \left(\frac{r}{2}\right)\mathbf{k}$. For $0 \le z \le 3$, $0 \le \frac{r}{2} \le 3 \Rightarrow 0 \le r \le 6$; to get only the first octant, let $0 \le \theta \le \frac{\pi}{2}$.

5. In cylindrical coordinates, let $x = r\cos\theta$, $y = r\sin\theta$ since $x^2 + y^2 = 9 \Rightarrow z^2 = 9 - \left(x^2 + y^2\right) = 9 - r^2$ $\Rightarrow z = \sqrt{9 - r^2}$, $z \ge 0$. Then $\mathbf{r}(r,\theta) = (r\cos\theta)\mathbf{i} + (r\sin\theta)\mathbf{j} + \sqrt{9 - r^2}\mathbf{k}$. Let $0 \le \theta \le 2\pi$. For the domain of r: $z = \sqrt{x^2 + y^2}$ and $x^2 + y^2 + z^2 = 9 \Rightarrow x^2 + y^2 + \left(\sqrt{x^2 + y^2}\right)^2 = 9 \Rightarrow 2\left(x^2 + y^2\right) = 9 \Rightarrow 2r^2 = 9$ $\Rightarrow r = \frac{3}{\sqrt{2}} \Rightarrow 0 \le r \le \frac{3}{\sqrt{2}}$.

7. In spherical coordinates, $x = \rho\sin\phi\cos\theta$, $y = \rho\sin\phi\sin\theta$, $\rho = \sqrt{x^2 + y^2 + z^2} \Rightarrow \rho^2 = 3 \Rightarrow \rho = \sqrt{3}$ $\Rightarrow z = \sqrt{3}\cos\phi$ for the sphere; $z = \frac{\sqrt{3}}{2} = \sqrt{3}\cos\phi \Rightarrow \cos\phi = \frac{1}{2} \Rightarrow \phi = \frac{\pi}{3}$; $z = -\frac{\sqrt{3}}{2} \Rightarrow -\frac{\sqrt{3}}{2} = \sqrt{3}\cos\phi$ $\Rightarrow \cos\phi = -\frac{1}{2} \Rightarrow \phi = \frac{2\pi}{3}$. Then $\mathbf{r}(r,\theta) = \left(\sqrt{3}\sin\phi\cos\theta\right)\mathbf{i} + \left(\sqrt{3}\sin\phi\sin\theta\right)\mathbf{j} + \left(\sqrt{3}\cos\phi\right)\mathbf{k}$, $\frac{\pi}{3} \le \phi \le \frac{2\pi}{3}$ and $0 \le \theta \le 2\pi$.

9. Since $z = 4 - y^2$, we can let $\mathbf{r}$ be a function of x and y $\Rightarrow \mathbf{r}(x,y) = x\mathbf{i} + y\mathbf{j} + (4 - y^2)\mathbf{k}$. Then $z = 0$ $\Rightarrow 0 = 4 - y^2 \Rightarrow y = \pm 2$. Thus, let $-2 \le y \le 2$ and $0 \le x \le 2$.

11. When $x = 0$, let $y^2 + z^2 = 9$ be the circular section in the yz-plane. Use polar coordinates in the yz-plane $\Rightarrow y = 3 \cos \theta$ and $z = 3 \sin \theta$. Thus let $x = u$ and $\theta = v \Rightarrow \mathbf{r}(u,v) = x\mathbf{i} + (3 \cos v)\mathbf{j} + (3 \sin v)\mathbf{k}$ where $0 \le u \le 3$, and $0 \le v \le 2\pi$.

13. (a) $x + y + z = 1 \Rightarrow z = 1 - x - y$. In cylindrical coordinates, let $x = r \cos \theta$ and $y = r \sin \theta$ $\Rightarrow z = 1 - r \cos \theta - r \sin \theta \Rightarrow \mathbf{r}(r,\theta) = (r \cos \theta)\mathbf{i} + (r \sin \theta)\mathbf{j} + (1 - r \cos \theta - r \sin \theta)\mathbf{k}$, $0 \le \theta \le 2\pi$ and $0 \le r \le 3$.

(b) In a fashion similar to cylindrical coordinates, but working in the yz-plane instead of the xy-plane, let $y = u \cos v$, $z = u \sin v$ where $u = \sqrt{y^2 + z^2}$ and v is the angle formed by (x,y,z), $(x,0,0)$, and $(x,y,0)$ with $(x,0,0)$ as vertex. Since $x + y + z = 1 \Rightarrow x = 1 - y - z \Rightarrow x = 1 - u \cos v - u \sin v$, then $\mathbf{r}$ is a function of u and v $\Rightarrow \mathbf{r}(u,v) = (1 - u \cos v - u \sin v)\mathbf{i} + (u \cos v)\mathbf{j} + (u \sin v)\mathbf{k}$, $0 \le u \le 3$ and $0 \le v \le 2\pi$.

15. Let $x = w \cos v$ and $z = w \sin v$. Then $(x - 2)^2 + z^2 = 4 \Rightarrow x^2 - 4x + z^2 = 0 \Rightarrow w^2 \cos^2 v - 4w \cos v + w^2 \sin^2 v$ $= 0 \Rightarrow w^2 - 4w \cos v = 0 \Rightarrow w = 0$ or $w - 4 \cos v = 0 \Rightarrow w = 0$ or $w = 4 \cos v$. Now $w = 0 \Rightarrow x = 0$ and $y = 0$, which is a line not a cylinder. Therefore, let $w = 4 \cos v \Rightarrow x = (4 \cos v)(\cos v) = 4 \cos^2 v$ and $z = 4 \cos v \sin v$. Finally, let $y = u$. Then $\mathbf{r}(u,v) = (4 \cos^2 v)\mathbf{i} + u\mathbf{j} + (4 \cos v \sin v)\mathbf{k}$, $-\frac{\pi}{2} \le v \le \frac{\pi}{2}$ and $0 \le u \le 3$.

17. Let $x = r \cos \theta$ and $y = r \sin \theta$. Then $\mathbf{r}(r,\theta) = (r \cos \theta)\mathbf{i} + (r \sin \theta)\mathbf{j} + \left(\frac{2 - r \sin \theta}{2}\right)\mathbf{k}$, $0 \le r \le 1$ and $0 \le \theta \le 2\pi$ $\Rightarrow \mathbf{r}_r = (\cos \theta)\mathbf{i} + (\sin \theta)\mathbf{j} - \left(\frac{\sin \theta}{2}\right)\mathbf{k}$ and $\mathbf{r}_\theta = (-r \sin \theta)\mathbf{i} + (r \cos \theta)\mathbf{j} - \left(\frac{r \cos \theta}{2}\right)\mathbf{k}$

$$\Rightarrow \mathbf{r}_r \times \mathbf{r}_\theta = \begin{vmatrix} \mathbf{i} & \mathbf{j} & \mathbf{k} \\ \cos \theta & \sin \theta & -\frac{\sin \theta}{2} \\ -r \sin \theta & r \cos \theta & -\frac{r \cos \theta}{2} \end{vmatrix}$$

$$= \left(\frac{-r \sin \theta \cos \theta}{2} + \frac{(\sin \theta)(r \cos \theta)}{2}\right)\mathbf{i} + \left(\frac{r \sin^2 \theta}{2} + \frac{r \cos^2 \theta}{2}\right)\mathbf{j} + (r \cos^2 \theta + r \sin^2 \theta)\mathbf{k} = \frac{r}{2}\mathbf{j} + r\mathbf{k}$$

$$\Rightarrow |\mathbf{r}_r \times \mathbf{r}_\theta| = \sqrt{\frac{r^2}{4} + r^2} = \frac{\sqrt{5}\,r}{2} \Rightarrow A = \int_0^{2\pi}\int_0^1 \frac{\sqrt{5}\,r}{2}\,dr\,d\theta = \int_0^{2\pi} \left[\frac{\sqrt{5}\,r^2}{4}\right]_0^1 d\theta = \int_0^{2\pi} \frac{\sqrt{5}}{4}\,d\theta = \frac{\pi\sqrt{5}}{2}$$

19. Let $x = r \cos \theta$ and $y = r \sin \theta \Rightarrow z = 2\sqrt{x^2 + y^2} = 2r$, $1 \le r \le 3$ and $0 \le \theta \le 2\pi$. Then $\mathbf{r}(r,\theta) = (r \cos \theta)\mathbf{i} + (r \sin \theta)\mathbf{j} + 2r\mathbf{k} \Rightarrow \mathbf{r}_r = (\cos \theta)\mathbf{i} + (\sin \theta)\mathbf{j} + 2\mathbf{k}$ and $\mathbf{r}_\theta = (-r \sin \theta)\mathbf{i} + (r \cos \theta)\mathbf{j}$

$$\Rightarrow \mathbf{r}_r \times \mathbf{r}_\theta = \begin{vmatrix} \mathbf{i} & \mathbf{j} & \mathbf{k} \\ \cos\theta & \sin\theta & 2 \\ -r\sin\theta & r\cos\theta & 0 \end{vmatrix} = (-2r\cos\theta)\mathbf{i} - (2r\sin\theta)\mathbf{j} + \left(r\cos^2\theta + r\sin^2\theta\right)\mathbf{k}$$

$$= (-2r\cos\theta)\mathbf{i} - (2r\sin\theta)\mathbf{j} + r\mathbf{k} \Rightarrow |\mathbf{r}_r \times \mathbf{r}_\theta| = \sqrt{4r^2\cos^2\theta + 4r^2\sin^2\theta + r^2} = \sqrt{5r^2} = r\sqrt{5}$$

$$\Rightarrow A = \int_0^{2\pi}\int_1^3 r\sqrt{5}\, dr\, d\theta = \int_0^{2\pi}\left[\frac{r^2\sqrt{5}}{2}\right]_1^3 d\theta = \int_0^{2\pi} 4\sqrt{5}\, d\theta = 8\pi\sqrt{5}$$

21. Let $x = r\cos\theta$ and $y = r\sin\theta \Rightarrow r^2 = x^2 + y^2 = 1$, $1 \le z \le 4$ and $0 \le \theta \le 2\pi$. Then $\mathbf{r}(z,\theta) = (\cos\theta)\mathbf{i} + (\sin\theta)\mathbf{j} + z\mathbf{k} \Rightarrow \mathbf{r}_z = \mathbf{k}$ and $\mathbf{r}_\theta = (-\sin\theta)\mathbf{i} + (\cos\theta)\mathbf{j}$

$$\Rightarrow \mathbf{r}_\theta \times \mathbf{r}_z = \begin{vmatrix} \mathbf{i} & \mathbf{j} & \mathbf{k} \\ -\sin\theta & \cos\theta & 0 \\ 0 & 0 & 1 \end{vmatrix} = (\cos\theta)\mathbf{i} + (\sin\theta)\,\mathbf{j} = |\mathbf{r}_\theta \times \mathbf{r}_z| = \sqrt{\cos^2\theta + \sin^2\theta} = 1$$

$$\Rightarrow A = \int_0^{2\pi}\int_1^4 1\, dr\, d\theta = \int_0^{2\pi} 3\, d\theta = 6\pi$$

23. $z = 2 - x^2 - y^2$ and $z = \sqrt{x^2+y^2} \Rightarrow z = 2 - z^2 \Rightarrow z^2 + z - 2 = 0 \Rightarrow z = -2$ or $z = 1$. Since $z = \sqrt{x^2+y^2} \ge 0$, we get $z = 1$ where the cone intersects the paraboloid. When $x = 0$ and $y = 0$, $z = 2 \Rightarrow$ the vertex of the paraboloid is $(0,0,2)$. Therefore, z ranges from 1 to 2 on the "cap" $\Rightarrow$ r ranges from 1 (when $x^2 + y^2 = 1$) to 0 (when $x = 0$ and $y = 0$ at the vertex). Let $x = r\cos\theta$, $y = r\sin\theta$, and $z = 2 - r^2$. Then $\mathbf{r}(r,\theta) = (r\cos\theta)\mathbf{i} + (r\sin\theta)\mathbf{j} + \left(2 - r^2\right)\mathbf{k}$, $0 \le r \le 1$, $0 \le \theta \le 2\pi \Rightarrow \mathbf{r}_r = (\cos\theta)\mathbf{i} + (\sin\theta)\mathbf{j} - 2r\mathbf{k}$ and

$$\mathbf{r}_\theta = (-r\sin\theta)\mathbf{i} + (r\cos\theta)\mathbf{j} \Rightarrow \mathbf{r}_r \times \mathbf{r}_\theta = \begin{vmatrix} \mathbf{i} & \mathbf{j} & \mathbf{k} \\ \cos\theta & \sin\theta & -2r \\ -r\sin\theta & r\cos\theta & 0 \end{vmatrix}$$

$$= \left(2r^2\cos\theta\right)\mathbf{i} + \left(2r^2\sin\theta\right)\mathbf{j} + r\mathbf{k} \Rightarrow |\mathbf{r}_r \times \mathbf{r}_\theta| = \sqrt{4r^4\cos^2\theta + 4r^4\sin^2\theta + r^2} = r\sqrt{4r^2+1}$$

$$\Rightarrow A = \int_0^{2\pi}\int_0^1 r\sqrt{4r^2+1}\, dr\, d\theta = \int_0^{2\pi}\left[\tfrac{1}{12}\left(4r^2+1\right)^{3/2}\right]_0^1 d\theta = \int_0^{2\pi}\left(\frac{5\sqrt{5}-1}{12}\right) d\theta = \tfrac{\pi}{6}\left(5\sqrt{5}-1\right)$$

25. Let $x = \rho\sin\phi\cos\theta$, $y = \rho\sin\phi\sin\theta$, and $z = \rho\cos\phi \Rightarrow \rho = \sqrt{x^2+y^2+z^2} = \sqrt{2}$ on the sphere. Next, $x^2 + y^2 + z^2 = 2$ and $z = \sqrt{x^2+y^2} \Rightarrow z^2 + z^2 = 2 \Rightarrow z^2 = 1 \Rightarrow z = 1$ since $z \ge 0 \Rightarrow \phi = \frac{\pi}{4}$. For the lower portion of the sphere cut by the cone, we get $\phi = \pi$. Then
$\mathbf{r}(\phi,\theta) = \left(\sqrt{2}\sin\phi\cos\theta\right)\mathbf{i} + \left(\sqrt{2}\sin\phi\sin\theta\right)\mathbf{j} + \left(\sqrt{2}\cos\phi\right)\mathbf{k}$, $\frac{\pi}{4} \le \phi \le \pi$, $0 \le \theta \le 2\pi$
$\Rightarrow \mathbf{r}_\phi = \left(\sqrt{2}\cos\phi\cos\theta\right)\mathbf{i} + \left(\sqrt{2}\cos\phi\sin\theta\right)\mathbf{j} - \left(\sqrt{2}\sin\phi\right)\mathbf{k}$ and $\mathbf{r}_\theta = \left(-\sqrt{2}\sin\phi\sin\theta\right)\mathbf{i} + \left(\sqrt{2}\sin\phi\cos\theta\right)\mathbf{j}$

$$\Rightarrow \mathbf{r}_\phi \times \mathbf{r}_\theta = \begin{vmatrix} \mathbf{i} & \mathbf{j} & \mathbf{k} \\ \sqrt{2}\cos\phi\cos\theta & \sqrt{2}\cos\phi\sin\theta & -\sqrt{2}\sin\phi \\ -\sqrt{2}\sin\phi\sin\theta & \sqrt{2}\sin\phi\cos\theta & 0 \end{vmatrix}$$

$$= (2\sin^2\phi\cos\theta)\mathbf{i} + (2\sin^2\phi\sin\theta)\mathbf{j} + (2\sin\phi\cos\phi)\mathbf{k}$$

$$\Rightarrow |\mathbf{r}_\phi \times \mathbf{r}_\theta| = \sqrt{4\sin^4\phi\cos^2\theta + 4\sin^4\phi\sin^2\theta + 4\sin^2\phi\cos^2\phi} = \sqrt{4\sin^2\phi} = 2|\sin\phi| = 2\sin\phi$$

$$\Rightarrow A = \int_0^{2\pi}\int_{\pi/4}^{\pi} 2\sin\phi\, d\phi\, d\theta = \int_0^{2\pi} (2+\sqrt{2})\, d\theta = (4+2\sqrt{2})\pi$$

27. Let the parametrization be $\mathbf{r}(x,z) = x\mathbf{i} + x^2\mathbf{j} + z\mathbf{k} \Rightarrow \mathbf{r}_x = \mathbf{i} + 2x\mathbf{j}$ and $\mathbf{r}_z = \mathbf{k} \Rightarrow \mathbf{r}_x \times \mathbf{r}_z = \begin{vmatrix} \mathbf{i} & \mathbf{j} & \mathbf{k} \\ 1 & 2x & 0 \\ 0 & 0 & 1 \end{vmatrix}$

$$= 2x\mathbf{i} + \mathbf{j} \Rightarrow |\mathbf{r}_x \times \mathbf{r}_z| = \sqrt{4x^2+1} \Rightarrow \iint_S G(x,y,z)\, d\sigma = \int_0^3\int_0^2 x\sqrt{4x^2+1}\, dx\, dz = \int_0^3 \left[\tfrac{1}{12}(4x^2+1)^{3/2}\right]_0^2 dz$$

$$= \int_0^3 \tfrac{1}{12}(17\sqrt{17}-1)\, dz = \frac{17\sqrt{17}-1}{4}$$

29. Let the parametrization be $\mathbf{r}(\phi,\theta) = (\sin\phi\cos\theta)\mathbf{i} + (\sin\phi\sin\theta)\mathbf{j} + (\cos\phi)\mathbf{k}$ (spherical coordinates with $\rho = 1$ on the sphere), $0 \le \phi \le \pi$, $0 \le \theta \le 2\pi \Rightarrow \mathbf{r}_\phi = (\cos\phi\cos\theta)\mathbf{i} + (\cos\phi\sin\theta)\mathbf{j} - (\sin\phi)\mathbf{k}$ and

$$\mathbf{r}_\theta = (-\sin\phi\sin\theta)\mathbf{i} + (\sin\phi\cos\theta)\mathbf{j} \Rightarrow \mathbf{r}_\phi \times \mathbf{r}_\theta = \begin{vmatrix} \mathbf{i} & \mathbf{j} & \mathbf{k} \\ \cos\phi\cos\theta & \cos\phi\sin\theta & -\sin\phi \\ -\sin\phi\sin\theta & \sin\phi\cos\theta & 0 \end{vmatrix}$$

$$= (\sin^2\phi\cos\theta)\mathbf{i} + (\sin^2\phi\sin\theta)\mathbf{j} + (\sin\phi\cos\phi)\mathbf{k} \Rightarrow |\mathbf{r}_\phi \times \mathbf{r}_\theta| = \sqrt{\sin^4\phi\cos^2\theta + \sin^4\phi\sin^2\theta + \sin^2\phi\cos^2\phi}$$

$$= \sin\phi;\ x = \sin\phi\cos\theta \Rightarrow G(x,y,z) = \cos^2\theta\sin^2\phi \Rightarrow \iint_S G(x,y,z)\, d\sigma = \int_0^{2\pi}\int_0^{\pi} (\cos^2\theta\sin^2\phi)(\sin\phi)\, d\phi\, d\theta$$

$$= \int_0^{2\pi}\int_0^{\pi} (\cos^2\theta)(1-\cos^2\phi)(\sin\phi)\, d\phi\, d\theta; \begin{bmatrix} u = \cos\phi \\ du = -\sin\phi\, d\phi \end{bmatrix} \to \int_0^{2\pi}\int_{-1}^{1} (\cos^2\theta)(u^2-1)\, du\, d\theta$$

$$= \int_0^{2\pi} (\cos^2\theta)\left[\frac{u^3}{3} - u\right]_1^{-1} d\theta = \frac{4}{3}\int_0^{2\pi} \cos^2\theta\, d\theta = \frac{4}{3}\left[\frac{\theta}{2} + \frac{\sin 2\theta}{4}\right]_0^{2\pi} = \frac{4\pi}{3}$$

31. Let the parametrization be $\mathbf{r}(x,y) = x\mathbf{i} + y\mathbf{j} + (4 - x - y)\mathbf{k} \Rightarrow \mathbf{r}_x = \mathbf{i} - \mathbf{k}$ and $\mathbf{r}_y = \mathbf{j} - \mathbf{k}$

$$\Rightarrow \mathbf{r}_x \times \mathbf{r}_y = \begin{vmatrix} \mathbf{i} & \mathbf{j} & \mathbf{k} \\ 1 & 0 & -1 \\ 0 & 1 & -1 \end{vmatrix} = \mathbf{i} + \mathbf{j} + \mathbf{k} \Rightarrow |\mathbf{r}_x \times \mathbf{r}_y| = \sqrt{3} \Rightarrow \iint_S F(x,y,z)\, d\sigma = \int_0^1 \int_0^1 (4 - x - y)\sqrt{3}\, dy\, dx$$

$$= \int_0^1 \sqrt{3}\left[4y - xy - \frac{y^2}{2}\right]_0^1 dx = \int_0^1 \sqrt{3}\left(\frac{7}{2} - x\right) dx = \sqrt{3}\left[\frac{7}{2}x - \frac{x^2}{2}\right]_0^1 = 3\sqrt{3}$$

33. Let the parametrization be $\mathbf{r}(r,\theta) = (r\cos\theta)\mathbf{i} + (r\sin\theta)\mathbf{j} + (1 - r^2)\mathbf{k}$, $0 \le r \le 1$ (since $0 \le z \le 1$) and $0 \le \theta \le 2\pi$

$$\Rightarrow \mathbf{r}_r = (\cos\theta)\mathbf{i} + (\sin\theta)\mathbf{j} - 2r\mathbf{k} \text{ and } \mathbf{r}_\theta = (-r\sin\theta)\mathbf{i} + (r\cos\theta)\mathbf{j} \Rightarrow \mathbf{r}_r \times \mathbf{r}_\theta = \begin{vmatrix} \mathbf{i} & \mathbf{j} & \mathbf{k} \\ \cos\theta & \sin\theta & -2r \\ -r\sin\theta & r\cos\theta & 0 \end{vmatrix}$$

$$= (2r^2\cos\theta)\mathbf{i} + (2r^2\sin\theta)\mathbf{j} + r\mathbf{k} \Rightarrow |\mathbf{r}_r \times \mathbf{r}_\theta| = \sqrt{(2r^2\cos\theta)^2 + (2r^2\sin\theta) + r^2} = r\sqrt{1 + 4r^2};\ z = 1 - r^2 \text{ and}$$

$$x = r\cos\theta \Rightarrow H(x,y,z) = (r^2\cos^2\theta)\sqrt{1 + 4r^2} \Rightarrow \iint_S H(x,y,z)\, d\sigma$$

$$= \int_0^{2\pi}\int_0^1 (r^2\cos^2\theta)\left(\sqrt{1+4r^2}\right)\left(r\sqrt{1+4r^2}\right) dr\, d\theta = \int_0^{2\pi}\int_0^1 r^3(1 + 4r^2)\cos^2\theta\, dr\, d\theta = \frac{11\pi}{12}$$

35. Let the parametrization be $\mathbf{r}(x,y) = x\mathbf{i} + y\mathbf{j} + (4 - y^2)\mathbf{k}$, $0 \le x \le 1$, $-2 \le y \le 2$; $z = 0 \Rightarrow 0 = 4 - y^2$

$$\Rightarrow y = \pm 2;\ \mathbf{r}_x = \mathbf{i} \text{ and } \mathbf{r}_y = \mathbf{j} - 2y\mathbf{k} \Rightarrow \mathbf{r}_x \times \mathbf{r}_y = \begin{vmatrix} \mathbf{i} & \mathbf{j} & \mathbf{k} \\ 1 & 0 & 0 \\ 0 & 1 & -2y \end{vmatrix} = 2y\mathbf{j} + \mathbf{k} \Rightarrow \mathbf{F}\cdot\mathbf{n}\, d\sigma$$

$$= \mathbf{F}\cdot\frac{\mathbf{r}_x \times \mathbf{r}_y}{|\mathbf{r}_x \times \mathbf{r}_y|}|\mathbf{r}_x \times \mathbf{r}_y|\, dy\, dx = (2xy - 3z)\, dy\, dx = \left[2xy - 3(4 - y^2)\right] dy\, dx \Rightarrow \iint_S \mathbf{F}\cdot\mathbf{n}\, d\sigma$$

$$= \int_0^1\int_{-2}^2 (2xy + 3y^2 - 12)\, dy\, dx = \int_0^1 \left[xy^2 + y^3 - 12y\right]_{-2}^2 dx = \int_0^1 -32\, dx = -32$$

37. Let the parametrization be $\mathbf{r}(\phi,\theta) = (a\sin\phi\cos\theta)\mathbf{i} + (a\sin\phi\sin\theta)\mathbf{j} + (a\cos\phi)\mathbf{k}$ (spherical coordinates with $\rho = a$, $a \ge 0$, on the sphere), $0 \le \phi \le \frac{\pi}{2}$ (for the first octant)

$$\Rightarrow \mathbf{r}_\phi = (a\cos\phi\cos\theta)\mathbf{i} + (a\cos\phi\sin\theta)\mathbf{j} - (a\sin\phi)\mathbf{k} \text{ and } \mathbf{r}_\theta = (-a\sin\phi\sin\theta)\mathbf{i} + (a\sin\phi\cos\theta)\mathbf{j}$$

$$\Rightarrow \mathbf{r}_\phi \times \mathbf{r}_\theta = \begin{vmatrix} \mathbf{i} & \mathbf{j} & \mathbf{k} \\ a\cos\phi\cos\theta & a\cos\phi\sin\theta & -a\sin\phi \\ -a\sin\phi\sin\theta & a\sin\phi\cos\theta & 0 \end{vmatrix}$$

$= (a^2 \sin^2\phi \cos\theta)\mathbf{i} + (a^2 \sin^2\phi \sin\theta)\mathbf{j} + (a^2 \sin\phi \cos\phi)\mathbf{k} \Rightarrow \mathbf{F}\cdot\mathbf{n}\, d\sigma = \mathbf{F}\cdot\frac{\mathbf{r}_\phi \times \mathbf{r}_\theta}{|\mathbf{r}_\phi \times \mathbf{r}_\theta|}|\mathbf{r}_\phi \times \mathbf{r}_\theta|\, d\theta\, d\phi$

$= a^3 \cos^2\phi \sin\phi\, d\theta\, d\phi$ since $\mathbf{F} = z\mathbf{k} = (a\cos\phi)\mathbf{k} \Rightarrow \iint_S \mathbf{F}\cdot\mathbf{n}\, d\sigma = \int_0^{\pi/2}\int_0^{\pi/2} a^3 \cos^2\phi \sin\phi\, d\phi\, d\theta = \frac{\pi a^3}{6}$

39. Let the parametrization be $\mathbf{r}(x,y) = x\mathbf{i} + y\mathbf{j} + (2a - x - y)\mathbf{k}$, $0 \le x \le a$, $0 \le y \le a \Rightarrow \mathbf{r}_x = \mathbf{i} - \mathbf{k}$ and $\mathbf{r}_y = \mathbf{j} - \mathbf{k}$

$$\Rightarrow \mathbf{r}_x \times \mathbf{r}_y = \begin{vmatrix} \mathbf{i} & \mathbf{j} & \mathbf{k} \\ 1 & 0 & -1 \\ 0 & 1 & -1 \end{vmatrix} = \mathbf{i} + \mathbf{j} + \mathbf{k} \Rightarrow \mathbf{F}\cdot\mathbf{n}\, d\sigma = \mathbf{F}\cdot\frac{\mathbf{r}_x \times \mathbf{r}_y}{|\mathbf{r}_x \times \mathbf{r}_y|}|\mathbf{r}_x \times \mathbf{r}_y|\, dy\, dx$$

$= [2xy + 2y(2a - x - y) + 2x(2a - x - y)]\, dy\, dx$ since $\mathbf{F} = 2xy\mathbf{i} + 2yz\mathbf{j} + 2xz\mathbf{k}$

$= 2xy\mathbf{i} + 2y(2a - x - y)\mathbf{j} + 2x(2a - x - y)\mathbf{k} \Rightarrow \iint_S \mathbf{F}\cdot\mathbf{n}\, d\sigma$

$$= \int_0^a\int_0^a [2xy + 2y(2a - x - y) + 2x(2a - x - y)]\, dy\, dx = \int_0^a\int_0^a (4ay - 2y^2 + 4ax - 2x^2 - 2xy)\, dy\, dx$$

$$= \int_0^a \left(\tfrac{4}{3}a^3 + 3a^2x - 2ax^2\right) dx = \left(\tfrac{4}{3} + \tfrac{3}{2} - \tfrac{2}{3}\right)a^4 = \frac{13a^4}{6}$$

41. Let the parametrization be $\mathbf{r}(r,\theta) = (r\cos\theta)\mathbf{i} + (r\sin\theta)\mathbf{j} + r\mathbf{k}$, $0 \le r \le 1$ (since $0 \le z \le 1$) and $0 \le \theta \le 2\pi$

$$\Rightarrow \mathbf{r}_r = (\cos\theta)\mathbf{i} + (\sin\theta)\mathbf{j} + \mathbf{k} \text{ and } \mathbf{r}_\theta = (-r\sin\theta)\mathbf{i} + (r\cos\theta)\mathbf{j} \Rightarrow \mathbf{r}_\theta \times \mathbf{r}_r = \begin{vmatrix} \mathbf{i} & \mathbf{j} & \mathbf{k} \\ -r\sin\theta & r\cos\theta & 0 \\ \cos\theta & \sin\theta & 1 \end{vmatrix}$$

$= (r\cos\theta)\mathbf{i} + (r\sin\theta)\mathbf{j} - r\mathbf{k} \Rightarrow \mathbf{F}\cdot\mathbf{n}\, d\sigma = \mathbf{F}\cdot\frac{\mathbf{r}_\theta \times \mathbf{r}_r}{|\mathbf{r}_\theta \times \mathbf{r}_r|}|\mathbf{r}_\theta \times \mathbf{r}_r|\, d\theta\, dr = (r^3 \sin\theta \cos^2\theta + r^2)\, d\theta\, dr$ since

$\mathbf{F} = (r^2 \sin\theta \cos\theta)\mathbf{i} - r\mathbf{k} \Rightarrow \iint_S \mathbf{F}\cdot\mathbf{n}\, d\sigma = \int_0^{2\pi}\int_0^1 (r^3 \sin\theta \cos^2\theta + r^2)\, dr\, d\theta = \int_0^{2\pi} \left(\frac{1}{4}\sin\theta \cos^2\theta + \frac{1}{3}\right) d\theta$

$$= \left[-\frac{1}{12}\cos^3\theta + \frac{\theta}{3}\right]_0^{2\pi} = \frac{2\pi}{3}$$

43. Let the parametrization be $\mathbf{r}(r,\theta) = (r\cos\theta)\mathbf{i} + (r\sin\theta)\mathbf{j} + r\mathbf{k}$, $1 \le r \le 2$ (since $1 \le z \le 2$) and $0 \le \theta \le 2\pi$

$$\Rightarrow \mathbf{r}_r = (\cos\theta)\mathbf{i} + (\sin\theta)\mathbf{j} + \mathbf{k} \text{ and } \mathbf{r}_\theta = (-r\sin\theta)\mathbf{i} + (r\cos\theta)\mathbf{j} \Rightarrow \mathbf{r}_\theta \times \mathbf{r}_r = \begin{vmatrix} \mathbf{i} & \mathbf{j} & \mathbf{k} \\ -r\sin\theta & r\cos\theta & 0 \\ \cos\theta & \sin\theta & 1 \end{vmatrix}$$

$= (r\cos\theta)\mathbf{i} + (r\sin\theta)\mathbf{j} - r\mathbf{k} \Rightarrow \mathbf{F}\cdot\mathbf{n}\, d\sigma = \mathbf{F}\cdot\frac{\mathbf{r}_\theta \times \mathbf{r}_r}{|\mathbf{r}_\theta \times \mathbf{r}_r|}|\mathbf{r}_\theta \times \mathbf{r}_r|\, d\theta\, dr = (-r^2 \cos^2\theta - r^2 \sin^2\theta - r^3)\, d\theta\, dr$

$= (-r^2 - r^3)\,d\theta\,dr$ since $\mathbf{F} = (-r\cos\theta)\mathbf{i} - (r\sin\theta)\mathbf{j} + r^2\mathbf{k} \Rightarrow \iint_S \mathbf{F}\cdot\mathbf{n}\,d\sigma = \int_0^{2\pi}\int_1^2 (-r^2 - r^3)\,dr\,d\theta = -\frac{73\pi}{6}$

45. Let the parametrization be $\mathbf{r}(\phi,\theta) = (a\sin\phi\cos\theta)\mathbf{i} + (a\sin\phi\sin\theta)\mathbf{j} + (a\cos\phi)\mathbf{k}$, $0 \le \phi \le \frac{\pi}{2}$, $0 \le \theta \le \frac{\pi}{2}$
$\Rightarrow \mathbf{r}_\phi = (a\cos\phi\cos\theta)\mathbf{i} + (a\cos\phi\sin\theta)\mathbf{j} - (a\sin\phi)\mathbf{k}$ and $\mathbf{r}_\theta = (-a\sin\phi\sin\theta)\mathbf{i} + (a\sin\phi\cos\theta)\mathbf{j}$

$$\Rightarrow \mathbf{r}_\phi \times \mathbf{r}_\theta = \begin{vmatrix} \mathbf{i} & \mathbf{j} & \mathbf{k} \\ a\cos\phi\cos\theta & a\cos\phi\sin\theta & -a\sin\phi \\ -a\sin\phi\sin\theta & a\sin\phi\cos\theta & 0 \end{vmatrix}$$

$= (a^2\sin^2\phi\cos\theta)\,\mathbf{i} + (a^2\sin^2\phi\sin\theta)\mathbf{j} + (a^2\sin\phi\cos\phi)\mathbf{k}$

$\Rightarrow |\mathbf{r}_\phi \times \mathbf{r}_\theta| = \sqrt{a^4\sin^4\phi\cos^2\theta + a^4\sin^4\phi\sin^2\theta + a^4\sin^2\phi\cos^2\phi} = \sqrt{a^4\sin^2\phi} = a^2\sin\phi$. The mass is

$M = \iint_S d\sigma = \int_0^{\pi/2}\int_0^{\pi/2} (a^2\sin\phi)\,d\phi\,d\theta = \frac{a^2\pi}{2}$; the first moment is $M_{yz} = \iint_S x\,d\sigma$

$= \int_0^{\pi/2}\int_0^{\pi/2} (a\sin\phi\cos\theta)(a^2\sin\phi)\,d\phi\,d\theta = \frac{a^3\pi}{4} \Rightarrow \bar{x} = \frac{\left(\frac{a^3\pi}{4}\right)}{\left(\frac{a^2\pi}{2}\right)} = \frac{a}{2} \Rightarrow$ the centroid is located at $\left(\frac{a}{2}, \frac{a}{2}, \frac{a}{2}\right)$ by symmetry

47. Let the parametrization be $\mathbf{r}(\phi,\theta) = (a\sin\phi\cos\theta)\mathbf{i} + (a\sin\phi\sin\theta)\mathbf{j} + (a\cos\phi)\mathbf{k}$, $0 \le \phi \le \pi$, $0 \le \theta \le 2\pi$
$\Rightarrow \mathbf{r}_\phi = (a\cos\phi\cos\theta)\mathbf{i} + (a\cos\phi\sin\theta)\mathbf{j} - (a\sin\phi)\mathbf{k}$ and $\mathbf{r}_\theta = (-a\sin\phi\sin\theta)\mathbf{i} + (a\sin\phi\cos\theta)\mathbf{j}$

$$\Rightarrow \mathbf{r}_\phi \times \mathbf{r}_\theta = \begin{vmatrix} \mathbf{i} & \mathbf{j} & \mathbf{k} \\ a\cos\phi\cos\theta & a\cos\phi\sin\theta & -a\sin\phi \\ -a\sin\phi\sin\theta & a\sin\phi\cos\theta & 0 \end{vmatrix}$$

$= (a^2\sin^2\phi\cos\theta)\,\mathbf{i} + (a^2\sin^2\phi\sin\theta)\mathbf{j} + (a^2\sin\phi\cos\phi)\mathbf{k}$

$\Rightarrow |\mathbf{r}_\phi \times \mathbf{r}_\theta| = \sqrt{a^4\sin^4\phi\cos^2\theta + a^4\sin^4\phi\sin^2\theta + a^4\sin^2\phi\cos^2\phi} = \sqrt{a^4\sin^2\phi} = a^2\sin\phi$. The moment of inertia is $I_z = \iint_S \delta(x^2+y^2)\,d\sigma = \int_0^{2\pi}\int_0^{\pi} \delta[(a\sin\phi\cos\theta)^2 + (a\sin\phi\sin\theta)^2](a^2\sin\phi)\,d\phi\,d\theta$

$= \int_0^{2\pi}\int_0^{\pi} \delta(a^2\sin^2\phi)(a^2\sin\phi)\,d\phi\,d\theta = \int_0^{2\pi}\int_0^{\pi} \delta a^4\sin^3\phi\,d\phi\,d\theta = \int_0^{2\pi} \delta a^4\left[\left(-\frac{1}{3}\cos\phi\right)(\sin^2\phi + 2)\right]_0^{\pi} d\theta = \frac{8\delta\pi a^4}{3}$

49. The parametrization $\mathbf{r}(r,\theta) = (r\cos\theta)\mathbf{i} + (r\sin\theta)\mathbf{j} + r\mathbf{k}$ at $P_0 = (\sqrt{2}, \sqrt{2}, 2) \Rightarrow \theta = \frac{\pi}{4},\ r = 2,$

$\mathbf{r}_r = (\cos\theta)\mathbf{i} + (\sin\theta)\mathbf{j} + \mathbf{k} = \frac{\sqrt{2}}{2}\mathbf{i} + \frac{\sqrt{2}}{2}\mathbf{j} + \mathbf{k}$ and

$\mathbf{r}_\theta = (-r\sin\theta)\mathbf{i} + (r\cos\theta)\mathbf{j} = -\sqrt{2}\mathbf{i} + \sqrt{2}\mathbf{j}$

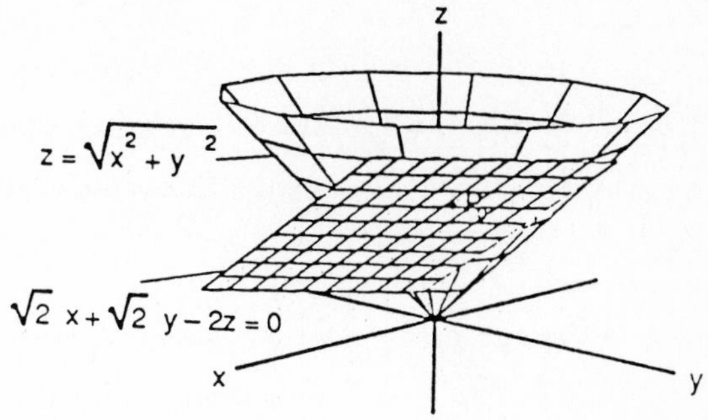

$$\Rightarrow \mathbf{r}_r \times \mathbf{r}_\theta = \begin{vmatrix} \mathbf{i} & \mathbf{j} & \mathbf{k} \\ \sqrt{2}/2 & \sqrt{2}/2 & 1 \\ -\sqrt{2} & \sqrt{2} & 0 \end{vmatrix}$$

$= -\sqrt{2}\mathbf{i} - \sqrt{2}\mathbf{j} + 2\mathbf{k} \Rightarrow$ the tangent plane is

$\left(-\sqrt{2}\mathbf{i} - \sqrt{2}\mathbf{j} + 2\mathbf{k}\right)\cdot\left[\left(x - \sqrt{2}\right)\mathbf{i} + \left(y - \sqrt{2}\right)\mathbf{j} + (z-2)\mathbf{k}\right] = \sqrt{2}x + \sqrt{2}y - 2z = 0$, or $x + y - \sqrt{2}z = 0$. The parametrization $\mathbf{r}(r,\theta) \Rightarrow x = r\cos\theta,\ y = r\sin\theta$ and $z = r \Rightarrow x^2 + y^2 = r^2 = z^2 \Rightarrow$ the surface is $z = \sqrt{x^2 + y^2}$.

51. The parametrization $\mathbf{r}(\theta, z) = (3\sin 2\theta)\mathbf{i} + \left(6\sin^2\theta\right)\mathbf{j} + z\mathbf{k}$ at $P_0 = \left(\frac{3\sqrt{3}}{2}, \frac{9}{2}, 0\right) \Rightarrow \theta = \frac{\pi}{3}$ and $z = 0$. Then

$\mathbf{r}_\theta = (6\cos 2\theta)\mathbf{i} + (12\sin\theta\cos\theta)\mathbf{j}$
$= -3\mathbf{i} + 3\sqrt{3}\mathbf{j}$ and $\mathbf{r}_z = \mathbf{k}$ at P_0

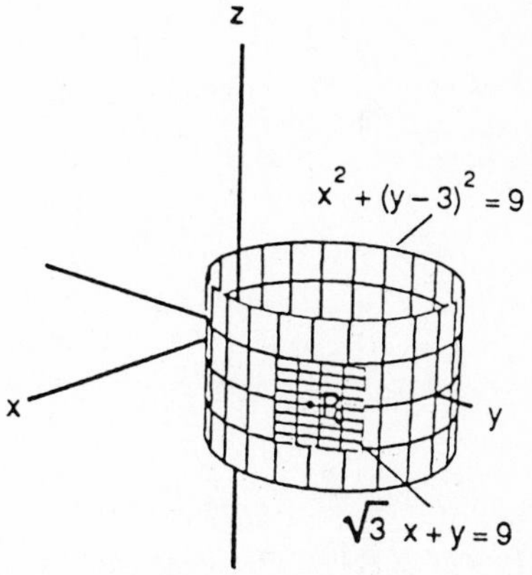

$$\Rightarrow \mathbf{r}_\theta \times \mathbf{r}_z = \begin{vmatrix} \mathbf{i} & \mathbf{j} & \mathbf{k} \\ -3 & 3\sqrt{3} & 0 \\ 0 & 0 & 1 \end{vmatrix} = 3\sqrt{3}\mathbf{i} + 3\mathbf{j} \Rightarrow \text{the tangent}$$

plane is $\left(3\sqrt{3}\mathbf{i} + 3\mathbf{j}\right)\cdot\left[\left(x - \frac{3\sqrt{3}}{2}\right)\mathbf{i} + \left(y - \frac{9}{2}\right)\mathbf{j} + (z-0)\mathbf{k}\right] = 0$

$\Rightarrow \sqrt{3}x + y = 9$. The parametrization $\Rightarrow x = 3\sin 2\theta$ and $y = 6\sin^2\theta \Rightarrow x^2 + y^2 = 9\sin^2 2\theta + \left(6\sin^2\theta\right)^2$
$= 9\left(4\sin^2\theta\cos^2\theta\right) + 36\sin^4\theta = 6\left(6\sin^2\theta\right) = 6y \Rightarrow x^2 + y^2 - 6y + 9 = 9 \Rightarrow x^2 + (y-3)^2 = 9$

53. (a) An arbitrary point on the circle C is $(x,z) = (R + r\cos u,\ r\sin u) \Rightarrow (x,y,z)$ is on the torus with $x = (R + r\cos u)\cos v,\ y = (R + r\cos u)\sin v$, and $z = r\sin u,\ 0 \le u \le 2\pi,\ 0 \le v \le 2\pi$

(b) $\mathbf{r}_u = (-r\sin u\cos v)\mathbf{i} - (r\sin u\sin v)\mathbf{j} + (r\cos u)\mathbf{k}$ and $\mathbf{r}_v = (-(R + r\cos u)\sin v)\mathbf{i} + ((R + r\cos u)\cos v)\mathbf{j}$

$$\Rightarrow \mathbf{r}_u \times \mathbf{r}_v = \begin{vmatrix} \mathbf{i} & \mathbf{j} & \mathbf{k} \\ -r\sin u\cos v & -r\sin u\sin v & r\cos u \\ -(R + r\cos u)\sin v & (R + r\cos u)\cos v & 0 \end{vmatrix}$$

$= -(R + r\cos u)(r\cos v\cos u)\mathbf{i} - (R + r\cos u)(r\sin v\cos u)\mathbf{j} + (-r\sin u)(R + r\cos u)\mathbf{k}$

$\Rightarrow |\mathbf{r}_u \times \mathbf{r}_v|^2 = (R + r\cos u)^2\left(r^2\cos^2 v\cos^2 u + r^2\sin^2 v\cos^2 u + r^2\sin^2 u\right) \Rightarrow |\mathbf{r}_u \times \mathbf{r}_v| = r(R + r\cos u)$

$$\Rightarrow A = \int_0^{2\pi}\int_0^{2\pi}\left(rR + r^2\cos u\right)du\,dv = \int_0^{2\pi} 2\pi rR\,dv = 4\pi^2 rR$$

55. (a) Let $w^2 + \frac{z^2}{c^2} = 1$ where $w = \cos\phi$ and $\frac{z}{c} = \sin\phi \Rightarrow \frac{x^2}{a^2} + \frac{y^2}{b^2} = \cos^2\phi \Rightarrow \frac{x}{a} = \cos\phi\cos\theta$ and $\frac{y}{b} = \cos\phi\sin\theta$

$\Rightarrow x = a\cos\theta\cos\phi,\ y = b\sin\theta\cos\phi$, and $z = c\sin\phi$

$\Rightarrow \mathbf{r}(\theta,\phi) = (a\cos\theta\cos\phi)\mathbf{i} + (b\sin\theta\cos\phi)\mathbf{j} + (c\sin\phi)\mathbf{k}$

(b) $\mathbf{r}_\theta = (-a\sin\theta\cos\phi)\mathbf{i} + (b\cos\theta\cos\phi)\mathbf{j}$ and $\mathbf{r}_\phi = (-a\cos\theta\sin\phi)\mathbf{i} - (b\sin\theta\sin\phi)\mathbf{j} + (c\cos\phi)\mathbf{k}$

$$\Rightarrow \mathbf{r}_\theta \times \mathbf{r}_\phi = \begin{vmatrix} \mathbf{i} & \mathbf{j} & \mathbf{k} \\ -a\sin\theta\cos\phi & b\cos\theta\cos\phi & 0 \\ -a\cos\theta\sin\phi & -b\sin\theta\sin\phi & c\cos\phi \end{vmatrix}$$

$= \left(bc\cos\theta\cos^2\phi\right)\mathbf{i} + \left(ac\sin\theta\cos^2\phi\right)\mathbf{j} + (ab\sin\phi\cos\phi)\mathbf{k}$

$\Rightarrow \left|\mathbf{r}_\theta \times \mathbf{r}_\phi\right|^2 = b^2c^2\cos^2\theta\cos^4\phi + a^2c^2\sin^2\theta\cos^4\phi + a^2b^2\sin^2\phi\cos^2\phi$, and the result follows.

57. $\mathbf{r}(\theta,u) = (5\cosh u\cos\theta)\mathbf{i} + (5\cosh u\sin\theta)\mathbf{j} + (5\sinh u)\mathbf{k} \Rightarrow \mathbf{r}_\theta = (-5\cosh u\sin\theta)\mathbf{i} + (5\cosh u\cos\theta)\mathbf{j}$ and $\mathbf{r}_u = (5\sinh u\cos\theta)\mathbf{i} + (5\sinh u\sin\theta)\mathbf{j} + (5\cosh u)\mathbf{k}$

$$\Rightarrow \mathbf{r}_\theta \times \mathbf{r}_u = \begin{vmatrix} \mathbf{i} & \mathbf{j} & \mathbf{k} \\ -5\cosh u\sin\theta & 5\cosh u\cos\theta & 0 \\ 5\sinh u\cos\theta & 5\sinh u\sin\theta & 5\cosh u \end{vmatrix}$$

$= \left(25\cosh^2 u\cos\theta\right)\mathbf{i} + \left(25\cosh^2 u\sin\theta\right)\mathbf{j} - (25\cosh u\sinh u)\mathbf{k}$. At the point $(x_0, y_0, 0)$, where $x_0^2 + y_0^2 = 25$ we have $5\sinh u = 0 \Rightarrow u = 0$ and $x_0 = 25\cos\theta,\ y_0 = 25\sin\theta \Rightarrow$ the tangent plane is $(x_0\mathbf{i} + y_0\mathbf{j}) \cdot [(x - x_0)\mathbf{i} + (y - y_0)\mathbf{j} + z\mathbf{k}] = 0 \Rightarrow x_0x - x_0^2 + y_0y - y_0^2 = 0 \Rightarrow x_0x + y_0y = 25$

14.7 STOKES'S THEOREM

1. $\text{curl}\,\mathbf{F} = \nabla \times \mathbf{F} = \begin{vmatrix} \mathbf{i} & \mathbf{j} & \mathbf{k} \\ \frac{\partial}{\partial x} & \frac{\partial}{\partial y} & \frac{\partial}{\partial z} \\ x^2 & 2x & z^2 \end{vmatrix} = 0\mathbf{i} + 0\mathbf{j} + (2 - 0)\mathbf{k} = 2\mathbf{k}$ and $\mathbf{n} = \mathbf{k} \Rightarrow \text{curl}\,\mathbf{F}\cdot\mathbf{n} = 2 \Rightarrow d\sigma = dx\,dy$

$\Rightarrow \oint_C \mathbf{F}\cdot d\mathbf{r} = \iint_R 2\,dA = 2(\text{Area of the ellipse}) = 4\pi$

3. $\text{curl}\,\mathbf{F} = \nabla \times \mathbf{F} = \begin{vmatrix} \mathbf{i} & \mathbf{j} & \mathbf{k} \\ \frac{\partial}{\partial x} & \frac{\partial}{\partial y} & \frac{\partial}{\partial z} \\ y & xz & x^2 \end{vmatrix} = -x\mathbf{i} - 2x\mathbf{j} + (z - 1)\mathbf{k}$ and $\mathbf{n} = \frac{\mathbf{i} + \mathbf{j} + \mathbf{k}}{\sqrt{3}} \Rightarrow \text{curl}\,\mathbf{F}\cdot\mathbf{n}$

$= \frac{1}{\sqrt{3}}(-3x + z - 1) \Rightarrow d\sigma = \frac{\sqrt{3}}{1}\,dA \Rightarrow \oint_C \mathbf{F}\cdot d\mathbf{r} = \iint_R \frac{1}{\sqrt{3}}(-3x + z - 1)\sqrt{3}\,dA$

$= \int_0^1 \int_0^{1-x} [-3x + (1 - x - y) - 1]\,dy\,dx = \int_0^1 \int_0^{1-x} (-4x - y)\,dy\,dx = \int_0^1 -\left[4x(1 - x) + \frac{1}{2}(1 - x)^2\right] dx$

$$= -\int_0^1 \left(\tfrac{1}{2} + 3x - \tfrac{7}{2}x^2\right) dx = -\tfrac{5}{6}$$

5. $\text{curl } \mathbf{F} = \nabla \times \mathbf{F} = \begin{vmatrix} \mathbf{i} & \mathbf{j} & \mathbf{k} \\ \frac{\partial}{\partial x} & \frac{\partial}{\partial y} & \frac{\partial}{\partial z} \\ y^2+z^2 & x^2+y^2 & x^2+y^2 \end{vmatrix} = 2y\mathbf{i} + (2z-2x)\mathbf{j} + (2x-2y)\mathbf{k}$ and $\mathbf{n} = \mathbf{k}$

$$\Rightarrow \text{curl } \mathbf{F} \cdot \mathbf{n} = 2x - 2y \Rightarrow d\sigma = dx\,dy \Rightarrow \oint_C \mathbf{F} \cdot d\mathbf{r} = \int_{-1}^{1}\int_{-1}^{1} (2x-2y)\,dx\,dy = \int_{-1}^{1} \left[x^2 - 2xy\right]_{-1}^{1} dy$$

$$= \int_{-1}^{1} -4y\,dy = 0$$

7. $x = 3\cos t$ and $y = 2\sin t \Rightarrow \mathbf{F} = (2\sin t)\mathbf{i} + \left(9\cos^2 t\right)\mathbf{j} + \left(9\cos^2 t + 16\sin^4 t\right)\sin e^{\sqrt{(6\sin t\cos t)(0)}}\mathbf{k}$ at the base of the shell; $\mathbf{r} = (3\cos t)\mathbf{i} + (2\sin t)\mathbf{j} \Rightarrow d\mathbf{r} = (-3\sin t)\mathbf{i} + (2\cos t)\mathbf{j} \Rightarrow \mathbf{F} \cdot \frac{d\mathbf{r}}{dt} = -6\sin^2 t + 18\cos^3 t$

$$\Rightarrow \iint_S \nabla \times \mathbf{F} \cdot \mathbf{n}\,d\sigma = \int_0^{2\pi} \left(-6\sin^2 t + 18\cos^3 t\right) dt = \left[-3t + \tfrac{3}{2}\sin 2t + 6(\sin t)\left(\cos^2 t + 2\right)\right]_0^{2\pi} = -6\pi$$

9. Flux of $\nabla \times \mathbf{F} = \iint_S \nabla \times \mathbf{F} \cdot \mathbf{n}\,d\sigma = \oint_C \mathbf{F} \cdot d\mathbf{r}$, so let C be parametrized by $\mathbf{r} = (a\cos t)\mathbf{i} + (a\sin t)\mathbf{j}$, $0 \le t \le 2\pi \Rightarrow \frac{d\mathbf{r}}{dt} = (-a\sin t)\mathbf{i} + (a\cos t)\mathbf{j} \Rightarrow \mathbf{F} \cdot \frac{d\mathbf{r}}{dt} = ay\sin t + ax\cos t = a^2\sin^2 t + a^2\cos^2 t = a^2$

$$\Rightarrow \text{Flux of } \nabla \times \mathbf{F} = \oint_C \mathbf{F} \cdot d\mathbf{r} = \int_0^{2\pi} a^2\,dt = 2\pi a^2$$

11. Let S_1 and S_2 be oriented surfaces that span C and that induce the same positive direction on C. Then

$$\iint_{S_1} \nabla \times \mathbf{F} \cdot \mathbf{n}_1\,d\sigma_1 = \oint_C \mathbf{F} \cdot d\mathbf{r} = \iint_{S_2} \nabla \times \mathbf{F} \cdot \mathbf{n}_2\,d\sigma_2$$

13. $\nabla \times \mathbf{F} = \begin{vmatrix} \mathbf{i} & \mathbf{j} & \mathbf{k} \\ \frac{\partial}{\partial x} & \frac{\partial}{\partial y} & \frac{\partial}{\partial z} \\ 2z & 3x & 5y \end{vmatrix} = 5\mathbf{i} + 2\mathbf{j} + 3\mathbf{k}$; $\mathbf{r}_r = (\cos\theta)\mathbf{i} + (\sin\theta)\mathbf{j} - 2r\mathbf{k}$ and $\mathbf{r}_\theta = (-r\sin\theta)\mathbf{i} + (r\cos\theta)\mathbf{j}$

$$\Rightarrow \mathbf{r}_r \times \mathbf{r}_\theta = \begin{vmatrix} \mathbf{i} & \mathbf{j} & \mathbf{k} \\ \cos\theta & \sin\theta & -2r \\ -r\sin\theta & r\cos\theta & 0 \end{vmatrix} = \left(2r^2\cos\theta\right)\mathbf{i} + \left(2r^2\sin\theta\right)\mathbf{j} + r\mathbf{k};\ \mathbf{n} = \frac{\mathbf{r}_r \times \mathbf{r}_\theta}{|\mathbf{r}_r \times \mathbf{r}_\theta|} \text{ and } d\sigma = |\mathbf{r}_r \times \mathbf{r}_\theta|\,dr\,d\theta$$

$$\Rightarrow \nabla \times \mathbf{F} \cdot \mathbf{n}\,d\sigma = (\nabla \times \mathbf{F}) \cdot (\mathbf{r}_r \times \mathbf{r}_\theta)\,dr\,d\theta = \left(10r^2\cos\theta + 4r^2\sin\theta + 3r\right) dr\,d\theta \Rightarrow \iint_S \nabla \times \mathbf{F} \cdot \mathbf{n}\,d\sigma$$

$$= \int_0^{2\pi}\int_0^2 \left(10r^2 \cos\theta + 4r^2 \sin\theta + 3r\right) dr\, d\theta = \int_0^{2\pi} \left[\tfrac{10}{3}r^3 \cos\theta + \tfrac{4}{3}r^3 \sin\theta + \tfrac{3}{2}r^2\right]_0^2 d\theta$$

$$= \int_0^{2\pi} \left(\tfrac{80}{3}\cos\theta + \tfrac{32}{3}\sin\theta + 6\right) d\theta = 6(2\pi) = 12\pi$$

15. $\nabla \times \mathbf{F} = \begin{vmatrix} \mathbf{i} & \mathbf{j} & \mathbf{k} \\ \frac{\partial}{\partial x} & \frac{\partial}{\partial y} & \frac{\partial}{\partial z} \\ x^2y & 2y^3z & 3z \end{vmatrix} = -2y^3\mathbf{i} + 0\mathbf{j} - x^2\mathbf{k}$; $\mathbf{r}_r \times \mathbf{r}_\theta = \begin{vmatrix} \mathbf{i} & \mathbf{j} & \mathbf{k} \\ \cos\theta & \sin\theta & 1 \\ -r\sin\theta & r\cos\theta & 0 \end{vmatrix}$

$= (-r\cos\theta)\mathbf{i} - (r\sin\theta)\mathbf{j} + r\mathbf{k}$ and $\nabla \times \mathbf{F} \cdot \mathbf{n}\, d\sigma = (\nabla \times \mathbf{F}) \cdot (\mathbf{r}_r \times \mathbf{r}_\theta)\, dr\, d\theta$ (see Exercise 13 above)

$$\Rightarrow \iint_S \nabla \times \mathbf{F} \cdot \mathbf{n}\, d\sigma = \iint_R \left(2ry^3 \cos\theta - rx^2\right) dr\, d\theta = \int_0^{2\pi}\int_0^1 \left(2r^4 \sin\theta \cos\theta - r^3 \cos^2\theta\right) dr\, d\theta$$

$$= \int_0^{2\pi} \left(\tfrac{2}{5}\sin\theta\cos\theta - \tfrac{1}{4}\cos^2\theta\right) d\theta = \left[\tfrac{1}{5}\sin^2\theta - \tfrac{1}{4}\left(\tfrac{\theta}{2} + \tfrac{\sin 2\theta}{4}\right)\right]_0^{2\pi} = -\tfrac{\pi}{4}$$

17. $\nabla \times \mathbf{F} = \begin{vmatrix} \mathbf{i} & \mathbf{j} & \mathbf{k} \\ \frac{\partial}{\partial x} & \frac{\partial}{\partial y} & \frac{\partial}{\partial z} \\ 3y & 5-2x & z^2-2 \end{vmatrix} = 0\mathbf{i} + 0\mathbf{j} - 5\mathbf{k}$;

$$\mathbf{r}_\phi \times \mathbf{r}_\theta = \begin{vmatrix} \mathbf{i} & \mathbf{j} & \mathbf{k} \\ \sqrt{3}\cos\phi\cos\theta & \sqrt{3}\cos\phi\sin\theta & -\sqrt{3}\sin\phi \\ -\sqrt{3}\sin\phi\sin\theta & \sqrt{3}\sin\phi\cos\theta & 0 \end{vmatrix}$$

$= (3\sin^2\phi\cos\theta)\mathbf{i} + (3\sin^2\phi\sin\theta)\mathbf{j} + (3\sin\phi\cos\phi)\mathbf{k}$; $\nabla \times \mathbf{F} \cdot \mathbf{n}\, d\sigma = (\nabla \times \mathbf{F}) \cdot (\mathbf{r}_\phi \times \mathbf{r}_\theta)\, d\phi\, d\theta$ (see Exercise 13 above) $\Rightarrow \iint_S \nabla \times \mathbf{F} \cdot \mathbf{n}\, d\sigma = \int_0^{2\pi}\int_0^{\pi/2} -15\cos\phi\sin\phi\, d\phi\, d\theta = \int_0^{2\pi} \left[\tfrac{15}{2}\cos^2\phi\right]_0^{\pi/2} d\theta = \int_0^{2\pi} -\tfrac{15}{2}\, d\theta = -15\pi$

19. (a) $\mathbf{F} = 2x\mathbf{i} + 2y\mathbf{j} + 2z\mathbf{k} \Rightarrow \text{curl } \mathbf{F} = \mathbf{0} \Rightarrow \oint_C \mathbf{F} \cdot d\mathbf{r} = \iint_S \nabla \times \mathbf{F} \cdot \mathbf{n}\, d\sigma = \iint_S d\sigma = 0$

(b) Let $f(x,y,z) = x^2y^2z^3 \Rightarrow \nabla \times \mathbf{F} = \nabla \times \nabla f = \mathbf{0} \Rightarrow \text{curl } \mathbf{F} = \mathbf{0} \Rightarrow \oint_C \mathbf{F} \cdot d\mathbf{r} = \iint_S \nabla \times \mathbf{F} \cdot \mathbf{n}\, d\sigma$

$$= \iint_S 0\, d\sigma = 0$$

(c) $\mathbf{F} = \nabla \times (x\mathbf{i} + y\mathbf{j} + z\mathbf{k}) = \mathbf{0} \Rightarrow \nabla \times \mathbf{F} = \mathbf{0} \Rightarrow \oint_C \mathbf{F} \cdot d\mathbf{r} = \iint_S \nabla \times \mathbf{F} \cdot \mathbf{n}\, d\sigma = \iint_S 0\, d\sigma = 0$

(d) $\mathbf{F} = \nabla f \Rightarrow \nabla \times \mathbf{F} = \nabla \times \nabla f = \mathbf{0} \Rightarrow \oint_C \mathbf{F} \cdot d\mathbf{r} = \iint_S \nabla \times \mathbf{F} \cdot \mathbf{n}\, d\sigma = \iint_S 0\, d\sigma = 0$

21. Let $\mathbf{F} = 2y\mathbf{i} + 3z\mathbf{j} - x\mathbf{k} \Rightarrow \nabla \times \mathbf{F} = \begin{vmatrix} \mathbf{i} & \mathbf{j} & \mathbf{k} \\ \frac{\partial}{\partial x} & \frac{\partial}{\partial y} & \frac{\partial}{\partial z} \\ 2y & 3z & -x \end{vmatrix} = -3\mathbf{i} + \mathbf{j} - 2\mathbf{k}$; $\mathbf{n} = \frac{2\mathbf{i} + 2\mathbf{j} + \mathbf{k}}{3}$

$$\Rightarrow \nabla \times \mathbf{F} \cdot \mathbf{n} = -2 \Rightarrow \oint_C 2y\,dx + 3z\,dy - x\,dz = \oint_C \mathbf{F} \cdot d\mathbf{r} = \iint_S \nabla \times \mathbf{F} \cdot \mathbf{n}\,d\sigma = \iint_S -2\,d\sigma$$

$= -2 \iint_S d\sigma$, where $\iint_S d\sigma$ is the area of the region enclosed by C on the plane S: $2x + 2y + z = 2$

23. Suppose $\mathbf{F} = M\mathbf{i} + N\mathbf{j} + P\mathbf{k}$ exists such that $\nabla \times \mathbf{F} = \left(\frac{\partial P}{\partial y} - \frac{\partial N}{\partial z}\right)\mathbf{i} + \left(\frac{\partial M}{\partial z} - \frac{\partial P}{\partial x}\right)\mathbf{j} + \left(\frac{\partial N}{\partial x} - \frac{\partial M}{\partial y}\right)\mathbf{k}$

$= x\mathbf{i} + y\mathbf{j} + z\mathbf{k}$. Then $\frac{\partial}{\partial x}\left(\frac{\partial P}{\partial y} - \frac{\partial N}{\partial z}\right) = \frac{\partial}{\partial x}(x) \Rightarrow \frac{\partial^2 P}{\partial x \partial y} - \frac{\partial^2 N}{\partial x \partial z} = 1$. Likewise, $\frac{\partial}{\partial y}\left(\frac{\partial M}{\partial z} - \frac{\partial P}{\partial x}\right) = \frac{\partial}{\partial y}(y)$

$\Rightarrow \frac{\partial^2 M}{\partial y \partial z} - \frac{\partial^2 P}{\partial y \partial x} = 1$ and $\frac{\partial}{\partial z}\left(\frac{\partial N}{\partial x} - \frac{\partial M}{\partial y}\right) = \frac{\partial}{\partial z}(z) \Rightarrow \frac{\partial^2 N}{\partial z \partial x} - \frac{\partial^2 M}{\partial z \partial y} = 1$. Summing the calculated equations

$\Rightarrow \left(\frac{\partial^2 P}{\partial x \partial y} - \frac{\partial^2 P}{\partial y \partial x}\right) + \left(\frac{\partial^2 N}{\partial z \partial x} - \frac{\partial^2 N}{\partial y \partial z}\right) + \left(\frac{\partial^2 M}{\partial y \partial z} - \frac{\partial^2 M}{\partial z \partial y}\right) = 3$ or $0 = 3$ (assuming the second mixed partials are equal). This result is a contradiction, so there is no field $\mathbf{F}$ such that curl $\mathbf{F} = x\mathbf{i} + y\mathbf{j} + z\mathbf{k}$.

25. $r = \sqrt{x^2 + y^2} \Rightarrow r^4 = (x^2 + y^2)^2 \Rightarrow \mathbf{F} = \nabla(r^4) = 4x(x^2 + y^2)\mathbf{i} + 4y(x^2 + y^2)\mathbf{j} = M\mathbf{i} + N\mathbf{j}$

$$\Rightarrow \oint_C \nabla(r^4) \cdot \mathbf{n}\,ds = \oint_C \mathbf{F} \cdot \mathbf{n}\,ds = \oint_C M\,dy - N\,dx = \iint_R \left(\frac{\partial M}{\partial x} + \frac{\partial N}{\partial y}\right) dx\,dy$$

$$= \iint_R \left[4(x^2 + y^2) + 8x^2 + 4(x^2 + y^2) + 8y^2\right] dA = \iint_R 16(x^2 + y^2)\,dA = 16 \iint_R x^2\,dA + 16 \iint_R y^2\,dA$$

$= 16I_y + 16I_x$.

14.8 THE DIVERGENCE THEOREM AND A UNIFIED THEORY

1. $\mathbf{F} = \frac{-y\mathbf{i} + x\mathbf{j}}{\sqrt{x^2 + y^2}} \Rightarrow \text{div } \mathbf{F} = \frac{xy - xy}{(x^2 + y^2)^{3/2}} = 0$

3. $\mathbf{F} = -\frac{GM(x\mathbf{i} + y\mathbf{j} + z\mathbf{k})}{(x^2 + y^2 + z^2)^{3/2}} \Rightarrow \text{div } \mathbf{F} = -GM\left[\frac{(x^2 + y^2 + z^2)^{3/2} - 3x^2(x^2 + y^2 + z^2)^{1/2}}{(x^2 + y^2 + z^2)^3}\right]$

$$- GM\left[\frac{(x^2 + y^2 + z^2)^{3/2} - 3y^2(x^2 + y^2 + z^2)^{1/2}}{(x^2 + y^2 + z^2)^3}\right] - GM\left[\frac{(x^2 + y^2 + z^2)^{3/2} - 3z^2(x^2 + y^2 + z^2)^{1/2}}{(x^2 + y^2 + z^2)^3}\right]$$

$$= -GM\left[\frac{3(x^2 + y^2 + z^2)^2 - 3(x^2 + y^2 + z^2)(x^2 + y^2 + z^2)}{(x^2 + y^2 + z^2)^{7/2}}\right] = 0$$

5. $\frac{\partial}{\partial x}(y-x)=-1,\ \frac{\partial}{\partial y}(z-y)=-1,\ \frac{\partial}{\partial z}(y-x)=0 \Rightarrow \nabla\cdot\mathbf{F}=-2 \Rightarrow \text{Flux}=\int_{-1}^{1}\int_{-1}^{1}\int_{-1}^{1} -2\,dx\,dy\,dz=-2(2^3)$

$=-16$

7. $\frac{\partial}{\partial x}(y)=0,\ \frac{\partial}{\partial y}(xy)=x,\ \frac{\partial}{\partial z}(-z)=-1 \Rightarrow \nabla\cdot\mathbf{F}=x-1;\ z=x^2+y^2 \Rightarrow z=r^2$ in cylindrical coordinates

$\Rightarrow \text{Flux}=\iiint_D (x-1)\,dz\,dy\,dx=\int_0^{2\pi}\int_0^2\int_0^{r^2}(r\cos\theta-1)\,dz\,r\,dr\,d\theta=\int_0^{2\pi}\int_0^2\left(r^3\cos\theta-r^2\right)r\,dr\,d\theta$

$=\int_0^{2\pi}\left[\frac{r^5}{5}\cos\theta-\frac{r^4}{4}\right]_0^2 d\theta=\int_0^{2\pi}\left(\frac{32}{5}\cos\theta-4\right)d\theta=\left[\frac{32}{5}\sin\theta-4\theta\right]_0^{2\pi}=-8\pi$

9. $\frac{\partial}{\partial x}(x^2)=2x,\ \frac{\partial}{\partial y}(-2xy)=-2x,\ \frac{\partial}{\partial z}(3xz)=3x \Rightarrow \text{Flux}=\iiint_D 3x\,dx\,dy\,dz$

$=\int_0^{\pi/2}\int_0^{\pi/2}\int_0^2(3\rho\sin\phi\cos\theta)(\rho^2\sin\phi)\,d\rho\,d\phi\,d\theta=\int_0^{\pi/2}\int_0^{\pi/2}12\sin^2\phi\cos\theta\,d\phi\,d\theta=\int_0^{\pi/2}3\pi\cos\theta\,d\theta=3\pi$

11. $\frac{\partial}{\partial x}(2xz)=2z,\ \frac{\partial}{\partial y}(-xy)=-x,\ \frac{\partial}{\partial z}(-z^2)=-2z \Rightarrow \nabla\cdot\mathbf{F}=-x \Rightarrow \text{Flux}=\iiint_D -x\,dV$

$=\int_0^2\int_0^{\sqrt{16-4x^2}}\int_0^{4-y}-x\,dz\,dy\,dx=\int_0^2\int_0^{\sqrt{16-4x^2}}(xy-4x)\,dy\,dx=\int_0^2\left[\frac{1}{2}x\left(16-4x^2\right)-4x\sqrt{16-4x^2}\right]dx$

$=\left[4x^2-\frac{1}{2}x^4+\frac{1}{3}\left(16-4x^2\right)^{3/2}\right]_0^2=-\frac{40}{3}$

13. Let $\rho=\sqrt{x^2+y^2+z^2}$. Then $\frac{\partial\rho}{\partial x}=\frac{x}{\rho},\ \frac{\partial\rho}{\partial y}=\frac{y}{\rho},\ \frac{\partial\rho}{\partial z}=\frac{z}{\rho} \Rightarrow \frac{\partial}{\partial x}(\rho x)=\left(\frac{\partial\rho}{\partial x}\right)x+\rho=\frac{x^2}{\rho}+\rho,\ \frac{\partial}{\partial y}(\rho y)=\left(\frac{\partial\rho}{\partial y}\right)y+\rho$

$=\frac{y^2}{\rho}+\rho,\ \frac{\partial}{\partial z}(\rho z)=\left(\frac{\partial\rho}{\partial z}\right)z+\rho=\frac{z^2}{\rho}+\rho \Rightarrow \nabla\cdot\mathbf{F}=\frac{x^2+y^2+z^2}{\rho}+3\rho=4\rho$, since $\rho=\sqrt{x^2+y^2+z^2}$

$\Rightarrow \text{Flux}=\iiint_D 4\rho\,dV=\int_0^{2\pi}\int_0^{\pi}\int_1^{\sqrt{2}}(4\rho)\left(\rho^2\sin\phi\right)d\rho\,d\phi\,d\theta=\int_0^{2\pi}\int_0^{\pi}3\sin\phi\,d\phi\,d\theta=\int_0^{2\pi}6\,d\theta=12\pi$

15. $\frac{\partial}{\partial x}\left(5x^3+12xy^2\right)=15x^2+12y^2,\ \frac{\partial}{\partial y}\left(y^3+e^y\sin z\right)=3y^2+e^y\sin z,\ \frac{\partial}{\partial z}\left(5z^3+e^y\cos z\right)=15z^2-e^y\sin z$

$\Rightarrow \nabla\cdot\mathbf{F}=15x^2+15y^2+15z^2=15\rho^2 \Rightarrow \text{Flux}=\iiint_D 15\rho^2\,dV=\int_0^{2\pi}\int_0^{\pi}\int_1^{\sqrt{2}}\left(15\rho^2\right)\left(\rho^2\sin\phi\right)d\rho\,d\phi\,d\theta$

$$= \int_0^{2\pi}\int_0^{\pi} (12\sqrt{2}-3)\sin\phi\, d\phi\, d\theta = \int_0^{2\pi} (24\sqrt{2}-6)\, d\theta = (48\sqrt{2}-12)\pi$$

17. (a) $\mathbf{G} = M\mathbf{i} + N\mathbf{j} + P\mathbf{k} \Rightarrow \nabla\times\mathbf{G} = \text{curl } \mathbf{G} = \left(\frac{\partial P}{\partial y} - \frac{\partial N}{\partial z}\right)\mathbf{i} + \left(\frac{\partial M}{\partial z} - \frac{\partial P}{\partial x}\right)\mathbf{k} + \left(\frac{\partial N}{\partial x} - \frac{\partial M}{\partial y}\right)\mathbf{k} \Rightarrow \nabla\cdot\nabla\times\mathbf{G}$

$$= \text{div}(\text{curl } \mathbf{G}) = \frac{\partial}{\partial x}\left(\frac{\partial P}{\partial y} - \frac{\partial N}{\partial z}\right) + \frac{\partial}{\partial y}\left(\frac{\partial M}{\partial z} - \frac{\partial P}{\partial x}\right) + \frac{\partial}{\partial z}\left(\frac{\partial N}{\partial x} - \frac{\partial M}{\partial y}\right)$$

$$= \frac{\partial^2 P}{\partial x\partial y} - \frac{\partial^2 N}{\partial x\partial z} + \frac{\partial^2 M}{\partial y\partial z} - \frac{\partial^2 P}{\partial y\partial x} + \frac{\partial^2 N}{\partial z\partial x} - \frac{\partial^2 M}{\partial z\partial y} = 0$$ if all first and second partial derivatives are continuous

(b) By the Divergence Theorem, the outward flux of $\nabla\times\mathbf{G}$ across a closed surface is zero because

outward flux of $\nabla\times\mathbf{G} = \iint_S (\nabla\times\mathbf{G})\cdot\mathbf{n}\, d\sigma$

$= \iiint_D \nabla\cdot\nabla\times\mathbf{G}\, dV$ [Divergence Theorem with $\mathbf{F} = \nabla\times\mathbf{G}$]

$= \iiint_D (0)\, dV = 0$ [by part (a)]

19. (a) $\text{div}(g\mathbf{F}) = \nabla\cdot g\mathbf{F} = \frac{\partial}{\partial x}(gM) + \frac{\partial}{\partial y}(gN) + \frac{\partial}{\partial z}(gP) = \left(g\frac{\partial M}{\partial x} + M\frac{\partial g}{\partial x}\right) + \left(g\frac{\partial N}{\partial y} + N\frac{\partial g}{\partial y}\right) + \left(g\frac{\partial P}{\partial z} + P\frac{\partial g}{\partial x}\right)$

$$= \left(M\frac{\partial g}{\partial x} + N\frac{\partial g}{\partial y} + P\frac{\partial g}{\partial z}\right) + g\left(\frac{\partial M}{\partial x} + \frac{\partial N}{\partial y} + \frac{\partial P}{\partial z}\right) = g\nabla\cdot\mathbf{F} + \nabla g\cdot\mathbf{F}$$

(b) $\nabla\times(g\mathbf{F}) = \left[\frac{\partial}{\partial y}(gP) - \frac{\partial}{\partial z}(gN)\right]\mathbf{i} + \left[\frac{\partial}{\partial z}(gM) - \frac{\partial}{\partial x}(gP)\right]\mathbf{j} + \left[\frac{\partial}{\partial x}(gN) - \frac{\partial}{\partial y}(gM)\right]\mathbf{k}$

$$= \left(P\frac{\partial g}{\partial y} + g\frac{\partial P}{\partial y} - N\frac{\partial g}{\partial z} - g\frac{\partial N}{\partial z}\right)\mathbf{i} + \left(M\frac{\partial g}{\partial z} + g\frac{\partial M}{\partial z} - P\frac{\partial g}{\partial x} - g\frac{\partial P}{\partial x}\right)\mathbf{j} + \left(N\frac{\partial g}{\partial x} + g\frac{\partial N}{\partial x} - M\frac{\partial g}{\partial y} - g\frac{\partial M}{\partial y}\right)\mathbf{k}$$

$$= \left(P\frac{\partial g}{\partial y} - N\frac{\partial g}{\partial z}\right)\mathbf{i} + \left(g\frac{\partial P}{\partial y} - g\frac{\partial N}{\partial z}\right)\mathbf{i} + \left(M\frac{\partial g}{\partial z} - P\frac{\partial g}{\partial x}\right)\mathbf{j} + \left(g\frac{\partial M}{\partial z} - g\frac{\partial P}{\partial x}\right)\mathbf{j} + \left(N\frac{\partial g}{\partial x} - M\frac{\partial g}{\partial y}\right)\mathbf{k}$$

$$+ \left(g\frac{\partial N}{\partial x} - g\frac{\partial M}{\partial y}\right)\mathbf{k} = g\nabla\times\mathbf{F} + \nabla g\times\mathbf{F}$$

21. The integral's value never exceeds the surface area of S. Since $|\mathbf{F}| \le 1$, we have $|\mathbf{F}\cdot\mathbf{n}| = |\mathbf{F}|\,|\mathbf{n}| \le (1)(1) = 1$ and

$\iiint_D \nabla\cdot\mathbf{F}\, d\sigma = \iint_S \mathbf{F}\cdot\mathbf{n}\, d\sigma$ [Divergence Theorem]

$\le \iint_S |\mathbf{F}\cdot\mathbf{n}|\, d\sigma$ [A property of integrals]

$\le \iint_S (1)\, d\sigma$ [$|\mathbf{F}\cdot\mathbf{n}| \le 1$]

= Area of S.

23. (a) $\frac{\partial}{\partial x}(x) = 1$, $\frac{\partial}{\partial y}(y) = 1$, $\frac{\partial}{\partial z}(z) = 1 \Rightarrow \nabla \cdot \mathbf{F} = 3 \Rightarrow \text{Flux} = \iiint\limits_D 3\, dV = 3 \iiint\limits_D dV$

$= 3(\text{Volume of the solid})$

(b) If **F** is orthogonal to **n** at every point of S, then $\mathbf{F}\cdot\mathbf{n} = 0$ everywhere $\Rightarrow \text{Flux} = \iint\limits_S \mathbf{F}\cdot\mathbf{n}\, d\sigma = 0$.

But the flux is 3(Volume of the solid) $\neq 0$, so **F** is not orthogonal to **n** at every point.

25. $\iint\limits_S \mathbf{F}\cdot\mathbf{n}\, d\sigma = \iiint\limits_D \nabla\cdot\mathbf{F}\, dV = \iiint\limits_D 3\, dV \Rightarrow \frac{1}{3}\iint\limits_S \mathbf{F}\cdot\mathbf{n}\, d\sigma = \iiint\limits_D dV = \text{Volume of D}$

27. (a) From the Divergence Theorem, $\iint\limits_S \nabla f\cdot\mathbf{n}\, d\sigma = \iiint\limits_D \nabla\cdot\nabla f\, dV = \iiint\limits_D \nabla^2 f\, dV = \iiint\limits_D 0\, dV = 0$

(b) From the Divergence Theorem, $\iint\limits_S f\nabla f\cdot\mathbf{n}\, d\sigma = \iiint\limits_D \nabla\cdot f\nabla f\, dV$. Now,

$$f\nabla f = \left(f\frac{\partial f}{\partial x}\right)\mathbf{i} + \left(f\frac{\partial f}{\partial y}\right)\mathbf{j} + \left(f\frac{\partial f}{\partial z}\right)\mathbf{k} \Rightarrow \nabla\cdot f\nabla f = \left[f\frac{\partial^2 f}{\partial x^2} + \left(\frac{\partial f}{\partial x}\right)^2\right] + \left[f\frac{\partial^2 f}{\partial y^2} + \left(\frac{\partial f}{\partial y}\right)^2\right] + \left[f\frac{\partial^2 f}{\partial z^2} + \left(\frac{\partial f}{\partial z}\right)^2\right]$$

$= f\nabla^2 f + |\nabla f|^2 = 0 + |\nabla f|^2$ since f is harmonic $\Rightarrow \iint\limits_S f\nabla f\cdot\mathbf{n}\, d\sigma = \iiint\limits_D |\nabla f|^2\, dV$, as claimed.

29. $\iint\limits_S f\nabla g\cdot\mathbf{n}\, d\sigma = \iiint\limits_D \nabla\cdot f\nabla g\, dV = \iiint\limits_D \nabla\cdot\left(f\frac{\partial g}{\partial x}\mathbf{i} + f\frac{\partial g}{\partial y}\mathbf{j} + f\frac{\partial g}{\partial z}\mathbf{k}\right) dV$

$$= \iiint\limits_D \left(f\frac{\partial^2 g}{\partial x^2} + \frac{\partial f}{\partial x}\frac{\partial g}{\partial x} + f\frac{\partial^2 g}{\partial y^2} + \frac{\partial f}{\partial y}\frac{\partial g}{\partial y} + f\frac{\partial^2 g}{\partial z^2} + \frac{\partial f}{\partial z}\frac{\partial g}{\partial z}\right) dV$$

$$= \iiint\limits_D \left[f\left(\frac{\partial^2 g}{\partial x^2} + \frac{\partial^2 g}{\partial y^2} + \frac{\partial^2 g}{\partial z^2}\right) + \left(\frac{\partial f}{\partial x}\frac{\partial g}{\partial x} + \frac{\partial f}{\partial y}\frac{\partial g}{\partial y} + \frac{\partial f}{\partial z}\frac{\partial g}{\partial z}\right)\right] dV = \iiint\limits_D (f\nabla^2 g + \nabla f\cdot\nabla g)\, dV$$

31. (a) The integral $\iiint\limits_D p(t,x,y,z)\, dV$ represents the mass of the fluid at any time t. The equation says that the instantaneous rate of change of mass is flux of the fluid through the surface S enclosing the region D: the mass decreases if the flux is outward (so the fluid flows out of D), and increases if the flow is inward (interpreting **n** as the outward pointing unit normal to the surface).

(b) $\iiint\limits_D \frac{\partial p}{\partial t}\, dV = \frac{d}{dt}\iiint\limits_D p\, dV = -\iint\limits_S p\mathbf{v}\cdot\mathbf{n}\, d\sigma = -\iiint\limits_D \nabla\cdot p\mathbf{v}\, dV \Rightarrow \frac{\partial \rho}{\partial t} = -\nabla\cdot p\mathbf{v}$

$\Rightarrow \nabla\cdot p\mathbf{v} + \frac{\partial p}{\partial t} = 0$, as claimed

CHAPTER 14 PRACTICE EXERCISES

1. Path 1: $\mathbf{r} = t\mathbf{i} + t\mathbf{j} + t\mathbf{k} \Rightarrow x = t,\ y = t,\ z = t,\ 0 \le t \le 1 \Rightarrow f(g(t), h(t), k(t)) = 3 - 3t^2$ and $\frac{dx}{dt} = 1,\ \frac{dy}{dt} = 1,$

$$\frac{dz}{dt} = 1 \Rightarrow \sqrt{\left(\frac{dx}{dt}\right)^2 + \left(\frac{dy}{dt}\right)^2 + \left(\frac{dz}{dt}\right)^2}\,dt = \sqrt{3}\,dt \Rightarrow \int_C f(x,y,z)\,ds = \int_0^1 \sqrt{3}(3 - 3t^2)\,dt = 2\sqrt{3}$$

Path 2: $\mathbf{r}_1 = t\mathbf{i} + t\mathbf{j},\ 0 \le t \le 1 \Rightarrow x = t,\ y = t,\ z = 0 \Rightarrow f(g(t), h(t), k(t)) = 2t - 3t^2 + 3$ and $\frac{dx}{dt} = 1,\ \frac{dy}{dt} = 1,$

$$\frac{dz}{dt} = 0 \Rightarrow \sqrt{\left(\frac{dx}{dt}\right)^2 + \left(\frac{dy}{dt}\right)^2 + \left(\frac{dz}{dt}\right)^2}\,dt = \sqrt{2}\,dt \Rightarrow \int_{C_1} f(x,y,z)\,ds = \int_0^1 \sqrt{2}(2t - 3t^2 + 3)\,dt = 3\sqrt{2};$$

$\mathbf{r}_2 = \mathbf{i} + \mathbf{j} + t\mathbf{k} \Rightarrow x = 1,\ y = 1,\ z = t \Rightarrow f(g(t), h(t), k(t)) = 2 - 2t$ and $\frac{dx}{dt} = 0,\ \frac{dy}{dt} = 0,\ \frac{dz}{dt} = 1$

$$\Rightarrow \sqrt{\left(\frac{dx}{dt}\right)^2 + \left(\frac{dy}{dt}\right)^2 + \left(\frac{dz}{dt}\right)^2}\,dt = dt \Rightarrow \int_{C_2} f(x,y,z)\,ds = \int_0^1 (2 - 2t)\,dt = 1$$

$$\Rightarrow \int_C f(x,y,z)\,ds = \int_{C_1} f(x,y,z)\,ds + \int_{C_2} f(x,y,z) = 3\sqrt{2} + 1$$

3. $\mathbf{r} = (a\cos t)\mathbf{j} + (a\sin t)\mathbf{k} \Rightarrow x = 0,\ y = a\cos t,\ z = a\sin t \Rightarrow f(g(t), h(t), k(t)) = \sqrt{a^2\sin^2 t} = a\,|\sin t|$ and

$$\frac{dx}{dt} = 0,\ \frac{dy}{dt} = -a\sin t,\ \frac{dz}{dt} = a\cos t \Rightarrow \sqrt{\left(\frac{dx}{dt}\right)^2 + \left(\frac{dy}{dt}\right)^2 + \left(\frac{dz}{dt}\right)^2}\,dt = a\,dt$$

$$\Rightarrow \int_C f(x,y,z)\,ds = \int_0^{2\pi} a^2|\sin t|\,dt = \int_0^{\pi} a^2\sin t\,dt + \int_{\pi}^{2\pi} -a^2\sin t\,dt = 4a^2$$

5. $\frac{\partial P}{\partial y} = -\frac{1}{2}(x+y+z)^{-3/2} = \frac{\partial N}{\partial z},\ \frac{\partial M}{\partial z} = -\frac{1}{2}(x+y+z)^{-3/2} = \frac{\partial P}{\partial x},\ \frac{\partial N}{\partial x} = -\frac{1}{2}(x+y+z)^{-3/2} = \frac{\partial M}{\partial y}$

$$\Rightarrow M\,dx + N\,dy + P\,dz \text{ is exact; } \frac{\partial f}{\partial x} = \frac{1}{\sqrt{x+y+z}} \Rightarrow f(x,y,z) = 2\sqrt{x+y+z} + g(y,z) \Rightarrow \frac{\partial f}{\partial y} = \frac{1}{\sqrt{x+y+z}} + \frac{\partial g}{\partial y}$$

$$= \frac{1}{\sqrt{x+y+z}} \Rightarrow \frac{\partial g}{\partial y} = 0 \Rightarrow g(y,z) = h(z) \Rightarrow f(x,y,z) = 2\sqrt{x+y+z} + h(z) \Rightarrow \frac{\partial f}{\partial z} = \frac{1}{\sqrt{x+y+z}} + h'(z)$$

$$= \frac{1}{\sqrt{x+y+z}} \Rightarrow h'(x) = 0 \Rightarrow h(z) = C \Rightarrow f(x,y,z) = 2\sqrt{x+y+z} + C \Rightarrow \int_{(-1,1,1)}^{(4,-3,0)} \frac{dx + dy + dz}{\sqrt{x+y+z}}$$

$$= f(4,-3,0) - f(-1,1,1) = 2\sqrt{1} - 2\sqrt{1} = 0$$

7. $\frac{\partial P}{\partial y} = x\cos z = \frac{\partial N}{\partial z},\ \frac{\partial M}{\partial z} = y\cos z = \frac{\partial P}{\partial x},\ \frac{\partial N}{\partial x} = \sin z = \frac{\partial M}{\partial y} \Rightarrow \mathbf{F}$ is conservative $\Rightarrow \int_C \mathbf{F}\cdot d\mathbf{r} = 0$

9. Let $M = 8x \sin y$ and $N = -8y \cos x \Rightarrow \frac{\partial M}{\partial y} = 8x \cos y$ and $\frac{\partial N}{\partial x} = 8y \sin x \Rightarrow \int_C 8x \sin y\, dx - 8y \cos x\, dy$

$$= \iint_R (8y \sin x - 8x \cos y)\, dy\, dx = \int_0^{\pi/2}\int_0^{\pi/2} (8y \sin x - 8x \cos y)\, dy\, dx = \int_0^{\pi/2} \left(\pi^2 \sin x - 8x\right) dx$$

$$= -\pi^2 + \pi^2 = 0$$

11. Let $z = 1 - x - y \Rightarrow f_x(x,y) = -1$ and $f_y(x,y) = -1 \Rightarrow \sqrt{f_x^2 + f_y^2 + 1} = \sqrt{3} \Rightarrow$ Surface Area $= \iint_R \sqrt{3}\, dx\, dy$

$= \sqrt{3}$(Area of the circular region in the xy-plane) $= \pi\sqrt{3}$

13. $\nabla f = 2x\mathbf{i} + 2y\mathbf{j} + 2z\mathbf{k}$, $\mathbf{p} = \mathbf{k} \Rightarrow |\nabla f| = \sqrt{4x^2 + 4y^2 + 4z^2} = 2\sqrt{x^2 + y^2 + z^2} = 2$ and $|\nabla f \cdot \mathbf{p}| = |2z| = 2z$ since

$z \ge 0 \Rightarrow$ Surface Area $= \iint_R \frac{2}{2z}\, dA = \iint_R \frac{1}{z}\, dA = \iint_R \frac{1}{\sqrt{1 - x^2 - y^2}}\, dx\, dy = \int_0^{2\pi}\int_0^{1/\sqrt{2}} \frac{1}{\sqrt{1 - r^2}}\, r\, dr\, d\theta$

$$\int_0^{2\pi} \left[-\sqrt{1 - r^2}\right]_0^{1/\sqrt{2}} d\theta = \int_0^{2\pi} \left(1 - \frac{1}{\sqrt{2}}\right) d\theta = 2\pi\left(1 - \frac{1}{\sqrt{2}}\right)$$

15. $f(x,y,z) = \frac{x}{a} + \frac{y}{b} + \frac{z}{c} = 1 \Rightarrow \nabla f = \left(\frac{1}{a}\right)\mathbf{i} + \left(\frac{1}{b}\right)\mathbf{j} + \left(\frac{1}{c}\right)\mathbf{k} \Rightarrow |\nabla f| = \sqrt{\frac{1}{a^2} + \frac{1}{b^2} + \frac{1}{c^2}}$ and $\mathbf{p} = \mathbf{k} \Rightarrow |\nabla f \cdot \mathbf{p}| = \frac{1}{c}$

since $c > 0 \Rightarrow$ Surface Area $= \iint_R \frac{\sqrt{\frac{1}{a^2} + \frac{1}{b^2} + \frac{1}{c^2}}}{\left(\frac{1}{c}\right)}\, dA = c\sqrt{\frac{1}{a^2} + \frac{1}{b^2} + \frac{1}{c^2}} \iint_R dA = \frac{1}{2}abc\sqrt{\frac{1}{a^2} + \frac{1}{b^2} + \frac{1}{c^2}}$,

since the area of the triangular region R is $\frac{1}{2}ab$

17. $\nabla f = 2y\mathbf{j} + 2z\mathbf{k}$, $\mathbf{p} = \mathbf{k} \Rightarrow |\nabla f| = \sqrt{4y^2 + 4z^2} = 2\sqrt{y^2 + z^2} = 10$ and $|\nabla f \cdot \mathbf{p}| = 2z$ since $z \ge 0$

$\Rightarrow d\sigma = \frac{10}{2z}\, dx\, dy = \frac{5}{z}\, dx\, dy = \iint_S g(x,y,z)\, d\sigma = \iint_R (x^4 y)(y^2 + z^2)\left(\frac{5}{z}\right) dx\, dy$

$$= \iint_R (x^4 y)(25)\left(\frac{5}{\sqrt{25 - y^2}}\right) dx\, dy = \int_0^4\int_0^1 \frac{125y}{\sqrt{25 - y^2}}\, x^4\, dx\, dy = \int_0^4 \frac{25y}{\sqrt{25 - y^2}}\, dy = 50$$

19. A possible parametrization is $\mathbf{r}(\phi,\theta) = (6 \sin\phi \cos\theta)\mathbf{i} + (6 \sin\phi \sin\theta)\mathbf{j} + (6 \cos\phi)\mathbf{k}$ (spherical coordinates); now $\rho = 6$ and $z = -3 \Rightarrow -3 = 6 \cos\phi \Rightarrow \cos\phi = -\frac{1}{2} \Rightarrow \phi = \frac{2\pi}{3}$ and $z = 3\sqrt{3} \Rightarrow 3\sqrt{3} = 6 \cos\phi$

$\Rightarrow \cos\phi = \frac{\sqrt{3}}{2} \Rightarrow \phi = \frac{\pi}{6} \Rightarrow \frac{\pi}{6} \le \phi \le \frac{2\pi}{3}$; also $0 \le \theta \le 2\pi$

21. A possible parametrization is $\mathbf{r}(r,\theta) = (r\cos\theta)\mathbf{i} + (r\sin\theta)\mathbf{j} + (1+r)\mathbf{k}$ (cylindrical coordinates);

now $r = \sqrt{x^2+y^2} \Rightarrow z = 1 + r$ and $1 \le z \le 3 \Rightarrow 1 \le 1 + r \le 3 \Rightarrow 0 \le r \le 2$; also $0 \le \theta \le 2\pi$

23. Let $x = u\cos v$ and $z = u\sin v$, where $u = \sqrt{x^2+z^2}$ and v is the angle in the xz-plane with the x-axis $\Rightarrow \mathbf{r}(u,v) = (u\cos v)\mathbf{i} + 2u^2\mathbf{j} + (u\sin v)\mathbf{k}$ is a possible parametrization; $0 \le y \le 2 \Rightarrow 2u^2 \le 2 \Rightarrow u^2 \le 1$ $\Rightarrow 0 \le u \le 1$ since $u \ge 0$; also, for just the upper half of the paraboloid, $0 \le v \le \pi$

25. $\mathbf{r}_u = \mathbf{i}+\mathbf{j},\ \mathbf{r}_v = \mathbf{i}-\mathbf{j}+\mathbf{k} \Rightarrow \mathbf{r}_u \times \mathbf{r}_v = \begin{vmatrix} \mathbf{i} & \mathbf{j} & \mathbf{k} \\ 1 & 1 & 0 \\ 1 & -1 & 1 \end{vmatrix} = \mathbf{i} - \mathbf{j} - 2\mathbf{k} \Rightarrow |\mathbf{r}_u \times \mathbf{r}_v| = \sqrt{6}$

$$\Rightarrow \text{Surface Area} = \iint_{R_{uv}} |\mathbf{r}_u \times \mathbf{r}_v|\,du\,dv = \int_0^1\int_0^1 \sqrt{6}\,du\,dv = \sqrt{6}$$

27. $\mathbf{r}_r = (\cos\theta)\mathbf{i} + (\sin\theta)\mathbf{j},\ \mathbf{r}_\theta = (-r\sin\theta)\mathbf{i} + (r\cos\theta)\mathbf{j} + \mathbf{k} \Rightarrow \mathbf{r}_r \times \mathbf{r}_\theta = \begin{vmatrix} \mathbf{i} & \mathbf{j} & \mathbf{k} \\ \cos\theta & \sin\theta & 0 \\ -r\sin\theta & r\cos\theta & 1 \end{vmatrix}$

$= (\sin\theta)\mathbf{i} - (\cos\theta)\mathbf{j} + r\mathbf{k} \Rightarrow |\mathbf{r}_r \times \mathbf{r}_\theta| = \sqrt{\sin^2\theta + \cos^2\theta + r^2} = \sqrt{1+r^2} \Rightarrow \text{Surface Area} = \iint_{R_{r\theta}} |\mathbf{r}_r \times \mathbf{r}_\theta|\,dr\,d\theta$

$$= \int_0^{2\pi}\int_0^1 \sqrt{1+r^2}\,dr\,d\theta = \int_0^{2\pi} \left[\frac{r}{2}\sqrt{1+r^2} + \frac{1}{2}\ln\left(r+\sqrt{1+r^2}\right)\right]_0^1 d\theta = \int_0^{2\pi} \left[\frac{1}{2}\sqrt{2} + \frac{1}{2}\ln\left(1+\sqrt{2}\right)\right] d\theta$$

$$= \pi\left[\sqrt{2} + \ln\left(1+\sqrt{2}\right)\right]$$

29. $\frac{\partial P}{\partial y} = 0 = \frac{\partial N}{\partial z},\ \frac{\partial M}{\partial z} = 0 = \frac{\partial P}{\partial x},\ \frac{\partial N}{\partial x} = 0 = \frac{\partial M}{\partial y} \Rightarrow$ Conservative

31. $\frac{\partial P}{\partial y} = 0 \ne ye^z = \frac{\partial N}{\partial z} \Rightarrow$ Not Conservative

33. $\frac{\partial f}{\partial x} = 2 \Rightarrow f(x,y,z) = 2x + g(y,z) \Rightarrow \frac{\partial f}{\partial y} = \frac{\partial g}{\partial y} = 2y + z \Rightarrow g(y,z) = y^2 + zy + h(z)$

$\Rightarrow f(x,y,z) = 2x + y^2 + zy + h(z) \Rightarrow \frac{\partial f}{\partial z} = y + h'(z) = y + 1 \Rightarrow h'(z) = 1 \Rightarrow h(z) = z + C$

$\Rightarrow f(x,y,z) = 2x + y^2 + zy + z$

35. Over Path 1: $\mathbf{r} = t\mathbf{i} + t\mathbf{j} + t\mathbf{k},\ 0 \le t \le 1 \Rightarrow x = t,\ y = t,\ z = t$ and $d\mathbf{r} = (\mathbf{i}+\mathbf{j}+\mathbf{k})\,dt \Rightarrow \mathbf{F} = 2t^2\mathbf{i} + \mathbf{j} + t^2\mathbf{k}$

$\Rightarrow \mathbf{F}\cdot d\mathbf{r} = (3t^2+1)\,dt \Rightarrow \text{Work} = \int_0^1 (3t^2+1)\,dt = 2;$

Over Path 2: $\mathbf{r}_1 = t\mathbf{i} + t\mathbf{j},\ 0 \le t \le 1 \Rightarrow x = t,\ y = t,\ z = 0$ and $d\mathbf{r}_1 = (\mathbf{i}+\mathbf{j})\,dt \Rightarrow \mathbf{F}_1 = 2t^2\mathbf{i} + \mathbf{j} + t^2\mathbf{k}$

$\Rightarrow \mathbf{F}_1 \cdot d\mathbf{r}_1 = (2t^2+1)\, dt \Rightarrow \text{Work}_1 = \int_0^1 (2t^2+1)\, dt = \frac{5}{3}$; $\mathbf{r}_2 = \mathbf{i}+\mathbf{j}+t\mathbf{k}$, $0 \le t \le 1 \Rightarrow x = 1,\ y = 1,\ z = t$ and

$d\mathbf{r}_2 = \mathbf{k}\, dt \Rightarrow \mathbf{F}_2 = 2\mathbf{i}+\mathbf{j}+\mathbf{k} \Rightarrow \mathbf{F}_2 \cdot d\mathbf{r}_2 = dt \Rightarrow \text{Work}_2 = \int_0^1 dt = 1 \Rightarrow \text{Work} = \text{Work}_1 + \text{Work}_2 = \frac{5}{3}+1 = \frac{8}{3}$

37. (a) $\mathbf{r} = (e^t \cos t)\mathbf{i} + (e^t \sin t)\mathbf{j} \Rightarrow x = e^t \cos t,\ y = e^t \sin t$ from $(1,0)$ to $(e^{2\pi},0) \Rightarrow 0 \le t \le 2\pi$

$\Rightarrow \frac{d\mathbf{r}}{dt} = (e^t \cos t - e^t \sin t)\,\mathbf{i} + (e^t \sin t + e^t \cos t)\,\mathbf{j}$ and $\mathbf{F} = \frac{x\mathbf{i}+y\mathbf{j}}{(x^2+y^2)^{3/2}} = \frac{(e^t \cos t)\mathbf{i} + (e^t \sin t)\mathbf{j}}{(e^{2t}\cos^2 t + e^{2t}\sin^2 t)^{3/2}}$

$= \left(\frac{\cos t}{e^{2t}}\right)\mathbf{i} + \left(\frac{\sin t}{e^{2t}}\right)\mathbf{j} \Rightarrow \mathbf{F} \cdot \frac{d\mathbf{r}}{dt} = \left(\frac{\cos^2 t}{e^t} - \frac{\sin t \cos t}{e^t} + \frac{\sin^2 t}{e^t} + \frac{\sin t \cos t}{e^t}\right) = e^{-t}$

$\Rightarrow \text{Work} = \int_0^{2\pi} e^{-t}\, dt = 1 - e^{-2\pi}$

(b) $\mathbf{F} = \frac{x\mathbf{i}+y\mathbf{j}}{(x^2+y^2)^{3/2}} \Rightarrow \frac{\partial f}{\partial x} = \frac{x}{(x^2+y^2)^{3/2}} \Rightarrow f(x,y,z) = -(x^2+y^2)^{-1/2} + g(y,z) \Rightarrow \frac{\partial f}{\partial y} = \frac{y}{(x^2+y^2)^{3/2}} + \frac{\partial g}{\partial y}$

$= \frac{y}{(x^2+y^2)^{3/2}} \Rightarrow g(y,z) = C \Rightarrow f(x,y,z) = -(x^2+y^2)^{-1/2}$ is a potential function for $\mathbf{F} \Rightarrow \int_C \mathbf{F} \cdot d\mathbf{r}$

$= f(e^{2\pi},0) - f(1,0) = 1 - e^{-2\pi}$

39. (a) $x^2+y^2 = 1 \Rightarrow \mathbf{r} = (\cos t)\mathbf{i} + (\sin t)\mathbf{j}$, $0 \le t \le \pi \Rightarrow x = \cos t$ and $y = \sin t \Rightarrow \mathbf{F} = (\cos t + \sin t)\mathbf{i} - \mathbf{j}$ and

$\frac{d\mathbf{r}}{dt} = (-\sin t)\mathbf{i} + (\cos t)\mathbf{j} \Rightarrow \mathbf{F} \cdot \frac{d\mathbf{r}}{dt} = -\sin t \cos t - \sin^2 t - \cos t$

$\Rightarrow \text{Flow} = \int_0^{\pi} (-\sin t \cos t - \sin^2 t - \cos t)\, dt = \left[\frac{1}{2}\cos^2 t - \frac{t}{2} + \frac{\sin 2t}{4} - \sin t\right]_0^{\pi} = -\frac{\pi}{2}$

(b) $\mathbf{r} = -t\mathbf{i}$, $-1 \le t \le 1 \Rightarrow x = -t$ and $y = 0 \Rightarrow \mathbf{F} = -t\mathbf{i} - t^2\mathbf{j}$ and $\frac{d\mathbf{r}}{dt} = -\mathbf{i} \Rightarrow \mathbf{F} \cdot \frac{d\mathbf{r}}{dt} = t \Rightarrow \text{Flow} = \int_{-1}^{1} t\, dt = 0$

(c) $\mathbf{r}_1 = (1-t)\mathbf{i} - t\mathbf{j}$, $0 \le t \le 1 \Rightarrow \mathbf{F}_1 = (1-2t)\mathbf{i} - (1-2t+2t^2)\mathbf{j}$ and $\frac{d\mathbf{r}_1}{dt} = -\mathbf{i} - \mathbf{j} \Rightarrow \mathbf{F}_1 \cdot \frac{d\mathbf{r}_1}{dt} = 2t^2$

$\Rightarrow \text{Flow}_1 = \int_0^1 2t^2\, dt = \frac{2}{3}$; $\mathbf{r}_2 = -t\mathbf{i} + (t-1)\mathbf{j}$, $0 \le t \le 1 \Rightarrow \mathbf{F}_2 = -\mathbf{i} - (2t^2 - 2t + 1)\mathbf{j}$ and $\frac{d\mathbf{r}_2}{dt} = -\mathbf{i} + \mathbf{j}$

$\Rightarrow \mathbf{F}_2 \cdot \frac{d\mathbf{r}_2}{dt} = -2t^2 + 2t \Rightarrow \text{Flow}_2 = \int_0^1 (-2t^2 + 2t)\, dt = \frac{1}{3} \Rightarrow \text{Flow} = \text{Flow}_1 + \text{Flow}_2 = \frac{2}{3} + \frac{1}{3} = 1$

41. $\nabla \times \mathbf{F} = \begin{vmatrix} \mathbf{i} & \mathbf{j} & \mathbf{k} \\ \frac{\partial}{\partial x} & \frac{\partial}{\partial y} & \frac{\partial}{\partial z} \\ y^2 & -y & 3z^2 \end{vmatrix} = -2y\mathbf{k}$; unit normal to the plane is $\mathbf{n} = \frac{2\mathbf{i} + 6\mathbf{j} - 3\mathbf{k}}{\sqrt{4+36+9}} = \frac{2}{7}\mathbf{i} + \frac{6}{7}\mathbf{j} - \frac{3}{7}\mathbf{k}$

$\Rightarrow \nabla \times \mathbf{F} \cdot \mathbf{n} = \frac{6}{7}y$; $\mathbf{p} = \mathbf{k}$ and $f(x,y,z) = 2x + 6y - 3z \Rightarrow |\nabla f \cdot \mathbf{p}| = 3 \Rightarrow d\sigma = \frac{|\nabla f|}{|\nabla f \cdot \mathbf{p}|}\, dA = \frac{7}{3}\, dA$

$\Rightarrow \oint_C \mathbf{F} \cdot d\mathbf{r} = \iint_R \frac{6}{7}y\, d\sigma = \iint_R \left(\frac{6}{7}y\right)\left(\frac{7}{3}\, dA\right) = \iint_R 2y\, dA = \int_0^{2\pi}\int_0^1 2r \sin\theta\, r\, dr\, d\theta = \int_0^{2\pi} \frac{2}{3} \sin\theta\, d\theta = 0$

43. (a) $\mathbf{r} = \sqrt{2}t\mathbf{i} + \sqrt{2}t\mathbf{j} + (4 - t^2)\mathbf{k}$, $0 \le t \le 1 \Rightarrow x = \sqrt{2}t$, $y = \sqrt{2}t$, $z = 4 - t^2 \Rightarrow \frac{dx}{dt} = \sqrt{2}$, $\frac{dy}{dt} = \sqrt{2}$, $\frac{dz}{dt} = -2t$

$\Rightarrow \sqrt{\left(\frac{dx}{dt}\right)^2 + \left(\frac{dy}{dt}\right)^2 + \left(\frac{dz}{dt}\right)^2}\, dt = \sqrt{4 + 4t^2}\, dt \Rightarrow M = \int_C \delta(x,y,z)\, ds = \int_0^1 3t\sqrt{4 + 4t^2}\, dt = \left[\frac{1}{4}(4+4t)^{3/2}\right]_0^1$

$= 4\sqrt{2} - 2$

(b) $M = \int_C \delta(x,y,z)\, ds = \int_0^1 \sqrt{4 + 4t^2}\, dt = \left[t\sqrt{1+t^2} + \ln\left(t + \sqrt{1+t^2}\right)\right]_0^1 = \sqrt{2} + \ln\left(1 + \sqrt{2}\right)$

45. $\mathbf{r} = t\mathbf{i} + \left(\frac{2\sqrt{2}}{3}t^{3/2}\right)\mathbf{j} + \left(\frac{t^2}{2}\right)\mathbf{k}$, $0 \le t \le 2 \Rightarrow x = t$, $y = \frac{2\sqrt{2}}{3}t^{3/2}$, $z = \frac{t^2}{2} \Rightarrow \frac{dx}{dt} = 1$, $\frac{dy}{dt} = \sqrt{2}\,t^{1/2}$, $\frac{dz}{dt} = t$

$\Rightarrow \sqrt{\left(\frac{dx}{dt}\right)^2 + \left(\frac{dy}{dt}\right)^2 + \left(\frac{dz}{dt}\right)^2}\, dt = \sqrt{1 + 2t + t^2}\, dt = \sqrt{(t+1)^2}\, dt = |t+1|\, dt = (t+1)\, dt$ on the domain given.

Then $M = \int_C \delta\, ds = \int_0^2 \left(\frac{1}{t+1}\right)(t+1)\, dt = \int_0^2 dt = 2$; $M_{yz} = \int_C x\delta\, ds = \int_0^2 t\left(\frac{1}{t+1}\right)(t+1)\, dt = \int_0^2 t\, dt = 2$;

$M_{xz} = \int_C y\delta\, ds = \int_0^2 \left(\frac{2\sqrt{2}}{3}t^{3/2}\right)\left(\frac{1}{t+1}\right)(t+1)\, dt = \int_0^2 \frac{2\sqrt{2}}{3}t^{3/2}\, dt = \frac{32}{15}$; $M_{xy} = \int_C z\delta\, ds$

$= \int_0^2 \left(\frac{t^2}{2}\right)\left(\frac{1}{t+1}\right)(t+1)\, dt = \int_0^2 \frac{t^2}{2}\, dt = \frac{4}{3} \Rightarrow \bar{x} = \frac{M_{yz}}{M} = \frac{2}{2} = 1$; $\bar{y} = \frac{M_{xz}}{M} = \frac{\left(\frac{32}{15}\right)}{2} = \frac{16}{15}$; $\bar{z} = \frac{M_{xy}}{M}$

$= \frac{\left(\frac{4}{3}\right)}{2} = \frac{2}{3}$; $I_x = \int_C (y^2 + z^2)\delta\, ds = \int_0^2 \left(\frac{8}{9}t^3 + \frac{t^4}{4}\right) dt = \frac{232}{45}$; $I_y = \int_C (x^2 + z^2)\delta\, ds = \int_0^2 \left(t^2 + \frac{t^4}{4}\right) dt = \frac{64}{15}$;

$I_z = \int_C (y^2 + x^2)\delta\, ds = \int_0^2 \left(t^2 + \frac{8}{9}t^3\right) dt = \frac{56}{9}$; $R_x = \sqrt{\frac{I_x}{M}} = \sqrt{\frac{\left(\frac{232}{45}\right)}{2}} = \sqrt{\frac{116}{45}}$; $R_y = \sqrt{\frac{I_y}{M}} = \sqrt{\frac{\left(\frac{64}{15}\right)}{2}} = \sqrt{\frac{32}{15}}$;

$R_z = \sqrt{\frac{I_z}{M}} = \sqrt{\frac{\left(\frac{56}{9}\right)}{2}} = \sqrt{\frac{28}{9}}$

47. $\mathbf{r}(t) = (e^t \cos t)\mathbf{i} + (e^t \sin t)\mathbf{j} + e^t\mathbf{k},\ 0 \le t \le \ln 2 \Rightarrow x = e^t \cos t,\ y = e^t \sin t,\ z = e^t \Rightarrow \frac{dx}{dt} = (e^t \cos t - e^t \sin t),$

$\frac{dy}{dt} = (e^t \sin t + e^t \cos t),\ \frac{dz}{dt} = e^t \Rightarrow \sqrt{\left(\frac{dx}{dt}\right)^2 + \left(\frac{dy}{dt}\right)^2 + \left(\frac{dz}{dt}\right)^2}\ dt$

$= \sqrt{(e^t \cos t - e^t \sin t)^2 + (e^t \sin t + e^t \cos t)^2 + (e^t)^2}\ dt = \sqrt{3e^{2t}}\ dt = \sqrt{3}\,e^t\ dt;\ M = \int_C \delta\ ds = \int_0^{\ln 2} \sqrt{3}\,e^t\ dt$

$= \sqrt{3};\ M_{xy} = \int_C z\delta\ ds = \int_0^{\ln 2} (\sqrt{3}\,e^t)(e^t)\ dt = \int_0^{\ln 2} \sqrt{3}\,e^{2t}\ dt = \frac{3\sqrt{3}}{2} \Rightarrow \bar{z} = \frac{M_{xy}}{M} = \frac{\left(\frac{3\sqrt{3}}{2}\right)}{\sqrt{3}} = \frac{3}{2};$

$I_z = \int_C (x^2 + y^2)\delta\ ds = \int_0^{\ln 2} (e^{2t} \cos^2 t + e^{2t} \sin^2 t)(\sqrt{3}\,e^t)\ dt = \int_0^{\ln 2} \sqrt{3}\,e^{3t}\ dt = \frac{7\sqrt{3}}{3} \Rightarrow R_z = \sqrt{\frac{I_z}{M}}$

$= \sqrt{\frac{7\sqrt{3}}{3\sqrt{3}}} = \sqrt{\frac{7}{3}}$

49. Because of symmetry $\bar{x} = \bar{y} = 0$. Let $f(x,y,z) = x^2 + y^2 + z^2 = 25 \Rightarrow \nabla f = 2x\mathbf{i} + 2y\mathbf{j} + 2z\mathbf{k}$

$\Rightarrow |\nabla f| = \sqrt{4x^2 + 4y^2 + 4z^2} = 10$ and $\mathbf{p} = \mathbf{k} \Rightarrow |\nabla f \cdot \mathbf{p}| = 2z$, since $z \ge 0 \Rightarrow M = \iint_R \delta(x,y,z)\ d\sigma$

$= \iint_R z\left(\frac{10}{2z}\right) dA = \iint_R 5\ dA = 5(\text{Area of the circular region}) = 80\pi;\ M_{xy} = \iint_R z\delta\ d\sigma = \iint_R 5z\ dA$

$= \iint_R 5\sqrt{25 - x^2 - y^2}\ dx\,dy = \int_0^{2\pi}\int_0^4 (5\sqrt{25 - r^2})\ r\ dr\,d\theta = \int_0^{2\pi} \frac{490}{3}\ d\theta = \frac{980}{3}\pi \Rightarrow \bar{z} = \frac{\left(\frac{980}{3}\pi\right)}{80\pi} = \frac{49}{12}$

$\Rightarrow (\bar{x},\bar{y},\bar{z}) = \left(0,0,\frac{49}{12}\right);\ I_z = \iint_R (x^2 + y^2)\delta\ d\sigma = \iint_R 5(x^2 + y^2)\ dx\,dy = \int_0^{2\pi}\int_0^4 5r^3\ dr\,d\theta = \int_0^{2\pi} 320\ d\theta = 640\pi;$

$R_z = \sqrt{\frac{I_z}{M}} = \sqrt{\frac{640\pi}{80\pi}} = 2\sqrt{2}$

51. $M = 2xy + x$ and $N = xy - y \Rightarrow \frac{\partial M}{\partial x} = 2y + 1,\ \frac{\partial M}{\partial y} = 2x,\ \frac{\partial N}{\partial x} = y,\ \frac{\partial N}{\partial y} = x - 1 \Rightarrow \text{Flux} = \iint_R \left(\frac{\partial M}{\partial x} + \frac{\partial N}{\partial y}\right) dx\,dy$

$= \iint_R (2y + 1 + x - 1)\ dy\,dx = \int_0^1\int_0^1 (2y + x)\ dy\,dx = \frac{3}{2};\ \text{Circ} = \iint_R \left(\frac{\partial N}{\partial x} - \frac{\partial M}{\partial y}\right) dx\,dy$

$= \iint_R (y - 2x)\ dy\,dx = \int_0^1\int_0^1 (y - 2x)\ dy\,dx = -\frac{1}{2}$

53. $M = -\frac{\cos y}{x}$ and $N = \ln x \sin y \Rightarrow \frac{\partial M}{\partial y} = \frac{\sin y}{x}$ and $\frac{\partial N}{\partial x} = \frac{\sin y}{x} \Rightarrow \oint_C \ln x \sin y\ dy - \frac{\cos y}{x}\ dx$

$= \iint_R \left(\frac{\partial N}{\partial x} - \frac{\partial M}{\partial y}\right) dx\,dy = \iint_R \left(\frac{\sin y}{x} - \frac{\sin y}{x}\right) dx\,dy = 0$

55. $\frac{\partial}{\partial x}(2xy) = 2y$, $\frac{\partial}{\partial y}(2yz) = 2z$, $\frac{\partial}{\partial z}(2xz) = 2x \Rightarrow \nabla \cdot \mathbf{F} = 2y + 2z + 2x \Rightarrow \text{Flux} = \iiint_D (2x + 2y + 2z)\, dV$

$= \int_0^1 \int_0^1 \int_0^1 (2x + 2y + 2z)\, dx\, dy\, dz = \int_0^1 \int_0^1 (1 + 2y + 2z)\, dy\, dz = \int_0^1 (2 + 2z)\, dz = 3$

57. $\frac{\partial}{\partial x}(-2x) = -2$, $\frac{\partial}{\partial y}(-3y) = -3$, $\frac{\partial}{\partial z}(z) = 1 \Rightarrow \nabla \cdot \mathbf{F} = -4$; $x^2 + y^2 + z^2 = 2$ and $x^2 + y^2 = z \Rightarrow z = 1$

$\Rightarrow x^2 + y^2 = 1 \Rightarrow \text{Flux} = \iiint_D -4\, dV = -4 \int_0^{2\pi} \int_0^1 \int_{r^2}^{\sqrt{2-r^2}} dz\, r\, dr\, d\theta = -4 \int_0^{2\pi} \int_0^1 \left(r\sqrt{2 - r^2} - r^3\right) dr\, d\theta$

$= -4 \int_0^{2\pi} \left(-\frac{7}{12} + \frac{2}{3}\sqrt{2}\right) d\theta = \frac{2}{3}\pi\left(7 - 8\sqrt{2}\right)$

59. $\mathbf{F} = y\mathbf{i} + z\mathbf{j} + x\mathbf{k} \Rightarrow \nabla \cdot \mathbf{F} = 0 \Rightarrow \text{Flux} = \iint_S \mathbf{F} \cdot \mathbf{n}\, d\sigma = \iiint_D \nabla \cdot \mathbf{F}\, dV = 0$

61. $\mathbf{F} = xy^2\mathbf{i} + x^2y\mathbf{j} + y\mathbf{k} \Rightarrow \nabla \cdot \mathbf{F} = y^2 + x^2 + 0 \Rightarrow \text{Flux} = \iint_S \mathbf{F} \cdot \mathbf{n}\, d\sigma = \iiint_D \nabla \cdot \mathbf{F}\, dV$

$= \iiint_D (x^2 + y^2)\, dV = \int_0^{2\pi} \int_0^1 \int_{-1}^1 r^2\, dz\, r\, dr\, d\theta = \int_0^{2\pi} \int_0^1 2r^3\, dr\, d\theta = \int_0^{2\pi} \frac{1}{2}\, d\theta = \pi$

CHAPTER 14 ADDITIONAL EXERCISES–THEORY, EXAMPLES, APPLICATIONS

1. $dx = (-2 \sin t + 2 \sin 2t)\, dt$ and $dy = (2 \cos t - 2 \cos 2t)\, dt$; $\text{Area} = \frac{1}{2} \oint_C x\, dy - y\, dx$

$= \frac{1}{2} \int_0^{2\pi} [(2 \cos t - \cos 2t)(2 \cos t - 2 \cos 2t) - (2 \sin t - \sin 2t)(-2 \sin t + 2 \sin 2t)]\, dt$

$= \frac{1}{2} \int_0^{2\pi} [6 - (6 \cos t \cos 2t + 6 \sin t \sin 2t)]\, dt = \frac{1}{2} \int_0^{2\pi} (6 - 6 \cos t)\, dt = 6\pi$

3. $dx = \cos 2t\, dt$ and $dy = \cos t\, dt$; $\text{Area} = \frac{1}{2} \oint_C x\, dy - y\, dx = \frac{1}{2} \int_0^{\pi} \left(\frac{1}{2} \sin 2t \cos t - \sin t \cos 2t\right) dt$

$= \frac{1}{2} \int_0^{\pi} \left[\sin t \cos^2 t - (\sin t)\left(2 \cos^2 t - 1\right)\right] dt = \frac{1}{2} \int_0^{\pi} \left(-\sin t \cos^2 t + \sin t\right) dt = \frac{1}{2}\left[\frac{1}{3} \cos^3 t - \cos t\right]_0^{\pi} = -\frac{1}{3} + 1 = \frac{2}{3}$

5. (a) $\mathbf{F}(x,y,z) = z\mathbf{i} + x\mathbf{j} + y\mathbf{k}$ is $\mathbf{0}$ only at the point $(0,0,0)$, and curl $\mathbf{F}(x,y,z) = \mathbf{i}+\mathbf{j}+\mathbf{k}$ is never $\mathbf{0}$.

(b) $\mathbf{F}(x,y,z) = z\mathbf{i} + y\mathbf{k}$ is $\mathbf{0}$ only on the line $x = t$, $y = 0$, $z = 0$ and curl $\mathbf{F}(x,y,z) = \mathbf{i}+\mathbf{j}$ is never $\mathbf{0}$.

(c) $\mathbf{F}(x,y,z) = z\mathbf{i}$ is $\mathbf{0}$ only when $z = 0$ (the xy-plane) and curl $\mathbf{F}(x,y,z) = \mathbf{j}$ is never $\mathbf{0}$.

7. Set up the coordinate system so that $(a,b,c) = (0,R,0) \Rightarrow \delta(x,y,z) = \sqrt{x^2+(y-R)^2+z^2}$

$= \sqrt{x^2+y^2+z^2-2Ry+R^2} = \sqrt{2R^2-2Ry}$; let $f(x,y,z) = x^2+y^2+z^2 = R^2$ and $\mathbf{p} = \mathbf{i}$

$\Rightarrow \nabla f = 2x\mathbf{i}+2y\mathbf{j}+2z\mathbf{k} \Rightarrow |\nabla f| = 2\sqrt{x^2+y^2+z^2} = 2R \Rightarrow d\sigma = \frac{|\nabla f|}{|\nabla f\cdot \mathbf{i}|}\,dz\,dy = \frac{2R}{2x}\,dz\,dy$

$$\Rightarrow \text{Mass} = \iint_S \delta(x,y,z)\,d\sigma = \iint_{R_{yz}} \sqrt{2R^2-2Ry}\left(\frac{R}{x}\right)dz\,dy = R\iint_{R_{yz}} \frac{\sqrt{2R^2-2Ry}}{\sqrt{R^2-y^2-z^2}}\,dz\,dy$$

$$= 4R\int_{-R}^{R}\int_0^{\sqrt{R^2-y^2}} \frac{\sqrt{2R^2-2Ry}}{\sqrt{R^2-y^2-z^2}}\,dz\,dy = \frac{16\pi R^3}{3} \quad \text{(we used a CAS integrator)}$$

9. $M = x^2+4xy$ and $N = -6y \Rightarrow \frac{\partial M}{\partial x} = 2x+4y$ and $\frac{\partial N}{\partial x} = -6 \Rightarrow \text{Flux} = \int_0^b\int_0^a (2x+4y-6)\,dx\,dy$

$= \int_0^b (a^2+4ay-6a)\,dy = a^2b+2ab^2-6ab$. We want to minimize $f(a,b) = a^2b+2ab^2-6ab = ab(a+2b-6)$.

Thus, $f_a(a,b) = 2ab+2b^2-6b = 0$ and $f_b(a,b) = a^2+4ab-6a = 0 \Rightarrow b(2a+2b-6) = 0 \Rightarrow b = 0$ or $b = -a+3$. Now $b = 0 \Rightarrow a^2-6a = 0 \Rightarrow a = 0$ or $a = 6 \Rightarrow (0,0)$ and $(6,0)$ are critical points. On the other hand, $b = -a+3 \Rightarrow a^2+4a(-a+3)-6a = 0 \Rightarrow -3a^2+6a = 0 \Rightarrow a = 0$ or $a = 2 \Rightarrow (0,3)$ and $(2,1)$ are also critical points. The flux at $(0,0) = 0$, the flux at $(6,0) = 0$, the flux at $(0,3) = 0$ and the flux at $(2,1) = -4$. Therefore, the flux is minimized at $(2,1)$ with value -4.

11. (a) Partition the string into small pieces. Let $\Delta_i s$ be the length of the ith piece. Let (x_i, y_i) be a point in the ith piece. The work done by gravity in moving the ith piece to the x-axis is approximately $W_i = (gx_iy_i\Delta_i s)y_i$ where $x_iy_i\Delta_i s$ is approximately the mass of the ith piece. The total work done by gravity in moving the string to the x-axis is $\sum_i W_i = \sum_i gx_iy_i^2\Delta_i s \Rightarrow \text{Work} = \int_C gxy^2\,ds$

(b) $\text{Work} = \int_C gxy^2\,ds = \int_0^{\pi/2} g(2\cos t)(4\sin^2 t)\sqrt{4\sin^2 t+4\cos^2 t}\,dt = 16g\int_0^{\pi/2} \cos t\sin^2 t\,dt$

$= \left[16g\left(\frac{\sin^3 t}{3}\right)\right]_0^{\pi/2} = \frac{16}{3}g$

(c) $\bar{x} = \dfrac{\int_C x(xy)\,ds}{\int_C xy\,ds}$ and $\bar{y} = \dfrac{\int_C y(xy)\,ds}{\int_C xy\,ds}$; the mass of the string is $\int_C xy\,ds$ and the weight of the string is

$g\int_C xy\,ds$. Therefore, the work done in moving the point mass at $(\bar{x},\bar{y})$ to the x-axis is

$$W = \left(g\int_C xy\,ds\right)\bar{y} = g\int_C xy^2\,ds = \frac{16}{3}g.$$

13. (a) Partition the sphere $x^2 + y^2 + (z-2)^2 = 1$ into small pieces. Let $\Delta_i\sigma$ be the surface area of the ith piece and let (x_i, y_i, z_i) be a point on the ith piece. The force due to pressure on the ith piece is approximately $w(4 - z_i)\Delta_i\sigma$. The total force on S is approximately $\sum_i w(4 - z_i)\Delta_i\sigma$. This gives the actual force to be $\iint_S w(4-z)\,d\sigma$.

(b) The upward buoyant force is a result of the **k**-component of the force on the ball due to liquid pressure. The force on the ball at (x,y,z) is $w(4-z)(-\mathbf{n}) = w(z-4)\mathbf{n}$, where $\mathbf{n}$ is the outer unit normal at (x,y,z). Hence the **k**-component of this force is $w(z-4)\mathbf{n}\cdot\mathbf{k} = w(z-4)\mathbf{k}\cdot\mathbf{n}$. The (magnitude of the) buoyant force on the ball is obtained by adding up all these **k**-components to obtain $\iint_S w(z-4)\mathbf{k}\cdot\mathbf{n}\,d\sigma$.

(c) The Divergence Theorem says $\iint_S w(z-4)\mathbf{k}\cdot\mathbf{n}\,d\sigma = \iiint_D \operatorname{div}(w(z-4)\mathbf{k})\,dV = \iiint_D w\,dV$, where D is $x^2 + y^2 + (z-2)^2 \le 1 \Rightarrow \iint_S w(z-4)\mathbf{k}\cdot\mathbf{n}\,d\sigma = w\iiint_D 1\,dV = \frac{4}{3}\pi w$, the weight of the fluid if it were to occupy the region D.

15. Assume that S is a surface to which Stokes's Theorem applies. Then $\oint_C \mathbf{E}\cdot d\mathbf{r} = \iint_S (\nabla\times\mathbf{E})\cdot\mathbf{n}\,d\sigma$ $= \iint_S -\frac{\partial \mathbf{B}}{\partial t}\cdot\mathbf{n}\,d\sigma = -\frac{\partial}{\partial t}\iint_S \mathbf{B}\cdot\mathbf{n}\,d\sigma$. Thus the voltage around a loop equals the negative of the rate of change of magnetic flux through the loop.

17. $\oint_C f\nabla g\cdot d\mathbf{r} = \iint_S \nabla\times(f\nabla g)\cdot\mathbf{n}\,d\sigma$ (Stokes's Theorem)

$= \iint_S (f\nabla\times\nabla g + \nabla f\times\nabla g)\cdot\mathbf{n}\,d\sigma$ (Section 14.8, Exercise 19a)

$= \iint_S [(f)(\mathbf{0}) + \nabla f \times \nabla g] \cdot \mathbf{n}\, d\sigma$ (Section 14.7, Equation 12)

$= \iint_S (\nabla f \times \nabla g) \cdot \mathbf{n}\, d\sigma$

19. False; let $\mathbf{F} = y\mathbf{i} + x\mathbf{j} \neq \mathbf{0} \Rightarrow \nabla \cdot \mathbf{F} = \frac{\partial}{\partial x}(y) + \frac{\partial}{\partial y}(x) = 0$ and $\nabla \times \mathbf{F} = \begin{vmatrix} \mathbf{i} & \mathbf{j} & \mathbf{k} \\ \frac{\partial}{\partial x} & \frac{\partial}{\partial y} & \frac{\partial}{\partial z} \\ x & y & 0 \end{vmatrix} = 0\mathbf{i} + 0\mathbf{j} + 0\mathbf{k} = \mathbf{0}$

21. $\mathbf{r} = x\mathbf{i} + y\mathbf{j} + z\mathbf{k} \Rightarrow \nabla \cdot \mathbf{r} = 1 + 1 + 1 = 3 \Rightarrow \iiint_D \nabla \cdot \mathbf{r}\, dV = 3 \iiint_D dV = 3V \Rightarrow V = \frac{1}{3} \iiint_D \nabla \cdot \mathbf{r}\, dV$

$= \frac{1}{3} \iint_S \mathbf{r} \cdot \mathbf{n}\, d\sigma$, by the Divergence Theorem

NOTES:

NOTES:

NOTES:

NOTES:

NOTES:

NOTES:

NOTES: